Representation Theory
and Mathematical Physics

Representation Theory and Mathematical Physics
Conference in honor of Gregg Zuckerman's 60th Birthday
Yale University, October 24-27, 2009

$$\Gamma^{\mathfrak{g},K}_{\mathfrak{g},L\cap K}(W) = Hom_{R(\mathfrak{g},L\cap K)}(R(\mathfrak{g},K),W)^K$$

$$\{u,v\} = (-1)^{|u|} Res_w Res_{z-w}[b(z-w)u](w)v$$

Speakers:

Barbasch, Dan (Cornell)
Frenkel, Igor (Yale)
Gaitsgory, Dennis (Harvard)
Garland, Howard (Yale)
Howe, Roger (Yale)
Kostant, Bert (MIT)
Kobayashi, Toshiyuki (Tokyo)
Lusztig, George (MIT)
Penkov, Ivan (Jacobs University, Bremen)
Sally, Paul (Chicago)
Schmid, Wilfried (Harvard)
Serganova, Vera (Berkeley)
Speh, Birgit (Cornell)
Stein, Elias (Princeton)
Trapa, Peter (University of Utah)
Vogan, David (MIT)
Willenbring, Jeb (University of Wisconsin, Milwaukee)

Organizers:

Adams, Jeffrey (University of Maryland)
Kapranov, Mikhail (Yale)
Lian, Bong (Brandeis)
Sahi, Siddhartha (Rutgers)

Sponsors:

National Science Foundation
Yale University

Contact:

Website: http://www.liegroups.org/zuckerman/
Local contact: Mary Ellen DelVecchio, *mary.delvecchio@yale.edu* 203-432-7058
Poster background: courtesy of http://media4.its.yale.edu/res/sites/OPA/

CONTEMPORARY MATHEMATICS

557

Representation Theory and Mathematical Physics

Conference in Honor of
Gregg Zuckerman's 60th Birthday
October 24–27, 2009
Yale University

Jeffrey Adams
Bong Lian
Siddhartha Sahi
Editors

American Mathematical Society
Providence, Rhode Island

2010 *Mathematics Subject Classification.* Primary 22E45, 22E46, 22E47, 17B65, 17B68, 17B69, 33D52.

Library of Congress Cataloging-in-Publication Data

Representation theory and mathematical physics : conference in honor of Gregg Zuckerman's 60th birthday, October 24–27, 2009, Yale University / Jeffrey Adams, Bong H. Lian, Siddhartha Sahi, editors.

    p. cm. — (Contemporary mathematics ; v. 557)
    Includes bibliographical references.
    ISBN 978-0-8218-5246-0 (alk. paper)
    1. Linear algebraic groups—Congresses. 2. Representations of Lie groups—Congresses. 3. Mathematical physics—Congresses. I. Zuckerman, Gregg. II. Adams, Jeffrey. III. Lian, Bong H., 1962- IV. Sahi, Siddhartha, 1958-

QA179.R47 2011
515′.7223—dc23

                  2011030808

Dedicated to Gregg Zuckerman on the occasion of his 60th birthday.

Contents

Preface

Lie groups and their representations are a fundamental area of mathematics, with connections to geometry, topology, number theory, physics, combinatorics, and many other areas. Gregg Zuckerman's work lies at the very heart of the modern theory of representations of Lie groups. His influential ideas on derived functors, the translation principle, and coherent continuation laid the groundwork of modern algebraic theory.

Zuckerman has long been active in the fruitful interplay between mathematics and physics. Developments in this area include work on chiral algebras, and the representation theory of affine Kac-Moody algebras. Recent progress on the geometric Langlands program points to exciting connections between automorphic representations and dual fibrations in geometric mirror symmetry.

These topics were the subject of a conference in honor of Gregg Zuckerman's 60th birthday, held at Yale, October 24-27, 2009.

Summary of Contributions

The classical Plancherel theorem is a statement about the Fourier transform on $L^2(\mathbb{R})$. It has generalizations to any locally compact group. *The Plancherel Formula, The Plancherel Theorem, and the Fourier Transform of Orbital Integrals* by Rebecca A. Herb and Paul J. Sally, Jr. surveys the history of this subject for non-abelian Lie groups and p-adic groups.

One of Zuckerman's major contributions to representation theory is the technique now known as cohomological induction or the derived functor construction of representations. An important special case of this construction are the so-called $A_q(\lambda)$ representations which are cohomologically induced from one-dimensional characters. The paper *Branching Problems of Zuckerman Derived Functor Modules* by Toshiyuki Kobayashi provides a comprehensive survey of known results on the restrictions of the $A_q(\lambda)$ to symmetric subgroups, along with sketches of the most important ideas of the proofs.

Chiral Equivariant Cohomology of Spheres, by Bong H. Lian, Andrew R. Linshaw, and Bailin Song, is a survey of their work on the theory of chiral equivariant cohomology. This is a new topological invariant which is vertex algebra valued and contains the Borel-Cartan equivariant cohomology theory of a G-manifold as a substructure. The paper describes some of the general structural features of the new invariant—a quasi-conformal structure, equivariant homotopy invariance, and the values of this cohomology on homogeneous spaces—as well as a class of group actions on spheres having the same classical equivariant cohomology, but which can all be distinguished by the new invariant.

An irreducible admissible representation of a Lie group G is determined by its global character, which is an invariant distribution. It is represented by a conjugation invariant function defined on a dense subset of the semisimple elements. In *Computing Global Characters* Jeffrey Adams gives an explicit algorithm for such characters, based on the Kazhdan-Lusztig-Vogan polynomials.

Part of Arthur's conjectures predict the existence of certain stable virtual characters associated to nilpotent orbits. These conjectures are known in the case of an Archimedean field. In *Stable Combinations of Special Unipotent Representations*, Dan M. Barbasch and Peter E. Trapa study the space of stable virtual representations associated certain nilpotent orbits $\mathcal{O}$, and relate this space to the *special piece* of $\mathcal{O}$.

Levi Components of Parabolic Subalgebras of Finitary Lie Algebras, by Elizabeth Dan-Cohen and Ivan Penkov, turns to understanding certain "standard" subalgebras of the finitary Lie algebras including $sl(\infty)$, $so(\infty)$ and $sp(\infty)$. A lot is known about their Cartan, Borel and parabolic subalgebras. This paper goes much further by giving a description of subalgebras which can appear as the Levi component of a simple finitary Lie algebra, and a characterization of all parabolic subalgebras of which a given subalgebra is a Levi component.

In *On Extending the Langlands-Shahidi Method to Arithmetic Quotients of Loop Groups*, Howard Garland describes an important new approach to proving analytic properties of L-functions on certain finite dimensional groups by using the theory of Eisenstein series on infinite dimensional loop groups. This theory of Eisenstein series on loop groups for minimal parabolics has been developed by him in a series of previous papers. In this paper, he shows how the theory changes when one works with Eisenstein series associated to cusp forms on maximal parabolics.

If a group K acts on a vector space V then one is often interested in understanding the algebra of invariant polynomials $S(V^*)^K$ as explicitly as possible. In the paper *The Measurement of Quantum Entanglement and Enumeration of Graph Coverings,* motivated by questions in quantum computing, Michael W. Hero, Jeb F. Willenbring, and Lauren Kelly Williams consider the case of the group $K = U(n_1) \times \cdots \times U(n_r)$ acting in the usual manner on the tensor product $V = \mathbb{C}^{n_1} \otimes \cdots \otimes \mathbb{C}^{n_r}$, and they determine the invariant polynomials in the limiting case where all the $n_i \to \infty$.

Roger Howe's theory of dual pairs typically refers to a commuting pair of subgroups of the symplectic group. Although Howe's original formulation was in terms of Lie superalgebras, and such algebras have a wide variety of applications, their dual pairs have not received much attention. In *The Dual Pairs* $(O_{(p,q)}, O\widetilde{Sp}_{2,2})$ *and Zuckerman Translation*, Dan Lu and Roger Howe study the representation theory of the dual pair consisting of the group $O(p, q)$ and the Lie superalgebra $O\widetilde{Sp}_{2,2}$.

The paper *On the algebraic set of singular elements in a complex simple Lie algebra* by Bertram Kostant and Nolan Wallach studies a space of defining equations, denoted M, for the cone of singular elements of a semisimple Lie algebra $\mathfrak{g}$. The main result of the paper is the explicit decomposition of M as a $\mathfrak{g}$-module.

In *An Explicit Embedding of Gravity and the Standard Model in E_8*, Garrett Lisi gives some preliminary steps in the construction of a unified theory of all interactions based on the gauge group E_8. The kinematic framework he proposes accounts for one of the three observed generations of matter, but also include many particles that are not yet observed. His description includes an explicit embedding of the standard model and gravitational gauge groups into E_8, and the action of the corresponding Lie algebra generators on fermions.

Harmonic analysis on a connected reductive algebraic group G is a special case of harmonic analysis on symmetric spaces: one writes $G = G \times G/G^\delta$ where G^δ is the fixed points of the involution interchanging the two factors. In *From Groups to Symmetric Spaces* George Lusztig studies various properties which are known in the group case, and to what extent they generalize to other symmetric spaces.

In *Study of Antiorbital Complexes* George Lusztig studies a problem on the support of the Fourier transform over a finite field. Special cases are related to cuspidal character sheaves and canonical bases.

A "classical" automorphic representation V of a group G corresponds to an imbedding $V \hookrightarrow L^2(G/\Gamma)$. Composing this with evaluation at identity, one obtains an automorphic distribution. The paper *Adelization of Automorphic Distributions and Mirabolic Eisenstein Series* by Stephen D. Miller and Wilfried Schmid analyzes the mirabolic Eisenstein series attached to a congruence subgroup of $GL(n, Z)$ and its associated automorphic distribution.

In an algebraic setting, Ivan Penkov and Vera Serganova propose a systematic approach to studying various categories of modules over the Lie algebras $sl(\infty)$, $o(\infty)$ and $sp(\infty)$. One of the main results of *Categories of Integrable $sl(\infty)$-, $o(\infty)$-, $sp(\infty)$-modules* states that an integrable module with finite dimensional weight spaces is semisimple. Another important result gives a description of the largest category of integrable modules which is closed under dualization and whose objects have finite Loewy lengths.

Macdonald polynomials are a far-reaching generalization of a number of important special functions in representation theory and combinatorics. The expansion coefficients of a product of two Macdonald polynomials may be regarded as generalized Littlewood-Richardson coefficients. The paper *Binomial Coefficients and Littlewood–Richardson Coefficients for Interpolation Polynomials and Macdonald Polynomials* by Siddhartha Sahi obtains explicit formulas for these coefficients by solving a more general problem involving the interpolation polynomials introduced by Knop and Sahi.

The paper *Restriction of some Representations of $U(p, q)$ to a Symmetric Subgroup* by Birgit Speh studies the restriction of derived functor modules of $U(p, q)$ to certain noncompact symmetric subgroups. It is shown that in various cases the decomposition is discrete with finite multiplicities.

Expository Papers

Contemporary Mathematics
Volume **557**, 2011

THE PLANCHEREL FORMULA, THE PLANCHEREL THEOREM, AND THE FOURIER TRANSFORM OF ORBITAL INTEGRALS

REBECCA A. HERB (UNIVERSITY OF MARYLAND) AND PAUL J. SALLY, JR. (UNIVERSITY OF CHICAGO)

ABSTRACT. We discuss various forms of the Plancherel Formula and the Plancherel Theorem on reductive groups over local fields.

Dedicated to Gregg Zuckerman on his 60th birthday

1. INTRODUCTION

The classical Plancherel Theorem proved in 1910 by Michel Plancherel can be stated as follows:

Theorem 1.1. *Let $f \in L^2(\mathbb{R})$ and define $\phi_n : \mathbb{R} \to \mathbb{C}$ for $n \in \mathbb{N}$ by*

$$\phi_n(y) = \frac{1}{\sqrt{2\pi}} \int_{-n}^{n} f(x)e^{iyx}dx.$$

The sequence ϕ_n is Cauchy in $L^2(\mathbb{R})$ and we write $\phi = \lim_{n\to\infty} \phi_n$ (in L^2). Define $\psi_n : \mathbb{R} \to \mathbb{C}$ for $n \in \mathbb{N}$ by

$$\psi_n(x) = \frac{1}{\sqrt{2\pi}} \int_{-n}^{n} \phi(y)e^{-iyx}dy.$$

The sequence ψ_n is Cauchy in $L^2(\mathbb{R})$ and we write $\psi = \lim_{n\to\infty} \psi_n$ (in L^2). Then,

$$\psi = f \text{ almost everywhere, and } \int_{\mathbb{R}} |f(x)|^2 dx = \int_{\mathbb{R}} |\phi(y)|^2 dy.$$

This theorem is true in various forms for any locally compact abelian group. It is often proved by starting with $f \in L^1(\mathbb{R}) \cap L^2(\mathbb{R})$, but it is really a theorem about square integrable functions.

There is also a "smooth" version of Fourier analysis on $\mathbb{R}$, motivated by the work of Laurent Schwartz, that leads to the Plancherel Theorem.

Definition 1.2 (The Schwartz Space). The *Schwartz space*, $\mathcal{S}(\mathbb{R})$, is the collection of complex-valued functions f on $\mathbb{R}$ satisfying:

(1) $f \in C^\infty(\mathbb{R})$.
(2) f and all its derivatives vanish at infinity faster than any polynomial. That is, $\lim_{|x|\to\infty} |x|^k f^{(m)}(x) = 0$ for all $k, m \in \mathbb{N}$.

Fact 1.3. *The Schwartz space has the following properties:*

Date: June 21, 2011.

(1) *The space $\mathcal{S}(\mathbb{R})$ is dense in $L^p(\mathbb{R})$ for $1 \le p < \infty$.*
(2) *The space $\mathcal{S}(\mathbb{R})$ is not dense in $L^\infty(\mathbb{R})$.*
(3) *The space $\mathcal{S}(\mathbb{R})$ is a vector space over $\mathbb{C}$.*
(4) *The space $\mathcal{S}(\mathbb{R})$ is an algebra under both pointwise multiplication and convolution.*
(5) *The space $\mathcal{S}(\mathbb{R})$ is invariant under translation.*

For $f \in \mathcal{S}(\mathbb{R})$, we define the Fourier transform as usual by

$$\widehat{f}(y) = \frac{1}{\sqrt{2\pi}} \int_{\mathbb{R}} f(x) e^{iyx} dx.$$

Of course, there are no convergence problems here, and we have

$$f(x) = \frac{1}{\sqrt{2\pi}} \int_{\mathbb{R}} \widehat{f}(y) e^{-iyx} dy.$$

This leads to the Plancherel Theorem for functions in $\mathcal{S}(\mathbb{R})$ by setting $\widetilde{f}(x) = \overline{f(-x)}$ and considering $f * \widetilde{f}$ at 0. Using the fact that the Fourier transform carries convolution product to function product, we have

$$\|f\|^2 = \left[f * \widetilde{f}\right](0) = \frac{1}{\sqrt{2\pi}} \int_{\mathbb{R}} \widehat{f * \widetilde{f}}(y) dy = \left\|\widehat{f}\right\|^2.$$

It is often simpler to work on the space $C_c^\infty(\mathbb{R})$ of complex-valued, compactly supported, infinitely differentiable functions on $\mathbb{R}$. However, nonzero functions in $C_c^\infty(\mathbb{R})$ do not have Fourier transforms in $C_c^\infty(\mathbb{R})$. On the other hand, the Fourier transform is an isometric isomorphism from $\mathcal{S}(\mathbb{R})$ to $\mathcal{S}(\mathbb{R})$.

The spaces $C_c^\infty(\mathbb{R})$ and $\mathcal{S}(\mathbb{R})$ can be turned into topological vector spaces so that the embedding from $C_c^\infty(\mathbb{R})$ into $\mathcal{S}(\mathbb{R})$ is continuous. However, the topology on $C_c^\infty(\mathbb{R})$ is not the relative topology from $\mathcal{S}(\mathbb{R})$. A continuous linear functional on $C_c^\infty(\mathbb{R})$ is a *distribution* on $\mathbb{R}$, and this distribution is *tempered* if it can be extended to a continuous linear functional on $\mathcal{S}(\mathbb{R})$ with the appropriate topology. This situation will arise again in our discussion of the Plancherel Formula on reductive groups.

Work on the Plancherel Formula for non-abelian groups began in earnest in the late 1940s. There were two distinct approaches. The first, for separable, locally compact, unimodular groups, was pursued by Mautner [65], Segal [84], and others. The second, for semisimple Lie groups, was followed by Gel'fand–Naimark [23], and Harish–Chandra [24], along with others. Segal's paper [84] and Mautner's paper [65] led eventually to the following statement (see [21], Theorem 7.44).

Theorem 1.4. *Let G be a separable, unimodular, type I group, and let dx be a fixed Haar measure on G. There exists a positive measure μ on $\widehat{G}$ (determined uniquely up to a constant that depends only on dx) such that, for $f \in L^1(G) \cap L^2(G)$, $\pi(f)$ is a Hilbert–Schmidt operator for μ-almost all $\pi \in \widehat{G}$, and*

$$\int_G |f(x)|^2 dx = \int_{\widehat{G}} \|\pi(f)\|_{HS}^2 d\mu(\pi).$$

Here, of course, $\widehat{G}$ denotes the set of equivalence classes of irreducible unitary representations of G.

At about the same time, Harish-Chandra stated the following theorem in his paper *Plancherel Formula for Complex Semisimple Lie Groups*.

Theorem 1.5. *Let G be a connected, complex, semisimple Lie group. Then, for $f \in C_c^\infty(G)$,*

$$f(1) = \lim_{H \to 0} \prod_{\alpha \in P} D_\alpha \overline{D_\alpha} \left[e^{\rho(H) + \overline{\rho(H)}} \int_{K \times N} f\left(u \exp(H) n u^{-1}\right) du\, dn \right].$$

An explanation of the notation here can be found in [24]. We do note two things. First of all, f is taken to be in $C_c^\infty(G)$, and the formula for $f(1)$ is the limit of a differential operator applied to what may be regarded as a Fourier inversion formula for the orbital integral over a conjugacy class of $\exp(H)$ in G. It should also be mentioned that not all irreducible unitary representations are contained in the support of the Plancherel measure for complex semisimple Lie groups. In particular, the complementary series are omitted.

In this note, we will trace the evolution of the Plancherel Formula over the past sixty years. For real groups, we observe that the original Plancherel Formula and the Fourier inversion formula ultimately became a decomposition of the Schwartz space into orthogonal components indexed by conjugacy classes of Cartan subgroups. While the distinction between the Fourier inversion formula and the decomposition of the Schwartz space might not have been clear for real semisimple Lie groups, it certainly appeared in the development of the Plancherel Theorem for reductive p-adic groups by Harish-Chandra in his paper *The Plancherel Formula for Reductive p-adic Groups* in [40]. See also the papers of Waldspurger [95] and Silberger [91], [92]. For p-adic groups, the lack of information about irreducible characters and suitable techniques for Fourier inversion has made the derivation of an explicit Plancherel Formula very difficult.

In this paper, the authors have drawn extensively on the perceptive description of Harish-Chandra's work by R. Howe, V. S. Varadarajan, and N. Wallach (see [39]). The authors would like to thank Jonathan Gleason and Nick Ramsey for their assistance in preparing this paper. We also thank David Vogan for his valuable comments on the first draft.

2. ORBITAL INTEGRALS AND THE PLANCHEREL FORMULA

Let G be a reductive group over a local field. For $\gamma \in G$, let G_γ be the centralizer of γ in G. Assume G_γ is unimodular. For f "smooth" on G, define

$$\Lambda_\gamma(f) = \int_{G/G_\gamma} f\left(x\gamma x^{-1}\right) d\dot{x},$$

with $d\dot{x}$ a G-invariant measure on G/G_γ.

Then Λ_γ is an invariant distribution on G, that is, $\Lambda_\gamma(f) = \Lambda_\gamma({}^y f)$ where ${}^y f(x) = f\left(yxy^{-1}\right)$ for $y \in G$. A major problem in harmonic analysis on reductive groups is to find the Fourier transform of the invariant distribution Λ_γ. That is, find a linear functional $\widehat{\Lambda_\gamma}$ such that

$$\Lambda_\gamma(f) = \widehat{\Lambda_\gamma}\left(\hat{f}\right),$$

where $\hat{f}$ is a function defined on the space of tempered invariant "eigendistributions" on G. This space includes the tempered irreducible characters of G along with other invariant distributions. For example, if Π is an admissible representation of G with character Θ_Π, then

$$\hat{f}(\Pi) = \mathrm{tr}(\Pi(f)) = \int_G f(x)\Theta_\Pi(x)dx.$$

The nature of the other distributions is an intriguing problem. The hope is that the Plancherel Formula for G can be obtained through some limiting process for Λ_γ.

For example, if $G = SU(1,1) \cong SL(2,\mathbb{R})$, we let

$$\gamma = \begin{bmatrix} e^{i\theta_0} & 0 \\ 0 & e^{-i\theta_0} \end{bmatrix}, \theta_0 \neq 0, \pi.$$

Then γ is a regular element in G, and $G_\gamma = \mathbb{T}$, where

$$\mathbb{T} = \left\{ \begin{bmatrix} e^{i\theta} & 0 \\ 0 & e^{-i\theta} \end{bmatrix} \Big| 0 \leq \theta < 2\pi \right\}.$$

After a simple computation, we get

$$F_f^{\mathbb{T}}(\gamma) := \left| e^{i\theta_0} - e^{-i\theta_0} \right| \Lambda_\gamma(f)$$
$$= -\frac{1}{2}\left(\pi^{(+,+)}(f) - \pi^{(+,-)}(f) \right) - \sum_{n \neq 0} \mathrm{sgn}(n)\chi_{\omega(n)}(f)e^{-in\theta_0}$$
$$+ \frac{i}{4}\left[\int_{\mathbb{R}} \pi^{(+,\nu)}(f)\frac{\sinh(\nu(\theta_0 - \pi/2))}{\sinh(\nu\pi/2)}d\nu - \int_{\mathbb{R}} \pi^{(-,\nu)}(f)\frac{\cosh(\nu(\theta_0 - \pi/2))}{\cosh(\nu\pi/2)}d\nu \right].$$

The parameter $n \neq 0$ indexes the discrete series and the parameter ν indexes the principal series representations of G. The terms $\pi^{(+,+)}(f)$ and $\pi^{(+,-)}(f)$ represent the characters of the irreducible components of the reducible principal series, and we obtain a "singular invariant eigendistribution" on G by subtracting one from the other and dividing by 2. This is exactly the invariant distribution that makes harmonic analysis work. It is called a *supertempered distribution* by Harish-Chandra.

This leads directly to the Plancherel Formula. By a theorem of Harish-Chandra, it follows that

$$\lim_{\theta \to 0}\left[\frac{1}{i}\frac{d}{d\theta}\left[F_f^{\mathbb{T}}(\gamma) \right] \right] = 8\pi f(1)$$
$$= \sum_{n \in \mathbb{Z}} |n|\chi_{\omega(n)}(f) + 1/2 \int_0^\infty \pi^{(+,\nu)}(f)\nu\coth(\pi/2\nu)d\nu$$
$$+ 1/2 \int_0^\infty \pi^{(-,\nu)}(f)\nu\tanh(\pi/2\nu)d\nu.$$

The representations of $SL(2,\mathbb{R})$ were first determined by Bargmann [8]. In his 1952 paper [25], Harish-Chandra gave hints to the entire picture for Fourier analysis on real groups. He constructed the unitary representations, computed their characters, found the Fourier transform of orbital integrals, and deduced the Plancherel Formula. This was done in about four and one-half pages.

We mention again that the support of the Fourier transform of the tempered invariant distribution Λ_γ contains not only the characters of the principal series and the discrete series, but also the tempered invariant distribution

$$\frac{1}{2}\left(\pi^{(+,+)} - \pi^{(+,-)} \right).$$

This singular invariant eigendistribution (appropriately normalized) is equal to 1 on the elliptic set and 0 off the elliptic set, thereby having no effect on harmonic analysis of the principal series.

Through the 1950s, along with an intensive study of harmonic analysis on semisimple Lie groups, Harish-Chandra analyzed invariant distributions, their Fourier transforms, and limit formulas related to these. This was mainly with reference to distributions on $C_c^\infty(G)$. He showed that G has discrete series iff G has a compact Cartan subgroup. For the rest of this section, we will assume that G has discrete series. He also suspected quite early that the irreducible unitary representations that occurred in the Plancherel Formula would be indexed by a series of representations parameterized by characters of conjugacy classes of Cartan subgroups.

In the 1960s, Harish-Chandra proved deep results about the character theory of semisimple Lie groups, in particular, the discrete series characters. In developing the Fourier analysis on a semisimple Lie group, he had to work with the smooth matrix coefficients of the discrete series. These matrix coefficients vanish rapidly at infinity, but are not compactly supported. This led to the definition of the Schwartz space $\mathcal{C}(G)$ [27]. The Schwartz space was designed to include matrix coefficients of the discrete series and slightly more. The Schwartz space is dense in $L^2(G)$, but is not contained in $L^1(G)$. Moreover, the Schwartz space $\mathcal{C}(G)$ does not contain the smooth matrix coefficients of parabolically induced representations. Nonetheless, the matrix coefficients of these parabolically induced representations are tempered distributions, that is, if m is such a matrix coefficient and $f \in \mathcal{C}(G)$, then $\int_G fm$ converges. Hence, one can consider the orthogonal complement of these matrix coefficients in $\mathcal{C}(G)$.

The collection of parabolically induced representations is indexed by non-compact Cartan subgroups of G. If H is a Cartan subgroup of G with split component A, then the centralizer L of A is a Levi subgroup of G. Now the representations corresponding to H are induced from parabolic subgroups with Levi component L, and the subspace $\mathcal{C}_H(G)$ is generated by so called wave packets associated to these induced representations. Thus, we have an orthogonal decomposition

$$\mathcal{C}(G) = \bigoplus \mathcal{C}_H(G),$$

where H runs over conjugacy classes of Cartan subgroups. When H is the compact Cartan subgroup of G, $\mathcal{C}_H(G)$ is the space of cusp forms in $\mathcal{C}(G)$. This decomposition of the Schwartz space is a version of the Plancherel Theorem for G, and it is in this form that the Plancherel Theorem appears for reductive p-adic groups.

As he approached his final version of the Plancherel Theorem and Formula for real semisimple Lie groups, Harish-Chandra presented a development of the Plancherel Formula for functions in $C_c^\infty(G)$ in his paper *Two Theorems on Semisimple Lie Groups* [28]. Here, he shows exactly how irreducible tempered characters decompose the δ distribution. In particular, for G of real rank 1, he gives an explicit formula for the Fourier transform of an elliptic orbital integral, and derives the Plancherel Formula from this. To understand the Plancherel Theorem for real groups in complete detail, one should consult the three papers [34], [35], [36], and the expository renditions of this material [30], [31], [32].

3. The Fourier Transform of Orbital Integrals, the Plancherel Formula, and Supertempered Distributions

In a paper in *Acta Mathematica* in 1973 [81], Sally and Warner re-derived, by somewhat different methods, the inversion formula that Harish-Chandra proved in his "Two Theorems" paper [28]. The purpose of the Sally–Warner paper was to explore the support of the Fourier transform of an elliptic orbital integral. To quote: "In this paper, we give explicit formulas for the Fourier transform of Λ_y, that is, we determine a linear functional $\widehat{\Lambda_y}$ such that

$$\Lambda_y(f) = \widehat{\Lambda_y}\left(\hat{f}\right), f \in C_c^\infty(G).$$

Here, we regard $\hat{f}$ as being defined on the space of tempered invariant eigendistributions on G. This space contains the characters of the principal series and the discrete series for G along with some 'singular' invariant eigendistributions whose character-theoretic nature has not yet been completely determined."

In fact, the character theoretic nature of these singular invariant eigendistributions was determined in a paper [44] by Herb and Sally in 1977. In this paper, the present authors used results of Hirai [55], Knapp–Zuckerman [60], Schmid [83], and Zuckerman [98] to show that, as in the case of $SU(1,1)$, these distributions are alternating sums of characters of limits of discrete series representations which can be embedded as the irreducible components of certain reducible principal series. In his final published paper [38], Harish-Chandra developed a comprehensive version of these singular invariant eigendistributions, and he called them "supertempered distributions." These supertempered distributions include the characters of discrete series along with some finite linear combinations of irreducible tempered elliptic characters that arise from components of reducible generalized principal series. This situation has already been illustrated for $SL(2,\mathbb{R})$ in Section 2 of this paper. One notable fact about supertempered distributions is that they appear discretely in the Fourier transforms of elliptic orbital integrals; hence they play an essential role in the study of invariant harmonic analysis. For the remainder of this section, we present a collection of results of the first author related to Fourier inversion and the Plancherel Theorem for real groups.

In order to explain the steps needed to derive the Fourier transform for orbital integrals in general, we first look in more detail at the case that G has real rank one. In this case G has at most two non-conjugate Cartan subgroups: a non-compact Cartan subgroup H with vector part A of dimension one, and possibly a compact Cartan subgroup T. We assume for simplicity that G is acceptable, that is, the half-sum of positive roots (denoted ρ) exponentiates to give a well defined character on T. The characters Θ_τ^T of the discrete series representations are indexed by $\tau \in \hat{T}'$, the set of regular characters of T, and the characters Θ_χ^H of the principal series are indexed by characters $\chi \in \hat{H}$. In addition, for $f \in C_c^\infty(G)$ we have invariant integrals $F_f^T(t), t \in T$, and $F_f^H(a), a \in H$. These are normalized versions of the orbital integrals $\Lambda_\gamma(f), \gamma \in G$, which have better properties as functions on the Cartan subgroups.

The analysis on the non-compact Cartan subgroup is elementary. First, as functions on G', the set of regular elements of G, the principal series characters are supported on conjugates of H. In addition, for $\chi \in \hat{H}$ and $a \in H' = H \cap G'$, $\Theta_\chi^H(a)$ is given by a simple formula in terms of $\chi(a)$. As a result it is easy to show that the abelian Fourier transform $\hat{F}_f^H(\chi), \chi \in \hat{H}$, is equal up to a constant to $\Theta_\chi^H(f)$, the principal series character evaluated at f. Finally, $F_f^H \in C_c^\infty(H)$, and so the abelian Fourier inversion formula on H yields an expansion

$$(3.1) \qquad F_f^H(a) = c_H \int_{\hat{H}} \epsilon(\chi)\overline{\chi(a)}\Theta_\chi^H(f)d\chi, a \in H,$$

where c_H is a constant depending on normalizations of measures and $\epsilon(\chi) = \pm 1$.

The situation on the compact Cartan subgroup is more complicated. There are three main differences. First, for $\tau \in \hat{T}'$ and $t \in T' = T \cap G'$, $\Theta_\tau^T(t)$ is given by a simple formula in terms of the character $\tau(t)$. However, Θ_τ^T is also non-zero on H'. Thus for $\tau \in \hat{T}'$ and $f \in C_c^\infty(G)$, the abelian Fourier coefficient $\hat{F}_f^T(\tau)$ is equal up to a constant to $\Theta_\tau^T(f)$ plus an error term which is an integral over H of F_f^H times the numerator of Θ_τ^T. Second, the singular characters $\tau_0 \in \hat{T}$ do not correspond to discrete series characters. They do however parameterize singular invariant eigendistributions $\Theta_{\tau_0}^T$, and $\hat{F}_f^T(\tau_0)$ can be given in terms of $\Theta_{\tau_0}^T(f)$. Finally, F_f^T is smooth on T', but has jump discontinuities at singular elements. Because of this there are convergence issues when the abelian Fourier inversion formula is used to expand F_f^T in terms of its Fourier coefficients.

Sally and Warner were able to compute the explicit Fourier transform of F_f^T in the rank one situation where discrete series character formulas on the non-compact Cartan subgroup were known. The resulting formula is very similar to the one for the special case of $SU(1,1)$ given in the previous section. The discrete series characters and singular invariant eigendistributions occur discretely in a sum over $\hat{T}$ and the principal series characters occur in an integral over $\hat{A}$ with hyperbolic sine and cosine factors. They were also able to differentiate the resulting formula to obtain the Plancherel Formula.

The key to computing an explicit Fourier transform for orbital integrals in the general case is an understanding of discrete series character formulas on non-compact Cartan subgroups. Thus we briefly review some of these formulas. The results are valid for any connected reductive Lie group, but we assume for simplicity of notation that G is acceptable. A detailed expository account of all results about discrete series characters presented in this section is given in [53].

Assume that G has discrete series representations, and hence a compact Cartan subgroup T, and identify the character group of T with a lattice $L \subset E = i\mathbf{t}^*$. For each $\lambda \in E$, let $W(\lambda) = \{w \in W : w\lambda = \lambda\}$ where W is the full complex Weyl group, and let $E' = \{\lambda \in E : W(\lambda) = \{1\}\}$. Then each $\lambda \in L' = L \cap E'$ is regular, and corresponds to a discrete series character Θ_λ^T. For $t \in T'$, we have the simple character formula

$$(3.2) \qquad \Theta_\lambda^T(t) = \epsilon(E^+)\Delta(t)^{-1} \sum_{w \in W_K} \det(w)e^{w\lambda}(t),$$

where Δ is the Weyl denominator, W_K is the subgroup of W generated by reflections in the compact roots, and $\epsilon(E^+) = \pm 1$ depends only on the connected component (Weyl chamber) E^+ of E' containing λ.

Now assume that H is a non-compact Cartan subgroup of G, and let H^+ be a connected component of H'. Then for $h \in H^+$,

$$(3.3) \qquad \Theta_\lambda^T(h) = c(H^+)\epsilon(E^+)\Delta(h)^{-1} \sum_{w \in W} \det(w)c(w\colon E^+\colon H^+)\xi_{w,\lambda}(h),$$

where $c(H^+)$ is an explicit constant given as a quotient of certain Weyl groups and the $c(w\colon E^+\colon H^+)$ are integer constants depending only on the data shown in the notation. The sum is over the full complex Weyl group W, and for w such that $c(w\colon E^+\colon H^+)$ is potentially non-zero, $\xi_{w,\lambda}$ is a character of H obtained from w and λ using a Cayley transform. This formula is a restatement of results of Harish-Chandra in [26]. In that paper, Harish-Chandra gave properties of the constants $c(w\colon E^+\colon H^+)$ which characterize them completely. These properties can in theory be used to determine the constants by induction on the dimension of the vector component of H. This easily yields formulas when this dimension is one or two, but quickly becomes cumbersome for higher dimensions.

With the above notation, it is easy to describe the singular invariant eigendistributions corresponding to $\lambda \in L^s = L \backslash L'$. Let $\lambda_0 \in L^s$, and let E^+ be a chamber whose closure contains λ_0. The exponential terms $\xi_{w,\lambda_0}(h), h \in H^+$, still make sense, and the "limit of discrete series" $\Theta^T_{\lambda_0, E^+} = \lim_{\lambda \to \lambda_0, \lambda \in L \cap E^+} \Theta^T_\lambda$ is given by (3.3) using the constants $c(w : E^+ : H^+)$. Zuckerman [98] showed that the limits of discrete series are the characters of tempered unitary representations of G. The singular invariant eigendistribution corresponding to λ_0 is the alternating sum of the limits of discrete series taken over all chambers with closures containing λ_0:

$$(3.4) \qquad \Theta^T_{\lambda_0} = [W(\lambda_0)]^{-1} \sum_{w \in W(\lambda_0)} \det w \; \Theta^T_{\lambda_0, wE^+}.$$

The main results of [45] are as follows. Let $\Phi(\lambda_0)$ denote the roots of T which are orthogonal to λ_0. Then $\Theta^T_{\lambda_0}$ vanishes if $\Phi(\lambda_0)$ contains any compact roots. Thus we may as well assume that all roots in $\Phi(\lambda_0)$ are non-compact. By using Cayley transforms with respect to the roots of $\Phi(\lambda_0)$ we obtain a Cartan subgroup H and corresponding cuspidal Levi subgroup M. Because the Cayley transform of λ_0 is regular with respect to the roots of H in M, it determines a discrete series character of M, which can then be parabolically induced to obtain a unitary principal series character $\Theta^H_{\lambda_0}$ of G.

Theorem 3.5 (Herb–Sally). $\Theta^H_{\lambda_0} = \sum_{w \in W(\lambda_0)} \Theta^T_{\lambda_0, wE^+}.$

It follows from Knapp [59] that $\Theta^H_{\lambda_0}$ has at most $[W(\lambda_0)]$ irreducible components. Thus each limit of discrete series character is irreducible, and $\Theta^T_{\lambda_0}$ is the alternating sum of the characters of the irreducible constituents of $\Theta^H_{\lambda_0}$.

In [46], Herb used the methods of Sally and Warner, and the discrete series character formulas of Harish-Chandra, to obtain a Fourier inversion formula for orbital integrals for groups of arbitrary real rank. As in the rank one case, for any Cartan subgroup H of G we have normalized orbital integrals $F^H_f(h), h \in H, f \in C_c^\infty(G)$. We also have characters $\Theta^H_\chi, \chi \in \hat{H}$. If H is compact, these are discrete series characters for regular χ and singular invariant eigendistributions for singular χ. If H is non-compact, corresponding to the Levi subgroup M, then they are parabolically induced from discrete series or singular invariant eigendistributions on M. Using standard character formulas for parabolic induction, these characters can also be written using Harish-Chandra's discrete series formulas for M.

Fix a Cartan subgroup H_0. The goal is to find a formula

$$(3.6) \qquad F^{H_0}_f(h_0) = \sum_H \int_{\hat{H}} \Theta^H_\chi(f) K^H(h_0, \chi) d\chi, \; h_0 \in H_0',$$

where H runs over a set of representatives of conjugacy classes of Cartan subgroups of G, $d\chi$ is Haar measure on $\hat{H}$, and $K^H(h_0, \chi)$ is a function depending on h_0 and χ. The problem is to compute the functions $K^H(h_0, \chi)$, or at least show they exist.

As in the rank one case, for $\chi_0 \in \hat{H}_0$ and $f \in C_c^\infty(G)$, the abelian Fourier coefficient $\hat{F}_f^{H_0}(\chi_0)$ is equal up to a constant to $\Theta_{\chi_0}^{H_0}(f)$ plus an error term for each of the other Cartan subgroups. The error term corresponding to H is an integral over H of the numerator of $\Theta_{\chi_0}^{H_0}$ times F_f^H. Because $\Theta_{\chi_0}^{H_0}$ is parabolically induced, its character is non-zero only on Cartan subgroups of G which are conjugate to Cartan subgroups of M_0, the corresponding Levi subgroup. Thus the error term will be identically zero unless H can be conjugated into M_0, but is not conjugate to H_0. This implies in particular that the vector dimension of H is strictly greater than that of H_0. Thus if H_0 is maximally split in G there are no error terms. However, if $H_0 = T$ is compact, then $M_0 = G$ and all non-compact Cartan subgroups contribute error terms.

Let H be a Cartan subgroup of M_0 which is not conjugate to H_0 and let M be the corresponding Levi subgroup. In analyzing the error term corresponding to H, we obtain a primary term involving the characters $\Theta_\chi^H(f), \chi \in \hat{H}$, plus secondary error terms, one for each Cartan subgroup of M not conjugate to H. This leads to messy bookkeeping, but the process eventually terminates since the vector dimension of the Cartan subgroups with non-zero error terms increases strictly at each step.

In particular, if H is a Cartan subgroup of G not conjugate to a Cartan subgroup of M_0, then it never occurs in a non-zero error term and K^H is identically zero. Our original Cartan subgroup H_0 also is not involved in any error term, and we have

$$(3.7) \qquad K^{H_0}(h_0, \chi_0) = c_{H_0} \epsilon(\chi_0) \overline{\chi_0(h_0)}, h_0 \in H_0', \chi_0 \in \hat{H}_0.$$

The formulas for K^H become progressively more complicated as the vector dimension of H increases. In particular, if H is maximally split in G, then K^H has contributions from error terms at many different steps.

Aside from the proliferation of error terms, the analysis which will lead to the functions $K^H(h_0, \chi)$ involves two main problems that do not occur in real rank one. The main problem is that the final formulas contain the unknown integer constants $c(w: E^+: H^+)$ appearing in discrete series character formulas. These occur in complicated expressions which can be interpreted as Fourier series in several variables. These series are not absolutely convergent and have no obvious closed form. Thus although [45] showed the existence of the functions $K^H(h_0, \chi)$, it does not result in a formula which is suitable for applications. In particular, it cannot be differentiated to obtain the Plancherel Formula for G. Second, in the rank one case the analysis can be carried out for any $h \in H'$. However there are cases in higher rank, for example the real symplectic group of real rank three, in which certain integrals diverge for some elements $h \in H'$. However, the analysis is valid on a dense open subset of H'.

In order to improve these results and obtain a satisfactory Fourier inversion formula similar to that of Sally and Warner for rank one groups, it was necessary to have more information about the discrete series constants. The first of these improvements came from a consideration of stable discrete series characters and stable orbital integrals.

Assume that G has a compact Cartan subgroup T, and use the notation from the earlier discussion of discrete series characters. For $\lambda \in L$ we define

$$(3.8) \qquad \Theta_\lambda^{T,st} = [W_K]^{-1} \sum_{w \in W} \Theta_{w\lambda}^T.$$

If $\lambda \in L'$, then $\Theta_\lambda^{T,st}$ is called a stable discrete series character. For $\lambda \in L^s$, we have $\Theta_\lambda^{T,st} = 0$. Similarly we define the stable orbital integral

$$(3.9) \qquad \Lambda_t^{st}(f) = \sum_{w \in W} \Lambda_{wt}(f), f \in C_c^\infty(G), t \in T'.$$

If we normalize the orbital integral as usual, we have

$$(3.10) \qquad F_f^{T,st}(t) = \Delta(t)\Lambda_t^{st}(f) = \sum_{w \in W} \det(w) F_f^T(wt).$$

Similarly, for any Cartan subgroup H with corresponding Levi subgroup M there is a series of stable characters $\Theta_\chi^{H,st}, \chi \in \hat{H}$, induced from stable discrete series characters of M. We also obtain stable orbital integrals by averaging over the complex Weyl group of H in M.

Recall that there is a differential operator Π such that

$$(3.11) \qquad f(1) = \lim_{t \to 1, t \in T'} \Pi F_f^T(t).$$

Since the differential operator Π transforms by the sign character of W, it follows immediately that we also have

$$(3.12) \qquad f(1) = [W]^{-1} \lim_{t \to 1, t \in T'} \Pi F_f^{T,st}(t).$$

The advantage of stabilizing is that the formulas for the stable discrete series characters on the non-compact Cartan subgroups are simpler than those of the individual discrete series characters. The Fourier inversion formula for stable orbital integrals involves only these stable characters and has the general form

$$(3.13) \qquad F_f^{T,st}(t) = \sum_H \int_{\hat{H}} \Theta_\chi^{H,st}(f) K^{H,st}(t,\chi) d\chi, \ t \in T'.$$

When G has real rank one the Fourier inversion formulas for the stable orbital integrals are no simpler than those obtained by Sally and Warner. However when G has real rank two there is already significant simplification, and Sally's student Chao [13] was able to obtain expressions for the functions $K^{H,st}(t,\chi)$ in closed form and differentiate them to obtain the Plancherel Formula.

Herb [47], [48] then developed the theory of two-structures and showed that the constants occurring in stable discrete series character formulas for any group can be expressed in terms of constants occurring in the stable discrete series character formulas for the group $SL(2,\mathbb{R})$ and the rank two symplectic group $Sp(4,\mathbb{R})$. As a consequence she was able to write each function $K^{H,st}(t,\chi)$ occurring in (3.13) as a product of factors which occur in the corresponding formulas for $SL(2,\mathbb{R})$ and $Sp(4,\mathbb{R})$.

This formula can be differentiated to yield the Plancherel Formula. However, the Fourier inversion formulas for stable orbital integrals are of independent interest, and much of the complexity of these distributions is lost when they are differentiated and evaluated at $t = 1$. In particular the functions occurring in the Plancherel Formula, which had already been obtained by different methods by Harish-Chandra

[36], reduce to a product of rank one factors which occur in the Plancherel Formula for $SL(2,\mathbb{R})$. The discrete series character formulas and Fourier inversion formula for $F_f^{T,st}(t)$ require both $SL(2,\mathbb{R})$ and $Sp(4,\mathbb{R})$ type factors coming from the theory of two-structures.

In [49] Herb was able to use Shelstad's ideas on endoscopy to obtain explicit Fourier inversion formulas for the individual (not stabilized) orbital integrals. The idea is that certain weighted sums of orbital integrals, $\Lambda_\gamma^\kappa(f)$, correspond to stable orbital integrals on endoscopic groups. Thus their Fourier inversion formulas can be computed as in [48]. This is done for sufficiently many weights κ that the original orbital integrals $\Lambda_\gamma(f)$ can be recovered. Again, the theory of two-structures was important, and the functions $K^H(h_0, \chi)$ occurring in (3.6) can be given in closed form using products of terms coming from the groups $SL(2,\mathbb{R})$ and $Sp(4,\mathbb{R})$.

Although this gave a satisfactory Fourier inversion formula, the derivation is complicated by the use of stability and endoscopy. Stability and endoscopy also combined to yield explicit, but cumbersome, formulas for the discrete series constants $c(w\colon E^+\colon H^+)$ occurring in (3.3). In [52], Herb found simpler formulas for these constants that bypass the theories of stability and endoscopy, and are easier to prove independently of these results. Using special two-structures called two-structures of non-compact type, she obtained a formula for the constants $c(w\colon E^+\colon H^+)$ directly in terms of constants occurring in discrete series character formulas for $SL(2,\mathbb{R})$ and $Sp(4,\mathbb{R})$. These formulas could be used to give a direct and simpler proof of the Fourier inversion formulas for orbital integrals given in [49].

4. The p-adic Case

We now focus on the representation theory and harmonic analysis of reductive p-adic groups. Since the 1960s, there has been a flurry of activity related to these groups. Some of this has been generated by the so-called "Langlands Program" (see Jacquet–Langlands [57] and Langlands [63]). However, a number of results in representation theory and harmonic analysis were completed well before this activity related to the Langlands Program by Bruhat [9], Satake [82], Gel'fand–Graev [22], and Macdonald [64]. Of particular interest were the results of Mautner [66] that gave the first construction of supercuspidal representations. Here, a supercuspidal representation is an infinite-dimensional, irreducible, unitary representation with compactly supported matrix coefficients (mod the center). In the mid-1960s, for a p-adic field F with odd residual characteristic, all supercuspidal representations for $SL(2,F)$ were constructed by Shalika [88], and for $PGL(2,F)$ by Silberger [90]. These two were Mautner's Ph.D. students. At roughly the same time, Shintani [89] constructed some supercuspidal representations for the group of $n \times n$ matrices over F whose determinant is a unit in the ring of integers of F. Shintani also proved the existence of a Frobenius-type formula for computing supercuspidal characters as induced characters. Incidentally, in 1967–1968, the name "supercuspidal" had not emerged, and these representations were called "absolutely cuspidal," "compactly supported discrete series," and other illustrative titles.

We also note that, in this same period, Sally and Shalika computed the characters of the discrete series of $SL(2,F)$ as induced characters [77] (see also [2]), derived the Plancherel Formula for $SL(2,F)$ [78], and developed an explicit Fourier transform for elliptic orbital integrals in $SL(2,F)$ [80]. This Fourier transform led directly to the Plancherel Formula through the use of the Shalika germ expansion [87].

The guide for this progression of results was the 1952 paper of Harish-Chandra on $SL(2,\mathbb{R})$ [25].

In the autumn of 1969, Harish-Chandra presented his first complete set of notes on reductive p-adic groups [29]. These are known as the "van Dijk Notes". These notes appear to be the origin of the terms "supercusp form" and "supercuspidal representation". They present a wealth of information about supercusp forms, discrete series characters, and other related topics. At the end of the introduction, Harish-Chandra states the following: "Of course the main goal here is the Plancherel Formula. However, I hope that a correct understanding of this question would lead us in a natural way to the discrete series for G. (This is exactly what happens in the real case. But the p-adic case seems to be much more difficult here.)" It seems that that Harish-Chandra favored the prefix "super" as in "supercusp form," "supertempered distribution," etc.

We now proceed to the description of Harish-Chandra's Plancherel Theorem (see [40]) and Waldspurger's exposition of Harish-Chandra's ideas [95]. We then give an outline of the current state of knowledge of the discrete series of reductive p-adic groups and their characters. Finally, we give details (as currently known) of the Plancherel Formula and the Fourier transform of orbital integrals.

The background for Harish-Chandra's Plancherel Theorem was developed in his Williamstown lectures [33]. He showed that, using the philosophy of cusp forms, one could prove a formula similar to that for real groups that we outlined in Section 2. He was able to do this despite the lack of information about the discrete series and their characters.

Following the model of real groups, for each special torus A, Harish-Chandra constructed a subspace $\mathcal{C}_A(G)$ from the matrix coefficients of representations corresponding to A. These representations are parabolically induced from relative discrete series representations of M, the centralizer of A. There are two notable differences between the real case and the p-adic case. First of all, because, in the p-adic case, there are discrete series that are not supercuspidal (for example, the Steinberg representation of $SL(2,F)$), the theory of the constant term must be modified. Second, because of a compactness condition on the dual of A, it is not necessary to consider the asymptotics of the Plancherel measure that are required in the real case because of non-compactness.

Thus, even though the understanding of the discrete series and their characters for p-adic groups is quite rudimentary, Harish-Chandra succeeded in proving a version of the Plancherel Theorem. This version, as stated by Howe [39], is: "The (Schwartz) space $\mathcal{C}(G)$ is the orthogonal direct sum of wave packets formed from series of representations induced unitarily from discrete series of (the Levi components of) parabolic subgroups P. Moreover if two such series of induced representations yield the same subspace of $\mathcal{C}(G)$, then the parabolics from which they are induced are associate, and the representations of the Levi components are conjugate." Equivalently, as stated by Harish-Chandra (Lemma 5 of *The Plancherel Formula for Reductive p-adic Groups* in [40]), if G is a connected reductive p-adic group and $\mathcal{C}(G)$ is the Schwartz space of G, then

$$\mathcal{C}(G) = \sum_{A \in S} \mathcal{C}_A(G)$$

where S is the set of conjugacy classes of special tori in G and the sum is orthogonal.

In 2002, Waldspurger produced a carefully designed version of Harish-Chandra's Plancherel Theorem. This work is executed with remarkable precision, and we quote here from Waldspurger's introduction (the translation here is that of the authors of the present article).

"The Plancherel formula is an essential tool of invariant harmonic analysis on real or p-adic reductive groups. Harish-Chandra dedicated several articles to it. He first treated the case of real groups, his last article on this subject being [36]. A little later, he proved the formula in the p-adic case. But he published only a summary of these results [40]. The complete proof was to be found in a hand-written manuscript that was hardly publishable in that state. Several years ago, L. Clozel and the present author conceived of a project to publish these notes. This project was not realized, but the preparatory work done on that occasion has now become the text that follows. It is a redaction of Harish-Chandra's proof, based on the unpublished manuscript.

. . .

As this article is appearing more than fifteen years after Harish-Chandra's manuscript, we had the choice between scrupulously respecting the original or introducing several modifications taking account of the evolution of the subject in the meantime. We have chosen the latter option. As this choice is debatable and the fashion in which we observe the subject to have evolved is rather subjective, let us attempt to explain the modifications that we have wrought.

There are several changes of notation: we have used those which seemed to us to be the most common and which have been used since Arthur's work on the trace formula. We work on a base field of any characteristic, positive characteristic causing only the slightest disturbance. We have eliminated the notion of the Eisenstein integral in favor of the equivalent and more popular coefficient of the induced representation. We have used the algebraic methods introduced by Bernstein. They allow us to demonstrate more naturally that certain functions are polynomial or rational, where Harish-Chandra proved their holomorphy or meromorphy. At the end of the article, we have slightly modified the method of extending the results obtained for semi-simple groups to reductive groups, in particular, the manner in which one treats the center. In fact, the principal change concerns the 'constant terms' and the intertwining operators. Harish-Chandra began with the study of the 'constant terms' of the coefficients of the induced representations and deduced from this study the properties of the intertwining operators. These latter having seemed to us more popular than the 'constant terms,' we have inverted the order, first studying the intertwining operators, in particular their rational extension, and having deduced from this the properties of the 'constant terms.' All of these modifications remain, nevertheless, minor and concern above all the preliminaries. The proof of the Plancherel formula itself (sections VI, VII and VIII below) has not been altered and is exactly that of Harish-Chandra."

It remains to address the current status of the three central problems of harmonic analysis on reductive p-adic groups. These are the construction of the discrete series, the determination of the characters of the discrete series, and the derivation of the

Fourier transform of orbital integrals as linear functionals on the space of tempered irreducible characters and supertempered distributions.

There is a long list of authors who have attacked the construction of discrete series of p-adic groups over the past forty years. We limit ourselves to a few of the major stepping stones. The work of Howe [56] on $GL(n)$ in the tame case set the stage for a great deal of the future work. Howe's supercuspidal representations for $GL(n)$ were proved to be exhaustive by Moy in [70]. Further work in the direction of tame supercuspidals may be found in the papers [68] and [69] of L. Morris.

In the mid 1980s, Bushnell and Kutzko attacked $GL(n)$ in the wild case. Their main weapon was the theory of types, and the definitive results for $GL(n)$ and $SL(n)$ were published in [10], [11], and [12]. While in the tame case, one gets a reasonable parameterization in terms of characters of tori, it does not seem that such a parameterization can be expected in the wild case. It is difficult to associate certain characters with any particular torus, as well as to tell when representations constructed from different tori are distinct. We also mention the work of Corwin on division algebras in both the tame [14] and the wild [15] case.

A big breakthrough came in J.-K. Yu's construction of tame supercuspidal representations for a wide class of groups in [97]. In this paper, Yu points to the fact that he was guided by the results of Adler [1] at the beginning of this undertaking. Under certain restrictions on p, Yu's supercuspidal representations were proved to be exhaustive by Ju-Lee Kim [58] using tools from harmonic analysis in a remarkable way. Throughout this period, the work of Moy–Prasad [71], [72] was quite influential. Also, Stevens [94] succeeded in applying the Bushnell–Kutzko methods to the classical groups to obtain all their supercuspidal representations as induced representations when the underlying field has odd residual characteristic. Finally, major results have been obtained by Mœglin and Tadic for non-supercuspidal discrete series in [67]. There is still much work to be done, but considerable progress has been made.

The theory of characters has been slower in its development. There are two avenues of approach that have been cultivated. The first is the local character expansion of Harish-Chandra. If $\mathcal{O}$ is a G-orbit in $\mathfrak{g}$, then $\mathcal{O}$ carries a G-invariant measure denoted by $\mu_{\mathcal{O}}$ (see, for example, [76]). The Fourier transform of the distribution $f \mapsto \mu_{\mathcal{O}}(f)$ is represented by a function $\widehat{\mu_{\mathcal{O}}}$ on $\mathfrak{g}$ that is locally summable on the set of regular elements $\mathfrak{g}'$ in $\mathfrak{g}$. The local character expansion is:

Theorem 4.1. *Let π be an irreducible smooth representation of G. There are complex numbers $c_{\mathcal{O}}(\pi)$, indexed by nilpotent orbits $\mathcal{O}$, such that*

$$\Theta_\pi(\exp Y) = \sum_{\mathcal{O}} c_{\mathcal{O}}(\pi)\widehat{\mu_{\mathcal{O}}}(Y)$$

for Y sufficiently near 0 in $\mathfrak{g}'$.

This result is presented in Harish-Chandra's Queen's Notes [37] and is fully explicated in [41]. The local character expansion could be a very valuable tool if three problems are overcome. These are: (1) determine the functions $\widehat{\mu_{\mathcal{O}}}$, (2) find the constants $c_{\mathcal{O}}(\pi)$, and (3) determine the domain of validity of the expansion. For progress in these directions, see Murnaghan [73], [74], Waldspurger [96], DeBacker–Sally [19], and DeBacker [17].

The second approach is the direct use of the Frobenius formula for induced characters to produce full character formulas on the regular elements in G. See

Harish-Chandra [29] (p. 94), Sally [79], and Rader–Silberger [75]. This approach has been used by DeBacker for $GL(\ell)$, ℓ a prime [20], and Spice for $SL(\ell)$, ℓ a prime [93]. Recent work of Adler and Spice [3] and DeBacker and Reeder [18] shows some promise in this direction, but their results are still quite limited. The paper [3] of Adler and Spice gives an interesting report on the development and current status of character theory on reductive p-adic groups. For additional results on the theory of characters, consult the papers of Cunningham and Gordon [16] and Kutzko and Pantoja [61].

We finish this paper with an update on the Plancherel Theorem, the Plancherel Formula, and the Fourier transform of orbital integrals in the p-adic case. As regards the Plancherel Theorem, it seems that some flesh is beginning to appear on the bones. Thus, for some special cases, an explicit Plancherel measure related to the components in the Schwartz space decomposition has been found (see Shahidi [85], [86], Kutzko–Morris [62], and Aubert–Plymen [6], [7]). The results seem to be applicable mainly to $GL(n)$ and $SL(n)$. In some cases, restrictions on the residual characteristic have been completely avoided. These methods seem to a great extent to be independent of explicit character formulas. It would be interesting to determine how far these techniques can be carried for general reductive p-adic groups. For this, one should consult the papers [42] and [43] of Heiermann.

It is one of the purposes of this paper to point out the nature of the Plancherel Formula in the theory of harmonic analysis on reductive p-adic groups. As was the case originally with Harish-Chandra, the Plancherel Formula should be considered as the Fourier transform of the δ distribution regarded as an invariant distribution on a space of smooth functions on the underlying group. This is achieved in the real case by determining the Fourier transform of an elliptic orbital integral and applying a limit formula involving differential operators to deduce an expression for $f(1)$ as a linear functional on the space of tempered invariant distributions. This space is directly connected to the space of tempered irreducible characters of G along with some additional supertempered virtual characters. It appears to be the case that, to accomplish this goal, one has to have a full understanding of the irreducible tempered characters of G. This, of course, requires a detailed knowledge of the discrete series. This is exactly the approach that was detailed in Section 3.

As pointed out by Harish-Chandra, a complete knowledge of the discrete series and their characters would yield the Plancherel measure for p-adic groups exactly as in the real case. In the p-adic case, the role of differential operators in the limit formula to obtain $f(1)$ is assumed by the *Shalika germ expansion*.

For a connected semi-simple p-adic group G, Shalika defines in [87]

$$I_f(x) = \int_{G(x)} f d\mu,$$

where x is a regular element in G, $G(x)$ is its conjugacy class, μ is a G-invariant measure on $G(x)$, and $f \in C_c^\infty(G)$. Shalika shows that $I_f(x)$ has an asymptotic expansion in terms of the integrals

$$\Lambda_{\mathcal{O}}(f) = \int_{\mathcal{O}} f d\mu$$

of f over the unipotent conjugacy classes $\mathcal{O}$. Here, for $\mathcal{O} = \{1\}$, we take $\Lambda_{\mathcal{O}}(f) = f(1)$. The coefficients $C_{\mathcal{O}}(x)$ occurring in this expansion are called the *Shalika germs*.

We start with $G = SL(2, F)$ where F has odd residual characteristic, and then use Shalika germs to produce the Plancherel Formula for G. This result of Sally and Shalika was proved in 1969 and is presented in detail in [80]. We repeat it here to indicate the role that such a formula can play in the harmonic analysis on a reductive p-adic group.

Let T be a compact Cartan subgroup of G. For each nontrivial unipotent orbit $\mathcal{O}$, there is a subset $T_{\mathcal{O}}$ of the set of regular elements in T such that the following asymptotic expansion holds.

$$F_f^T(t) = |D(t)|^{1/2} I_f(t) \sim -A_T |D(t)|^{1/2} f(1) + B_T \sum_{\dim \mathcal{O} > 0} C_{\mathcal{O}}(t) \Lambda_{\mathcal{O}}(f)$$

where the Shalika germ $C_{\mathcal{O}}(t)$ is the characteristic function of $T_{\mathcal{O}}$. The constants A_T and B_T depend on normalization of measures and whether T is ramified or unramified.

By summing products of characters, we are led to the following expression.

$$\begin{aligned}
\mu(T) I_f(t) &= \sum_{\Pi \in D} \overline{\chi_{\Pi}(t)} \hat{f}(\Pi) + \frac{1}{2} \sum_{\Pi \in \mathrm{RPS}_V} \overline{\chi_{\Pi}(t)} \hat{f}(\Pi) \\
&\quad - \frac{q+1}{2q} \mu(A_1) \int_{\substack{\xi \in \widehat{F^{\times}} \\ \xi | A_{h_0+1} = 1}} |\Gamma(\xi)|^{-2} \hat{f}(\xi) d\xi \\
&\quad + \frac{q}{2} \mu(A_1) \kappa_T |D(t)|^{-1/2} \int_{\substack{\xi \in \widehat{F^{\times}} \\ \xi | A_{h_0+1} = 1}} \hat{f}(\xi) d\xi
\end{aligned}$$

This is the Fourier transform of the elliptic orbital integral corresponding to the regular element t. Note the occurrence of the characters of the reducible principal series, denoted RPS_V, corresponding to the three sgn characters on $F^{\times}$. As in the case of $SL(2, \mathbb{R})$, each represents the difference of two characters divided by 2, and that difference is 0 except on the compact Cartan subgroups corresponding to the sgn character associated to the quadratic extension V. So again, these singular tempered invariant distributions (see [54]) appear in the Fourier transform of an elliptic orbit.

Using the Shalika germ expansion, we are led directly to the Plancherel Formula for $SL(2, F)$:

$$\mu(K) f(1) = \sum_{\Pi \in D} \hat{f}(\Pi) d(\Pi) + \frac{1}{2} \left(\frac{q^2 - 1}{q} \right) \mu(A_1) \int_{\xi \in \widehat{F^{\times}}} |\Gamma(\xi)|^{-2} \hat{f}(\xi) d\xi$$

The results above for $SL(2)$ indicate the major themes of this paper. We have attempted to illustrate the distinction among three basic objects in harmonic analysis on reductive groups over a local field. First of all, there is the Plancherel Theorem that gives a decomposition of the Schwartz space into a collection of pairwise-orthogonal components parameterized by characters of conjugacy classes of Cartan subgroups in G. This was carried out by Harish-Chandra for both real and p-adic groups, along with a careful rendering by Waldspurger in the p-adic case. Secondly, there is the Plancherel Formula, that is, the computation of the Fourier transform of the δ distribution as a linear functional on the space of irreducible tempered characters. With knowledge of the discrete-series characters, this can be accomplished directly as in [78]. Finally, there is the derivation of the Fourier transform of elliptic orbital integrals as linear functionals on the space of tempered irreducible characters and supertempered distributions. For real groups, this is carried out in Section 3. With this in hand, we can use a suitable "limit formula" to obtain the Plancherel Formula.

It is clear that, in the p-adic case, a complete theory of the Fourier transform of orbital integrals would lead to direct results about lifting, matching, and transferring orbital integrals. Such a theory would entail a deep understanding of discrete series characters and their properties. A start in this direction may be found in papers of Arthur [4], [5] and Herb [50], [51]. We expect to return to this subject in the near future.

REFERENCES

[1] Jeffrey D. Adler. Refined anisotropic K-types and supercuspidal representations. *Pacific J. Math.*, 185(1):1–32, 1998.

[2] Jeffrey D. Adler, Stephen DeBacker, Paul J. Sally, Jr., and Loren Spice. Supercuspidal characters of SL_2 over a p-adic field. *To appear in Harmonic Analysis on reductive, p-adic groups (Contemp. Math.).*

[3] Jeffrey D. Adler and Loren Spice. Supercuspidal characters of reductive p-adic groups. *Amer. J. Math.*, 131(4):1137–1210, 2009.

[4] James Arthur. On elliptic tempered characters. *Acta Math.*, 171(1):73–138, 1993.

[5] James Arthur. On the Fourier transforms of weighted orbital integrals. *J. Reine Angew. Math.*, 452:163–217, 1994.

[6] Anne-Marie Aubert and Roger Plymen. Explicit Plancherel formula for the p-adic group $GL(n)$. *C. R. Math. Acad. Sci. Paris*, 338(11):843–848, 2004.

[7] Anne-Marie Aubert and Roger Plymen. Plancherel measure for $GL(n, F)$ and $GL(m, D)$: explicit formulas and Bernstein decomposition. *J. Number Theory*, 112(1):26–66, 2005.

[8] V. Bargmann. Irreducible unitary representations of the Lorentz group. *Ann. of Math. (2)*, 48:568–640, 1947.

[9] François Bruhat. Sur les réprésentations des groupes classiques P-adiques. I, II. *Amer. J. Math.*, 83:321–338, 343–368, 1961.

[10] Colin J. Bushnell and Philip C. Kutzko. *The admissible dual of* $GL(N)$ *via compact open subgroups*, volume 129 of *Annals of Mathematics Studies*. Princeton University Press, Princeton, NJ, 1993.

[11] Colin J. Bushnell and Philip C. Kutzko. The admissible dual of $SL(N)$. I. *Ann. Sci. École Norm. Sup. (4)*, 26(2):261–280, 1993.

[12] Colin J. Bushnell and Philip C. Kutzko. The admissible dual of $SL(N)$. II. *Proc. London Math. Soc. (3)*, 68(2):317–379, 1994.

[13] Wen-Min Chao. Fourier inversion and Plancherel formula for semisimple Lie groups of real rank two. *University of Chicago Thesis*, 1977.

[14] Lawrence Corwin. Representations of division algebras over local fields. *Advances in Math.*, 13:259–267, 1974.

[15] Lawrence Corwin. The unitary dual for the multiplicative group of arbitrary division algebras over local fields. *J. Amer. Math. Soc.*, 2(3):565–598, 1989.

[16] Clifton Cunningham and Julia Gordon. Motivic proof of a character formula for SL(2). *Experiment. Math.*, 18(1):11–44, 2009.

[17] Stephen Debacker. Homogeneity results for invariant distributions of a reductive p-adic group. *Ann. Sci. École Norm. Sup. (4)*, 35(3):391–422, 2002.

[18] Stephen DeBacker and Mark Reeder. Depth-zero supercuspidal L-packets and their stability. *Ann. of Math. (2)*, 169(3):795–901, 2009.

[19] Stephen DeBacker and Paul J. Sally, Jr. Germs, characters, and the Fourier transforms of nilpotent orbits. In *The mathematical legacy of Harish-Chandra (Baltimore, MD, 1998)*, volume 68 of *Proc. Sympos. Pure Math.*, pages 191–221. Amer. Math. Soc., Providence, RI, 2000.

[20] Stephen M. DeBacker. On supercuspidal characters of GL_ℓ, ℓ a prime. *University of Chicago Thesis*, 1997.

[21] Gerald B. Folland. *A course in abstract harmonic analysis*. Studies in Advanced Mathematics. CRC Press.

[22] I. M. Gel'fand and M. I. Graev. Representations of the group of second-order matrices with elements in a locally compact field and special functions on locally compact fields. *Uspehi Mat. Nauk*, 18(4 (112)):29–99, 1963.

[23] I. M. Gel'fand and M. A. Naĭmark. *Unitarnye predstavleniya klassičeskih grupp*. Trudy Mat. Inst. Steklov., vol. 36. Izdat. Nauk SSSR, Moscow-Leningrad, 1950.

[24] Harish-Chandra. Plancherel formula for complex semi-simple Lie groups. *Proc. Nat. Acad. Sci. U. S. A.*, 37:813–818, 1951.

[25] Harish-Chandra. Plancherel formula for the 2×2 real unimodular group. *Proc. Nat. Acad. Sci. U. S. A.*, 38:337–342, 1952.

[26] Harish-Chandra. Discrete series for semisimple Lie groups. I. Construction of invariant eigendistributions. *Acta Math.*, 113:241–318, 1965.

[27] Harish-Chandra. Discrete series for semisimple Lie groups. II. Explicit determination of the characters. *Acta Math.*, 116:1–111, 1966.

[28] Harish-Chandra. Two theorems on semi-simple Lie groups. *Ann. of Math. (2)*, 83:74–128, 1966.

[29] Harish-Chandra. *Harmonic analysis on reductive p-adic groups*. Lecture Notes in Mathematics, Vol. 162. Springer-Verlag, Berlin, 1970. Notes by G. van Dijk.

[30] Harish-Chandra. Harmonic analysis on semisimple Lie groups. *Bull. Amer. Math. Soc.*, 76:529–551, 1970.

[31] Harish-Chandra. Some applications of the Schwartz space of a semisimple Lie group. In *Lectures in Modern Analysis and Applications. II*, Lecture Notes in Mathematics, Vol. 140, pages 1–7. Springer, Berlin, 1970.

[32] Harish-Chandra. On the theory of the Eisenstein integral. In *Conference on Harmonic Analysis (Univ. Maryland, College Park, Md., 1971)*, pages 123–149. Lecture Notes in Math., Vol. 266. Springer, Berlin, 1972.

[33] Harish-Chandra. Harmonic analysis on reductive p-adic groups. In *Harmonic analysis on homogeneous spaces (Proc. Sympos. Pure Math., Vol. XXVI, Williams Coll., Williamstown, Mass., 1972)*, pages 167–192. Amer. Math. Soc., Providence, R.I., 1973.

[34] Harish-Chandra. Harmonic analysis on real reductive groups. I. The theory of the constant term. *J. Functional Analysis*, 19:104–204, 1975.

[35] Harish-Chandra. Harmonic analysis on real reductive groups. II. Wavepackets in the Schwartz space. *Invent. Math.*, 36:1–55, 1976.

[36] Harish-Chandra. Harmonic analysis on real reductive groups. III. The Maass-Selberg relations and the Plancherel formula. *Ann. of Math. (2)*, 104(1):117–201, 1976.

[37] Harish-Chandra. Admissible invariant distributions on reductive p-adic groups. In *Lie theories and their applications (Proc. Ann. Sem. Canad. Math. Congr., Queen's Univ., Kingston, Ont., 1977)*, pages 281–347. Queen's Papers in Pure Appl. Math., No. 48. Queen's Univ., Kingston, Ont., 1978.

[38] Harish-Chandra. Supertempered distributions on real reductive groups. In *Studies in applied mathematics*, volume 8 of *Adv. Math. Suppl. Stud.*, pages 139–153. Academic Press, New York, 1983.

[39] Harish-Chandra. *Collected papers. Vol. I*. Springer-Verlag, New York, 1984. 1944–1954, Edited and with an introduction by V. S. Varadarajan, With introductory essays by Nolan R. Wallach and Roger Howe.

[40] Harish-Chandra. *Collected papers. Vol. IV*. Springer-Verlag, New York, 1984. 1970–1983, Edited by V. S. Varadarajan.

[41] Harish-Chandra. *Admissible invariant distributions on reductive p-adic groups*, volume 16 of *University Lecture Series*. American Mathematical Society, Providence, RI, 1999. Preface and notes by Stephen DeBacker and Paul J. Sally, Jr.

[42] Volker Heiermann. Une formule de Plancherel pour l'algèbre de Hecke d'un groupe réductif p-adique. *Comment. Math. Helv.*, 76(3):388–415, 2001.

[43] Volker Heiermann. Décomposition spectrale et représentations spéciales d'un groupe réductif p-adique. *J. Inst. Math. Jussieu*, 3(3):327–395, 2004.

[44] R. A. Herb and P. J. Sally, Jr. Singular invariant eigendistributions as characters. *Bull. Amer. Math. Soc.*, 83(2):252–254, 1977.

[45] R. A. Herb and P. J. Sally, Jr. Singular invariant eigendistributions as characters in the Fourier transform of invariant distributions. *J. Funct. Anal.*, 33(2):195–210, 1979.

[46] Rebecca A. Herb. Fourier inversion of invariant integrals on semisimple real Lie groups. *Trans. Amer. Math. Soc.*, 249(2):281–302, 1979.

[47] Rebecca A. Herb. Fourier inversion and the Plancherel theorem. In *Noncommutative harmonic analysis and Lie groups (Marseille, 1980)*, volume 880 of *Lecture Notes in Math.*, pages 197–210. Springer, Berlin, 1981.

[48] Rebecca A. Herb. Fourier inversion and the Plancherel theorem for semisimple real Lie groups. *Amer. J. Math.*, 104(1):9–58, 1982.

[49] Rebecca A. Herb. Discrete series characters and Fourier inversion on semisimple real Lie groups. *Trans. Amer. Math. Soc.*, 277(1):241–262, 1983.

[50] Rebecca A. Herb. Elliptic representations for $Sp(2n)$ and $SO(n)$. *Pacific J. Math.*, 161(2):347–358, 1993.

[51] Rebecca A. Herb. Supertempered virtual characters. *Compositio Math.*, 93(2):139–154, 1994.

[52] Rebecca A. Herb. Discrete series characters and two-structures. *Trans. Amer. Math. Soc.*, 350(8):3341–3369, 1998.

[53] Rebecca A. Herb. Two-structures and discrete series character formulas. In *The mathematical legacy of Harish-Chandra (Baltimore, MD, 1998)*, volume 68 of *Proc. Sympos. Pure Math.*, pages 285–319. Amer. Math. Soc., Providence, RI, 2000.

[54] Rebecca A. Herb, Nick Ramsey, and Paul J. Sally, Jr. Some remarks on the representations of p-adic SL_2. *To appear.*

[55] Takeshi Hirai. Invariant eigendistributions of Laplace operators on real simple Lie groups. III. Methods of construction for semisimple Lie groups. *Japan. J. Math. (N.S.)*, 2(2):269–341, 1976.

[56] Roger E. Howe. Tamely ramified supercuspidal representations of Gl_n. *Pacific J. Math.*, 73(2):437–460, 1977.

[57] H. Jacquet and R. P. Langlands. *Automorphic forms on* GL(2). Lecture Notes in Mathematics, Vol. 114. Springer-Verlag, Berlin, 1970.

[58] Ju-Lee Kim. Supercuspidal representations: an exhaustion theorem. *J. Amer. Math. Soc.*, 20(2):273–320 (electronic), 2007.

[59] A. W. Knapp. Commutativity of intertwining operators. II. *Bull. Amer. Math. Soc.*, 82(2):271–273, 1976.

[60] A. W. Knapp and Gregg Zuckerman. Classification of irreducible tempered representations of semi-simple Lie groups. *Proc. Nat. Acad. Sci. U.S.A.*, 73(7):2178–2180, 1976.

[61] Phil Kutzko and José Pantoja. Character formulas for supercuspidal representations of the groups GL_2, SL_2. *Comm. Algebra*, 26(6):1679–1697, 1998.

[62] Philip Kutzko and Lawrence Morris. Explicit Plancherel theorems for $\mathcal{H}(q_1, q_2)$ and $\mathbb{SL}_2(F)$. *Pure Appl. Math. Q.*, 5(1):435–467, 2009.

[63] R. P. Langlands. Problems in the theory of automorphic forms. In *Lectures in modern analysis and applications, III*, pages 18–61. Lecture Notes in Math., Vol. 170. Springer, Berlin, 1970.

[64] I. G. Macdonald. Spherical functions on a p-adic Chevalley group. *Bull. Amer. Math. Soc.*, 74:520–525, 1968.

[65] F. I. Mautner. Unitary representations of locally compact groups. II. *Ann. of Math. (2)*, 52:528–556, 1950.

[66] F. I. Mautner. Spherical functions over p-adic fields. II. *Amer. J. Math.*, 86:171–200, 1964.

[67] Colette Mœglin and Marko Tadić. Construction of discrete series for classical p-adic groups. *J. Amer. Math. Soc.*, 15(3):715–786 (electronic), 2002.

[68] Lawrence Morris. Tamely ramified supercuspidal representations of classical groups. I. Filtrations. *Ann. Sci. École Norm. Sup. (4)*, 24(6):705–738, 1991.

[69] Lawrence Morris. Tamely ramified supercuspidal representations of classical groups. II. Representation theory. *Ann. Sci. École Norm. Sup. (4)*, 25(3):233–274, 1992.

[70] Allen Moy. Local constants and the tame Langlands correspondence. *Amer. J. Math.*, 108(4):863–930, 1986.

[71] Allen Moy and Gopal Prasad. Unrefined minimal K-types for p-adic groups. *Invent. Math.*, 116(1-3):393–408, 1994.

[72] Allen Moy and Gopal Prasad. Jacquet functors and unrefined minimal K-types. *Comment. Math. Helv.*, 71(1):98–121, 1996.

[73] Fiona Murnaghan. Characters of supercuspidal representations of classical groups. *Ann. Sci. École Norm. Sup. (4)*, 29(1):49–105, 1996.

[74] Fiona Murnaghan. Local character expansions and Shalika germs for GL(n). *Math. Ann.*, 304(3):423–455, 1996.

[75] Cary Rader and Allan Silberger. Some consequences of Harish-Chandra's submersion principle. *Proc. Amer. Math. Soc.*, 118(4):1271–1279, 1993.

[76] R. Ranga Rao. Orbital integrals in reductive groups. *Ann. of Math.*, 96:505–510, 1972.

[77] P. J. Sally, Jr. and J. A. Shalika. Characters of the discrete series of representations of SL(2) over a local field. *Proc. Nat. Acad. Sci. U.S.A.*, 61:1231–1237, 1968.

[78] P. J. Sally, Jr. and J. A. Shalika. The Plancherel formula for SL(2) over a local field. *Proc. Nat. Acad. Sci. U.S.A.*, 63:661–667, 1969.

[79] Paul J. Sally, Jr. Some remarks on discrete series characters for reductive p-adic groups. In *Representations of Lie groups, Kyoto, Hiroshima, 1986*, volume 14 of *Adv. Stud. Pure Math.*, pages 337–348. Academic Press, Boston, MA, 1988.

[80] Paul J. Sally, Jr. and Joseph A. Shalika. The Fourier transform of orbital integrals on SL$_2$ over a p-adic field. In *Lie group representations, II (College Park, Md., 1982/1983)*, volume 1041 of *Lecture Notes in Math.*, pages 303–340. Springer, Berlin, 1984.

[81] Paul J. Sally, Jr. and Garth Warner. The Fourier transform on semisimple Lie groups of real rank one. *Acta Math.*, 131:1–26, 1973.

[82] Ichirô Satake. Theory of spherical functions on reductive algebraic groups over p-adic fields. *Inst. Hautes Études Sci. Publ. Math.*, (18):5–69, 1963.

[83] Wilfried Schmid. On the characters of the discrete series. The Hermitian symmetric case. *Invent. Math.*, 30(1):47–144, 1975.

[84] I. E. Segal. An extension of Plancherel's formula to separable unimodular groups. *Ann. of Math. (2)*, 52:272–292, 1950.

[85] Freydoon Shahidi. Fourier transforms of intertwining operators and Plancherel measures for GL(n). *Amer. J. Math.*, 106(1):67–111, 1984.

[86] Freydoon Shahidi. A proof of Langlands' conjecture on Plancherel measures; complementary series for p-adic groups. *Ann. of Math. (2)*, 132(2):273–330, 1990.

[87] J. A. Shalika. A theorem on semi-simple P-adic groups. *Ann. of Math. (2)*, 95:226–242, 1972.

[88] Joseph A. Shalika. Representation of the two by two unimodular group over local fields. In *Contributions to automorphic forms, geometry, and number theory*, pages 1–38. Johns Hopkins Univ. Press, Baltimore, MD, 2004.

[89] Takuro Shintani. On certain square-integrable irreducible unitary representations of some p-adic linear groups. *J. Math. Soc. Japan*, 20:522–565, 1968.

[90] Allan J. Silberger. PGL$_2$ over the p-adics: its representations, spherical functions, and Fourier analysis. Lecture Notes in Mathematics, Vol. 166. Springer-Verlag, Berlin, 1970.

[91] Allan J. Silberger. Harish-Chandra's Plancherel theorem for p-adic groups. *Trans. Amer. Math. Soc.*, 348(11):4673–4686, 1996.

[92] Allan J. Silberger. Correction to: "Harish-Chandra's Plancherel theorem for p-adic groups" [Trans. Amer. Math. Soc. **348** (1996), no. 11, 4673–4686; MR1370652 (99c:22026)]. *Trans. Amer. Math. Soc.*, 352(4):1947–1949, 2000.

[93] Loren Spice. Supercuspidal characters of SL$_l$ over a p-adic field, l a prime. *Amer. J. Math.*, 127(1):51–100, 2005.

[94] Shaun Stevens. The supercuspidal representations of p-adic classical groups. *Invent. Math.*, 172(2):289–352, 2008.

[95] J.-L. Waldspurger. La formule de Plancherel pour les groupes p-adiques (d'après Harish-Chandra). *J. Inst. Math. Jussieu*, 2(2):235–333, 2003.

[96] Jean-Loup Waldspurger. Intégrales orbitales nilpotentes et endoscopie pour les groupes classiques non ramifiés. *Astérisque*, (269):vi+449, 2001.

[97] Jiu-Kang Yu. Construction of tame supercuspidal representations. *J. Amer. Math. Soc.*, 14(3):579–622 (electronic), 2001.

[98] Gregg Zuckerman. Tensor products of finite and infinite dimensional representations of semisimple Lie groups. *Ann. Math. (2)*, 106(2):295–308, 1977.

Contemporary Mathematics
Volume **557**, 2011

Branching problems of Zuckerman derived functor modules

Toshiyuki Kobayashi

Dedicated to Gregg Zuckerman on the occasion of his 60th birthday

ABSTRACT. We discuss recent developments on branching problems of irreducible unitary representations π of real reductive groups when restricted to reductive subgroups. Highlighting the case where the underlying $(\mathfrak{g}, K)$-modules of π are isomorphic to Zuckerman derived functor modules $A_{\mathfrak{q}}(\lambda)$, we show various and rich features of branching laws such as infinite multiplicities, irreducible restrictions, multiplicity-free restrictions, and discrete decomposable restrictions. We also formulate a number of conjectures.

1. Introduction

Zuckerman derived functor is powerful algebraic machinery to construct irreducible unitary representations by cohomological parabolic induction. The $(\mathfrak{g}, K)$-modules $A_{\mathfrak{q}}(\lambda)$, referred to as Zuckerman derived functor modules, give a far reaching generalization of the Borel–Weil–Bott construction of irreducible finite dimensional representations of compact Lie groups. They include Harish-Chandra's discrete series representations of real reductive Lie groups as a special case, and may be thought of as a geometric quantization of elliptic orbits (see Fact 6.1).

Branching problems in representation theory ask how irreducible representations π of a group G decompose when restricted to a subgroup G'.

The subject of our study is branching problems with emphasis on the setting when (G, G') is a reductive symmetric pair (Subsection 2.3), and when π is the unitarization of a Zuckerman derived functor module $A_{\mathfrak{q}}(\lambda)$. We see that branching problems in this setting include a wide range of examples: a very special case is equivalent to finding the Plancherel formula for homogeneous spaces (e.g. Proposition 2.4 and Example 4.8) and another special case is of combinatorial nature (e.g. the Blattner formula).

In this article, we give new perspectives on branching problems by revealing the following surprisingly rich and various features:

2010 *Mathematics Subject Classification.* Primary 22E46; Secondary 53C35.

Key words and phrases. branching law, symmetric pair, Zuckerman derived functor module, unitary representation, multiplicity-free representation.

Partially supported by Institut des Hautes Études Scientifiques, France and Grant-in-Aid for Scientific Research (B) (22340026), Japan Society for the Promotion of Science.

- The multiplicities may be infinite (Section 2) and may be one (Section 4).
- The restriction may stay irreducible (Section 3).
- The spectrum may be purely continuous and may be discretely decomposable (Section 5).

Finally, we present a number of open problems that might be interesting for further study (see Conjectures 4.2, 4.3, 5.4, and 5.11).

This article is based on the talk presented at the conference "Representation Theory and Mathematical Physics" in honor of Gregg Zuckerman's 60th birthday at Yale University on October 2009. The author is one of those who have been inspired by Zuckerman's work, and would like to express his sincere gratitude to the organizers of the stimulating conference, Professors J. Adams, M. Kapranov, B. Lian, and S. Sahi for their hospitality.

2. Wild aspects of branching laws

2.1. Analysis and synthesis. One of the most distinguished feature of *unitary* representations is that they are always built up from the smallest objects, namely, irreducible ones. For a locally compact group G, we denote by $\widehat{G}$ the set of equivalence classes of irreducible unitary representations of G, endowed with the Fell topology.

FACT 2.1 (Mautner–Teleman). *Every unitary representation π of a locally compact group G is unitarily equivalent to a direct integral of irreducible unitary representations:*

$$(2.1) \qquad\qquad \pi \simeq \int_{\widehat{G}}^{\oplus} n_\pi(\sigma)\sigma \, d\mu(\sigma).$$

Here, $d\mu$ is a Borel measure on $\widehat{G}$, $n_\pi : \widehat{G} \to \mathbb{N} \cup \{\infty\}$ is a measurable function, and $n_\pi(\sigma)\sigma$ stands for the multiple of an irreducible unitary representation σ with multiplicity $n_\pi(\sigma)$.

The decomposition (2.1) is unique if G is of type I in the sense of von Neumann algebras. Reductive Lie groups are of type I. Then the *multiplicity function* n_π is well-defined up to a measure zero set with respect to $d\mu$. We say that π has a *uniformly bounded multiplicity* if there is $C > 0$ such that $n_\pi(\sigma) \leq C$ almost everywhere; π is *multiplicity-free* if $n_\pi(\sigma) \leq 1$ almost everywhere, or equivalently, if the ring of continuous G-endomorphisms of π is commutative.

2.2. Branching laws and Plancherel formulas. Suppose that G' is a closed subgroup of G. Here are two basic settings where the problem of decomposing unitary representations arises naturally.

1) (Induction $G' \uparrow G$) *Plancherel formula.*
    For simplicity, assume that there exists a G-invariant Borel measure on the homogeneous space G/G'. Then the group G acts unitarily on the Hilbert space $L^2(G/G')$ by translations. The irreducible decomposition of the regular representation of G on $L^2(G/G')$ is called the *Plancherel formula* for G/G'.
2) (Restriction $G \downarrow G'$) *Branching laws.*
    Given an irreducible unitary representation π of G. By the symbol $\pi|_{G'}$, we think of π as a representation of the subgroup G'. The *branching law* of the restriction $\pi|_{G'}$ means the formula of decomposing π into irreducible

representations of G'. Special cases of branching laws include the classical Clebsch–Gordan formula, or more generally, the decomposition of the tensor product of two irreducible representations (*fusion rule*), and the Blattner formula, etc.

2.3. Symmetric pairs. We are particularly interested in the branching laws with respect to reductive symmetric pairs. Let us fix some notation.

Suppose σ is an involutive automorphism of a Lie group G. We denote by $G^\sigma := \{g \in G : \sigma g = g\}$, the group of fixed points by σ. We say that (G, G') is a *symmetric pair* if G' is an open subgroup of G^σ. Then the homogeneous space G/G' becomes an affine symmetric space with respect to the canonical G-invariant affine connection. The pair (G, G') is said to be a *reductive symmetric pair* if G is reductive. Further, if G' is compact then G/G' becomes a Riemannian symmetric space.

EXAMPLE 2.2. 1) (group case) Let $`G$ be a Lie group, $G := `G \times `G$ the direct product group, and $\sigma \in \mathrm{Aut}(G)$ be defined as $\sigma(x, y) := (y, x)$. Then $G^\sigma \equiv \mathrm{diag}(`G) := \{(x, x) : x \in `G\}$. Since the homogeneous space $G/G^\sigma$ is diffeomorphic to $`G$, we refer to the symmetric pair $(G, G^\sigma) = (`G \times `G, \mathrm{diag}(`G))$ as a *group case*.

2) The followings are chains of reductive symmetric pairs:

$$GL(2n, \mathbb{H}) \supset GL(n, \mathbb{C}) \supset GL(n, \mathbb{R}) \supset GL(p, \mathbb{R}) \times GL(q, \mathbb{R}) \quad (p + q = n),$$

$$O(4p, 4q) \supset U(2p, 2q) \supset Sp(p, q) \supset U(p, q) \supset O(p, q).$$

2.4. Finite multiplicity theorem of van den Ban. Let (G, G') be a reductive symmetric pair.

The irreducible decomposition (2.1) is well-behaved for the induction $G' \uparrow G$, namely, for the Plancherel formula of the symmetric space G/G':

FACT 2.3 (van den Ban [**2**]). *Suppose (G, G') is a reductive symmetric pair. Then the regular representation π on $L^2(G/G')$ has a uniformly bounded multiplicity.*

2.5. Plancherel formulas v.s. branching laws. Fairly many cases of the Plancherel formula for $L^2(G/G')$ treated in Fact 2.3 can be realized as a special example of branching laws of the restriction of irreducible unitary representations of other groups. For example, we recall from [**21**, Propositions 6.1, 6.2] and [**29**, Theorem 36]:

PROPOSITION 2.4. *Let G/G' be a reductive symmetric space. Then the regular representation of G on $L^2(G/G')$ is unitarily equivalent to the restriction $\pi|_G$ for some irreducible unitary representation π of a reductive group $\widetilde{G}$ containing G as its subgroup if (G, G') fulfills one of the following conditions:*

(A) *G' is compact and the crown domain D of the Riemannian symmetric space G/G' is a Hermitian symmetric space,*

or

(B) *G'/Z_G has a split center. Here Z_G stands for the center of G.*

REMARK 2.5.   1) Most Riemannian symmetric pairs (G, G') satisfy the assumption (A) (see [**37**] for details).

2) As the proof below shows,

$$\widetilde{G} \supset G \supset G'$$

is a chain of reductive symmetric pairs.

3) We can take π to be the unitarization of some $A_{\mathfrak{q}}(\lambda)$ in (A) and also in (B) when G is a complex reductive Lie group.

4) There are some more cases other than (A) or (B) for which the conclusion of Proposition 2.4 holds. For instance, see Example 4.8 for the group case $L^2(GL(n,\mathbb{C}))$ and also for a more general case $L^2(GL(2n,\mathbb{R})/GL(n,\mathbb{C}))$.

.

OUTLINE OF THE PROOF. The choice of π and $\widetilde{G}$ depends on each case (A) and (B).

(A) We take $\widetilde{G}$ to be the automorphism group of D, and π to be any holomorphic discrete series representation of $\widetilde{G}$ of scalar type. Then π is realized in the Hilbert space consisting of square integrable, holomorphic sections of a G-equivariant holomorphic line bundle over D. Since holomorphic sections are determined uniquely by the restriction to the totally real submanifold G/G', we get a realization of the restriction $\pi|_G$ in a certain Hilbert subspace of $\mathcal{A}(G/G')$, which itself is not $L^2(G/G')$ but is unitarily equivalent to the regular representation on $L^2(G/G')$ (see [12]).

(B) Let P be a maximal parabolic subgroup of G whose Levi part is G'. Take $\widetilde{G}$ to be the direct product $G \times G$, and π to be the outer tensor product representation $\pi_1 \boxtimes \pi_2$ where π_1 is a degenerate unitary principal series representation induced from a unitary character of P and π_2 is the contragredient representation of π_1. Then apply the Mackey theory. $\square$

EXAMPLE 2.6. 1) The regular representation on $L^2(G/G') = L^2(GL(n,\mathbb{R})/O(n))$ is unitarily equivalent to the restriction of a holomorphic discrete series representation of $\widetilde{G} := Sp(n,\mathbb{R})$ to G.

2) The regular representation on $L^2(GL(n,\mathbb{R})/GL(p,\mathbb{R}) \times GL(q,\mathbb{R}))$ with $(p + q = n)$ is unitarily equivalent to the restriction of a degenerate principal representation of $\widetilde{G} := GL(n,\mathbb{R}) \times GL(n,\mathbb{R})$ to G (namely, to the tensor product representation).

2.6. Wild aspects of branching laws. Retain our assumption that (G,G') is a reductive symmetric pair.

Proposition 2.4 suggests that branching problems include a wide range of examples. In fact, while the 'good behavior' in Fact 2.3 for the Plancherel formula of the symmetric space G/G', the branching law of the restriction $\pi|_{G'}$ does not behave well in general. Even when π_K is a Zuckerman derived functor module $A_{\mathfrak{q}}(\lambda)$, we cannot expect:

'FALSE THEOREM' 2.7. *Let (G,G') be a reductive symmetric pair, and π an irreducible unitary representation of G. Then the multiplicities of the discrete spectrum in the branching laws $\pi|_{G'}$ are finite.*

REMARK 2.8. Such a multiplicity theorem holds for reductive symmetric pairs (G,G') under the assumption that the restriction $\pi|_{G'}$ is infinitesimally discretely decomposable in the sense of Definition 5.3 (cf. [22, 28]). A key to the proof is Theorem 5.6 on a criterion of K'-admissibility and Corollary 5.8 on an estimate of the associated variety. See Remark 5.14 for the case $\pi_K \simeq A_{\mathfrak{q}}(\lambda)$.

Before giving a counterexample to (false) 'Theorem' 2.7 about the discrete spectrum, we discuss an easier case, namely, an example of infinite multiplicities in the continuous spectrum of the branching law:

PROPOSITION 2.9 ($G \times G \downarrow \mathrm{diag}\, G$). (Gelfand–Graev [8].) *If π_1 and π_2 are two unitary principal series representations of $G = SL(n, \mathbb{C})$ ($n \geq 3$), then the multiplicities in the decomposition of the tensor product $\pi_1 \otimes \pi_2$ are infinite almost everywhere with respect to the measure $d\mu$ in the direct integral (2.1).*

We recall the underlying $(\mathfrak{g}, K)$-modules of unitary principal series representations of a complex reductive Lie group are obtained as a special case of Zuckerman derived functor modules $A_\mathfrak{q}(\lambda)$.

Hence we get

OBSERVATION 2.10. *The multiplicities of the continuous spectrum in the branching law of the restriction $\pi|_{G'}$ may be infinite even in the setting where $\pi_K \simeq A_\mathfrak{q}(\lambda)$ and (G, G') is a reductive symmetric pair.*

Here is a more delicate example, which yields a counterexample to (false) 'Theorem' 2.7 about the discrete spectrum.

PROPOSITION 2.11 ($G_{\mathbb{C}} \downarrow G_{\mathbb{R}}$). (see [26]) *There exist an irreducible unitary principal series representation π of $G = SO(5, \mathbb{C})$ and two irreducible unitary representations τ_1 (a holomorphic discrete series representation) and τ_2 (a non-holomorphic discrete series representation) of the subgroup $G' = SO(3, 2)$ such that*

$$0 < \dim \mathrm{Hom}_{G'}(\tau_1, \pi|_{G'}) < \infty \quad and \quad \dim \mathrm{Hom}_{G'}(\tau_2, \pi|_{G'}) = \infty.$$

Here, $\mathrm{Hom}_{G'}(\cdot, \cdot)$ denotes the space of continuous G'-intertwining operators.

3. Almost irreducible branching laws

Let G be a real reductive Lie group, G' a subgroup, and π an irreducible unitary representation of G.

We have seen some wild aspects of branching laws in the previous section. As its opposite extremal case, this section highlights especially nice cases, namely, where the restriction $\pi|_{G'}$ remains irreducible or almost irreducible in the following (obvious) sense:

DEFINITION 3.1. We say a unitary representation π is *almost irreducible* if π is a finite direct sum of irreducible representations.

It may well happen that the restriction $\pi|_{G'}$ is almost irreducible when G' is a maximal parabolic subgroup of G, but is a rare phenomenon when G' is a reductive subgroup. Nevertheless, we find in Subsections 3.2–3.3 that there exist a small number of examples where the restriction $\pi|_{G'}$ stays irreducible, or is almost irreducible in some cases.

We divide such irreducible unitary representations π of G into three cases, according as π_K are Zuckerman derived functor modules $A_\mathfrak{q}(\lambda)$ (see Theorem 3.5), principal series representations (see Theorem 3.8), and minimal representations (see Theorem 3.11). From the view point of the Kostant–Kirillov–Duflo orbit method, they may be thought of as the geometric quantization of elliptic, hyperbolic, and nilpotent orbits, respectively.

3.1. Restriction to compact subgroups. First of all, we observe that almost irreducible restrictions $\pi|_{G'}$ happen only when G' is non-compact if $\dim \pi = \infty$.

Let K be a maximal compact subgroup of a real reductive Lie group G.

OBSERVATION 3.2. *For any irreducible infinite dimensional unitary representation π of G, the branching law of the restriction $\pi|_K$ contains infinitely many irreducible representations of K.*

PROOF. Clear from Harish-Chandra's admissibility theorem (see Fact 3.4 below). □

For later purpose, we introduce the following terminology:

DEFINITION 3.3. Suppose K' is a compact group and π is a representation of K'. We say π is K'-admissible if $\dim \operatorname{Hom}_{K'}(\tau, \pi) < \infty$ for any $\tau \in \widehat{K'}$.

With this terminology, we state:

FACT 3.4 (Harish-Chandra's admissibility theorem). *Any irreducible unitary representation π of G is K-admissible.*

We shall apply the notion of K'-admissibility when K' is a subgroup of K, and see that it plays a crucial role in the theory of discretely decomposable restrictions in Section 5.

3.2. Irreducible restriction $\pi|_{G'}$ with $\pi_K = A_{\mathfrak{q}}(\lambda)$. This subsection discusses for which triple (G, G', π) the restriction $\pi|_{G'}$ is (almost) irreducible in the setting that the underlying $(\mathfrak{g}, K)$-module π_K is isomorphic to a Zuckerman derived functor module $A_{\mathfrak{q}}(\lambda)$.

Let $\mathfrak{q}$ be a θ-stable parabolic subalgebra of $\mathfrak{g}_{\mathbb{C}} = \mathfrak{g} \otimes_{\mathbb{R}} \mathbb{C}$, $L := N_G(\mathfrak{q}) \equiv \{g \in G : \operatorname{Ad}(g)\mathfrak{q} = \mathfrak{q}\}$, and $\overline{A_{\mathfrak{q}}(\lambda)}$ the unitary representation of G whose underlying $(\mathfrak{g}, K)$-module is $A_{\mathfrak{q}}(\lambda)$.

THEOREM 3.5 ([**19**]). *Suppose that (G, G', L) is one of the following triples:*

G	G'	L
$SU(n,n)$	$Sp(n, \mathbb{R})$	$U(n-1, n)$
$SU(2p, 2q)$	$Sp(p, q)$	$U(2p-1, 2q)$
$SO_0(2p, 2q)$	$SO_0(2p, 2q-1)$	$U(p, q)$
$SO_0(4, 3)$	$G_2(\mathbb{R})$	$SO_0(4,1) \times SO(2)$
$SO_0(4, 3)$	$G_2(\mathbb{R})$	$SO(2) \times SO_0(2,3)$
$SL(2n, \mathbb{C})$	$Sp(n, \mathbb{C})$	$GL(2n-1, \mathbb{C})$
$SO(2n, \mathbb{C})$	$SO(2n-1, \mathbb{C})$	$GL(n, \mathbb{C})$
$SO(7, \mathbb{C})$	$G_2(\mathbb{C})$	$\mathbb{C}^{\times} \times SO(5, \mathbb{C})$
$SU(2n)$	$Sp(n)$	$U(2n-1)$
$SO(2n)$	$SO(2n-1)$	$U(n)$
$SO(7)$	$G_{2, compact}$	$SO(2) \times SO(5)$

Then, the restriction $\overline{A_{\mathfrak{q}}(\lambda)}|_{G'}$ is almost irreducible for any λ satisfying the positivity and integrality condition (see Subsection 6.2). Further, the restriction $\overline{A_{\mathfrak{q}}(\lambda)}|_{G'}$ stays irreducible if the character $\lambda|_{\mathfrak{l}'}$ of $\mathfrak{l}' := \mathfrak{g}' \cap \mathfrak{l}$ is in the good range with respect to $\mathfrak{q}' := \mathfrak{g}'_\mathbb{C} \cap \mathfrak{q}$ *(see (6.2)). On the level of Harish-Chandra modules, we have an isomorphism*

$$A_\mathfrak{q}(\lambda) \simeq A_{\mathfrak{q}'}(\lambda|_{\mathfrak{l}'}),$$

as $(\mathfrak{g}', K')$-*modules.*

OUTLINE OF PROOF. We recall the following well-known representations of spheres:

$$Sp(n)/Sp(n-1) \overset{\sim}{\to} U(2n)/U(2n-1) \quad \simeq S^{4n-1},$$
$$U(n)/U(n-1) \overset{\sim}{\to} SO(2n)/SO(2n-1) \simeq S^{2n-1},$$
$$Spin(5)/Spin(3) \overset{\sim}{\to} Spin(7)/G_2 \quad\quad \simeq S^7.$$

Then the trick in [**20**, Lemma 5.1] shows that the natural inclusion map $G'/L' \hookrightarrow G/L$ is in fact surjective for any of the specific triples (G, G', L) in Theorem 3.5, where we set $L' := G' \cap L$. Further, we have $\mathfrak{g}'_\mathbb{C} + \mathfrak{q} = \mathfrak{g}_\mathbb{C}$ so that the inclusion $\mathfrak{g}'_\mathbb{C} \hookrightarrow \mathfrak{g}_\mathbb{C}$ induces the bijection $\mathfrak{g}'_\mathbb{C}/\mathfrak{q}' \overset{\sim}{\to} \mathfrak{g}_\mathbb{C}/\mathfrak{q}$ and L' coincides with $N_{G'}(\mathfrak{q}')$. Thus, the diffeomorphism $G'/L' \overset{\sim}{\to} G/L$ is biholomorphic. In turn, we get an isomorphism of canonical line bundles (see (6.1)):

$$
\begin{array}{ccc}
G' \times_{L'} \mathbb{C}_{2\rho(\mathfrak{u}')} & \overset{\sim}{\to} & G \times_L \mathbb{C}_{2\rho(\mathfrak{u})} \\
\downarrow & & \downarrow \\
G'/L' & \overset{\sim}{\to} & G/L
\end{array}
$$

This implies

$$\rho(\mathfrak{u})|_{\mathfrak{l}'} = \rho(\mathfrak{u}')$$

in the setting of Theorem 3.5. Let $\mathcal{L}_\lambda$ be a G-equivariant holomorphic line bundle over G/L for $\lambda \in \sqrt{-1}\mathfrak{l}^*$. Then the pull-back of $\mathcal{L}_{\lambda+2\rho(\mathfrak{u})}$ to G'/L' yields a G'-equivariant holomorphic line bundle $\mathcal{L}_{\lambda|_{\mathfrak{l}'}+2\rho(\mathfrak{u}')}$ over G'/L'. Hence, we have natural isomorphisms

$$H^*_{\bar{\partial}}(G/L, \mathcal{L}_{\lambda+2\rho(\mathfrak{u})}) \overset{\sim}{\to} H^*_{\bar{\partial}}(G'/L', \mathcal{L}_{\lambda|_{\mathfrak{l}'}+2\rho(\mathfrak{u}')})$$

between Dolbeault cohomology groups. Thus, we get Theorem 3.5 in view of the geometric interpretation of Zuckerman derived functor modules (see Section 6). □

REMARK 3.6. 1) The pairs (G, G') in Theorem 3.5 are reductive symmetric pairs except for the case $(G, G') = (SO_0(4, 3), G_2(\mathbb{R}))$.

2) The pair $(\mathfrak{g}, \mathfrak{l})$ is a reductive symmetric pair in all the cases of Theorem 3.5 ($\mathfrak{q}$ is of symmetric type in the sense of Definition 4.1). Correspondingly there are two choices of θ-stable parabolic subalgebras $\mathfrak{q}$ of $\mathfrak{g}_\mathbb{C}$ with $N_G(\mathfrak{q}) \simeq L$. In either case, $\overline{A_\mathfrak{q}(\lambda)}|_{G'}$ is almost irreducible.

3) In the compact case (i.e. the last three rows), the restriction is irreducible for all λ.

EXAMPLE 3.7. 1) In [**20**] we gave a different proof of Theorem 3.5 for the pair $SO_0(4, 3) \downarrow G_2(\mathbb{R})$ based on the Beilinson–Bernstein localization theory, and then applied it to construct (all) discrete series representations for non-symmetric homogeneous spaces $G_2(\mathbb{R})/SL(3, \mathbb{R})$ and $G_2(\mathbb{R})/SU(2, 1)$.

2) H. Sekiguchi applied the restriction of $A_\mathfrak{q}(\lambda)$ with respect to the symmetric pair $U(n, n) \downarrow Sp(n, \mathbb{R})$ for more general $\mathfrak{q}$ to get a range characterization theorem of the Penrose transform (see [**41**]). Following the notation in [**41**, Proposition 1.5], we see that the unitary character $\mathbb{C}_\lambda$ is in the weakly fair range for the θ-stable maximal parabolic subalgebra $\mathfrak{q}$ considered in Theorem 3.5 if and only if $\lambda = \lambda_1 e_1$

with $\lambda_1 \geq -n$. Further, $A_{\mathfrak{q}}(\lambda)$ is irreducible as a $\mathfrak{u}(n,n)$-module for all $\lambda_1 \geq -n$. Its restriction to $\mathfrak{sp}(n,\mathbb{R})$ stays irreducible for $\lambda_1 > -n$, but splits into two irreducible modules $(W(n,1)_+)_K \oplus (W(n,1)_-)_K$.

3) Dunne and Zierau [6] determined the automorphism groups of elliptic orbits. It follows from their results that our list in Theorem 3.5 exhausts all the cases where $\overline{A_{\mathfrak{q}}(\lambda)}|_{G'}$ stays irreducible for sufficiently positive λ.

3.3. Irreducible restriction $\pi|_{G'}$ with $\pi = \operatorname{Ind}_P^G(\tau)$. This subsection discusses for which triples (G, G', π) the restriction $\pi|_{G'}$ is (almost) irreducible in the setting that π is a (degenerate) principal series representation $\pi = \operatorname{Ind}_P^G(\tau)$ of G.

Let P be a parabolic subgroup of G with Levi decomposition $P = LN$. For an irreducible unitary representation τ of L, we extend it to P by letting N act trivially, and denote by $\operatorname{Ind}_P^G(\tau)$ the unitarily induced representation of G.

THEOREM 3.8. *Suppose that (G, G', L) is one of the following triples:*

G	G'	L
$SL(2n,\mathbb{C})$	$Sp(n,\mathbb{C})$	$GL(2n-1,\mathbb{C})$
$SO(2n,\mathbb{C})$	$SO(2n-1,\mathbb{C})$	$GL(n,\mathbb{C})$
$SO(7,\mathbb{C})$	$G_2(\mathbb{C})$	$\mathbb{C}^\times \times SO(5,\mathbb{C})$
$SL(2n,\mathbb{R})$	$Sp(n,\mathbb{R})$	$GL(2n-1,\mathbb{R})$
$SO(2n,2n)$	$SO(2n,2n-1)$	$GL(2n,\mathbb{R})$
$SO(4,3)$	$G_2(\mathbb{R})$	$SO(1,1) \times SO(3,2)$

Then, the degenerate unitary principal series representations $\pi = \operatorname{Ind}_P^G(\tau)$ of G are almost irreducible when restricted to the subgroup G' for any one dimensional unitary representation τ of any parabolic subgroup P having L as its Levi part.

OUTLINE OF THE PROOF. The subgroup G' acts transitively on the (real) flag variety G/P in the setting of Theorem 3.8, and the isotropy subgroup $P' := G' \cap P$ becomes a parabolic subgroup of G'. Then we get an isomorphism $G'/P' \xrightarrow{\sim} G/P$, and hence the conclusion follows. □

We note that the parabolic subgroup P in Theorem 3.8 is maximal.

EXAMPLE 3.9. For simplicity, we use $GL(2n,\mathbb{R})$ instead of the semisimple group $SL(2n,\mathbb{R})$ in the fourth row, and consider the reductive symmetric pair $(G, G') = (GL(2n,\mathbb{R}), Sp(n,\mathbb{R}))$. Let P be a maximal parabolic subgroup P of G with Levi subgroup $L = GL(2n-1,\mathbb{R}) \times GL(1,\mathbb{R})$. Then P has an abelian unipotent radical $\mathbb{R}^{2n-1}$ and $P' = G' \cap P$ has a non-abelian unipotent radical which is isomorphic to the Heisenberg group H^{2n-1}. In this case the unitary representation $\pi = \operatorname{Ind}_P^G(\tau)$ is irreducible as a representation of G for any unitary character τ of P. On the other hand, the restriction of π to G' is more delicate. It stays irreducible for generic τ (i.e. $d\tau \neq 0$) and splits into two irreducible representations of G' for singular τ, giving rise to a 'special unipotent representation' of $G' = Sp(n,\mathbb{R})$. See [35] for a detailed analysis in connection with the Weyl operator calculus.

REMARK 3.10. For a complex reductive group, the underlying $(\mathfrak{g}, K)$-modules of (degenerate) principal series representations are isomorphic to some $A_{\mathfrak{q}}(\lambda)$. Thus the first three cases in Theorem 3.8 have already appeared in Theorem 3.5 in the context of $A_{\mathfrak{q}}(\lambda)$.

3.4. Irreducible restriction of minimal representation. Thirdly, we present an example of almost irreducible branching laws for representations π which are supposed to be attached to minimal nilpotent coadjoint orbits.

Let ϖ be the irreducible unitary representation of the indefinite orthogonal group $G = O(p, q)$ for $p, q \geq 2$, $(p, q) \neq (2, 2)$ and $p + q$ even, constructed in [**3**] or [**34**, Part I]. It is a representation of Gelfand–Kirillov dimension $p + q - 3$, and is *minimal* in the sense that its annihilator in the enveloping algebra $U(\mathfrak{g})$ is the Joseph ideal if $p + q > 6$.

THEOREM 3.11 $(O(p, q) \downarrow O(p, q - 1))$.

$$\varpi|_{O(p, q-1)} \simeq V_+ + V_-$$

where $V_\pm$ are irreducible representations of $O(p, q - 1)$.

PROOF. See [**34**, Corollary 7.2.1]. $\qquad\qquad\qquad\qquad\qquad\qquad\qquad\square$

REMARK 3.12. The irreducible decomposition $V_+ + V_-$ has a geometric meaning in connection to the smallest L^2-eigenvalues of the (ultra-hyperbolic) Laplacian on pseudo-Riemannian space forms.

4. Multiplicity-free conjecture

Irreducible restrictions to reductive subgroups are a somewhat rare phenomenon, as we have seen in the previous section. On the other hand, it happens more often that the restriction is multiplicity-free with respect to reductive symmetric pairs (G, G') (see [**29**] for examples). In this section, we propose a conjectural sufficient condition for the restriction $\pi|_{G'}$ to be multiplicity-free in the setting where π_K is a Zuckerman derived functor module $A_{\mathfrak{q}}(\lambda)$. Our conjecture is motivated by the propagation theorem of multiplicity-free property under 'visible actions' [**31**].

DEFINITION 4.1. 1) We say a θ-stable parabolic subalgebra $\mathfrak{q} = \mathfrak{l} + \mathfrak{u}$ is of *symmetric type* if $(\mathfrak{g}, \mathfrak{l})$ forms a symmetric pair.

2) We say that $\mathfrak{q}$ is of *virtually symmetric type* if there exists a θ-stable parabolic subalgebra $\widetilde{\mathfrak{q}}$ of symmetric type such that $\widetilde{L}/L \equiv N_G(\widetilde{\mathfrak{q}})/N_G(\mathfrak{q})$ is compact.

REMARK. 1) If $\mathfrak{q}$ is of virtually symmetric type, then we have a fibration $\widetilde{L}/L \to G/L \to G/\widetilde{L}$ with compact fiber $\widetilde{L}/L$.

2) If $\mathfrak{q}$ is of symmetric type, then $\mathfrak{q}$ is obviously of virtually symmetric type.

3) Any parabolic subalgebra is of virtually symmetric type if G is compact.

Let $\mathfrak{q}$ be a θ-stable parabolic subalgebra of $\mathfrak{g}_{\mathbb{C}}$, and $\overline{A_{\mathfrak{q}}(\lambda)}$ be the unitarization of $A_{\mathfrak{q}}(\lambda)$. Suppose $(\mathfrak{g}, \mathfrak{g}')$ is a reductive symmetric pair. We then propose the following two conjectures:

CONJECTURE 4.2. *If a θ-stable parabolic subalgebra $\mathfrak{q}$ is of symmetric type, then the restriction $\overline{A_{\mathfrak{q}}(\lambda)}|_{G'}$ is multiplicity-free for sufficiently regular λ.*

CONJECTURE 4.3. *If $\mathfrak{q}$ is of virtually symmetric type, then the restriction $\overline{A_{\mathfrak{q}}(\lambda)}|_{G'}$ has a uniformly bounded multiplicity.*

Here are some affirmative cases:

EXAMPLE 4.4. Suppose G is a non-compact simple Lie group such that G/K is a Hermitian symmetric space. We write $\mathfrak{g} = \mathfrak{k} + \mathfrak{p}$ for the Cartan decomposition.

Then $\mathfrak{p}_{\mathbb{C}} := \mathfrak{p} \otimes_{\mathbb{R}} \mathbb{C}$ decomposes into a direct sum of two irreducible representations of K, say $\mathfrak{p}_{\mathbb{C}} = \mathfrak{p}_+ \oplus \mathfrak{p}_-$. Then $\mathfrak{q} := \mathfrak{k}_{\mathbb{C}} + \mathfrak{p}_+$ is a θ-stable parabolic subalgebra of symmetric type. If λ is in the good range, then $A_{\mathfrak{q}}(\lambda)$ is the underlying $(\mathfrak{g}, K)$-module of a holomorphic discrete series representation of scalar type. In this case, we see Conjecture 4.2 holds by the explicit branching law:

$$G' = K \qquad\qquad \cdots \text{ Hua } [\mathbf{13}], \text{ Kostant, Schmid } [\mathbf{40}],$$

$$G' : \text{ non-compact } \cdots \text{ Kobayashi } [\mathbf{29}].$$

EXAMPLE 4.5. As a generalization of Example 4.4, we retain that G/K is a Hermitian symmetric space, and assume that $\mathfrak{q}$ is of holomorphic type in the sense that $\mathfrak{q} \cap \mathfrak{p}_{\mathbb{C}} \supset \mathfrak{p}_+$. Then $A_{\mathfrak{q}}(\lambda)$ is at most a finite direct sum of irreducible unitary highest weight modules if λ is in the weakly fair range (see [$\mathbf{1}$]). In this case, Conjecture 4.3 is true for any $A_{\mathfrak{q}}(\lambda)$ (see [$\mathbf{30}$, Theorem B]). Further, it was proved in [$\mathbf{30}$, Theorems A, C] as a special case of the propagation theorem of multiplicity-free property that the restriction $\pi|_{G'}$ is multiplicity-free if π is an irreducible unitary highest weight module of scalar type.

EXAMPLE 4.6. For $(G, G') = (O(p,q), O(r) \times O(p-r,q))$ and for a θ-stable parabolic subalgebra $\mathfrak{q}$ of maximal dimension, we see from explicit branching laws [$\mathbf{18}$] that Conjecture 4.2 holds in this case. Likewise, Conjecture 4.2 holds for the restriction $O(2p, 2q) \downarrow U(p,q)$ again by explicit branching laws [$\mathbf{20}$].

EXAMPLE 4.7. For any compact group G, the restriction $\pi_\lambda|_{G'}$ is always multiplicity-free if $\mathfrak{q}$ is of symmetric type ([$\mathbf{30}$, Theorems E, F]) and hence, Conjecture 4.2 is true.

EXAMPLE 4.8. Let $(G, G') = (GL(2n, \mathbb{C}), GL(n, \mathbb{C}) \times GL(n, \mathbb{C}))$, and $\mathfrak{q}$ a θ-stable parabolic subalgebra such that $N_G(\mathfrak{q}) \simeq G'$. Then $\mathfrak{q}$ is of symmetric type. Further, we have the following unitary equivalence:

$$\overline{A_{\mathfrak{q}}(\lambda)}|_{GL(n,\mathbb{C}) \times GL(n,\mathbb{C})} \simeq L^2(GL(n, \mathbb{C})).$$

Thanks to the Plancherel formula of the group $GL(n, \mathbb{C})$ due to the Gelfand school and Harish-Chandra, we see that Conjecture 4.2 holds also in this case.

Let us retain the same θ-stable parabolic subalgebra $\mathfrak{q}$ and consider another reductive symmetric pair $(G, G'') = (GL(2n, \mathbb{C}), GL(2n, \mathbb{R}))$. Then, we get the following unitary isomorphism:

$$\overline{A_{\mathfrak{q}}(\lambda)}|_{GL(2n,\mathbb{R})} \simeq L^2(GL(2n, \mathbb{R})/GL(n, \mathbb{C})).$$

Again, the right-hand side is multiplicity-free by the Plancherel formula for reductive symmetric space due to Oshima, van den Ban, Schlichtkrull, and Delorme [$\mathbf{4}$] among others. (It should be noted that the Plancherel formula for a reductive symmetric space is not multiplicity-free in general.)

REMARK 4.9. As we have seen in Example 4.8, Conjectures 4.2 and 4.3 refer to the multiplicities in both discrete and continuous spectrum in the branching law $\overline{A_{\mathfrak{q}}(\lambda)}|_{G'}$.

5. Discretely decomposable branching laws

This section highlights another nice class of branching problems, namely, when the restriction $\pi|_{G'}$ splits discretely without continuous spectrum.

An obvious case is when $\dim \pi < \infty$ or when G' is compact. One of the advantages of discretely decomposable restrictions is that we can expect a combinatorial and detailed study of branching laws by purely algebraic methods because we do not have analytic difficulties arising from continuous spectrum.

Prior to [18], discretely decomposable restrictions $\pi|_{G'}$ were known in some specific settings, e.g. the θ-correspondence for the Weil representation with respect to compact dual pair [11], or when π is a holomorphic discrete series representation and G' is a Hermitian Lie group [14]. A systematic study in the general case including Zuckerman derived functor modules $A_{\mathfrak{q}}(\lambda)$ was initiated by the author in a series of papers [20, 21, 22, 25, 27]. See [9, 18, 20, 34, 39] for a number of concrete examples of branching laws $\pi|_{G'}$ in this framework, [33] for some application to modular symbols, [24] for the construction of new discrete series representations on non-symmetric spaces. See also the lecture notes [28] for a survey on representation theoretic aspects, and [25, 27] for some applications.

In this section, we give a brief overview of discretely decomposable restrictions including some recent developments and open problems.

5.1. Infinitesimally discretely decomposable restrictions. Let us begin with an algebraic formulation. Suppose $\mathfrak{g}'$ is a Lie algebra.

DEFINITION 5.1. A $\mathfrak{g}'$-module V is said to be *discretely decomposable* if there exists an increasing filtration $\{V_n\}$ such that $V = \bigcup_{n=0}^{\infty} V_n$ and each V_n is of finite length as a $\mathfrak{g}'$-module.

In the setting where G' is a real reductive Lie group with maximal compact subgroup K', the terminology 'discretely decomposable' fits well if V is a unitarizable $(\mathfrak{g}', K')$-module, namely, if V is the underlying $(\mathfrak{g}', K')$-module of a unitary representation of G':

REMARK 5.2 ([22, Lemma 1.3]). Suppose V is a unitarizable $(\mathfrak{g}', K')$-module. Then V is discretely decomposable as a $\mathfrak{g}'$-module if and only if V is decomposed into an algebraic direct sum of irreducible $(\mathfrak{g}', K')$-modules.

We apply Definition 5.1 to branching problems. Let G be a real reductive Lie group, and G' a reductive subgroup of G. We may and do assume that K is a maximal compact subgroup of G and $K' := K \cap G'$ is that of G'.

DEFINITION 5.3. Let π be a unitary representation of G of finite length. We say the restriction $\pi|_{G'}$ is *infinitesimally discretely decomposable* if the underlying $(\mathfrak{g}, K)$-module π_K is discretely decomposable as a $\mathfrak{g}'$-module.

Here is a comparison between the category of unitary representations and that of $(\mathfrak{g}, K)$-modules:

CONJECTURE 5.4. *Let π be an irreducible unitary representation of G, and G' a reductive subgroup of G. Then the following two conditions on (G, G', π) are equivalent:*

(i) *The restriction $\pi|_{G'}$ is infinitesimally discretely decomposable.*
(ii) *The unitary representation π decomposes discretely into a direct sum of irreducible unitary representations of G'.*

In general, the implication (i) $\Rightarrow$ (ii) holds. Moreover, the branching law for the restriction of the unitary representation π to G' and that for the restriction of the

$(\mathfrak{g}, K)$-module π_K to $(\mathfrak{g}', K')$ are essentially the same under the assumption (i) (see [**26**, Theorem 2.7]). The converse statement (ii) $\Rightarrow$ (i) remains open; affirmative results have been partially obtained by Duflo and Vargas [**5**] for discrete series representations π, see also [**26**, Conjecture D] and [**45**].

For the study of discretely decomposable restrictions, the concept of K'-admissible restrictions is useful:

PROPOSITION 5.5. *If the restriction $\pi|_{K'}$ is K'-admissible then both the conditions (i) and (ii) in Conjecture 5.4 hold.*

PROOF. See [**22**, Proposition 1.6] and [**20**, Theorem 1.2], respectively. $\square$

5.2. Analytic approach. We now consider a criterion for the K'-admissibility of a representation π.

Let K' be a closed subgroup of K. Associated to the Hamiltonian K-action on the cotangent bundle $T^*(K/K')$, we consider the momentum map

$$\mu : T^*(K/K') \to \sqrt{-1}\mathfrak{k}^*.$$

Then its image equals $\sqrt{-1}\,\mathrm{Ad}^*(K)(\mathfrak{k}')^\perp$, where $(\mathfrak{k}')^\perp$ is the kernel of the projection $\mathrm{pr}_{\mathfrak{k}\to\mathfrak{k}'} : \mathfrak{k}^* \to (\mathfrak{k}')^*$, the dual to the inclusion $\mathfrak{k}' \subset \mathfrak{k}$ of Lie algebras. The momentum set $C_K(K')$ is defined as the intersection of Image μ with a dominant Weyl chamber C_+ ($\subset \sqrt{-1}\mathfrak{t}^*$) with respect to a fixed positive system $\Delta^+(\mathfrak{k}, \mathfrak{t})$ and a Cartan subalgebra $\mathfrak{t}$ of $\mathfrak{k}$:

(5.1) $C_K(K') := C_+ \cap \sqrt{-1}\,\mathrm{Ad}^*(K)(\mathfrak{k}')^\perp.$

Here we regard $\mathfrak{t}^*$ as a subspace of $\mathfrak{k}^*$ via a K-invariant non-degenerate bilinear form on $\mathfrak{k}$.

Next, let π be a K-module. We write $\mathrm{AS}_K(\pi)$ for the asymptotic K-support introduced by Kashiwara and Vergne [**15**], that is, the limit cone of the set of highest weights of K-types in π. $\mathrm{AS}_K(\pi)$ is a closed cone in C_+.

We are ready to state a criterion for admissible restrictions.

THEOREM 5.6. *Let $G \supset G'$ be a pair of reductive Lie groups, and take maximal compact subgroups $K \supset K'$, respectively. Suppose π is an irreducible unitary representation of G.*

1) *Then the following two conditions are equivalent:*
   (i) $C_K(K') \cap \mathrm{AS}_K(\pi) = \{0\}$.
   (ii) *The restriction $\pi|_{K'}$ is K'-admissible.*
2) *If one of the equivalent conditions (i) or (ii) is fulfilled, then the restriction $\pi|_{G'}$ is infinitesimally discretely decomposable (see Definition 5.3), and the restriction $\pi|_{G'}$ is unitarily equivalent to the Hilbert direct sum:*

$$\pi|_{G'} \simeq \sum_{\tau \in \widehat{G'}}^{\oplus} n_\pi(\tau)\tau \qquad \text{with } n_\pi(\tau) < \infty \quad \text{for any} \quad \tau \in \widehat{G'}.$$

OUTLINE OF PROOF. The proof of the implication (i) $\Rightarrow$ (ii) was proved first by the author [**21**, Theorem 2.8] by using the singularity spectrum of hyperfunction characters in a more general setting where π is just a K-module such that the multiplicity

$$m_\pi(\tau) := \dim \mathrm{Hom}_K(\tau, \pi)$$

is of infra-exponential growth. In the same spirit, Hansen, Hilgert, and Keliny [**10**] gave an alternative proof by using the wave front set of distribution characters under the assumption that $m_\pi(\tau)$ is at most of polynomial growth. The last statement was proved in [**21**, Theorem 2.9] as a consequence of Proposition 5.5. See also [**28**]. □

The condition (i) in Theorem 5.6 is obviously fulfilled if $C_K(K') = \{0\}$ or if $\mathrm{AS}_K(\pi) = \{0\}$. We pin down the meanings of these extremal cases:

1) $C_K(K') = \{0\} \Leftrightarrow K' = K$. Then the conclusion in Theorem 5.6 2) is nothing but Harish-Chandra's admissibility theorem (see Fact 3.4).

2) $\mathrm{AS}_K(\pi) = \{0\} \Leftrightarrow \dim \pi < \infty$.

5.3. Algebraic approach. For a finitely generated $\mathfrak{g}$-module X, the associated variety $\mathcal{V}_{\mathfrak{g}_{\mathbb{C}}}(X)$ is a subvariety in the nilpotent cone $\mathcal{N}_{\mathfrak{g}_{\mathbb{C}}}$ of $\mathfrak{g}_{\mathbb{C}}^*$ (see [**42**]). In what follows, let X be the underlying $(\mathfrak{g}, K)$-module of $\pi \in \widehat{G}$ and Y the underlying $(\mathfrak{g}', K')$-module of $\tau(\in \widehat{G'})$.

We write $\mathrm{pr}_{\mathfrak{g} \to \mathfrak{g}'} : \mathfrak{g}_{\mathbb{C}}^* \to (\mathfrak{g}_{\mathbb{C}}')^*$ for the natural projection dual to $\mathfrak{g}_{\mathbb{C}}' \hookrightarrow \mathfrak{g}_{\mathbb{C}}$.

THEOREM 5.7 (see [**22**, Theorem 3.1]). *If* $\mathrm{Hom}_{\mathfrak{g}'}(Y, X) \neq \{0\}$, *then*

$$(5.2) \qquad \mathrm{pr}_{\mathfrak{g} \to \mathfrak{g}'}(\mathcal{V}_{\mathfrak{g}_{\mathbb{C}}}(X)) \subset \mathcal{V}_{\mathfrak{g}_{\mathbb{C}}'}(Y).$$

Theorem 5.7 leads us to a useful criterion for discrete decomposability by means of associated varieties:

COROLLARY 5.8. *If the restriction X is infinitesimally discretely decomposable as a $\mathfrak{g}'$-module, then $\mathrm{pr}_{\mathfrak{g} \to \mathfrak{g}'}(\mathcal{V}_{\mathfrak{g}_{\mathbb{C}}}(X))$ is contained in the nilpotent cone of $\mathfrak{g}_{\mathbb{C}}'$.*

REMARK 5.9. An analogous statement to Theorem 5.7 fails if we replace $\mathrm{Hom}_{\mathfrak{g}'}(Y, X) \neq \{0\}$ by $\mathrm{Hom}_{G'}(\tau, \pi|_{G'}) \neq \{0\}$.

REMARK 5.10. Analogous results to Theorem 5.7 and Corollary 5.8 hold in the category $\mathcal{O}$. See [**32**].

It is plausible that the following holds:

CONJECTURE 5.11. *The inclusion (5.2) in Theorem 5.7 is equality.*

Here are some affirmative results to Conjecture 5.11.

PROPOSITION 5.12.
1) X *is the Segal–Shale–Weil representation, and $\mathfrak{g}' = \mathfrak{g}_1' \oplus \mathfrak{g}_2'$ is the compact dual pair in $\mathfrak{g} = \mathfrak{sp}(n, \mathbb{R})$.*
2) X *is the underlying $(\mathfrak{g}, K)$-module of the minimal representation of $O(p, q)$ $(p + q$ even), and $(\mathfrak{g}, \mathfrak{g}')$ is a symmetric pair.*
3) X *is a (generalized) Verma module, and $(\mathfrak{g}, \mathfrak{g}')$ is a symmetric pair.*
4) $X = A_{\mathfrak{q}}(\lambda)$ *and $(\mathfrak{g}, \mathfrak{g}')$ is a symmetric pair.*

PROOF. The first statement could be read off from the results in [**7**, **38**] by case-by-case argument though they were not formulated by means of Theorem 5.7. See [**34**] for the proof of the second, and [**32**] for that of the third statement, respectively. The fourth statement is proved recently by Y. Oshima by using a $\mathcal{D}$-module argument. □

5.4. Restriction of $A_{\mathfrak{q}}(\lambda)$ to symmetric pair. For the restriction of $A_{\mathfrak{q}}(\lambda)$ to a reductive symmetric pair, our criterion is computable. Let us have a closer look.

Suppose that (G, G') is a symmetric pair defined by an involutive automorphism σ of G. As usual, the differential of σ will be denoted by the same letter. By taking a conjugation by G if necessary, we may and do assume that σ stabilizes K and that $\mathfrak{t}$ and $\Delta^+(\mathfrak{k}, \mathfrak{t})$ are chosen so that

1) $\mathfrak{t}^{-\sigma} := \mathfrak{t} \cap \mathfrak{k}^{-\sigma}$ is a maximal abelian subspace of $\mathfrak{k}^{-\sigma}$,
2) $\sum^+(\mathfrak{k}, \mathfrak{t}^{-\sigma}) := \{\lambda|_{\mathfrak{t}^{-\sigma}} : \lambda \in \Delta^+(\mathfrak{k}, \mathfrak{t})\} \setminus \{0\}$ is a positive system of the restricted root system $\sum(\mathfrak{k}, \mathfrak{t}^{-\sigma})$.

Then the momentum set $C_K(K')$ coincides with the dominant Weyl chamber ($\subset \sqrt{-1}(\mathfrak{t}^{-\sigma})^*$) with respect to $\Sigma^+(\mathfrak{k}, \mathfrak{t}^{-\sigma})$.

Let $\Delta(\mathfrak{u} \cap \mathfrak{p}) \subset \sqrt{-1}\mathfrak{t}^*$ be the set of weights in $\mathfrak{u} \cap \mathfrak{p}$, and $\mathbb{R}_+\Delta(\mathfrak{u} \cap \mathfrak{p})$ the closed cone spanned by $\Delta(\mathfrak{u} \cap \mathfrak{p})$. Then the asymptotic support $AS_K(A_{\mathfrak{q}}(\lambda))$ is contained in $\mathbb{R}_+\Delta(\mathfrak{u} \cap \mathfrak{p})$.

THEOREM 5.13. *The following six conditions on $(\mathfrak{g}, \mathfrak{g}^\sigma, \mathfrak{q})$ are equivalent:*

(i) *$A_{\mathfrak{q}}(\lambda)$ is non-zero and discretely decomposable as a $\mathfrak{g}'$-module for some λ in the weakly fair range.*

(i)' *$A_{\mathfrak{q}}(\lambda)$ is discretely decomposable as a $\mathfrak{g}'$-module for any λ in the weakly fair range.*

(ii) *$\mathbb{R}_+\Delta(\mathfrak{u} \cap \mathfrak{p}) \cap \sqrt{-1}\mathfrak{t}^{-\sigma} = \{0\}$.*

(iii) *$A_{\mathfrak{q}}(\lambda)$ is non-zero and K'-admissible for some λ in the weakly fair range.*

(iii)' *$A_{\mathfrak{q}}(\lambda)$ is K'-admissible for any λ in the weakly fair range.*

(iv) *$\mathrm{pr}_{\mathfrak{g}\to\mathfrak{g}'}(\mathcal{V}_{\mathfrak{g}c}(A_{\mathfrak{q}}(\lambda)))$ is contained in the nilpotent cone of $\mathfrak{g}'_{\mathbb{C}}$.*

PROOF. The equivalences (i) $\Leftrightarrow$ (i)' and (iii) $\Leftrightarrow$ (iii)' are easy. The implication (ii) $\Rightarrow$ (iii) was first proved in [**20**]. Alternatively, we can use Theorem 5.6 and the inclusive relation $AS_K(A_{\mathfrak{q}}(\lambda)) \subset \mathbb{R}_+\Delta(\mathfrak{u} \cap \mathfrak{p})$. This was the approach taken in [**21**]. Other implications are proved in [**23**] based on Theorem 5.7.   □

See [**36**] for the list of all such triples $(\mathfrak{g}, \mathfrak{g}^\sigma, \mathfrak{q})$.

REMARK 5.14. The implication (i) $\Rightarrow$ (iii)' and Theorem 5.6 show that

$$\dim \mathrm{Hom}_{G'}(\tau, \pi|_{G'}) < \infty \quad \text{for any } \tau \in \widehat{G'}$$

if the restriction $\pi|_{G'}$ is infinitesimally discretely decomposable for any $\pi_K \simeq A_{\mathfrak{q}}(\lambda)$.

6. Appendix – basic properties of $A_{\mathfrak{q}}(\lambda)$

This section gives a quick summary of basic properties on Zuckeman's derived functor modules and the "geometric quantization" of elliptic coadjoint orbits $\mathcal{O}_\lambda$ in the following scheme:

$\lambda \in \sqrt{-1}\mathfrak{g}^*$ an elliptic and integral element

$\mathcal{L}_{\lambda+\rho_\lambda} \to \mathcal{O}_\lambda$ a G-equivariant holomorphic line bundle

$H^*_{\bar\partial}(\mathcal{O}_\lambda, \mathcal{L}_{\lambda+\rho_\lambda})$ a Fréchet representation of G

π_λ a unitary representation of G

There is no new result in this section, and the normalization of the parameters and formulation follows the expository notes [23, 28]. See [16] for a more complete treatment and references therein.

6.1. Zuckerman derived functor modules.

Let G be a connected real reductive Lie group, $\mathfrak{g} = \mathfrak{k} + \mathfrak{p}$ a Cartan decomposition of the Lie algebra of $\mathfrak{g}$, and θ the corresponding Cartan involution.

Let $\mathfrak{q}$ be a θ-stable parabolic subalgebra of $\mathfrak{g}_{\mathbb{C}}$. Then the normalizer $L = N_G(\mathfrak{q})$ is a connected reductive subgroup of G, and the homogeneous space G/L carries a G-invariant complex structure such that the holomorphic tangent bundle $T(G/L)$ is given as a homogeneous bundle $G \times_L (\mathfrak{g}_{\mathbb{C}}/\mathfrak{q})$. Let $\mathfrak{l}_{\mathbb{C}}$ be the complexification of the Lie algebra $\mathfrak{l}$ of L, and $\mathfrak{u}$ the unipotent radical of $\mathfrak{q}$. Then we have a Levi decomposition $\mathfrak{q} = \mathfrak{l}_{\mathbb{C}} + \mathfrak{u}$. We set $\rho(\mathfrak{u})(X) := \frac{1}{2}\operatorname{Trace}(\operatorname{ad}(X) : \mathfrak{u} \to \mathfrak{u})$ for $X \in \mathfrak{l}$.

We say a Lie algebra homomorphism $\lambda : \mathfrak{l} \to \mathbb{C}$ is *integral* if λ lifts to a character of the connected group L, denoted by $\mathbb{C}_\lambda$. Then $\mathcal{L}_\lambda := G \times_{G_\lambda} \mathbb{C}_\lambda$ is a G-equivariant holomorphic line bundle over G/L. For example, $2\rho(\mathfrak{u})$ is integral, and the canonical bundle $\Omega(G/L) := \Lambda^{top}(T^*(G/L))$ is isomorphic to

$$(6.1) \qquad\qquad \Omega(G/L) \simeq \mathcal{L}_{2\rho(\mathfrak{u})}$$

as a G-equivariant holomorphic line bundle. The Zuckerman derived functor $W \mapsto \mathcal{R}^j_{\mathfrak{q}}(W \otimes \mathbb{C}_{\rho(\mathfrak{u})})$ is a covariant functor from the category of $(\mathfrak{l}, L \cap K)$-modules to the category of $(\mathfrak{g}, K)$-modules. We note that L is not necessarily compact. In this generality, H. Wong proved in [44] that the Dolbeault cohomology groups

$$H^j_{\bar{\partial}}(G/L, \mathcal{L}_\lambda \otimes \Omega(G/L)) \simeq H^j_{\bar{\partial}}(G/L, \mathcal{L}_{\lambda + 2\rho(\mathfrak{u})})$$

carry a Fréchet topology on which G acts continuously and that $\mathcal{R}^j_{\mathfrak{q}}(\mathbb{C}_{\lambda + \rho(\mathfrak{u})})$ are isomorphic to their underlying $(\mathfrak{g}, K)$-modules. We set $S := \dim_{\mathbb{C}}(\mathfrak{u} \cap \mathfrak{k}_{\mathbb{C}})$, and

$$A_{\mathfrak{q}}(\lambda) := \mathcal{R}^S_{\mathfrak{q}}(\mathbb{C}_{\lambda + \rho(\mathfrak{u})}).$$

In our normalization, $A_{\mathfrak{q}}(0)$ is an irreducible and unitarizable $(\mathfrak{g}, K)$-module with non-zero $(\mathfrak{g}, K)$-cohomology [43], and in particular, has the same infinitesimal character with that of the trivial one dimensional representation $\mathbb{C}$ of G.

6.2. Geometric quantization of elliptic coadjoint orbit.

Let $\lambda \in \sqrt{-1}\mathfrak{g}^*$. We say that the coadjoint orbit $\mathcal{O}_\lambda := \operatorname{Ad}^*(G) \cdot \lambda$ is *elliptic* if $\lambda|_{\mathfrak{p}} \equiv 0$. We identify $\mathfrak{g}$ with the dual space $\mathfrak{g}^*$ by a non-degenerate G-invariant bilinear form, and write $X_\lambda \in \sqrt{-1}\mathfrak{g}$ for the corresponding element to λ. Then $\operatorname{ad}(X_\lambda)$ is semisimple and all the eigenvalues are pure imaginary. The sum of the eigenspaces for non-negative eigenvalues of $-\sqrt{-1}\operatorname{ad}(X_\lambda)$ defines a θ-stable parabolic subalgebra $\mathfrak{q} = \mathfrak{l}_{\mathbb{C}} + \mathfrak{u}$, and consequently, the elliptic orbit $\mathcal{O}_\lambda$ carries a G-invariant complex structure such that the holomorphic tangent bundle is given by $G \times_L (\mathfrak{g}_{\mathbb{C}}/\mathfrak{q})$.

We set $\rho_\lambda := \rho(\mathfrak{u})$. If $\lambda + \rho_\lambda$ is integral, namely, if $\lambda + \rho_\lambda$ lifts to a character of L, then we can define a G-equivariant holomorphic line bundle $\mathcal{L}_{\lambda + \rho_\lambda} := G \times_L \mathbb{C}_{\lambda + \rho_\lambda}$ over $\mathcal{O}_\lambda$.

Here is a brief summary of the important achievements on unitary representation theory in 1980s and 1990s on the geometric quantization of elliptic orbits due to Parthasarathy, Zuckerman, Vogan and Wallach (algebraic construction, unitarizability of Zuckerman derived functor modules $A_{\mathfrak{q}}(\lambda)$), and Schmid and Wong (realization in Dolbeault cohomology, in particular, the closed range property of the $\bar{\partial}$-operator) among others. See [16, 28] for the original references therein.

FACT 6.1. *Let* $\lambda \in \sqrt{-1}\mathfrak{g}^*$ *be elliptic such that* $\lambda + \rho_\lambda$ *is integral.*

1) *(vanishing theorem)* $H^j_{\bar{\partial}}(\mathcal{O}_\lambda, \mathcal{L}_{\lambda+\rho_\lambda}) = 0$ *if* $j \neq S$.

2) *The Dolbeault cohomology group* $H^S_{\bar{\partial}}(\mathcal{O}_\lambda, \mathcal{L}_{\lambda+\rho_\lambda})$ *carries a Fréchet topology, on which* G *acts continuously. It is the maximal globalization of* $\mathcal{R}^S_{\mathfrak{q}}(\mathbb{C}_\lambda) = A_{\mathfrak{q}}(\lambda - \rho_\lambda)$ *in the sense of Schmid.*

3) *(unitarizability) There is a dense subspace* $\mathcal{H}$ *in* $H^S_{\bar{\partial}}(\mathcal{O}_\lambda, \mathcal{L}_{\lambda+\rho_\lambda})$ *on which a* G*-invariant Hilbert structure exists. We denote by* π_λ *the resulting unitary representation on* $\mathcal{H}$.

4) *If* λ *is in the good range in the sense of Vogan, then the unitary representation of* G *on* $\mathcal{H}$ *is irreducible and non-zero.*

Here, by 'good range', we mean that λ satisfies

$$(6.2) \qquad \langle \lambda + \rho_{\mathfrak{l}}, \alpha \rangle > 0 \quad \text{for any} \quad \alpha \in \Delta(\mathfrak{u}, \mathfrak{h}_{\mathbb{C}}),$$

where $\mathfrak{h}$ is a fundamental Cartan subalgebra containing X_λ and $\rho_{\mathfrak{l}}$ is half the sum of positive roots for $\Delta(\mathfrak{l}_{\mathbb{C}}, \mathfrak{h}_{\mathbb{C}})$. (This condition is independent of the choice of $\mathfrak{h}$ and $\Delta^+(\mathfrak{l}_{\mathbb{C}}, \mathfrak{h}_{\mathbb{C}})$.)

References

[1] J. Adams, Unitary highest weight modules, Adv. in Math. **63** (1987), 113–137.

[2] E. van den Ban, Invariant differential operators on a semisimple symmetric space and finite multiplicities in a Plancherel formula, Arkiv Mat. **25** (1987), 175–187.

[3] B. Binegar and R. Zierau, Unitarization of a singular representation of $SO(p,q)$, Commun. Math. Phys. **138** (1991), 245–258.

[4] P. Delorme, Formule de Plancherel pour les espaces symétriques réductifs, Ann. of Math. (2), **147** (1998), 417–452.

[5] M. Duflo and J. A. Vargas, Branching laws for square integrable representations, Proc. Japan Acad. Ser. A, Math. Sci. **86** (2010), 49–54.

[6] E. Dunne and R. Zierau, The automorphism groups of complex homogeneous spaces. Math. Ann. **307** (1997), 489–503.

[7] T. Enright and J. Willenbring, Hilbert series, Howe duality and branching for classical groups, Ann. of Math. (2) **159** (2004), 337–375.

[8] I. M. Gelfand and M. I. Graev, Geometry of homogeneous spaces, representations of groups in homogeneous spaces and related questions of integral geometry. I, Trudy Moskov. Mat. Obšč. 8 (1959), 321–390.

[9] B. Gross and N. Wallach, Restriction of small discrete series representations to symmetric subgroups, *Proc. Sympos. Pure Math.*, **68** (2000), Amer. Math. Soc., 255–272.

[10] S. Hansen, J. Hilgert and S. Keliny, Asymptotic K-support and restrictions of representations, Represent. Theory **13** (2009), 460–469.

[11] R. Howe, θ-series and invariant theory, Proc. Symp. Pure Math. **33** (1979), Amer. Math. Soc., 275–285.

[12] R. Howe, Reciprocity laws in the theory of dual pairs, Progr. in ath. Birkhäuser, 40 (1983), 159–175

[13] L. K. Hua, Harmonic Analysis of Functions of Several Complex Variables in the Classical Domains, Amer. Math. Soc., 1963.

[14] H. P. Jakobsen and M. Vergne, Restrictions and expansions of holomorphic representations, J. Funct. Anal. **34** (1979), 29–53.

[15] M. Kashiwara and M. Vergne, K-types and singular spectrum, In: Lect. Notes in Math. **728**, 1979, Springer-Verlag, 177–200.

[16] A. W. Knapp and D. Vogan, Jr., Cohomological Induction and Unitary Representations, Princeton U.P., 1995.

[17] T. Kobayashi, Unitary representations realized in L^2-sections of vector bundles over semisimple symmetric spaces, Proceedings at the 27-28th Symp. of Functional Analysis and Real Analysis (1989), Math. Soc. Japan, 39–54.

[18] T. Kobayashi, The restriction of $A_q(\lambda)$ to reductive subgroups, *Proc. Japan Acad.*, **69** (1993), 262–267.

[19] T. Kobayashi, Irreducible restriction of $A_q(\lambda)$ to reductive subgroups, Lecture at Summer workshop on representation theory, Polytechnic University, August 24, 1993.

[20] T. Kobayashi, Discrete decomposability of the restriction of $A_q(\lambda)$ with respect to reductive subgroups and its application, *Invent. Math.*, **117** (1994), 181–205.

[21] T. Kobayashi, Discrete decomposability of the restriction of $A_q(\lambda)$, II. —micro-local analysis and asymptotic K-support, *Ann. of Math.*, **147** (1998), 709–729.

[22] T. Kobayashi, Discrete decomposability of the restriction of $A_q(\lambda)$, III. —restriction of Harish-Chandra modules and associated varieties, *Invent. Math.*, **131** (1998), 229–256.

[23] T. Kobayashi, Harmonic analysis on homogeneous manifolds of reductive type and unitary representation theory, *Sugaku*, **46** (1994), Math. Soc. Japan (in Japanese), 124–143; *Translations, Series II*, Selected Papers on Harmonic Analysis, Groups, and Invariants (K. Nomizu, ed.), **183** (1998), Amer. Math. Soc., 1–31.

[24] T. Kobayashi, Discrete series representations for the orbit spaces arising from two involutions of real reductive Lie groups, *J. Funct. Anal.*, **152** (1998), 100–135.

[25] T. Kobayashi, Theory of discrete decomposable branching laws of unitary representations of semisimple Lie groups and some applications, *Sugaku*, **51** (1999), Math. Soc. Japan (in Japanese), 337–356; English translation, *Sugaku Exposition*, **18** (2005), Amer. Math. Soc. 1–37.

[26] T. Kobayashi, Discretely decomposable restrictions of unitary representations of reductive Lie groups — examples and conjectures, *Advanced Study in Pure Math.*, **26** (2000), 98–126.

[27] T. Kobayashi, Unitary representations and branching laws, *Proceedings of the I.C.M. 2002 at Beijing*, **2** (2002), 615–627.

[28] T. Kobayashi, Restrictions of unitary representations of real reductive groups, *Progr. in Math.* **229**, pages 139–207, Birkhäuser, 2005.

[29] T. Kobayashi, Multiplicity-free representations and visible actions on complex manifolds, *Publ. Res. Inst. Math. Sci.* **41**(2005), 497–549 (a special issue of Publications of the Research Institute for Mathematical Sciences commemorating the fortieth anniversary of the founding of the Research Institute for Mathematical Sciences).

[30] T. Kobayashi, Multiplicity-free theorems of the restrictions of unitary highest weight modules with respect to reductive symmetric pairs. *Progr. Math.* **255**, pages 45–109. Birkhäuser, 2007.

[31] T. Kobayashi, Visible actions on symmetric spaces. *Transformation Groups*, **12** (2007), 671–694.

[32] T. Kobayashi, Restrictions of generalized Verma modules to symmetric pairs, submitted, arXiv:1008.4544

[33] T. Kobayashi and T. Oda, Vanishing theorem of modular symbols on locally symmetric spaces, *Comment. Math. Helvetici*, **73** (1998), 45–70.

[34] T. Kobayashi and B. Ørsted, Analysis on minimal representations of $O(p,q)$, Part II. Branching Laws, *Adv. in Math.*, **180** (2003), 513–550.

[35] T. Kobayashi, B. Ørsted, and M. Pevzner, Geometric analysis on small unitary representations of $GL(n, \mathbb{R})$, *J. Funct. Anal.*, **260** (2011), 1682–1720.

[36] T. Kobayashi and Y. Oshima, Classification of discretely decomposable $A_q(\lambda)$ with respect to reductive symmetric pairs, submitted, arXiv:1104.4400

[37] B. Krötz and R. J. Stanton, Holomorphic extensions of representations. I. Automorphic functions, *Ann. of Math.* (2) **159** (2004), 641–724.

[38] K. Nishiyama, H. Ochiai, and K. Taniguchi, Bernstein degree and associated cycles of Harish-Chandra modules — Hermitian symmetric case —, *Astérisque*, **273** (2001), 13–80.

[39] B. Ørsted and B. Speh, Branching laws for some unitary representations of $SL(4, \mathbb{R})$, SIGMA 4 (2008) doi:10.3842/SIGMA.2008.017.

[40] W. Schmid, Die Randwerte holomorphe Funktionen auf hermetisch symmetrischen Raumen. Invent. Math. **9** (1969–70), 61–80.

[41] H. Sekiguchi, The Penrose transform for $Sp(n, \mathbb{R})$ and singular unitary representations, J. Math. Soc. Japan **54** (2002), 215–253.

[42] D. A. Vogan, Jr., Associated varieties and unipotent representations, Harmonic Analysis on Reductive Lie Groups, *Progress in Math.*, **101** (1991), Birkhäuser, 315–388.

[43] D. A. Vogan, Jr. and G. J. Zuckerman, Unitary representations with nonzero cohomology, Compositio Math. **53** (1984), 51–90.

[44] H. Wong, Dolbeault cohomological realization of Zuckerman modules associated with finite rank representations, J. Funct. Anal. **129** (1995), 428–454.
[45] F. Zhu and K. Liang, On a branching law of unitary representations and a conjecture of Kobayashi, C. R. Acad. Sci. Paris, Ser. I, **348** (2010), 959–962.

GRADUATE SCHOOL OF MATHEMATICAL SCIENCES, IPMU,, THE UNIVERSITY OF TOKYO, KOMABA, MEGURO, TOKYO, 153-8914 JAPAN
    E-mail address: toshi@ms.u-tokyo.ac.jp

Contemporary Mathematics
Volume **557**, 2011

Chiral Equivariant Cohomology
of Spheres

Bong H. Lian, Andrew R. Linshaw, and Bailin Song

*In honor of our friend and teacher, Professor Gregg J. Zuckerman,
on the occasion of his 60th birthday*

ABSTRACT. The chiral equivariant cohomology of a smooth manifold with a Lie group action is a vertex algebra valued cohomology which contains and generalizes the classical equivariant cohomology à la H. Cartan. This paper is a survey of some of the highlights of this theory. We describe some general structural features, including a quasi-conformal structure, equivariant homotopy invariance, and the values of this cohomology on homogeneous spaces. We show that for any simple, connected group G, there is a sphere with infinitely many actions of G with the same classical equivariant cohomology, which can all be distinguished by our new invariant. We conclude with a discussion of the chiral equivariant cohomology of a point, whose structure is still mysterious in the case of nonabelian groups.

Contents

MSC 57S25

1. Introduction

Let G be a compact, connected Lie group with complexified Lie algebra $\mathfrak{g}$, and let M be a topological G-space. The equivariant cohomology $H_G^*(M)$ is defined to be

$$H^*((M \times E)/G)),$$

where E is any contractible space on which G acts freely. This is known as the *Borel construction*. If M is a smooth manifold on which G acts by diffeomorphisms, there is a de Rham model of $H_G^*(M)$ due to H. Cartan [1][2], and developed further by Duflo-Kumar-Vergne [3] and Guillemin-Sternberg [7]. Following [7], one can define the equivariant cohomology $H_G^*(A)$ of any G^*-algebra A. A G^*-algebra is a commutative superalgebra equipped with an action of G, together with a compatible action of a certain differential Lie superalgebra $(\mathfrak{sg}, d)$ associated to the Lie algebra $\mathfrak{g}$ of G. A G^*-algebra (A, d) is a cochain complex, and the subalgebra of A which is both G-invariant and killed by $\mathfrak{sg}$ forms a subcomplex known as the basic subcomplex. $H_G^*(A)$ is defined to be $H_{bas}^*(A \otimes W(\mathfrak{g}))$, where $W(\mathfrak{g}) = \Lambda(\mathfrak{g}^*) \otimes S(\mathfrak{g}^*)$ is the Weil complex of $\mathfrak{g}$. The de Rham model of $H_G^*(M)$ is obtained by taking A to be the algebra $\Omega(M)$ of differential forms on M, and $H_G^*(\Omega(M)) \cong H_G^*(M)$ by an equivariant version of the de Rham theorem.

In [9], the chiral equivariant cohomology $\mathbf{H}_G^*(\mathcal{A})$ of an $O(\mathfrak{sg})$-algebra $\mathcal{A}$ was introduced as a vertex algebra analogue of the equivariant cohomology of G^*-algebras. The main idea is to replace the key ingredients in Cartan's definition of $H_G^*(A)$ with their vertex algebra counterparts, using an appropriate notion of invariant theory. Examples of $O(\mathfrak{sg})$-algebras include the semi-infinite Weil complex $\mathcal{W}(\mathfrak{g})$, which was introduced by Feigin-Frenkel in [4], and the chiral de Rham complex $\mathcal{Q}(M)$ of a smooth G-manifold M, which was introduced by Malikov-Schechtman-Vaintrob in [14]. In [11], the chiral equivariant cohomology functor was extended to the larger categories of $\mathfrak{sg}[t]$-algebras and $\mathfrak{sg}[t]$-modules. The main example of an $\mathfrak{sg}[t]$-algebra which is *not* an $O(\mathfrak{sg})$-algebra is the subalgebra $\mathcal{Q}'(M) \subset \mathcal{Q}(M)$ generated by the weight-zero subspace. Both $\mathbf{H}_G^*(\mathcal{Q}(M))$ and $\mathbf{H}_G^*(\mathcal{Q}'(M))$ are "chiralizations" of $H_G^*(M)$, that is, vertex algebras equipped with weight gradings

$$\mathbf{H}_G^*(\mathcal{Q}(M)) = \bigoplus_{m \geq 0} \mathbf{H}_G^*(\mathcal{Q}(M))[m],$$

$$\mathbf{H}_G^*(\mathcal{Q}'(M)) = \bigoplus_{m \geq 0} \mathbf{H}_G^*(\mathcal{Q}'(M))[m],$$

such that $\mathbf{H}_G^*(\mathcal{Q}(M))[0] \cong H_G^*(M) \cong \mathbf{H}_G^*(\mathcal{Q}'(M))[0]$.

We briefly recall the construction of chiral equivariant cohomology, following the notation in [9][11][12]. We will assume that the reader is familiar with the basic notions in vertex algebra theory. For a list of references, see page 102 of [9]. A *differential vertex algebra* (DVA) is a degree graded vertex algebra $\mathcal{A}^* = \oplus_{p \in \mathbf{Z}} \mathcal{A}^p$ equipped with a vertex algebra derivation d of degree 1 such that $d^2 = 0$. A DVA will be called *degree-weight graded* if it has an additional $\mathbf{Z}_{\geq 0}$-grading by weight, which is compatible with the degree in the sense that $\mathcal{A}^p = \oplus_{n \geq 0} \mathcal{A}^p[n]$. There is an auxiliary structure on a DVA which is analogous to the structure of a G^*-algebra in [7]. Associated to $\mathfrak{g}$ is a Lie superalgebra $\mathfrak{sg} := \mathfrak{g} \triangleright \mathfrak{g}^{-1}$ with bracket $[(\xi, \eta), (x, y)] = ([\xi, x], [\xi, y] - [x, \eta])$, which is equipped with a differential $d : (\xi, \eta) \mapsto (\eta, 0)$. This differential extends to the loop algebra $\mathfrak{sg}[t, t^{-1}]$, and gives rise to a vertex algebra derivation on the corresponding current algebra $O(\mathfrak{sg}) := O(\mathfrak{sg}, 0)$. Here 0 denotes the zero bilinear form on $\mathfrak{sg}$. An $O(\mathfrak{sg})$-algebra is a degree-weight graded DVA $\mathcal{A}$ equipped with a DVA homomorphism $\rho : O(\mathfrak{sg}) \to \mathcal{A}$, which we denote by $(\xi, \eta) \to L_\xi + \iota_\eta$.

In [11], it was observed that the chiral equivariant cohomology functor can be defined on the larger class of spaces which carry only a representation of the Lie subalgebra $\mathfrak{sg}[t]$ of $\mathfrak{sg}[t, t^{-1}]$. An $\mathfrak{sg}[t]$-module is a degree-weight graded complex $(\mathcal{A}, d_{\mathcal{A}})$ equipped with a Lie algebra homomorphism $\rho : \mathfrak{sg}[t] \rightarrow End\ \mathcal{A}$, which we denote by $(\xi, \eta)t^n \rightarrow L_\xi(n) + \iota_\xi(n)$, $n \geq 0$. We also require that for all $x \in \mathfrak{sg}[t]$ we have $\rho(dx) = [d_{\mathcal{A}}, \rho(x)]$, and $\rho(x)$ has degree 0 whenever x is even in $\mathfrak{sg}[t]$, and degree -1 whenever x is odd, and has weight $-n$ if $x \in \mathfrak{sg}t^n$. Finally, we require $\mathcal{A}$ to admit a compatible action $\hat{\rho} : G \rightarrow Aut(\mathcal{A})$ of G satisfying:

$$\frac{d}{dt}\hat{\rho}(exp(t\xi))|_{t=0} = L_\xi(0), \tag{1.1}$$

$$\hat{\rho}(g)L_\xi(n)\hat{\rho}(g^{-1}) = L_{Ad(g)(\xi)}(n), \tag{1.2}$$

$$\hat{\rho}(g)\iota_\xi(n)\hat{\rho}(g^{-1}) = \iota_{Ad(g)(\xi)}(n), \tag{1.3}$$

$$\hat{\rho}(g)d\hat{\rho}(g^{-1}) = d, \tag{1.4}$$

for all $\xi \in \mathfrak{g}$, $g \in G$, and $n \geq 0$. These conditions are analogous to Equations (2.23)-(2.26) of [7]. In order for (1.1) to make sense, we must be able to differentiate along appropriate curves in $\mathcal{A}$, which is the case in our main examples. Given an $\mathfrak{sg}[t]$-module $(\mathcal{A}, d)$, we define the chiral horizontal, invariant and basic subspaces of $\mathcal{A}$ to be respectively

$$\mathcal{A}_{hor} = \{a \in \mathcal{A} | \rho(x)a = 0\ \forall x \in \mathfrak{g}^{-1}[t]\},$$

$$\mathcal{A}_{inv} = \{a \in \mathcal{A} | \rho(x)a = 0\ \forall x \in \mathfrak{g}[t],\ \hat{\rho}(g)(a) = a\ \forall g \in G\},$$

$$\mathcal{A}_{bas} = \mathcal{A}_{hor} \cap \mathcal{A}_{inv}.$$

Both $\mathcal{A}_{inv}$ and $\mathcal{A}_{bas}$ are subcomplexes of $\mathcal{A}$, but $\mathcal{A}_{hor}$ is not a subcomplex of $\mathcal{A}$ in general.

An $O(\mathfrak{sg})$-algebra which plays an important role in our theory is the semi-infinite Weil complex $\mathcal{W}(\mathfrak{g})$. As a vertex algebra, $\mathcal{W}(\mathfrak{g})$ is just the $bc\beta\gamma$-system $\mathcal{E}(\mathfrak{g}) \otimes \mathcal{S}(\mathfrak{g})$. In this notation, $\mathcal{E}(\mathfrak{g})$ is the vertex algebra with odd generators $b^\xi, c^{\xi'}$, which are linear in $\xi \in \mathfrak{g}$ and $\xi' \in \mathfrak{g}^*$, and satisfy the OPE relations

$$b^\xi(z)c^{\xi'}(w) \sim \langle \xi', \xi \rangle (z - w)^{-1},$$

$$c^{\xi'}(z)b^\xi(w) \sim \langle \xi', \xi \rangle (z - w)^{-1},$$

$$b^\xi(z)b^\eta(w) \sim 0, \qquad c^{\xi'}(z)c^{\eta'}(w) \sim 0.$$

Here $\langle , \rangle$ denotes the natural pairing between $\mathfrak{g}^*$ and $\mathfrak{g}$. Similarly, $\mathcal{S}(\mathfrak{g})$ is the vertex algebra with even generators $\beta^\xi, \gamma^{\xi'}$, which are linear in $\xi \in \mathfrak{g}$ and $\xi' \in \mathfrak{g}^*$, and satisfy

$$\beta^\xi(z)\gamma^{\xi'}(w) \sim \langle \xi', \xi \rangle (z - w)^{-1},$$

$$\gamma^{\xi'}(z)\beta^\xi(w) \sim -\langle \xi', \xi \rangle (z - w)^{-1},$$

$$\beta^\xi(z)\beta^\eta(w) \sim 0, \qquad \gamma^{\xi'}(z)\gamma^{\eta'}(w) \sim 0.$$

$\mathcal{W}(\mathfrak{g})$ possesses a Virasoro element $\omega_{\mathcal{W}}$ given by

$$\omega_{\mathcal{W}} = \omega_{\mathcal{E}} + \omega_{\mathcal{S}},$$

$$\omega_{\mathcal{E}} = -\sum_{i=1}^{n} : b^{\xi_i} \partial c^{\xi_i'} :, \qquad \omega_{\mathcal{S}} = \sum_{i=1}^{n} : \beta^{\xi_i} \partial \gamma^{\xi_i'} :, \tag{1.5}$$

where $n = dim(\mathfrak{g})$. The generators $\beta^\xi, \gamma^{\xi'}, b^\xi, c^{\xi'}$ are primary of weights $1, 0, 1, 0$ with respect to $\omega_{\mathcal{W}}$. There is an additional grading by degree, in which $\beta^\xi, \gamma^{\xi'}, b^\xi, c^{\xi'}$ (and their respective derivatives) have degrees $-2, 2, -1, 1$. Note that the weight-zero component is isomorphic to the classical Weil algebra $W(\mathfrak{g}) = \bigwedge(\mathfrak{g}^*) \otimes Sym(\mathfrak{g}^*)$, where the degree 1 and degree 2 generators $c^{\xi'}, \gamma^{\xi'}$ play the role of connection 1-forms and curvature 2-forms, respectively.

Next, we recall the $O(\mathfrak{sg})$-algebra structure on $\mathcal{W}(\mathfrak{g})$. Define vertex operators

$$\Theta_{\mathcal{W}}^\xi = \Theta_{\mathcal{E}}^\xi + \Theta_{\mathcal{S}}^\xi, \qquad D = J + K$$

$$\Theta_{\mathcal{E}}^{\xi} = \sum_{i=1}^{n} : b^{[\xi, \xi_i]} c^{\xi_i'} :, \qquad \Theta_{\mathcal{S}}^{\xi} = -\sum_{i=1}^{n} : \beta^{[\xi, \xi_i]} \gamma^{\xi_i'} : .$$

$$J = \sum_{i=1}^{n} : (\Theta_{\mathcal{S}}^{\xi_i} + \frac{1}{2} \Theta_{\mathcal{E}}^{\xi_i}) c^{\xi_i'} :, \qquad K = \sum_{i=1}^{n} : \gamma^{\xi_i} b^{\xi_i} : .$$

The Fourier modes $J(0)$, $K(0)$, and $D(0)$ are called the BRST, chiral Koszul, and chiral Weil differentials, respectively. They satisfy $J(0)^2 = K(0)^2 = D(0)^2 = [K(0), J(0)] = 0$. Moreover, we have

$$\Theta_{\mathcal{W}}^{\xi}(z) \Theta_{\mathcal{W}}^{\eta}(w) \sim \Theta_{\mathcal{W}}^{[\xi, \eta]}(w)(z - w)^{-1},$$

$$\Theta^{\xi}(z) b^{\eta}(w) \sim b^{[\xi, \eta]}(w)(z - w)^{-1},$$

$$[D(0), b^{\xi}(z)] = \Theta_{\mathcal{W}}^{\xi}(z), \qquad [D(0), \Theta_{\mathcal{W}}^{\xi}(z)] = 0.$$

In other words, the map $O(\mathfrak{sg}) \to \mathcal{W}(\mathfrak{g})$ sending $(\xi, \eta)(z) \mapsto \Theta^{\xi}(z) + b^{\eta}(z)$ and sending $d \mapsto [D(0), -]$, is a homomorphism of DVAs.

Definition 1.1. *For any $\mathfrak{sg}[t]$-module $(\mathcal{A}, d_{\mathcal{A}})$, we define its chiral basic cohomology $\mathbf{H}_{bas}^*(\mathcal{A})$ to be $H^*(\mathcal{A}_{bas}, d_{\mathcal{A}})$. We define its chiral equivariant cohomology $\mathbf{H}_G^*(\mathcal{A})$ to be $\mathbf{H}_{bas}^*(\mathcal{W}(\mathfrak{g}) \otimes \mathcal{A})$.*

In the case where $\mathcal{A}$ is the trivial $\mathfrak{sg}[t]$-module $\mathbf{C}$, we will refer to $\mathbf{H}_G^*(\mathbf{C}) = \mathbf{H}_{bas}^*(\mathcal{W}(\mathfrak{g}))$ as the *chiral point algebra*. It plays the role of $H_G^*(pt) \cong Sym(\mathfrak{g}^*)^G$ in the classical theory, and $\mathbf{H}_G^*(\mathbf{C})[0] \cong H_G^*(pt)$. For any $\mathfrak{sg}[t]$-module $\mathcal{A}$, there is a chiral Chern-Weil map

$$\kappa_G : \mathbf{H}_G^*(\mathbf{C}) \to \mathbf{H}_G^*(\mathcal{A}), \tag{1.6}$$

and we regard the elements of $\mathbf{H}_G^*(\mathbf{C})$ as universal characteristic classes.

In their seminal paper [14], Malikov-Schechtman-Vaintrob introduced a sheaf Ω_M^{ch} of vertex algebras on any nonsingular algebraic variety

M, which they call the *chiral de Rham sheaf*. As the authors pointed out, a similar construction can be done in the analytic and smooth categories. In this paper we work exclusively in the smooth setting, and we use the notation $\mathcal{Q}_M$ instead of Ω_M^{ch}. It is a *weak sheaf* of vertex algebras, meaning that it is a presheaf in which

$$0 \to \mathcal{Q}_M(U) \to \prod_i \mathcal{Q}_M(U_i) \rightrightarrows \prod_{i,j} \mathcal{Q}_M(U_i \cap U_j) \qquad (1.7)$$

is exact for *finite* open covers $\{U_i\}$ of an open set U (see Section 1.1 of [11]). The space $\mathcal{Q}(M) = \mathcal{Q}_M(M)$ of global sections is a degree-weight graded vertex algebra which contains the ordinary de Rham algebra $\Omega(M)$ as the weight-zero subspace. There is a square-zero derivation $d_{\mathcal{Q}}$ on $\mathcal{Q}(M)$ whose restriction to $\Omega(M)$ is the ordinary de Rham differential d, and the inclusion of complexes $(\Omega(M), d) \hookrightarrow (\mathcal{Q}(M), d_{\mathcal{Q}})$ induces an isomorphism in cohomology. When M is a G-manifold, $\mathcal{Q}(M)$ has the structure of an $O(\mathfrak{sg})$-algebra, and we may consider its chiral equivariant cohomology. By Theorem 1.5 of [12], for any compact, connected group G and any G-manifold M we have an isomorphism of vertex algebras

$$\mathbf{H}_G^*(\mathcal{Q}(M)) \cong \mathbf{H}_{K_0}^*(\mathbf{C}) \otimes H_{G'}^*(M),$$

where K_0 is the identity component of $K = Ker(G \to Diff(M))$ and $G' = G/K_0$. Here $H_{G'}^*(M)$ is regarded as a vertex algebra in which all circle products except $\circ_{-1}$ are trivial. In particular, the positive-weight subspace $\mathbf{H}_G^*(\mathcal{Q}(M))_+ = \bigoplus_{m>0} \mathbf{H}_G^*(\mathcal{Q}(M))[m]$ vanishes whenever K is finite. This shows that $\mathbf{H}_G^*(\mathcal{Q}(M))_+$ only depends on K, and therefore carries no nontrivial information about M.

The sheaf $\mathcal{Q}_M$ contains a subsheaf $\mathcal{Q}'_M$ of abelian vertex algebras which was studied in [11]. For each open set $U \subset M$, $\mathcal{Q}'(U) = \mathcal{Q}'_M(U)$ is defined to be the vertex subalgebra of $\mathcal{Q}(U)$ generated by the weight-zero subspace $\mathcal{Q}(U)[0] \cong \Omega(U)$. This sheaf has better functorial properties than $\mathcal{Q}_M$. It is contravariant in M since it is built from the de Rham sheaf, whereas $\mathcal{Q}_M$ is not functorial in M since its construction depends on both vector fields and differential forms on M. If M is a G-manifold, $\mathcal{Q}'(M)$ is not an $O(\mathfrak{sg})$-algebra, but it retains the structure of an $\mathfrak{sg}[t]$-algebra, so its chiral equivariant cohomology is well-defined. In this paper, our main focus is on $\mathbf{H}^*_G(\mathcal{Q}'(M))$, which we denote by $\mathbf{H}^*_G(M)$ for simplicity. Unlike the case of $\mathcal{Q}(M)$, the positive weight subspace $\mathbf{H}^*_G(M)_+ = \bigoplus_{m>0} \mathbf{H}^*_G(M)[m]$ contains nontrivial information about M.

Using a G-invariant partition of unity on M, it is easy to check (see Section 2 of [12]) that for G-invariant open sets $U, V \subset M$, there exists a Mayer-Vietoris sequence

$$\cdots \to \mathbf{H}^{k-1}_G(U \cap V) \to \mathbf{H}^k_G(U \cup V) \to \mathbf{H}^k_G(U) \oplus \mathbf{H}^k_G(V)$$
$$\to \mathbf{H}^k_G(U \cap V) \to \cdots .$$

There are three foundational results from [12] that we present in this paper. First, $\mathbf{H}^*_G(-)$ is invariant under G-equivariant homotopy. That is, if M and N are G-manifolds and $\phi_0, \phi_1 : M \to N$ are equivariantly homotopic G-maps, the induced maps $\phi_0^*, \phi_1^* : \mathbf{H}^*_G(N) \to \mathbf{H}^*_G(M)$ are the same. Second, for any G and M, $\mathbf{H}^*_G(M)$ possesses a *quasi-conformal structure*, i.e., an action of the subalgebra of the Virasoro algebra generated by $\{L_n | \ n \geq -1\}$, such that L_{-1} acts by ∂ and L_0 acts by $n \cdot id$ on the subspace of weight n. The quasi-conformal structure provides a powerful vanishing criterion for $\mathbf{H}^*_G(M)_+$; it suffices to show that L_0 acts by zero. Third, we give a description of $\mathbf{H}^*_G(G/H)$ for any closed subgroup $H \subset G$, relative to the vertex algebra $\mathbf{H}^*_{K_0}(\mathbf{C})$, where K_0 is the identity component of $K = Ker(G \to Diff(G/H))$. Using these results, we show that for compact M, the degree p and weight n subspace $\mathbf{H}^p_G(M)[n]$ is finite-dimensional for all $p \in \mathbf{Z}$ and $n \geq 0$, which extends a classical result in the case $n = 0$. Finally, using results of R. Oliver [15][16] which describe the fixed-point subsets of group actions on contractible spaces, we show that for any simple group G, there is a sphere with infinitely many smooth actions of G with the same classical equivariant cohomology, which can all be distinguished by $\mathbf{H}^*_G(-)$. One can even construct a morphism $f : M \to N$ in the category of compact G-manifolds which induces a ring isomorphism $H^*_G(N) \to H^*_G(M)$ with $\mathbf{Z}$-coefficients, such that $\mathbf{H}^*_G(M) \neq \mathbf{H}^*_G(N)$. Hence $\mathbf{H}^*_G(-)$ *is a strictly stronger invariant than* $H^*_G(-)$ *on the category of compact G-manifolds.*

We conclude with a few remarks about the chiral point algebra $\mathbf{H}^*_G(\mathbf{C})$. If G is abelian, $\mathbf{H}^*_G(\mathbf{C})$ is the abelian vertex algebra generated by $H^*_G(pt)$, but for nonabelian G, $\mathbf{H}^*_G(\mathbf{C})$ possesses a rich algebraic structure. For semisimple G, $\mathbf{H}^*_G(\mathbf{C})$ has a conformal structure of central charge zero, and the Virasoro class $\mathbf{L}$ has no classical analogue in $H^*_G(pt)$. In [13], the structure of $\mathbf{H}^*_G(\mathbf{C})$ was studied in the simplest nontrivial case $G = SU(2)$, and we briefly recall the main results. In particular, we construct an injective linear map

$$\Psi : U(\mathfrak{sl}_2) \to \mathbf{H}^*_{SU(2)}(\mathbf{C})$$

whose image consists entirely of nonclassical elements. The Virasoro element $\mathbf{L}$ is precisely $\Psi(1)$, and we conjecture that the image of Ψ, together with the classical generator of $H^*_{SU(2)}(pt)$, forms a strong generating set for $\mathbf{H}^*_{SU(2)}(\mathbf{C})$. We expect that there exists an alternative construction of $\mathbf{H}^*_G(M)$ which is suitable for any topological G-space M, and gives the same cohomology as our theory when M is a manifold. Such a construction would necessarily include a topological realization of the chiral point algebra. We hope that the structure we describe in the case $G = SU(2)$ may give some hint about where to look for such a construction.

2. Homotopy invariance

Let M and N be G-manifolds, and let $\phi_0, \phi_1 : M \to N$ be G-equivariant maps. Let I denote the interval $[0, 1]$, which we regard as a

G-manifold equipped with the trivial action. A *G-equivariant homotopy* from ϕ_0 to ϕ_1 is a smooth G-equivariant map $\Phi : M \times I \to N$ such that for all $x \in M$, $\Phi(x, 0) = \phi_0(x)$ and $\Phi(x, 1) = \phi_1(x)$. For each $t \in I$, $\phi_t : M \to N$ will denote the map $\phi_t(x) = \Phi(x, t)$.

Theorem 2.1. *Let M and N be G-manifolds, and let $\phi_0, \phi_1 : M \to N$ be G-equivariant maps. If there exists a G-equivariant homotopy Φ from ϕ_0 to ϕ_1, the induced maps $\phi_0^*, \phi_1^* : \mathbf{H}_G^*(N) \to \mathbf{H}_G^*(M)$ are the same.*

The proof is modeled on the proof of the corresponding statement in [7] in the classical setting. Suppose first that $\mathcal{A}$ and $\mathcal{B}$ are $\mathfrak{sg}[t]$-algebras. Define a *chiral chain homotopy* to be a linear map $P : \mathcal{A} \to \mathcal{B}$, homogeneous of weight 0 and degree -1, which is G-equivariant and satisfies

$$P\iota_\xi^{\mathcal{A}}(k) + \iota_\xi^{\mathcal{B}}(k)P = 0, \quad PL_\xi^{\mathcal{A}}(k) - L_\xi^{\mathcal{B}}(k)P = 0, \tag{2.1}$$

for all $\xi \in \mathfrak{g}$ and $k \geq 0$.

If $P : \mathcal{A} \to \mathcal{B}$ is a chiral chain homotopy, the map $\tau = Pd_{\mathcal{A}} + d_{\mathcal{B}}P$ is a morphism of $\mathfrak{sg}[t]$-modules. Two $\mathfrak{sg}[t]$-module homomorphisms $\phi_0, \phi_1 : \mathcal{A} \to \mathcal{B}$ are said to be chiral chain homotopic if there is a chiral chain homotopy $P : \mathcal{A} \to \mathcal{B}$ such that $\phi_1 - \phi_0 = \tau$. This clearly implies that ϕ_0, ϕ_1 induce the same map from $\mathbf{H}_G^*(\mathcal{A}) \to \mathbf{H}_G^*(\mathcal{B})$, which is analogous to Proposition 2.4.1 of [7].

Suppose that $\phi_0, \phi_1 : M \to N$ are G-equivariantly homotopic via $\Phi : M \times I \to N$. We recall the classical construction of a chain homotopy $P : \Omega^*(N) \to \Omega^{*-1}(M)$ between $\phi_0^*, \phi_1^* : \Omega^*(N) \to \Omega^*(M)$, following [7]. For fixed $x \in M$, consider the curve in N given by $s \mapsto \phi_s(x)$, and let

$\xi_t : M \to TN$ be the map which assigns to x the tangent vector to this curve at $s = t$. Consider the map

$$f_t : \Omega^*(N) \to \Omega^{*-1}(M), \qquad \sigma \mapsto \phi_t^*(\iota_{\xi_t}(\sigma)) \qquad (2.2)$$

defined by $\phi_t^*(\iota_{\xi_t}(\sigma))(\eta_1, \ldots, \eta_k) = \sigma(\xi_t(x), d\phi_t(\eta_1), \ldots, d\phi_t(\eta_k))$, given vectors $\eta_1, \ldots, \eta_k \in TM_x$. It is well-known that

$$\frac{d}{dt}\phi_t^*\sigma = \phi_t^*(\iota_{\xi_t}(d\sigma) + d(\phi_t^*(\iota_{\xi_t}(\sigma)). \qquad (2.3)$$

Define $P : \Omega^*(N) \to \Omega^{*-1}(M)$ by $P\sigma = \int_0^1 \phi_t^*(\iota_{\xi_t}(\sigma))\ dt$. Integrating (2.3) over I shows that $Pd + dP = \phi_1^* - \phi_0^*$, so P is a chain homotopy. Clearly f_t is G-equivariant and satisfies $f_t\iota_\xi^N + \iota_\xi^M f_t = 0$, for all $t \in I$. It follows that P is also G-equivariant and satisfies $P\iota_\xi^N + \iota_\xi^M P = 0$. Hence P is a chain homotopy, and ϕ_0, ϕ_1 induce the same maps in equivariant cohomology.

We need to show that P extends to a linear map $\mathbf{P} : \mathcal{Q}'(N) \to \mathcal{Q}'(M)$ which is a chiral chain homotopy between $\phi_0^*, \phi_1^* : \mathcal{Q}'(N) \to \mathcal{Q}'(M)$. By Lemma 3.2 of [11], for each $m \geq 0$, we may regard $\mathcal{Q}'_M[m]$ as a smooth vector bundle over M of finite rank. Let $\pi : M \times I \to M$ be the projection onto the first factor. Pulling back $\mathcal{Q}'_M[m]$ to a vector bundle on $M \times I$, let $\Gamma[m] = \Gamma(M \times I, \pi^*(\mathcal{Q}'_M[m]))$ denote the space of smooth sections. Note that $\Gamma = \oplus_{m \geq 0}\Gamma[m]$ is an $\mathfrak{sg}[t]$-algebra, and that $\frac{d}{dt}$ and ∂ are commuting derivations on Γ. Furthermore, the (fiberwise) integral $\int_0^1 \sigma(x,t)\ dt$ is a well-defined map from $\Gamma \to \Gamma(M, \mathcal{Q}'_M)$, and $\int_0^1 \partial\sigma(x,t)\ dt = \partial\int_0^1 \sigma(x,t)\ dt$.

Suppose that $\phi : \mathcal{A} \to \mathcal{B}$ is a morphism of $\mathfrak{sg}[t]$-algebras. A degree-weight homogeneous linear map $f : \mathcal{A} \to \mathcal{B}$ will be called a ϕ-derivation if

$$f(a \circ_n b) = f(a) \circ_n \phi(b) + (-1)^{(deg\ f)(deg\ a)}\phi(a) \circ_n f(b), \qquad (2.4)$$

for all homogeneous $a, b \in \mathcal{A}$ and $n \in \mathbf{Z}$. Clearly $f(1) = 0$ and $f(\partial a) = \partial f(a)$ for all $a \in \mathcal{A}$. If $\mathcal{A}, \mathcal{B}$ are abelian vertex algebras, to check that f is a ϕ-derivation, it is enough to show that for all $a, b \in \mathcal{A}$,

$$f(: ab :) = \, : f(a)\phi(b) : \, + (-1)^{(deg\ f)(deg\ a)} : \phi(a)f(b) :,$$
$$f(\partial a) = \partial f(a). \tag{2.5}$$

A ϕ-derivation f is determined by its values on a set of generators of $\mathcal{A}$. In the case $\mathcal{A} = \mathcal{Q}'(N)$, $\mathcal{B} = \mathcal{Q}'(M)$, $\phi = \phi_t^*$, since $\mathcal{Q}'(N)$ is generated by $\Omega(N)$, any two ϕ_t^*-derivations which agree on $\Omega(N)$ must agree on all of $\mathcal{Q}'(N)$. It is easy to check that there is a unique extension of the map $f_t : \Omega^*(N) \to \Omega^{*-1}(M)$ defined by (2.2), to a linear map $F_t : \mathcal{Q}'^*(N) \to \mathcal{Q}'^{*-1}(M)$, which is a ϕ_t^*-derivation. Moreover, (2.3) holds for any σ in $\mathcal{Q}'(N)$, not just $\Omega(N)$.

We now define $\mathbf{P} : \mathcal{Q}'(N) \to \mathcal{Q}'(M)$ by $\mathbf{P}(\sigma) = \int_0^1 F_t(\sigma)\ dt$, which coincides with P at weight zero. Integration of (2.3) over I shows that $d\mathbf{P} + \mathbf{P}d = \phi_1^* - \phi_0^*$. Finally, we need to show that $\mathbf{P}$ is a chiral chain homotopy. For all $\sigma \in \Omega(N)$, $\xi \in \mathfrak{g}$, and $t \in I$, f_t satisfies

$$f_t L_\xi^N - L_\xi^M f_t = 0, \qquad\qquad f_t \iota_\xi^N + \iota_\xi^M f_t = 0. \tag{2.6}$$

For $\xi \in \mathfrak{g}$ and $k \geq 0$, consider the maps

$$R_{t,\xi,k} = F_t L_\xi^N(k) - L_\xi^M(k)F_t, \qquad S_{t,\xi,k} = F_t \iota_\xi^N(k) + \iota_\xi^M(k)F_t.$$

Clearly $R_{t,\xi,k}$ and $S_{t,\xi,k}$ are ϕ_t^*-derivations from $\mathcal{Q}'(N) \to \mathcal{Q}'(M)$, which are homogeneous of weight $-k$ and degree -1 and -2, respectively. For $k > 0$, $R_{t,\xi,k}$ and $S_{t,\xi,k}$ both act by zero on $\mathcal{Q}'(N)[0]$ by weight considerations. For $k = 0$, $R_{t,\xi,k}$ and $S_{t,\xi,k}$ act by zero on $\mathcal{Q}'(N)[0]$ by (2.6). Since $R_{t,\xi,k}$ and $S_{t,\xi,k}$ are ϕ_t^*-derivations, it follows that they act by zero on all of $\mathcal{Q}'(N)$. Finally, since this holds for each $t \in I$, it is immediate that

$$\mathbf{P}L_\xi^N(k)(\sigma) - L_\xi^M(k)\mathbf{P}(\sigma) = \int_0^1 R_{t,\xi,k}(\sigma)\ dt = 0,$$

$$\mathbf{P}\iota_\xi^N(k)(\sigma) + \iota_\xi^M(k)\mathbf{P}(\sigma) = \int_0^1 S_{t,\xi,k}(\sigma)\, dt = 0,$$

for all $\xi \in \mathfrak{g}$, $k \geq 0$ and $\sigma \in \mathcal{Q}'(N)$. Hence $\mathbf{P}$ is a chiral chain homotopy, as desired. $\square$

3. Conformal and quasi-conformal structures

First we recall the conformal element $\mathbf{L}$ of $\mathbf{H}_G^*(\mathbf{C})$ in the case of semisimple G. Let $\{\xi_1, \ldots, \xi_n\}$ be an orthonormal basis for $\mathfrak{g}$ relative to the Killing form. Recall from [9] that the Virasoro element $\mathbf{L}$ is represented by

$$L = \omega_{\mathcal{S}} - L_{\mathcal{S}} + C$$

where $L_{\mathcal{S}} = -\sum_{i=1}^n\, :\Theta_{\mathcal{S}}^{\xi_i}\Theta_{\mathcal{S}}^{\xi_i}:$ and $C = \sum_{i,j=1}^n\, :b^{\xi_i}b^{\xi_j}\gamma^{ad^*(\xi_i)(\xi_j')}: =:$ $(K(0)\Theta_{\mathcal{S}}^{\xi_i})b^{\xi_i}:$. The term $\omega_{\mathcal{S}} - L_{\mathcal{S}}$ lies in $\mathcal{W}(\mathfrak{g})_{bas}$ and satisfies the Virasoro OPE with central charge zero, but it is not $D(0)$-closed. The purpose of the term C is to correct this flaw, and L still satisfies $L(z)L(w) \sim 2L(w)(z-w)^{-2} + \partial L(w)(z-w)^{-1}$. By Corollary 7.17 of [9], L represents a nontrivial class $\mathbf{L}$ in $\mathbf{H}_G^*(\mathbf{C})$, and by Corollary 4.18 of [11], $\mathbf{L}$ is a conformal structure on $\mathbf{H}_G^*(\mathbf{C})$. Finally, Theorem 4.17 of [11] shows that for any G-manifold M, $\kappa_G(\mathbf{L})$ is a conformal structure on $\mathbf{H}_G^*(M)$, where κ_G is the chiral Chern-Weil map.

In this section we show that for any G and M, $\mathbf{H}_G^*(M)$ has a *quasiconformal* structure, that is, an action of the subalgebra of the Virasoro algebra generated by $\{L_n|\ n \geq -1\}$, such that $L_{-1} = L\circ_0$ acts by ∂ and $L_0\omega = L\circ_1 \omega = n\omega$ for $\omega \in \mathbf{H}_G^*(M)[n]$. When G is semisimple, this quasiconformal structure extends to the above conformal structure $\kappa_G(\mathbf{L})$. This provides a vanishing criterion for $\mathbf{H}_G^*(M)_+$; it suffices to show that $L\circ_1$ acts by zero.

We work in the setting of a general $O(\mathfrak{sg})$ topological vertex algebra (TVA) [11]. Recall that an $O(\mathfrak{sg})$ TVA is a degree-weight graded DVA $(\mathcal{A}, d)$ equipped with an $O(\mathfrak{sg})$-structure $(\xi, \eta) \mapsto L_\xi^{\mathcal{A}} + \iota_\eta^{\mathcal{A}}$, a chiral horizontal element $g^{\mathcal{A}}$, such that $L^{\mathcal{A}} = dg^{\mathcal{A}}$ is a conformal structure, with respect to which the $L_\xi^{\mathcal{A}}$ and $\iota_\eta^{\mathcal{A}}$ are primary of weight one. We call $g^{\mathcal{A}}$ a chiral contracting homotopy of $\mathcal{A}$. Given an $O(\mathfrak{sg})$ TVA $(\mathcal{A}, d)$, a differential vertex subalgebra $\mathcal{B}$ is called a half $O(\mathfrak{sg})$ TVA if the non-negative Fourier modes of the vertex operators $\iota_\xi^{\mathcal{A}}$ and $g^{\mathcal{A}}$ preserve $\mathcal{B}$. Note that the non-negative Fourier modes of $L_\xi^{\mathcal{A}} = d\iota_\xi^{\mathcal{A}}$ and $L^{\mathcal{A}} = dg^{\mathcal{A}}$ automatically preserve $\mathcal{B}$ as well. In particular, the action of $\{L^{\mathcal{A}} \circ_n \mid n \geq 0\}$ is a quasi-conformal structure on $\mathcal{B}$. Since $[d, g^{\mathcal{A}} \circ_1] = L^{\mathcal{A}} \circ_1$ and $g^{\mathcal{A}} \circ_1$ act on $\mathcal{B}_{bas}$, Theorem 4.8 of [11] shows that $\mathbf{H}_{bas}^*(\mathcal{B})$ vanishes beyond weight zero.

For a G-manifold M, $\mathcal{Q}(M)$ is our main example of an $O(\mathfrak{sg})$ TVA. In local coordinates,

$$g = g^M = b^i \partial \gamma^i, \qquad L = L^M = \beta^i \partial \gamma^i - b^i \partial c^i. \qquad (3.1)$$

The subalgebra $\mathcal{Q}'(M)$ is then a half $O(\mathfrak{sg})$ TVA as above.

Let $\mathcal{B}$ be a half $O(\mathfrak{sg})$ TVA inside some $O(\mathfrak{sg})$ TVA $\mathcal{A}$ as above. Then the non-negative Fourier modes of

$$L^{tot} = L^{\mathcal{W}} \otimes 1 + 1 \otimes L^{\mathcal{A}} \in \mathcal{W} \otimes \mathcal{A}$$

act on $\mathcal{W} \otimes \mathcal{B}$, giving $\mathcal{W} \otimes \mathcal{B}$ a quasi-conformal structure. Moreover,

$$L_\xi^{tot} = L_\xi^{\mathcal{W}} \otimes 1 + 1 \otimes L_\xi^{\mathcal{A}}, \qquad \iota_\xi^{tot} = \iota_\xi^{\mathcal{W}} \otimes 1 + 1 \otimes \iota_\xi^{\mathcal{A}}$$

are primary of weight one with respect to L^{tot}, and $dL^{tot} = 0$.

Theorem 3.1. $L^{tot}\circ_n$ *operates on* $\mathbf{H}_G^*(\mathcal{B})$ *for* $n \geq 0$, *and gives* $\mathbf{H}_G^*(\mathcal{B})$ *a quasi-conformal structure. If* G *is semisimple,* $\kappa_G(\mathbf{L})\circ_n = L^{tot}\circ_n$ *as operators on* $\mathbf{H}_G^*(\mathcal{B})$ *for all* $n \geq 0$.

Proof: This is immediate from Theorem 4.8 and Theorem 4.17 of [11]. $\square$

Lemma 3.2. *Suppose that* $\alpha \in \mathcal{W} \otimes \mathcal{B}$ *is homogeneous of weight* 2 *and degree* -1, *is* G-*invariant, chiral horizontal, and satisfies*

$$L_\xi^{tot} \circ_1 \alpha = \beta^\xi \otimes 1, \tag{3.2}$$

for all $\xi \in \mathfrak{g}$. *Then* $L^{tot}\circ_1$ *acts by zero on* $\mathbf{H}_G^*(\mathcal{B})$, *and we have* $\mathbf{H}_G^*(\mathcal{B})_+ = 0$.

Proof: Recall from [9] that $d(\beta^{\xi_i}\partial c^{\xi_i'} \otimes 1) = L^{\mathcal{W}} \otimes 1$, but $\beta^{\xi_i}\partial c^{\xi_i'} \otimes 1$ is not chiral horizontal since $\iota_\xi^{\mathcal{W}} \circ_n (\beta^{\xi_i}\partial c^{\xi_i'} \otimes 1) = -\delta_{n,1}\beta^\xi \otimes 1$ for $n \geq 0$. Let $\omega_0 = \beta^{\xi_i}\partial c^{\xi_i'} \otimes 1 + d\alpha$. Clearly $d\omega_0 = L^{\mathcal{W}} \otimes 1$ and

$$\iota_\xi^{tot} \circ_n (d\alpha) = L_\xi^{tot} \circ_n \alpha = \delta_{n,1}\beta^\xi \otimes 1 = -\iota_\xi^{tot} \circ_n (\beta^{\xi_i}\partial c^{\xi_i'} \otimes 1)$$

for $n \geq 0$, since α is chiral horizontal. It follows that $\iota_\xi^{tot} \circ_n \omega_0 = 0$ for all $n \geq 0$, so ω_0 is chiral horizontal. In particular, the non-negative Fourier modes of ω_0 act on $(\mathcal{W} \otimes \mathcal{B})_{bas}$. Finally, let $\omega = \omega_0 + g^{\mathcal{A}} \in \mathcal{W} \otimes \mathcal{A}$. Since $dg^{\mathcal{A}} = L^{\mathcal{A}}$ we have $d\omega = L^{tot}$. The non-negative Fourier modes of ω clearly preserve $(\mathcal{W} \otimes \mathcal{B})_{bas}$ since both ω_0 and $g^{\mathcal{A}}$ have this property. In particular, $[d, \omega\circ_1] = L^{tot}\circ_1$, so $\omega\circ_1$ is a contracting homotopy for $L^{tot}\circ_1$, as desired. $\square$

Note that when G is semisimple, the existence of α as above guarantees that $\kappa_G(\mathbf{L}) = 0$; take $\omega = \beta^{\xi_i}\partial c^{\xi_i'} \otimes 1 + d\alpha + \theta_S^{\xi_i} b^{\xi_i} \otimes 1$. An OPE calculation shows that ω is chiral basic and $d\omega = \kappa_G(\mathbf{L})$.

For example, in the case where G acts locally freely on M, we constructed α as above in [11]. If G acts locally freely on M, we have a map $\mathfrak{g}^* \to \Omega^1(M)$ sending $\xi' \to \theta^{\xi'}$, such that $\iota_\xi \theta^{\eta'} = \langle \eta', \xi \rangle$. The $\theta^{\xi'}$ are known as *connection one-forms*. Choose an orthonormal basis $\{\xi_i\}$ for $\mathfrak{g}$ relative to the Killing form, and let $\Gamma^{\xi_i'} = g \circ_0 \theta^{\xi_i'}$. Then

$$\alpha = \beta^{\xi_i} \otimes \Gamma^{\xi_i'} \in \mathcal{W} \otimes \mathcal{Q}'(M)$$

is G-invariant, chiral horizontal, and satisfies (3.2). This shows that $\mathbf{H}_G^*(M)_+ = 0$.

The next lemma shows that locally defined vertex operators α satisfying the conditions of Lemma 3.2 can be glued together.

Lemma 3.3. *Let M be a G-manifold and let $\{U_i|\ i \in I\}$ be a cover of M by G-invariant open sets. Suppose that $\alpha_i \in \mathcal{W}(\mathfrak{g}) \otimes \mathcal{Q}'(U_i)$ satisfies the conditions of Lemma 3.2. Then $\mathbf{H}_G^*(M)_+ = 0$.*

Proof: Let $\{\phi_i|\ i \in I\}$ be a G-invariant partition of unity subordinate to the cover. Let $\alpha = \sum_i \phi_i \alpha_i$, which is a well-defined global section of $\mathcal{W}(\mathfrak{g}) \otimes \mathcal{Q}'(M)$. Moreover, since ϕ_i is basic, it follows that α remains G-invariant, G-chiral horizontal and satisfies (3.2), as desired. $\square$

4. Chiral equivariant cohomology of homogeneous spaces

Theorem 4.1. *For any compact, connected G and closed subgroup $H \subset G$,*

$$\mathbf{H}_G^*(G/H) \cong \mathbf{H}_{K_0}^*(\mathbf{C}) \otimes H_{G'}^*(G/H), \qquad (4.1)$$

where K_0 is the identity component of $K = Ker(G \to Diff(G/H))$, and $G' = G/K_0$. Here $H^*_{G'}(G/H)$ is regarded as a vertex algebra in which all circle products are trivial except $\circ_{-1}$, and (4.1) is a vertex algebra isomorphism.

Proof: We prove Theorem 4.1 only in the case where G is simple. The general case requires a somewhat more detailed analysis and can found in [12]. We may assume without loss of generality that H has postitive codimension in G. Since G is simple, K is finite and $\mathbf{H}^*_{K_0}(\mathbf{C}) = \mathbf{C}$, so it suffices to prove that $\mathbf{H}^*_G(G/H)_+ = 0$. Fix a basis $\{\xi_1, \ldots, \xi_n\}$ of $\mathfrak{g}$ and a corresponding dual basis $\{\xi'_1, \ldots, \xi'_n\}$ of $\mathfrak{g}^*$ (relative to the Killing form), such that $\xi_1, \ldots, \xi_h$ is a basis of $\mathfrak{h}$. By Lemma 3.2, it suffices to construct a G-invariant, G-chiral horizontal element $\alpha \in \mathcal{W}(\mathfrak{g}) \otimes \mathcal{Q}'(G/H)$ satisfying $L^{tot}_\xi \circ_1 \alpha = \beta^\xi \otimes 1$ for all $\xi \in \mathfrak{g}$.

In order to study G/H as a G-space under left multiplication, it is convenient to regard G as a $G \times H$-space, on which G acts on the left and H acts on the right. The right H-action induces compatible actions of H and $\mathfrak{sh}[t]$ on $\mathcal{Q}'(G)$ which commute with the actions of G and $\mathfrak{sg}[t]$ coming from the left G-action. By Lemma 3.9 of [11], the projection $\pi : G \to G/H$ induces an isomorphism of vertex algebras

$$\pi^* : \mathcal{Q}'(G/H)) \to \mathcal{Q}'(G)_{H-bas}.$$

Moreover, by declaring that H and $\mathfrak{sh}[t]$ act trivially on $\mathcal{W}(\mathfrak{g})$, we may extend the actions of H and $\mathfrak{sh}[t]$ to $\mathcal{W}(\mathfrak{g}) \otimes \mathcal{Q}'(G)$. We identify the complexes $\mathcal{W}(\mathfrak{g}) \otimes \mathcal{Q}'(G/H)$ and $\mathcal{W}(\mathfrak{g}) \otimes \mathcal{Q}'(G)_{H-bas}$ and regard $\mathcal{W}(\mathfrak{g}) \otimes \mathcal{Q}'(G/H)_{H-bas}$ as a subcomplex of $\mathcal{W}(\mathfrak{g}) \otimes \mathcal{Q}'(G)$. Thus in order to prove Theorem 4.1, we need to find a G-invariant, G-chiral horizontal element $\alpha \in \mathcal{W}(\mathfrak{g}) \otimes \mathcal{Q}'(G)_{H-bas}$ satisfying $L^{tot}_\xi \circ_1 \alpha = \beta^\xi \otimes 1$ for all $\xi \in \mathfrak{g}$. Define a new operator

$$\mathcal{L} : \mathcal{W}(\mathfrak{g}) \otimes \mathcal{Q}'(G) \to \mathfrak{g} \otimes \mathcal{W}(\mathfrak{g}) \otimes \mathcal{Q}'(G)$$

sending $\omega \mapsto \xi_k \otimes L^{tot}_{\xi_k} \circ_1 \omega$. Clearly $\mathcal{L}$ is G-equivariant, and the condition $L^{tot}_{\xi} \circ_1 \alpha = \beta^{\xi} \otimes 1$ for all $\xi \in \mathfrak{g}$ is equivalent to $\mathcal{L}(\alpha) = \xi_k \otimes \beta^{\xi_k} \otimes 1$.

Let $f \in \mathcal{C}^{\infty}(G)$ be a smooth function, and fix $\zeta \in \mathfrak{g}$, $\xi \in \mathfrak{g}$, and $\eta' \in \mathfrak{g}^*$. Then for $k = 1, \ldots, n$ we have

$$
\begin{aligned}
L^{tot}_{\xi_k} \circ_1 (\beta^{\zeta} b^{\xi} c^{\eta'} \otimes f) &= L^{W}_{\xi_k} \circ_1 (\beta^{\zeta} b^{\xi} c^{\eta'} \otimes f) \\
&= \beta^{\zeta} \otimes \langle [\xi_k, \xi], \eta' \rangle f \\
&= \beta^{\zeta} \otimes \langle \xi_k, ad^*_{\xi} \eta' \rangle f.
\end{aligned}
\tag{4.2}
$$

Next, we identify $\mathfrak{g}$ with $\mathfrak{g}^*$ via the Killing form, and in particular we identify $\mathfrak{h}^{\perp}$ with $(\mathfrak{g}/\mathfrak{h})^*$. It is easy to check that under the coadjoint action of $\mathfrak{g}$ on the subspace $\mathfrak{h}^{\perp} \subset \mathfrak{g}^*$ we have

$$
ad^*(\mathfrak{g}/\mathfrak{h})^* = \{ad^*_{\xi}(\eta') | \xi \in \mathfrak{g}, \eta' \in (\mathfrak{g}/\mathfrak{h})^* \} = \mathfrak{g}^*.
\tag{4.3}
$$

Hence there exist elements $\chi_i \in \mathfrak{g}$ and $\eta'_i \in (\mathfrak{g}/\mathfrak{h})^*$ for which $ad^*_{\chi_i} \eta'_i = \xi'_i$, for $i = 1, \ldots, n$. Then

$$
L^{tot}_{\xi_k} \circ_1 (\sum_i \beta^{\xi_i} b^{\chi_i} c^{\eta'_i} \otimes 1) = \beta^{\xi_i} \otimes \langle \xi_k, \xi'_i \rangle = \beta^{\xi_k} \otimes 1,
$$

so that

$$
\mathcal{L}(\sum_i \beta^{\xi_i} b^{\chi_i} c^{\eta'_i} \otimes 1) = \xi_k \otimes \beta^{\xi_k} \otimes 1.
\tag{4.4}
$$

However, $\sum_i \beta^{\xi_i} b^{\chi_i} c^{\eta'_i} \otimes 1$ is not G-invariant. We seek a G-invariant element

$$
\alpha_0 = \sum_{j,k,l} \beta^{\xi_j} b^{\xi_k} c^{\xi'_l} \otimes f_{jkl}
$$

which also satisfies (4.4).

We will construct α_0 using the connections one-forms coming from both the left and right actions of G on itself, which we denote by $\theta^{\xi'}$, $\bar{\theta}^{\xi'}$, respectively, for $\xi' \in \mathfrak{g}^*$. We denote the $\mathfrak{sg}[t]$-algebra structure on

$\mathcal{Q}'(G)$ coming from the *right* G-action by $(\xi, \eta) \mapsto \bar{L}_\xi + \bar{\iota}_\eta$. Evaluating the functions $\iota_\xi \circ_0 \bar{\theta}^{\xi'}$ and $\bar{\iota}_\xi \circ_0 \theta^{\xi'}$ at the identity $e \in G$, we have

$$\iota_\xi \circ_0 \bar{\theta}^{\xi'}|_e = \langle \xi, \xi' \rangle = \bar{\iota}_\xi \circ_0 \theta^{\xi'}|_e. \tag{4.5}$$

Define

$$\alpha_0 = \sum_{i,j,k,l} \beta^{\xi_j} b^{\xi_k} c^{\xi_l'} \otimes \bar{\iota}_{\xi_i}(\theta^{\xi_j'}) \bar{\iota}_{\chi_i}(\theta^{\xi_k'}) \iota_{\xi_l}(\bar{\theta}^{\eta_i'}).$$

Clearly α_0 is G-invariant, and $\alpha_0|_e = \sum_i \beta^{\xi_i} b^{\chi_i} c^{\eta_i'} \otimes 1$, by (4.5). Acting by $\mathcal{L}$ we see that $(\mathcal{L}(\alpha_0))|_e = \xi_k \otimes \beta^{\xi_k} \otimes 1$. Finally, since α_0 is G-invariant and the operator $\mathcal{L}$ is G-equivariant, it follows that $\mathcal{L}(\alpha_0) = \xi_k \otimes \beta^{\xi_k} \otimes 1$ at every point of G, as desired.

Next, we will correct α_0 to make it G-chiral horizontal without destroying G-invariance or condition (4.4). Note that for $r \geq 0$,

$$\begin{aligned} \iota_{\xi_t}^{tot} \circ_r \alpha_0 &= b^{\xi_t} \circ_r \alpha_0 \\ &= \delta_{r,0} \sum_{i,j,k} \beta^{\xi_j} b^{\xi_k} \otimes \bar{\iota}_{\xi_i}(\theta^{\xi_j'}) \bar{\iota}_{\chi_i}(\theta^{\xi_k'}) \iota_{\xi_t}(\bar{\theta}^{\eta_i'}). \end{aligned}$$

Let

$$\alpha_1 = - \sum_{i,j,k,l} \beta^{\xi_j} b^{\xi_k} \otimes \bar{\iota}_{\xi_i}(\theta^{\xi_j'}) \bar{\iota}_{\chi_i}(\theta^{\xi_k'}) \iota_{\xi_l}(\bar{\theta}^{\eta_i'}) \theta^{\xi_l'}.$$

An OPE calculation shows that for $r \geq 0$

$$\begin{aligned} L_{\xi_t}^{tot} \circ_r \alpha_1 &= 0, \\ \iota_{\xi_t}^{tot} \circ_r \alpha_1 &= -\delta_{r,0} \sum_{i,j,k} \beta^{\xi_j} b^{\xi_k} \otimes \bar{\iota}_{\xi_i}(\theta^{\xi_j'}) \bar{\iota}_{\chi_i}(\theta^{\xi_k'}) \iota_{\xi_t}(\bar{\theta}^{\eta_i'}). \end{aligned} \tag{4.6}$$

Let $\alpha = \alpha_0 + \alpha_1$. It follows from (4.6) that α is G-invariant, G-chiral horizontal, and satisfies $\mathcal{L}(\alpha) = \xi_k \otimes \beta^{\xi_k} \otimes 1$.

We need to correct α so that it lies in $\mathcal{W}(\mathfrak{g}) \otimes \mathcal{Q}'(G)_{H-bas}$, without destroying the above properties. First, α_0 is already H-chiral horizontal.

This is clear since α is a sum of terms of the form $\beta^{\xi_j} b^{\xi_k} c^{\xi_l} \otimes f_{jkl}$ where $f_{jkl} \in \mathcal{C}^\infty(G)$, and $\bar{\iota}_\xi \circ_r$ lowers degree and only acts on the second factor of $\mathcal{W}(\mathfrak{g}) \otimes \mathcal{Q}'(G)$ for $\xi \in \mathfrak{h}$.

Second, we claim that α_1 is H-chiral horizontal as well. First note that for $\xi \in \mathfrak{h}$, $\bar{\iota}_\xi \circ_r$ acts by derivations on $\mathcal{Q}'(G)$ for all $r \geq 0$, so it suffices to show that it acts by zero on each term of the form $\bar{\iota}_{\xi_i}(\theta^{\xi_j'})$, $\bar{\iota}_{\chi_i}(\theta^{\xi_k'})$ and $\iota_{\xi_l}(\bar{\theta}^{\eta_i'})\theta^{\xi_l'}$. Clearly $\bar{\iota}_\xi \circ_r$ acts by zero on $\bar{\iota}_{\xi_i}(\theta^{\xi_j'})$ and $\bar{\iota}_{\chi_i}(\theta^{\xi_k'})$ since these terms have degree 0 and $\bar{\iota}_\xi \circ_r$ has degree -1. Next, note that for each $\eta_i' \in (\mathfrak{g}/\mathfrak{h})^*$ we have

$$\iota_{\xi_l}(\bar{\theta}^{\eta_i'})\theta^{\xi_l'} = \bar{\theta}^{\eta_i'}.$$

which can be checked by applying ι_η, $\eta \in \mathfrak{g}/\mathfrak{h}$ to both sides. Since $\bar{\iota}_\xi \circ_r \bar{\theta}^{\eta_i'} = 0$ for all $\xi \in \mathfrak{h}$ and $\eta_i' \in (\mathfrak{g}/\mathfrak{h})^*$, the claim follows.

Finally, if α is not H-invariant, we can make it H-invariant by averaging it over H, that is, we take $\alpha' = \frac{1}{|H|}\int_H h\alpha \, d\mu$, where $d\mu$ is the Haar measure on H. Since the G and H actions commute, α' is G-invariant and G-chiral horizontal, and $\mathcal{L}(\alpha') = \xi_k \otimes \beta^{\xi_k} \otimes 1$. Moreover, since α is H-chiral horizontal, it follows that $h\alpha$ is still H-chiral horizontal for all $h \in H$, so α' is H-chiral horizontal as well. Since α' is H-invariant and lives in $\mathcal{W}(\mathfrak{g})[2] \otimes \mathcal{Q}'(G)[0]$, α' is in fact H-chiral invariant, so that $\alpha' \in \mathcal{W}(\mathfrak{g}) \otimes \mathcal{Q}'(G)_{H-bas}$, as desired. $\square$

5. Finite-dimensionality of $\mathbf{H}_G^*(M)$ for compact M

The subspace $\mathcal{W}(\mathfrak{g})^p[n] \subset \mathcal{W}(\mathfrak{g})$ of degree p and weight n is finite-dimensional, so $\mathbf{H}_G^p(\mathbf{C})[n]$ is finite-dimensional for all $p \in \mathbf{Z}$ and $n \geq 0$. Similarly, for any closed subgroup $H \subset G$, Theorem 4.1 implies that $\mathbf{H}_G^p(G/H)[n]$ is finite-dimensional for all p, n. We will show that for an arbitrary compact M, $\mathbf{H}_G^p(M)[n]$ is finite-dimensional for all p, n as well, which generalizes a well-known classical result in the case $n = 0$.

Lemma 5.1. *If M has a finite cover $\{U_1, \ldots, U_m\}$ of G-invariant open sets, such that*

$$dim \ \mathbf{H}_G^p(U_{i_1} \cap \cdots \cap U_{i_k})[n] < \infty,$$

for each p, n and for fixed indices $i_1, \ldots, i_k$, then $\mathbf{H}^p(M)[n]$ is finite-dimensional.

Proof: This is the standard generalized Mayer-Vietoris argument by induction on m. $\square$

Lemma 5.2. *Suppose the G-manifold M is a fiber bundle whose general fiber is G/H. If M is compact, then $\mathbf{H}_G^p(M)[n]$ is finite-dimensional.*

Proof: We can cover M by finitely many open sets whose multiple intersections are equivariantly contractible to G/H. The claim then follows by the preceding lemma. $\square$

Given a closed subgroup $H \subset G$, let $M_{(H)}$ denote the subset of M consisting of points with isotropy group conjugate to H. $M_{(H)}$ is a closed submanifold of M, which may be regarded as a G/H-fiber bundle over the manifold $M_{(H)}/G$. By the preceding lemma, $\mathbf{H}_G^p(M_{(H)})[n]$ is finite-dimensional for each p, n.

Theorem 5.3. *Suppose M is compact. Then $\mathbf{H}_G^p(M)[n]$ is finite-dimensional.*

Proof: Since M is compact, there are only finitely many conjugacy classes (H) for which $M_{(H)}$ is nonempty. For $dim \ H > 0$, each such $M_{(H)}$ has

a G-invariant tubular neighborhood $U_{(H)}$ which is equivariantly contractible to $M_{(H)}$. M has a finite cover consisting of the $U_{(H)}$ together
with the open set U of points with finite isotropy group. Without loss of
generality we can shrink each $U_{(H)}$ so it contains only two orbit types:
(H) and (e).

By homotopy invariance, $\mathbf{H}_G^*(U_{(H)}) = \mathbf{H}_G^*(M_{(H)})$, which is finite-
dimemsional at each p, n. The action of G on U is locally free so
$\mathbf{H}_G^*(U))_+ = 0$. Since the multiple intersections of the $U_{(H)}$ all lie in
U, they also have no higher-weight cohomology. It follows from Lemma
5.1 that for $n > 0$, $\mathbf{H}_G^p(M)[n]$ is finite-dimensional for all p. For $n = 0$,
the finite-dimensionality of $\mathbf{H}_G^*(M)[0] = H_G^p(M)$ is classical. $\square$

6. Chiral equivariant cohomology of spheres

In this section, we show that Theorem 4.1 has the following remarkable consequence, which shows that $\mathbf{H}_G^*(-)$ is a strictly stronger
invariant on the category of compact G-manifolds than $H_G^*(-)$.

Theorem 6.1. *For any simple G, there is a sphere with infinitely many
smooth actions of G, which have pairwise distinct chiral equivariant cohomology, but identical classical equivariant cohomology.*

First, we need a result which describes the postitive weight subspace
$\mathbf{H}_G^*(M)_+$ for an arbitrary G-manifold M in the case where G is simple.

Lemma 6.2. *For any simple G and G-manifold M, the map*

$$\mathbf{H}_G^*(M)_+ \to \mathbf{H}_G^*(M^G)_+ \tag{6.1}$$

induced by the inclusion $M^G \hookrightarrow M$, is a linear isomorphism. Hence $\mathbf{H}_G(M)_+ \cong \mathbf{H}_G^*(\mathbf{C})_+ \otimes H^*(M^G)$.

Proof: We may assume that M^G is non-empty. Let U_0 be a G-invariant tubular neighborhood of M^G and let $U_1 = M \setminus M^G$. It suffices to show that $\mathbf{H}_G^*(U_1)_+ = 0$ and $\mathbf{H}_G^*(U_0 \cap U_1)_+ = 0$. In this case, we have $\mathbf{H}_G^*(M)_+ = \mathbf{H}_G^*(U_0 \cup U_1)_+$, and since $\mathbf{H}_G^*(U_0)_+ \cong \mathbf{H}_G^*(M^G)_+$ by homotopy invariance, the claim follows from a Mayer-Vietoris argument.

For each point $x \in U_1$, the isotropy group G_x has positive codimension in G since G is connected. Let U_x be a G-invariant neighborhood of the orbit Gx, which we may take to be a vector bundle of the form $G \times_{G_x} V$ whose zero-section is Gx.

By Theorem 4.1, there exists a G-invariant, G-chiral horizontal element $\alpha_x \in \mathcal{W}(\mathfrak{g}) \otimes \mathcal{Q}'(G/G_x)$ satisfying (3.2). Via the projection $U_x \to Gx$, this pulls back to an element $\alpha_{U_x} \in \mathcal{W}(\mathfrak{g}) \otimes \mathcal{Q}'(U_x)$ satisfying the same conditions. Using a G-invariant partition of unity as in Lemma 3.3, we can glue the α_{U_x} together to obtain $\alpha \in \mathcal{W}(\mathfrak{g}) \otimes \mathcal{Q}'(U_1)$ satisfying these conditions as well. It follows that $\mathbf{H}_G^*(U_1)_+ = 0$. Finally, the same argument shows that $\mathbf{H}_G^*(U_0 \cap U_1)_+ = 0$. $\square$

Remark 6.3. *For any G and M, $\mathbf{H}_G^*(M)_+$ is a vertex algebra ideal, by Theorem 6.2 of [12]. In other words, $\mathbf{H}_G^*(M)_+$ is closed under $\alpha \circ_n$ and $\circ_n \alpha$ for all $n \in \mathbf{Z}$ and $\alpha \in \mathbf{H}_G^*(M)$. The map (6.1) is in fact an isomorphism of vertex algebra ideals. Both the ring structure of $H^*(M^G)$ and the map $H_G^*(M) \to H_G^*(M^G)$ are encoded in the vertex algebra structure of $\mathbf{H}_G^*(M)$; see the proof of Theorem 1.6 of [12].*

The next theorem we will need is an immediate consequence of results in [15][16] which describe the fixed-point sets of group actions on disks.

Theorem 6.4. *(Oliver) Let F be a finite CW-complex. If G is semisimple, there exists a smooth action of G on a closed disk D with fixed point set D^G having the homotopy type of F. If G is a torus, there exists such an action if and only if F is $\mathbf{Z}$-acyclic.*

Proof of Theorem 6.1: If G acts smoothly on an n-dimensional disk D, we may glue together two copies of D along their boundaries to obtain a smooth action of G on the sphere S^n. Let F be a CW-complex with 3 zero-cells and no higher-dimensional cells. Choose an n-dimensional disk D with a smooth G-action such that D^G has the homotopy type of F. Let S_0 be the copy of S^n obtained by gluing together two copies of D along their boundaries. Note that each connected component $\mathcal{C}$ of D^G gives rise to one component of S_0^G (if $\mathcal{C} \cap \partial D \neq \varnothing$), or two components of S_0^G (if $\mathcal{C} \cap \partial D_0 = \varnothing$). Hence $3 \leq c_0 \leq 6$, where c_0 is the number of components of S_0^G.

Given $x \in S_0^G$, we can find a G-invariant ball $B_0 \subset S_0$ containing x, which intersects exactly one component of S_0^G. By removing B_0 from two copies of S_0 and then gluing them together along their boundaries, we obtain an n-dimensional G-sphere S_1 such that S_1^G has either $2c_0 - 2$ components (if $\partial B \cap S_0^G = \varnothing$) or $2c_0 - 1$ components (if $\partial B \cap S_0^G \neq \varnothing$). We continue this process as follows. Assume that n-dimensional G-spheres $S_0, S_1, \ldots, S_{i-1}$ have been defined. Let $B_{i-1} \subset S_{i-1}$ be a G-invariant ball intersecting exactly one component of S_{i-1}^G. Define $D_i = S_{i-1} \backslash B_{i-1}$, and let S_i be the sphere obtained by gluing two copies of D_i along their boundaries. We thus obtain a sequence of n-dimensional spheres $S_0, S_1, S_2, \cdots$ with smooth G-actions, such that the number of fixed-point components $c_0, c_1, c_2, \ldots$ are all distinct. Since $c_i = dim\, H^0(S_i^G)$, it is immediate from Lemma 6.2 that the vertex algebras $\mathbf{H}_G^*(S_i)$ are all distinct.

It remains to show that the classical equivariant cohomology rings $H_G^*(S_i)$ are all isomorphic. Recall that a G-manifold M is said to be *equivariantly formal* if the spectral sequence of the fibration

$$M \hookrightarrow (M \times E)/G \to E/G$$

collapses. If M is equivariantly formal, the map $i^* : H_G^*(M) \to H_G^*(M^G)$ is injective, and $H_G^*(M) \cong H^*(M) \otimes H_G^*(pt)$ as a module over $H_G^*(pt)$. It is easy to check that a G-sphere S is equivariantly formal if and only if it has a G-fixed point, and in this case $H_G^*(S) \cong H_G^*(pt)[\omega]/(\omega^2)$. $\square$

In fact, it is possible to construct compact G-manifolds M and N and a smooth G-equivariant map $f : M \to N$ which is a homotopy equivalence (and hence induces an isomorphism $f^* : H_G^*(N) \to H_G^*(M)$ with $\mathbf{Z}$-coefficients), such that $\mathbf{H}_G^*(M) \neq \mathbf{H}_G^*(N)$; see Theorem 7.5 of [12]. Unlike the classical equivariant cohomology, the functor $\mathbf{H}_G^*(-)$ can distinguish G-manifolds M and N which admit a G-equivariant map which is a homotopy equivalence, as long as M and N are not equivariantly homotopic.

7. The chiral point algebra

As we have seen, for any simple group G and G-manifold M, $\mathbf{H}_G^*(M)$ can be described in terms of $\mathbf{H}_G^*(\mathbf{C})$ together with the classical ring $H^*(M^G)$. An important problem in this theory is to describe $\mathbf{H}_G^*(\mathbf{C})$ for any G. In this section, we make a few general remarks about the chiral point algebra, and recall some results from [13] describing certain features of this algebra in the simplest nontrivial case $G = SU(2)$.

Since any compact G can be written as a quotient $(G_1 \times \cdots \times G_r \times T)/\Gamma$ where the G_i are simple, T is a torus, and Γ is finite,

$\mathbf{H}_G^*(\mathbf{C}) = \mathbf{H}_{G_1}^*(\mathbf{C}) \otimes \cdots \otimes \mathbf{H}_{G_r}^*(\mathbf{C}) \otimes \mathbf{H}_T^*(\mathbf{C})$. By Theorem 6.1 of [9], $\mathbf{H}_T^*(\mathbf{C})$ is the free abelian vertex algebra with generators $\gamma^{\xi_1'}, \ldots, \gamma^{\xi_n'}$, where $\{\xi_1, \ldots, \xi_n\}$ is a basis for the Lie algebra of T. In other words, $\mathbf{H}_T^*(\mathbf{C})$ is the polynomial algebra generated by $\partial^k \gamma^{\xi_i'}$ for $k \geq 0$. So we may assume without loss of generality that G is simple. There is a "classical sector" of $\mathbf{H}_G^*(\mathbf{C})$, which is the abelian subalgebra generated by the weight-zero component $\mathbf{H}_G^*(\mathbf{C})[0] = Sym(\mathfrak{g}^*)^G$. Unlike the case where G is abelian, $\mathbf{H}_G^*(\mathbf{C})$ contains additional elements that have no classical analogues, most notably a Virasoro element $\mathbf{L}$, which is represented by

$$L = \omega_{\mathcal{S}} - L_{\mathcal{S}} + C. \tag{7.1}$$

Here $\omega_{\mathcal{S}} = \sum_{i=1}^n : \beta^{\xi_i} \partial \gamma^{\xi_i'} :$, $L_{\mathcal{S}} = -\sum_{i=1}^n : \Theta_{\mathcal{S}}^{\xi_i} \Theta_{\mathcal{S}}^{\xi_i} :$, where $\xi_1, \ldots, \xi_n$ is an orthonormal basis for $\mathfrak{g}$ (relative to the Killing form), and

$$C = \sum_{i,j=1}^n : b^{\xi_i} b^{\xi_j} \gamma^{ad^*(\xi_i)(\xi_j')} : =: (K(0)\Theta_{\mathcal{S}}^{\xi_i}) b^{\xi_i} :, \tag{7.2}$$

Recall that $\mathcal{W}_{bas}$ is a complex under the commuting differentials $K(0)$ and $J(0)$. It is convenient to begin with the $K(0)$-cohomology $H^*(\mathcal{W}_{bas}, K(0))$, and then construct elements of $\mathbf{H}_G^*(\mathbf{C}) = H^*(\mathcal{W}_{bas}, D(0))$ using the spectral sequence of the double complex. From this point of view, the Virasoro class $\mathbf{L}$ arises in a very simple way from the element C given by (7.2), which represents a nontrivial class in $H^*(\mathcal{W}_{bas}, K(0))$. The main technical difficulty in studying $H^*(\mathcal{W}_{bas}, K(0))$ is that $\mathcal{W}_{bas}$ is a commutant subalgebra of a $bc\beta\gamma$-system, and commutant vertex algebras of this kind are not well understood. Given a vertex algebra $\mathcal{V}$ and a subalgebra $\mathcal{A}$, recall that the commutant $Com(\mathcal{A}, \mathcal{V})$ is defined to be

$$\{v \in \mathcal{V} | [a(z), v(w)] = 0, \forall a \in \mathcal{A}\}.$$

If $\mathcal{A}$ is a homomorphic image of a current algebra $O(\mathfrak{g}, B)$ of some Lie algebra $\mathfrak{g}$, we have $Com(\mathcal{A}, \mathcal{V}) = \mathcal{V}^{\mathfrak{g}[t]}$. In this notation, $\mathcal{W}(\mathfrak{g})_{bas} =$

$Com(\mathcal{O}, \mathcal{W}(\mathfrak{g}))$, where $\mathcal{O}$ is the copy of $O(\mathfrak{sg})$ generated by $\Theta_{\mathcal{W}}^{\xi}$ and b^{ξ} for $\xi \in \mathfrak{g}$. This algebra is difficult to study because $O(\mathfrak{sg})$ does not act completely reducibly on $\mathcal{W}(\mathfrak{g})$. Moreover, it is difficult to describe $\mathbf{H}_G^*(\mathbf{C})$ without first giving a reasonable description of $\mathcal{W}(\mathfrak{g})_{bas}$.

A *good increasing filtration* on a vertex algebra $\mathcal{A}$ is a $\mathbf{Z}_{\geq 0}$-filtration

$$\mathcal{A}_{(0)} \subset \mathcal{A}_{(1)} \subset \mathcal{A}_{(2)} \subset \cdots, \quad \mathcal{A} = \bigcup_{k \geq 0} \mathcal{A}_{(k)}$$

such that $\mathcal{A}_{(0)} = \mathbf{C}$, and for all $a \in \mathcal{A}_{(k)}$, $b \in \mathcal{A}_{(l)}$, we have

$$a \circ_n b \in \mathcal{A}_{(k+l)}, \quad for \ n < 0,$$

$$a \circ_n b \in \mathcal{A}_{(k+l-1)}, \quad for \ n \geq 0.$$

Filtrations satisfying these conditions were introduced in [8], and were used in [10] to study the vertex algebra commutant problem. If $\mathcal{A}$ possesses such a filtration, the associated graded object $gr(\mathcal{A}) = \bigoplus_{k>0} \mathcal{A}_{(k)}/\mathcal{A}_{(k-1)}$ is a $\mathbf{Z}_{\geq 0}$-graded associative, supercommutative algebra with a unit 1 under a product induced by the Wick product on $\mathcal{A}$. For example, $\mathcal{W}(\mathfrak{g})$ admits a good increasing filtration where $\mathcal{W}(\mathfrak{g})_{(k)}$ is defined to be the vector space spanned by iterated Wick products of the generators $b^{\xi}, c^{\xi'}, \beta^{\xi}, \gamma^{\xi'}$ and their derivatives, of length at most k. We say that elements of $\mathcal{W}(\mathfrak{g})_{(k)} \setminus \mathcal{W}(\mathfrak{g})_{(k-1)}$ have *polynomial degree* k. This filtration is $\mathfrak{sg}[t]$-invariant, and we have an isomorphism of supercommutative rings

$$gr(\mathcal{W}(\mathfrak{g})) \cong Sym(\bigoplus_{k \geq 0} (V_k \oplus V_k^*) \bigotimes \wedge (\bigoplus_{k \geq 0} (U_k \oplus U_k^*)).$$

This filtration is inherited by any subalgebra of $\mathcal{W}(\mathfrak{g})$, such as $\mathcal{W}(\mathfrak{g})_{hor}$ and $\mathcal{W}(\mathfrak{g})_{bas}$.

Recall that the horizontal subalgebra $\mathcal{W}(\mathfrak{g})_{hor} \subset \mathcal{W}(\mathfrak{g})$ is generated by $\beta^{\xi}, \gamma^{\xi'}, b^{\xi}$. Clearly $\mathcal{W}(\mathfrak{g})_{hor}$ (but not $\mathcal{W}(\mathfrak{g})$) has a $\mathbf{Z}_{\geq 0}$ grading by

b-number, which is the eigenvalue of the Fourier mode $-F(0)$, where $F = -\sum_{i=1}^{n} : b^{\xi_i} c^{\xi'_i} :$. For all $\xi \in \mathfrak{g}$, β^{ξ_i} and γ^{ξ_i} have b-number 0, and b^{ξ_i} and its derivatives have b-number 1. This grading is inherited by $\mathcal{W}(\mathfrak{g})_{bas}$, since the action of $\Theta^{\xi}_{\mathcal{W}}$ on $\mathcal{W}(\mathfrak{g})_{hor}$ preserves b-number, for all $\xi \in \mathfrak{g}$. Let us introduce the notation $\mathcal{W}(\mathfrak{g})^{(k)}_{bas}$ to denote the subspace of b-number k; in this notation, $\mathcal{S}(\mathfrak{g})^{\mathfrak{g}[t]} = \mathcal{W}(\mathfrak{g})^{(0)}_{bas}$.

For the rest of this section, we focus on the case $G = SU(2)$. The complexified Lie algebra of $SU(2)$ is $\mathfrak{sl}_2$, and we work in the standard root basis x, y, h, satisfying

$$[x, y] = h, \qquad [h, x] = 2x, \qquad [h, y] = -2y.$$

For simplicity of notation, we denote $\mathcal{W}(\mathfrak{sl}_2)$ and $\mathcal{S}(\mathfrak{sl}_2)$ by $\mathcal{W}$ and $\mathcal{S}$, respectively. Even in this case, we are unable to describe $\mathcal{W}_{bas}$ completely. However, there are eight obvious elements of $\mathcal{W}_{bas}$ that can be written down using Weyl's first fundamental theorem of invariant theory for the adjoint representation of $\mathfrak{sl}_2$:

$$v^h =: \beta^h \gamma^{h'} : + : \beta^x \gamma^{x'} : + : \beta^y \gamma^{y'} :, \tag{7.3}$$

$$v^x = \frac{1}{2} \left(: \gamma^{h'} \gamma^{h'} : + : \gamma^{x'} \gamma^{y'} : \right), \tag{7.4}$$

$$v^y = -\frac{1}{2} \left(: \beta^h \beta^h : +4 : \beta^x \beta^y : \right), \tag{7.5}$$

$$K =: \gamma^{h'} b^h : + : \gamma^{x'} b^x : + : \gamma^{y'} b^y :, \tag{7.6}$$

$$Q^{\beta b} =: \beta^h b^h : +2 : \beta^x b^y : +2 : \beta^y b^x :, \tag{7.7}$$

$$C^{\gamma bb} = - : \gamma^{h'} b^x b^y : + \frac{1}{2} : \gamma^{x'} b^x b^h : - \frac{1}{2} : \gamma^{y'} b^y b^h :, \tag{7.8}$$

$$C^{\beta bb} =: \beta^h b^x b^y : + : \beta^x b^y b^h : - : \beta^y b^x b^h :, \tag{7.9}$$

$$C^{bbb} =: b^x b^y b^h : . \tag{7.10}$$

Note that K is the same vertex operator whose zero-mode $K(0)$ is the chiral Koszul differential, and $C^{\gamma bb}$ coincides with the element C given by (7.2). An OPE calculation shows that v^x, v^y, v^h generate a copy of the current algebra $O(\mathfrak{sl}_2, -\frac{3}{8}\kappa)$, which has level $-\frac{3}{2}$ in the standard normalization. Let $\mathcal{C}$ denote the subalgebra of $\mathcal{W}_{bas}$ generated by the elements (7.3)-(7.10). It is easy to check that $\mathcal{C}$ is a super current algebra associated to a certain 8-dimensional Lie superalgebra which is an extension of $\mathfrak{sl}_2$ (see Equations (32)-(38) of [13]), and in particular $\mathcal{C}$ is *strongly* generated by the elements (7.3)-(7.10). In fact, $\mathcal{C}$ is a subcomplex of $\mathcal{W}_{bas}$ under the both differentials $K(0)$ and $J(0)$.

In [13], it was conjectured that $\mathcal{W}_{bas} = \mathcal{C}$, and using some techniques from commutative algebra (in particular the theory of jet schemes), the following partial result was obtained.

Theorem 7.1. *The subalgebra $\mathcal{S}^{\mathfrak{sl}_2[t]} \subset \mathcal{W}_{bas}$ is strongly generated by v^x, v^y, v^h, and hence is isomorphic to $O(\mathfrak{sl}_2, -\frac{3}{8}\kappa)$. The subspace $\mathcal{W}_{bas}^{(1)}$ has a basis consisting of normally ordered polynomials in $v^x, v^h, v^y, Q^{\beta b}, K$ and their derivatives, which are linear in $Q^{\beta b}$ and K, and their derivatives.*

This shows that the equality $\mathcal{W}_{bas} = \mathcal{C}$ holds for b-numbers 0 and 1. In particular, $\mathcal{W}_{bas}^{(0)}$ and $\mathcal{W}_{bas}^{(1)}$ are both homogeneous of even polynomial degree. The differential $K(0)$ is easily seen to preserve polynomial degree and raise the b-number by one. It follows that any element of $\mathcal{W}_{bas}^{(2)} \cap Ker(K(0))$ which has odd polynomial degree, must represent a nontrivial class in $H^*(\mathcal{W}_{bas}, K(0))$. For example, define

$$h_4 = \; : v^y C^{\gamma bb} : \; -\frac{1}{4} : v^h C^{\beta bb} : \; -\frac{5}{12} \partial C^{\beta bb},$$

which is homogeneous of weight 4 and degree -4. An OPE calculation shows that h_4 lies in the kernel of $K(0)$, and since h_4 has polynomial degree 5, it represents a nontrivial class in $H^{-4}(\mathcal{W}_{bas}, K(0))$. Similarly, for $n \geq 2$ define

$$
\begin{aligned}
h_{2n+2} =: & (v^y)^n C^{\gamma bb} : - \frac{n}{2n+2} : (v^y)^{n-1} v^h C^{\beta bb} : \\
& - \frac{n^2 - n}{2n^2 + 3n + 1} : \partial v^y (v^y)^{n-2} C^{\beta bb} : \\
& - \frac{2n^2 + 3n}{4n^2 + 6n + 2} : (v^y)^{n-1} \partial C^{\beta bb} : .
\end{aligned}
$$

Clearly h_{2n+2} has weight $2n + 2$, degree $-4n$, and polynomial degree $2n+3$. An OPE calculation shows that $K(0)(h_{2n+2}) = 0$ (see Lemma 7.3 of [13]). Hence h_{2n+2} represents a nontrivial class in $H^{-4n}(\mathcal{W}_{bas}, K(0))$ of conformal weight $2n + 2$.

In fact, we can construct many more elements of $H^*(\mathcal{W}_{bas}, K(0))$ in a similar way. Consider the standard monomial basis for the universal enveloping algebra $U(\mathfrak{sl}_2)$ given by

$$
\{x^r y^s h^t \in U(\mathfrak{sl}_2) | r, s, t \geq 0\}.
$$

To the monomial $\mu = x^r y^s h^t \in U(\mathfrak{sl}_2)$, we can associate the normally ordered monomial

$$
: (v^x)^r (v^y)^s (v^h)^t :\in \mathcal{S}^{\mathfrak{sl}_2[t]},
$$

which we also denote by μ. Note that μ has polynomial degree $2(r+s+t)$, weight $2s+t$, and degree $4r-4s$. As shown in [13], there exists an element $h_\mu \in \mathcal{W}_{bas} \cap Ker(K(0))$ of polynomial degree $2(r + s + t) + 3$, degree $4r - 4s$ and weight $2s + t + 2$, of the form

$$
: \mu C^{\gamma bb} : + : g C^{\beta bb} : + \cdots.
$$

Here $g \in \mathcal{S}^{\mathfrak{sl}_2[t]}$ has polynomial degree at most $2(r + s + t)$, and $(\cdots)$ has polynomial degree at most $2(r + s + t) + 1$. In particular, for $\mu = 1$, we have $h_\mu = C^{\gamma bb}$. The assignment $\mu \mapsto h_\mu$ extends to a linear map $\phi : U(\mathfrak{sl}_2) \to \mathcal{W}^{(2)}_{bas} \cap Ker(K(0))$, which induces an injective linear map $\Phi : U(\mathfrak{sl}_2) \to H^*(\mathcal{W}_{bas}, K(0))$. Finally, each nontrivial element of $H^*(\mathcal{W}_{bas}, K(0))$ in the image of Φ gives rise to a nontrivial element of $\mathbf{H}^*_{SU(2)}(\mathbf{C})$ via the spectral sequence of $\mathcal{W}_{bas}$, regarded as a double complex with respect to the commuting differentials $K(0)$ and $J(0)$. More precisely, we have

Theorem 7.2. *There exists a linear map $\psi : U(\mathfrak{sl}_2) \to \mathcal{S}^{\mathfrak{sl}_2[t]}$ such that for any $f \in U(\mathfrak{sl}_2)$, $\phi(f) + \psi(f)$ lies in $Ker(D(0))$. Moreover, the linear map*

$$\Psi : U(\mathfrak{sl}_2) \to \mathbf{H}^*_{SU(2)}(\mathbf{C})$$

sending f to the class of $\phi(f) + \psi(f)$, is injective.

For the sake of illustration, we write down explicit representatives for a few of the classes in $\mathbf{H}^*_{SU(2)}(\mathbf{C})$ given by Theorem 7.2. First, for the identity element $1 \in U(\mathfrak{sl}_2)$, we have

$$\phi(1) + \psi(1) = C^{\gamma bb} + 2 : v^x v^y : + \frac{1}{2} : v^h v^h : - \frac{1}{2} \partial v^h.$$

This coincides with the element L given by (7.1), expressed in terms of the generators of $\mathcal{C}$. Hence $\Psi(1)$ is precisely the Virasoro class $\mathbf{L} \in \mathbf{H}^*_{SU(2)}(\mathbf{C})$. Next, for $n \geq 1$, let H_{2n+2} denote $\phi(f) + \psi(f)$ for $f = y^n \in U(\mathfrak{sl}_2)$, which represents a class of degree $-4n$ and weight $2n + 2$. Similarly, let F_{n+2} denote $\phi(f) + \psi(f)$ for $f = h^n$, which represents a class of degree 0 and weight $n + 2$. We have

$$F_3 =: v^h C^{\gamma bb} : + \frac{2}{3} : v^x C^{\beta bb} : - \frac{5}{3} \partial C^{\gamma bb}$$
$$+ \frac{4}{3} : v^y v^x v^h : + \frac{1}{3} : v^h v^h v^h : - \frac{1}{3} : v^h \partial v^h :$$
$$- \frac{16}{3} : (\partial v^y) v^x : + \frac{2}{3} : v^y \partial v^x : - \frac{5}{3} \partial^2 v^h,$$

$$F_4 = : v^h v^h C^{\gamma bb} : + : (\partial v^h) C^{\gamma bb} : + : v^h v^x C^{\beta bb} :$$

$$+ \frac{2}{3} : (\partial v^x) C^{\beta bb} : + \frac{1}{3} : v^x \partial C^{\beta bb} :$$

$$+ \frac{1}{4} : v^h v^h v^h v^h : + : v^h v^h v^x v^y :$$

$$+ \frac{1}{2} : v^h v^h \partial v^h : + \frac{4}{3} : (\partial v^x) v^y v^h : + \frac{2}{3} v^x (\partial v^y) v^h :$$

$$+ \frac{5}{3} : v^x v^y \partial v^h : + 4 : (\partial^2 v^x) v^y : -2 : v^x \partial^2 v^y :$$

$$- \frac{1}{4} : (\partial v^h)(\partial v^h) : - \frac{1}{12} \partial^3 v^h,$$

$$H_4 = : v^y C^{\gamma bb} : - \frac{1}{4} : v^h C^{\beta bb} : - \frac{5}{12} \partial C^{\beta bb}$$

$$+ : v^x v^y v^y : + \frac{1}{4} : v^h v^h v^y : + \frac{7}{6} : v^h \partial v^y :$$

$$- \frac{19}{12} : (\partial v^h) v^y : + \frac{1}{12} \partial^2 v^y,$$

$$H_6 = : v^y v^y C^{\gamma bb} : - \frac{1}{3} : v^y v^h C^{\beta bb} : - \frac{2}{15} : (\partial v^y) C^{\beta bb} :$$

$$- \frac{7}{15} : v^y \partial C^{\beta bb} : + \frac{1}{6} : v^y v^y v^h v^h : + \frac{2}{3} : v^x v^y v^y v^y :$$

$$+ \frac{2}{15} : (\partial v^y) v^y v^h : - \frac{53}{30} : v^y v^y \partial v^h : +2 : v^y \partial^2 v^y :,$$

$$H_8 = : v^y v^y v^y v^y C^{\gamma bb} - \frac{3}{8} : v^y v^y v^h C^{\beta bb} : - \frac{3}{14} : (\partial v^y) v^y C^{\beta bb} :$$

$$- \frac{27}{56} : v^y v^y \partial C^{\beta bb} : + \frac{1}{2} : v^x v^y v^y v^y v^y : + \frac{1}{8} : v^y v^y v^y v^h v^h :$$

$$- \frac{103}{56} : v^y v^y v^y \partial v^h : + \frac{3}{28} : (\partial v^y) v^y v^y v^h : +3 : (\partial^2 v^y) v^y v^y : .$$

In [13], it was conjectured that the elements $\{\Psi(f) | \ f \in U(\mathfrak{sl}_2)\}$, together with the classical element $[v^x] \in \mathbf{H}^*_{SU(2)}(\mathbf{C})[0] = H^*_{SU(2)}(pt)$

represented by v^x, form a strong generating set for $\mathbf{H}^*_{SU(2)}(\mathbf{C})$. Finally, we expect that there exists an alternative and more geometric construction of the chiral equivariant cohomology, and in particular of $\mathbf{H}^*_G(\mathbf{C})$. We hope that the structure described above in the case $G = SU(2)$ may give some hint about where to look for such a construction.

References

[1] H. Cartan, Notions d'algèbre différentielle; application aux groupes de Lie et aux variétiés où opère un groupe de Lie, Colloque de Topologie, C.B.R.M., Bruxelles 15-27 (1950).

[2] H. Cartan, La Transgression dans un groupe de Lie et dans un espace fibré principal, Colloque de Topologie, C.B.R.M., Bruxelles 57-71 (1950).

[3] M. Duflo, S. Kumar, and M. Vergne, Sur la Cohomologie Équivariante des Variétés Différentiables, Astérisque 215 (1993).

[4] B. Feigin and E. Frenkel, Semi-infinite Weil complex and the Virasoro algebra, Comm. Math. Phys. 137 (1991), 617-639.

[5] I.B. Frenkel, H. Garland, and G.J. Zuckerman, Semi-Infinite Cohomology and String Theory, Proc. Natl. Acad. Sci. USA Vol. 83, No. 22 (1986) 8442-8446.

[6] M. Goresky, R. Kottwitz, and R. MacPherson, Equivariant cohomology, Koszul duality, and the localization theorem, Invent. Math. 131, 25-83 (1998).

[7] V. Guillemin and S. Sternberg, Supersymmetry and Equivariant de Rham Theory, Springer, 1999.

[8] H. Li, Vertex algebras and vertex Poisson algebras, Commun. Contemp. Math. 6 (2004) 61-110.

[9] B. Lian and A. Linshaw, Chiral equivariant cohomology I, Adv. Math. 209, 99-161 (2007).

[10] B. Lian and A. Linshaw, Howe pairs in the theory of vertex algebras, J. Algebra 317, 111-152 (2007).

[11] B. Lian, A. Linshaw, and B. Song, Chiral equivariant cohomology II, Trans. Amer. Math. Soc. 360 (2008), 4739-4776.

[12] B. Lian, A. Linshaw, and B. Song, Chiral equivariant cohomology III, to appear in Amer. J. Math.

[13] A. Linshaw, Chiral equivariant cohomology of a point: a first look, arXiv:1007.3015.

[14] F. Malikov, V. Schechtman, and A. Vaintrob, Chiral de Rham complex, Comm. Math. Phys. 204 (1999), no. 2, 439–473.

[15] R. Oliver, Compact Lie group actions on disks, Math. Z. 149, 79-97 (1976).

[16] R. Oliver, Fixed points of disk actions, Bull. Am. Mat. Soc, Vol.85, No. 2, 279-280 (1976).

Bong H. Lian, Department of Mathematics, Brandeis University, Waltham MA 02454, USA. lian@brandeis.edu

Andrew R. Linshaw, Fachbereich Mathematik, Technische Universität at Armstadt, 64289 Darmstadt, Germany. alinshaw@math.ucsd.edu

Bailin Song, Department of Mathematics, The University of Science and Technology of China, Hefei, China. bailinso@ustc.edu.cn

Research Papers

Contemporary Mathematics
Volume **557**, 2011

Computing Global Characters

Jeffrey Adams

ABSTRACT. The character of an irreducible admissible representation π of a real reductive group is a function on the regular semisimple elements. We give an algorithm for computing the character of an arbitrary π. The presentation is as self-contained and explicit as possible, The calculation is made as explicit as possible, and will implement in the Atlas of Lie Groups and Representations software.

Dedicated to my advisor, Gregg Zuckerman, on his 60th birthday

1. Introduction

Let π be an irreducible representation of a real reductive group G. The global character θ_π of π may be considered as a function on the regular semisimple elements of G. The global character determines π, and it is of great interest to compute it. For example Harish-Chandra found the discrete series representations of G by computing their characters, and characters play an important role in the Langlands program.

Fix a Cartan subgroup H of G and let D be the Weyl denominator. Let λ be the infinitesimal character of π. The function θ_π restricted to the regular elements of H is roughly of the form

$$(1.1) \qquad \theta_\pi = \frac{\sum_{w \in W} a_w(\pi) e^{w\lambda}}{D}$$

for certain integers $a_w(\pi)$. We would like to compute these integers.

There are several methods for doing this in the literature. The character of any irreducible representation is an integer combination of characters of standard modules, i.e., induced from (limits of) discrete series representations. Together with the induced character formula this reduces the problem in principle to computing characters of the discrete series.

The most conceptual algorithm for this case is due to Rebecca Herb [**11**]. It is based on the theory of endoscopy. First one computes the character of a stable sum of discrete series representations. Then one computes the character of an individual discrete series representation in terms of the characters of stable sums of discrete series for G and for various endoscopic groups of lower dimension.

Alternatively, properties of the discrete series characters, including the Hecht-Schmid character identities, give recursive formulas which determine these constants

---

2010 *Mathematics Subject Classification.* 22E30, 22E46.

Supported in part by NSF grant #DMS0968275

[**15**, Chapter 13, §4]. In [**14**] Hirai uses this approach to give a very different formula from that of Herb.

Another very different formula is [**6**]. Other special cases include [**17**], [**18**], [**8**], [**16**], [**19**], [**5**], and unpublished work of Zuckerman. Also see the references in [**11**] and [**14**] and [**15**, Chaper 13].

We follow a more computational approach, based on the Kazhdan-Lusztig-Vogan (KLV) polynomials. Certain terms $e^{w\lambda}$ only appear with nonzero coefficient in the character formula for a single standard module I. It follows that for any irreducible representation π, $a_w(\pi) = \pm M(I, \pi)$, where $M(I, \pi)$ is the coefficient of I in the expression of π as a sum of standard modules, which is given by a KLV polynomial. We then compute other coefficients $a_w(\pi)$ by using coherent continuation. The KLV polynomials are computed by the atlas software (available at www.liegroups.org). A version of this approach appears in an unpublished manuscript of Vogan [**20**].

The main ideas here are all present, in one form or another, in the literature, especially [**10**], as well as [**20**]. The main point here is to formulate the result in as self-contained and clean a form as possible, with an emphasis on the KLV polynomials for use in computations. These calculations will eventually be incorporated in the atlas software, which is a primary motivation for this paper.

One noteworthy aspect of this presentation is our use of covers of Cartan subgroups, which simplifies many of the formulas. In particular we give an elementary proof of the Zuckerman character formula. In these terms we write the trivial representation as a linear combination of standard modules parametrized by *holomorphic* characters of (covers of) Cartan subgroups.

An excellent reference for the basics of character theory is [**10**]. Also see [**15**], especially Chapters 10-13. We have tried to keep the presentation as elementary as possible. Everything is based on Harish-Chandra's theory, and most of what is required was available in the early 1980s. A notable exception is the theory of $\mathcal{D}$-modules, which are not discussed explicitly here, but which play an important role behind the scenes, and a fundamental one in the theory of the KLV polynomials.

We thank David Vogan for some helpful discussions, and Dragan Miličić and Becky Herb for assistance with Section 9.

2. Weyl denominators and related functions

Fix a connected, complex reductive group G and a Cartan subgroup H. Let $\Delta = \Delta(G, H)$ be the set of roots of H in G, and let $W = W(G, H) = W(\Delta)$ be the Weyl group. If Δ^+ is a set of positive roots, let $\rho = \rho(\Delta^+) = \frac{1}{2} \sum_{\Delta^+} \alpha$ as usual. The ρ-cover of H is defined as in [**4**, Section 5]:

$$(2.1) \qquad H_\rho = \{(h, z) \in H \times \mathbb{C}^* \mid e^{2\rho}(h) = z^2\}.$$

Projection on the first factor is a two-to-one cover of H. Projection on the second factor, denoted e^ρ, is a genuine character of H_ρ (one not factoring to H) satisfying $e^\rho(h, z)^2 = e^{2\rho}(h)$. If $w\Delta^+$ is another choice of positive roots then H_ρ is canonically isomorphic to $H_{w\rho}$, via the map $(h, z) \to (h, e^{w\rho - \rho}(h)z)$. It is convenient to eliminate the dependence on Δ^+: define $\widetilde{H}$ to be the inverse limit of $\{H_{w\rho} \mid w \in W\}$.

The Weyl group acts on H_ρ by $w : (h, z) \to (wh, e^{w^{-1}\rho - \rho}(h)z)$, and hence on $\widetilde{H}$.

Now assume H is defined over $\mathbb{R}$, with real points $H(\mathbb{R})$. Let $H(\mathbb{R})_\rho$ be the inverse image of $H(\mathbb{R})$ in H_ρ, and define $\widetilde{H(\mathbb{R})}$ to be the inverse limit of the $H(\mathbb{R})_{w\rho}$.

Let $\mathfrak{h}$ be the Lie algebra of H, $\mathfrak{h}^* = \mathrm{Hom}_\mathbb{C}(\mathfrak{h}, \mathbb{C})$, and write $\langle, \rangle$ for the pairing $\mathfrak{h}^* \times \mathfrak{h} \to \mathbb{C}$. If $\alpha \in \Delta \subset \mathfrak{h}^*$, $\alpha^\vee \in \mathfrak{h}$ denotes the corresponding coroot.

The cross action on characters is defined as follows. For $\lambda \in \mathfrak{h}^*$ let $W(\lambda) = \{w \in W \mid w\lambda - \lambda \text{ is a sum of roots}\}$. For Λ a genuine character of $H(\mathbb{R})_\rho$, $w \in W(d\Lambda)$, define

$$(2.2) \qquad w \times \Lambda = \Lambda \otimes (wd\Lambda - d\Lambda).$$

The expression $wd\Lambda - d\Lambda$ is a sum of roots, which we consider as a (non-genuine) character of $H(\mathbb{R})_\rho$; the right hand side is naturally a genuine character of $H(\mathbb{R})_\rho$.

This definition of the cross action is simpler than that of [**21**, Definition 8.3.1], thanks to the fact that we are using the ρ-cover of H.

Now suppose θ is a Cartan involution of G corresponding to $G(\mathbb{R})$, i.e. $G(\mathbb{R})^\theta$ is a maximal compact subgroup of $G(\mathbb{R})$. For H a θ-stable Cartan subgroup set

$$(2.3) \qquad \begin{aligned} \Delta_i &= \{\alpha \in \Delta \mid \theta(\alpha) = \alpha\} \\ \Delta_r &= \{\alpha \in \Delta \mid \theta(\alpha) = -\alpha\} \\ \Delta_{cx} &= \{\alpha \in \Delta \mid \theta(\alpha) \neq \pm\alpha\}. \end{aligned}$$

Then Δ_i, Δ_r are root systems (Δ_{cx} typically is not), and we denote their Weyl groups W_i, W_r.

Fix a set of positive roots Δ^+. For $* = i, r$ or cx let $\Delta_*^+ = \Delta^+ \cap \Delta_*$. Then Δ_i^+, Δ_r^+ are positive roots for Δ_i, Δ_r, respectively. Define $\rho_* = \frac{1}{2}\sum_{\Delta_*^+} \alpha$.

We need to allow the positive systems for Δ_i, Δ_r to vary independently of Δ^+. We will write $\Psi_i \subset \Delta_i, \Psi_r \subset \Delta_r$ for choices of positive imaginary and real roots respectively, with ρ_i, ρ_r defined accordingly. The covers $H(\mathbb{R})_{\rho_i}$ and $H(\mathbb{R})_{\rho_r}$ are defined as above.

DEFINITION 2.1. If Δ^+ is a set of positive roots define

$$(2.4)(a) \qquad D^0(\Delta^+, g) = \prod_{\alpha \in \Delta^+} (1 - e^{-\alpha}(g)) \quad (g \in H),$$

$$(2.4)(b) \qquad D(\Delta^+, \widetilde{g}) = D^0(\Delta^+, g)e^\rho(\widetilde{g}) \quad (\widetilde{g} \in \widetilde{H})$$

where g is the image of $\widetilde{g}$ in H, and

$$(2.4)(c) \qquad |D(\Delta^+, g)| = |D^0(\Delta^+, g)e^\rho(\widetilde{g})| \quad (g \in H)$$

where $\widetilde{g}$ is any inverse image of g in $\widetilde{H}$.

The dependence of D on Δ^+ is obvious (modulo chasing the covers a bit):

$$(2.5) \qquad D(\Delta^+, w^{-1}\widetilde{g}) = D(w\Delta^+, \widetilde{g}) = \mathrm{sgn}(w)D(\Delta^+, \widetilde{g}) \quad (w \in W).$$

It is also easy to see

$$(2.6) \qquad D(\Delta^+, \widetilde{g}) = \sum_{w \in W} \mathrm{sgn}(w)(w \times e^\rho)(\widetilde{g}) \quad (\widetilde{g} \in H_\rho).$$

Suppose H is defined over $\mathbb{R}$. After conjugating by $G(\mathbb{R})$ we may assume H is θ-stable, which we often do without further comment. The *real Weyl group* is $W(G(\mathbb{R}), H(\mathbb{R})) = \mathrm{Norm}_{G(\mathbb{R})}(H(\mathbb{R}))/H(\mathbb{R})$. This is a subgroup of W^θ, the elements of W commuting with the action of θ on H.

Fix positive imaginary and real roots Ψ_i, Ψ_r. For $w \in W^\theta$ define $w_i \in W(\Delta_i), w_r \in W(\Delta_r)$ by $w\Psi_i = w_i\Psi_i$ and $w\Psi_r = w_r\Psi_r$. Define w_{cx} by $w = w_{cx}w_iw_r$. Then $w_{cx} \in W^\theta$, w_{cx} fixes Ψ_i, Ψ_r, and $\mathrm{sgn}(w_{cx}) = 1$. See [**22**, Section 3]. Define $\mathrm{sgn}_i(w) = \mathrm{sgn}(w_i)$. Although w_i depends on the choice of Ψ_i, $\mathrm{sgn}(w_i)$ is independent of this choice.

If H is θ-stable let A be the identity component of $\{h \in H \,|\, \theta(h) = h^{-1}\}$. Let $M = \mathrm{Cent}_G(A)$; this is a connected complex reductive group. If H is defined over $\mathbb{R}$ then so are A and M, and M^θ is a complexified maximal compact subgroup of $M(\mathbb{R})$. With d denoting the derived group define

$$(2.7) \qquad \begin{aligned} q_H &= \frac{1}{2}\dim_{\mathbb{C}}(M_d/(M_d)^\theta) \\ &= \frac{1}{2}|\{\alpha \in \Delta_i \,|\, \alpha \text{ is noncompact}\}|. \end{aligned}$$

Fix a set of positive real roots Ψ_r. For later use we define functions $\epsilon(\Psi_r, g) = \pm 1$, $\gamma(\Psi_r)(\tilde{g}) = \pm 1, \pm i$ and $\tau(\Psi_r, w)(g) = \pm 1$, and give some elementary properties.

For $g \in H(\mathbb{R})$ define

$$(2.8) \qquad \epsilon(\Psi_r, g) = \mathrm{sgn} \prod_{\alpha \in \Psi_r} (1 - e^{-\alpha}(g)).$$

View e^{ρ_r} as a genuine character of $H(\mathbb{R})_{\rho_r}$ and define

$$(2.9) \qquad \gamma(\Psi_r)(\tilde{g}) = \frac{e^{\rho_r}(\tilde{g})}{|e^{\rho_r}(\tilde{g})|} \quad (\tilde{g} \in H(\mathbb{R})_{\rho_r}).$$

Since $|e^{\rho_r}(\tilde{g})|$ factors to $H(\mathbb{R})$, $\gamma(\Psi_r)$ is a genuine character of $H(\mathbb{R})_{\rho_r}$, and $e^{2\rho_r}(g) \in \mathbb{R}^\times$ implies $\gamma(\Psi_r)(\tilde{g})^4 = 1$. For $w \in W^\theta$ define

$$(2.10) \qquad \tau(\Psi_r, w)(g) = \frac{\gamma(\Psi_r)}{\gamma(w\Psi_r)}(g) = \mathrm{sgn}(e^{\rho_r - w\rho_r}(g)) \quad (g \in H(\mathbb{R})).$$

As indicated this factors to a character on $H(\mathbb{R})$, with values in ± 1.

LEMMA 2.2. *For all* $w, x, y \in W^\theta$:
(1) $\tau(\Psi_r, w) = \tau(\Psi_r, w_r)$.
(2) $\tau(\Psi_r, xy) = \tau(\Psi_r, x)\, x\tau(\Psi_r, y)$;
(3) $\epsilon(\Psi_r, wg) = sgn(w_r)\epsilon(\Psi_r, g)\tau(\Psi_r, w_r^{-1})(g)$.

We omit the straightforward proofs. For (2) see [**4**, (8.26)(b)].

DEFINITION 2.3. Suppose H is θ-stable and Δ^+ is a set of positive roots. Let

$$(2.11) \qquad B(\Delta^+) = \frac{1}{2}|\{\alpha \in \Delta^+ \text{ complex} \,|\, \theta\alpha \in \Delta^+\}|.$$

We make repeated use of the condition

$$(2.12) \qquad \alpha \in \Delta^+ \text{ complex} \Rightarrow \theta(\alpha) \notin \Delta^+,$$

i.e. $B(\Delta^+) = 0$. See [**21**, Lemma 6.7.1].

Suppose α is a complex root. Then $e^{\alpha + \bar{\alpha}}$ takes positive real values on $H(\mathbb{R})$, and has a natural positive square-root. Accordingly for $h \in H(\mathbb{R})$ we define

$$(2.13) \qquad \begin{aligned} e^{\frac{1}{2}(\alpha + \bar{\alpha})}(h) &= \sqrt{e^{\alpha + \bar{\alpha}}(h)} \\ e^{\frac{1}{2}(\alpha - \bar{\alpha})}(h) &= e^{\alpha}(h)e^{-\frac{1}{2}(\alpha + \bar{\alpha})}(h). \end{aligned}$$

Since ρ_{cx} is a sum of terms $\frac{1}{2}(\alpha \pm \overline{\alpha})$, $e^{\rho_{cx}}$ is a well defined character of $H(\mathbb{R})$. In particular if Δ^+ satisfies (2.12) then $\widetilde{e^{\rho_{cx}}}(h) > 0$ for $h \in H(\mathbb{R})$.

Suppose Λ is a character of $\widetilde{H(\mathbb{R})}$, and $\langle d\Lambda, \alpha^\vee \rangle \in \mathbb{R}_{\neq 0}$ for all roots. We define

$$(2.14) \qquad \epsilon(\Lambda, \Delta^+) = \text{sgn}(w) \quad \text{where } \langle wd\Lambda, \alpha^\vee \rangle > 0 \text{ for all } \alpha \in \Delta^+.$$

We often apply this to Δ_i^+ in which case it is enough to assume $\langle d\Lambda, \alpha^\vee \rangle \neq 0$ for all imaginary roots.

3. Limits of Relative Discrete Series

We say a Cartan subgroup H (defined over $\mathbb{R}$) is *relatively compact* if $H(\mathbb{R})$ is compact modulo the center of $G(\mathbb{R})$. By a result of Harish-Chandra $G(\mathbb{R})$ has relative discrete series representations (i.e. discrete series modulo the center) if and only if it has a relatively compact Cartan subgroup. It also has limits of such representations, which are obtained by translating to singular infinitesimal character.

Let $q = q_H = \frac{1}{2} \dim_\mathbb{C}(G_d/(G_d)^\theta)$ (cf. 2.7).

DEFINITION 3.1. A relative discrete series parameter is a pair $\gamma = (H, \Lambda)$ where H is relatively compact, Λ is a genuine character of $\widetilde{H(\mathbb{R})}$ and

$$(3.1) \qquad \langle d\Lambda, \alpha^\vee \rangle \neq 0 \quad \text{for all } \alpha \in \Delta(G, H).$$

Let $\Delta^+ = \Delta^+(d\Lambda) = \{\alpha \,|\, \langle d\Lambda, \alpha^\vee \rangle > 0\}$.

Let $I(\gamma)$ be the unique relative discrete series representation whose character restricted to the regular elements of $H(\mathbb{R})$ is

$$(3.2) \qquad \Theta_{I(\gamma)}(g) = (-1)^q D(\Delta^+, \widetilde{g})^{-1} \sum_{w \in W(G(\mathbb{R}), H(\mathbb{R}))} \text{sgn}(w)(w\Lambda)(\widetilde{g})$$

where $\widetilde{g} \in \widetilde{H(\mathbb{R})}$ maps to g.

More generally let $\gamma = (H, \Delta^+, \Lambda)$ where Λ is a genuine character of $\widetilde{H(\mathbb{R})}$ satisfying

$$(3.3)(a) \qquad \langle d\Lambda, \alpha^\vee \rangle \neq 0 \quad \text{if } \alpha \text{ is compact and simple for } \Delta^+,$$

and Δ^+ satisfies

$$(3.3)(b) \qquad \{\alpha \,|\, \langle d\Lambda, \alpha^\vee \rangle > 0\} \subset \Delta^+.$$

Choose Λ' regular so that $\Lambda' - \Lambda$ is a sum of roots, and satisfying $\Delta^+ = \Delta^+(\Lambda')$. Define a discrete series representation $I(H, \Lambda')$ by the preceding construction. Let $I(H, \Delta^+, \Lambda) = \Psi(I(H, \Lambda'))$ where Ψ is the Jantzen-Zuckerman translation functor taking infinitesimal character $d\Lambda'$ to infinitesimal character $d\Lambda$. See [**15**, Chapter 12, §7].

Every relative limit of discrete series representation is obtained this way, and $I(\gamma) \simeq I(\gamma')$ if and only if γ is $G(\mathbb{R})$ conjugate to γ'.

See [**4**, Example 8.14].

Formula (3.2) holds for limits of discrete series. However the uniqueness statement does not. For example for $SO(2, 1)$ the character of the (unique) limit of discrete series representation vanishes on the compact Cartan subgroup (since $s_\alpha \in W(G(\mathbb{R}), H(\mathbb{R}))$).

4. Regular Integral Standard Modules

In this section we consider standard modules with regular integral infinitesimal
character. The basic reference for this section and the next is [**4**, Section 8]. Also
see [**3**, Theorem 11.4] and [**1**, Theorem 6.1].

We first dispense with a technical issue about the covers. Suppose H is θ-stable,
Δ^+ is a set of positive roots, and define $\rho_i, \rho_r, \rho_{cx}$ with respect to Δ_i^+, Δ_r^+ and Δ_{cx}^+
as in Section 2. Recall (end of Section 2) $e^{\rho_{cx}}$ is a well defined character of $H(\mathbb{R})$.

DEFINITION 4.1. Fix H (defined over $\mathbb{R}$) and Δ^+. Suppose χ, τ are genuine
characters of $H(\mathbb{R})_\rho, H(\mathbb{R})_{\rho_r}$, respectively. Define:

$$(4.1) \qquad (\chi \otimes \tau)(\overline{h}) = (\chi e^{-\rho})(h)(\tau e^{\rho_r})(h) e^{\rho_{cx}}(h) e^{\rho_i}(\overline{h}) \quad (\overline{h} \in H(\mathbb{R})_{\rho_i})$$

where h is the image of $\overline{h}$ in $H(\mathbb{R})$. This is a genuine character of $H(\mathbb{R})_{\rho_i}$, satisfying
$d(\chi \otimes \tau) = d\chi + d\tau$.

DEFINITION 4.2. A *regular character* for $G(\mathbb{R})$ is a pair $\gamma = (H, \Lambda)$ where H is
a Cartan subgroup of G, defined over $\mathbb{R}$, Λ is a genuine character of $\widetilde{H(\mathbb{R})}$, and $d\Lambda$
is regular and integral, i.e. $\langle d\Lambda, \alpha^\vee \rangle \in \mathbb{Z}_{\neq 0}$ for all roots.

We say $\gamma = (H, \Lambda)$ is *based on* H. The group $G(\mathbb{R})$ acts on regular characters
by conjugation.

Now suppose Λ is a genuine character of $\widetilde{H(\mathbb{R})}$. Fix positive imaginary and real
roots Ψ_i, Ψ_r. Choose positive roots Δ^+ containing Ψ_i, Ψ_r, and apply Definition 4.1
to define (cf. (2.9))

$$(4.2) \qquad \Lambda_M(\Psi_r) = \Lambda \otimes \gamma(\Psi_r)^{-1},$$

a genuine character of $H(\mathbb{R})_{\rho_i}$. It is easy to see this is independent of the choice of
Δ^+, and the dependence on Ψ_r is given by

$$(4.3) \qquad \Lambda_M(w\Psi_r) = \tau(\Psi_r, w)\Lambda_M(\Psi_r) \quad (w \in W_r).$$

Given $\gamma = (H, \Lambda)$, conjugate by $G(\mathbb{R})$ as usual to assume $H(\mathbb{R})$ is θ-stable.
Define A and $M = \text{Cent}_G(A)$ as in Section 2. Let $\mathfrak{a} = \text{Lie}(A)$. Recall M is defined
over $\mathbb{R}$. Let $P = MN$ be a parabolic subgroup such that

$$(4.4) \qquad \{\alpha \mid \langle d\Lambda|_\mathfrak{a}, \alpha^\vee \rangle \geq 0\} \quad \text{for all weights } \alpha \text{ of } \mathfrak{h} \text{ in Lie}(N).$$

It is easy to see that N and P are defined over $\mathbb{R}$.

DEFINITION 4.3. Suppose $\gamma = (H, \Lambda)$ is a regular character. Let

$$(4.5) \qquad \Psi_r = \{\alpha \in \Delta_r \mid \langle d\Lambda, \alpha^\vee \rangle > 0\},$$

define Λ_M by (4.2), and the relative discrete series representation $I_M(H, \Lambda_M)$ by
Definition 3.1 applied to $M(\mathbb{R})$. Choose P as above, and define

$$(4.6) \qquad I(\gamma) = \text{Ind}_{P(\mathbb{R})}^{G(\mathbb{R})}(I_M(H, \Lambda_M)).$$

THEOREM 4.4. $I(\gamma)$ *is non-zero, independent of the choice of* P, *and has a
unique irreducible quotient, denoted* $\pi(\gamma)$. *Every irreducible representation of* $G(\mathbb{R})$
with regular integral infinitesimal character is isomorphic to $\pi(\gamma)$ *for some regular
character* γ, *unique up to conjugacy by* $G(\mathbb{R})$.

EXAMPLE 4.5. Let $G(\mathbb{R}) = SL(2, \mathbb{R})$ and let $H(\mathbb{R})$ be a split Cartan subgroup. Choose an isomorphism $h : \mathbb{R}^* \simeq H(\mathbb{R})$. The cover $H(\mathbb{R})_\rho$ is trivial, so we may drop it from the notation. Let $\Psi_r = \{\alpha\}$ where $\alpha(h(x)) = \mathrm{sgn}(x)$. Then $e^{\rho_r}(h(x)) = x$ and $\gamma(\Psi_r)(h(x)) = \mathrm{sgn}(x)$. If Λ is a character of $H(\mathbb{R})$ then $\Lambda_M(h(x)) = \Lambda(h(x))\mathrm{sgn}(x)$. In particular if $\Lambda = e^\rho$ then $\Lambda_M(h(x)) = |x|$.

EXAMPLE 4.6. Consider the split Cartan subgroup $H(\mathbb{R}) \simeq \mathbb{R}^*$ of $PGL(2, \mathbb{R})$. Now some care is required with the covers. Choose an isomorphism $h : \mathbb{R}^* \simeq H(\mathbb{R})$, and choose $\Psi_r = \{\alpha\}$ where $\alpha(h(x)) = x$. Then $H(\mathbb{R})_\rho = H(\mathbb{R})_{\rho_r} = \{(h(x), z) \mid z^2 = x\}$. On the other hand $H(\mathbb{R})_{\rho_i} = \{(h(x), \epsilon) \mid \epsilon^2 = 1\}$ (trivial cover).

Suppose Λ is a genuine character of $H(\mathbb{R})_\rho$. Write $\Lambda(h(x), z) = \mu(x)z$ for μ a character of $\mathbb{R}^*$. Note that $\gamma(\Psi_r)(h, z) = z/|z|$. According to Definition 4.1 if $(h, \epsilon) \in H(\mathbb{R})_{\rho_i}$:

$$\Lambda_M(h(x), \epsilon) = (\Lambda e^{-\rho})(h(x))(\gamma(\Psi_r)^{-1} e^{\rho_r})(h(x))\epsilon$$

$$= \mu(x)|z|\epsilon = \mu(x)|x|^{\frac{1}{2}}\epsilon.$$

We may ignore the trivial cover $H(\mathbb{R})_{\rho_i}$ and write $\Lambda_M(h(x)) = \mu(x)|x|^{\frac{1}{2}}$. In particular $\mu = 1$ gives $\Lambda = e^\rho$ and $\Lambda_M(h(x)) = |x|^{\frac{1}{2}}$.

5. General Standard Modules

Although our computation of characters of irreducible representations (see Section 12) is limited to the case of regular integral infinitesimal character, it is only a little extra effort to prove various intermediate results in greater generality.

In this section we consider general standard modules, where are obtained by dropping the assumption that $d\Lambda$ is regular and integral. This necessitates some further choices. First of all we need to choose Ψ_i to define limits of discrete series for $M(\mathbb{R})$. Secondly Ψ_r is not determined by (4.5), so we need to choose Ψ_r to define Λ_M. For references see the beginning of the previous section.

DEFINITION 5.1. Standard limit data is a set $\gamma = (H, \Psi_i, \Psi_r, \Lambda)$ where H is defined over $\mathbb{R}$, Ψ_i, Ψ_r are sets of positive imaginary and real roots, respectively, and Λ is a genuine character of $\widetilde{H(\mathbb{R})}$. We require:

(5.1)(a) $\{\alpha \in \Delta_i \mid \langle d\Lambda, \alpha^\vee \rangle > 0\} \subset \Psi_i$

(5.1)(b)   if $\alpha \in \Delta_i$ is compact and simple for Ψ_i then $\langle d\Lambda, \alpha^\vee \rangle \neq 0$.

We say $\gamma = (H, \Psi_i, \Psi_r, \Lambda)$ is based on H. The group $G(\mathbb{R})$ acts on standard limit data by conjugation.

DEFINITION 5.2. Suppose $\gamma = (H, \Psi_i, \Psi_r, \Lambda)$ is standard limit data. Define Λ_M by (4.2) and $I_M(H, \Psi_i, \Lambda_M)$ by Definition 3.1 applied to M. Choose a parabolic subgroup $P = MN$ such that

(5.2) $Re\langle d\Lambda|_{\mathfrak{a}}, \alpha^\vee \rangle \geq 0$ for all roots α of $\mathfrak{h}$ in $\mathrm{Lie}(N)$

and define

(5.3) $$I(\gamma) = \mathrm{Ind}_{P(\mathbb{R})}^{G(\mathbb{R})}(I_M(H, \Psi_i, \Lambda_M(\Psi_r))).$$

As before $I(\gamma)$ is independent of the choice of P. However it may not have a unique irreducible quotient, and we let $\pi(\gamma)$ be the co-socle of $I(\gamma)$, the direct sum of all of the irreducible quotients.

If $d\Lambda$ is regular and integral we recover Definition 4.2:

$$(5.4)(a) \qquad\qquad I(H, \Lambda) = I(H, \Delta_i^+, \Delta_r^+, \Lambda)$$

where $\Delta^+ = \{\alpha \mid \langle d\Lambda, \alpha^\vee \rangle > 0\}$. More generally if $\langle d\Lambda, \alpha^\vee \rangle \neq 0$ for all imaginary roots we define

$$(5.4)(b) \qquad\qquad I(H, \Psi_r, \Lambda) = I(H, \Psi_i, \Psi_r, \Lambda)$$

where $\Psi_i = \{\alpha \in \Delta_i \mid \langle d\Lambda, \alpha^\vee \rangle > 0\}$.

The choice of Ψ_r introduces a twist in the notion of equivalence of data. By (2.10) and (4.2) if $w \in W_r$, then $I(H, \Psi_i, \Psi_r, \Lambda) \simeq I(H, \Psi_i, w\Psi_r, \tau(\Psi_r, w)\Lambda)$. We therefore define

$$(H, \Psi_i, \Psi_r, \Lambda) \sim (H, \Psi_i, w\Psi_r, \tau(\Psi_r, w)\Lambda) \quad (w \in W_r).$$

Incorporating conjugation by $G(\mathbb{R})$ we make the following definition.

DEFINITION 5.3. Fix $\gamma = (H, \Psi_i, \Psi_r, \Lambda)$. We say standard limit data γ' is equivalent to γ if there exists $g \in G(\mathbb{R})$, $w \in W_r$ such that

$$(5.5) \qquad\qquad g\gamma' g^{-1} = (H, \Psi_i, w\Psi_r, \tau(\Psi_r, w)\Lambda).$$

In particular $(H, \Psi_i, \Psi_r, \Lambda) \sim (H, \Psi_i', \Psi_r', \Lambda')$ if and only if there exists $y \in W(G(\mathbb{R}), H(\mathbb{R}))$ such that

  (1) $\Psi_i = y\Psi_i'$,
  (2) $y\Lambda' = \tau(\Psi_r, w)\Lambda$ where $w \in W_r$ satisfies $w\Psi_r = y\Psi_r'$.

THEOREM 5.4.
(1) Every irreducible representation is a summand of $I(\gamma)$ for some γ.
(2) Suppose γ, γ' are standard limit data. Then $I(\gamma) \simeq I(\gamma')$ if $\gamma \sim \gamma'$;

If the infinitesimal character is regular then the converse holds in (2) (Theorem 4.4). In general an additional *final* condition is necessary to make this hold, and to make $\pi(\gamma)$ irreducible. See [**3**, Definition 11.13, Condition (b)] or [**1**, Theorem 6.1, Condition 5].

A virtual character is an element of the Grothendieck group, i.e. a finite linear combination of irreducible representations, with integral coefficients. A basic result is that the representations $I(\gamma)$ span the virtual characters. At regular infinitesimal character these form a basis (at singular infinitesimal character this is not true without the final condition of the previous paragraph). We restrict to regular infinitesimal character.

DEFINITION 5.5. For π a virtual representation with regular infinitesimal character define integers $M(I(\gamma), \pi)$ by

$$(5.6) \qquad\qquad \pi = \sum_\gamma M(I(\gamma), \pi) I(\gamma)$$

in the Grothendieck group.

For π irreducible the integers $M(I(\gamma), \pi)$ are computed by the Kazhdan-Lusztig-Vogan algorithm.

6. Character Formulas

Fix a Cartan subgroup H defined over $\mathbb{R}$ and write $H(\mathbb{R})_{reg}$ for the regular elements of $H(\mathbb{R})$. Fix a set of positive real roots Ψ_r. Let

$$(6.1) \qquad H(\mathbb{R})_+ = \{g \in H(\mathbb{R})_{reg} \,|\, |e^\alpha(g)| > 1 \text{ for all } \alpha \in \Psi_r\}.$$

Every element of $H(\mathbb{R})_{reg}$ is conjugate via $W_r \subset W(G(\mathbb{R}), H(\mathbb{R}))$ to a unique element of $H(\mathbb{R})_+$.

We parametrize infinitesimal characters by Weyl group orbits of elements of $\mathfrak{h}^*$. Let

$$(6.2) \qquad \mathcal{P}(H, \lambda) = \{\Lambda \,|\, \Lambda \text{ is a genuine character of } \widetilde{H(\mathbb{R})},\ d\Lambda \in W\lambda\}.$$

Since we make frequent use of the trivial representation, let ρ be one-half the sum of any set of positive roots and define

$$(6.3) \qquad \mathcal{P}(H, \mathbb{C}) = \mathcal{P}(H, \rho).$$

Recall if Δ^+ is any set of positive roots then $\Delta_r^+ = \Delta^+ \cap \Delta_r$ is a set of positive roots of Δ_r.

PROPOSITION 6.1. Suppose π is an admissible representation. Fix a Cartan subgroup H, defined over $\mathbb{R}$, and a set of positive roots Δ^+. Suppose π has infinitesimal character $\lambda \in \mathfrak{h}^*$. Use Δ_r^+ to define $H(\mathbb{R})_+$. Then the restriction of Θ_π to $H(\mathbb{R})_+$ may be written

$$(6.4) \qquad \theta_\pi(g) = \frac{\sum_{\Lambda \in \mathcal{P}(H, \lambda)} a(\pi, \Delta^+, \Lambda)\Lambda(\widetilde{g})}{D(\Delta^+, \widetilde{g})}$$

for some unique integers $a(\pi, \Delta^+, \Lambda)$.

Here $\widetilde{g}$ is an inverse image of g in $\widetilde{H(\mathbb{R})}$; the right hand side is independent of this choice.

This is essentially due to Harish-Chandra [7], see [10, 3.41], except that the set on which this expansion is valid is larger. Also see see [1, Theorem 5.8] and [20].

We want to compute the integers $a(\pi, \Delta^+, \Lambda)$.

From invariance of θ_π it is easy to see that if $w \in W(G(\mathbb{R}), H(\mathbb{R}))$ then

$$(6.5)(a) \qquad a(\pi, w\Delta^+, w\Lambda) = a(\pi, \Delta^+, \Lambda)$$

and if, furthermore, $w\Delta_r^+ = \Delta_r^+$, then (cf. (2.5))

$$(6.5)(b) \qquad a(\pi, \Delta^+, w\Lambda) = \operatorname{sgn}(w)a(\pi, \Delta^+, \Lambda).$$

However if $w \in W_r$ there is often no relationship between $a(\pi, \Delta^+, \Lambda)$ and $a(\pi, \Delta^+, w\Lambda)$.

The dependence of $a(\pi, \Delta^+, \Lambda)$ on Δ^+ is fairly innocuous, and we record it here.

LEMMA 6.2. Fix a θ-stable Cartan subgroup H, Δ^+, and suppose $w \in W$. There exist unique $x, y \in W$ such that

$$(6.6)(a) \qquad x \in W_r \text{ and } (w\Delta^+)_r = x(\Delta_r^+),$$

$$(6.6)(b) \qquad (y\Delta^+)_r = \Delta_r^+,$$

$$(6.6)(c) \qquad w = xy.$$

If $w \in W^\theta$ then $(w\Delta^+)_r = w(\Delta_r^+)$ and $x = w_r$ (cf. Section 2).

PROOF. The element x satisfying (a) exists and is unique, so let $y = x^{-1}w$. It is enough to show y satisfies (b). Note that $(x^{-1}w\Delta^+)_r = x^{-1}[(w\Delta^+)_r]$ since $x \in W_r$. Thus

$$(6.7) \qquad (y\Delta^+)_r = (x^{-1}w\Delta^+)_r = x^{-1}[(w\Delta^+)_r] = x^{-1}[x(\Delta_r^+)] = \Delta_r^+.$$

$\square$

LEMMA 6.3. *In the setting of Proposition 6.1 suppose $w \in W$, and write $w = xy$ as in the Lemma. Then*

$$(6.8)(a) \qquad\qquad a(\pi, w\Delta^+, \Lambda) = sgn(y)a(\pi, \Delta^+, x^{-1}\Lambda).$$

In particular

$$(6.8)(b) \qquad\quad a(\pi, w\Delta^+, \Lambda) = sgn(w)a(\pi, \Delta^+, \Lambda) \quad (w\Delta^+)_r = w(\Delta_r^+)$$

$$(6.8)(c) \qquad\quad a(\pi, w\Delta^+, \Lambda) = a(\pi, \Delta^+, w^{-1}\Lambda) \quad (w \in W_r).$$

In particular (b) holds for $w \in W_i$.

PROOF. Assuming $|e^\alpha(g)| > 1$ for all $\alpha \in \Delta_r^+$, we have

$$(6.9) \qquad\qquad \Theta_\pi(g) = D(\Delta^+, \widetilde{g})^{-1} \sum a(\pi, \Delta^+, \Lambda)\Lambda(\widetilde{g}).$$

Since $(w\Delta^+)_r = x(\Delta_r^+)$, g satisfies the previous condition if and only if $|e^\beta(xg)| > 1$ for all $\beta \in (w\Delta^+)_r$. Therefore

$$
\begin{aligned}
\Theta_\pi(xg) &= D(w\Delta^+, x\widetilde{g})^{-1} \sum a(\pi, w\Delta^+, \Gamma)\Gamma(x\widetilde{g}) \\
(6.10) \qquad &= sgn(w)sgn(x)D(\Delta^+, \widetilde{g})^{-1} \sum a(\pi, w\Delta^+, \Gamma)(x^{-1}\Gamma)(\widetilde{g}) \\
&= sgn(y)D(\Delta^+, \widetilde{g})^{-1} \sum a(\pi, w\Delta^+, x\Lambda)\Lambda(\widetilde{g})
\end{aligned}
$$

Setting $\theta_\pi(g) = \theta_\pi(xg)$ gives the result.

$\square$

It makes sense to use (6.8)(b) to move Λ to the dominant Δ_i^+-chamber using W_i, which (provided $d\Lambda$ is regular) introduces a term $\epsilon(\Lambda, \Delta_i^+)$ (see the end of Section 2).

It is also useful to record how $a(\pi, \Delta^+, \Lambda)$ depends on Δ^+ provided Δ_i^+, Δ_r^+ are fixed. Recall (Definition 2.3) $B(\Delta^+) = \frac{1}{2}|\{\alpha \in \Delta^+ \text{complex} \mid \theta\alpha \in \Delta^+\}|$.

LEMMA 6.4. *Let Δ_1^+, Δ_2^+ be sets of positive roots, containing the same real and imaginary roots. Then*

$$(6.11) \qquad\qquad a(\pi, \Delta_1^+, \Lambda) = (-1)^{B(\Delta_1^+)-B(\Delta_2^+)}a(\pi, \Delta_2^+, \Lambda).$$

PROOF. Choose w such that $w\Delta_1^+ = \Delta_2^+$. By Lemma 6.3 $a(\pi, \Delta_1^+, \Lambda) = (-1)^{\ell(w)}a(\pi, \Delta_2^+, \Lambda)$. It is easy to see $\ell(w) = B(\Delta_1^+) - B(\Delta_2^+) \pmod 2$. $\square$

EXAMPLE 6.5. Suppose H is a relatively compact Cartan subgroup, and $I(H, \Lambda)$ is a relative discrete series representation (Section 3). Then with q_H as in (2.7),

$$a(I(H, \Lambda), \Delta^+, \Gamma) = \begin{cases} (-1)^{q_H}sgn(w) & \Gamma = w\Lambda \quad (w \in W(G(\mathbb{R}), H(\mathbb{R}))) \\ 0 & \text{otherwise.} \end{cases}$$

Equivalently,

$$a(I(H,\Lambda),\Delta^+,\Gamma) = \begin{cases} (-1)^{q_H}\epsilon(\Gamma,\Delta^+) & \Gamma \text{ is } G(\mathbb{R}) - \text{conjugate to } \Lambda \\ 0 & \text{otherwise.} \end{cases}$$

7. Example: The Trivial Representation

The character of the trivial representation $\mathbb{C}$ is the identity function. By (2.6):

$$(7.1) \qquad \theta_{\mathbb{C}}(g) = D(\Delta^+,\tilde{g})^{-1} \sum_{w\in W} \text{sgn}(w)(w \times e^\rho)(\tilde{g}) \quad (\tilde{g} \in H(\mathbb{R})_\rho)$$

(where $\rho = \rho(\Delta^+)$).

We reinterpret this formula in terms of holomorphic characters. The group H_ρ inherits from H a holomorphic structure. Therefore it makes sense to talk about holomorphic characters of H_ρ. The character ρ is holomorphic, so the genuine holomorphic characters of H_ρ are precisely the holomorphic characters of H, (pulled back to H_ρ and) tensored with ρ.

It is clear from the definition of the cross action that if Λ is holomorphic then $w \times \Lambda$ is (defined and) holomorphic for all $w \in W$. It is also clear that for $w \in W$, $w \times e^\rho$ is the unique genuine holomorphic character of $H(\mathbb{R})_\rho$ with differential $w\rho$. Thus Λ occurs in the character formula of the trivial representation if and only if $d\Lambda \in W\rho$ and Λ is holomorphic. Recall (6.3) $\mathcal{P}(H,\mathbb{C})$ is the set of genuine characters of with differential in $W\rho$. So if we let

$$\mathcal{P}^{\text{hol}}(H,\mathbb{C}) = \{\Lambda \in \mathcal{P}(H,\mathbb{C}) \,|\, \Lambda \text{ is holomorphic}\}$$

we may reformulate (7.1) as

$$(7.2) \qquad \theta_{\mathbb{C}} = D(\Delta^+)^{-1} \sum_{\Lambda\in\mathcal{P}^{hol}(H,\mathbb{C})} \epsilon(\Lambda,\Delta^+)\Lambda$$

(for ϵ see (2.14)). In other words:

COROLLARY 7.1. Fix H and Δ^+. Then

$$(7.3) \qquad a(\mathbb{C},\Delta^+,\Lambda) = \begin{cases} \epsilon(\Lambda,\Delta^+) & \Lambda \in \mathcal{P}^{\text{hol}}(H,\mathbb{C}) \\ 0 & \text{otherwise.} \end{cases}$$

Equivalently

$$(7.4) \qquad a(\mathbb{C},\Delta^+,w \times e^\rho) = \text{sgn}(w)$$

and all other $a(\mathbb{C},\Delta^+,\Lambda)$ are 0.

8. Character formula for standard modules

Suppose γ is standard limit data based on H. We give a formula for the character of $I(\gamma)$ on $H(\mathbb{R})$. This combines the character formula for (relative) discrete series on a (relatively) compact Cartan subgroup (3.2) with a special case of the induced character formula, which we now state. See [**10**, Theorem 5.7].

PROPOSITION 8.1. Suppose H is a Cartan subgroup, defined over $\mathbb{R}$, and $P = MN$ is a parabolic subgroup defined over $\mathbb{R}$ as usual. Suppose σ is an admissible representation of $M(\mathbb{R})$, pulled back to $P(\mathbb{R})$. Let $\pi = \mathrm{Ind}_{P(\mathbb{R})}^{G(\mathbb{R})}(\sigma \otimes 1)$ (normalized induction). Then (for any choice of positive roots Δ^+)

$$(8.1) \qquad \Theta_\pi(g) = |D(\Delta^+, g)|^{-1} \sum_w |D(\Delta_i^+, wg)| \Theta_\sigma(wg) \quad (g \in H(\mathbb{R})_{\mathrm{reg}})$$

where the sum is over $W(M(\mathbb{R}), H(\mathbb{R})) \backslash W(G(\mathbb{R}), H(\mathbb{R}))$.

Here is the main result of this section. Recall q_H is given in (2.7), $B(\Delta^+)$ is given by (2.11), and (see Section 2) for $w \in W^\theta$, $\mathrm{sgn}_i(w)$ is defined.

PROPOSITION 8.2. Suppose $\gamma = (H, \Psi_i, \Psi_r, \Lambda)$ is standard limit data. Choose positive roots Δ^+ containing Ψ_i and Ψ_r and set $\delta = (-1)^{q_H + B(\Delta^+)}$. Then for $g \in H(\mathbb{R})_{\mathrm{reg}}$

$$(8.2) \qquad \Theta_{I(\gamma)}(g) = \delta \frac{\epsilon(\Psi_r, g)}{D(\Delta^+, \widetilde{g})} \sum_{w \in W(G(\mathbb{R}), H(\mathbb{R}))} \mathrm{sgn}_i(w) \tau(\Psi_r, w)(g)(w\Lambda)(\widetilde{g})$$

where $\widetilde{g}$ is an inverse image of g in $\widetilde{H(\mathbb{R})}$. In other words for Λ a genuine character of $\widetilde{H(\mathbb{R})}$:

$$a(I(\gamma), \Delta^+, \Gamma) = \begin{cases} \delta \mathrm{sgn}_i(w) & \Gamma = \tau(\Psi_r, w)(w\Lambda) \quad (w \in W(G(\mathbb{R}), H(\mathbb{R}))) \\ 0 & \text{otherwise.} \end{cases}$$

REMARK 8.3. The fact that the right hand side of (8.2) is invariant under $g \to wg$ for $w \in W(G(\mathbb{R}), H(\mathbb{R}))$ follows from (and is essentially equivalent to) Lemma 2.2.

PROOF OF PROPOSITION 8.2. Write $W_G = W(G(\mathbb{R}), H(\mathbb{R}))$ and $W_M = W(M(\mathbb{R}), H(\mathbb{R}))$. Recall $I(H, \Psi_i, \Psi_r, \Lambda)$ is induced from $I_M(H, \Psi_i, \Lambda_M(\Psi_r))$ where $\Lambda_M(\Psi_r)$ is given by (4.2). To save a little space write $\Lambda_M = \Lambda_M(\Psi_r)$. By Proposition 8.2 and (3.2):

$$\Theta_{I(\gamma)}(g) = |D(\Delta^+, g)|^{-1} \sum_{w \in W_M \backslash W_G} |D(\Psi_i, wg)| \Theta_{I_M(H, \Psi_i, \Lambda_M)}(wg)$$

$$(8.3) \qquad = |D(\Delta^+, g)|^{-1} \sum_{w \in W_M \backslash W_G} |D(\Psi_i, wg)| \times$$

$$(-1)^{q_H} D(\Psi_i, \overline{wg})^{-1} \sum_{y \in W_M} \mathrm{sgn}(y)(y\Lambda_M)(\overline{wg}).$$

Here $\overline{wg}$ is an inverse image of wg in $H(\mathbb{R})_{\rho_i}$. Also choose an inverse image $\widetilde{g}$ of g in $H(\mathbb{R})_\rho$. This gives

$$(8.4) \qquad (-1)^{q_H} D(\Delta^+, \widetilde{g})^{-1} \sum_{y, w} \frac{D(\Delta^+, \widetilde{g})}{|D(\Delta^+, g)|} \frac{|D(\Psi_i, wg)|}{D(\Psi_i, \overline{wg})} \mathrm{sgn}(y) \Lambda_M(y^{-1}\overline{wg}).$$

Using $D(\Psi_i, y^{-1}\overline{wg}) = \mathrm{sgn}(y) D(\Psi_i, \overline{wg})$ and $D(\Delta^+, \widetilde{g}) = \mathrm{sgn}(w) D(\Delta^+, w\widetilde{g})$ we can combine the sums:

$$(8.5) \qquad (-1)^{q_H} D(\Delta^+, \widetilde{g})^{-1} \sum_{w \in W(G(\mathbb{R}), H(\mathbb{R}))} \mathrm{sgn}(w) \frac{D(\Delta^+, w\widetilde{g})}{|D(\Delta^+, wg)|} \frac{|D(\Psi_i, wg)|}{D(\Psi_i, \overline{wg})} \Lambda_M(\overline{wg})$$

The final terms equal

$$(8.6) \qquad \frac{e^{\rho}(w\widetilde{g})/e^{\rho_i}(\overline{wg})}{|e^{\rho}(w\widetilde{g})|/|e^{\rho_i}(\overline{wg})|} \prod_{\Delta^+ \setminus \Psi_i} \frac{(1 - e^{-\alpha}(wg))}{|(1 - e^{-\alpha}(wg))|} \Lambda_M(\overline{wg})$$

Assume Δ^+ satisfies (2.12). Since $\theta\alpha = -\overline{\alpha}$, this says the positive complex roots come in pairs $\alpha, \overline{\alpha}$. Then the product becomes the product over Ψ_r, i.e. $\epsilon(\Psi_r, wg)$, giving

$$(8.7) \qquad \epsilon(\Psi_r, wg) \frac{e^{\rho}(w\widetilde{g})/e^{\rho_i}(\overline{wg})}{|e^{\rho}(w\widetilde{g})|/|e^{\rho_i}(\overline{wg})|} \Lambda_M(\overline{wg}).$$

By Definition 4.1 and (4.2)

$$\Lambda_M(\overline{wg}) = (\Lambda e^{-\rho})(wg)(\gamma(\Psi_r)^{-1} e^{\rho_r})(wg) e^{\rho_{cx}}(wg) e^{\rho_i}(\overline{wg})$$

Using

$$(\Lambda e^{-\rho})(wg) = \Lambda(w\widetilde{g}) e^{-\rho}(w\widetilde{g})$$

$$(\gamma(\Psi_r)^{-1} e^{\rho_r})(wg) = |e^{\rho_r}(wg)|$$

$$|e^{\rho}(w\widetilde{g})|/|e^{\rho_i}(\overline{wg})| = |e^{\rho_r}(wg)| e^{\rho_{cx}}(wg) \quad \text{(since (2.12) holds)}$$

it is easy to see the final two terms of (8.7) are equal to $\Lambda(w\widetilde{g})$. Therefore

$$(8.8) \qquad \Theta_{I(\gamma)}(g) = (-1)^{q_H} D(\Delta^+, \widetilde{g})^{-1} \sum_{w \in W(G(\mathbb{R}), H(\mathbb{R}))} \text{sgn}(w)\epsilon(\Psi_r, wg)\Lambda(w\widetilde{g}).$$

By Lemma 2.2(3) $\epsilon(\Psi_r, wg) = \text{sgn}(w_r)\tau(\Psi_r, w_r^{-1})(g)\epsilon(\Psi_r, g)$. Also (cf. Section 2) $\text{sgn}(w)\text{sgn}(w_r) = \text{sgn}(w_i)$. Inserting these, and replacing w with w^{-1}, gives (8.2).

This completes the proof provided Δ^+ satisfies (2.12), i.e. $B(\Delta^+) = 0$. The general case follows from Lemma 6.4. $\qquad \square$

EXAMPLE 8.4. Consider principal series for $SL(2, \mathbb{R})$. We use the notation of Example 4.5.

Let $s_\alpha \in W$ be the simple reflection. Then $\tau(\Psi_r, s_\alpha)(h(x)) = \text{sgn}(\alpha(h(x))) = 1$, and (8.2) gives

$$(8.9) \qquad \begin{aligned} \Theta_{I(\gamma)}(h(x)) &= \frac{\text{sgn}(1 - \frac{1}{x^2})}{(1 - \frac{1}{x^2})x}(\Lambda(x) + \Lambda(x^{-1})) \\ &= \frac{\text{sgn}(x - \frac{1}{x})}{x - \frac{1}{x}}\text{sgn}(x)(\Lambda(x) + \Lambda(x^{-1})) \\ &= \frac{(\Lambda \otimes \text{sgn})(x) + (\Lambda \otimes \text{sgn})(x^{-1})}{|x - \frac{1}{x}|} \end{aligned}$$

This is the familiar character formula for $\text{Ind}_{B(\mathbb{R})}^{SL(2,\mathbb{R})}(\Lambda \otimes \text{sgn})$ ($B(\mathbb{R})$ is a Borel subgroup). For example if $\Lambda = e^{\rho}$ then $e^{\rho}(h(x)) = x$ and $(\Lambda \otimes \text{sgn})(h(x)) = |x|$, so

$$(8.10) \qquad \theta_{I(\gamma)}(h(x)) = \frac{|x| + |x|^{-1}}{|x - \frac{1}{x}|}$$

This is the character formula of the spherical principal series with infinitesimal character ρ, with the trivial representation as a quotient.

EXAMPLE 8.5. Consider $PGL(2, \mathbb{R})$ and use notation of Example 4.6. We compute

$$s_\alpha(h(x), z) = (h(\frac{1}{x}), e^{-2\rho_r}(h(x))z) = (h(\frac{1}{x}), \frac{z}{x}) = (h(\frac{1}{x}), \frac{1}{z})$$

and

$$\tau(\Psi_r, s_\alpha)(h(x)) = \text{sgn}(\alpha(h(x))) = \text{sgn}(x).$$

Write $\Lambda(h(x), z) = \mu(x)z$, so (8.2) gives

$$
\begin{aligned}
\Theta_{I(\gamma)}(h(x)) &= \Theta_{I(\gamma)}(h(x), z) \\
&= \frac{\text{sgn}(1 - \frac{1}{x})}{(1 - \frac{1}{x})z}(\mu(x)z + \mu(x^{-1})\text{sgn}(x)/z) \\
&= \frac{\mu(x) + \mu(x^{-1})\frac{1}{|x|}}{|1 - \frac{1}{x}|} \\
&= \frac{\mu(x)|x|^{\frac{1}{2}} + \mu(x^{-1})|x|^{-\frac{1}{2}}}{|1 - \frac{1}{x}||x|^{\frac{1}{2}}}
\end{aligned}
$$

(8.11)

This is the formula for $\text{Ind}_{B(\mathbb{R})}^{PGL(2,\mathbb{R})}(\mu||^{\frac{1}{2}})$. For example $\Lambda = e^\rho$ is given by $\mu = 1$, and

$$\theta_{I(\gamma)}(h(x)) = \frac{|x|^{\frac{1}{2}} + |x|^{-\frac{1}{2}}}{|1 - \frac{1}{x}||x|^{\frac{1}{2}}}.$$

(8.12)

This is the spherical principal series with infinitesimal character ρ, with the trivial representation as a quotient. Note that $\theta_{I(\gamma)}(h(x)) = 1$ for $x < 0$.

EXAMPLE 8.6. This example generalizes (parts of) the previous two. Suppose $H(\mathbb{R})$ is split, and take $\Lambda = e^\rho$. Let $\pi = I(H, \Lambda)$. A short calculation gives

$$\Theta_{I(\gamma)}(g) = |D(\Delta^+, g)|^{-1} \sum_{w \in W} |e^{w\rho}(g)| \quad (g \in H(\mathbb{R})).$$

(8.13)

This is the character of the spherical principal series representation with infinitesimal character ρ, with the trivial representation as quotient.

COROLLARY 8.7. Fix H, Δ^+ and Λ. Let γ be standard limit data based on H. Then

$$a(I(\gamma), \Delta^+, \Lambda) = \begin{cases} (-1)^{q_H + B(\Delta^+)}\text{sgn}(w) & \gamma \sim (H, w\Delta_i^+, \Delta_r^+, \Lambda) \quad (w \in W_i) \\ \text{otherwise.} \end{cases}$$

PROOF. Write $\gamma = (H, \Psi_i, \Psi_r, \Gamma)$. Define $x \in W_i, y \in W_r$ by $x\Delta_i^+ = \Psi_i, y\Delta_r^+ = \Psi_r$. Apply Lemma 6.3 or (6.5):

$$a(I(\gamma), \Delta^+, \Lambda) = a(I(\gamma), xy\Delta^+, y\Lambda)\text{sgn}(x)$$

(8.14)

and we now have $\Psi_i, \Psi_r \subset xy\Delta^+$. By Proposition 8.2 this is non-zero if and only if $y\Lambda = \tau(\Psi_r, u)u\Gamma$ for some $u \in W(G(\mathbb{R}), H(\mathbb{R}))$, i.e. (using $u^{-1}\tau(\Psi_r, u) = \tau(\Psi_r, u^{-1})$) $\Gamma = \tau(\Psi_r, u^{-1})u^{-1}y\Lambda$. So

$$
\begin{aligned}
\gamma = (H, \Psi_i, \Psi_r, \Gamma) &= (H, \Psi_i, \Psi_r, \tau(\Psi_r, u^{-1})u^{-1}y\Lambda) \\
&\sim (H, \Psi_i, u^{-1}\Psi_r, u^{-1}y\Lambda) \quad \text{(by Definition 5.3)} \\
&\sim (H, u\Psi_i, \Psi_r, y\Lambda) \\
&\sim (H, u_i x\Delta_i^+, \Delta_r^+, \Lambda).
\end{aligned}
$$

(8.15)

If this holds the sign is $(-1)^{q_H + B(\Delta^+)}\operatorname{sgn}(u_i x)$, so set $w = u_i x \in W_i$. $\qquad\square$

If $d\Lambda$ is regular with respect to the imaginary roots this has an important consequence – a character Λ determines a unique standard module based on H:

COROLLARY 8.8. In the setting of the previous Corollary, assume $\langle d\Lambda, \alpha^\vee \rangle \neq 0$ for all imaginary roots. Then

$$(8.16) \qquad a(I(\gamma), \Delta^+, \Lambda) = \begin{cases} (-1)^{B(\Delta^+)}(-1)^{q_H}\epsilon(\Lambda, \Delta_i^+) & \gamma \sim (H, \Delta_r^+, \Lambda) \\ 0 & \text{otherwise.} \end{cases}$$

In particular if γ, γ' are based on H, and Λ occurs in the character formulas for both $I(\gamma)$ and $I(\gamma')$, then $\gamma \sim \gamma'$.

The conclusion of the Corollary does not hold if we drop the assumption that γ, γ' are based on H. For example let $G = SL(2, \mathbb{R})$ and consider the principal series representations (cf. Example 8.4). The only principal series representation with the term $\Lambda(h(x)) = x^{-1}$ is the spherical one. However this also occurs in the character formula for both of the discrete series representations (with trivial infinitesimal character), based on the compact Cartan subgroup. We will see in the next section that such a result does hold for certain characters. See Section 14, Example 1.

9. Computation of $a(\pi, \Delta^+, \Lambda)$ for leading terms

If Λ satisfies a certain positivity condition with respect to Δ^+ then the conclusion of Corollary holds 8.7 without assuming γ is based on H:

PROPOSITION 9.1. Fix H, Δ^+ and a genuine character Λ of $H(\mathbb{R})_\rho$. Assume

(a) $\langle d\Lambda, \alpha^\vee \rangle \neq 0$ for all $\alpha \in \Delta$,
(b) $\operatorname{Re}\langle d\Lambda, \alpha^\vee \rangle \geq 0$ for all $\alpha \in \Delta_r^+$.

Suppose γ is standard limit data. Then

$$(9.1) \qquad a(I(\gamma), \Delta^+, \Lambda) = \begin{cases} (-1)^{q_H + B(\Delta^+)}\epsilon(\Lambda, \Delta_i^+) & \gamma \sim (H, \Delta_r^+, \Lambda) \\ 0 & \text{otherwise.} \end{cases}$$

The key point is that $a(I(\gamma), \Delta^+, \Lambda) = 0$ if γ is not based on H. If $I(\gamma)$ is a (relative) discrete series representation this follows from the fact that $I(\gamma)$ is tempered, and growth conditions on matrix coefficients [7]. For example see [12, 3.3] or [15, 13.26]. The general case follows from this and the induced character formula. See [10, Section 3]. It also follows from the theory of $\mathcal{D}$-modules.

Using this we see that for certain Λ, computing $a(\pi, \Delta^+, \Lambda)$ is equivalent to computing the multiplicity of a standard module in the character formula for π:

COROLLARY 9.2. Let π be a virtual representation with regular infinitesimal character. Fix H, Δ^+, and Λ such that $\operatorname{Re}\langle d\Lambda, \alpha^\vee \rangle \geq 0$ for all $\alpha \in \Delta_r^+$. Then

$$(9.2) \qquad a(\pi, \Delta^+, \Lambda) = (-1)^{q_H + B(\Delta^+)}\epsilon(\Lambda, \Delta_i^+)M(I(H, \Delta_r^+, \Lambda), \pi).$$

10. The Zuckerman Character formula for the Trivial Representation

Recall

$$(10.1) \qquad \mathbb{C} = \sum_{\gamma} M(I(\gamma), \mathbb{C}) I(\gamma)$$

for certain integers $M(I(\gamma), \mathbb{C})$. These are computed by the Kazhdan-Lusztig-Vogan (KLV) polynomials, but in this case this goes back to a result of Zuckerman [**21**, Proposition 9.4.16]. We give an elementary proof here. Since we are working at regular integral infinitesimal character we may write $\gamma = (H, \Lambda)$ for a regular character and $I(H, \Lambda)$ for the corresponding standard module (cf. Section 4).

Fix H and consider a standard module $I(H, \Lambda)$ with $\Lambda \in \mathcal{P}(H, \mathbb{C})$ (cf. (6.3)). Let

$$(10.2) \qquad \Delta^+ = \Delta^+(\Lambda) = \{\alpha \mid \langle d\Lambda, \alpha^\vee \rangle > 0\}.$$

By definition $I(H, \Lambda) = I(H, \Delta_r^+, \Lambda)$ (see (5.4)). Therefore we may apply Corollary 9.2 to conclude

$$(10.3) \qquad M(I(H, \Lambda), \mathbb{C}) = a(\mathbb{C}, \Delta^+(\Lambda), \Lambda)(-1)^{q_H + B(\Delta^+(\Lambda))}$$

(using $\epsilon(\Lambda, \Delta^+(\Lambda)_i) = 1$). On the other hand by Corollary 7.1

$$(10.4) \qquad a(\mathbb{C}, \Delta^+(\Lambda), \Lambda) = \begin{cases} 1 & \Lambda \in \mathcal{P}^{hol}(H, \mathbb{C}) \\ 0 & else. \end{cases}$$

Therefore

$$(10.5) \qquad M(I(H, \Lambda), \mathbb{C}) = \begin{cases} (-1)^{q_H + B(\Delta^+(\Lambda))} & \Lambda \in \mathcal{P}^{hol}(H, \mathbb{C}) \\ 0 & else. \end{cases}$$

In other words:

LEMMA 10.1. *There is an identity in the Grothendieck group*

$$(10.6) \qquad \mathbb{C} = \sum_{H} \sum_{\Lambda \in \mathcal{P}^{hol}(H, \mathbb{C})} (-1)^{q_H + B(\Delta^+(\Lambda))} I(H, \Lambda).$$

(the first sum is over $G(\mathbb{R})$-conjugacy classes of Cartan subgroups defined over $\mathbb{R}$).

Subtracting (a) and (c) in [**21**, Lemma 9.4.15] shows that

$$(10.7) \qquad (-1)^{q_H + B(\Delta^+(\Lambda))} = (-1)^{\ell(\gamma_0) - \ell(\gamma)}$$

where ℓ is the length function of [**21**, Definition 8.1.4] and γ_0, γ are the parameters for the trivial representation and $I(H, \Lambda)$, respectively. Thus the Lemma is a version of Zuckerman's character formula for the trivial representation [**21**, Proposition 9.4.16]. A nice feature of these parameters is that it is precisely the *holomorphic* characters which appear.

11. Coherent Continuation

We give the definition of coherent continuation [**10**, (3.38)], [**21**, Definitions 7.2.5, 7.2.28] in our terms.

Fix a Cartan subgroup H, with Weyl group W, and $\lambda \in \mathfrak{h}^*$. Let $X^*(H)$ be the algebraic (holomorphic) characters of H. If F is a finite dimensional representation of $G(\mathbb{C})$ write $\Delta(F) \subset X^*(H)$ for its weights.

We say a family $\{\pi[\lambda + \mu] \mid \mu \in X^*(H)\}$ of virtual Harish-Chandra modules is a *coherent family* if for all $\mu \in X^*(H)$:

(1) $\pi[\lambda + \mu]$ has infinitesimal character $\lambda + \mu$,

(2) If F is any finite dimensional representation of $G(\mathbb{C})$ then

$$\pi[\lambda + \mu] \otimes F = \sum_{\mu' \in \Delta(F)} \pi[\lambda + \mu + \mu']$$

where $\Delta(F)$ is the set of weights of F.

We consider only finite dimensional representations of $G(\mathbb{C})$ rather than of $G(\mathbb{R})$ as in [**21**, Chapter 7] – this is sufficient for our purposes.

This has a direct interpretation in terms of characters. Fix Δ^+ and drop it from the notation, so write $a(\pi[\lambda + \mu], \Lambda)$ for the coefficients in the character formula for $\pi[\lambda + \mu]$.

LEMMA 11.1. *(a) Suppose π is a virtual character with regular infinitesimal character λ. There is a unique coherent family $\{\pi[\lambda + \mu] \mid \mu \in X^*(H)]\}$ such that $\pi[\lambda] = \pi$.*

(b) Suppose $\{\pi[\lambda + \mu]\}$ is a coherent family. Assume $d\Lambda = w\lambda$ for some $w \in W$, and $\mu \in X^*(H)$. Then

(11.1)
$$a(\pi[\lambda + \mu], \Lambda \otimes w\mu) = a(\pi[\lambda], \Lambda).$$

For (a) see [**10**, Lemma 3.39] or [**21**, Theorem 7.2.7 and Corollary 7.2.27]. Part (b) is the first statement of [**10**, Lemma 3.44] carried over to our setting. This can be made into a necessary condition by considering all Cartan subgroups, but we won't need this.

LEMMA 11.2. *Suppose $\{\pi[\lambda + \mu]\}$ is a coherent family, and fix $w \in W(\lambda)$. Suppose $d\Lambda \in W\lambda$, and write $d\Lambda = yw\lambda$ for $y \in W$. Then*

(11.2)
$$a(\pi[w\lambda], \Lambda) = a(\pi[\lambda], (yw^{-1}y^{-1}) \times \Lambda).$$

PROOF. This is a simple change of variables. Let $\mu = w\lambda - \lambda$ and consider $\Lambda \otimes (-y\mu)$. Then $d(\Lambda \otimes (-y\mu)) = yw\lambda - y\mu = y(w\lambda - \mu) = y\lambda$. Apply (11.1):

(11.3)
$$a(\pi[w\lambda], \Lambda) = a(\pi[\lambda + \mu], (\Lambda \otimes (-y\mu)) \otimes y\mu)$$
$$= a(\pi[\lambda], \Lambda \otimes (-y\mu)).$$

It is enough to show $(yw^{-1}y^{-1}) \times \Lambda = \Lambda - y\mu$, i.e. $(yw^{-1}y^{-1})d\Lambda - d\Lambda = -y\mu$. This follows from $d\Lambda = yw\lambda$ and $\lambda - w\lambda = -\mu$.   □

Fix once and for all an *abstract* Cartan subalgebra $\mathfrak{h}_a$, a choice of positive roots Δ_a^+ of $\mathfrak{h}_a$, and let W_a be the abstract Weyl group, as in [**22**, 2.6]. If $\mathfrak{h}$ is any Cartan subalgebra, and $\lambda \in \mathfrak{h}^*$, there is a unique $\lambda_a \in \mathfrak{h}_a^*$, dominant for Δ_a^+, and a unique inner isomorphism ϕ_λ taking λ_a to λ. This induces an isomorphism $\phi_\lambda : W_a \to W$. We need an identity describing the dependence on λ, which is immediate:

(11.4)
$$\phi_{y\lambda}(w_a) = y\phi_\lambda(w_a)y^{-1} \quad (y \in W, w_a \in W_a).$$

DEFINITION 11.3. Suppose π is a virtual character, with regular infinitesimal character $\lambda_a \in \mathfrak{h}_a^*$ (dominant for Δ_a^+), and $w_a \in W_a(\lambda_a)$. Fix a Cartan subgroup H and $\lambda \in \mathfrak{h}^*$ conjugate to λ_a. Let $\{\pi[\lambda + \gamma]\}$ be the coherent family with $\pi = \pi[\lambda]$.

For $w_a \in W_a(\lambda_a)$ define

(11.5)
$$w_a \cdot \pi = \pi[\phi_\lambda(w_a^{-1})\lambda].$$

This is independent of the choice of H and λ.

This is the *coherent continuation* action on virtual characters, due to Zuckerman. See [**21**, Definition 7.2.28].

Here is the formulation in terms of characters. We first define the cross action of the abstract Weyl group on genuine characters (cf. (11.6)). For $w_a \in W(\lambda_a)$, and Λ such that $d\Lambda$ is conjugate to λ_a, define

$$(11.6) \qquad w_a \times \Lambda = \phi_{d\Lambda}(w_a^{-1}) \times \Lambda.$$

Note that the differential of the right hand side is $d(w_a^{-1} \times \Lambda) = \phi_{d\Lambda}(w_a \lambda_a)$, and this implies $d(w_a \times \Lambda) - d\Lambda = \phi_{d\Lambda}(w_a^{-1}\lambda_a - \lambda_a)$, which is a sum of roots since $w_a \in W(\lambda_a)$, so the cross action is well defined. It is straightforward to check that $(x_a y_a) \times \Lambda = x_a \times (y_a \times \Lambda)$.

PROPOSITION 11.4. Suppose π is a virtual character, with regular infinitesimal character $\lambda_a \in \mathfrak{h}_a^*$ (dominant for Δ_a^+), and $w_a \in W_a(\lambda_a)$. Suppose H is a Cartan subgroup, defined over $\mathbb{R}$, and Λ is a genuine character of $\widetilde{H(\mathbb{R})}$ with $d\Lambda$ conjugate to λ_a. Then

$$(11.7) \qquad a(w_a \cdot \pi, \Lambda) = a(\pi, w_a^{-1} \times \Lambda).$$

PROOF. This is a straightforward unwinding of the definitions. First of all choose $\lambda \in \mathfrak{h}^*$ conjugate to λ_a. Let $\{\pi[\lambda+\mu]\}$ be the coherent family with $\pi[\lambda] = \pi$. Then

$$
\begin{aligned}
a(w_a \cdot \pi, \Lambda) &= a(w_a \cdot \pi[\lambda], \Lambda) \\
&= a(\pi[\phi_\lambda(w_a^{-1})\lambda], \Lambda) \quad \text{(by Definition 11.3)} \\
&= a(\pi, (y\phi_\lambda(w_a)y^{-1}) \times \Lambda) \quad \text{(by (11.2))},
\end{aligned}
$$

where $y \in W$ satisfies $y\phi_\lambda(w_a^{-1})\lambda = d\Lambda$.

On the other hand

$$
\begin{aligned}
a(\pi, w_a^{-1} \times \Lambda) &= a(\pi, \phi_{d\Lambda}(w_a) \times \Lambda) \\
&= a(\pi, \phi_{y\phi_\lambda(w_a^{-1})\lambda}(w_a) \times \Lambda \\
&= a(\pi, (y\phi_\lambda(w_a^{-1})\phi_\lambda(w_a)\phi_\lambda(w_a)y^{-1}) \times \Lambda) \quad (\text{by (11.4)}) \\
&= a(\pi, (y\phi_\lambda(w_a)y^{-1}) \times \Lambda)
\end{aligned}
$$

$\square$

REMARK 11.5. Note that for $w_a \in W_a$, $w_a \times \Lambda = w \times \Lambda$ for some $w \in W$ *depending on* Λ, so some care is needed when using this formula when Λ varies. Fix Λ_0, and let $w_0 = \phi_{d\Lambda_0}(w_a^{-1}) \in W$, so

$$w_a^{-1} \times \Lambda = w_0^{-1} \times \Lambda \quad \text{(if } d\Lambda_0 = d\Lambda).$$

Then

$$(11.8) \qquad w_a^{-1} \times \Lambda = (yw_0^{-1}y^{-1}) \times \Lambda \quad \text{(if } yd\Lambda_0 = d\Lambda).$$

EXAMPLE 11.6. Consider the trivial representation $\mathbb{C}$:

$$(11.9) \qquad w_a \cdot \mathbb{C} = \text{sgn}(w_a)\mathbb{C} \quad (w_a \in W_a).$$

This is immediate from (7.4) and (11.7).

12. Computation of general $a(\pi, \Delta^+, \Lambda)$

As in the previous section fix $\mathfrak{h}_a$, W_a, and define the cross action of W_a accordingly. We use coherent continuation to give a formula for $a(\pi, \Delta^+, \Lambda)$ for arbitrary Λ.

PROPOSITION 12.1. Let π be a virtual representation with regular infinitesimal character $\lambda_a \in \mathfrak{h}_a^*$. Fix H, Δ^+ and a genuine character Λ of $H(\mathbb{R})_\rho$ with $d\Lambda$ conjugate to λ_a. Suppose $w_a \in W_a(\lambda_a)$ satisfies

(12.1) $\mathrm{Re}\langle d(w_a \times \Lambda), \alpha^\vee \rangle \geq 0$ for all $\alpha \in \Delta_r^+$.

Then

$$a(\pi, \Delta^+, \Lambda) = (-1)^{q_H + B(\Delta^+)} \epsilon(w_a \times \Lambda, \Delta_i^+) M(I(H, \Delta_r^+, w_a \times \Lambda), w_a \cdot \pi).$$

PROOF. By Proposition 11.4 and Corollary 9.2

$$a(\pi, \Delta^+, \Lambda) = a(w_a \cdot \pi, \Delta^+, w_a \times \Lambda)$$
$$= (-1)^{q_H + B(\Delta^+)} \epsilon(w_a \times \Lambda, \Delta_i^+) M(I(H, \Delta_r^+, w_a \times \Lambda), w_a \cdot \pi).$$

$\square$

If λ_a is integral we can always find $w_a \in W_a(\lambda_a) = W_a$ satisfying the conditions, so this determines all coefficients. In fact there is some flexibility in choosing w_a. If λ_a is not integral a little more work is required to determine all coefficients.

Here is one convenient reformulation in the integral case.

COROLLARY 12.2. Assume λ_a is regular and integral. Suppose π is a virtual character with infinitesimal character λ_a, and having a central character. Choose Δ^+ so that (2.12) holds, and let λ be Δ^+-dominant and conjugate to λ_a. Let $\Lambda_1, \ldots, \Lambda_n$ be the genuine characters of $\widetilde{H(\mathbb{R})}$ such that $d\Lambda_i = \lambda$, and such that $\Lambda_i e^{-\rho}$ and π have the same restriction to the center of $G(\mathbb{R})$.

Then for any $w_a \in W_a$ and $i \leq n$:

(12.2) $a(\pi, \Delta^+, w_a^{-1} \times \Lambda_i) = (-1)^{q_H} M(I(H, \Lambda_i), w_a \cdot \pi)$.

Every nonzero $a(\pi, \Delta^+, \Lambda)$ is of this form.

We now sketch how to use this to compute $a(\pi, \Delta^+, \Gamma)$ using the atlas software, which is freely available from www.liegroups.org/software. We assume the infinitesimal character is regular and integral.

First assume $I(\gamma)$ is a standard module and H is an arbitrary (θ-stable) Cartan subgroup. With Δ^+ and Λ_i as in Corollary 12.2 we conclude

(12.3) $a(I(\gamma), \Delta^+, w_a^{-1} \times \Lambda_i) = (-1)^{q_H} M(I(H, \Lambda_i), w_a \cdot I(\gamma))$.

The list of parameters with regular integral infinitesimal character is given by the output of the block command in the atlas software. First identify the modules $I(H, \Lambda_i)$ for $1 \leq i \leq n$ in the output of block.

Next compute $w_a \cdot I(\gamma)$ for all $w_a \in W_a$ in the basis of standard modules. This is elementary; it requires only the output of the block command, and not the KLV polynomials. In particular if s_α is a simple reflection then $s_\alpha \cdot I(\gamma)$ is a single standard module unless α is noncompact imaginary, in which case it is given by a Hecht-Schmid identity with 2 or 3 terms. See [9], [21, Corollary 8.4.6] or [23]. The application coherentContinuation, available at www.liegroups.org/software/helpers, is useful for computing coherent continuation. See [2].

Then (12.3) gives all nonzero terms $a(I(\gamma), \Delta^+, \Lambda)$. To vary Δ^+ use Lemma 6.3. This can be used to compute the character of a discrete series representation on an arbitrary Cartan subgroup.

For example for the split real form of E_8, it take about 14 seconds on a small computer to compute the `block` command, which produces a list of $452,690$ standard modules. The output of this command takes up about 165 megabytes of disk space.

Now suppose π is an irreducible representation. There are at least two ways to compute $a(\pi, \Delta^+, \Lambda)$. First of all use the KLV polynomials to write π as a sum of standard modules

$$(12.4) \qquad \pi = \sum_{\gamma} M(I(\gamma), \pi) I(\gamma).$$

Then

$$(12.5) \qquad a(\pi, \Delta^+, \Lambda) = \sum_{\gamma} M(I(\gamma), \pi) a(I(\gamma), \Delta^+, \Lambda),$$

and $a(I(\gamma), \Delta^+, \Lambda)$ are computed as above. The KLV polynomials, and hence $M(I(\gamma), \pi)$, are provided by the `klbasis` command of `atlas`. These are readily available for groups of rank ≤ 7. However, in contrast with the `block` command, these are difficult to compute in higher rank. For example to compute the KLV polynomials for the split real form of E_8 takes about 5.5 hours on a very large machine, and storing them takes 60 gigabytes of disk space.

Alternatively, the `wgraph` command of `atlas` computes the coherent continuation action in the basis of irreducible modules. Note, however, that computing the `wgraph` command is the same order of difficulty as computing `klbasis`. In any event one can compute

$$(12.6) \qquad a(\pi, \Delta^+, w_a^{-1} \times \Lambda_i) = (-1)^{q_H} M(I(H, \Lambda_i), w_a \cdot \pi)$$

directly, although this uses the KLV polynomials twice, once to compute $w_a \cdot \pi$, and again to compute $M(I(H, \Lambda_i), w_a \cdot \pi)$.

13. Alternative version

While the parametrization of Section 5 using the ρ-cover of H has many advantages, the extra choice of Ψ_r is unappealing. For this reason it worthwhile to write Langlands parameters in terms of the ρ_i cover of H. This is done in [1]. It is useful to give the translation between the two versions.

DEFINITION 13.1. ρ_i-standard limit data is a triple (H, Ψ_i, Γ) where H is a Cartan subgroup (defined over $\mathbb{R}$ as usual), Ψ_i is a set of positive imaginary roots, and Γ is a genuine character of $H(\mathbb{R})_{\rho_i}$ (where ρ_i is defined by Δ_i^+). We assume these satisfy

$$(13.1) \qquad \langle d\Gamma, \alpha^\vee \rangle \geq 0 \quad (\alpha \in \Psi_i)$$

$$(13.2) \qquad \langle d\Gamma, \alpha^\vee \rangle \neq 0 \quad (\alpha \in \Psi_i \text{ simple and compact})$$

We define equivalence by conjugation by $G(\mathbb{R})$.

See [1, Theorem 6.1].

LEMMA 13.2. *Suppose* $(H, \Psi_i, \Psi_r, \Lambda)$ *is standard limit data. Define* ρ_i-*standard limit data* $(H, \Psi_i, \Lambda \otimes \gamma(\Psi_r)^{-1})$ *(see Section 4).*

Conversely suppose (H, Ψ_i, Γ) *is* ρ_i-*standard limit data. Choose a set of positive roots* Ψ_r, *and define standard limit data* $(H, \Psi_i, \Psi_r, \Gamma \otimes \gamma(\Psi_r))$.

These define a bijection between equivalence classes of standard limit data and ρ_i-*standard limit data.*

The proof is immediate (using (4.3)).

To write a character formula using ρ_i-standard limit data we need a version of the Weyl denominator defined on $H(\mathbb{R})_{\rho_i}$.

DEFINITION 13.3. Fix a set Δ^+ of positive roots, and define ρ_i, ρ_r and ρ_{cx} as usual. Recall (end of Section 2) $e^{\rho_{cx}}$ is well defined on $H(\mathbb{R})$. Define a genuine character $e^{\rho'}$ of $H(\mathbb{R})_{\rho_i}$:

$$(13.3) \qquad e^{\rho'}(\overline{g}) = e^{\rho_{cx}}(g)|e^{\rho_r}(g)|e^{\rho_i}(\overline{g}) \quad (g \in H(\mathbb{R})_{\rho_i})$$

where g is the image of $\overline{g}$ in $H(\mathbb{R})$, and

$$(13.4) \qquad D'(\Delta^+, \overline{g}) = \prod_{\alpha \in \Delta^+} (1 - e^{-\alpha}(g))e^{\rho'}(\overline{g}).$$

We then define constants $a'(\pi, \Delta^+, \Lambda)$ for Λ a genuine character of $H(\mathbb{R})_{\rho_i}$ by analogy with (6.4), using D' in place of D.

It follows easily from Definition (4.1) that

$$(13.5) \qquad e^{\rho'} = e^{\rho} \otimes \gamma(\Delta_r^+)^{-1}$$

Suppose Λ is a genuine character of $H(\mathbb{R})_\rho$. It follows immediately from this and Definition (4.2) that

$$(13.6) \qquad (\Lambda e^{-\rho})(g) = (\Lambda_M e^{-\rho'})(g).$$

Also an elementary calculation gives

$$(13.7) \qquad (w\Lambda)_M = \tau(\Psi_r, w)\Lambda_M \quad (w \in W(G(\mathbb{R}), H(\mathbb{R})).$$

The character formula of Proposition 8.2 takes the following form.

PROPOSITION 13.4. Suppose $\gamma = (H, \Psi_i, \Lambda)$ is ρ_i-standard limit data. Choose positive roots Δ^+ containing Ψ_i, and let $\Psi_r = \Delta_r^+$. Let $\delta = (-1)^{q_H + B(\Delta^+)}$. Then for $g \in H(\mathbb{R})_{\mathrm{reg}}$

$$\Theta_{I(\gamma)}(g) = \delta \frac{\epsilon(\Psi_r, g)}{D'(\Delta^+, \overline{g})} \sum_{w \in W(G(\mathbb{R}), H(\mathbb{R}))} \mathrm{sgn}_i(w)(w\Lambda)(\overline{g}).$$

where $\overline{g}$ is an inverse image of g in $H(\mathbb{R})_{\rho_i}$. In other words for Γ a genuine character of $H(\mathbb{R})_{\rho_i}$:

$$a'(I(\gamma), \Delta^+, \Gamma) = \begin{cases} \delta \mathrm{sgn}_i(w) & \Gamma = w\Lambda \quad (w \in W(G(\mathbb{R}), H(\mathbb{R})) \\ 0 & \text{otherwise.} \end{cases}$$

The remaining results, including Proposition 12.1, hold with minor changes.

14. Examples

We compute some character formulas for $SL(2, \mathbb{R}), PGL(2, \mathbb{R})$ and $Sp(4, \mathbb{R})$, using Corollary (12.2).

14.1. Example 1: $SL(2,\mathbb{R})$. Here is the character table for $SL(2,\mathbb{R})$ at infinitesimal character ρ. See Examples 4.5 and 8.4.

Since ρ exponentiates to H we may ignore the ρ-cover. Identify compact and split Cartan subgroup with $\mathbb{R}^*, S^1$ respectively. Choose $\Delta^+ = \{\alpha\}$ for $\mathbb{R}^*$ with $\alpha(x) = x^2$, and for S^1 with $\alpha(e^{i\theta}) = e^{2i\theta}$.

With the obvious notation there are 4 standard modules:

(14.1)
$$DS_+ = I(S^1, e^{i\theta}) : \text{holomorphic discrete series}$$
$$DS_- = I(S^1, e^{-i\theta}) : \text{anti-holomorphic discrete series}$$
$$PS_+ = I(\mathbb{R}^*, x) : \text{reducible principal series with even } K\text{-types}$$
$$PS_- = I(\mathbb{R}^*, |x|) : \text{irreducible principal series with odd } K\text{-types}$$

The only reducible standard module is PS_+; in the Grothendieck group

(14.2)
$$PS_+ = DS_+ + DS_- + \mathbb{C}.$$

Since Δ^+ is fixed we drop it from the notation. The coefficients $a(I(\gamma), \Lambda)$ and $a(\pi, \Lambda)$ are given by the following table.

$SL(2,\mathbb{R})$						
	$\mathbb{R}^*$				S^1	
	x	$\lvert x\rvert$	$\frac{1}{x}$	$\frac{1}{\lvert x\rvert}$	$e^{i\theta}$	$e^{-i\theta}$
DS_+	0	0	1	0	-1	0
DS_-	0	0	1	0	0	1
PS_+	1	0	1	0	0	0
PS_-	0	1	0	1	0	0
$\mathbb{C}$	1	0	-1	0	1	-1

We show how to compute some of these coefficients. The main coherent continuation identity we need is the basic Hecht-Schmid identity:

(14.3)(a)
$$s_\alpha \cdot DS_\pm = PS_+ - DS_\mp$$

and we'll also use

(14.3)(b)
$$s_\alpha \cdot PS_\pm = PS_\pm$$

which follows from (a) and (14.2).

First we consider the easy character formulas for PS_+ on $H = \mathbb{R}^*$. Using (12.2), with $q_H = 0 = B(\Delta^+) = 0$, we have:

$$a(PS_+, x) = M(I(H,x), PS_+) = M(PS_+, PS_+) = 1$$
$$a(PS_+, |x|) = M(I(H,|x|), PS_+) = M(PS_-, PS_+) = 0$$
$$a(PS_+, x^{-1}) = a(PS_+, s_\alpha \times x) = M(I(H,x), s_\alpha \cdot PS_+) = M(PS_+, PS_+) = 1$$
$$a(PS_+, |x|^{-1}) = a(PS_+, s_\alpha \times |x|) = M(I(H,|x|), s_\alpha \cdot PS_+) = M(PS_-, PS_+) = 0$$

The formulas for the characters of the principal series are given in Example 8.4.

Next, here are the elementary character formulas for $DS_\pm$ on $H = S^1$. In this case $q_H = 1$ and $B(\Delta^+) = 0$. Using $I(H, e^{i\theta}) = DS_+$ and (12.2) we compute:

$$a(DS_+, e^{i\theta}) = -M(DS_+, DS_+) = -1$$
$$a(DS_+, e^{-i\theta}) = a(DS_+, s_\alpha \times e^{i\theta}) = -M(DS_+, s_\alpha \cdot DS_+)$$
$$= -M(DS_+, PS - DS_-) = 0$$
$$a(DS_-, e^{i\theta}) = -M(DS_+, DS_-) = 0$$
$$a(DS_-, e^{-i\theta}) = a(DS_-, s_\alpha \times e^{i\theta}) = -M(DS_+, s_\alpha \cdot DS_-)$$
$$= -M(DS_+, PS - DS_+) = 1$$

These give the well known formulas (and special cases of (3.2)):

(14.4)
$$\theta_{DS_+}(e^{i\theta}) = \frac{-e^{i\theta}}{e^{i\theta} - e^{-i\theta}}$$

$$\theta_{DS_-}(e^{i\theta}) = \frac{e^{-i\theta}}{e^{i\theta} - e^{-i\theta}}$$

Finally the most interesting case, the discrete series characters on $\mathbb{R}^*$:

$$a(DS_\pm, x) = M(I(\mathbb{R}^*, x), DS_\pm) = M(PS_+, DS_\pm) = 0$$
$$a(DS_\pm, |x|) = M(I(\mathbb{R}^*, |x|), DS_\pm) = M(PS_-, DS_\pm) = 0$$
$$a(DS_\pm, \frac{1}{x}) = a(DS_\pm, s_\alpha \times x) = M(I(\mathbb{R}^*, x), s_\alpha \cdot DS_\pm)$$
$$= M(PS_+, PS_+ - DS_\mp) = 1$$
$$a(DS_\pm, \frac{1}{|x|}) = a(DS_\pm, s_\alpha \times |x|) = M(I(\mathbb{R}^*, |x|), s_\alpha \cdot DS_\pm)$$
$$= M(PS_-, PS_+ - DS_\mp) = 0$$

Therefore

(14.5)
$$\theta_{DS_\pm}(x) = \frac{x^{-1}}{x - \frac{1}{x}} \quad (|x| > 1).$$

This implies

(14.6)
$$\theta_{DS_\pm}(x) = \frac{-x}{x - \frac{1}{x}} \quad (|x| < 1).$$

14.2. Example 2: $PGL(2, \mathbb{R})$. We now give the character table for $PGL(2, \mathbb{R})$ at infinitesimal character ρ. We cannot ignore the ρ-cover in this case. See Examples 4.6 and 8.5.

Again identify compact and split Cartan subgroupos with $\mathbb{R}^*, S^1$ respectively. Choose $\Delta^+ = \{\alpha\}$ for $\mathbb{R}^*$ with $\alpha(x) = x$, and for S^1 with $\alpha(e^{i\theta}) = e^{i\theta}$.

Now the cover of $\mathbb{R}^*$ is $\{(x, z) \mid z^2 = x\}$, and $e^\rho(x, z) = z$. Similarly the cover of S^1 is $\{(w, z) \mid |w| = |z| = 1, z^2 = w\}$, and $e^\rho(w, z) = z$.

There are 3 standard modules at infinitesimal character ρ:

$$DS = I(S^1, e^\rho) : \text{discrete series}$$

(14.7)
$$PS_\mathbb{C} = I(\mathbb{R}^*, e^\rho) : \text{reducible spherical principal series}$$

$$PS_{\text{sgn}} = I(\mathbb{R}^*, \text{sgn } e^\rho) : \text{reducible non-spherical principal series}$$

Both standard modules are reducible; in the Grothendieck group

(14.8)(a) $$PS_{\mathbb{C}} = DS + \mathbb{C}$$

(14.8)(b) $$PS_{\text{sgn}} = DS + \text{sgn}$$

There is no irreducible principal series representation at infinitesimal character ρ (there are two irreducible principal series at infinitesimal character 2ρ, which we don't consider).

The coefficients $a(I(\gamma), \Lambda)$ and $a(\pi, \Lambda)$ are given by the following table.

	$PGL(2, \mathbb{R})$					
	$\mathbb{R}^*$				S^1	
	e^ρ	$\text{sgn}\, e^\rho$	$e^{-\rho}$	$\text{sgn}\, e^{-\rho}$	e^ρ	$e^{-\rho}$
DS	0	0	1	1	-1	1
$PS_{\mathbb{C}}$	1	0	0	1	0	0
PS_{sgn}	0	1	1	0	0	0
$\mathbb{C}$	1	0	-1	0	1	-1
sgn	0	1	0	-1	1	-1

In this case the relevant Hecht-Schmid identity is:

(14.9)(a) $$s_\alpha \cdot DS = PS_{\mathbb{C}} + PS_{\text{sgn}} - DS$$

We also use that fact that

(14.9)(b) $$s_\alpha \cdot PS_{\mathbb{C}} = PS_{\text{sgn}}$$

which follows from (a), (14.8)(a) and Example 11.6.

Here is the easy character formula for $PS_{\mathbb{C}}$ on $H = \mathbb{R}^*$. In this case $q_H = 1$ and $B(\Delta^+) = 0$.

$$a(PS_{\mathbb{C}}, e^\rho) = M(I(H, e^\rho), PS_{\mathbb{C}}) = M(PS_{\mathbb{C}}, PS_{\mathbb{C}}) = 1$$
$$a(PS_{\mathbb{C}}, \text{sgn}\, e^\rho) = M(I(H, e^\rho \text{sgn}), PS_{\mathbb{C}}) = M(PS_{\text{sgn}}, PS_{\mathbb{C}}) = 0$$
$$a(PS_{\mathbb{C}}, e^{-\rho}) = a(PS_{\mathbb{C}}, s_\alpha \times e^\rho) = M(I(H, e^\rho), s_\alpha \cdot PS_{\mathbb{C}})$$
$$= M(PS_{\mathbb{C}}, PS_{\text{sgn}}) = 0$$
$$a(PS_{\mathbb{C}}, \text{sgn}\, e^{-\rho}) = a(PS_{\mathbb{C}}, s_\alpha \times \text{sgn}\, e^\rho) = M(I(H, \text{sgn}\, e^\rho), s_\alpha \cdot PS_{\mathbb{C}})$$
$$= M(PS_{\text{sgn}}, PS_{\text{sgn}}) = 1$$
$$a(PS_{\mathbb{C}}, \text{sgn}\, e^{-\rho}) = M(I(H, s_\alpha \times \text{sgn}\, e^\rho), PS_{\mathbb{C}}) = M(PS_{\text{sgn}}, PS_{\text{sgn}}) = 1$$

The formulas for the characters of the principal series are given in Example 8.5.

Next consider the elementary character formula for the discrete series representation on $H = S^1$, with $q_H = 1$, $B(\Delta^+) = 0$, and $I(H, e^\rho) = DS$:

$$a(DS, e^\rho) = -M(DS, DS) = -1$$
$$a(DS, e^{-\rho}) = -M(DS, s_\alpha \cdot DS) = -M(DS, PS_{\mathbb{C}} + PS_{\text{sgn}} - DS) = 1.$$

So:

(14.10) $$\theta_{DS}(e^{i\theta}) = \frac{-e^{i\theta} + e^{i\theta}}{e^{i\theta} - e^{-i\theta}} = -1.$$

Note that $\theta_{DS}(e^{i\theta}) = -\theta_{\mathbb{C}}(e^{i\theta})$, which follows from (14.8)(a).

The most interesting case is the discrete series on $H = \mathbb{R}^*$, with $q_H = B(\Delta^+) = 0$, $I(\mathbb{R}^*, e^\rho) = PS_{\mathbb{C}}$, $I(\mathbb{R}^*, \text{sgn } e^\rho) = PS_-$:

$$a(DS, e^\rho) = M(I(\mathbb{R}^*, e^\rho), DS) = M(PS_{\mathbb{C}}, DS) = 0$$
$$a(DS, \text{sgn } e^\rho) = M(I(\mathbb{R}^*, \text{sgn } e^\rho), DS) = M(PS_{\text{sgn}}, DS) = 0$$
$$a(DS, e^{-\rho}) = a(DS, s_\alpha \times e^\rho) = M(I(\mathbb{R}^*, e^\rho), s_\alpha \cdot DS)$$
$$= M(PS_{\mathbb{C}}, PS_{\mathbb{C}} + PS_{\text{sgn}} - DS) = 1$$
$$a(DS, \text{sgn } e^{-\rho}) = a(DS, s_\alpha \text{sgn } e^{-\rho}) = M(I(\mathbb{R}^*, \text{sgn } e^\rho), s_\alpha \cdot DS)$$
$$= M(PS_{\text{sgn}}, PS_{\mathbb{C}} + PS_{\text{sgn}} - DS) = 1$$

Therefore for $|x| > 1$ we have:

$$\theta_{DS}(x) = \frac{z^{-1} + \text{sgn}(x)z^{-1}}{(1 - \frac{1}{x})z} \quad \text{(where } z^2 = x\text{)}$$
$$= \frac{|x|^{-\frac{1}{2}} + \text{sgn}(x)|x|^{-\frac{1}{2}}}{|x|^{\frac{1}{2}} - |x|^{-\frac{1}{2}}}$$
$$= \begin{cases} \frac{2x^{-\frac{1}{2}}}{|x|^{\frac{1}{2}} - |x|^{-\frac{1}{2}}} & x > 1 \\ 0 & x < -1. \end{cases}$$

Another way to write this is

$$\theta_{DS}(x) = \frac{1 + \text{sgn}(x)}{x - 1} \quad (|x| > 1).$$

Also note that

$$(14.11) \qquad \theta_{DS}(x) = \frac{-|x|^{\frac{1}{2}} - \text{sgn}(x)|x|^{\frac{1}{2}}}{|x|^{\frac{1}{2}} - |x|^{-\frac{1}{2}}} \quad (|x| < 1).$$

14.3. Discrete series representations of $Sp(4, \mathbb{R})$. We now calculate the characters of the discrete series representations of $Sp(4, \mathbb{R})$. We make some use of the atlas software; see [2] for an introduction. The formulas we obtain can be readily shown to agree with those of [13, 6.2], and (with a little more work) with [15, page 499] and [12, (4.9)].

There are 4 discrete series representations of $Sp(4, \mathbb{R})$ with infinitesimal character ρ. Two of these are holomorphic/anti-holomorphic (in chambers with a compact simple root), and the other two are large (in a chamber in which both simple roots are noncompact). The only representations we need to consider are in the *block* of these representations, since this is preserved by the coherent continuation action. This block is the span in the Grothendieck group of 12 standard representations, or the corresponding irreducible representations. These can be read off from the output of the block command in atlas, and are numbered 0 to 11. Representations $0 - 3$ are in the discrete series; $0, 1$ are large, and $2, 3$ are holomorphic/anti-holomorphic. For more information on blocks, and the atlas software see [2, Example 10.3].

Here is the output of the block command:

```
0( 0,6):  0  [i1,i1]  1  2  ( 4, *)  ( 5, *)  0  e
1( 1,6):  0  [i1,i1]  0  3  ( 4, *)  ( 6, *)  0  e
2( 2,6):  0  [ic,i1]  2  0  ( *, *)  ( 5, *)  0  e
```

3(3,6):	0	[ic,i1]	3	1	(*, *)	(6, *)	0	e
4(4,5):	1	[r1,C+]	4	9	(0, 1)	(*, *)	1	1
5(5,4):	1	[C+,r1]	7	5	(*, *)	(0, 2)	2	2
6(6,4):	1	[C+,r1]	8	6	(*, *)	(1, 3)	2	2
7(7,3):	2	[C-,i1]	5	8	(*, *)	(10, *)	2	1,2,1
8(8,3):	2	[C-,i1]	6	7	(*, *)	(10, *)	2	1,2,1
9(9,2):	2	[i2,C-]	9	4	(10,11)	(*, *)	1	2,1,2
10(10,0):	3	[r2,r1]	11	10	(9, *)	(7, 8)	3	2,1,2,1
11(10,1):	3	[r2,rn]	10	11	(9, *)	(*, *)	3	2,1,2,1

Write $I(k)$ and $\pi(k)$ for the standard and irreducible modules with number $0 \leq k \leq 11$ from the table.

This block has 2 principal series representations, numbers 10 and 11. Standard representation $I(10)$ is the spherical principal series representation. Its irreducible quotient $\pi(10)$ is the trivial representation. Standard module $I(11)$ is the unique nonspherical principal series representation with trivial infinitesimal and central characters. This is reducible, and $\pi(11)$ is infinite dimensional.

(There are 6 other standard representations of $Sp(4,\mathbb{R})$ with infinitesimal character ρ. These are not in the previous block, have nontrivial central character, and include two minimal principle series representations, one of which is irreducible.)

14.3.1. *Split Cartan subgroup.* Let $H(\mathbb{R})$ be a split Cartan subgroup. Identify $H(\mathbb{R})$ with $\mathbb{R}^{*2}$ such that the roots are $(x,y) \to x^{\pm 2}, y^{\pm 2}, x^{\pm 1}y^{\pm 1}$. Write (a,b) for the character $(x,y) \to x^a y^b$. Choose simple roots $\alpha = (1,-1)$ and $\beta = (0,2)$, which define Δ^+, and give $\rho = (2,1)$. Write D for the Weyl denominator defined by Δ^+. Since ρ exponentiates we may ignore the cover. Finally let $\chi(x,y) = \mathrm{sgn}(xy)$.

In the setting of Corollary (12.2) and the subsequent discussion we only need to consider $\Lambda = e^\rho$ or χe^ρ. In this terminology

$$I(H, e^\rho) = I(10)$$
$$\text{(14.12)} \qquad I(H, \chi e^\rho) = I(11).$$

First we take π to be the large discrete series representation $\pi(0) = I(0)$. With a little care we may ignore the difference between W_a and W. Using (14.12) and (12.2) we have

$$\text{(14.13)} \qquad a(\pi, e^{w^{-1}\rho}) = M(I(10), w \cdot \pi)$$

$$\text{(14.14)} \qquad a(\pi, \chi e^{w^{-1}\rho}) = M(I(11), w \cdot \pi)$$

The first column of the next table gives $w \in W$ as a product of simple reflections s_α, s_β, which we label $1, 2$, respectively. The second column gives $w^{-1}\rho$. There is a subtlety here: because of (2.2), if you view the first column as an element w' of W (not W_a), then the second column is $w'\rho$ (not $w^{-1}\rho$).

We compute $w \cdot \pi$, as a sum of standard modules, using the output of the `block` command. The application `coherentContinuation`, available at `www.liegroups.org/software/helpers` is very useful for expediting this calculation. We omit the details how to do this, but the result is listed in column 3 of the next table. For example the second row indicates $s_\alpha \cdot \pi = -I(1) + I(4)$ (the coefficients are all ± 1).

By (14.13) $a(\pi, e^{w^{-1}\rho})$ (respectively $a(\pi, \chi e^{w^{-1}\rho})$) is computed as the multiplicity of $I(10)$ (respectively $I(11)$) in column 3, which is given in columns 4 and 5.

w	$w^{-1}\rho$	$w \cdot \pi$	$a(\pi, e^{w^{-1}\rho})$	$a(\pi, \chi e^{w^{-1}\rho})$
e	$(2,1)$	0	0	0
1	$(1,2)$	$-1+4$	0	0
2	$(2,-1)$	$-2+5$	0	0
12	$(-1,2)$	$2+7$	0	0
21	$(1,-2)$	$3-6+9$	0	0
121	$(-2,1)$	$-3-8-9+10+11$	1	1
212	$(-1,-2)$	$-0+5-8+10$	1	0
1212	$(-2,-1)$	$1-4-6+7+11$	0	1

Let

(14.15) $$H(\mathbb{R})_+ = \{(x,y) \mid |x| > |y| > 1\}.$$

From the table we conclude that on $H(\mathbb{R})_+$ we have

(14.16) $$(D\theta_\pi)(x,y)) = x^{-2}y + \text{sgn}(xy)x^{-2}y + x^{-1}y^{-2} + \text{sgn}(xy)x^{-2}y^{-1}$$

Perhaps a more familiar way to write this is as follows. Let $t = (\epsilon_1, \epsilon_2)$ ($\epsilon_i = \pm 1$). Write (a,b) for the differential of the character (a,b) of H. Suppose $X \in \mathfrak{h}_0$, and assume $e^X \in H(\mathbb{R})_+$. Let $H(\mathbb{R})^0$ identity component of $H(\mathbb{R})$. Then the character formula on $H(\mathbb{R})^0 \cap H(\mathbb{R})_+$ is

(14.17)(a) $$(D\theta_\pi)(\exp(X)) = 2e^{(-2,1)X} + e^{(-1,-2)X} + e^{(-2,-1)X}.$$

It is worth pointing out that all nonzero KLV polynomials are 1, in spite of the 2 appearing in this formula (but this is explained by the fact that $1+1 = 2$). We also see, as is evident from the central character and the fact that $e^\rho(-1,-1) = -1$, that

$$(D\theta_\pi)(-\exp(X)) = -(D\theta_\pi)(\exp X).$$

On the other hand if $t = (\epsilon, -\epsilon)$ ($\epsilon = \pm 1$) then on $tH(\mathbb{R})^0 \cap H(\mathbb{R})_+$ we have:

(14.17)(b) $$(D\theta_\pi)(t\exp(X)) = \epsilon(e^{(-1,-2)X} + e^{(-2,-1)X}).$$

Formula (14.16) has the advantage that it holds on all of $H(\mathbb{R})_+$.

More succinctly, here are character formulas for the other discrete series representations $\pi(1)$ (large) and $\pi(2), \pi(3)$ (holomorphic/antiholomorphic). In the column $w \cdot \pi(i)$ we only list the terms $10, 11$ which occur.

In this table the entry in column $6,7,8$ in the multiplicity of $I(10)$ in column $3,4,5$, respectively. The entry in column $9,10,11$ is the multiplicity of $I(11)$ in column $3,4,5$, respectively.

		$w \cdot \pi(i)$			$a(\pi(i), e^{w^{-1}\rho})$			$a(\pi(i), \chi e^{w^{-1}\rho})$		
w	$w^{-1}\rho$	$\pi(1)$	$\pi(2)$	$\pi(3)$	$\pi(1)$	$\pi(2)$	$\pi(3)$	$\pi(1)$	$\pi(2)$	$\pi(3)$
121	$(-2,1)$	$10+11$	$*$	$*$	1	0	0	1	0	0
212	$(-1,-2)$	10	10	10	1	1	1	0	0	0
1212	$(-2,-1)$	11	-10	-10	0	-1	-1	1	0	0

We conclude that the characters of $\pi(0)$ and $\pi(1)$ are equal on $H(\mathbb{R})$. Moreover (still on $H(\mathbb{R})_+$) we have

(14.18)(a) $$(D\theta_{\pi(k)})(x,y) = x^{-1}y^{-2} - x^{-2}y^{-1} \quad (k = 2,3).$$

Alternatively, with $t = (\epsilon_1, \epsilon_2)$:

(14.18)(b) $\qquad\qquad (D\theta_{\pi(k)})(t\exp(X)) = \epsilon_1 e^{(-1,-2)X} - \epsilon_2 e^{(-2,-1)X}$

14.3.2. *Cartan #2:* $H(\mathbb{R}) \simeq \mathbb{R}^* \times S^1$. We briefly consider a Cartan subgroup with $H(\mathbb{R}) \simeq \mathbb{R}^* \times S^1$; this is Cartan #2 in `atlas`. Identify $H(\mathbb{R})$ with $\mathbb{R}^* \times S^1$, and choose simple roots α, β so that $\alpha(x, e^{i\theta}) = xe^{-i\theta}$, $\beta(x, e^{i\theta}) = e^{2i\theta}$. Write (a, b) for the character $(x, e^{i\theta}) \to x^a e^{ib\theta}$, so $e^\rho = (2, 1)$.

In this case $q_H = 1$, $B(\Delta^+) = 0$, and $\epsilon(\Delta_i^+, e^\rho) = 1$. We can check that $I(H, e^\rho) = I(8)$, and conclude

(14.19) $\qquad\qquad a(\pi(i), e^{w^{-1}\rho}) = -M(I(8), w \cdot \pi(i)).$

Using `atlas` we compute the following table, showing only the occurences of 8:

w	$w^{-1}\rho$	$w \cdot \pi(i)$				$a(\pi(i), e^{w^{-1}\rho})$			
		$\pi(0)$	$\pi(1)$	$\pi(2)$	$\pi(3)$	$\pi(0)$	$\pi(1)$	$\pi(2)$	$\pi(3)$
12	$(-1, 2)$	*	8	*	8	0	-1	0	-1
121	$(-2, 1)$	-8	*	*	-8	1	0	0	1
212	$(-1, -2)$	-8	*	-8	*	1	0	1	0
1212	$(-2, -1)$	*	8	8	*	0	-1	-1	0

Therefore on

$$H(\mathbb{R})_+ = \{(x, e^{i\theta}) \,|\, |x| > 1\}$$

we have formulas

(14.20)(a) $\qquad\qquad (D\theta_{\pi(0)})(xe^{i\theta}) = x^{-1}e^{-2i\theta} + x^{-2}e^{i\theta}$

(14.20)(b) $\qquad\qquad (D\theta_{\pi(1)})(xe^{i\theta}) = -x^{-1}e^{2i\theta} - x^{-2}e^{-i\theta}$

(14.20)(c) $\qquad\qquad (D\theta_{\pi(2)})(xe^{i\theta}) = x^{-1}e^{-2i\theta} - x^{-2}e^{-i\theta}$

(14.20)(d) $\qquad\qquad (D\theta_{\pi(3)})(xe^{i\theta}) = -x^{-1}e^{2i\theta} + x^{-2}e^{i\theta}.$

14.3.3. *Cartan #1:* $H(\mathbb{R}) \simeq \mathbb{C}^*$. Finally suppose $H(\mathbb{R}) \simeq \mathbb{C}^*$ (Cartan #3 in `atlas`). Choose the isomorphism, and simple roots α, β, so that

$$\alpha(z) = z/\overline{z}, \quad \beta(z) = \overline{z}^2.$$

Write (a, b) for the character $z \to z^a \overline{z}^b$, i.e. $e^{x+iy} \to e^{(a+b)x+(a-b)iy}$. In particular $e^\rho = (2, 1)$, and $e^\rho(e^{x+iy}) = e^{3x+iy}$.

In this case $I(H, e^\rho) = I(9)$, $q_H = 1$, $\epsilon(\Delta_i^+, e^\rho) = 1$, $B(\Delta^+) = 0$, and

(14.21) $\qquad\qquad a(\pi(i), e^{w^{-1}\rho}) = -M(I(9), w \cdot \pi(i)).$

Using `atlas` we compute this table, showing only the occurences of 9:

w	$w^{-1}\rho$	$w \cdot \pi(i)$				$a(\pi(i), e^{w^{-1}\rho})$			
		$\pi(0)$	$\pi(1)$	$\pi(2)$	$\pi(3)$	$\pi(0)$	$\pi(1)$	$\pi(2)$	$\pi(3)$
21	$(1, -2)$	9	9	*	*	1	1	0	0
121	$(-2, 1)$	-9	-9	*	*	-1	-1	0	0
212	$(-1, -2)$	*	*	-9	-9	0	0	-1	-1
1212	$(-2, -1)$	*	*	9	9	0	0	1	1

Therefore on

$$H(\mathbb{R})_+ = \{e^{x+iy} \,|\, x > 0\}$$

we have

$$(14.22) \qquad (D\theta_{\pi(i)})(e^{x+iy}) = \begin{cases} -e^{-x+3iy} + e^{-x-3iy} & i = 0,1 \\ e^{-3x+iy} - e^{-3x-iy} & i = 1,2. \end{cases}$$

14.4. Character Table for $Sp(4,\mathbb{R})$ at ρ. Without giving any more details of the calculations, here is complete information about the characters of the irreducible representations of $Sp(4,\mathbb{R})$ in the block of the trivial representation.

We use notation for the four Cartan subgroups $S^1 \times S^1, \mathbb{C}^*, \mathbb{R}^* \times S^1, \mathbb{R}^* \times \mathbb{R}^*$ as in Section 14.3.1, 14.3.2 and 14.3.3. (We haven't considered the compact Cartan subgroup yet; the notation is obvious here, and we choose Δ^+ so that $\rho = (2,1)$ as usual.) For each Cartan subgroup we have fixed a choice of Δ^+, and D is the corresponding Weyl denominator. The characters Λ of $H(\mathbb{R})$ with $d\Lambda \in W\rho$ are parametrized in each case by $\{(\pm a, \pm b)\}$ with $(a,b) = (2,1)$ or $(1,2)$. In the case of the split Cartan subgroup there are two characters (a,b) and $\chi(a,b)$ (see Section 14.3.1).

As discussed in Section 14.3 the block of the trivial representation consists of 12 standard modules $I(i)$, and their corresponding irreducible representations $\pi(i)$, with $0 \le i \le 11$. The character formulas for the irreducible representations, in terms of standard modules are as follows. This was computed using the `klbasis` command; and may also be found in [**23**].

$$\pi(0) = I(0)$$
$$\pi(1) = I(1)$$
$$\pi(2) = I(2)$$
$$\pi(3) = I(3)$$
$$\pi(4) = -I(0) - I(1) + I(4)$$
$$\pi(5) = -I(0) - I(2) + I(5)$$
$$\pi(6) = -I(1) - I(3) + I(6)$$
$$\pi(7) = I(0) + I(1) + I(2) - I(4) - I(5)) + I(7)$$
$$\pi(8) = I(0) + I(1) + I(3) - I(4) - I(6) + I(8)$$
$$\pi(9) = I(0) + I(1) + I(2) + I(3) - I(4) - I(5) - I(6) + I(9)$$
$$\pi(10) = -I(0) - I(1) - I(2) - I(3) + I(4) + I(5) + I(6)$$
$$\qquad - I(7) - I(8) - I(9) + I(10)$$
$$\pi(11) = -I(2) - I(3) - I(9) + I(11)$$

Each row in the following tables gives the character formula for a single standard or irreducible module, on the given Cartan subgroup.

For example the first row in the first table below says the formula for the large discrete series representation $I(0) = \pi(0)$ on the compact Cartan subgroup is:

$$(D\theta_{I(0)})(e^{i\theta_1}, e^{i\theta_2}) = -e^{2i\theta_1 - i\theta_2} + e^{-i\theta_1 + 2i\theta_2}$$

The last entry in the second table gives

$$(D\theta_{\pi(11)})(e^{i\theta_1}, e^{i\theta_2}) = -e^{2i\theta_1 + i\theta_2} + e^{i\theta_1 + 2i\theta_2} + e^{-i\theta_1 - 2i\theta_2} - e^{-2i\theta_1 - i\theta_2}$$

Compact Cartan subgroup

$$(a,b): (e^{i\theta_1}, e^{i\theta_2}) \to e^{ai\theta_1 + bi\theta_2}$$

$$H(\mathbb{R})_+ = H(\mathbb{R})$$

	$S^1 \times S^1$: Standard Modules							
I	(2,1)	(1,2)	(2,-1)	(-1,2)	(1,-2)	(-2,1)	(-1,-2)	(-2,-1)
$I(0)$			-1	1				
$I(1)$					1	-1		
$I(2)$	1	-1						
$I(3)$							-1	1

	$S^1 \times S^1$: Irreducible Modules							
π	(2,1)	(1,2)	(2,-1)	(-1,2)	(1,-2)	(-2,1)	(-1,-2)	(-2,-1)
$\pi(0)$			-1	1				
$\pi(1)$					1	-1		
$\pi(2)$	1	-1						
$\pi(3)$							-1	1
$\pi(4)$			1	-1	-1	1		
$\pi(5)$	-1	1	1	-1				
$\pi(6)$					-1	1	1	-1
$\pi(7)$	1	-1	-1	1	1	-1		
$\pi(8)$			-1	1	1	-1	-1	1
$\pi(9)$	1	-1	-1	1	1	-1	-1	1
$\pi(10)$	-1	1	1	-1	-1	1	1	-1
$\pi(11)$	-1	1					1	-1

The identity (6.5)(b) is clear in the table, for w the short simple reflection, which is in $W(G(\mathbb{R}), H(\mathbb{R})) \subset W_i = W$.

Cartan #2: $\mathbb{R}^* \times S^1$

$$(a,b) : (x, e^{i\theta}) \to x^a e^{bi\theta}$$

$$H(\mathbb{R})_+ = \{(x, e^{i\theta}) \,|\, |x| > 1\}$$

See Section 14.3.2.

I	$\mathbb{R}^* \times S^1$: Standard Modules							
	(2,1)	(1,2)	(2,-1)	(-1,2)	(1,-2)	(-2,1)	(-1,-2)	(-2,-1)
$I(0)$						1	1	
$I(1)$				-1				-1
$I(2)$							1	-1
$I(3)$				-1	1			
$I(5)$					1		1	
$I(6)$		-1		-1				
$I(7)$			1					1
$I(8)$	-1					-1		

π	$\mathbb{R}^* \times S^1$: Irreducible Modules							
	(2,1)	(1,2)	(2,-1)	(-1,2)	(1,-2)	(-2,1)	(-1,-2)	(-2,-1)
$\pi(0)$						1	1	
$\pi(1)$				-1				-1
$\pi(2)$							1	-1
$\pi(3)$				-1	1			
$\pi(4)$				1		-1	-1	1
$\pi(5)$					1	-1	-1	1
$\pi(6)$		-1		1		-1		1
$\pi(7)$			1	-1	-1	1	1	-1
$\pi(8)$	-1	1		-1		1	1	-1
$\pi(9)$		1		-1	-1	2	1	-2
$\pi(10)$	1	-1	-1	1	1	-1	-1	1
$\pi(11)$				1	0	-1	-1	1

The first four lines of the tables are equivalent to (14.20)(a-d).

Cartan #1: $\mathbb{C}^*$

$$(a,b): e^{x+iy} \to e^{(a+b)x+(a-b)iy}$$

$$H(\mathbb{R})_+ = \{(e^{x+iy} \mid x > 0\}$$

See Section 14.3.3.

	$\mathbb{C}^*$: Standard Modules							
I	$(2,1)$	$(1,2)$	$(2,-1)$	$(-1,2)$	$(1,-2)$	$(-2,1)$	$(-1,-2)$	$(-2,-1)$
$I(0)$					-1	1		
$I(1)$					-1	1		
$I(2)$							1	-1
$I(3)$							1	-1
$I(4)$			-1	1	-1	1		
$I(9)$	-1	1					-1	1

	$\mathbb{C}^*$: Irreducible Modules							
π	$(2,1)$	$(1,2)$	$(2,-1)$	$(-1,2)$	$(1,-2)$	$(-2,1)$	$(-1,-2)$	$(-2,-1)$
$\pi(0)$					-1	1		
$\pi(1)$					-1	1		
$\pi(2)$							1	-1
$\pi(3)$							1	-1
$\pi(4)$			-1	1	1	-1		
$\pi(5)$					1	-1	-1	1
$\pi(6)$					1	-1	-1	1
$\pi(7)$			1	-1	-1	1	1	-1
$\pi(8)$			1	-1	-1	1	1	-1
$\pi(9)$	-1	1	1	-1	-1	1	1	-1
$\pi(10)$	1	-1	-1	1	1	-1	-1	1
$\pi(11)$	1	-1					-1	1

The first four lines of the tables are equivalent to (14.22).

As in the case of the compact Cartan subgroup the identity (6.5)(b) is clear; the short simple reflection is in $W(G(\mathbb{R}), H(\mathbb{R})) \cap W_i$.

Split Cartan subgroup

In the next two tables the two entries in an ordered pair in a column labelled (a,b) gives the multipicity of the characters

$$(a,b): (x,y) \to x^a y^b, \quad (x,y) \to \mathrm{sgn}(xy) x^a y^b$$

respectively. For example the last entry in the row for $\pi(4)$ is $1,-1$ in a column labelled $(-2,-1)$; this means the function $(x,y) \to x^{-2}y^{-1} - \mathrm{sgn}(xy)x^{-2}y^{-1}$. Also

$$H(\mathbb{R})_+ = \{(x,y) \mid |x| > |y| > 0\}.$$

See Section 14.3.3.

I	$(2,1)$	$(1,2)$	$(2,-1)$	$(-1,2)$	$(1,-2)$	$(-2,1)$	$(-1,-2)$	$(-2,-1)$
	$\mathbb{R}^* \times \mathbb{R}^*$: Standard Modules							
$I(0)$						$1,1$	$1,0$	$0,1$
$I(1)$						$1,1$	$1,0$	$0,1$
$I(2)$							$1,0$	$-1,0$
$I(3)$							$1,0$	$-1,0$
$I(4)$				$1,1$		$1,1$	$1,1$	$1,1$
$I(5)$					$1,0$	$0,1$	$1,0$	$0,1$
$I(6)$					$1,0$	$0,1$	$1,0$	$0,1$
$I(7)$			$1,0$	$0,1$			$0,1$	$1,0$
$I(8)$			$1,0$	$0,1$			$0,1$	$1,0$
$I(9)$		$1,1$		$1,1$	$1,1$			$1,1$
$I(10)$	$1,0$	$0,1$	$1,0$	$0,1$	$0,1$	$1,0$	$0,1$	$1,0$
$I(11)$	$0,1$	$1,0$	$0,1$	$1,0$	$1,0$	$0,1$	$1,0$	$0,1$

π	$(2,1)$	$(1,2)$	$(2,-1)$	$(-1,2)$	$(1,-2)$	$(-2,1)$	$(-1,-2)$	$(-2,-1)$
	$\mathbb{R}^* \times \mathbb{R}^*$: Irreducible Modules							
$\pi(0)$						$1,1$	$1,0$	$0,1$
$\pi(1)$						$1,1$	$1,0$	$0,1$
$\pi(2)$							$1,0$	$-1,0$
$\pi(3)$							$1,0$	$-1,0$
$\pi(4)$				$1,1$		-1,-1	-1,1	1,-1
$\pi(5)$					$1,0$	-1,0	-1,0	$1,0$
$\pi(6)$					$1,0$	-1,0	-1,0	$1,0$
$\pi(7)$			$1,0$	-1,0	-1,0	$1,0$	$1,0$	-1,0
$\pi(8)$			$1,0$	-1,0	-1,0	$1,0$	$1,0$	-1,0
$\pi(9)$		$1,1$		-1,-1	-1,1	2,0	1,-1	-2,0
$\pi(10)$	$1,0$	-1,0	-1,0	$1,0$	$1,0$	-1,0	-1,0	$1,0$
$\pi(11)$	$0,1$	0,-1	$0,1$	$1,0$	0,-1	-1,0	-1,0	$1,0$

The first four lines of the tables are equivalent to (14.16) and (14.18).

References

1. J. Adams, M. van Leeuwen, A. Paul, P. Trapa, D. A. Jr. Vogan, and W.-L. Yee. Unitary dual of real reductive groups. preprint.

2. Jeffrey Adams. Guide to the atlas software: Computational representation theory of real reductive groups. In *Representation Theory of Real Reductive Groups, Proceedings of Conference at Snowbird, July 2006*, Contemp. Math. Amer. Math. Soc., 2008.

3. Jeffrey Adams, Dan Barbasch, and David A. Vogan, Jr. *The Langlands classification and irreducible characters for real reductive groups*, volume 104 of *Progress in Mathematics*. Birkhäuser Boston Inc., Boston, MA, 1992.

4. Jeffrey Adams and David A. Vogan, Jr. *L*-groups, projective representations, and the Langlands classification. *Amer. J. Math.*, 114(1):45–138, 1992.

5. Juan Bigeon and Jorge Vargas. A new formula for discrete series characters on split groups. *J. Lie Theory*, 16(2):329–349, 2006.

6. M. Goresky, R. Kottwitz, and R. MacPherson. Discrete series characters and the Lefschetz formula for Hecke operators. *Duke Math. J.*, 89(3):477–554, 1997.

7. Harish-Chandra. The characters of semisimple Lie groups. *Trans. Amer. Math. Soc.*, 83:98–163, 1956.

8. Henryk Hecht. The characters of some representations of Harish-Chandra. *Math. Ann.*, 219(3):213–226, 1976.

9. Henryk Hecht and Wilfried Schmid. A proof of Blattner's conjecture. *Invent. Math.*, 31(2):129–154, 1975.

10. Henryk Hecht and Wilfried Schmid. Characters, asymptotics and n-homology of Harish-Chandra modules. *Acta Math.*, 151(1-2):49–151, 1983.

11. Rebecca A. Herb. Discrete series characters and Fourier inversion on semisimple real Lie groups. *Trans. Amer. Math. Soc.*, 277(1):241–262, 1983.

12. Rebecca A. Herb. Two-structures and discrete series character formulas. In *The mathematical legacy of Harish-Chandra (Baltimore, MD, 1998)*, pages 285–319. Amer. Math. Soc., Providence, RI, 2000.

13. Takeshi Hirai. Invariant eigendistributions of Laplace operators on real simple Lie groups. IV. Explicit forms of the characters of discrete series representations for Sp(n, **R**). *Japan. J. Math. (N.S.)*, 3(1):1–48, 1977.

14. Takeshi Hirai. The characters of the discrete series for semisimple Lie groups. *J. Math. Kyoto Univ.*, 21(3):417–500, 1981.

15. Anthony Knapp. *Representation Theory of Semisimple Groups. An overview based on.* Princeton University Press, Princeton, NJ, 1986.

16. Susan Martens. The characters of the holomorphic discrete series. *Proc. Nat. Acad. Sci. U.S.A.*, 72(9):3275–3276, 1975.

17. Wilfried Schmid. On the characters of the discrete series. The Hermitian symmetric case. *Invent. Math.*, 30(1):47–144, 1975.

18. Wilfried Schmid. Some remarks about the discrete series characters of Sp(n, **R**). In *Noncommutative harmonic analysis (Actes Colloq., Marseille-Luminy, 1974)*, pages 172–194. Lecture Notes in Math., Vol. 466. Springer, Berlin, 1975.

19. Jorge Vargas. A character formula for the discrete series of a semisimple Lie group. *Bull. Amer. Math. Soc. (N.S.)*, 2(3):465–467, 1980.

20. David A. Vogan, Jr. Complex geometry and representations of reductive groups. preprint.

21. David A. Vogan, Jr. *Representations of Real Reductive Lie Groups*, volume 15 of *Progress in mathematics*. Birkhäuser, Boston, 1981.

22. David A. Vogan, Jr. Irreducible characters of semisimple Lie groups IV. character-multiplicity duality. *Duke Math. J.*, 49, No. 4:943–1073, 1982.

23. David A. Vogan, Jr. The Kazhdan-Lusztig conjecture for real reductive groups. In *Representation theory of reductive groups (Park City, Utah, 1982)*, volume 40 of *Progr. Math.*, pages 223–264. Birkhäuser Boston, Boston, MA, 1983.

DEPARTMENT OF MATHEMATICS, UNIVERSITY OF MARYLAND, COLLEGE PARK, MD 21210
E-mail address: jdamath.umd.edu

Contemporary Mathematics
Volume **557**, 2011

Stable Combinations of Special Unipotent Representations

Dan M. Barbasch and Peter E. Trapa

To Gregg Zuckerman, with respect and admiration

ABSTRACT. We define and study a class of virtual characters which are stable in the sense of Langlands and Shelstad. These combinations are associated to nonspecial nilpotent orbits in certain even "special pieces" of the Langlands dual, and are defined in terms of characteristic cycles of perverse sheaves on dual partial flag varieties. Our results generalize earlier work of Adams, Barbasch, and Vogan.

1. Introduction

In [**Ar1**]–[**Ar2**], Arthur outlined a set of conjectures describing the automorphic spectrum of semisimple Lie group over a local field. He suggested that the set of automorphic representations is arranged into (possibly overlapping) packets satisfying a number of properties. In particular, each packet was predicted to give rise to a canonical linear combination of its elements whose character was stable in the sense of Langlands and Shelstad [**La**], [**LaSh**].

In the real case, Arthur's predictions are made precise, refined, and in many cases established in [**BV3**] and, most completely, in [**ABV**]. Many of Arthur's conjectures can be reduced to the case of a certain (precisely defined) set of special unipotent representations. This set is a union of Arthur packets, and since each Arthur packet gives rise to a stable virtual representation, one thus obtains a collection of stable linear combinations of special unipotent representations. One is naturally led to ask if their span exhausts the space of stable virtual special unipotent representations. Simple examples show this is too naive. For example, in the complex case (where stability is empty) Arthur packets are typically not singletons. So the question becomes: can one give a canonical basis of stable virtual special unipotent representations which accounts for these "extra" stable sums?

Under certain natural hypotheses we give a positive answer in terms of the geometry of "special pieces" of nilpotent cone of the Langlands dual Lie algebra. Recall (from [**Sp**]) that if $\mathcal{O}'$ is a nilpotent adjoint orbit for a complex reductive Lie algebra, there is a unique special orbit $\mathcal{O}$ of smallest dimension which contains $\mathcal{O}'$ in its closure. The collection of all $\mathcal{O}'$ for which $\mathcal{O}$ is this unique orbit is called

---

2000 *Mathematics Subject Classification.* Primary 22E47, Secondary 11F70.

The first author was partially supported by NSF grants DMS-0967386 and DMS-0901104.

The second author was partially supported by NSF grants DMS-0554118 and DMS-0968060.

the special piece of nilpotent cone parametrized by $\mathcal{O}$. We denote it $\mathcal{SP}(\mathcal{O})$. The special pieces form a partition of the set of nilpotent adjoint orbits indexed by the special orbits.

In order to formulate our main results, like Theorem 2.1 below, a number of technicalities must be treated with care. A significant complication, as in [**ABV**], is that one cannot work with a single real form individually, but instead must work with an inner class of them simultaneously. (We begin recalling the relevant details in Section 2.) In spite of these technicalities, some consequences of our results are easy to state and have nothing to do with real groups. For example, suppose $\mathfrak{g}$ is a complex semisimple Lie algebra with adjoint group G. Fix a Cartan subalgebra $\mathfrak{h}$ and a system of positive roots $\Delta^+ = \Delta^+(\mathfrak{h}, \mathfrak{g})$, and write $\mathfrak{n} = \bigoplus_{\alpha \in \Delta^+} \mathfrak{g}_\alpha$. Write W for the Weyl group of $\mathfrak{h}$ in $\mathfrak{g}$. For $w \in W$, write $\mathfrak{n}^w = \bigoplus_{\alpha \in \Delta^+} \mathfrak{g}_{w\alpha}$. Then for each $w \in W$ there is always a dense nilpotent adjoint orbit $\mathcal{O}(w)$ contained in

$$G \cdot (\mathfrak{n} \cap \mathfrak{n}^w).$$

A closely related variant of the map $w \mapsto \mathcal{O}(w)$ was studied by Steinberg in [**St**]. It is natural to ask if the map admits a canonical section. That is, given a nilpotent adjoint orbit $\mathcal{O}$, can one canonically define a Weyl group element $w \in W$ such that $\mathcal{O}$ is dense in $G \cdot (\mathfrak{n} \cap \mathfrak{n}^w)$?

For example, suppose $\mathcal{O}$ is even in the sense all of the labels on the associated weighted Dynkin diagram are even. Let $\mathfrak{l} = \mathfrak{l}(\mathcal{O})$ denote the subalgebra of $\mathfrak{g}$ corresponding to the roots labeled zero. If $w_{\mathfrak{l}}$ denotes the long element of the Weyl group of $\mathfrak{l}$, $W(\mathfrak{l}) \subset W$, then indeed $\mathcal{O}$ is dense in $G \cdot (\mathfrak{n} \cap \mathfrak{n}^{w_{\mathfrak{l}}})$. As a consequence of Corollary 4.4 below (applied to the diagonal symmetric subgroup in $G \times G$), we have the following generalization.

THEOREM 1.1. *Suppose $\mathcal{O}$ is an even nilpotent adjoint orbit for $\mathfrak{g}$. Let $W^{\mathfrak{l}}$ denote the set of maximal length coset representatives of $W(\mathfrak{l})\backslash W/W(\mathfrak{l})$ where $\mathfrak{l} = \mathfrak{l}(\mathcal{O})$ corresponds to the nodes labeled zero in the weighted Dynkin diagram for $\mathcal{O}$. Let $d(\mathcal{O})$ denote the Spaltenstein dual of $\mathcal{O}$ (e.g. [**BV3**, Appendix B] or [**CM**, Section 6.3]). Assume that both $\mathcal{O}$ and $d(\mathcal{O})$ are even, and fix an adjoint orbit $\mathcal{O}'$ in $\mathcal{SP}(\mathcal{O})$.*

(a) *There exists a unique element $w(\mathcal{O}') \in W^{\mathfrak{l}}$ such that $\mathcal{O}'$ is dense in $G \cdot (\mathfrak{n} \cap \mathfrak{n}^{w(\mathcal{O}')})$. (For example, if $\mathcal{O}' = \mathcal{O}$, then $w(\mathcal{O}') = w_{\mathfrak{l}}$, the longest element in the identity coset.)*

(b) *Let π' denote the Springer representation associated to the trivial local system on $\mathcal{O}'$, and let sgn denote the sign representation of $W(\mathfrak{l})$. Then*

$$\dim \mathrm{Hom}_{W(\mathfrak{l})}(\mathrm{sgn}, \pi') = 1.$$

Under the conditions of the theorem, the map $\mathcal{O}' \mapsto w(\mathcal{O}')$ in part (a) is thus a natural section of the map $w \mapsto \mathcal{O}(w)$. (The equivalence of statements (a) and (b) goes back to Borho-MacPherson. A more general statement is given in Proposition 4.2 below.) It would be interesting to investigate how to relax the evenness hypotheses in the theorem.

The paper is organized as follows. After recalling the machinery of [**ABV**] in Section 2, we state our main result in Theorem 2.1. We prove it in the final two sections. Examples 2.2, 2.3, and 2.4 illustrate many of the main ideas.

Acknowledgements. We thank Jeffrey Adams for drawing our attention to the problem considered in this paper. In particular, using the software package `atlas` he computed a basis for the space of stable virtual special unipotent representations in many exceptional examples; see www.liegroups.org/tables/unipotent. These examples led us to the formulation of Theorem 2.1.

Finally, it is a pleasure to dedicate this paper to Gregg Zuckerman. His revolutionary ideas, particularly the construction of cohomological induction and his approach to the character theory of real reductive groups (and its relation with tensoring with finite-dimensional representations), are the foundations on which the results in this paper are built.

2. Statement and Examples of the Main Results

Let G be a connected reductive complex algebraic. We begin by fixing a weak extend group G^Γ for G as in [**ABV**, Definition 2.13]. This means that there is an exact sequence of real Lie groups

$$1 \longrightarrow G \longrightarrow G^\Gamma \longrightarrow \Gamma := \mathrm{Gal}(\mathbb{C}/\mathbb{R}) \longrightarrow 1$$

and each $\delta \in G^\Gamma - G$ acts by conjugation as an antiholomorphic automorphism of G. If $\delta \in G^\Gamma - G$ is such that $\delta^2 \in Z(G)$ — that is if δ is a strong real form for G^Γ in the language of [**ABV**] — then conjugation by δ defines an antiholomorphic involution of G. In this case, we write $G(\mathbb{R}, \delta)$ for the corresponding fixed points, a real form of G. It follows from [**ABV**, Proposition 2.14] that the set of real forms which arise in this way constitute exactly one inner class of real forms, and moreover every such inner class arises in this way. In particular, by fixing G^Γ we have fixed an inner class of real forms of G.

Recall (again from [**ABV**, Definition 2.13]) that a representation of a strong real form for G^Γ is a pair (π, δ) where δ is a strong real form of G^Γ and π is an admissible representation of $G(\mathbb{R}, \delta)$. Two representations (π, δ) and (π', δ') are said to be equivalent if there is an element $g \in G$ such that $\delta' = g\delta g^{-1}$ and π' is infinitesimal equivalent to $\pi \circ Ad(g^{-1})$. Write $\Pi(G/\mathbb{R})$ for the set of infinitesimal equivalence classes of irreducible representations of strong real forms for G^Γ.

Fix a maximal ideal I in the center of the enveloping algebra $U(\mathfrak{g})$ of the Lie algebra $\mathfrak{g}$ of G. Choose a Cartan subalgebra $\mathfrak{h} \in \mathfrak{g}$ and write W for the Weyl group of $\mathfrak{h}$ in $\mathfrak{g}$. According to the Harish-Chandra isomorphism we may attach an element $\lambda \in \mathfrak{h}^*/W$ to I. Let $\Pi^\lambda(G/\mathbb{R})$ denote the subset of $\Pi(G/\mathbb{R})$ consisting of those representations whose associated Harish-Chandra modules are annihilated by I. Write $\mathbb{Z}\Pi^\lambda(G/\mathbb{R})$ for the (finite rank) free $\mathbb{Z}$ module with basis indexed by $\Pi^\lambda(G/\mathbb{R})$.

We next introduce various objects on the dual side. Let $G^\vee$ denote the Langlands dual group corresponding to G, and write $\mathfrak{g}^\vee$ for its Lie algebra. Let $\widetilde{G}^\vee_{\mathrm{alg}}$ denote the algebraic universal cover of $G^\vee$ (e.g. [**ABV**, Definition 1.18]). For later use, recall that the construction of the dual group specifies a Cartan subalgebra $\mathfrak{h}^\vee$ of $\mathfrak{g}^\vee$ which is canonically isomorphic to $\mathfrak{h}^*$.

Definition 1.8 and Lemma 1.9 of [**ABV**] introduce a smooth complex algebraic variety $X = X(G^\Gamma)$ attached to the extended group fixed above, and provide an action of $\widetilde{G}^\vee_{\mathrm{alg}}$ on X which factors to an action of $G^\vee$. (To be more precise, [**ABV**, Definition 1.8] explains how to define X from an L-group, and the discussion around [**ABV**, Proposition 4.14] explains how to build an L-group from a fixed inner class of real forms, in particular the class specified by our fixed weak extended group G^Γ.)

The variety X is a disjoint union of smooth (possibly empty) finite-dimensional varieties X^λ indexed by $\lambda \in \mathfrak{h}^*/W$. The action of $G^\vee$ (and $\widetilde{G}^\vee_{\mathrm{alg}}$) on X preserves each X^λ. The orbits for both actions on X^λ are the same and are finite in number. We do not recall the general structure of X^λ here, but instead describe certain special cases in detail below.

Let $\mathcal{P}(X^\lambda, \widetilde{G}^\vee_{\mathrm{alg}})$ denote the category of $\widetilde{G}^\vee_{\mathrm{alg}}$-equivariant perverse sheaves on X^λ, and write $\mathbb{Z}\mathcal{P}(X^\lambda, \widetilde{G}^\vee_{\mathrm{alg}})$ for its integral Grothendieck group. Let $T^*_{G^\vee}(X^\lambda)$ denote the conormal variety for the action of $G^\vee$ on X^λ, namely the subvariety of $T^*(X^\lambda)$ consisting of the unions of the various conormal bundles $T^*_Q(X^\lambda)$ to $G^\vee$ orbits Q on X^λ. (Recall that the orbits of $G^\vee$ and $\widetilde{G}^\vee_{\mathrm{alg}}$ are the same.) The characteristic cycle functor gives a map

$$\mathrm{CC} \ : \ \mathbb{Z}\mathcal{P}(X^\lambda, \widetilde{G}^\vee_{\mathrm{alg}}) \longrightarrow H_{\mathrm{top}}\left(T^*_{G^\vee}(X^\lambda), \mathbb{Z}\right) \simeq \bigoplus_{Q \in G^\vee \backslash X^\lambda} \mathbb{Z}\left[\overline{T^*_Q(X^\lambda)}\right].$$

The right-hand side is the top-dimensional integral Borel-Moore homology group of $T^*_{G^\vee}(X^\lambda)$ which, as indicated, is isomorphic to the direct sum of the $\mathbb{Z}$ span of the fundamental classes of closures of the individual conormal bundles.

The ABV interpretation of the Local Langlands Conjecture, summarized in [**ABV**, Corollary 1.26], provides a $\mathbb{Z}$-module isomorphism

$$\Phi \ : \ \mathbb{Z}\Pi^\lambda(G/\mathbb{R}) \simeq \left(\mathbb{Z}\mathcal{P}(X^\lambda, \widetilde{G}^\vee_{\mathrm{alg}})\right)^\star$$

for each $\lambda \in \mathfrak{h}^*/W$; here and elsewhere $(\ \cdot\ )^\star$ applied to a $\mathbb{Z}$-module denotes $\mathrm{Hom}_{\mathbb{Z}}(\ \cdot\ , \mathbb{Z})$. The isomorphism Φ depends on more data than just the weak extended group G^Γ fixed above. It requires fixing a (strong) extended group $(G^\Gamma, \mathcal{W})$ as in [**ABV**, Definition 1.12] and a strong real form [**ABV**, Definition 1.13]. We define

$$(2.1) \qquad \mathbb{Z}_{\mathrm{st}}\Pi^\lambda(G/\mathbb{R}) := \Phi^{-1}\left(\mathbb{Z}\mathcal{P}(X^\lambda, \widetilde{G}^\vee_{\mathrm{alg}})/\ker(\mathrm{CC})\right)^\star.$$

This is a space of integral linear combinations of irreducible representations of G^Γ, that is virtual representations. (This space depends only on the weak extended group G^Γ.)

For the purpose of this paper, we may take (2.1) as the definition of the subspace of stable virtual characters in $\mathbb{Z}\Pi^\lambda(G/\mathbb{R})$. The equivalence with Langlands' original formulation of stability is given in [**ABV**, Chapter 18].

The main aim of this paper is to define a canonical basis of $\mathbb{Z}_{\mathrm{st}}\Pi^\lambda(G/\mathbb{R})$ (in certain special cases) indexed by rational forms of special pieces of the nilpotent cone of $\mathfrak{g}^\vee$. We now specify the special cases of interest. Begin by fixing a nilpotent adjoint orbit $\mathcal{O}^\vee$ for $\mathfrak{g}^\vee$. Choose a Jacobson-Morozov triple $\{e^\vee, f^\vee, h^\vee\}$ with $h^\vee \in \mathfrak{h}^\vee$ ($\mathfrak{h}^\vee$ as defined above). Set

$$(2.2) \qquad \lambda(\mathcal{O}^\vee) = \frac{1}{2}h^\vee \in \mathfrak{h}^\vee \simeq \mathfrak{h}^*.$$

Define

$$(2.3) \qquad \mathfrak{l}^\vee(\mathcal{O}^\vee) = \text{ the centralizer in } \mathfrak{g}^\vee \text{ of } \lambda(\mathcal{O}^\vee);$$

equivalently $\mathfrak{l}^\vee(\mathcal{O}^\vee)$ is the sum of the zero eigenspaces of $\mathrm{ad}(h^\vee)$. Set

$$(2.4) \qquad \mathfrak{p}^\vee(\mathcal{O}^\vee) = \text{ the sum of the non-negative eigenspaces of } \mathrm{ad}(h^\vee).$$

Let $I(\mathcal{O}^\vee)$ denote the maximal ideal in the center of $U(\mathfrak{g})$ corresponding to $\lambda(\mathcal{O}^\vee)$ under the Harish-Chandra isomorphism. According to a result of Dixmier [**Di**], there is a unique maximal primitive ideal $J(\mathcal{O}^\vee)$ in $U(\mathfrak{g})$ containing $I(\mathcal{O}^\vee)$. We say a representation (δ, π) of G^Γ is *special unipotent attached to* $\mathcal{O}^\vee$ if the Harish-Chandra module of π is annihilated by $J(\mathcal{O}^\vee)$. We write

$$\Pi(\mathcal{O}^\vee) \subset \Pi^{\lambda(\mathcal{O}^\vee)}(G/\mathbb{R})$$

for the subset of irreducible special unipotent representations of G^Γ attached to $\mathcal{O}^\vee$, write $\mathbb{Z}\Pi(\mathcal{O}^\vee)$ for their span, and define

$$\mathbb{Z}_{\mathrm{st}}\Pi(\mathcal{O}^\vee) := \mathbb{Z}\Pi(\mathcal{O}^\vee) \cap \mathbb{Z}_{\mathrm{st}}\Pi^\lambda(G/\mathbb{R}).$$

It is this space for which we will find a canonical basis under certain natural hypotheses.

To state our main results, we need more detailed information about the structure of the $G^\vee$ action on X^λ assuming λ is integral. Let

(2.5) $Y^\vee =$ the variety of parabolic subalgebras of $\mathfrak{g}^\vee$ conjugate to $\mathfrak{p}^\vee(\mathcal{O}^\vee)$

with notation as in (2.4). Proposition 6.16 of [**ABV**] provides a collection of symmetric subgroups $K_1^\vee, \ldots, K_k^\vee$ of $G^\vee$. Each $K_i^\vee$ acts on $Y^\vee$ with finitely many orbits. Furthermore, [**ABV**, Proposition 7.14] implies the existence of an isomorphism

(2.6) $\mathcal{P}(X^\lambda, G^\vee) \simeq \mathcal{P}(Y^\vee, K_1^\vee) \oplus \cdots \oplus \mathcal{P}(Y^\vee, K_k^\vee),$

where $\mathcal{P}(Y^\vee, K_i^\vee)$ once again denotes the category of $K_i^\vee$ equivariant perverse sheaves on $Y^\vee$. Moreover, if we let CC_i denote the characteristic cycle functor for $\mathcal{P}(Y^\vee, K_i^\vee)$, then the isomorphism in (2.6) descends to an isomorphism
(2.7)
$$\mathcal{P}(X^\lambda, G^\vee)/\ker(\mathrm{CC}) \simeq \mathcal{P}(Y^\vee, K_1^\vee)/\ker(\mathrm{CC}_1) \oplus \cdots \oplus \mathcal{P}(Y^\vee, K_k^\vee)/\ker(\mathrm{CC}_k).$$

General properties of the characteristic cycle construction imply that it is insensitive to central extensions of the group acting in the sense that

$$\mathcal{P}(X^\lambda, \widetilde{G}_{\mathrm{alg}}^\vee)/\ker(\mathrm{CC}) \simeq \mathcal{P}(X^\lambda, G^\vee)/\ker(\mathrm{CC});$$

see [**Ch**, Proposition 2.6.2], for example. Thus (2.7) in fact gives

(2.8) $\mathcal{P}(X^\lambda, \widetilde{G}_{\mathrm{alg}}^\vee)/\ker(\mathrm{CC}) \simeq \mathcal{P}(Y^\vee, K_1^\vee)/\ker(\mathrm{CC}_1) \oplus \cdots \oplus \mathcal{P}(Y^\vee, K_k^\vee)/\ker(\mathrm{CC}_k).$

As a matter of notation, we let $\mathfrak{k}_i^\vee$ denote the Lie algebra of $K_i^\vee$ and write

(2.9) $\mathfrak{g}^\vee = \mathfrak{k}_i^\vee \oplus \mathfrak{s}_i^\vee$

for the corresponding Cartan decomposition. According to [**KR**], if $\mathcal{O}^\vee$ is any nilpotent adjoint orbit in $\mathfrak{g}^\vee$, then each $K_i^\vee$ acts with finitely many orbits on $\mathcal{O}^\vee \cap \mathfrak{s}_i^\vee$,

$$\#(K_i^\vee \backslash (\mathcal{O}^\vee \cap \mathfrak{s}_i^\vee)) < \infty.$$

Recall that an orbit $\mathcal{O}^\vee$ for $\mathfrak{g}^\vee$ is said to be even, if the eigenvalues of $\mathrm{ad}(h^\vee)$ acting on $\mathfrak{g}^\vee$ are all even integers; equivalently if $\lambda(\mathcal{O}^\vee)$ is integral. Assume this is the case and fix an orbit

$$\mathcal{O}_K^\vee \in \bigcup_{i=1}^k K_i^\vee \backslash (\mathcal{O}^\vee \cap \mathfrak{s}_i^\vee).$$

Chapter 27 of [**ABV**] defines an Arthur packet parametrized by $\mathcal{O}_K^\vee$,

$$\mathrm{A}(\mathcal{O}_K^\vee) \subset \Pi(\mathcal{O}^\vee).$$

The union of the various Arthur packets (over all possible orbits $\mathcal{O}_K^\vee$) exhausts $\Pi(\mathcal{O}^\vee)$ (but the union is not in general disjoint). Moreover, for each $\mathcal{O}_K^\vee$, the discussion around [**ABV**, (1.34c)] defines a stable integral linear combination of elements of $\Pi(\mathcal{O}_K^\vee)$,

$$(2.10) \qquad\qquad \pi(\mathcal{O}_K^\vee) \in \mathbb{Z}_{\mathrm{st}}\Pi(\mathcal{O}^\vee).$$

These virtual representations are all linearly independent, so in particular one has

$$(2.11) \qquad\qquad \dim_{\mathbb{Z}} \mathbb{Z}_{\mathrm{st}}\Pi(\mathcal{O}^\vee) \geq \sum_i \#\left(K_i^\vee \backslash (\mathcal{O}^\vee \cap \mathfrak{s}_i^\vee)\right).$$

Our main result finds other interesting stable representations attached to $K^\vee$ orbits on the special piece parametrized by $\mathcal{O}^\vee$ and (in favorable instances) proves they are a basis of $\mathbb{Z}_{\mathrm{st}}(\mathcal{O}^\vee)$.

THEOREM 2.1. *Let G be a connected reductive algebraic group with dual group $G^\vee$. Fix a weak extended group G^Γ for G (in particular, an inner class of real forms for G). Fix an even nilpotent adjoint orbit $\mathcal{O}^\vee$ for $\mathfrak{g}^\vee$. Assume further that the Spaltenstein dual $\mathcal{O} := d(\mathcal{O}^\vee)$, a nilpotent adjoint orbit for $\mathfrak{g}$ (cf. [**BV3**, Appendix B]), is also even.*

Recall the Cartan decompositions of (2.9) and the corresponding symmetric subgroups $K_i^\vee$ introduced above. Write $\mathcal{SP}(\mathcal{O}^\vee)$ for the special piece of the nilpotent cone of $\mathfrak{g}^\vee$ containing $\mathcal{O}^\vee$. Then

$$(2.12) \qquad \dim_{\mathbb{Z}} \mathbb{Z}_{\mathrm{st}}\Pi(\mathcal{O}^\vee) = \sum_i \#\left(K_i^\vee \backslash (\mathcal{SP}(\mathcal{O}^\vee) \cap \mathfrak{s}_i^\vee)\right);$$

cf. (2.11). In fact, for each element $\mathcal{O}_K^\vee$ on the right-hand side, equation (4.9) below defines an element $\pi(\mathcal{O}_K^\vee) \in \mathbb{Z}_{\mathrm{st}}\Pi(\mathcal{O}^\vee)$ so that

$$\left\{ \pi(\mathcal{O}_K^\vee) \mid \mathcal{O}_K^\vee \in \bigcup_i K_i^\vee \backslash (\mathcal{SP}(\mathcal{O}^\vee) \cap \mathfrak{s}_i^\vee) \right\}$$

is a basis of $\mathbb{Z}_{\mathrm{st}}\Pi(\mathcal{O}^\vee)$. (When $G^\vee \cdot \mathcal{O}_K^\vee = \mathcal{O}^\vee$, $\pi(\mathcal{O}_K^\vee)$ coincides with the stable virtual representation in (2.10).)

EXAMPLE 2.2. Suppose $G = \mathrm{Sp}(4,\mathbb{C})$ and G^Γ gives rise to the inner class of G containing the split form. There are four equivalence classes of strong real forms for G^Γ, $\{\delta_s, \delta_{2,0}, \delta_{1,1}, \delta_{0,2}\}$. The labeling is arranged so that $G(\mathbb{R}, \delta_s) = \mathrm{Sp}(4,\mathbb{R})$ and $G(\mathbb{R}, \delta_{p,q}) = \mathrm{Sp}(p,q)$.

Let $\mathcal{O}^\vee$ denote the (even) nilpotent orbit for $G^\vee = \mathrm{SO}(5,\mathbb{C})$ whose Jordan type is given by the partition 311. Then $d(\mathcal{O}^\vee)$ is the orbit for G with Jordan type corresponding to the partition 22, which is also even, so Theorem 2.1 applies.

In [**ABV**, Example 27.14], the elements of $\Pi(\mathcal{O}^\vee)$ are enumerated. Among them are eight representations of $\mathrm{Sp}(4,\mathbb{R})$ and one of $\mathrm{Sp}(1,1)$. The representations of $\mathrm{Sp}(4,\mathbb{R})$ are the three irreducible constituents of $\mathrm{Ind}_{GL(2,\mathbb{R})}^{\mathrm{Sp}(4,\mathbb{R})}(\det)$; the three irreducible constituents of $\mathrm{Ind}_{GL(2,\mathbb{R})}^{\mathrm{Sp}(4,\mathbb{R})}(|\det|)$; and the two irreducible constituents of $\mathrm{Ind}_{GL(1,\mathbb{R})\times\mathrm{Sp}(2,\mathbb{R})}^{\mathrm{Sp}(4,\mathbb{R})}(\mathrm{sgn}(\det) \otimes 1)$. These eight representations are distinguished by their lowest $\mathrm{U}(2)$ types which in the respective cases are $(2,0)$, $(0,2)$, and $(0,0)$; $(1,1)$, $(-1,-1)$, and $(1,-1)$; and $(1,0)$ and $(0,-1)$. Write $\pi_s(m,n)$ for the corresponding special unipotent representation of $\mathrm{Sp}(4,\mathbb{R})$ with lowest $\mathrm{U}(2)$ type (m,n). Meanwhile the unique special unipotent representation of $\mathrm{Sp}(1,1)$ attached to $\mathcal{O}^\vee$

is the irreducible spherical representation with infinitesimal character $\lambda(\mathcal{O}^\vee)$ which we denote by $\pi_{(1,1)}(0)$.

The symmetric subgroups $K_i^\vee$ above in this case are

$$K_i^\vee = \mathrm{S}(\mathrm{O}(5-i) \times \mathrm{O}(i)) \text{ for } i = 0, 1, 2.$$

In terms of the signed tableau parametrization (for example, [**CM**, Chapter 9]), $\bigcup_i K_i^\vee \backslash (\mathcal{O}^\vee \cap \mathfrak{s}^\vee)$ consists of three elements

$$
\begin{array}{ccc}
\begin{array}{ccc} + & - & + \\ + & & \\ + & & \end{array}
&
\begin{array}{ccc} - & + & - \\ + & & \\ + & & \end{array}
&
\begin{array}{ccc} + & - & + \\ + & & \\ - & & \end{array}
\\
, & , & ;
\end{array}
$$

the first arises for $i = 1$, the second and third for $i = 2$. This means there are three Arthur packets in $\Pi(\mathcal{O}^\vee)$. They are listed in [**ABV**, (27.17)]. They give rise, respectively, to the following three stable virtual representations in $\mathbb{Z}_{\mathrm{st}}(\mathcal{O}^\vee)$,

$$\pi_s(1, 0) + \pi_s(0, -),$$
$$\pi_s(0, 0) + \pi_s(1, -1),$$
$$\pi_s(1, 1) + \pi_s(-1, -1) + \pi_s(2, 2) + \pi_s(-2, -2) + \pi_{(1,1)}(0).$$

Meanwhile there is another orbit $\mathcal{O}^{\vee'}$ (besides $\mathcal{O}^\vee$) in $\mathcal{SP}(\mathcal{O}^\vee)$, namely the orbit with Jordan type corresponding to the partition 221. This time $\bigcup_i K_i^\vee \backslash (\mathcal{O}^{\vee'} \cap \mathfrak{s}^\vee)$ consists of one element

(2.13)
$$
\begin{array}{cc}
+ & - \\
- & + \\
+ &
\end{array}
$$

arising for $K_2^\vee$. Theorem 2.1 thus implies $\dim_{\mathbb{Z}} \mathbb{Z}_{\mathrm{st}}(\mathcal{O}^\vee) = 3 + 1$, and gives an additional stable virtual representation parametrized by the orbit in (2.13). This extra stable sum is

(2.14) $$\pi_s(1, 1) + \pi_s(-1, -1) - \pi_s(2, 2) - \pi_s(-2, -2).$$

EXAMPLE 2.3. Let G be of type F4, and let $\mathcal{O}^\vee$ be the orbit labeled $F4(a3)$ in the Bala-Carter classification (e.g. [**CM**, Section 8.4]). If we orient the Dynkin diagram of F4 so that the first two roots are long, the weighted Dynkin diagram for $\mathcal{O}^\vee$ is 0200. In particular, the orbit is even. In fact $\mathcal{O}^\vee$ is equal to its own Spaltenstein dual, and thus Theorem 2.1 applies.

The special piece $\mathcal{SP}(\mathcal{O}^\vee)$ consists of four other orbits besides $F4(a3)$. In the Bala-Carter classification, they are labeled $C3(a1)$, $\widetilde{A2} + A1$, $B2$, and $A2 + \widetilde{A1}$. The respective weighted Dynkin diagrams are $1010, 0101, 2001$, and 0010.

There is a unique inner class of real forms for G; it contains the split, rank one, and compact forms. (In fact it is easy to see (from the singularity of the infinitesimal character $\lambda(\mathcal{O}^\vee)$ that $\Pi(\mathcal{O}^\vee)$ can consist of representations only of the split form.) The only symmetric subgroup $K^\vee$ appearing above in this case is the quotient of $Sp(6, \mathbb{C}) \times SL(2, \mathbb{C})$ by the diagonal copy of a central $\mathbb{Z}/2$. From the tables in [**CM**, Section 9.5], we deduce that

$$\#K^\vee \backslash (F4(a3) \cap \mathfrak{s}^\vee) = 3$$

and so there are three Arthur packets in $\Pi(\mathcal{O}^\vee)$. Meanwhile we have

$$\#K^\vee\backslash(C3(a1) \cap \mathfrak{s}^\vee) = 2$$

$$\#K^\vee\backslash((\widetilde{A2} + A1) \cap \mathfrak{s}^\vee) = 1$$

$$\#K^\vee\backslash(B2 \cap \mathfrak{s}^\vee) = 2$$

$$\#K^\vee\backslash((A2 + \widetilde{A1}) \cap \mathfrak{s}^\vee) = 1.$$

Thus Theorem 2.1 says

$$\dim_\mathbb{Z} \mathbb{Z}_{\text{st}}\Pi(\mathcal{O}^\vee) = 3 + 2 + 1 + 2 + 1.$$

The definition in (4.9) gives a canonical basis for the space. To write the basis down explicitly requires computing characteristic cycles of irreducible objects in $\mathcal{P}(K^\vee, Y^\vee)$. We have not performed the calculations required to do this.

EXAMPLE 2.4. Let G be of type E8, and let $\mathcal{O}^\vee$ denote the orbit labeled $E8(a7)$ in the Bala-Carter classification. It is even and self-dual, and thus Theorem 2.1 applies. The special piece it parametrizes consists of the additional orbits $E7(a5), E6(a3) + A1, D6(a2), D5(a1) + A2, A5 + A1$, and $A4 + A3$.

There is a unique inner class of real forms for G, and (arguing as in the previous example), $\Pi(\mathcal{O}^\vee)$ can consist of representations only of the split form. The only symmetric subgroup $K^\vee$ appearing above in this case is a quotient of $\text{Spin}(16, \mathbb{C})$ by a central $\mathbb{Z}/2$ (but $K^\vee$ is not isomorphic to $\text{SO}(16, \mathbb{C})$). Again using the tables in [**CM**, Section 9.5], we deduce that

$$\#K^\vee\backslash(E8(a7) \cap \mathfrak{s}^\vee) = 3$$

and so there are three Arthur packets in $\Pi(\mathcal{O}^\vee)$. Meanwhile we have

$$\#K^\vee\backslash(E7(a5) \cap \mathfrak{s}^\vee) = 2$$

$$\#K^\vee\backslash((E6(a3) + A1) \cap \mathfrak{s}^\vee) = 2$$

$$\#K^\vee\backslash(D6(a2) \cap \mathfrak{s}^\vee) = 2$$

$$\#K^\vee\backslash((A5 + A1) \cap \mathfrak{s}^\vee) = 1$$

$$\#K^\vee\backslash((D5(a1) + A2) \cap \mathfrak{s}^\vee) = 1$$

$$\#K^\vee\backslash((A4 + A3) \cap \mathfrak{s}^\vee) = 1.$$

Thus Theorem 2.1 implies

$$\dim_\mathbb{Z} \mathbb{Z}_{\text{st}}\Pi(\mathcal{O}^\vee) = 3 + 2 + 2 + 2 + 1 + 1 + 1.$$

3. Proof of Equality in (2.12)

Our main technique allows us to compute the numbers in (2.12) in terms of certain Weyl group calculations. The full Weyl group does not act at singular infinitesimal character, and so we must instead translate to regular infinitesimal character and work there.

Retain the setting of Theorem 2.1. Temporarily choose a system of simple roots for $\mathfrak{h}$ in $\mathfrak{g}$ and a representative $\lambda_\circ$ of $\lambda(\mathcal{O}^\vee)$ which is dominant. Let μ be the highest weight of a finite-dimensional representation of G such that $\nu_\circ := \lambda_\circ + \mu \in \mathfrak{h}^*$ is dominant and regular. Let ν denote the image of $\nu_\circ$ in $\mathfrak{h}^*/W$. As above, we can consider the set $\Pi^\nu(G/\mathbb{R})$ and its $\mathbb{Z}$ span $\mathbb{Z}\Pi^\nu(G/\mathbb{R})$. This space identifies with an appropriate Grothendieck group of representations at regular integral infinitesimal character which admits a coherent continuation action of W.

Recall the symmetric subgroups $K_1^\vee, \ldots, K_k^\vee$ of the previous section. Let $X^\vee$ denote the full flag variety for $\mathfrak{g}^\vee$. There is an action of W on each Grothendieck group $\mathbb{Z}\mathcal{P}(X^\vee, K_i^\vee)$. (One way to see this is to use the Riemann-Hilbert correspondence to identify $\mathcal{P}(X^\vee, K_i^\vee)$ with a category of $K_i^\vee$ equivariant holonomic $\mathcal{D}$ modules on $X^\vee$. In turn, by localization, this category is a equivalent to a category of $\mathfrak{g}^\vee$ modules which admits a coherent continuation action of W.)

Meanwhile Corollary 1.26 and Proposition 7.14 of [**ABV**] give an isomorphism (depending on the extended group $(G^\Gamma, \mathcal{W})$)

$$\Psi \ : \ \mathbb{Z}\Pi^\nu(G/\mathbb{R}) \longrightarrow \bigoplus_i \mathbb{Z}\mathcal{P}(X^\vee, K_i^\vee)^\star$$

which intertwines the W action on both sides. Once again we have characteristic cycle functors

$$\mathrm{CC}_i \ : \ \mathbb{Z}\mathcal{P}(X^\vee, K_i^\vee) \longrightarrow \mathrm{H}_{\mathrm{top}}\left(T_{K_i^\vee}^*(X^\vee), \mathbb{Z}\right) \simeq \bigoplus_{Q \in K_i^\vee \backslash X^\vee} \left[\overline{T_{K_i^\vee}^*(X^\vee)}\right].$$

We have remarked that the domain of CC_i carries an action of W. The range does as well, and according to results of Tanisaki [**Ta**], each CC_i is W-equivariant. Thus

$$\mathbb{Z}\mathcal{P}(X^\vee, K_i^\vee)/\ker(\mathrm{CC}_i) \simeq \mathrm{H}_{\mathrm{top}}\left(T_{K_i^\vee}^*(X^\vee), \mathbb{Z}\right)$$

as representations of W. Once again we define

$$\mathbb{Z}_{\mathrm{st}}\Pi^\nu(G/\mathbb{R}) := \Psi^{-1}\left(\bigoplus_i \mathbb{Z}\mathcal{P}(X^\vee, K_i^\vee)/\ker(\mathrm{CC}_i)\right)^\star.$$

and we have an isomorphism

$$(3.1) \qquad\qquad \mathbb{Z}_{\mathrm{st}}\Pi^\nu(G/\mathbb{R}) \simeq \bigoplus_i \mathrm{H}_{\mathrm{top}}\left(T_{K_i^\vee}^*(X^\vee), \mathbb{Z}\right)^\star$$

of representations of W.

For our counting argument, we need to specify a particular left cell representation. Let $\mathfrak{l}^\vee(\mathcal{O}^\vee)$ denote the centralizer in $\mathfrak{g}^\vee$ of $\lambda(\mathcal{O}^\vee) \in \mathfrak{h}^\vee$, and let $w(\mathcal{O}^\vee)$ denote the long element of the Weyl group of $\mathfrak{l}^\vee(\mathcal{O}^\vee)$ viewed as an element of $W(\mathfrak{h}^\vee, \mathfrak{g}^\vee) = W$. Let $V(\mathcal{O}^\vee)$ denote the representation of W afforded by the integral linear combinations of elements of the Kazhdan-Lusztig left cell containing $w(\mathcal{O}^\vee)$.

PROPOSITION 3.1 ([**BV3**, Section 5]). *Retain the setting above. In particular, assume $\mathcal{O}^\vee$ is even. We have*

$$\dim_\mathbb{Z} \mathbb{Z}\Pi(\mathcal{O}^\vee) = \dim \mathrm{Hom}_W(V(\mathcal{O}^\vee) \otimes \mathrm{sgn}, \mathbb{Z}\Pi^\nu(G/\mathbb{R})).$$

and

$$\dim_\mathbb{Z} \mathbb{Z}_{\mathrm{st}}\Pi(\mathcal{O}^\vee) = \dim \mathrm{Hom}_W(V(\mathcal{O}^\vee) \otimes \mathrm{sgn}, \mathbb{Z}\Pi_{\mathrm{st}}^\nu(G/\mathbb{R})).$$

The following result brings the role of special pieces into play. To state it, we need to introduce some notation for the Springer correspondence. Fix any nilpotent adjoint orbit $\mathcal{O}$ for $\mathfrak{g}$ and a representative x of $\mathcal{O}$. Let $A_G(\xi)$ denote the component group of the centralizer of x in G. We let $\mathrm{Sp}(x)$ denote the Springer representation of $W \times A_G(x)$ on the top homology of the Springer fiber over x (normalized so that $\mathrm{Sp}(x)$ is the sign representation of W if x is zero). As usual, we set

$$\mathrm{Sp}(x)^{A_G(x)} = \mathrm{Hom}_{A_G(x)}\left(\mathbb{1}, \mathrm{Sp}(\xi)\right).$$

This is a a representation of W.

PROPOSITION 3.2. *Suppose $\mathcal{O}^\vee$ is a an even nilpotent adjoint orbit for $\mathfrak{g}^\vee$. Let $\mathcal{O}$ denote special nilpotent orbit for $\mathfrak{g}$ obtained as the Spaltenstein dual of $\mathcal{O}^\vee$. Enumerate representative for the adjoint orbits in the special piece parametrized by $\mathcal{O}$ as $x_1, \cdots, x_l$. Then*

$$V(\mathcal{O}^\vee) \otimes \mathrm{sgn} \simeq \bigoplus_i \mathrm{Sp}(x_i)^{A_G(x_i)}.$$

PROOF. This follows by combining [**BV3**, Proposition 5.28] and [**Lu2**, Theorem 0.4]. □

The proposition involves special pieces on the group side, while the statement of Theorem 2.1 involves special pieces on the dual side. If we make the additional hypothesis that $\mathcal{O}$ is even, then we can match up the two sides.

PROPOSITION 3.3. *Suppose $\mathcal{O}^\vee$ is a an even nilpotent adjoint orbit for $\mathfrak{g}^\vee$. Let $\mathcal{O}$ denote its Spaltenstein dual, and further assume that $\mathcal{O}$ is even. Enumerate representative in the special piece parametrized by $\mathcal{O}^\vee$ as $x_1^\vee, \cdots, x_l^\vee$. Then*

$$V(\mathcal{O}^\vee) \simeq \bigoplus_i \mathrm{Sp}(x_i^\vee)^{A_{G^\vee}(x_i^\vee)}.$$

That is, $V(\mathcal{O}^\vee)$ is the sum over the orbits in $\mathcal{SP}(\mathcal{O}^\vee)$ of the Weyl group representations attached to the trivial local system on them.

PROOF. This follows from the previous proposition and Lusztig's classification of cells [**Lu1**] (the relevant details of which are recalled in [**BV3**, Theorem 4.7d]). □

Proposition 3.1 gives a way to compute the dimension of the left-hand side of (2.12) in terms of Weyl group representations. We need a way to do the same for the right-hand side.

PROPOSITION 3.4. *Suppose $K^\vee$ is a symmetric subgroup of $G^\vee$ and write $\mathfrak{g}^\vee = \mathfrak{k}^\vee \oplus \mathfrak{s}^\vee$ for the corresponding Cartan decomposition. Let $x_1^\vee, x_2^\vee, \ldots,$ denote representatives of the nilpotent $K^\vee$ orbits on $\mathfrak{s}^\vee$. Let $A_{K^\vee}(x_i^\vee)$ denote the component group of the centralizer in $K^\vee$ of $x_i^\vee$. (Since this group maps to $A_{G^\vee}(x_i^\vee)$, it makes sense to consider invariants $\mathrm{Sp}(x_i^\vee)^{A_{K^\vee}(x_i^\vee)}$ in $\mathrm{Sp}(x_i^\vee)$ of the image of $A_{K^\vee}(x_i^\vee)$ in $A_{G^\vee}(x_i^\vee)$.) As W representations, we have*

$$\mathrm{H}_{\mathrm{top}}(T_{K^\vee}^*(X^\vee), \mathbb{Z}) \simeq \sum_i \mathrm{Sp}(x_i^\vee)^{A_{K^\vee}(x_i^\vee)}.$$

In particular, each representation of W attached to the trivial local system on a complex nilpotent orbit appears with multiplicity equal to the number of $K^\vee$ orbits on its intersection with $\mathfrak{s}^\vee$.

Proof. This follows from [**Ro**, Theorem 3.3]. (Rossmann works with the conormal variety of orbits of real forms of $G^\vee$ on $X^\vee$. To translate to the conormal variety of orbits for a symmetric subgroup, one can use [**MUV**], for example.) □

Proof of equality in (2.12). From Proposition 3.1 and (3.1), we have

$$\dim_\mathbb{Z} \mathbb{Z}_{st}\Pi(\mathcal{O}^\vee) = \dim \operatorname{Hom}_W\left(V(\mathcal{O}^\vee) \otimes \operatorname{sgn}, \mathbb{Z}_{st}\Pi^\nu(G/\mathbb{R})\right)$$

$$= \dim \operatorname{Hom}_W\left(V(\mathcal{O}^\vee) \otimes \operatorname{sgn}, \bigoplus_i \operatorname{H}_{top}(T^*_{K_i^\vee}\mathcal{B}, \mathbb{Z})^*\right)$$

$$= \dim \operatorname{Hom}_W\left(V(\mathcal{O}^\vee), \bigoplus_i \operatorname{H}_{top}(T^*_{K_i^\vee}\mathcal{B}, \mathbb{Z})\right).$$

The concluding sentences of Propositions 3.3 and 3.4 show that the right-hand side equals

$$\sum_i \#\left(K_i^\vee \backslash (\mathcal{SP}(\mathcal{O}^\vee) \cap \mathfrak{s}_i^\vee)\right)$$

as claimed. □

4. Proof of Theorem 2.1

In this section we prove the last assertion of Theorem 2.1. According to (2.1),

$$(4.1) \qquad \mathbb{Z}_{st}\Pi(\mathcal{O}^\vee) \subset \mathbb{Z}_{st}\Pi^{\lambda(\mathcal{O}^\vee)}(G/\mathbb{R}) \simeq \bigoplus_i \left[\operatorname{H}_{top}(T^*_{K_i^\vee}(Y^\vee), \mathbb{Z})\right]^*.$$

Our main task is to determine which linear functionals on

$$(4.2) \qquad \operatorname{H}_{top}(T^*_{K_i^\vee}(Y^\vee), \mathbb{Z}) = \bigoplus_{Q \in K_i^\vee \backslash Y^\vee} \left[\overline{T^*_Q(Y^\vee)}\right]$$

correspond to elements of $\mathbb{Z}_{st}\Pi(\mathcal{O}^\vee)$ in (4.1). This is the content of part (2) of the next proposition. To formulate it, we recall the $G^\vee$ equivariant moment map μ from $T^*(Y^\vee)$ to $(\mathfrak{g}^\vee)^*$. We use an invariant form to identify $\mathfrak{g}^\vee$ and $(\mathfrak{g}^\vee)^*$, and view the image of the moment map in $\mathfrak{g}^\vee$ itself.

PROPOSITION 4.1. *Retain the setting of Theorem 2.1. For each orbit Q of some $K_i^\vee$ on $Y^\vee$, define*

$$m_Q \in \left[\operatorname{H}_{top}(T^*_{K_i^\vee}(Y^\vee), \mathbb{Z})\right]^*$$

as the multiplicity of the fundamental class corresponding to the closure of the conormal bundle to Q (c.f. (4.2)). Recall the isomorphism

$$(4.3) \qquad \mathbb{Z}_{st}\Pi^{\lambda(\mathcal{O}^\vee)}(G/\mathbb{R}) \simeq \bigoplus_i \left[\operatorname{H}_{top}(T^*_{K_i^\vee}(Y^\vee), \mathbb{Z})\right]^*$$

and write $\pi(Q)$ for the element of $\mathbb{Z}_{st}\Pi^{\lambda(\mathcal{O}^\vee)}(G/\mathbb{R})$ corresponding to m_Q.

(1) *The set*

$$(4.4) \qquad \{\pi(Q) \mid Q \in K_i^\vee \backslash Y^\vee \text{ for some } i\}$$

is a basis for

$$\mathbb{Z}_{st}\Pi^{\lambda(\mathcal{O}^\vee)}(G/\mathbb{R}).$$

(2) *The set*

(4.5) $$\{\pi(Q) \mid \mu\left(T_Q^*(Y^\vee)\right) \cap \mathcal{SP}(\mathcal{O}^\vee) \text{ is nonempty}\}$$

is a basis for the subspace

$$\mathbb{Z}_{\mathrm{st}}\Pi(\mathcal{O}^\vee) \subset \mathbb{Z}_{\mathrm{st}}\Pi^{\lambda(\mathcal{O}^\vee)}(G/\mathbb{R}).$$

PROOF. Since

$$\bigcup_i \{m_Q \mid Q \in K_i^\vee \backslash Y^\vee\}$$

is obviously a basis for the left-hand side of (4.3) (in light of (4.2)), statement (1) of the proposition is clear.

For the second statement, we begin by proving

$$\{\pi(Q) \mid \mu\left(T_Q^*(Y^\vee)\right) \cap \mathcal{SP}(\mathcal{O}^\vee) \text{ is nonempty}\}$$

are linearly independent elements of $\mathbb{Z}_{\mathrm{st}}\Pi(\mathcal{O}^\vee)$. We need some additional notation. For an object $\mathcal{S}$ in $\mathcal{P}(Y^\vee, K_i^\vee)$, write

$$\mathrm{CC}_i(\mathcal{S}) = a_1[\overline{T_{Q_1}^*(Y^\vee)}] + \cdots + a_r[\overline{T_{Q_r}^*(Y^\vee)}],$$

and define

$$\mathrm{AV}_{\mathbb{C}}(\mathcal{S}) = G^\vee \cdot \bigcup_i \mu\left(\overline{T_{Q_i}^*(Y^\vee)}\right) \subset \overline{\mathcal{O}^\vee}.$$

According to the irreducibility theorem of [**BB1**], if $\mathcal{S}$ is irreducible, then $\mathrm{AV}_{\mathbb{C}}(\mathcal{S})$ is the closure of a single adjoint orbit. The results of [**BV1, BV2**] show that the orbit must be special.

Next recall the isomorphism

$$\Phi \; : \; \mathbb{Z}_{\mathrm{st}}\Pi^{\lambda(\mathcal{O}^\vee)}(G/\mathbb{R}) \longrightarrow \bigoplus_i \left(\mathbb{Z}\mathcal{P}(Y^\vee, K_i^\vee)/\ker(\mathrm{CC}_i)\right)^\star$$

obtained from (2.1) and (2.8). It follows from [**ABV**, Theorem 27.12] that $\pi \in \mathbb{Z}_{\mathrm{st}}\Pi^{\lambda(\mathcal{O}^\vee)}(G/\mathbb{R})$ belongs to the subspace $\mathbb{Z}_{\mathrm{st}}\Pi(\mathcal{O}^\vee)$ if and only if there is an irreducible object $\mathcal{S}$ in some $\mathcal{P}(Y^\vee, K_i^\vee)$ with $\mathrm{AV}_{\mathbb{C}}(\mathcal{S}) = \overline{\mathcal{O}^\vee}$ such that $\Phi(\pi)(\mathcal{S}) \neq 0$. As a consequence, suppose Q is an orbit of $K_i^\vee$ on $Y^\vee$ such that

$$\mu\left(\overline{T_Q^*(Y^\vee)}\right) \cap \mathcal{SP}(\mathcal{O}^\vee) \neq \emptyset.$$

Set

$$m_Q' = m_Q \circ \mathrm{CC}_i \in \left(\mathbb{Z}\mathcal{P}(Y^\vee, K_i^\vee)/\ker(\mathrm{CC}_i)\right)^\star.$$

There exists at least one irreducible object $\mathcal{S}$ in $\mathcal{P}(Y^\vee, K_i^\vee)$ whose support is the closure of Q. (For example, let $\mathcal{S}$ be the DGM extension of the trivial local system on Q. The fundamental class of the closure of $T_Q^*(Y^\vee)$ appears with multiplicity one in $\mathrm{CC}_i(\mathcal{S})$.) Since $\mu\left(\overline{T_Q^*(Y^\vee)}\right) \cap \mathcal{SP}(\mathcal{O}^\vee)$ is nonempty by hypothesis, and since $\mathrm{AV}_{\mathbb{C}}(\mathcal{S})$ must be the closure of a special orbit, it follows that $\mathrm{AV}_{\mathbb{C}}(\mathcal{S}) = \overline{\mathcal{O}^\vee}$. Therefore, by the discussion above, the element $\pi(Q) \in \mathbb{Z}_{\mathrm{st}}\Pi^{\lambda(\mathcal{O}^\vee)}(G/\mathbb{R})$ corresponding to m_Q is nonzero and belongs to $\mathbb{Z}_{\mathrm{st}}\Pi(\mathcal{O}^\vee)$. In other words,

$$\{\pi(Q) \mid \mu\left(T_Q^*(Y^\vee)\right) \cap \mathcal{SP}(\mathcal{O}^\vee) \text{ is nonempty}\} \subset \mathbb{Z}_{\mathrm{st}}\Pi(\mathcal{O}^\vee).$$

Because the m_Q are clearly linearly independent, so are the elements $\pi(Q)$ on the left-hand side above.

It remains to show the elements in (4.5) are indeed a basis of $\mathbb{Z}_{\mathrm{st}}\Pi(\mathcal{O}^\vee)$. Because of the linear independence just established, it suffices to check

$$(4.6) \qquad \sum_i \#\left\{ Q \in K_i^\vee \backslash Y^\vee \mid \mu\left(T_Q^*(Y^\vee)\right) \cap \mathcal{SP}(\mathcal{O}^\vee) \right\} \geq \dim_{\mathbb{Z}} \mathbb{Z}_{\mathrm{st}}\Pi(\mathcal{O}^\vee).$$

The following general result will be the main tool we use for counting the left-hand side.

PROPOSITION 4.2. *As above, assume $\mathcal{O}^\vee$ is even (but do not necessarily assume the Spaltenstein dual of $\mathcal{O}^\vee$ is even). With notation as in (2.2)–(2.5), fix a symmetric subgroup $K^\vee$ of $G^\vee$ and write $\mathfrak{g}^\vee = \mathfrak{k}^\vee \oplus \mathfrak{s}^\vee$ for the corresponding Cartan decomposition. Fix a nilpotent $K^\vee$ orbit $\mathcal{O}_K^\vee$ on $\mathfrak{s}^\vee$. Let $c(\mathcal{O}_K^\vee)$ denote the number of $K^\vee$ orbits Q on $Y^\vee$ such that $\mu(T_Q^* Y^\vee)$ meets $\mathcal{O}_K^\vee$ in a dense open set. Then, for $x^\vee \in \mathcal{O}_K^\vee$,*

$$c(\mathcal{O}_K^\vee) = \dim \mathrm{Hom}_{W(\mathfrak{l}^\vee(\mathcal{O}^\vee))}\left(\mathrm{sgn}, \mathrm{Sp}(x)^{A_{K^\vee}(x^\vee)}\right)$$

with notation for the Springer correspondence in Proposition 3.4.

Proof. This is a general result (and doesn't have anything to do with the dual group). It follows from Rossmann's theory applied to the partial flag setting. See [**CNT**, Section 2] for a proof. □

PROPOSITION 4.3. *In the setting of Proposition 4.2, assume further that the Spaltenstein dual of $\mathcal{O}^\vee$ is even, and*

$$G^\vee \cdot \mathcal{O}_K^\vee \subset \mathcal{SP}(\mathcal{O}^\vee).$$

Then numbers $c(\mathcal{O}_{K^\vee}^\vee)$ appearing in Proposition 4.2 are all nonzero. More precisely,

$$\dim \mathrm{Hom}_{W(\mathfrak{l}^\vee(\mathcal{O}^\vee))}\left(\mathrm{sgn}, \mathrm{Sp}(x^\vee)^{A_{G^\vee}(x^\vee)}\right) = 1,$$

and since $A_{K^\vee}(\xi) \to A_{\vee G}(\xi)$,

$$(4.7) \qquad \dim \mathrm{Hom}_{W(\mathfrak{l}^\vee(\mathcal{O}^\vee))}\left(\mathrm{sgn}, \mathrm{Sp}(x^\vee)^{A_{K^\vee}(x^\vee)}\right) \geq 1.$$

Proof. Section 5 of [**BV3**] shows that

$$\dim \mathrm{Hom}_{W(\mathfrak{l}^\vee(\mathcal{O}^\vee))}(\mathrm{sgn}, U) = 1$$

for an irreducible representation U in the left cell representation $V(\mathcal{O}^\vee)$. So the current proposition follows from Proposition 3.3. □

We now return to (4.6). By Proposition 4.3,
$$(4.8)$$
$$\sum_i \#\left\{ Q \in K_i^\vee \backslash Y^\vee \mid \mu\left(T_Q^*(Y^\vee)\right) \cap \mathcal{SP}(\mathcal{O}^\vee) \right\} \geq \sum_i \#\left(K_i^\vee \backslash (\mathcal{SP}(\mathcal{O}^\vee) \cap \mathfrak{s}_i^\vee)\right).$$

By (2.12) (which was proved in the previous section), the right-hand side equals $\dim_{\mathbb{Z}} \mathbb{Z}_{\mathrm{st}}\Pi(\mathcal{O}^\vee)$. This proves (4.6), and hence completes the proof of Proposition 4.1. □

COROLLARY 4.4. *In the setting of Proposition 4.3, there exists a unique orbit $Q = Q(\mathcal{O}_K^\vee)$ of $K^\vee$ on $Y^\vee$ such that*

$$\mu\left(\overline{T_Q^*(Y^\vee)}\right) = \overline{\mathcal{O}_K^\vee}.$$

PROOF. The proof of Proposition 4.1 shows that the inequality in (4.8) must be an equality. Hence the inequality in (4.7) in an equality. Hence the number $c(\mathcal{O}_K^\vee)$ in Proposition 4.2 must be 1. This proves the corollary. □

Proof of Theorem 2.1. In the setting of Theorem 2.1, given $\mathcal{O}_K^\vee \in K_i^\vee \backslash (\mathcal{SP}(\mathcal{O}^\vee) \cap \mathfrak{s}_i^\vee)$ define $Q(\mathcal{O}_K^\vee)$ as in Corollary 4.4. In the notation of Proposition 4.1 set

$$(4.9) \qquad\qquad \pi(\mathcal{O}_K^\vee) := \pi(Q(\mathcal{O}_K^\vee)).$$

Then by Proposition 4.1(2),

$$\left\{ \pi(\mathcal{O}_K^\vee) \mid \mathcal{O}_K^\vee \in \bigcup_i K_i^\vee \backslash (\mathcal{SP}(\mathcal{O}^\vee) \cap \mathfrak{s}_i^\vee) \right\}$$

is a basis for $\mathbb{Z}_{\mathrm{st}} \Pi(\mathcal{O}^\vee)$. □

References

[ABV] J. Adams, D. Barbasch, and D. A. Vogan, Jr., *The Langlands Classification and Irreducible Characters for Real Reductive Groups*, Progress in Math, Birkhäuser (Boston), **104** (1992).

[Ar1] J. Arthur, "On some problems suggested by the trace formula," in *Proceedings of the Special Year in Harmonic Analysis, University of Maryland*, R. Herb, R. Lipsman, and J. Rosenberg, eds., Lecture Notes in Mathematics **1024** (1983), Springer-Verlag (Berlin-Heidelberg-New York).

[Ar2] J. Arthur, "Unipotent automorphic representations: conjectures," in *Orbites Unipotentes et Represésentations II. Groups p-adiques et Réels*, Astésque, **171–172** (1989), 13–71.

[BV1] D. Barbasch and D. A. Vogan, Jr., Primitive ideals and orbital integrals in complex classical groups, *Math. Ann*, **259** (1982), 350–382.

[BV2] D. Barbasch and D. A. Vogan, Jr., Primitive ideals and orbital integrals in complex exceptional groups, *J. Algebra*, **80** (1983), 350–382.

[BV3] D. Barbasch and D. A. Vogan, Jr., Unipotent representations of complex semisimple Lie groups, *Annals of Math. (2)*, **121** (1985), no. 1, 41–110.

[BB1] W. Borho, J.-L. Brylinski, Differential operators on homogeneous spaces I. Irreducibility of the associated variety for annihilators of induced modules, *Invent. Math,* **69** (1982), no. 3, 437–476.

[BB3] W. Borho, J.-L. Brylinski, Differential operators on homogeneous spaces III. Characteristic varieties of Harish-Chandra modules and of primitive ideals, *Invent. Math,* **80** (1985), no. 1, 1–68.

[Ch] J.-T. Chang, Asymptotics and characteristic cycles for representations of complex groups, *Compositio Math.,* **88** (1993), no. 3, 265–283.

[CNT] D. Ciubotaru, K. Nishiyama, P. E. Trapa, Orbits of symmetric subgroups on partial flag varieties, to appear in *Representation Theory, Complex Analysis, and Integral Geometry Vol. 2,* (Eds. B. Kroetz, O. Offen and E. Sayag), Progress in Math, Birkhäuser (Boston).

[CM] D. Collingwood and W. M. McGovern, *Nilpotent orbits in semisimple Lie algebras*, Van Nostrand Reinhold (New York), 1993.

[Di] J. Dixmier, Idéaux primitifs dans l'algébre enveloppante d'une de Lie semisimple complexe, *C. R. Acad. Sci. Paris (A)*, **271** (1970), 134–136.

[KR] B. Kostant, Stephen Rallis, On representations associated with symmetric spaces, *Bull. Amer. Math. Soc.,* **75** (1969), 884-888.

[La] R. P. Langlands, Stable conjugacy: definitions and lemmas, *Canad. J. Math.,* **31** (1979), 700–725.

[LaSh] R. P. Langlands and D. Shelstad, On the definition of transfer factors, *Math. Ann.,* **278** (1987), 219-271.

[Lu1] G. Lusztig, *Characters of Reductive Groups over a Finite Field*, Annals of Mathematics Studies, **107** (19874), Princeton University Press (Princeton, NJ).

[Lu2] G. Lusztig, Notes on unipotent classes, *Asian Journal of Math.,* **1** (1997), no. 1, 194-207.

[MUV] I. Mirković, T. Uzawa, K. Vilonen, The Matsuki correspondence for sheaves, *Inventiones Math.,* **109** (1992), no. 1, 231–245.

[Ro]   W. Rossmann, Nilpotent orbital integrals and characters, in Operator algebras, unitary representations, enveloping algebras and invariant theory; in honor of Jacques Dixmier, Progress in Math. **92** (1990), Birkhäuser.

[Sp]   N. Spaltenstein, Classes unipotentes et sous-groupes de Borel, Lecture Notes in Math **946** (1982), Springer Verlag (Berlin-Heidelberg-New York).

[St]   R. Steinberg, On the desingularisation of the unipotent variety, *Inventiones Math.*, **36** (1976), pp. 209–224.

[Ta]   T. Tanisaki, Hodge modules, equivariant K-theory and Hecke algebras, *Publ. RIMS*, **23** (1987).

DEPARTMENT OF MATHEMATICS, CORNELL UNIVERSITY, ITHACA, NY 14853, USA
E-mail address: dmb14@cornell.edu

DEPARTMENT OF MATHEMATICS, UNIVERSITY OF UTAH, SALT LAKE CITY, UT 84112, USA
E-mail address: ptrapa@math.utah.edu

Contemporary Mathematics
Volume **557**, 2011

Levi components of parabolic subalgebras of finitary Lie algebras

Elizabeth Dan-Cohen and Ivan Penkov

ABSTRACT. We characterize locally semisimple subalgebras $\mathfrak{l}$ of $\mathfrak{sl}_\infty$, $\mathfrak{so}_\infty$, and $\mathfrak{sp}_\infty$ which are Levi components of parabolic subalgebras. Given $\mathfrak{l}$, we describe the parabolic subalgebras $\mathfrak{p}$ such that $\mathfrak{l}$ is a Levi component of $\mathfrak{p}$. We also prove that not every maximal locally semisimple subalgebra of a finitary Lie algebra is a Levi component.

When the set of self-normalizing parabolic subalgebras $\mathfrak{p}$ with fixed Levi component $\mathfrak{l}$ is finite, we prove an estimate on its cardinality. We consider various examples which highlight the differences from the case of parabolic subalgebras of finite-dimensional simple Lie algebras.

1. Introduction

The foundations of the theory of finitary Lie algebras have been laid in [**B, B2, BS, PS**]. This has made possible the development of a more detailed structure theory for the finitary Lie algebras [**NP, DPSn, DiP, D, DP, DPW, DiP2**]. In particular, the notions of Levi components and parabolic subalgebras were developed for finitary Lie algebras in [**DP**]. Nevertheless, the problem of an explicit description of all Levi component of parabolic subalgebras was not addressed there. This is the purpose of the present paper. More precisely, we identify the subalgebras which occur as the Levi component of a simple finitary Lie algebra, and we characterize all parabolic subalgebras of which a given subalgebra is a Levi component. In addition, we provide criteria for the number of self-normalizing parabolic subalgebra with a prescribed Levi component to be finite; note that the finite numbers which occur can be quite unlike those in the finite-dimensional case.

Along the way we present examples to highlight the many differences between the finitary and finite-dimensional situations. One phenomenon seen here for the first time is a maximal locally semisimple subalgebra of a parabolic subalgebra which is not a Levi component of the parabolic subalgebra. This answers a question left open in [**DP**]. It follows immediately that a maximal locally reductive subalgebra of a parabolic subalgebra is not in general a locally reductive part of

2010 *Mathematics Subject Classification*. Primary 17B65; Secondary 17B05.

Key words and phrases. simple finitary Lie algebra, parabolic subalgebra, Levi component.

Both authors acknowledge the partial support of the DFG through Grants PE 980/2-1 and PE 980/3-1.

the parabolic subalgebra. Moreover, we give examples in which a maximal locally reductive subalgebra of a parabolic subalgebra, despite containing a Levi component, nevertheless is not a locally reductive part of the parabolic subalgebra.

2. Preliminaries

2.1. Background on locally finite Lie algebras. Let V and V_* be countable-dimensional vector spaces over the complex numbers, together with a nondegenerate pairing $\langle \cdot, \cdot \rangle : V \times V_* \to \mathbb{C}$. A subspace $F \subset V$ is said to be *closed* (in the Mackey topology) if $F = F^{\perp\perp}$. By a result of Mackey [**M**], the vector spaces V and V_* admit dual bases: that is, there are bases $\{v_i \mid i \in I\}$ and $\{v^i \mid i \in I\}$ of V and V_*, respectively, such that $\langle v_i, v^j \rangle = \delta_{ij}$. We denote by $\mathfrak{gl}(V, V_*)$ the Lie algebra associated to the associative algebra $V \otimes V_*$ with multiplication

$$V \otimes V_* \times V \otimes V_* \to V \otimes V_*$$
$$(v \otimes w, v' \otimes w') \mapsto \langle v', w \rangle v \otimes w'.$$

The vectors $v_i \otimes v^j$ form a basis for $\mathfrak{gl}(V, V_*)$, with commutation relations $[v_i \otimes v^j, v_k \otimes v^l] = \delta_{jk} v_i \otimes v^l - \delta_{il} v_k \otimes v^j$. Thus $\mathfrak{gl}(V, V_*) \cong \mathfrak{gl}_\infty$, where $\mathfrak{gl}_\infty$ is the direct limit of the system

$$\mathfrak{gl}_n \to \mathfrak{gl}_{n+1} \qquad A \mapsto \begin{pmatrix} A & 0 \\ 0 & 0 \end{pmatrix}.$$

The Lie algebras $\mathfrak{sl}_\infty$, $\mathfrak{so}_\infty$, and $\mathfrak{sp}_\infty$ are similarly defined; that is, they are direct limits of systems of finite-dimensional simple Lie algebras where the natural representation of each successive Lie algebra considered as a representation of the previous Lie algebra decomposes as a direct sum of the natural representation plus a trivial representation. We denote by $\mathfrak{sl}(V, V_*)$ the commutator subalgebra of $\mathfrak{gl}(V, V_*)$, so $\mathfrak{sl}(V, V_*) \cong \mathfrak{sl}_\infty$.

Suppose $V = V_*$. When the pairing $V \times V \to \mathbb{C}$ is symmetric, then we denote by $\mathfrak{so}(V)$ the Lie algebra $\Lambda^2 V \cong \mathfrak{so}_\infty$. When the pairing $V \times V \to \mathbb{C}$ is antisymmetric, then we denote by $\mathfrak{sp}(V)$ the Lie algebra $\mathrm{Sym}^2 V \cong \mathfrak{sp}_\infty$. A subspace $F \subset V$ is called *isotropic* if $\langle F, F \rangle = 0$ and *coisotropic* if $F^\perp \subset F$.

We fix notation for maps

$$\Lambda : \mathfrak{gl}(V, V) \to \mathfrak{so}(V)$$
$$v \otimes w \mapsto v \otimes w - w \otimes v$$

and

$$S : \mathfrak{gl}(V, V) \to \mathfrak{sp}(V)$$
$$v \otimes w \mapsto v \otimes w + w \otimes v,$$

and note that they give homomorphisms of Lie algebras when restricted to $\mathfrak{gl}(X, Y)$ for any isotropic subspaces $X, Y \subset V$ such that the restriction $\langle \cdot, \cdot \rangle|_{X \times Y}$ is nondegenerate.

We call $\mathfrak{s}$ a *standard special linear subalgebra* of $\mathfrak{gl}(V, V_*)$ or $\mathfrak{sl}(V, V_*)$ if

$$\mathfrak{s} = \mathfrak{sl}(X, Y)$$

for some subspaces $X \subset V$ and $Y \subset V_*$ such that the restriction $\langle \cdot, \cdot \rangle|_{X \times Y}$ is nondegenerate. We call $\mathfrak{s}$ a *standard special linear subalgebra* of $\mathfrak{so}(V)$ (resp., of

$\mathfrak{sp}(V))$ if

$$\mathfrak{s} = \Lambda(\mathfrak{sl}(X, Y))$$

(resp., if $\mathfrak{s} = S(\mathfrak{sl}(X,Y)))$ for some isotropic subspaces X, $Y \subset V$ such that the restriction $\langle \cdot, \cdot \rangle|_{X \times Y}$ is nondegenerate.

A Lie algebra $\mathfrak{g}$ is said to be *finitary* if there exists a faithful countable-dimensional representation $\mathfrak{g} \hookrightarrow \operatorname{End} W$ where W has a basis in which the matrix of each endomorphism in the image of $\mathfrak{g}$ has only finitely many nonzero entries. Suppose $\mathfrak{g}$ is a finitary Lie algebra, and let $\{w_i \mid i \in I\}$ be a basis of a representation W of $\mathfrak{g}$ as in the definition of finitary. For each i, define $w^i \in W^*$ by $w^i(w_j) := \delta_{ij}$; let $W_* := \operatorname{Span}\{w^i \mid i \in I\}$, so that W_* is a countable-dimensional subspace of the full algebraic dual space W^*. Then the map $\mathfrak{g} \hookrightarrow \operatorname{End} W$ factors through $\mathfrak{gl}(W, W_*)$, yielding an injective homomorphism $\mathfrak{g} \hookrightarrow \mathfrak{gl}(W, W_*)$.

A Lie algebra $\mathfrak{g}$ is *locally finite* if every finite subset of $\mathfrak{g}$ is contained in a finite-dimensional subalgebra. A countable-dimensional locally finite Lie algebra $\mathfrak{g}$ is therefore the direct limit of a system of injective homomorphisms of finite-dimensional Lie algebras $\mathfrak{g}_n \hookrightarrow \mathfrak{g}_{n+1}$ for $n \in \mathbb{Z}_{>0}$. Observe that any finitary Lie algebra, being isomorphic to a subalgebra of $\mathfrak{gl}_\infty$, is itself locally finite.

A locally finite Lie algebra $\mathfrak{g}$ is *locally solvable* (respectively, *locally nilpotent*) if every finite-dimensional subalgebra of $\mathfrak{g}$ is solvable (resp., nilpotent). A locally finite Lie algebra $\mathfrak{g}$ is *locally simple* (resp., *locally semisimple*) if there exists a system of finite-dimensional simple (resp., semisimple) subalgebras $\mathfrak{g}_i$ of which $\mathfrak{g}$ is the direct limit. A locally finite Lie algebra $\mathfrak{g}$ is *locally reductive* if it is the direct limit of some system of finite-dimensional reductive subalgebras $\mathfrak{g}_n \hookrightarrow \mathfrak{g}_m$ where $\mathfrak{g}_m$ is a semisimple $\mathfrak{g}_n$-module for all $n < m$.

A subalgebra $\mathfrak{g}$ of a locally finite Lie algebra is called *parabolic* if it contains a maximal locally solvable subalgebra of $\mathfrak{g}$.

Let $\mathfrak{q}$ be a subalgebra of a locally reductive Lie algebra $\mathfrak{g}$, and let $\mathfrak{r}$ denote the locally solvable radical of $\mathfrak{q}$. The *linear nilradical* $\mathfrak{n}_\mathfrak{q}$ of $\mathfrak{q}$ is the set of Jordan nilpotent elements of $\mathfrak{r}$ (see [**DPSn**] for the details of Jordan decomposition in a locally reductive Lie algebra). One may check that $\mathfrak{n}_\mathfrak{q}$ is a locally nilpotent ideal of $\mathfrak{q}$; see [**DP**] where the proof is given in the finitary case, and it generalizes.

A subalgebra $\mathfrak{l} \subset \mathfrak{q}$ is a *Levi component* of $\mathfrak{q}$ if $[\mathfrak{q}, \mathfrak{q}] = (\mathfrak{r} \cap [\mathfrak{q}, \mathfrak{q}]) \subsetplus \mathfrak{l}$. A locally reductive subalgebra $\mathfrak{q}_{red}$ of $\mathfrak{q}$ is a *locally reductive part* of $\mathfrak{q}$ if $\mathfrak{q} = \mathfrak{n}_\mathfrak{q} \subsetplus \mathfrak{q}_{red}$.

A subalgebra $\mathfrak{q} \subset \mathfrak{g}$ is *splittable* if the nilpotent and semisimple parts of each element of $\mathfrak{q}$ are also in $\mathfrak{q}$. The fact that every splittable subalgebra of $\mathfrak{gl}_\infty$ has a locally reductive part was shown in [**DP**]. If $\mathfrak{q} \subset \mathfrak{gl}_\infty$ is splittable, then any subalgebra of $\mathfrak{q}$ containing $\mathfrak{n}_\mathfrak{q} + [\mathfrak{q}, \mathfrak{q}]$ is said to be *defined by trace conditions* on $\mathfrak{q}$ [**DP**]. (Note that any vector space containing $\mathfrak{n}_\mathfrak{q} + [\mathfrak{q}, \mathfrak{q}]$ and contained in $\mathfrak{q}$ is a subalgebra of $\mathfrak{g}$.)

Let X be a vector space, and let $\mathcal{C}$ be a set of subspaces of X on which inclusion gives a total order. Suppose $F' \subsetneq F''$ are subspaces of X in $\mathcal{C}$. We call F' the *immediate predecessor* of F'' in $\mathcal{C}$ if for all $C \in \mathcal{C}$ either $C \subset F'$ or $F'' \subset C$. When F' is the immediate predecessor of F'' in $\mathcal{C}$, we also say that F'' is the *immediate successor* of F' in $\mathcal{C}$, and that $F' \subset F''$ are an *immediate predecessor-successor pair* in $\mathcal{C}$.

DEFINITION 2.1. [**DiP**] A set $\mathfrak{F}$ of subspaces of X for which inclusion gives a total order is called a *generalized flag* if the following two conditions hold:

(1) For all $F \in \mathfrak{F}$, there is an immediate predecessor-successor pair $F' \subset F''$ in $\mathfrak{F}$ such that $F \in \{F', F''\}$;
(2) For all nonzero $v \in X$, there is an immediate predecessor-successor pair $F' \subset F''$ in $\mathfrak{F}$ such that $v \in F'' \setminus F'$.

For any generalized flag $\mathfrak{F}$, we denote by A the set of immediate predecessor-successor pairs of $\mathfrak{F}$. Then by definition we have $\mathfrak{F} = \{F'_\alpha, F''_\alpha\}_{\alpha \in A}$, where F'_α is the immediate predecessor of F''_α in the inclusion order, and the two subspaces are the pair $\alpha \in A$. Similarly, we denote by B the set of immediate predecessor-successor pairs of any generalized flag denoted by $\mathfrak{G}$, so that $\mathfrak{G} = \{G'_\beta, G''_\beta\}_{\beta \in B}$.

For any generalized flag $\mathfrak{F}$ in V, the stabilizer of $\mathfrak{F}$ in $\mathfrak{gl}(V, V_*)$ is denoted by $\mathrm{St}_{\mathfrak{F}}$ and is given by the formula $\mathrm{St}_{\mathfrak{F}} = \sum_{\alpha \in A} F''_\alpha \otimes (F'_\alpha)^\perp$ [**DiP**].

2.2. Background on parabolic subalgebras of finitary Lie algebras.
Recall that there is a nondegenerate pairing $\langle \cdot, \cdot \rangle : V \times V_* \to \mathbb{C}$. We say that a generalized flag $\mathfrak{F}$ in V is *semiclosed* if $(F')^{\perp\perp} \in \{F', F''\}$ for each immediate predecessor-successor pair $F' \subset F''$ in $\mathfrak{F}$.

PROPOSITION 2.2. *Let $\mathcal{C}$ be a set of subspaces of V totally ordered by inclusion. Then the following exist:*
(1) *a generalized flag in V with the same $\mathfrak{gl}(V, V_*)$-stabilizer as $\mathcal{C}$* [**DiP**];
(2) *a unique semiclosed generalized flag in V with the same $\mathfrak{gl}(V, V_*)$-stabilizer as $\mathcal{C}$, if each nonclosed subspace in $\mathcal{C}$ is the immediate successor in $\mathcal{C}$ of a closed subspace.*

PROOF. We recall the construction from [**DiP**] that produces a generalized flag with the same stabilizer as a given set $\mathcal{C}$ of subspaces totally ordered by inclusion. For a fixed nonzero vector x in V, consider the subspace $F'(x)$ which is the union of the subspaces in $\mathcal{C}$ not containing x; it is properly contained in $F''(x)$, the intersection of the subspaces in $\mathcal{C}$ containing x. The set of subspaces of the form $F'(x)$ or $F''(x)$ as x runs over the nonzero vectors in V is a generalized flag with the same stabilizer as $\mathcal{C}$.

Assume that each nonclosed subspace in $\mathcal{C}$ has an immediate predecessor in $\mathcal{C}$, and the latter is closed. Then the set $\mathcal{D} := \mathcal{C} \cup \{X^{\perp\perp} \mid X \in \mathcal{C}\}$ is totally ordered by inclusion. To prove this, it suffices to show that $X_1 \subsetneq X_2$ implies $X_1^{\perp\perp} \subset X_2$ for all $X_1, X_2 \in \mathcal{C}$. If X_2 is closed, then this is clear. If X_2 is not closed, then by assumption X_2 has an immediate predecessor $X_3 \in \mathcal{C}$, and X_3 is closed; then $X_1 \subset X_3 \subset X_2$, so $X_1^{\perp\perp} \subset X_3 \subset X_2$.

For each nonzero vector $x \in V$, we define
$$F_1(x) := \bigcup_{Y \in \mathcal{D}, \, x \notin Y} Y \qquad F_2(x) := F_1(x)^{\perp\perp} \qquad F_3(x) := \bigcap_{Y \in \mathcal{D}, \, x \in Y} Y.$$

Applying the general construction from [**DiP**] to $\mathcal{D}$ yields the generalized flag $\{F_1(x), F_3(x) \mid 0 \neq x \in V\}$. We claim that the refinement
$$\mathfrak{F} := \{F_1(x), F_2(x), F_3(x) \mid 0 \neq x \in V\}$$
is a semiclosed generalized flag with the same stabilizer as $\mathcal{C}$.

To see that $\mathfrak{F}$ is a generalized flag, it suffices to show that $F_2(x) \subset F_3(x)$ for all nonzero $x \in V$. If $F_3(x)$ is closed, then this is clear. If $F_3(x)$ is not closed, then $F_3(x) \in \mathcal{C}$ (otherwise it would be the intersection of the closed subspaces containing it, since each nonclosed subspace in $\mathcal{C}$ is assumed to have an immediate predecessor in $\mathcal{C}$ which is closed); hence $F_3(x)$ is the immediate successor in $\mathcal{C}$ of a closed subspace, and the latter is then $F_1(x) = F_2(x)$.

Now each immediate predecessor-successor pair in $\mathfrak{F}$ has the form $F_1(x) \subset F_2(x)$ or $F_2(x) \subset F_3(x)$ for some nonzero $x \in V$. In either case the condition defining a semiclosed generalized flag is satisfied. By construction $\mathcal{C}$ and $\mathfrak{F}$ have the same stabilizer. Finally, the uniqueness of $\mathfrak{F}$ follows from [**DP**, Proposition 3.8] $\qquad\square$

We say that semiclosed generalized flags $\mathfrak{F}$ in V and $\mathfrak{G}$ in V_* form a *taut couple* if $F^\perp$ is stable under $\mathrm{St}_\mathfrak{G}$ for all $F \in \mathfrak{F}$ and $G^\perp$ is stable under $\mathrm{St}_\mathfrak{F}$ for all $G \in \mathfrak{G}$. If $V = V_*$, then a semiclosed generalized flag $\mathfrak{F}$ in V is called *self-taut* if $F^\perp$ is stable under $\mathrm{St}_\mathfrak{F}$ for all $F \in \mathfrak{F}$. In the interest of clarity, we should emphasize that $\mathrm{St}_\mathfrak{F}$ means the $\mathfrak{gl}(V, V)$-stabilizer of $\mathfrak{F}$ in the case $V = V_*$.

We now summarize Theorem 5.6 in [**DP**]. For any taut couple $\mathfrak{F}$, $\mathfrak{G}$, the subalgebra $\mathrm{St}_\mathfrak{F} \cap \mathrm{St}_\mathfrak{G}$ is a self-normalizing parabolic subalgebra of $\mathfrak{gl}(V, V_*)$. Moreover, the self-normalizing parabolic subalgebras of $\mathfrak{gl}(V, V_*)$ are in bijection with the taut couples in V and V_* [**DP**, Corollary 5.7]. If $\mathfrak{p}$ is any parabolic subalgebra of $\mathfrak{gl}(V, V_*)$, then the normalizer of $\mathfrak{p}$ is a self-normalizing parabolic subalgebra, which we denote $\mathfrak{p}_+$; furthermore, $\mathfrak{p}$ is defined by trace conditions on $\mathfrak{p}_+$. We call the (unique) taut couple $\mathfrak{F}$, $\mathfrak{G}$ such that $\mathfrak{p}_+ = \mathrm{St}_\mathfrak{F} \cap \mathrm{St}_\mathfrak{G}$ the *taut couple associated to* $\mathfrak{p}$. The smallest parabolic subalgebra with the associated taut couple $\mathfrak{F}$, $\mathfrak{G}$ is denoted by $\mathfrak{p}_-$, and it is the set of elements of $\mathrm{St}_\mathfrak{F} \cap \mathrm{St}_\mathfrak{G}$ such that each component in each infinite-dimensional block of a locally reductive part is traceless.

The situation for $\mathfrak{sl}(V, V_*)$, $\mathfrak{so}(V)$, and $\mathfrak{sp}(V)$ is quite similar to the above, in that every parabolic subalgebra is defined by trace conditions on its normalizer, which is a self-normalizing parabolic subalgebra. Theorem 5.6 in [**DP**] also characterizes the parabolic subalgebras of $\mathfrak{sl}(V, V_*)$, as follows. The self-normalizing parabolic subalgebras of $\mathfrak{sl}(V, V_*)$ are also in bijection with the taut couples in V and V_*, where the joint stabilizer $\mathrm{St}_\mathfrak{F} \cap \mathrm{St}_\mathfrak{G} \cap \mathfrak{sl}(V, V_*)$ is the self-normalizing parabolic subalgebra of $\mathfrak{sl}(V, V_*)$ corresponding to the taut couple $\mathfrak{F}$, $\mathfrak{G}$. The parabolic subalgebras of $\mathfrak{so}(V)$ and $\mathfrak{sp}(V)$ are described in Theorem 6.6 in [**DP**]. In the case of $\mathfrak{sp}(V)$, taking the stabilizer gives yet again a bijection between the self-taut generalized flags in V and the self-normalizing parabolic subalgebras of $\mathfrak{sp}(V)$. In the case of $\mathfrak{so}(V)$, the analogous map surjects onto the self-normalizing parabolic subalgebras of $\mathfrak{so}(V)$, but by contrast it is not injective. The fibers of size different from 1 are all of size 3 [**DPW**]. Note that the claim in [**DP**, Theorem 6.6] regarding the uniqueness in the $\mathfrak{so}_\infty$ case is erroneous; the correct statement is [**DPW**, Theorem 2.8].

2.3. Locally reductive parts of parabolic subalgebras of $\mathfrak{sl}_\infty$ and $\mathfrak{gl}_\infty$.
We denote by C the ordered subset

$$C := \{\alpha \in A \mid (F'_\alpha)^{\perp\perp} = F'_\alpha\}$$

for any semiclosed generalized flag $\mathfrak{F}$ in V with $V \neq V_*$. For any taut couple $\mathfrak{F}$, $\mathfrak{G}$ there is a natural bijection between C and the set $\{\beta \in B \mid (G'_\beta)^{\perp\perp} = G'_\beta\}$, under which $F'_\gamma = (G''_\gamma)^\perp$ for all $\gamma \in C$ [**DP**, Proposition 3.4]. This enables us to consider C as a subset of B, as well; note that the inclusion of C into B is order reversing.

In the case $V = V_*$ we denote by C the analogous subset

$$C := \{\alpha \in A \mid (F'_\alpha)^{\perp\perp} = F'_\alpha, \, F''_\alpha \subset (F''_\alpha)^\perp\}.$$

In this case, there is a natural order-reversing bijection between C and the set $\{\alpha \in A \mid (F'_\alpha)^{\perp\perp} = F'_\alpha \supset (F'_\alpha)^\perp\}$ [**DP**, Proposition 6.1]. For $\gamma \in C$, we denote by $G'_\gamma \subset G''_\gamma$ the corresponding pair where G'_γ is closed and coisotropic, and thus obtain the analogous statements $F'_\gamma = (G''_\gamma)^\perp$ and $G'_\gamma = (F''_\gamma)^\perp$ for all $\gamma \in C$.

The next theorem is slightly more general than Proposition 3.6 (ii) in [**DP**]. Note that subspaces X_γ and Y_γ satisfying the hypotheses necessarily exist [**DP**, Prop. 3.6 (ii)].

THEOREM 2.3. *Let $\mathfrak{p}$ be a parabolic subalgebra of $\mathfrak{sl}(V, V_*)$ or $\mathfrak{gl}(V, V_*)$, with the associated taut couple $\mathfrak{F}$, $\mathfrak{G}$. Let $X_\gamma \subset V$ and $Y_\gamma \subset V_*$ be any subspaces with*

$$F''_\gamma = F'_\gamma \oplus X_\gamma \text{ and } G''_\gamma = G'_\gamma \oplus Y_\gamma$$

for all $\gamma \in C$, such that $\langle X_\gamma, Y_\eta \rangle = 0$ for $\gamma \neq \eta$. Then $\mathfrak{p} \cap \bigoplus_{\gamma \in C} \mathfrak{gl}(X_\gamma, Y_\gamma)$ is a locally reductive part of $\mathfrak{p}$.

PROOF. The proof of Proposition 3.6 (ii) in [**DP**] shows that $\bigoplus_{\gamma \in C} \mathfrak{gl}(X_\gamma, Y_\gamma)$ is a locally reductive part of $\mathrm{St}_{\mathfrak{F}} \cap \mathrm{St}_{\mathfrak{G}}$. Note that there are additional hypotheses in [**DP**] which were not used in the proof. Intersecting a locally reductive part of $\mathrm{St}_{\mathfrak{F}} \cap \mathrm{St}_{\mathfrak{G}}$ with $\mathfrak{p}$ imposes the same trace conditions defining $\mathfrak{p}$, yielding a locally reductive part of $\mathfrak{p}$. $\qquad\square$

3. Parabolic subalgebras of $\mathfrak{sl}_\infty$ and $\mathfrak{gl}_\infty$ with given Levi component

In this section we prove the two main theorems of this paper. Theorem 3.4 identifies the subalgebras of $\mathfrak{sl}(V, V_*)$ and $\mathfrak{gl}(V, V_*)$ which can be realized as the Levi component of a parabolic subalgebra. Theorem 3.5 characterizes all parabolic subalgebras of which a given subalgebra $\mathfrak{l}$ is a Levi component.

Every parabolic subalgebra of $\mathfrak{sl}(V, V_*)$ or $\mathfrak{gl}(V, V_*)$ has a Levi component of the form $\bigoplus_{i \in I} \mathfrak{sl}_{n_i}$ for some $n_i \in \mathbb{Z}_{\geq 2} \cup \{\infty\}$, by Theorem 2.3. We therefore consider whether every such subalgebra of $\mathfrak{sl}(V, V_*)$ or $\mathfrak{gl}(V, V_*)$ is a Levi component of some parabolic subalgebra. An obstruction presents itself immediately, in consequence of the following lemma.

LEMMA 3.1. *Let $\mathfrak{p}$ be a parabolic subalgebra of $\mathfrak{gl}(V, V_*)$, and $\mathfrak{l}$ a Levi component of $\mathfrak{p}$. Then $\mathfrak{l}$ is a direct sum of standard special linear subalgebras.*

Furthermore, the order of the generalized flag $\mathfrak{F}$ in V associated to $\mathfrak{p}$ induces an order on the simple direct summands of $\mathfrak{l}$.

PROOF. The Levi component $\mathfrak{l}$ of $\mathfrak{p}$ is a maximal locally semisimple subalgebra of $\mathfrak{p}$ by [**DP**, Theorem 4.3]. Since $\mathfrak{l}$ is a locally semisimple subalgebra of $\mathfrak{g}$, it is a direct sum of simple subalgebras [**DiP2**]. Let $\mathfrak{s}$ denote one of the simple direct summands of $\mathfrak{l}$, and take $\mathfrak{l}_0$ to be the direct sum of all the other simple direct summands of $\mathfrak{l}$, so $\mathfrak{l} = \mathfrak{s} \oplus \mathfrak{l}_0$. We will show that $\mathfrak{s}$ is a standard special linear subalgebra.

When $\mathfrak{s}$ is finite dimensional, there exist nontrivial simple $\mathfrak{s}$-submodules X_1, $X_2, \ldots, X_k$ of V and $Y_1, Y_2, \ldots, Y_k$ of V_* such that $\langle X_i, Y_j \rangle = 0$ for $i \neq j$ and

$$\mathfrak{s} \subset \mathfrak{sl}(X_1, Y_1) \oplus \mathfrak{sl}(X_2, Y_2) \oplus \cdots \oplus \mathfrak{sl}(X_k, Y_k).$$

When $\mathfrak{s}$ is infinite dimensional, the same statement follows from [**DiP2**], where the authors characterize arbitrary subalgebras of $\mathfrak{gl}(V, V_*)$ isomorphic to $\mathfrak{sl}_\infty$, $\mathfrak{so}_\infty$, and $\mathfrak{sp}_\infty$.

We will show that $\mathfrak{sl}(X_1, Y_1) \oplus \mathfrak{sl}(X_2, Y_2) \oplus \cdots \oplus \mathfrak{sl}(X_k, Y_k) \subset \mathfrak{p}$. As the labeling is arbitrary, it is enough to show that $\mathfrak{sl}(X_1, Y_1) \subset \mathfrak{p}$. Moreover, it suffices to show that $\mathfrak{F}$ is stable under $\mathfrak{sl}(X_1, Y_1)$, where $\mathfrak{F}$, $\mathfrak{G}$ is the taut couple associated to $\mathfrak{p}$. Indeed, it then follows by symmetry that $\mathfrak{G}$ is also stable under $\mathfrak{sl}(X_1, Y_1)$, and hence $\mathfrak{sl}(X_1, Y_1) \subset [\mathrm{St}_{\mathfrak{F}} \cap \mathrm{St}_{\mathfrak{G}}, \mathrm{St}_{\mathfrak{F}} \cap \mathrm{St}_{\mathfrak{G}}] \subset \mathfrak{p}$.

Fix a nonzero vector $x_i \in X_i$ for $i = 1, \ldots, k$. By the definition of a generalized flag $x_i \in F_{\alpha_i}'' \setminus F_{\alpha_i}'$ for some $\alpha_i \in A$. Consider that $X_i = \mathfrak{s} \cdot x_i \subset \mathrm{St}_{\mathfrak{F}} \cdot x_i \subset F_{\alpha_i}''$. If there were a nonzero vector in the intersection $X_i \cap F_{\alpha_i}'$, then one would have similarly that $X_i \subset F_{\beta_i}''$ for some $\beta_i < \alpha_i$, contradicting the fact that $x_i \notin F_{\alpha_i}'$. Thus we conclude $X_i \cap F_{\alpha_i}' = 0$, and we have shown $F_{\alpha_i}' \oplus X_i \subset F_{\alpha_i}''$.

Observe that $\mathfrak{s} \cdot F_{\alpha_1}' \subset \mathrm{St}_{\mathfrak{F}} \cdot F_{\alpha_1}' \subset F_{\alpha_1}'$. Let π_i denote the ith projection for the decomposition $\mathfrak{sl}(X_1, Y_1) \oplus \mathfrak{sl}(X_2, Y_2) \oplus \cdots \oplus \mathfrak{sl}(X_k, Y_k)$. One has $\pi_i(\mathfrak{s}) \cdot F_{\alpha_1}' \subset (X_i \otimes Y_i) \cdot V \subset X_i$ for each i, and hence $\pi_1(\mathfrak{s}) \cdot F_{\alpha_1}' = 0$. Since $(Y_1)^{\perp}$ is the largest trivial $\pi_1(\mathfrak{s})$-submodule of V, we see that $F_{\alpha_1}' \subset (Y_1)^{\perp}$. As a result

$$\mathfrak{sl}(X_1, Y_1) \subset X_1 \otimes Y_1 \subset F_{\alpha_1}'' \otimes (F_{\alpha_1}')^{\perp} \subset \mathrm{St}_{\mathfrak{F}}.$$

Therefore $\mathfrak{sl}(X_1, Y_1) \oplus \mathfrak{sl}(X_2, Y_2) \oplus \cdots \oplus \mathfrak{sl}(X_k, Y_k) \oplus \mathfrak{l}_0$ is a locally semisimple subalgebra containing $\mathfrak{l}$ and contained in $\mathfrak{p}$. By the maximality of $\mathfrak{l}$, we obtain $\mathfrak{s} = \mathfrak{sl}(X_1, Y_1) \oplus \mathfrak{sl}(X_2, Y_2) \oplus \cdots \oplus \mathfrak{sl}(X_k, Y_k)$ and in particular $k = 1$, so $\mathfrak{s}$ is a standard special linear subalgebra.

This shows that $\mathfrak{l} = \bigoplus_{i \in I} \mathfrak{sl}(X_i, Y_i)$ for some subspaces $X_i \subset V$ and $Y_i \subset V_*$. Fix $i \in I$. As shown above, there exists $\alpha_i \in A$ such that $F_{\alpha_i}' \oplus X_i \subset F_{\alpha_i}''$ and $\langle F_{\alpha_i}', Y_i \rangle = 0$. Similarly, there exists $\beta_i \in B$ such that $G_{\beta_i}' \oplus Y_i \subset G_{\beta_i}''$ and $\langle X_i, G_{\beta_i}' \rangle = 0$. Since $\mathfrak{F}$, $\mathfrak{G}$ form a taut couple, it follows that $F_{\alpha_i}' = (G_{\beta_i}'')^{\perp}$ and $G_{\beta_i}' = (F_{\alpha_i}'')^{\perp}$, and hence $\alpha_i \in D$. Thus the rule $i \mapsto \alpha_i$ gives a well-defined map $\kappa : I \to D$. We claim that κ is an injective map. To see this, suppose $\kappa(j) = \kappa(k)$ for some $j, k \in I$. Then $\mathfrak{l} \subset \mathfrak{sl}(X_j + X_k, Y_j + Y_k) \oplus \bigoplus_{i \neq j,k} \mathfrak{s}_i \subset \mathfrak{p}$, so the maximality of $\mathfrak{l}$ implies that $j = k$. Since D is an ordered set, there is an induced order on I, the set of direct summands of $\mathfrak{l}$. $\quad\square$

Theorem 3.4 below shows that there are no further obstructions to finding a parabolic subalgebra such that a given subalgebra is a Levi component. We first prove a lemma and a proposition.

LEMMA 3.2. *Fix subspaces $X \subset V$ and $Y \subset V_*$ such that the restriction $\langle \cdot, \cdot \rangle|_{X \times Y}$ is nondegenerate. Let $T \subset V$, and define $U := ((T + X)^{\perp} \oplus Y)^{\perp}$. Then*

$$U = ((U \oplus X)^{\perp} \oplus Y)^{\perp}.$$

PROOF. To see that $U \subset ((U \oplus X)^{\perp} \oplus Y)^{\perp} = (U \oplus X)^{\perp\perp} \cap Y^{\perp}$, consider that $U = ((T+X)^{\perp} \oplus Y)^{\perp} = (T+X)^{\perp\perp} \cap Y^{\perp} \subset Y^{\perp}$, while the inclusion $U \subset (U+X)^{\perp\perp}$ is automatic. For the reverse containment, we observe that

$$\langle ((T+X)^{\perp} \oplus Y)^{\perp} \oplus X, (T+X)^{\perp} \rangle = \langle (T+X)^{\perp\perp} \cap Y^{\perp} \oplus X, (T+X)^{\perp} \rangle$$
$$\subset \langle (T+X)^{\perp\perp}, (T+X)^{\perp} \rangle = 0.$$

This shows that $(T + X)^{\perp} \subset (((T + X)^{\perp} \oplus Y)^{\perp} \oplus X)^{\perp}$. Hence

$$((((T + X)^{\perp} \oplus Y)^{\perp} \oplus X)^{\perp} \oplus Y)^{\perp} \subset ((T + X)^{\perp} \oplus Y)^{\perp},$$

i.e. $((U \oplus X)^{\perp} \oplus Y)^{\perp} \subset U$. $\quad\square$

For any semiclosed generalized flag $\mathfrak{F}$ we set

$$D := \{\gamma \in C \mid \dim F''_\gamma / F'_\gamma > 1\}.$$

Note that $D = \{\gamma \in C \mid \dim G''_\gamma / G'_\gamma > 1\}$, as the pairing $V \times V_* \to \mathbb{C}$ induces a nondegenerate pairing of F''_γ / F'_γ and G''_γ / G'_γ for all $\gamma \in C$.

PROPOSITION 3.3. *Let $\mathfrak{p}$ be a parabolic subalgebra of $\mathfrak{sl}(V, V_*)$ or $\mathfrak{gl}(V, V_*)$, with the associated taut couple $\mathfrak{F}$, $\mathfrak{G}$. Let $X_\gamma \subset V$ and $Y_\gamma \subset V_*$ be any subspaces with*

$$F''_\gamma = F'_\gamma \oplus X_\gamma \text{ and } G''_\gamma = G'_\gamma \oplus Y_\gamma$$

for all $\gamma \in D$, such that $\langle X_\gamma, Y_\eta \rangle = 0$ for $\gamma \neq \eta$. Then $\bigoplus_{\gamma \in D} \mathfrak{sl}(X_\gamma, Y_\gamma)$ is a Levi component of $\mathfrak{p}$.

PROOF. Let X_γ and Y_γ for $\gamma \in D$ be as in the statement, and let $\tilde{X}_\gamma$ and $\tilde{Y}_\gamma$ for $\gamma \in C$ be as in Theorem 2.3, so that $\mathrm{St}_{\mathfrak{F}} \cap \mathrm{St}_{\mathfrak{G}} = \mathfrak{n_p} \oplus \bigoplus_{\gamma \in C} \mathfrak{gl}(\tilde{X}_\gamma, \tilde{Y}_\gamma)$. Clearly the subalgebra $\mathfrak{n_p} \oplus \bigoplus_{\gamma \in D} \mathfrak{gl}(\tilde{X}_\gamma, \tilde{Y}_\gamma)$ is defined by trace conditions on $\mathrm{St}_{\mathfrak{F}} \cap \mathrm{St}_{\mathfrak{G}}$. A subalgebra has the same set of Levi components as any subalgebra defined by trace conditions on it, by [**DP**, Proposition 4.9]. Since $\mathfrak{p}$ is also defined by trace conditions on $\mathrm{St}_{\mathfrak{F}} \cap \mathrm{St}_{\mathfrak{G}}$, all three have the same set of Levi components, and it suffices to show that $\bigoplus_{\gamma \in D} \mathfrak{sl}(X_\gamma, Y_\gamma)$ is a Levi component of $\mathfrak{n_p} \oplus \bigoplus_{\gamma \in D} \mathfrak{gl}(\tilde{X}_\gamma, \tilde{Y}_\gamma)$.

Clearly $\bigoplus_{\gamma \in D} \mathfrak{sl}(X_\gamma, Y_\gamma)$ is a Levi component of $\mathfrak{n_p} \oplus \bigoplus_{\gamma \in D} \mathfrak{gl}(X_\gamma, Y_\gamma)$. We claim that

$$\mathfrak{n_p} \oplus \bigoplus_{\gamma \in D} \mathfrak{gl}(\tilde{X}_\gamma, \tilde{Y}_\gamma) = \mathfrak{n_p} \oplus \bigoplus_{\gamma \in D} \mathfrak{gl}(X_\gamma, Y_\gamma).$$

To see this, consider that for each $\gamma \in C$,

$$\begin{aligned}
\tilde{X}_\gamma \otimes \tilde{Y}_\gamma \subset F''_\gamma \otimes G''_\gamma &= (F'_\gamma \oplus X_\gamma) \otimes (G'_\gamma \oplus Y_\gamma) \\
&= (F'_\gamma \otimes G''_\gamma + F''_\gamma \otimes G'_\gamma) \oplus X_\gamma \otimes Y_\gamma.
\end{aligned}$$

Since $F'_\gamma \otimes G''_\gamma + F''_\gamma \otimes G'_\gamma \subset \mathfrak{n_p}$, we have shown that $\mathfrak{gl}(\tilde{X}_\gamma, \tilde{Y}_\gamma) \subset \mathfrak{n_p} \oplus \mathfrak{gl}(X_\gamma, Y_\gamma)$. One has symmetrically that $\mathfrak{gl}(X_\gamma, Y_\gamma) \subset \mathfrak{n_p} \oplus \mathfrak{gl}(\tilde{X}_\gamma, \tilde{Y}_\gamma)$ for each $\gamma \in C$. □

Let $\mathfrak{g}$ denote either $\mathfrak{sl}(V, V_*)$ or $\mathfrak{gl}(V, V_*)$.

THEOREM 3.4. *Let $\mathfrak{l}$ be a subalgebra of $\mathfrak{g}$. There exists a parabolic subalgebra $\mathfrak{p}$ of $\mathfrak{g}$ such that $\mathfrak{l}$ is a Levi component of $\mathfrak{p}$ if and only if $\mathfrak{l}$ is a direct sum of standard special linear subalgebras of $\mathfrak{g}$.*

Moreover, given a subalgebra $\mathfrak{l}$ for which such parabolic subalgebras exist, one exists that induces an arbitrary order on the simple direct summands of $\mathfrak{l}$ (see Lemma 3.1).

PROOF. In this and subsequent proofs, we assume (without loss of generality) that $\mathfrak{g} = \mathfrak{gl}(V, V_*)$. The only if direction was proved in Lemma 3.1.

Conversely, fix commuting standard special linear subalgebras $\mathfrak{s}_i \subset \mathfrak{gl}(V, V_*)$ for $i \in I$, as well as an order on I. We will construct a parabolic subalgebra $\mathfrak{p}$ such that $\mathfrak{l} := \bigoplus_{i \in I} \mathfrak{s}_i$ is a Levi component of $\mathfrak{p}$, and $\mathfrak{p}$ induces the given order on $\mathfrak{l}$. Each standard special linear subalgebra $\mathfrak{s}_i$ determines subspaces $X_i \subset V$ and $Y_i \subset V_*$ such that $\mathfrak{s}_i = \mathfrak{sl}(X_i, Y_i)$. As these direct summands commute, it must be that $\langle X_i, Y_j \rangle = 0$ for $i \neq j$.

For each i, we define

$$U_i := ((\bigoplus_{k \leq i} X_k)^\perp \oplus Y_i)^\perp.$$

One may check in an elementary fashion that $U_i \oplus X_i \subset U_j$ for all $i < j$. Since U_j is closed for all $j \in I$, we have moreover that $U_i \oplus X_i \subset (U_i \oplus X_i)^{\perp\perp} \subset U_j$ for all $i < j$. Furthermore, for each $i \in I$, an application of Lemma 3.2 using $T = \bigoplus_{k<i} X_k$ shows that $U_i = ((U_i \oplus X_i)^\perp \oplus Y_i)^\perp$.

We claim that there is a unique semiclosed generalized flag $\mathfrak{F}_0$ in V with the same stabilizer as the set $\{U_i, U_i \oplus X_i \mid i \in I\}$. This follows from Proposition 2.2 (2). Similarly, there is a unique semiclosed generalized flag in V_* with the same stabilizer as the set

$$\{(U_i \oplus X_i)^\perp, (U_i \oplus X_i)^\perp \oplus Y_i \mid i \in I\}.$$

One may check that $\mathfrak{F}_0$, $\mathfrak{G}_0$ form a taut couple using the identity $U_i = ((U_i \oplus X_i)^\perp \oplus Y_i)^\perp$. Indeed, $\mathfrak{F}_0$, $\mathfrak{G}_0$ is the minimal taut couple with immediate predecessor-successor pairs $U_i \subset U_i \oplus X_i$ in $\mathfrak{F}$ and $(U_i \oplus X_i)^\perp \subset (U_i \oplus X_i)^\perp \oplus Y_i$ in $\mathfrak{G}$ for all $i \in I$.

Let $\mathfrak{F}$, $\mathfrak{G}$ be maximal among the taut couples having immediate predecessor-successor pairs $U_i \subset U_i \oplus X_i$ in $\mathfrak{F}$ and $(U_i \oplus X_i)^\perp \subset (U_i \oplus X_i)^\perp \oplus Y_i$ in $\mathfrak{G}$ for all $i \in I$. Then there is a natural bijection between I and D. By [**DP**, Theorem 5.6], $\mathrm{St}_\mathfrak{F} \cap \mathrm{St}_\mathfrak{G}$ is a parabolic subalgebra of $\mathfrak{gl}(V, V_*)$. Moreover, it follows from Proposition 3.3 that $\mathfrak{l}$ is a Levi component of $\mathrm{St}_\mathfrak{F} \cap \mathrm{St}_\mathfrak{G}$. By construction, the induced order on the simple direct summands of $\mathfrak{l}$ is the given order on I. $\square$

We now characterize the parabolic subalgebras of which a given subalgebra is a Levi component.

THEOREM 3.5. *Let $\mathfrak{p}$ be a parabolic subalgebra of $\mathfrak{g}$, with the associated taut couple $\mathfrak{F}$, $\mathfrak{G}$. Then $\mathfrak{l}$ is a Levi component of $\mathfrak{p}$ if and only if there exist subspaces $X_\gamma \subset V$ and $Y_\gamma \subset V_*$ with*

$$(3.1) \qquad F_\gamma'' = F_\gamma' \oplus X_\gamma \ \text{and} \ G_\gamma'' = G_\gamma' \oplus Y_\gamma$$

for all $\gamma \in D$, such that $\mathfrak{l} = \bigoplus_{\gamma \in D} \mathfrak{sl}(X_\gamma, Y_\gamma)$.

PROOF. One direction is Proposition 3.3. Conversely, assume $\mathfrak{l}$ is an arbitrary Levi component of $\mathfrak{p}$. By Theorem 3.4, we have $\mathfrak{l} = \bigoplus_{i \in I} \mathfrak{s}_i$ for some standard special linear subalgebras $\mathfrak{s}_i \subset \mathfrak{gl}(V, V_*)$. Hence there exist subspaces $X_i \subset V$ and $Y_i \subset V_*$ such that $\mathfrak{s}_i = \mathfrak{sl}(X_i, Y_i)$; again, $\langle X_i, Y_j \rangle = 0$ for $i \neq j$. As shown in the proof of Lemma 3.1, there exists an injective map $\kappa : I \to D$, with the properties $F_{\kappa(i)}' \oplus X_i \subset F_{\kappa(i)}''$ and $\langle F_{\kappa(i)}', Y_i \rangle = 0$.

Let $\tilde{X}_\gamma$ and $\tilde{Y}_\gamma$ for $\gamma \in D$ be as in the statement of Proposition 3.3. Then since the span of the linear nilradical of $\mathfrak{p}$ and any Levi component equals $\mathfrak{n}_\mathfrak{p} + [\mathfrak{p}, \mathfrak{p}]$ [**DP**], we have

$$\mathfrak{n}_\mathfrak{p} \Subset \bigoplus_{\gamma \in D} \mathfrak{sl}(\tilde{X}_\gamma, \tilde{Y}_\gamma) = \mathfrak{n}_\mathfrak{p} \Subset \bigoplus_{i \in I} \mathfrak{sl}(X_i, Y_i).$$

Let $d \in D$ be arbitrary. Fix $v \in \tilde{X}_d$. Then

$$F_d' + \big(\mathfrak{n}_\mathfrak{p} \Subset \bigoplus_{\gamma \in D} \mathfrak{sl}(\tilde{X}_\gamma, \tilde{Y}_\gamma)\big) \cdot v = F_d' + \tilde{X}_d = F_d'',$$

since $\mathfrak{n}_\mathfrak{p} \cdot v \subset F'_d$. So

$$F''_d = F'_d + \left(\mathfrak{n}_\mathfrak{p} \Subset \bigoplus_{i \in I} \mathfrak{sl}(X_i, Y_i)\right) \cdot v = F'_d + \bigoplus_{i \in I} \langle v, Y_i \rangle X_i$$

$$= \begin{cases} F'_d \oplus X_i & \text{if } d = \kappa(i) \text{ for some } i \in I \\ F'_d & \text{if } \kappa(i) \neq d \text{ for all } i \in I, \end{cases}$$

since $\kappa(i) < d$ implies $X_i \subset F'_d$, while $\kappa(i) > d$ implies $\langle v, Y_i \rangle = 0$. As $F'_d \subsetneqq F''_d$, we conclude that $d = \kappa(i)$ for some $i \in I$. Hence κ is a bijection from I to D. Since we have shown that $F''_d = F'_d \oplus X_{\kappa^{-1}(d)}$ for all $d \in D$, we are done. $\qquad\square$

Here is an example notably different from the finite-dimensional case. Let V and V_* be vector spaces with bases $\{v\} \cup \{v_i \mid i \in \mathbb{Z}_{>0}\}$ and $\{v_i^* \mid i \in \mathbb{Z}_{>0}\}$, pairing according to the rules $\langle v_i, v_j^* \rangle = \delta_{ij}$ and

$$\langle v, v_j^* \rangle = 1 \text{ for all j.}$$

We will find all self-normalizing parabolic subalgebras of $\mathfrak{gl}(V, V_*)$ with Levi component $\mathfrak{sl}(X_1, Y_1) \oplus \mathfrak{sl}(X_2, Y_2)$, where

$$X_1 := \text{Span}\{v_{2i-1} \mid i \in \mathbb{Z}_{>0}\} \qquad Y_1 := \text{Span}\{v_{2i-1}^* \mid i \in \mathbb{Z}_{>0}\}$$
$$X_2 := \text{Span}\{v_{2i} \mid i \in \mathbb{Z}_{>0}\} \qquad Y_2 := \text{Span}\{v_{2i}^* \mid i \in \mathbb{Z}_{>0}\}.$$

By the above theorem, this is equivalent to finding all taut couples $\mathfrak{F}$, $\mathfrak{G}$ so that the given subspaces provide vector space complements for the pairs in D.

Since $Y_1 \oplus Y_2 = V_*$, the semiclosed generalized flag $\mathfrak{G}$ in V_* must be either $0 \subset Y_1 \subset V_*$ or $0 \subset Y_2 \subset V_*$. Then $\mathfrak{F}$ must be a refinement of the generalized flag $\{G^\perp \mid G \in \mathfrak{G}\}$; that is, $\mathfrak{F}$ is a refinement of $0 \subset X_2 \subset V$ or $0 \subset X_1 \subset V$. In either case, it is necessary to insert $X_1 \oplus X_2$ into $\mathfrak{F}$ in order to have the given subspaces X_1 and X_2 be vector space complements for the pairs in D. Thus the following is a complete list of the taut couples as desired:

$$0 \subset X_2 \subset X_1 \oplus X_2 \subset V$$
$$V_* \supset Y_1 \supset 0$$

and

$$0 \subset X_1 \subset X_1 \oplus X_2 \subset V$$
$$V_* \supset Y_2 \supset 0.$$

Note that the subspace $X_1 \oplus X_2$ appearing in both of the above taut couples has codimension 1 in V; nevertheless $(X_1 \oplus X_2)^{\perp\perp} = V$.

Let $\mathfrak{l} := \bigoplus_{i \in I} \mathfrak{s}_i$ for some commuting standard special linear subalgebras of $\mathfrak{g}$. By definition $\mathfrak{s}_i = \mathfrak{sl}(X_i, Y_i)$ for some subspaces $X_i \subset V$ and $Y_i \subset V_*$. The maximal trivial $\mathfrak{l}$-submodule of V is $(\bigoplus_{i \in I} Y_i)^\perp$, and $\mathfrak{l} \cdot V = \bigoplus_{i \in I} X_i$. Therefore the socle of V as an $\mathfrak{l}$-module (that is, the direct sum of all simple $\mathfrak{l}$-submodules of V) is

$$\bigoplus_{i \in I} X_i \oplus \left(\bigoplus_{i \in I} Y_i\right)^\perp,$$

and each nontrivial simple module in the socle of V has multiplicity 1. This shows that each subspace X_i for $i \in I$ is determined by $\mathfrak{l}$, and one can recover similarly the subspaces Y_i as the nontrivial simple submodules of V_*. This enables us to strengthen the above theorem as follows. Again, let $\mathfrak{p}$ be a parabolic subalgebra of $\mathfrak{g}$, with the associated taut couple $\mathfrak{F}$, $\mathfrak{G}$. The map which takes the subspaces

X_γ, Y_γ to the subalgebra $\bigoplus_{\gamma \in D} \mathfrak{sl}(X_\gamma, Y_\gamma)$ is a bijection from the sets of subspaces $X_\gamma \subset V$ and $Y_\gamma \subset V_*$ for $\gamma \in D$ such that (3.1) holds and $\langle X_\gamma, Y_\eta \rangle = 0$ for $\gamma \neq \eta$ to the Levi components of $\mathfrak{p}$.

Yet another restatement of Theorem 3.5 is in order. Let $\mathfrak{p}$ be a parabolic subalgebra of $\mathfrak{g}$, with the associated taut couple $\mathfrak{F}$, $\mathfrak{G}$. Then $\mathfrak{l} \subset \mathfrak{p}$ is a Levi component of $\mathfrak{p}$ if and only if the following conditions hold:

- The $\mathfrak{l}$-modules F_γ''/F_γ' and G_γ''/G_γ' are simple for all $\gamma \in D$;
- $\mathfrak{l} \cong \bigoplus_i \mathfrak{sl}_{n_i}$ for some $n_i \in \mathbb{Z}_{\geq 2} \cup \{\infty\}$;
- There is a unique nontrivial simple $\mathfrak{s}$-submodule of V for each simple direct summand $\mathfrak{s}$ of $\mathfrak{l}$;
- For each finite-dimensional simple direct summand $\mathfrak{s}$ of $\mathfrak{l}$, the nontrivial simple $\mathfrak{s}$-submodule of V is isomorphic to the natural or conatural $\mathfrak{s}$-module.

The last condition is automatic for the infinite-dimensional simple direct summands of $\mathfrak{l}$, as shown in [**DiP2**].

COROLLARY 3.6. *Not every maximal locally semisimple subalgebra of a finitary Lie algebra is a Levi component.*

PROOF. We will show that the finitary Lie algebra in question can be chosen to be a parabolic subalgebra of $\mathfrak{gl}_\infty$.

Let V and V_* be vector spaces with bases $\{\ldots, v_{-2}, v_{-1}, v_0, v_1, v_2, \ldots\}$ and $\{\ldots, v_{-2}^*, v_{-1}^*, v_0^*, v_1^*, v_2^*, \ldots\}$, respectively, and let the pairing be such that these are dual bases. Let

$$X_1 := \mathrm{Span}\{v_1, v_2, v_3, \ldots\} \qquad Y_1 := \mathrm{Span}\{v_0^* + v_1*, v_0^* + v_2^*, v_0^* + v_3^*, \ldots\}$$
$$X_2 := \mathrm{Span}\{v_{-1}, v_{-2}, v_3, \ldots\} \qquad Y_2 := \mathrm{Span}\{v_{-1}^*, v_{-2}^*, v_3^*, \ldots\}.$$

Let $\mathfrak{p}$ be the stabilizer of X_1. Then $\mathfrak{p}$ is the self-normalizing parabolic subalgebra of $\mathfrak{gl}(V, V_*)$ corresponding to the taut couple

$$0 \subset X_1 \subset V$$
$$V_* \supset (X_1)^\perp \supset 0.$$

Theorem 3.5 gives that $\mathfrak{sl}(X_1, Y_1) \oplus \mathfrak{sl}(X_2, Y_2)$ is not a Levi component of $\mathfrak{p}$, since

$$X_1 \oplus X_2 \subsetneq V.$$

We claim that $\mathfrak{sl}(X_1, Y_1) \oplus \mathfrak{sl}(X_2, Y_2)$ is nevertheless a maximal locally semisimple subalgebra of $\mathfrak{p}$.

To see this, let $\mathfrak{k}$ be any maximal locally semisimple subalgebra of $\mathfrak{p}$ containing $\mathfrak{sl}(X_1, Y_1) \oplus \mathfrak{sl}(X_2, Y_2)$. As one may check from the proof, Lemma 3.1 can be applied not just to Levi components but also to maximal locally semisimple subalgebras of $\mathfrak{p}$, and it implies that $\mathfrak{k}$ is a direct sum of standard special linear subalgebras. Thus $\mathfrak{k}$ must have direct summands $\mathfrak{k}_1$ and $\mathfrak{k}_2$ with $\mathfrak{sl}(X_i, Y_i) \subset \mathfrak{k}_i = \mathfrak{sl}(\tilde{X}_i, \tilde{Y}_i)$ for some subspaces $\tilde{X}_i \subset V$ and $\tilde{Y}_i \subset V_*$. Then one has necessarily that $X_i \subset \tilde{X}_i$ and $Y_i \subset \tilde{Y}_i$. Note that $\mathfrak{k}_1$ and $\mathfrak{k}_2$ must be distinct, as $\mathfrak{sl}(X_1 \oplus X_2, Y_1 \oplus Y_2)$ is not contained in $\mathfrak{p}$. Hence $[\mathfrak{k}_1, \mathfrak{k}_2] = 0$, which implies that $\langle \tilde{X}_1, \tilde{Y}_2 \rangle = \langle \tilde{X}_2, \tilde{Y}_1 \rangle = 0$. Consider that $X_2 \subset \tilde{X}_2 \subset (Y_1)^\perp = X_2$. Hence $\tilde{X}_2 = X_2$. Now

$$Y_2 \subset \tilde{Y}_2 \subset (X_1)^\perp = Y_2 \oplus \mathbb{C}v_0^*.$$

Since the restriction of the pairing on $V \times V_*$ to $\tilde{X}_2 \times \tilde{Y}_2$ is nondegenerate, we conclude that $\tilde{Y}_2 = Y_2$; that is, $\mathfrak{k}_2 = \mathfrak{sl}(X_2, Y_2)$.

Similarly, one sees that $X_1 \subsetneq \tilde{X}_1$ implies $\tilde{X}_1 = (Y_2)^\perp = X_1 \oplus \mathbb{C}v_0$, while $Y_1 \subsetneq \tilde{Y}_1$ implies $\tilde{Y}_1 = (X_2)^\perp = Y_1 \oplus \mathbb{C}v_0^*$. The only larger potential direct summand $\mathfrak{k}_1$ to consider based on nondegeneracy considerations is $\mathfrak{sl}(X_1 \oplus \mathbb{C}v_0, Y_1 \oplus \mathbb{C}v_0^*)$, but this does not stabilize X_1. Thus $\mathfrak{k} = \mathfrak{sl}(X_1, Y_1) \oplus \mathfrak{sl}(X_2, Y_2)$, and we have shown that $\mathfrak{sl}(X_1, Y_1) \oplus \mathfrak{sl}(X_2, Y_2)$ is a maximal locally semisimple subalgebra of $\mathfrak{p}$. □

4. Some further corollaries

We continue to take $\mathfrak{g}$ to be $\mathfrak{sl}(V, V_*)$ or $\mathfrak{gl}(V, V_*)$. Of the corollaries we present to Theorem 3.5, the first two in particular are useful when computing explicitly all parabolic subalgebras with a given Levi component.

COROLLARY 4.1. Fix a subalgebra $\mathfrak{l} = \bigoplus_{i \in I} \mathfrak{sl}(X_i, Y_i)$, where $X_i \subset V$ and $Y_i \subset V_*$. Assume that $\dim X_i \geq 2$ for all $i \in I$, and that I is an ordered set. Let $U_i \subset V$ be subspaces such that $U_i \oplus X_i \subset U_j$ for all $i < j$ and

$$U_i = ((U_i \oplus X_i)^\perp \oplus Y_i)^\perp$$

for each $i \in I$.

Let $\mathfrak{F}$ be a semiclosed generalized flag maximal among the semiclosed generalized flags in V in which $U_i \subset U_i \oplus X_i$ is an immediate predecessor-successor pair for all $i \in I$. Then there is a unique semiclosed generalized flag $\mathfrak{G}$ in V_* such that $\mathfrak{F}, \mathfrak{G}$ form a taut couple and $\mathfrak{l}$ is a Levi component of the self-normalizing parabolic subalgebra $\mathrm{St}_{\mathfrak{F}} \cap \mathrm{St}_{\mathfrak{G}}$.

PROOF. Let $\mathfrak{F}$ be maximal among the semiclosed generalized flags in V with immediate predecessor-successor pairs $U_i \subset U_i \oplus X_i$ for all $i \in I$. Let $\mathfrak{G}$ be any semiclosed generalized flag such that $\mathfrak{F}, \mathfrak{G}$ form a taut couple, and $\mathfrak{l}$ is a Levi component of $\mathrm{St}_{\mathfrak{F}} \cap \mathrm{St}_{\mathfrak{G}}$. By Theorem 3.5, for each $i \in I$ there is an immediate predecessor-successor pair $(U_i \oplus X_i)^\perp \subset (U_i \oplus X_i)^\perp \oplus Y_i$ in $\mathfrak{G}$. By [**DP**, Proposition 3.3] each closed subspace in $\mathfrak{G}$ is the union of some set of subspaces of the form $F^\perp$ for $F \in \mathfrak{F}$, and each closed subspace in $\mathfrak{F}$ is the union of some set of subspaces of the form $G^\perp$ for $G \in \mathfrak{G}$. Therefore $\mathfrak{G}$ if it exists must have the same stabilizer as the set

$$\{F^\perp \mid F \in \mathfrak{F}\} \cup \{(U_i \oplus X_i)^\perp \oplus Y_i \mid i \in I\}.$$

We show that the above set is totally ordered by inclusion, and $(U_i \oplus X_i)^\perp$ is the immediate predecessor of $(U_i \oplus X_i)^\perp \oplus Y_i$ for each $i \in I$. Indeed, consider that for each i there are no subspaces of the form $F^\perp$ for $F \in \mathfrak{F}$ properly between $(U_i \oplus X_i)^\perp$ and $(U_i)^\perp$, since $U_i \subset U_i \oplus X_i$ is an immediate predecessor-successor pair in $\mathfrak{F}$. Furthermore, one has $(U_i \oplus X_i)^\perp \subset (U_i \oplus X_i)^\perp \oplus Y_i \subset (U_i)^\perp$ because of the identity $U_i = ((U_i \oplus X_i)^\perp \oplus Y_i)^\perp$. Proposition 2.2 (2) gives the existence of a unique semiclosed generalized flag $\mathfrak{G}$ with the same stabilizer as the above set. Then $\mathfrak{F}, \mathfrak{G}$ form a taut couple by construction, and Theorem 3.5 implies that $\mathfrak{l}$ is a Levi component of the self-normalizing parabolic subalgebra $\mathrm{St}_{\mathfrak{F}} \cap \mathrm{St}_{\mathfrak{G}}$. □

The above corollary enables us to determine a self-normalizing parabolic subalgebra with a prescribed Levi component using only subspaces of V. The corollary below shows that any self-normalizing parabolic subalgebra of $\mathfrak{g}$ can be so described.

COROLLARY 4.2. Let $\mathfrak{p}$ be a parabolic subalgebra of $\mathfrak{g}$, with the associated taut couple $\mathfrak{F}$, $\mathfrak{G}$. Suppose the subalgebra $\bigoplus_{i \in I} \mathfrak{sl}(X_i, Y_i)$ is a Levi component of $\mathfrak{p}$, where $X_i \subset V$ and $Y_i \subset V_*$. Assume $\dim X_i \geq 2$ for all $i \in I$.

Then there exist subspaces $U_i \subset V$ for $i \in I$ with

$$U_i = ((U_i \oplus X_i)^\perp \oplus Y_i)^\perp$$

such that $\mathfrak{F}$ is maximal among the semiclosed generalized flags having immediate predecessor-successor pairs $U_i \subset U_i \oplus X_i$ for all $i \in I$.

PROOF. Fix $i \in I$. By Theorem 3.5, there exists some $\gamma \in D$ such that $F_\gamma'' = F_\gamma' \oplus X_i$ and $G_\gamma'' = G_\gamma' \oplus Y_i$. Then take $U_i := F_\gamma'$. Because $F_\gamma' = (G_\gamma'')^\perp$, and $G_\gamma' = (F_\gamma'')^\perp$, we see that $((U_i \oplus X_i)^\perp \oplus Y_i)^\perp = ((F_\gamma'')^\perp \oplus Y_i)^\perp = (G_\gamma' \oplus Y_i)^\perp = (G_\gamma'')^\perp = F_\gamma' = U_i$.

The maximality of $\mathfrak{F}$ as stated follows from Theorem 3.5, since all other immediate predecessor-successor pairs of $\mathfrak{F}$ (i.e. for $\alpha \notin D$) have $\dim F_\alpha''/F_\alpha' = 1$ or $(F_\alpha')^{\perp\perp} = F_\alpha''$, and in either case admit no further refinement. $\square$

The subspaces U_i of the above two corollaries already made an appearance in the proof of Theorem 3.4. There we constructed a parabolic subalgebra $\mathrm{St}_\mathfrak{F} \cap \mathrm{St}_\mathfrak{G}$ as in Corollary 4.1 by choosing $U_i := ((\bigoplus_{k \leq i} X_k)^\perp \oplus Y_i)^\perp$ for all $i \in I$. If $\mathfrak{p}$ is any such parabolic subalgebra of which, using the notation of Corollary 4.1, the subalgebra $\bigoplus_{i \in I} \mathfrak{sl}(X_i, Y_i)$ is a Levi component, then one may check that under the induced order on I

$$((\bigoplus_{k \leq i} X_k)^\perp \oplus Y_i)^\perp \subset U_i \subset (\bigoplus_{k \geq i} Y_k)^\perp$$

for each $i \in I$. We claim that it is also possible in general to construct such parabolic subalgebras by taking $U_i := (\bigoplus_{k \geq i} Y_k)^\perp$ for all $i \in I$. Indeed, in this case one may verify the property $U_i = ((U_i \oplus X_i)^\perp \oplus Y_i)^\perp$ using Lemma 3.2 (taking $T = (\bigoplus_{k > i} Y_k)^\perp$, with $X = X_i$ and $Y = Y_i$). That is, the largest possible subspaces U_i also satisfy the hypotheses of Corollary 4.1.

Let $\mathfrak{l}$ be a subalgebra of $\mathfrak{g}$. The next corollary shows that the parabolic subalgebras of which $\mathfrak{l}$ is a Levi component can be distinguished using only a single semiclosed generalized flag in V.

COROLLARY 4.3. Let $\mathfrak{p}_1$ and $\mathfrak{p}_2$ be parabolic subalgebras of $\mathfrak{g}$, with the associated taut couples $\mathfrak{F}_1$, $\mathfrak{G}_1$ and $\mathfrak{F}_2$, $\mathfrak{G}_2$, respectively. Suppose $\mathfrak{p}_1$ and $\mathfrak{p}_2$ have a Levi component in common. Then $\mathfrak{F}_1 = \mathfrak{F}_2$ implies $\mathfrak{G}_1 = \mathfrak{G}_2$.

PROOF. By Theorem 3.5, the common Levi component of $\mathfrak{p}_1$ and $\mathfrak{p}_2$ is of the form $\bigoplus_{\gamma \in D} \mathfrak{sl}(X_\gamma, Y_\gamma)$ for some subspaces $X_\gamma \subset V$ and $Y_\gamma \subset V_*$. As seen in Corollary 4.2, the generalized flag $\mathfrak{F}_1 = \mathfrak{F}_2$ is maximal among the semiclosed generalized flags in V having $F_\gamma' \subset F_\gamma' \oplus X_\gamma$ as immediate predecessor-successor pairs for all $\gamma \in D$. Evidently $F_\gamma' \oplus X_\gamma \subset F_\eta'$ for all $\gamma < \eta$ because $\mathfrak{F}$ is a generalized flag; the property $F_\gamma' = ((F_\gamma' \oplus X_\gamma)^\perp \oplus Y_\gamma)^\perp$ was shown in the proof of Corollary 4.2. Thus the uniqueness claim of Corollary 4.1 yields $\mathfrak{G}_1 = \mathfrak{G}_2$. $\square$

Consider the special case that a parabolic subalgebra $\mathfrak{p}$ of $\mathfrak{g}$ has 0 as a maximal locally semisimple subalgebra. Then $\mathfrak{p}$ is a *Borel* subalgebra (that is, a maximal locally solvable subalgebra) of $\mathfrak{g}$. Corollary 4.3 implies in this case that the associated taut couple of $\mathfrak{p}$ is determined by $\mathfrak{F}$, the part of the taut couple in V. Since maximal locally solvable subalgebras are minimal parabolic subalgebras, trace conditions are not relevant in this case. Hence a Borel subalgebra of $\mathfrak{g}$ is determined by a single (maximal) semiclosed generalized flag in V, as was proved in [**DiP**].

5. Counting parabolic subalgebras with given Levi component

In this section we address the question of how many parabolic subalgebras of $\mathfrak{g}$ have a given locally semisimple subalgebra $\mathfrak{l}$ as a Levi component. If $\mathfrak{l}$ is a Levi component of a parabolic subalgebra $\mathfrak{p}$ of a finitary Lie algebra, then $\mathfrak{l}$ is also a Levi component of $\mathfrak{p}_+$. Recall that $\mathfrak{p}_+$ is a self-normalizing parabolic subalgebra, and $\mathfrak{p}$ is defined by trace conditions on $\mathfrak{p}_+$. Therefore we will usually consider first the self-normalizing parabolic subalgebras of $\mathfrak{g}$ of which $\mathfrak{l}$ is a Levi component.

Fix for $i \in I$ commuting standard special linear subalgebras $\mathfrak{s}_i \subset \mathfrak{g}$. When $|I| = n < \infty$, Theorem 3.4 implies that there are at least $n!$ self-normalizing parabolic subalgebras of $\mathfrak{g}$ having $\bigoplus_i \mathfrak{s}_i$ as a Levi component; similarly there are uncountably many such parabolic subalgebras when I is a countable set. With Theorem 5.1 we find criteria for this number to be finite, and we also give an upper bound for this number when it is finite.

THEOREM 5.1. *Fix* $\mathfrak{l} = \mathfrak{sl}(X_1, Y_1) \oplus \cdots \oplus \mathfrak{sl}(X_n, Y_n) \subset \mathfrak{g}$ *for some subspaces* $X_i \subset V$ *and* $Y_i \subset V_*$ *with* $\dim X_i \geq 2$ *for all* i. *The number of self-normalizing parabolic subalgebras of* $\mathfrak{gl}(V, V_*)$ *with* $\mathfrak{l}$ *as a Levi component is finite if and only if*

$$\dim(\bigoplus_{i \notin J} Y_i)^{\perp}/(\bigoplus_{j \in J} X_j)^{\perp\perp} \leq 1$$

for all subsets $J \subset \{1, 2, \ldots, n\}$. *When finite, this number is at most* $3 \cdot 2^{n-2} \cdot n!$ *for* $n \geq 2$, *and at most* 2 *for* $n = 1$; *it is uncountable when infinite.*

PROOF. Suppose first that there exists a subset $J \subset \{1, 2, \ldots, n\}$ for which

$$\dim(\bigoplus_{i \notin J} Y_i)^{\perp}/(\bigoplus_{j \in J} X_j)^{\perp\perp} > 1.$$

Without loss of generality, suppose $J = \{1, 2, \ldots, k\}$. We define

$$U_j = \begin{cases} (X_1 \oplus X_2 \oplus \cdots \oplus X_j)^{\perp\perp} \cap Y_j^{\perp} & \text{if } 1 \leq j \leq k \\ (Y_j \oplus Y_{j+1} \oplus \ldots \oplus Y_n)^{\perp} & \text{if } k < j \leq n. \end{cases}$$

One may check that $U_1 \subset U_2 \subset \cdots \subset U_n$. As described in the paragraph after Corollary 4.2, one has $U_j = ((U_j \oplus X_j)^{\perp} \oplus Y_j)^{\perp}$ for $j = 1, \ldots n$.

Let $\mathfrak{F}_0$ be the semiclosed generalized flag

$$0 \subset \cdots \subset U_i \subset U_i \oplus X_i \subset (U_i \oplus X_i)^{\perp\perp} \subset U_{i+1} \subset \cdots \subset V.$$

By Corollary 4.1, any semiclosed generalized flag $\mathfrak{F}$ maximal among the refinements of $\mathfrak{F}_0$ retaining the immediacy of the pairs $U_i \subset U_i \oplus X_i$ for $i = 1, \ldots n$ determines a self-normalizing parabolic subalgebra of which $\mathfrak{l}$ is a Levi component.

Consider the following portion of $\mathfrak{F}_0$:

$$U_k \subset U_k \oplus X_k \subset (U_k \oplus X_k)^{\perp\perp} \subset U_{k+1}.$$

Since $X_1 \oplus X_2 \oplus \cdots \oplus X_{k-1} \subset U_k \subset (X_1 \oplus X_2 \oplus \cdots \oplus X_k)^{\perp\perp}$, we see that

$$(U_k \oplus X_k)^{\perp\perp} = (X_1 \oplus X_2 \oplus \cdots \oplus X_k)^{\perp\perp}.$$

We have assumed that $(X_1 \oplus X_2 \oplus \cdots \oplus X_k)^{\perp\perp}$ has codimension at least 2 in $(Y_{k+1} \oplus Y_{k+2} \oplus \cdots \oplus Y_n)^{\perp} = U_{k+1}$. Therefore there are uncountably many closed subspaces between them, and any such closed subspace can appear in a refinement $\mathfrak{F}$, $\mathfrak{G}$ as described above. Since different taut couples yield different self-normalizing parabolic subalgebras [**DP**, Proposition 3.8], we conclude that there are uncountably many self-normalizing parabolic subalgebras with $\mathfrak{l}$ as a Levi component.

Now suppose that

$$\dim(\bigoplus_{i \notin J} Y_i)^{\perp}/(\bigoplus_{j \in J} X_j)^{\perp\perp} \leq 1$$

for all subsets $J \subset \{1, 2, \ldots, n\}$. We show first that there are at most $2^n \cdot n!$ self-normalizing parabolic subalgebra of $\mathfrak{gl}(V, V_*)$ of which $\mathfrak{l}$ is a Levi component. Fix such a parabolic subalgebra $\mathfrak{p}$, and denote the associate taut couple by $\mathfrak{F}$, $\mathfrak{G}$. By Corollary 4.2, there exist subspaces U_i for $i = 1, \ldots n$ totally ordered by inclusion with the properties listed there. Without loss of generality, let us assume that $U_1 \subset U_2 \subset \cdots \subset U_n$. This reindexing produces the factor of $n!$.

Then $\mathfrak{F}$ is related to the semiclosed generalized flag

$$0 \subset U_1 \subset \cdots \subset U_i \subset U_i \oplus X_i \subset (U_i \oplus X_i)^{\perp\perp} \subset U_{i+1} \subset \cdots \subset (U_n \oplus X_n)^{\perp\perp} \subset V,$$

by maximally refining those pairs of the form $(U_i \oplus X_i)^{\perp\perp} \subset U_{i+1}$ for $i = 0, 1, \ldots n$, where we use the notation $U_0 = X_0 = 0$ and $U_{n+1} = V$.

For $i = 0, \ldots, n$, we have

$$(X_1 \oplus \cdots \oplus X_i)^{\perp\perp} \subset (U_i \oplus X_i)^{\perp\perp} \subset U_{i+1} \subset (Y_{i+1} \oplus \cdots \oplus Y_n)^{\perp}.$$

By hypothesis $\dim(Y_{i+1} \oplus \cdots \oplus Y_n)^{\perp}/(X_1 \oplus \cdots \oplus X_i)^{\perp\perp} \leq 1$ for each i, hence there are at most two possibilities for each U_i. Thus there at most 2^n possible choices of $U_1, \ldots, U_n$. Furthermore, since the pairs $(U_i \oplus X_i)^{\perp\perp} \subset U_{i+1}$ for each i have codimension at most 1, no further refinement of them is possible. Hence $\mathfrak{F}$ equals the semiclosed generalized flag given above, and by Corollary 4.1 $\mathfrak{p}$ is determined by the choice of $U_1, \ldots, U_n$. This shows that the number of self-normalizing parabolic subalgebras with $\mathfrak{l}$ as a Levi component is at most $2^n \cdot n!$. This completes the proof of the if and only if statement, as well as the estimate that the number is question is uncountable when infinite and at most $2^n \cdot n!$ when finite.

Now assume that $n \geq 2$, and that the number of self-normalizing parabolic subalgebras with $\mathfrak{l}$ as a Levi component is finite. To prove the stated upper bound of $3 \cdot 2^{n-2} \cdot n!$, we show that there are at most three possible combinations of U_1 and U_2.

Assume therefore that

$$(X_1)^{\perp\perp} \cap (Y_1)^{\perp} \subsetneq (Y_1 \oplus Y_2 \oplus \cdots \oplus Y_n)^{\perp} \subset (X_1 \oplus X_2)^{\perp\perp} \cap (Y_2)^{\perp}.$$

(If this assumption does not hold, it is clear that there are at most three possibilities for U_1 and U_2.) It suffices to show that

$$(X_1 \oplus X_2)^{\perp\perp} \cap (Y_2)^{\perp} = (Y_2 \oplus \cdots \oplus Y_n)^{\perp}.$$

We have already shown that $\dim(Y_1 \oplus Y_2 \oplus \cdots \oplus Y_n)^{\perp} \leq 1$. The condition

$$(X_1)^{\perp\perp} \cap (Y_1)^{\perp} \subsetneq (Y_1 \oplus Y_2 \oplus \cdots \oplus Y_n)^{\perp}$$

therefore implies that $(X_1)^{\perp\perp} \cap (Y_1)^{\perp} = 0$ and $\dim(Y_1 \oplus Y_2 \oplus \cdots \oplus Y_n)^{\perp} = 1$.

We have also assumed that $(Y_1 \oplus Y_2 \oplus \cdots \oplus Y_n)^\perp \subset (X_1 \oplus X_2)^{\perp\perp} \cap (Y_2)^\perp$, so indeed

$$(X_1)^{\perp\perp} + (Y_1 \oplus Y_2 \oplus \cdots \oplus Y_n)^\perp \subset (X_1 \oplus X_2)^{\perp\perp} \cap (Y_2)^\perp.$$

Since $(X_1)^{\perp\perp} \cap (Y_1 \oplus Y_2 \oplus \cdots \oplus Y_n)^\perp \subset (X_1)^{\perp\perp} \cap (Y_1)^\perp = 0$, we have the direct sum

$$(X_1)^{\perp\perp} \oplus (Y_1 \oplus Y_2 \oplus \cdots \oplus Y_n)^\perp \subset (X_1 \oplus X_2)^{\perp\perp} \cap (Y_2)^\perp \subset (Y_2 \oplus \cdots \oplus Y_n)^\perp.$$

We have already proved that the codimension of $(X_1)^{\perp\perp}$ in $(Y_2 \oplus \cdots \oplus Y_n)^\perp$ is at most 1; hence

$$(X_1)^{\perp\perp} \oplus (Y_1 \oplus Y_2 \oplus \cdots \oplus Y_n)^\perp = (X_1 \oplus X_2)^{\perp\perp} \cap (Y_2)^\perp = (Y_2 \oplus \cdots \oplus Y_n)^\perp,$$

and we are done.   □

Let us consider the finite numbers obtained as the number of self-normalizing parabolic subalgebras with a given subalgebra as a Levi component, subject to the restriction that the given subalgebra has n simple direct summands. In the case $n \geq 2$, Theorem 5.1 says that the number of self-normalizing parabolic subalgebras with Levi component $\mathfrak{s}_1 \oplus \cdots \oplus \mathfrak{s}_n$ is at most $3 \cdot 2^{n-2} \cdot n!$ if it is finite. (One can imagine that this upper bound is typically not sharp.) Considerations completely analogous to the finite-dimensional case show that the maximum of this set of finite numbers is at least $(n+1)!$. Nevertheless, the very last example in this section shows that, unlike in the finite-dimensional case, $(n+1)!$ is not an upper bound when $n = 5$.

As an illustration of Theorem 5.1, one might ask how many parabolic subalgebras of $\mathfrak{gl}(V, V_*)$ have Levi component $\mathfrak{sl}(X_1, Y_1) \oplus \mathfrak{sl}(X_2, Y_2)$, where

$$X_1 := \mathrm{Span}\{v_1 + v_{2i-1} \mid i \geq 2\} \qquad Y_1 := \mathrm{Span}\{v_{2i-1}^* \mid i \geq 2\}$$
$$X_2 := \mathrm{Span}\{v_1 + v_{2i} \mid i \geq 2\} \qquad Y_2 := \mathrm{Span}\{v_2^* + v_{2i}^* \mid i \geq 2\},$$

and V and V_* are vector spaces with dual bases $\{v_i \mid i \in \mathbb{Z}_{>0}\}$ and $\{v_i^* \mid i \in \mathbb{Z}_{>0}\}$. To answer the question, we compute the four quotient spaces in the hypotheses of Theorem 5.1:

- $(Y_1 \oplus Y_2)^\perp = \mathbb{C}v_1$,
- $(Y_1)^\perp / (X_2)^{\perp\perp} = \mathbb{C}v_1 \oplus \mathrm{Span}\{v_{2i} \mid i \geq 1\} / \mathbb{C}v_1 \oplus \mathrm{Span}\{v_{2i} \mid i \geq 2\}$,
- $(Y_2)^\perp / (X_1)^{\perp\perp} = \mathrm{Span}\{v_{2i-1} \mid i \geq 1\} / \mathrm{Span}\{v_{2i-1} \mid i \geq 1\}$,
- $V / (X_1 \oplus X_2)^{\perp\perp} = V / \mathrm{Span}\{v_i \mid i \neq 2\}$.

Since all the above have dimension no greater than 1, Theorem 5.1 implies that only a finite number of self-normalizing parabolic subalgebras of $\mathfrak{gl}(V, V_*)$ have $\mathfrak{sl}(X_1, Y_1) \oplus \mathfrak{sl}(X_2, Y_2)$ as a Levi component. Indeed, one may check that there are precisely three such self-normalizing parabolic subalgebras. Explicitly, they are the stabilizers of the following three taut couples. The order $1 < 2$ gives the taut couple

$$0 \subset \mathbb{C}v_1 \subset \mathbb{C}v_1 \oplus X_1 \subset \mathbb{C}v_1 \oplus X_1 \oplus X_2 \subset V$$
$$V_* \supset (v_1)^\perp \supset Y_2 \oplus \mathbb{C}v_2^* \supset \mathbb{C}v_2^* \supset 0,$$

while the other order gives the two taut couples

$$0 \subset \mathbb{C}v_1 \subset \mathbb{C}v_1 \oplus X_2 \subset \mathbb{C}v_1 \oplus X_2 \oplus X_1 \subset V$$
$$V_* \supset (v_1)^\perp \supset \mathbb{C}v_2^* \oplus Y_1 \supset \mathbb{C}v_2^* \supset 0$$

and

$$0 \subset \mathbb{C}v_1 \subset \mathbb{C}v_1 \oplus X_2 \subset \mathbb{C}v_1 \oplus X_2 \oplus \mathbb{C}v_2 \subset V$$
$$V_* \supset (v_1)^\perp \supset Y_1 \oplus \mathbb{C}v_2^* \supset Y_1 \supset 0.$$

Corollary 5.2 addresses the special case $n = 1$ of Theorem 5.1.

COROLLARY 5.2. Fix subspaces $X \subset V$ and $Y \subset V_*$ such that $\langle \cdot, \cdot \rangle|_{X \times Y}$ is nondegenerate. The number of self-normalizing parabolic subalgebras of $\mathfrak{g}$ with $\mathfrak{sl}(X, Y)$ as a Levi component is finite if and only if

$$\dim X^\perp \le 1 \text{ and } \dim Y^\perp \le 1.$$

When it is finite, this number is 1 if $\langle Y^\perp, X^\perp \rangle = 0$ and 2 if $\langle Y^\perp, X^\perp \rangle \ne 0$.

PROOF. Theorem 5.1 implies that the number of such parabolic subalgebras is finite if and only if $\dim Y^\perp/0 \le 1$ and $\dim V/X^{\perp\perp} \le 1$. Since $\dim V/X^{\perp\perp} = \dim X^\perp$, the if and only if statement is clear.

Now suppose $\dim X^\perp \le 1$ and $\dim Y^\perp \le 1$. No further refinement of the semiclosed generalized flag $0 \subset U \subset U \oplus X \subset V$ is possible for any subspace $U \subset V$ such that $U = ((U \oplus X)^\perp \oplus Y)^\perp$. Therefore by Corollaries 4.1 and 4.2, the parabolic subalgebras in question are in bijection with the subspaces $U \subset V$ such that $U = ((U \oplus X)^\perp \oplus Y)^\perp$.

Let $U \subset V$ be such that $U = ((U \oplus X)^\perp \oplus Y)^\perp$. It follows that $X^{\perp\perp} \cap Y^\perp \subset U \subset Y^\perp$. If $\langle Y^\perp, X^\perp \rangle = 0$, then $X^{\perp\perp} \cap Y^\perp = Y^\perp$, so $U = Y^\perp$, i.e. there is exactly one such parabolic subalgebra. If $\langle Y^\perp, X^\perp \rangle \ne 0$, then $X^{\perp\perp} \cap Y^\perp = 0$ and $Y^\perp \ne 0$, so $U = 0$ and $U = Y^\perp$ are the two possibilities. $\square$

The following example gives an illustration of Corollary 5.2. We will find all parabolic subalgebras of $\mathfrak{gl}(V, V_*)$ with the simple subalgebra $\mathfrak{sl}(X, Y)$ as a Levi component, where

$$X := \mathrm{Span}\{v_1 + v_i \mid i \ge 2\} \qquad Y := \mathrm{Span}\{v_i^* \mid i \ge 2\},$$

and $\{v_i \mid i \in \mathbb{Z}_{>0}\}$ and $\{v_i^* \mid i \in \mathbb{Z}_{>0}\}$ are dual bases of the vector spaces V and V_*.

We compute $X^\perp = 0$ and $Y^\perp = \mathbb{C}v_1$. By Corollary 5.2 there is exactly 1 self-normalizing parabolic subalgebra of $\mathfrak{gl}(V, V_*)$ having $\mathfrak{sl}(X, Y)$ as a Levi component. It is the stabilizer of the taut couple

$$0 \subset \quad \mathbb{C}v_1 \quad \subset V$$
$$V_* \supset \quad Y \quad \supset 0,$$

which by a computation is nothing but

$$\mathfrak{p}_+ = \left((v_1^*)^\perp \otimes v_1^*\right) \Subset \left((\mathbb{C}v_1 \otimes v_1^*) \oplus ((v_1^*)^\perp \otimes Y)\right).$$

Thus every parabolic subalgebra $\mathfrak{p}$ of $\mathfrak{gl}(V, V_*)$ with $\mathfrak{sl}(X, Y)$ as a Levi component is defined by trace conditions on the one infinite-dimensional block of a locally reductive part of $\mathfrak{p}_+$. There are precisely two such parabolic subalgebras, namely $\mathfrak{p}_+$ and $\mathfrak{p}_- = \left((v_1^*)^\perp \otimes v_1^*\right) \Subset \left((\mathbb{C}v_1 \otimes v_1^*) \oplus \mathfrak{sl}((v_1^*)^\perp, Y)\right)$.

Similarly, there are precisely two parabolic subalgebras of $\mathfrak{sl}(V, V_*)$ with Levi component $\mathfrak{sl}(X, Y)$. They are the traceless parts of the two parabolic subalgebras in the previous paragraph.

It is not hard to see now that it is impossible to extend $\mathfrak{sl}(X, Y)$ to a locally reductive part of $\mathfrak{p}_+$. Any locally reductive part of $\mathfrak{p}_+$ must be isomorphic to the locally reductive part $(\mathbb{C}v_1 \otimes v_1^*) \oplus ((v_1^*)^\perp \otimes Y)$, hence it must have two commuting blocks. However, the centralizer of $\mathfrak{sl}(X, Y)$ in $\mathfrak{gl}(V, V_*)$ is trivial. Relatedly, one may check that $\mathfrak{gl}(X, Y)$ is a maximal locally reductive subalgebra of $\mathfrak{p}_+$ which is not a locally reductive part of $\mathfrak{p}_+$.

We conclude this section with an example to demonstrate that $(n+1)!$ is not an upper bound for the finite numbers which occur as the number of self-normalizing parabolic subalgebras with prescribed Levi component when $n = 5$. We claim that there are precisely $8 \cdot 5!$ self-normalizing parabolic subalgebras of $\mathfrak{gl}(V, V_*)$ with Levi component $\mathfrak{sl}(X_1, Y_1) \oplus \cdots \oplus \mathfrak{sl}(X_5, Y_5)$, where

$$X_k := \mathrm{Span}\{v_i \mid i = k \bmod 5\} \qquad Y_k := \mathrm{Span}\{v_i^* \mid i = k \bmod 5\},$$

in the following notation. We take V and V_* to be the vector spaces with bases $\{z\} \cup \{w_1, w_2, \ldots, w_{15}\} \cup \{v_i \mid i \in \mathbb{Z}_{>0}\}$ and $\{\tilde{z}\} \cup \{\tilde{w}_1, \tilde{w}_2, \ldots, \tilde{w}_{15}\} \cup \{v_i^* \mid i \in \mathbb{Z}_{>0}\}$. Let $\langle \cdot, \cdot \rangle : V \times V_* \to \mathbb{C}$ be the nondegenerate pairing defined by setting

$$\langle v_i, v_j^* \rangle = \delta_{ij}$$
$$\langle z, \tilde{z} \rangle = 0$$
$$\langle w_k, \tilde{w}_l \rangle = 1$$
$$\langle v_i, \tilde{z} \rangle = \langle z, v_i^* \rangle = 0$$
$$\langle z, \tilde{w}_k \rangle = \langle w_k, \tilde{z} \rangle = \begin{cases} 0 & \text{if } 1 \leq k \leq 10 \\ 1 & \text{if } 11 \leq k \leq 15 \end{cases}$$
$$\langle v_i, \tilde{w}_k \rangle = \langle w_k, v_i^* \rangle = \begin{cases} 1 & \text{if } i \text{ is congruent} \bmod 40 \text{ to an element of } S_k \\ 0 & \text{otherwise} \end{cases}$$

for all $i, j \in \mathbb{Z}_{>0}$ and $k, l \in \{1, \ldots, 15\}$, where the sets S_k are the following:

$S_1 = \{1, 2\}$	$S_5 = \{9, 12\}$	$S_9 = \{15, 18\}$	$S_{13} = \{29, 30, 31, 32\}$
$S_2 = \{3, 6\}$	$S_6 = \{13, 14\}$	$S_{10} = \{19, 20\}$	$S_{14} = \{33, 34, 35, 36\}$
$S_3 = \{7, 8\}$	$S_7 = \{5, 16\}$	$S_{11} = \{21, 22, 23, 24\}$	$S_{15} = \{37, 38, 29, 0\}$.
$S_4 = \{4, 11\}$	$S_8 = \{10, 17\}$	$S_{12} = \{25, 26, 27, 28\}$	

It suffices to prove that there are 8 such parabolic subalgebras inducing the usual order $1 < 2 < 3 < 4 < 5$, due to symmetry. By Corollaries 4.1 and 4.2, these parabolic subalgebras are in correspondence with sets of subspaces $U_1, \ldots, U_5 \subset V$ such that $U_i = ((U_i \oplus X_i)^\perp \oplus Y_i)^\perp$ and $U_i \oplus X_i \subset U_{i+1}$. Indeed, the parabolic subalgebra associated to $U_1, \ldots, U_5$ is the stabilizer of the taut couple

$$0 \subset U_1 \subset U_1 \oplus X_1 \subset (U_1 \oplus X_1)^{\perp\perp} \subset U_2 \subset U_2 \oplus X_2 \subset \cdots$$
$$\cdots \subset (U_4 \oplus X_4)^{\perp\perp} \subset U_5 \subset U_5 \oplus X_5 \subset (U_5 \oplus X_5)^{\perp\perp} \subset V$$
$$V_* \supset (Y_1 \oplus T_1)^{\perp\perp} \supset Y_1 \oplus T_1 \supset T_1 \supset (Y_2 \oplus T_2)^{\perp\perp} \supset \cdots$$
$$\cdots \supset Y_4 \oplus T_4 \supset T_4 \supset (Y_5 \oplus T_5)^{\perp\perp} \supset Y_5 \oplus T_5 \supset T_5 \supset 0,$$

where $T_i := (U_i \oplus X_i)^\perp$.

By the proof of Theorem 5.1, there are at most 2 possibilities for each U_i. Thus each U_i must be either $\left(\left(\bigoplus_{j \leq i} X_j \right)^\perp \oplus Y_i \right)^\perp$ or $\left(\bigoplus_{j \geq i} Y_j \right)^\perp$. This enables us to list all the possibilities:

$$U_1 = 0 \text{ or } \operatorname{Span}\{z\}$$
$$U_2 = X_1 \oplus \operatorname{Span}\{z\}$$
$$U_3 = X_1 \oplus X_2 \oplus \operatorname{Span}\{z\} \text{ or } X_1 \oplus X_2 \oplus \operatorname{Span}\{z, w_1\}$$
$$U_4 = X_1 \oplus X_2 \oplus X_3 \oplus \operatorname{Span}\{z, w_1, w_2, w_3\}$$
$$U_5 = X_1 \oplus X_2 \oplus X_3 \oplus X_4 \oplus \operatorname{Span}\{z, w_1, w_2, w_3, w_4, w_5, w_6\} \text{ or }$$
$$X_1 \oplus X_2 \oplus X_3 \oplus X_4 \oplus \operatorname{Span}\{z, w_1, w_2, w_3, w_4, w_5, w_6, w_{11}\}.$$

Observe that for every combination of choices, the necessary inclusions remain, i.e. $U_i \oplus X_i \subset U_{i+1}$. The listed subspaces all satisfy $U_i = ((U_i \oplus X_i)^\perp \oplus Y_i)^\perp$, as noted immediately after Corollary 4.2. Hence there are exactly 8 self-normalizing parabolic subalgebras as desired, arising from the two possibilities each for U_1, U_3, and U_5.

6. Levi components of parabolic subalgebras of $\mathfrak{so}_\infty$ and $\mathfrak{sp}_\infty$

Assume $\mathfrak{g}$ is $\mathfrak{so}(V)$ or $\mathfrak{sp}(V)$. We omit the proofs, as they are similar to those given above.

THEOREM 6.1. *Let $\mathfrak{l}$ be a subalgebra of $\mathfrak{g}$. There exists a parabolic subalgebra $\mathfrak{p}$ of $\mathfrak{g}$ such that $\mathfrak{l}$ is a Levi component of $\mathfrak{p}$ if and only if $\mathfrak{l}$ is the direct sum of standard special linear subalgebras of $\mathfrak{g}$ and a subalgebra*

$$\mathfrak{k} = \begin{cases} \mathfrak{so}(W) & \text{if } \mathfrak{g} = \mathfrak{so}(V) \\ \mathfrak{sp}(W) & \text{if } \mathfrak{g} = \mathfrak{sp}(V) \end{cases}$$

for some subspace $W \subset V$ to which the restriction of the bilinear form on V is nondegenerate.

Moreover, given a subalgebra $\mathfrak{l}$ for which such parabolic subalgebras exist, one exists that induces an arbitrary order on the standard special linear direct summands of $\mathfrak{l}$.

THEOREM 6.2. *Let $\mathfrak{p}$ be a parabolic subalgebra of $\mathfrak{so}(V)$, with an associated self-taut generalized flag $\mathfrak{F}$. Let F denote the union of all isotropic subspaces F''_α for $\alpha \in A$, and let G denote the intersection of all coisotropic subspaces F'_α for $\alpha \in A$.*

Then $\mathfrak{l}$ is a Levi component of $\mathfrak{p}$ if and only if there exist isotropic subspaces $X_\gamma \subset V$ and $Y_\gamma \subset V$ for each $\gamma \in D$ with

$$F''_\gamma = F'_\gamma \oplus X_\gamma \text{ and } G''_\gamma = G'_\gamma \oplus Y_\gamma,$$

as well as a subspace W with $W = 0$ if $\dim G/F \leq 2$ and otherwise

$$G = F \oplus W$$

such that

$$\mathfrak{l} = \mathfrak{so}(W) \oplus \bigoplus_{\gamma \in D} \Lambda(\mathfrak{sl}(X_\gamma, Y_\gamma)).$$

THEOREM 6.3. *Let $\mathfrak{p}$ be a parabolic subalgebra of $\mathfrak{sp}(V)$, with the associated self-taut generalized flag $\mathfrak{F}$. Let F denote the union of all isotropic subspaces F''_α for $\alpha \in A$, and let G denote the intersection of all coisotropic subspaces F'_α for $\alpha \in A$.*

Then $\mathfrak{l}$ is a Levi component of $\mathfrak{p}$ if and only if there exist isotropic subspaces $X_\gamma \subset V$ and $Y_\gamma \subset V$ for each $\gamma \in D$ with

$$F_\gamma'' = F_\gamma' \oplus X_\gamma \text{ and } G_\gamma'' = G_\gamma' \oplus Y_\gamma,$$

as well as a subspace W with

$$G = F \oplus W$$

such that

$$\mathfrak{l} = \mathfrak{sp}(W) \oplus \bigoplus_{\gamma \in D} S(\mathfrak{sl}(X_\gamma, Y_\gamma)).$$

The parabolic subalgebra in the following example has Levi components isomorphic to $\mathfrak{g}$. Let V be the vector space with basis $\{v_i \mid i \in \mathbb{Z}_{\neq 0}\}$, and let $\langle \cdot, \cdot \rangle : V \times V \to \mathbb{C}$ be the nondegenerate pairing extending

$$\langle v_i, v_j \rangle = \begin{cases} 0 & \text{if } i \neq -j \\ 1 & \text{if } i = -j > 0 \end{cases}$$

symmetrically (or antisymmetrically). Take

$$W := \mathrm{Span}\{v_1 + v_i \mid i \neq \pm 1\},$$

and note that the restriction $\langle \cdot, \cdot \rangle|_{W \times W}$ is nondegenerate. We will show that there is a unique parabolic subalgebra of $\mathfrak{g}$ with $\mathfrak{so}(W)$ (resp. $\mathfrak{sp}(W)$) as a Levi component. In order to apply Theorem 6.2 (resp., Theorem 6.3), we consider self-taut generalized flags $\mathfrak{F}$ such that W provides a vector space complement for the single immediate predecessor-successor pair in D.

That is, any self-taut generalized flag $\mathfrak{F}$ in V such that $\mathrm{St}_\mathfrak{F} \cap \mathfrak{g}$ has the prescribed Levi component must have an immediate predecessor-successor pair of the form $U \subset U \oplus W$. Since W is neither isotropic nor coisotropic, U must be isotropic, with $U = (U \oplus W)^\perp$. This implies $W^{\perp\perp} \cap W^\perp \subset U$. We compute $W^{\perp\perp} \cap W^\perp = \mathbb{C}v_1$. As $\dim V/W = 2$, we conclude that $U = \mathbb{C}v_1$, so $\mathfrak{F}$ is the self-taut generalized flag

$$0 \subset \mathbb{C}v_1 \subset \mathbb{C}v_1 \oplus W \subset V.$$

This yields a single self-normalizing parabolic subalgebra $\mathfrak{p} := \mathrm{St}_\mathfrak{F} \cap \mathfrak{g}$ as desired. There are no nontrivial trace conditions on $\mathfrak{p}$, as there are no $\mathfrak{gl}_\infty$ blocks in a locally reductive part; hence $\mathfrak{p}$ is the unique parabolic subalgebra of $\mathfrak{g}$ with the prescribed Levi component.

Note that $\mathfrak{so}(W)$ (or $\mathfrak{sp}(W)$) is a maximal locally reductive subalgebra of $\mathfrak{p}$. On the other hand, any reductive part of $\mathfrak{p}$ is isomorphic to $\mathfrak{g} \oplus \mathbb{C}$. As in the example at the end of Section 5, we have found a maximal locally reductive subalgebra of $\mathfrak{p}$ which is not a locally reductive part of $\mathfrak{p}$.

Now that we have considered the three special cases of $\mathfrak{sl}_\infty$, $\mathfrak{so}_\infty$, and $\mathfrak{sp}_\infty$, the analogous statements hold for locally reductive finitary Lie algebras $\mathfrak{g}$. Then $[\mathfrak{g}, \mathfrak{g}]$ is locally simple, hence $[\mathfrak{g}, \mathfrak{g}] = \bigoplus_{i \in I} \mathfrak{s}_i$ for some simple finitary Lie algebras $\mathfrak{s}_i$. Let $\mathfrak{l}$ be a Levi component of a parabolic subalgebra $\mathfrak{p}$ of $\mathfrak{g}$. Then $\mathfrak{p} \cap \mathfrak{s}_i$ is a parabolic subalgebra of $\mathfrak{s}_i$ for each $i \in I$. It must be the case that $\mathfrak{l} = \bigoplus_{i \in I} \mathfrak{l} \cap \mathfrak{s}_i$, and moreover $\mathfrak{l} \cap \mathfrak{s}_i$ is a Levi component of $\mathfrak{p} \cap \mathfrak{s}_i$. Up to isomorphism the only infinite-dimensional simple finitary Lie algebras are $\mathfrak{sl}_\infty$, $\mathfrak{so}_\infty$, and $\mathfrak{sp}_\infty$ [**BS**]. Thus the results of this paper are enough to classify Levi components of parabolic subalgebras in this generality.

References

B. A. Baranov, Complex finitary simple Lie algebras, Arch. Math. 72 (1999), 101–106.

B2. A. Baranov, Finitary simple Lie algebras, J. Algebra 219 (1999), 299–329.

BS. A. Baranov, H. Strade, Finitary Lie algebras, J. Algebra 254 (2002), 173–211.

Bo. N. Bourbaki, Groupes et algèbres de Lie, Hermann, Paris, 1975.

D. E. Dan-Cohen, Borel subalgebras of root-reductive Lie algebras, J. Lie Theory 18 (2008), 215–241.

DP. E. Dan-Cohen, I. Penkov, Parabolic and Levi subalgebras of finitary Lie algebras, Internat. Math. Res. Notices 2010, No. 6 (2010), 1062–1101.

DPSn. E. Dan-Cohen, I. Penkov, N. Snyder, Cartan subalgebras of root-reductive Lie algebras, J. Algebra 308 (2007), 583–611.

DPW. E. Dan-Cohen, I. Penkov, J.A. Wolf, Parabolic subgroups of direct limit Lie groups, Cont. Math. 499 (2009) 47–59.

DiP. I. Dimitrov, I. Penkov, Borel subalgebras of $\mathfrak{gl}(\infty)$, Resenhas do Instituto de Matemática e Estatística da Universidade de São Paulo 6 (2004), No. 2/3, 153–163.

DiP2. I. Dimitrov, I. Penkov, Locally semisimple and maximal subalgebras of the finitary Lie algebras $\mathfrak{gl}(\infty)$, $\mathfrak{sl}(\infty)$, $\mathfrak{so}(\infty)$, and $\mathfrak{sp}(\infty)$, J. Algebra 322 (2009), 2069–2081.

M. G. Mackey, On infinite dimensional linear spaces, Trans. Amer. Math. Soc. 57 (1945) 155–207.

NP. K.-H. Neeb, I. Penkov, Cartan subalgebras of $\mathfrak{gl}_\infty$, Canad. Math. Bull. 46 (2003), 597–616.

PS. I. Penkov, H. Strade, Locally finite Lie algebras with root decomposition, Arch. Math. 80 (2004), 478–485.

JACOBS UNIVERSITY BREMEN, CAMPUS RING 1, 28759 BREMEN, GERMANY
E-mail address: elizabeth.dancohen@gmail.com

JACOBS UNIVERSITY BREMEN, CAMPUS RING 1, 28759 BREMEN, GERMANY
E-mail address: i.penkov@jacobs-university.de

Contemporary Mathematics
Volume **557**, 2011

ON EXTENDING THE LANGLANDS-SHAHIDI METHOD TO ARITHMETIC QUOTIENTS OF LOOP GROUPS

HOWARD GARLAND

To Gregg Zuckerman on his 60th birthday

ABSTRACT. We discuss certain Eisenstein series on arithmetic quotients of loop groups, $\hat{G}$, which are associated to cusp forms on finite-dimensional groups associated with maximal parabolics of $\hat{G}$.

INTRODUCTION

In his paper "Euler Products" ([L]), Langlands uses the meromorphic continuation of Eisenstein series associated with certain cusp forms φ on an arithmetic quotient of a Chevalley group H, to derive the meromorphic continuation of certain L -functions associated to φ and certain finite-dimensional representations π, of the L -group of H. His method was effective because he had already obtained the meromorphic continuation of the Eisenstein series ([L2]). However, there were limitations, one being that in order to apply the Langlands method, H must (up to local isomorphism) be realized as the semi-simple part of a maximal parabolic subgroup of a higher-dimensional group, since the method requires using the meromorphic continuation of an Eisenstein series associated with such a parabolic. There are then cases which are excluded; e.g., $H_1 = E_8$ and any π, and $H_2 = SL_3(\mathbb{R}) \times SL_3(\mathbb{R}) \times SL_3(\mathbb{R})$ and π being the representation $\tilde{\pi}$, the triple tensor product of the standard representation of $SL_3(\mathbb{R})$. The latter case has been of particular interest in the theory of automorphic L-functions.

Another limitation of the Langlands method has been that, by itself, it does not yield a holomorphic continuation in cases where that is expected and desired - only a meromorphic one. This limitation, and other issues (e.g., obtaining a functional equation) were dealt with to a significant extent by Shahidi and others, using the Langlands-Shahidi method.

In the present paper, we describe a possible alternative method for overcoming these limitations in a number of cases (including the case of $(H_2, \tilde{\pi})$, above). This method is based on the theory of Eisenstein series on arithmetic quotients of loop groups (see e.g., [LG], [R], [AC]), on a bold suggestion of A. Braverman and D. Kazhdan, which we will describe in more detail in §5 and later in this Introduction, and on a lemma of F. Shahidi (Lemma 4.1). Then for example, H_1, H_2 can each be realized as the semi-simple part of a maximal parabolic subgroup of $\hat{E}_8$ (= affine E_8) and $\hat{E}_6$ (= affine E_6), respectively. Remarkably, one can then obtain the

2000 *Mathematics Subject Classification.* Primary 11F99; Secondary 22E67.
Key words and phrases. automorphic L-functions, Eisenstein series, loop groups.

holomorphic continuation of appropriate Eisenstein series, with relative ease in a number of cases (again, including H_2).

Here we obtain the existence of the desired Eisenstein series (establishing a Godement criterion (3.11)), and then obtain the Maass-Selberg relations (see (4.3)). In fact, obtaining the latter proved remarkably easy, thanks to a result of F. Shahidi (Lemma 4.1). The proof of the Godement criterion (Cor. 1 to Theorem 3.2) depends on the convergence theorem in [AC] for minimal parabolics, and an argument in [GMRV] for extending convergence theorems for minimal parabolics to more general ones. It might seem then, that we are in a good position to extend the results of [L]. However, the same result of Shahidi, that simplifies the derivation of the Maass-Selberg relations for our Eisenstein series, also seems at first to prevent the extraction of the desired L-functions from the constant terms of such Eisenstein series: the problem is that when Shahidi's result does yield something like (4.3), the reason is precisely that the constant term is "elementary" and does not involve L-functions. This is what happens for example, for the pair $(\hat{E}_6, H_2)$.

A. Braverman and D.Kazhdan proposed a way out of this dilemma. Before discussing their idea, we note two things: *first*, thanks to (4.3) we obtain a holomorphic continuation of our Eisenstein series in some cases and *second*, the methods used in §§2-4 of this paper apply equally well to number fields and to function fields over finite fields (see e.g., [Lo]).

Braverman and Kazhdan proposed that instead of only computing the constant terms with respect to "upper triangular" parabolics (which are sufficient for obtaining convergence and the Maass-Selberg relations), that if possible, one also computes the constant terms with respect to "lower triangular" parabolics in order to obtain the L-functions (see §5, for the definition of upper and lower triangular parabolic subgroups). Of course in the finite-dimensional case there is no essential difference between lower and upper triangular parabolics: lower triangular parabolics are conjugate to upper triangular ones.

As in the finite-dimensional case treated in [L], the computation of the constant terms will depend on local computations, and in particular, on certain formulae of Gindikin-Karpelevich type for the lower triangular case. Such formulae have been conjectured in [BFK] in the non-archimedean case, and proved there for $F((t))$, F a finite field. A proof for all non-archimedean fields will be given in [BGKP]. We will discuss this in §5. For now let it suffice to say that the situation for lower triangular parabolics is more subtle than for upper triangular ones.

A striking feature of the loop case is that one has reproduced a significant portion of the theory of automorphic forms for finite-dimensional groups, and now, considering the two theories together, the loop case and the finite-dimensional case, one might obtain new results about the classical theory (e.g., holomorphic continuation of automorphic L-functions associated with (finite-dimensional)cusp forms). The situation is reminiscent of the proof of Bott periodicity, using the space of based loops of compact symmetric spaces: one obtains cell decompositions for generalized flag manifolds -Grassmannians for example (parameterized by coset spaces of finite Weyl groups), and also, cell decompositions of based loop spaces of compact symmetric spaces (parameterized by coset spaces of affine Weyl groups), and then, comparing the finite-dimensional cases and the loop cases, Bott periodicity falls out - a deep result about the homotopy of (finite-dimensional) compact, symmetric spaces. One might say, we are dealing here with a "Bott principle" for the theory of automorphic forms: comparing the loop and finite-dimensional theories, one might obtain new results about (finite-dimensional) automorphic forms and automorphic L-functions.

I am submitting this paper in honor of Gregg Zuckerman's 60th birthday. Over a period of many years, I have had the pleasure of collaborating with him on three papers, and of having an infinite number of discussions. These discussions ranged over a wide spectrum of mathematics and, as many of his colleagues know from their own experience, discussions with Gregg are memorable for his clear grasp of deep mathematical ideas and his uncanny ability to explain them with utter clarity.

1. THE SETTING

We let A be an irreducible, $l \times l$, classical, Cartan matrix, and we let $\tilde{A}$ be the corresponding affine, Cartan matrix. We let $\mathfrak{g} = \mathfrak{g}(A)$, $\hat{\mathfrak{g}} = \mathfrak{g}(\tilde{A})$ be the complex, Kac-Moody Lie algebras corresponding to A, $\tilde{A}$, respectively. We let $\mathfrak{g}_{\mathbb{Z}} \subseteq \mathfrak{g}$, $\hat{\mathfrak{g}}_{\mathbb{Z}} \subseteq \hat{\mathfrak{g}}$ denote the Chevalley $\mathbb{Z}$-forms, with $\hat{\mathfrak{g}}_{\mathbb{Z}}$ constructed from $\mathfrak{g}_{\mathbb{Z}}$, as in [LA]. We let

$$\hat{\mathfrak{g}}^e = \hat{\mathfrak{g}} \oplus \mathbb{C}D,$$

$$\hat{\mathfrak{g}}_{\mathbb{Z}}^e = \hat{\mathfrak{g}}_{\mathbb{Z}} \oplus \mathbb{Z}D$$

denote the extended, affine, Kac-Moody Lie algebra and $\mathbb{Z}$-form, respectively (D being the usual, homogeneous degree operator (see e.g., [LG], §3(after Prop. 3.3))).

We let $\{e_i, f_i, h_i\}_{i=1,...,l+1}$ be the Kac-Moody generators of $\hat{\mathfrak{g}}$, ordered so that $\{e_i, f_i, h_i\}_{i=1,...,l}$ generate $\mathfrak{g}$, which we may regard as a subalgebra of $\hat{\mathfrak{g}}$. We let $\mathfrak{h}$ (resp., $\hat{\mathfrak{h}}$) be the complex, linear span of the h_i, $i = 1,...,l$ (resp., $i = 1,....,l+1$), and set $\mathfrak{h}^e = \hat{\mathfrak{h}} \oplus \mathbb{C}D$, $\hat{\mathfrak{h}}_{\mathbb{Z}} = \mathbb{Z}$-span of the h_i, $i = 1,....,l+1$, $\hat{\mathfrak{h}}_{\mathbb{Z}}^e = \hat{\mathfrak{h}}_{\mathbb{Z}} \oplus \mathbb{Z}D$. Recall that $\lambda \in (\hat{\mathfrak{h}})^*$ is called *dominant integral*, in case

$$\lambda(h_i) \in \mathbb{Z}_{\geq 0}, \ i = 1,....,l+1.$$

We further adopt the convention that $\lambda(h_i)$ must be > 0, for at least one i.

Given $\lambda \in \hat{\mathfrak{h}}^*$ dominant integral, we let V^λ denote the corresponding irreducible highest weight module of $\hat{\mathfrak{g}}$, and we let $V_{\mathbb{Z}}^\lambda \subseteq V^\lambda$ be a Chevalley $\mathbb{Z}$-form, as constructed in [LA]. For a commutative ring with unit, we set $V_R^\lambda = R \otimes_{\mathbb{Z}} V_{\mathbb{Z}}^\lambda$ (we also set $\hat{\mathfrak{g}}_R = R \otimes_{\mathbb{Z}} \hat{\mathfrak{g}}_{\mathbb{Z}}$, $\mathfrak{g}_R = R \otimes_{\mathbb{Z}} \mathfrak{g}_{\mathbb{Z}}$, etc.).

For an algebraically closed field k, we let $\hat{G}_k^\lambda$ $(= \hat{G}_k^\lambda(\tilde{A}))$ be the Chevalley group contained in $Aut(V_k^\lambda)$, as defined in [LG], Definition (7.21). For an arbitrary field k with algebraic closure $\bar{k}$, we let

$$\hat{G}_k^\lambda(= \hat{G}_k^\lambda(\tilde{A}))$$

be the subgroup of $\hat{G}_{\bar{k}}^\lambda$ defined by

(1.1) $\hat{G}_k^\lambda = \{g \in \hat{G}_{\bar{k}}^\lambda | g(V_k^\lambda) = V_k^\lambda\}.$

In general, for k not algebraically closed, $\hat{G}_k^\lambda$ so defined, is larger than the corresponding group of [LG], Definition (7.21). This is essentially for the same reason that, e.g., $Ad(SL_2(\mathbb{Q}_p)) \subsetneq PSL_2(\mathbb{Q}_p)$, for the p-adic rationals $\mathbb{Q}_p$.

We let

(1.2) $\hat{G}_{\mathbb{Z}}^\lambda(= \hat{\Gamma}) = \{\gamma \in \hat{G}_{\mathbb{R}}^\lambda | \gamma(V_{\mathbb{Z}}^\lambda) = V_{\mathbb{Z}}^\lambda\}.$

We *adopt the notation* of [LG], [R], [AC]. For

$$\nu : \hat{h}_{\mathbb{R}} \to \mathbb{C} \text{ (real, linear)}$$

satisfying Godement's criterion

(1.3) $\operatorname{Re} \nu(h_i) < -2, \; i = 1, ..., l+1,$

we have from [R], Theorem 5.1, and [AC], Theorem 12.1:

Theorem 1.1. *The infinite sum*

(1.4) $$\sum_{\gamma \in \hat{\Gamma} / \hat{\Gamma} \cap \hat{B}} \Phi_{\nu}(g \exp(-rD)\gamma)$$

converges absolutely. Moreover, the convergence is uniform on sets $\hat{K}\Omega_A \eta(s)\hat{U}_{\mathcal{D}}$,*where* $s = e^{-r}$, $\Omega_A \subseteq \hat{A}$ *is compact,* $\hat{U}_{\mathcal{D}} \subseteq \hat{U}$, *as in [AC].*

The notation is as in [AC], but for the sake of completeness, we add a few words of explanation: As in [LG], $V_{\mathbb{C}}^{\lambda}$ admits a positive-definite, Hermitian inner product $\{,\}$,which is invariant with respect to a certain "compact form" $\hat{\mathfrak{k}} \subseteq \hat{\mathfrak{g}}_{\mathbb{C}}$ (as defined in [LA] ($\hat{\mathfrak{k}}$ being $\mathfrak{k}(\tilde{A})$ of [LA], §4). The form $\{,\}$ then restricts to a real, positive-definite inner product on $V_{\mathbb{R}}^{\lambda}$ and $\hat{K} \subseteq \hat{G}_{\mathbb{R}}^{\lambda}$ is defined by

$$\hat{K} = \{k \in \hat{G}_{\mathbb{R}}^{\lambda} | \{k\xi, k\eta\} = \{\xi, \eta\}, \; \xi, \eta \in V_{\mathbb{R}}^{\lambda}\}.$$

We fix a coherently ordered basis (see [LG], beginning of §12 for the definition) $\mathcal{B}$, say, of $V_{\mathbb{Z}}^{\lambda}$, and we let $\hat{A} \subseteq G_{\mathbb{R}}^{\lambda}$ be the subgroup of all diagonal (with respect to the basis $\mathcal{B}$)elements with positive entries. We let $\hat{U} \subseteq G_{\mathbb{R}}^{\lambda}$ be the subgroup of all upper triangular elements with diagonal elements all equal to one (again, with respect to $\mathcal{B}$).We then have the Iwasawa decomposition

(1.5) $$G_{\mathbb{R}}^{\lambda} = \hat{K}\hat{A}\hat{U}$$

(with uniqueness of expression)(see [LG], Lemma 16.14).

Now $\hat{\mathfrak{h}}_{\mathbb{R}}$ is the Lie algebra of $\hat{A}$ and ν defines a quasi-character

$$\nu : \hat{A} \to \mathbb{C}^{\times},$$

$$a \mapsto a^{\nu}, \; a \in \hat{A}.$$

Given $g \in \hat{G}_{\mathbb{R}}^{\lambda}$, g has a decomposition

$$g = k_g a_g u_g,$$

with respect to (1.5). We then set

$$\Phi_{\nu}(g) = a_g^{\nu}.$$

2. Extensions of the Convergence Theorem (Preliminaries).

For a field k with algebraic closure $\bar{k}$, we let $\hat{B}_{\bar{k}} \subseteq G_{\bar{k}}^{\lambda}$ be the upper triangular subgroup (with respect to the coherently ordered basis $\mathcal{B}$), and $\hat{B}_k = \hat{B}_{\bar{k}} \cap G_k^{\lambda}$. We let $\hat{P}_k \supseteq \hat{B}_k$ be a proper, parabolic subgroup of $\hat{G}_k$ (= $\hat{G}_k^{\lambda}$; we drop the superscript "λ" when there is no ambiguity about which λ we mean).

We consider various subgroups of $\hat{P}_k$. We first let $\alpha_1, ..., \alpha_{l+1} \in (\hat{\mathfrak{h}}^e)^*$, the complex dual of $\hat{\mathfrak{h}}^e$, be the simple roots:

$$\alpha_i(h_j) = \tilde{A}_{ij}, \; i, j = 1, ..., l+1,$$

where

$$\tilde{A} = (\tilde{A}_{ij})_{i,j=1,\dots,l+1}.$$

We let $\Xi = \{s_i\}_{i=1,\dots,l+1}$ denote the corresponding, simple root reflections (so s_i is the root reflection corresponding to α_i). Then the s_i generate the (affine) Weyl group $\hat{W}$ of $\hat{\mathfrak{g}}$ (with respect to $\hat{\mathfrak{h}}$) and for $\theta \subseteq \Xi$, we let $W_\theta \subseteq \hat{W}$ be the subgroup generated by the elements of θ. Then every subgroup $\hat{P}_k \supseteq \hat{B}_k$ is a group of the form

$$\hat{P}_k = \hat{P}_{\theta,k} = \hat{B}_k W_\theta \hat{B}_k,$$

and every proper, parabolic subgroup of $\hat{G}_k$ is a conjugate of a $\hat{P}_{\theta,k}$ for some $\theta \subsetneq \Xi$ (in fact, we take this as the definition of "proper parabolic").

We let

$$\hat{\Delta} \subseteq (\hat{\mathfrak{h}}^e)^*$$

be the affine roots of $\hat{\mathfrak{h}}$, and we let $\hat{\Delta}_+ \subseteq \hat{\Delta}$ be the positive roots determined by the choice of simple roots $\alpha_1, \dots, a_{l+1}$. When convenient, we identify Ξ with the set of simple roots. For $\theta \subseteq \Xi$ (considered then as the set of simple roots), we let $[\theta] \subseteq \hat{\Delta}$ denote the set of all roots in $\hat{\Delta}$ which are linear combinations of the elements of θ. We let $\hat{H}_k \subseteq \hat{B}_k$ be the diagonal subgroup (with respect to the coherently ordered basis $\mathcal{B}$), and for $\theta \subseteq \Xi$, we let

$$H_{\theta,k} = \{h \in \hat{H}_k | h^{\alpha_i} = 1, \ \alpha_i \in \theta\}.$$

For an algebraically closed field $\bar{k}$, we let $L_{\theta,\bar{k}} \subseteq \hat{G}_{\bar{k}}$ be the subgroup generated by elements $\{\chi_\alpha(u)\}_{\alpha \in [\theta], u \in \bar{k}}$. For an arbitrary field k with algebraic closure $\bar{k}$, we set

$$L_{\theta,k} = L_{\theta,\bar{k}} \cap \hat{G}_k.$$

One lets $\hat{U}_{\theta,k} \subseteq \hat{P}_{\theta,k}$ be the pro-unipotent radical; then

$$\hat{P}_{\theta,k} = M_{\theta,k} \hat{U}_{\theta,k},$$

with $\hat{U}_{\theta,k}$ normal, $M_{\theta,\bar{k}} = L_{\theta,\bar{k}} H_{\theta,\bar{k}}$, and $M_{\theta,k} = M_{\theta,\bar{k}} \cap \hat{G}_k$ (see [LG2], Theorem 6.1).

We now consider the case when $k = \mathbb{R}$. For $k = \mathbb{R}$, we set $\hat{G}^\lambda (= \hat{G}) = \hat{G}_\mathbb{R}^\lambda$, and

$$\hat{H} = \hat{H}_\mathbb{R},$$
$$H_\theta = H_{\theta,\mathbb{R}},$$
$$\hat{P}_\theta = \hat{P}_{\theta,\mathbb{R}},$$
$$L_\theta = L_{\theta,\mathbb{R}}, \ M_\theta = M_{\theta,\mathbb{R}},$$

etc. We let

$$Z \subseteq \hat{H}$$

be the subgroup of all elements whose diagonal elements are ± 1. We let

$$\lambda : \hat{H} \to \mathbb{C}^\times$$

be a quasi-character such that $\lambda|_Z$ is identically equal to 1. We may identify λ with a real, linear function

$$\lambda : \hat{\mathfrak{h}}_\mathbb{R} \to \mathbb{C},$$

where $\hat{\mathfrak{h}}_\mathbb{R}$ is the Lie algebra of $\hat{H}$.

We are assuming $\mathfrak{g} = \mathfrak{g}(A)$ is simple (for recall, we assumed at the beginning, that A is irreducible), and that we have ordered the h_i so that $h_1, \dots, h_l$ span $\mathfrak{h}$, the

Cartan subalgebra of $\mathfrak{g}(A) \subseteq \mathfrak{g}(\hat{A})$. We let α_0 be the corresponding highest root of $\mathfrak{g}(A)$ ($\alpha_0 \in \mathfrak{h}^*$, the complex dual of $\mathfrak{h}$) and we let

$$c = h_{\alpha_0} + h_{l+1} \in \hat{\mathfrak{h}},$$

(h_{α_0} denoting the coroot corresponding to α_0 - we will also denote this coroot by $\alpha_0^\vee$). Then c spans the center of $\hat{\mathfrak{g}}$. We have the extended Cartan

$$\hat{\mathfrak{h}}^e = \hat{\mathfrak{h}} \oplus \mathbb{C}D$$
$$= \mathfrak{h} \oplus \mathbb{C}c \oplus \mathbb{C}D,$$

and a corresponding decomposition of $(\hat{\mathfrak{h}}^e)^*$, the complex dual of $\hat{\mathfrak{h}}^e$,

$$(\hat{\mathfrak{h}}^e)^* = \mathfrak{h}^* \oplus \mathbb{C}\lambda_{l+1} \oplus \mathbb{C}\iota,$$

where, e.g.,

$$\lambda_{l+1}(c) = 1, \quad \iota(D) = 1.$$

Note that ι is the generating, imaginary root and λ_{l+1} is the $l+1^{st}$ fundamental weight defined by

$$\lambda_{l+1}(h_i) = \begin{cases} 0, & i \neq l+1 \\ 1, & i = l+1. \end{cases}$$

Now with these conventions and notations, assume

$$\theta = \theta_0 = \{\alpha_1, ..., \alpha_l\};$$

then (over $\mathbb{R}$)

$$\mathfrak{h}_\theta = \mathbb{R}c \ (\mathfrak{h}_\theta = \text{ Lie algebra of } H_\theta),$$
$$L_\theta = G,$$

a real, connected Lie group with Lie algebra $\mathfrak{g}_\mathbb{R} = \mathbb{R} \otimes_\mathbb{Z} \mathfrak{g}_\mathbb{Z}$.

3. EXTENSIONS OF THE CONVERGENCE THEOREM (CONTINUED).

We return to ν and to Φ_ν, as considered in (1.3) and (1.4) - see the exact definition of Φ_ν at the end of §1. We now further assume that ν is $\mathbb{R}$-valued; i.e., that

$$(3.1) \qquad \nu : \hat{\mathfrak{h}}_\mathbb{R} \to \mathbb{R}$$

(so now $\nu(h_i) < -2$, $i = 1, ..., l+1$). We let $\hat{P} = \hat{P}_\theta \supseteq \hat{B}$ be a proper, parabolic subgroup, and we consider the sum (1.4):

$$(3.2) \qquad \sum_{\gamma \in \hat{\Gamma}/\hat{\Gamma} \cap \hat{B}} \Phi_\nu(g \exp(-rD)\gamma)$$

$$= \sum_{\gamma \in \hat{\Gamma}/\hat{\Gamma} \cap \hat{P}} \sum_{\beta \in \hat{\Gamma} \cap \hat{P}/\hat{\Gamma} \cap \hat{B}} \Phi_\nu(g \exp(-rD)\gamma\beta).$$

Thanks to our present assumption (3.1), the series on either side of (3.2) are in fact series of positive terms, and since the left side is convergent (Theorem 1.1), so is the right side, and in particular, the series

$$(3.3) \qquad \sum_{\beta \in \hat{\Gamma} \cap \hat{P}/\hat{\Gamma} \cap \hat{B}} \Phi_\nu(g \exp(-rD)\gamma\beta)$$

($\gamma \in \hat{\Gamma}/\hat{\Gamma} \cap \hat{P}$ now fixed) is convergent (and of course, is absolutely convergent, since it is a sum of positive terms).

However, the series (3.3) is in fact a convergent Eisenstein series for the (finite-dimensional) reductive group

$$M_\theta$$

(where recall $\hat{P} = \hat{P}_\theta$). More precisely, let

$$\pi : \hat{P}_\theta \to M_\theta$$

be the projection. Let

$$K_\theta = \hat{K} \cap \hat{P}_\theta = \hat{K} \cap M_\theta,$$

where $\hat{K}$ is as in §1; then of course

$$\pi(\hat{K} \cap \hat{P}_\theta) = K_\theta.$$

Consider the elements g, γ appearing in (3.3). We have

$$g\exp(-rD)\gamma = k_{g\gamma}m_{g\gamma}\exp(-rD)u_{g\gamma}, \ k_{g\gamma} \in \hat{K}, \ m_{g\gamma} \in M_\theta, \ u_{g\gamma} \in \hat{U}_\theta;$$

then (β as in (3.3))

$$\Phi_\nu(g\exp(-rD)\gamma\beta) = \Phi_\nu(m_{g\gamma}\exp(-rD)\beta),$$

and the sum (3.3) becomes

(3.3′)
$$\sum_{\beta \in \hat{\Gamma}\cap\hat{P}/\hat{\Gamma}\cap\hat{B}} \Phi_\nu(m_{g\gamma}\exp(-rD)\beta).$$

Set

$$B_\theta = \pi(\hat{B}) \subseteq M_\theta,$$
$$\Gamma_\theta = \pi(\hat{\Gamma} \cap \hat{P}_\theta).$$

Then (3.3′) equals

(3.3″)
$$\sum_{\beta \in \Gamma_\theta/\Gamma_\theta\cap B_\theta} \Phi_\nu(m_{g\gamma}\exp(-rD)\beta),$$

where Γ_θ is an arithmetic subgroup of M_θ. For $m \in M_\theta$, we set

$$\Phi_\nu(m) = \Phi_\nu(m\exp(-rD)),$$

and we let

$$\Gamma_\theta^r =_{df} \exp(-rD)\Gamma_\theta\exp(rD);$$

Noting that $\exp(-rD)$ normalizes B_θ, we have that the sum (3.3″) equals

(3.4)
$$\sum_{\beta \in \Gamma_\theta^r/\Gamma_\theta^r\cap B_\theta} \Phi_\nu(m_{g\gamma}\beta),$$

which is an Eisenstein series for the pair $(M_\theta, \Gamma_\theta^r)$.

On the one hand, the absolute convergence of (3.4) follows from that of (3.2). On the other hand, our assumption that ν in (3.1) satisfies Godement's criterion ($\nu(h_i) < -2$, $i = 1, ..., l+1$) in fact implies that (3.4) is absolutely convergent, thanks to Godement's criterion for finite-dimensional groups.

It is useful to give an alternate description of the sum (3.4). Let

$$\mathfrak{h}(\theta) \subseteq \hat{\mathfrak{h}}_\mathbb{R},$$

be the (real) linear span of the h_i, ($s_i \in \theta$); then $\mathfrak{h}(\theta)$ may be regarded as the Cartan subalgebra of $\mathfrak{l}_\theta$, the Lie algebra of L_θ. We let

$$H(\theta) \subseteq L_\theta$$

denote the group generated by the elements

$$h_{\alpha_i}(s), \ \alpha_i \in \theta, s \in \mathbb{R}^\times,$$

so that $\mathfrak{h}(\theta)$ is the Lie algebra of $H(\theta)$.

We let

$$L'_\theta = M_\theta/Z(M_\theta), \ Z(M_\theta) = \text{center of } M_\theta.$$

We then have that

$$\Phi_\nu|_{L_\theta}$$

is the lift of a function Φ'_ν on L'_θ. If we let

$$\tilde{\omega} : M_\theta \to L'_\theta$$

denote the projection, if we let $H(\theta)' = \tilde{\omega}(H(\theta))$ and $A(\theta)'$ denote the identity component of $H(\theta)'$, $K'_\theta = \tilde{\omega}(K_\theta)$ and $U'_\theta = \tilde{\omega}(\hat{U} \cap M_\theta)$, then we have the Iwasawa decomposition

$$L'_\theta = K'_\theta A(\theta)' U'_\theta,$$

and

$$\Phi'_\nu(k'a'u') = (a')^\nu, \ k' \in K'_\theta, \ a' \in A(\theta)', \ u' \in U'_\theta,$$

(where we may identify $A(\theta)'$ with a subgroup of L_θ, in order to define $(a')^\nu$); then we have for $m \in M_\theta$,

$$\sum_{\beta \in \Gamma^r_\theta/\Gamma^r_\theta \cap B_\theta} \Phi_\nu(m\beta)$$

$$(3.5) \qquad\qquad = \sum_{\beta \in (\Gamma^r_\theta)'/(\Gamma^r_\theta)' \cap \tilde{\omega}(B_\theta)} \Phi'_\nu(\tilde{\omega}(m)\beta),$$

which is a convergent Eisenstein series on L'_θ (with respect to $(\Gamma^r_\theta)' =_{df} \tilde{\omega}(\Gamma^r_\theta)$ and the Borel subgroup $\tilde{\omega}(B_\theta) \subseteq L'_\theta$.) Let

$$E_\theta(m) = \tilde{E}_\theta(\tilde{\omega}(m)), \ m \in M_\theta,$$

denote the convergent sum (3.5) (as we noted, $\tilde{E}_\theta(\cdot)$ is then an Eisenstein series on L'_θ).

On the other hand, we consider

$$F_\theta(m) =_{df} \sum_{\beta \in \Gamma^r_\theta/\Gamma^r_\theta \cap B_\theta} \Phi_\nu(m\beta), \ m \in M_\theta;$$

then

$$F_\theta E_\theta^{-1}(\cdot)$$

is entirely determined by its restriction to $Z(M_\theta)$, and indeed

$$Z(M_\theta) \subseteq \hat{H},$$

and

$$F_\theta E_\theta^{-1}(z) = z^\nu, \ z \in Z(M_\theta).$$

Some explanation is required here, since, strictly speaking, the quasicharacter ν is only defined on the identity component $\hat{A} \subseteq \hat{H}$, and hence only on the identity component $A(M_\theta)$ of $Z(M_\theta)$. However, one has a homomorphism

$$\hat{H} \xrightarrow{\sigma} \hat{A}$$

given by absolute value (each $h \in \hat{H}$ is represented by a diagonal matrix, and $\sigma(h)$ is just the corresponding matrix of absolute values, and ν on $\hat{H}$ is just taken to be $\nu \circ \sigma$.

We set

$$\xi_\nu(\cdot) = F_\theta E_\theta^{-1}(\cdot);$$

then of course,

$$F_\theta = E_\theta \xi_\nu$$

(where $\xi_\nu|_{L_\theta} \equiv 1$).

But then, since (3.3) equals (3.4), we have

$$(3.6) \qquad \sum_{\beta \in \hat{\Gamma} \cap \hat{P}/\hat{\Gamma} \cap \hat{B}} \Phi_\nu(g\exp(-rD)\gamma\beta) = E_\theta(m_{g\gamma})\xi_\nu(m_{g\gamma}),$$

and so (see (3.2))

$$(3.7) \qquad \sum_{\gamma \in \hat{\Gamma}/\hat{\Gamma} \cap \hat{B}} \Phi_\nu(g\exp(-rD)\gamma) = \sum_{\gamma \in \hat{\Gamma}/\hat{\Gamma} \cap \hat{P}} E_\theta(m_{g\gamma})\xi_\nu(m_{g\gamma}).$$

Now $E_\theta(\cdot)$ is by definition the lift of an Eisenstein series on L'_θ, and hence is bounded below by some $\kappa > 0$. It follows that

$$(3.8) \qquad \sum_{\gamma \in \hat{\Gamma}/\hat{\Gamma} \cap \hat{P}} \xi_\nu(m_{g\gamma}) < \infty$$

since the series on either side of (3.7) is (absolutely) convergent. (We learned this method of deriving convergence for general parabolics from convergence for minimal ones, from [GMRV] - also see [MW], Proposition II.1.5.)

It is instructive to analyze the element $m_{g\gamma} \in M_\theta$. Recall (before (3.3')) the equation

$$g\exp(-rD)\gamma = k_{g\gamma} m_{g\gamma} \exp(-rD)u_{g\gamma}.$$

M_θ is then a direct product

$$M_\theta = \tilde{L}_\theta A(M_\theta),$$

where recall that $A(M_\theta)$ is the identity component of $Z(M_\theta)$, and where $\tilde{L}_\theta$ is a subgroup of $\hat{G}^\lambda$ containing L_θ with $\tilde{L}_\theta/L_\theta$ finite. Of course $m \in M_\theta$ then has a corresponding expression $m = la$, $l \in \tilde{L}_\theta$, $a \in A(M_\theta)$, and in particular, this is the case for $m_{g\gamma}$. Moreover

$$\xi_\nu(la) = a^\nu, \ l \in \tilde{L}_\theta, a \in A(M_\theta).$$

Given $g \in \hat{G}^\lambda$, and the decomposition

$$\hat{G}^\lambda = \hat{K}\hat{P}$$

$$= \hat{K}\tilde{L}_\theta A(M_\theta)\hat{U}_\theta,$$

we have that $g \in \hat{G}^\lambda$ has a corresponding decomposition

$$g = k_g l_g a_g u_g,$$

$$k_g \in \hat{K}, l_g \in \tilde{L}_\theta, a_g \in A(M_\theta), u_g \in \hat{U}_\theta.$$

One has:

Lemma 3.1. a_g *is uniquely determined by* g.

We briefly sketch the proof, which is a straightforward application of representation theory. Let μ be the sum of those fundamental weights λ_i such that α_i is not in θ. Then the group $\hat{G}^\lambda$ acts on $V^{m\mu}$ for some positive multiple $m\mu$ of μ.(see [LG], Prop. 20.2). We have a positive-definite, Hermitian inner product $\{,\}$, on $V^{m\mu}$, as in §1, and we let $v_{m\mu}$ be a highest weight vector of norm one, with respect to $\|\cdot\|$, the norm corresponding to $\{,\}$. Then $\|g \cdot v_{m\mu}\| = a_g^{m\mu}$, and the lemma follows.

We now specialize to the case when

$$\theta = \theta_{i_0} = \Xi - \{\alpha_{i_0}\},$$

for a single, simple root α_{i_0}. We have a relation

$$\sum_{i=1}^{l} n_i \alpha_i^\vee + h_{l+1} = c,$$

(c defined as in §2) where if α_0 is the highest root of $\mathfrak{g}(A)$,then

$$\alpha_0^\vee = \sum_{i=1}^{l} n_i \alpha_i^\vee,$$

with $\alpha_0^\vee$, $\alpha_i^\vee$ denoting the coroots corresponding to α_0, α_i, respectively.

We set $\alpha_{l+1}^\vee = h_{l+1}$, and note that $\alpha_i^\vee = h_i$, $i = 1, ..., l$. Now if

(3.9) $$\nu(\alpha_i^\vee) < -2, \ i = 1, ..., l+1,$$

then of course

$$\nu(c) < -2(1 + \sum_{i=1}^{l} n_i);$$

we set

$$h^\vee = 1 + \sum_{i=1}^{l} n_i$$

(which we will call the dual Coxeter number). Then

(3.10) $$\nu(c) < -2h^\vee.$$

On the other hand, if

$$\tilde{\nu} : \mathbb{R}c \to \mathbb{R}$$

satisfies (3.10), then $\tilde{\nu}$ is the restriction of some

$$\nu : \hat{\mathfrak{h}}_\mathbb{R} \to \mathbb{R}$$

satisfying (3.9).

If $g \in \hat{G}^\lambda$,then

$$g \exp(-rD)\gamma = k_{g\gamma} l_{g\gamma} a_{g\gamma} \exp(-rD) u_{g\gamma}, \ k_{g\gamma} \in \hat{K}, l_{g\gamma} \in \tilde{L}_\theta, a_{g\gamma} \in A(M_\theta), u_{g\gamma} \in \hat{U}_\theta,$$

and if we set

$$\xi_\nu(m_{g\gamma}) = \xi_\nu(a_{g\gamma}) = a_{g\gamma}^\nu,$$

then our above arguement shows

Theorem 3.2. *For* $\theta = \theta_{i_0}$, *as above and for*

$$\nu : \hat{\mathfrak{h}}_\mathbb{R} \to \mathbb{R}$$

a real, linear function such that

$$\nu(c) < -2h^\vee,$$

we have

$$\sum_{\gamma \in \hat{\Gamma}/\hat{\Gamma} \cap \hat{P}} a_{g\gamma}^{\nu} < \infty.$$

Corollary 1. *If $\tilde{\varphi}$ is a cusp form on $L'_{\theta_{i_0}}$ which is rapidly decreasing (e.g., an eigenfunction for the center of the universal enveloping algebra of $L'_{\theta_{i_0}}$), if $\varphi = \tilde{\varphi} \circ \tilde{\omega}$, and if*

$$\nu : \hat{\mathfrak{h}}_{\mathbb{R}} \to \mathbb{C}$$

is a real linear function, such that

(3.11) $$\mathrm{Re}(\nu)(c) < -2h^{\vee},$$

then

(3.12) $$\sum_{\gamma \in \hat{\Gamma}/\hat{\Gamma} \cap \hat{P}} \varphi(m_{g\gamma}) \xi_{\nu}(a_{g\gamma}), \; g \in \hat{G},$$

converges absolutely.

Proof. φ is of course bounded. If ν satisfies (3.11), we can dominate the series (3.12) by

$$\sum_{\gamma \in \hat{\Gamma}/\hat{\Gamma} \cap \hat{P}} \xi_{\mathrm{Re}\,\nu}(a_{g\gamma}),$$

which converges by Theorem 3.2. □

4. Shahidi's Argument

Thanks to the Corollary to Theorem 3.2, we have Eisenstein series on loop groups which are associated to certain cusp forms on finite-dimensional, semi-simple groups. For example, consider the affine Dynkin diagram associated with E_6, with the vertices numbered as in [Bourb] (assign the number 7 to the vertex corresponding to the negative of the highest root). Consider (with this numbering)

$$\theta_4 = \{\alpha_1, \alpha_2, \alpha_3, \alpha_5, \alpha_6, \alpha_7\},$$

and the subgroups

$$M_{\theta_4}, \; L_{\theta_4}$$

of $\hat{G}^{\lambda}$. Now take

$$\lambda = \lambda_4,$$

the fundamental weight corresponding to node 4 (in the numbering of [Bourb]). Then L_{θ_4} is locally isomorphic to

$$SL_3(\mathbb{R}) \times SL_3(\mathbb{R}) \times SL_3(\mathbb{R}).$$

Hence, starting with a cusp form φ on L'_{θ_4}, one obtains an Eisenstein series on $\hat{G}^{\lambda_4}$ (denote this group by $\hat{E}_6$).

But then, motivated by [L], one can ask to find the constant term for such an Eisenstein series, and then hope to obtain (for suitable φ) an expression involving L-functions associated with φ, and certain representations of the Langlands dual $\hat{E}_6^L$.

However, at first, this strategy seemed doomed to fail: the constant terms with respect to parabolics $\hat{P}_{\theta}$, as above, do not yield L-functions as in [L]. The problem is that M_{θ_4} is not self-associate, this being an instance of Shahidi's lemma (see [S]):

Lemma 4.1. *Let $\theta = \theta_i$, $i = 1, \ldots, l + 1$; then M_θ is not self-associate.*

Lemma 4.1 has an important implication for computing the constant term of Eisenstein series (see [S2], Theorem 6.2.1 and also Lemma 4.2, below).

As the proof in [S] is not terribly long, we include it here for the convenience of the reader: We have set $\mathfrak{h}(\theta)$ equal to the real linear span of the h_j, $j \neq i$ ($\theta = \theta_i$), and so $\mathfrak{h}(\theta)$ is the Lie algebra of $H(\theta) \subseteq L_\theta$, the subgroup generated by the $h_{\alpha_j}(s)$, $j \neq i$, $s \in \mathbb{R}^\times$. We let $\hat{W}_\theta$ denote the subgroup of the Weyl group generated by the s_j, $j \neq i$, and we set

$$w_0^\theta = \text{ longest element in } \hat{W}_\theta.$$

Assume then, that there is an element $w_0 \in \hat{W}$ such that

$$w_0(\theta) = \theta, \ w_0(\alpha_i) < 0$$

(this being the definition of M_θ being self-associate). Then

$$w_0 w_0^\theta(\theta) = -\theta,$$

while

$$w_0 w_0^\theta(\alpha_i) < 0.$$

To see this last assertion, we note that $w_0^\theta(\alpha_i)$ has an expression

$$w_0^\theta(\alpha_i) = \alpha_i + \sum_{j \neq i} k_j \alpha_j,$$

and then

$$(4.1) \qquad\qquad w_0 w_0^\theta(\alpha_i) = w_0(\alpha_i) + \sum_{j \neq i} k_j' \alpha_j,$$

(since $w_0(\theta) = \theta$). But, by assumption, $w_0(\alpha_i)$ is negative, and

$$w_0(\alpha_i) = \sum_{j=1}^{l+1} b_j \alpha_j, \text{ with } b_i \neq 0$$

(otherwise $w_0(\alpha_j) \in [\theta]$, the roots which are linear combinations of the elements of θ, for all j, and this is not possible). Hence

$$w_0 w_0^\theta(\alpha_i) < 0, \text{ by } (4.1).$$

Hence $w_0 w_0^\theta(\Delta_+) = \Delta_-$, and in particular, $w_0 w_0^\theta$ maps positive imaginary roots to negative roots. This is not possible, and so we obtain Lemma 4.1.

Now Lemma 4.1 seems to have an unfortunate consequence: At least for certain maximal parabolic subgroups, one can not obtain non-trivial constant terms from Eisenstein series associated with cusp forms of the reductive part.

But as Shahidi noted, there is also good news here: the theory of Eisenstein series associated to cusp forms for the reductive part of a maximal parabolic subgroup of a loop group does not depend on the knowledge of any new L-functions, and so might be more accesible than otherwise. In fact, the constant term of such Eisenstein series can be extremely simple. In the notation of the Corollary to Theorem 3.2 , if $E_\varphi(\nu)$ denotes the convergent sum (3.12), and if we consider the case of $\hat{E}_6$ and θ_4 (as described above) then $\hat{P}_{\theta_4}$ is not associate to *any* other parabolic and is not self associate by Lemma 4.1. We then have as a consequence of Lemma 4.1:

Lemma 4.2. *The constant term*

$$E_\varphi(\nu)_{\hat U_{\theta_i}}(g) =_{df} \int_{\hat U_{\theta_i}/\hat U_{\theta_i}\cap\hat\Gamma} E_\varphi(\nu)(g\exp(-rD)u)du$$

is equal to $\varphi(m_g)\xi_\nu(a_g)$, *for* $i = 4$, *and to zero otherwise.*

In our example of $(\hat E_6, \theta_4)$ then, this is the only non-zero contribution to the constant term. In [MS2], we extended Arthur's definition of truncation to loop groups ([MS2], Definition 3.2). In the notation of that paper, we have as a consequence of the simplicity of the constant terms we have just discussed, that for $(\hat E_6, \theta_4)$:

$$\wedge^{H_0} E_\varphi(\nu)(g\eta(s)) \ (s = \exp(-r), \ \eta(s) = \exp(-rD), \text{ as in [MS 2]})$$

$$= \sum_{\gamma\in\hat\Gamma/\hat\Gamma\cap\hat P_{\theta_4}} (1 - \hat T_{\theta_4, H_0}(g\eta(s)\gamma))\varphi(m_{g\gamma})\xi_\nu(a_{g\gamma}),$$

where $m_{g\gamma}$, $a_{g\gamma}$ are recall, defined by

$$g\eta(s)\gamma = k_{g\gamma}l_{g\gamma}a_{g\gamma}\eta(s)u_{g\gamma}, \ m_{g\gamma} = l_{g\gamma}a_{g\gamma},$$

as in §3, just before Theorem 3.2, and $\hat T_{\theta, H_0}(\cdot)$ is defined in [MS2], page 736.

Now let $\nu' : \mathbb{R}c \to \mathbb{C}$ be a second, real linear map satisfying (3.11) and let $\tilde\psi$ be a second cusp form on L'_θ (and set $\psi = \tilde\psi \circ \tilde\omega$); then

$$\{\wedge^{H_0} E_\varphi(\nu), \wedge^{H_0} E_\psi(\nu')\}$$

$$=_{df} \int_{\hat K\backslash\hat G^\lambda/\hat\Gamma} \wedge^{H_0} E_\varphi(\nu)(g\eta(s))\overline{\wedge^{H_0} E_\psi(\nu')}(g\eta(s))dg,$$

and one obtains that this last expression equals

$$= -\{\varphi, \psi\}_{L'_{\theta_4}} \frac{\exp((\sigma + \bar\sigma')(H_0))}{(\sigma + \bar\sigma')(c)}.$$

The notation here is as follows:

$$\sigma = \nu + \rho, \ \sigma' = \nu' + \rho,$$

$\{,\}_{L'_{\theta_4}}$ denotes the inner product induced from a suitable Haar measure on $L'_{\theta_4}/(\Gamma^r_{\theta_4})'$, and da is a suitable Haar measure on $A(M_{\theta_4})$. To obtain this result, one uses the methods of [MS3], [MS4]. As in [MS3], one first replaces the Eisenstein series $E_\varphi(\nu)$ (and similarly, $E_\psi(\nu')$) by a pseudo-Eisenstein series: let $\Phi = \Phi(a)$ on $A(M_{\theta_4})$ be a C^∞ function with compact support, and let

$$E_\varphi(\Phi)(g\eta(s)) = \sum_{\gamma\in\hat\Gamma/\hat\Gamma\cap\hat P_{\theta_4}} \varphi(m_{g\gamma})\Phi(a_{g\gamma});$$

then $E_\varphi(\Phi)$ is called a pseudo-Eisenstein series. One lets

$$\hat\Phi(\mu) = \int_{A(M_{\theta_4})} \Phi(a)\exp(-(\mu - \rho)(\log a))d\mu_I,$$

where the notation is as follows: $\mu : \mathbb{R}c = \mathfrak{h}_{\theta_4} \to \mathbb{C}$ is real linear, and μ_I denotes the imaginary part of μ. One can then define the truncation $\wedge^{H_0} E_\varphi(\Phi)$, just as we defined $\wedge^{H_0} E_\varphi(\nu)$, and then for Ψ a second C^∞ function with compact support on $A(M_{\theta_4})$, we have for $\mu_0, \mu'_0 : \mathbb{R}c \to \mathbb{R}$ with

$$\mu_0(c) < -h^\vee, \ \mu'_0(c) < -h^\vee,$$

that, similarly to [MS3],

$$\{\wedge^{H_0} E_\varphi(\Phi), \quad \wedge^{H_0} E_\psi(\Psi)\}$$

$$= -\int_{\operatorname{Re}\mu=\mu_0} \int_{\operatorname{Re}\mu'=\mu_0'} \hat{\Phi}(\mu)\overline{\hat{\Psi}(\mu')}\Xi(\mu,\bar{\mu}')d\mu_I d\mu_I',$$

where

$$\Xi(\mu,\bar{\mu}') = \{\varphi,\psi\}_{L_{\theta_4}'} \frac{\exp(\mu+\bar{\mu}')(H_0)}{(\mu+\bar{\mu}')(c)}.$$

The argument in the present setting is in fact simpler than that in [MS3]: one does not have to contend with the infinite sums over the affine Weyl group that appear in [MS3], and one does not need to use the functional equation for c-functions, in order to show that certain poles cancel, and so, as a result, that one can move the contours of certain integrals past these (non-existent) poles. In the present setting, there are no poles from the c-functions, since non-trivial c-functions don't even occur in the formula for $\Xi(\mu,\bar{\mu}')$!

Finally, we can pass from the inner product for truncated pseudo-Eisenstein series to that for truncated Eisenstein series, as in [MS4]. In particular, we obtain that the truncated Eisenstein series $\wedge^{H_0} E_\varphi(\nu)$ is square summable.

Now the above computation is valid for

$$\operatorname{Re}\sigma(c) < -h^\vee,$$
$$\operatorname{Re}\sigma'(c) < -h^\vee,$$

or equivalently

$$(4.2) \qquad\qquad \operatorname{Re}\nu(c) < -2h^\vee,$$
$$\operatorname{Re}\nu'(c) < -2h^\vee.$$

But clearly (in ν, ν') the right side of the equality (Maass-Selberg relation)

$$(4.3) \qquad\qquad \{\wedge^{H_0} E_\varphi(\nu), \wedge^{H_0} E_\psi(\nu')\}$$

$$= -\{\varphi,\psi\}_{L_\theta'} \frac{\exp((\sigma+\bar{\sigma}')(H_0))}{(\sigma+\bar{\sigma}')(c)}$$

is holomorphic in the region (4.2), and in fact, has a holomorphic extension to the region

$$(4.4) \qquad\qquad \operatorname{Re}\nu(c) < -h^\vee,$$
$$\operatorname{Re}\nu'(c) < -h^\vee;$$

i.e.,

$$(4.4') \qquad\qquad \operatorname{Re}\sigma(c) < 0, \quad \operatorname{Re}\sigma'(c) < 0.$$

From this one can deduce that the Eisenstein series has a *holomorphic* continuation (in ν), as a locally integrable function, to the region

$$\operatorname{Re}\nu(c) < -\rho(c) = -h^\vee.$$

We emphasize again: this is a *holomorphic* continuation! - *not* just a meromorphic one.

We note that the validity of (4.3) only depends on our assumption that $\hat{P}_\theta$ is not associate to any $\hat{P}_{\theta'}$, $\theta' \neq \theta$, by virtue of $\mathfrak{l}_\theta$, $\mathfrak{l}_{\theta'}$ not being isomorphic to one another ($\mathfrak{l}_\theta$, $\mathfrak{l}_{\theta'}$ being the Lie algebras of L_θ, $L_{\theta'}$, respectively). There are of course many instances other than the case of $(\hat{E}_6, \theta_4)$ considered earlier, where this assumption holds; e.g.,

$$(\hat{E}_7, \theta), \ \theta = \Xi - \{\alpha_4\},$$

$\Xi =$ set of simple roots, α_4 as in [Bourb].

5. LOCAL ISSUES: A SUMMARY OF WHERE THINGS STAND.

The question remains: Are there applications of (4.3) and the holomorphic continuation of loop Eisenstein series to the theory of L-functions, as in the finite-dimensional case treated in [L]? The seeming paradox here is that the argument for (4.3) (which is based on Shahidi's lemma 4.1) also seems to preclude obtaining new results on L-functions: For $(\hat{E}_6, \theta_4)$ for example, Lemma 4.2 implies that L-functions do not even occur in the constant term.

It was Braverman and Kazhdan who pointed to a possible way out of this dilemma: they argued that though the constant terms with respect to the "upper triangular" $\hat{P}_\theta$ are trivial, one could consider the constant terms with respect to "lower triangular" parabolics. By a "lower triangular" parabolic one means a proper subgroup of $\hat{G}_k^\lambda$, k a field, containing the group of elements in $\hat{G}_k^\lambda$ which are lower triangular with respect to the coherently ordered basis $\mathcal{B}$ (a "lower triangular" Borel subgroup). An "upper triangular" parabolic is simply a parabolic subgroup as defined in §2.

One expects that any computation of such constant terms would depend on local computations, and in particular, would depend on suitable Gindikin-Karpelevich formulae. These formulae would have to be established for the following three cases: (i). $k = \mathbb{R}$ or $\mathbb{C}$, (ii) $k = F((t))$, F a finite field, and (iii) $k = $ a finite algebraic extension $\mathcal{K}$ of a p-adic completion of the rational numbers. We note that the results of §§1-4, above, can all be developed equally well for function fields over finite fields.Concerning cases (ii) and (iii), a Gindikin-Karpelevich formula was conjectured in [BFK]. This formula was derived by assuming that a certain result in [BFG] for $F((t))$, F a field of characteristic 0, was also valid for F a finite field. Recently, A. Braverman informed me that this was in fact the case. The resulting formula for case (ii) then also suggested the formula for case (iii). In [BGKP], we prove this conjecture (with a small modification) for both cases (ii) and (iii). The proof in [BGKP] is based on a formula of A. Braverman, D. Kazhdan, and M. Patnaik, for spherical functions on p-adic loop groups and on loop groups over $F((t))$, F a finite field.

6. BIBLIOGRAPHY

[A] J. Arthur, A trace formula for reductive groups. II: applications of a truncation operator, Compos. Math. **40**(1980), 87-121.

[Bourb] N. Bourbaki, Groupes et algèbres de Lie, Chapitres 4, 5 et 6, Hermann, Paris(1968).

[BFG] A. Braverman, M. Finkelberg and D. Gaitsgory, Uhlenbeck spaces via affine Lie algebras, The unity of mathematics, 17-135, Progr. Math., **244**, Birkhäuser, Boston MA, 2006.

[BFK] A. Braverman, M. Finkelberg, D. Kazhdan, "Affine Gindikin-Karpelevich formula via Uhlenbeck spaces, arXiv: 0912.5132 v.2 [math.RT].

[BGKP] A. Braverman, H. Garland, D. Kazhdan and M. Patnaik, A Gindikin Karpelevich formula for loop groups over non-archimedian, local fields, in preparation.

[GMRV] M. B. Green, S. D. Miller, J. G. Russo, P. Vanhove, Eisenstein series for higher-rank groups and string theory amplitudes, arXiv: 1004.0163 v. 2 [hep-th].

[LA] H. Garland, The arithmetic theory of loop algebras, J. Algebra **53**(1978), 480-551.

[LG] H. Garland, The arithmetic theory of loop groups, Inst. Hautes Études Sci. Publ. Math. **52**(1980), 5-136.

[R] H. Garland, Certain Eisenstein series on loop groups: convergence and the constant term, Proceedings of the International Conference on Algebraic Groups and Arithmetic (in Honor of M. S. Raghunathan), December 2001 (S. G. Dani and Gopal Prasad, eds.), Tata Institute of Fundamental Research, Mumbai, India, 2004, 275-319.

[AC] H. Garland, Absolute convergence of Eisenstein series on loop groups, Duke Math. J. **135**(2006), 203-260.

[MS2] H. Garland, Eisenstein series on loop groups: Maass-Selberg relations 2, Amer. J. Math. **129**(2007), 723-784.

[MS3] H. Garland, Eisenstein series on loop groups: Maass-Selberg relations 3, Amer. J. Math. **129**(2007), 1277-1353.

[MS4] H. Garland, Eisenstein series on loop groups: Maass-Selberg relations 4, Contemp. Math. **442**(2007), 115-158, Proceedings of the conference "Lie Algebras, Vertex Operator Algebras and their Applications" in honor of James Lepowsky and Robert Wilson.

[L] R. P. Langlands, Euler products, Yale Mathematical Monographs **1**, Yale University Press, New Haven CT(1971).

[L2] R. P. Langlands, On the functional equations satisfied by Eisenstein series, Lecture Notes in Mathematics **544**, Springer-Verlag, New York (1976)

[Lo] P. J. Lombardo, The constant terms of Eisenstein series on affine Kac-Moody groups over function fields, Ph.D. thesis, University of Connecticut, Storrs, CT (2010).

[MW] C. Moeglin and J.-L. Waldspurger, Spectral decomposition and Eisenstein series, Cambridge Tracts in Mathematics 113, Cambridge University Press, Cambridge (1995).

[S] F. Shahidi, Infinite dimensional groups and automorphic L-functions, Pure Appl. Math. Q. **1**(2005), 683-699.

[S2] F. Shahidi, Eisenstein series and automorphic L-funtions, American Mathematical Society Colloquium Publications, Volume 58, Providence R.I., (2010).

DEPARTMENT OF MATHEMATICS, YALE UNIVERSITY
E-mail address: garland-howard@yale.edu

Contemporary Mathematics
Volume **557**, 2011

The measurement of quantum entanglement and enumeration of graph coverings

Michael W. Hero, Jeb F. Willenbring, and Lauren Kelly Williams

It is our honor to dedicate this article to Gregg Zuckerman.

ABSTRACT. We provide formulas for invariants defined on a tensor product of defining representations of unitary groups, under the action of the product group. This situation has a physical interpretation, as it is related to the quantum mechanical state space of a multi-particle system in which each particle has finitely many outcomes upon observation. Moreover, these invariant functions separate the entangled and unentangled states, and are therefore viewed as measurements of quantum entanglement.

When the ranks of the unitary groups are large, we provide a graph theoretic interpretation for the dimension of the invariants of a fixed degree. We also exhibit a bijection between isomorphism classes of finite coverings of connected simple graphs and a basis for the space of invariants. The graph coverings are related to branched coverings of surfaces.

1. Introduction

Understanding the orbit structure of a group action is among the central themes of mathematics. That is, if a group G acts on a set $\mathcal{X}$, one wishes to parameterize the set $\mathcal{X}/G = \{\mathcal{O}_G(x) | x \in \mathcal{X}\}$, where $\mathcal{O}_G(x) = \{g \cdot x | g \in G\}$, in a natural way. This paper addresses a specific situation in this broad theme: Let $\mathcal{H}_1, \cdots, \mathcal{H}_r$ denote finite dimensional Hilbert spaces with $\dim \mathcal{H}_i = n_i$. An unsolved problem is to explicitly parameterize the orbits in the tensor product, $\mathcal{H} = \mathcal{H}_1 \otimes \cdots \otimes \mathcal{H}_r$, under the action of the product of unitary groups, $U(\mathcal{H}_1) \times \cdots \times U(\mathcal{H}_r)$, given by

$$(u_1, \cdots, u_r) \cdot (v_1 \otimes \cdots \otimes v_r) = (u_1 v_1) \otimes \cdots \otimes (u_r v_r)$$

where $u_i \in U(\mathcal{H}_i)$ are unitary operators and $v_i \in \mathcal{H}_i$ for $i = 1, \cdots, r$.

A slightly simpler, but still open, problem is to describe a set $\mathcal{F}$ of functions $f : \mathcal{H} \to \mathbb{C}$, which are invariant under the group $U(\mathcal{H}_1) \times \cdots \times U(\mathcal{H}_r)$ and separate the orbits. That is, two tensors $x, y \in \mathcal{H}$ are in the same orbit if and only if

1991 *Mathematics Subject Classification.* 22E70, 81P15, 05C30.
Key words and phrases. Classical invariant theory, graph covering, Schur-Weyl duality, quantum entanglement.
This research was supported in part by the University of Wisconsin - Milwaukee, Research Growth Initiative Grant, and by National Security Agency grant # H98230-09-0054.

$f(x) = f(y)$ for all $f \in \mathcal{F}$. Such a set $\mathcal{F}$ does indeed exist in the algebra of polynomial functions on the underlying real vector space of $\mathcal{H}$ (see [**MW02**]).

The motivation for studying this particular group action goes back to [**EPR35**], and in the literature is often described in context with the physical effect known as "quantum entanglement", which has gained enormous popularity as the effect suggests vastly improved models of computation (see [**Fey81**]).

In line with this nomenclature, the invariant functions on $\mathcal{H}$ are called "measurements of quantum entanglement". The primary purpose of the present article is to point out some additional combinatorial/geometric structure related to an earlier work by the first two authors in [**HW09**]. Specifically, the enumeration problems addressed in [**HW09**] can be translated into enumeration problems addressed in [**KL01**]. We recall the situation briefly and then provide some examples illustrating a correspondence between coverings of simple graphs and the measurements of quantum entanglement.

1.1. General Setup. We recall a general situation which includes the above problem. Let K be a compact Lie group acting $\mathbb{C}$-linearly on a finite dimensional complex vector space V. It is a difficult problem in representation theory to provide a description of the K-orbits in V. One approach set out in [**MW02**] and [**Wal05**] is to use the invariant theory of K to separate orbits. More precisely: Set $\mathcal{P}_{\mathbb{R}}(V)$ to be the algebra of complex valued polynomial functions on the vector space V when viewed as a real vector space. The group K acts in the standard way on $\mathcal{P}_{\mathbb{R}}(V)$ by $g \cdot f(v) = f(g^{-1}v)$ for $g \in K$, $f \in \mathcal{P}_{\mathbb{R}}(V)$ and $v \in V$. Let the algebra of K-invariants in $\mathcal{P}_{\mathbb{R}}(V)$ be denoted by $\mathcal{P}_{\mathbb{R}}(V)^K$. We have:

THEOREM. *(c.f. Theorem 3.1 of* [**MW02**]*) If $v, w \in V$ then $f(v) = f(w)$ for all $f \in \mathcal{P}_{\mathbb{R}}(V)^K$ if and only if $\mathcal{O}_K(v) = \mathcal{O}_K(w)$; that is, v and w are in the same K-orbit.*

Fix a sequence of positive integers $\mathbf{n} = (n_1, \cdots, n_r)$. Let

$$V(\mathbf{n}) = \mathbb{C}^{n_1} \otimes \mathbb{C}^{n_2} \otimes \cdots \otimes \mathbb{C}^{n_r}$$

be the representation of $K(\mathbf{n}) = \prod_{i=1}^{r} U(n_i)$ under the standard action on each tensor[1] factor. (Here $U(n)$ denotes the group of $n \times n$ unitary[2] matrices.)

Well known results of Hilbert establish that the $K(\mathbf{n})$-invariant subalgebra of $\mathcal{P}_{\mathbb{R}}(V(\mathbf{n}))$ is finitely generated. In spite of this result, our situation lacks a complete description of such generators, except for certain small values of the parameter space $\mathbf{n} = (n_1, n_2, \cdots, n_r)$. We do not solve this problem here, but make an encouraging first step: We provide formulas for a set of polynomials that span the vector space of $K(\mathbf{n})$-invariants in $\mathcal{P}_{\mathbb{R}}(V(\mathbf{n}))$. Within a certain "stable range" this spanning set is linearly independent.

The $K(\mathbf{n})$-invariant subalgebra inherits a gradation from $\mathcal{P}_{\mathbb{R}}(V(\mathbf{n}))$. Thus, let $\mathcal{P}_{\mathbb{R}}^d(V(\mathbf{n}))$ denote the subspace of degree d homogeneous polynomial functions contained in $\mathcal{P}_{\mathbb{R}}(V(\mathbf{n}))$. We set $\mathcal{P}_{\mathbb{R}}^d(V(\mathbf{n}))^{K(\mathbf{n})} = \mathcal{P}_{\mathbb{R}}^d(V(\mathbf{n})) \cap \mathcal{P}_{\mathbb{R}}(V(\mathbf{n}))^{K(\mathbf{n})}$. One can see easily that $\mathcal{P}_{\mathbb{R}}^d(V(\mathbf{n}))^{K(\mathbf{n})} = (0)$ for d odd (see Lemma 2.1). However, the dimension of $\mathcal{P}_{\mathbb{R}}^d(V(\mathbf{n}))^{K(\mathbf{n})}$ for even d is more subtle. Set $h_m(\mathbf{n}) =$

---

[1]Here we tensor over $\mathbb{C}$.

[2]A *unitary* matrix, u, is an invertible complex matrix s.t. $\bar{u}^t = u^{-1}$.

$\dim \mathcal{P}_{\mathbb{R}}{}^{2m}(V(\mathbf{n}))^{K(\mathbf{n})}$. For fixed $\mathbf{n}$, the formal power series in q,

$$h_0(\mathbf{n}) + h_1(\mathbf{n})q + h_2(\mathbf{n})q^2 + \cdots$$

is called the *Hilbert series* of the $K(\mathbf{n})$-invariant subalgebra. As we shall see, calculating these coefficients is a step in the solution to the problem of finding a vector space basis of the invariants.

In [**HW09**], it is shown that for fixed $d = 2m$ and r the value of $h_m(\mathbf{n})$ stabilizes as the components of $\mathbf{n}$ grow large. Consequently, we can define

$$\widetilde{h}_{m,r} = \lim_{n_1 \to \infty} \lim_{n_2 \to \infty} \cdots \lim_{n_r \to \infty} h_m(n_1, \cdots, n_r).$$

Several papers in the recent literature investigate Hilbert series related to measurements of quantum entanglement. See, for example [**MW02, Wal05**]. Despite the fact that the value of $h_m(\mathbf{n})$ is not known in general, the value of $\widetilde{h}_{m,r}$ has a surprisingly simple description, which we present next.

We first set up the standard notation for partitions, which we define as weakly decreasing finite sequences of positive integers. We will always use lower case Greek letters to denote partitions. We will write $\lambda \vdash m$ to indicate that λ is a partition of size m. Lastly, if λ has a_1 ones, a_2 twos, a_3 threes etc., let

$$z_\lambda = 1^{a_1} 2^{a_2} 3^{a_3} \cdots a_1! a_2! a_3! \cdots$$

We have

THEOREM. *(c.f. Theorem 1.1 of [**HW09**]) For any integers $m \geq 0$ and $r \geq 1$,*

$$(1.1) \qquad \widetilde{h}_{m,r} = \sum_{\lambda \vdash m} z_\lambda^{r-2}$$

1.2. A Combinatorial Interpretation.

Let S_m denote the symmetric group on the set $\{1, \cdots, m\}$. The r-fold cartesian product, denoted,

$$S_m^r = \{\mathbf{s} = (\sigma_1, \sigma_2, \cdots, \sigma_r) | \sigma_i \in S_m \text{ for all } i\}$$

is acted upon by $S_m \times S_m$ under the action, $(\alpha, \beta) \cdot \mathbf{s} = \alpha \mathbf{s} \beta^{-1}$ where $\alpha \mathbf{s} \beta^{-1} = (\alpha \sigma_1 \beta^{-1}, \cdots, \alpha \sigma_r \beta^{-1})$. The orbits of this group action are the double cosets, $\Delta \backslash S_m^r / \Delta$, where $\Delta = \{(\sigma, \cdots, \sigma) | \sigma \in S_m\}$. Next, we shall see that the number of these double cosets is $\widetilde{h}_{m,r}$.

In [**HW09**], it is shown that $\widetilde{h}_{m,r}$ is the number of orbits under the S_m-action of "simultaneous conjugation",

$$\gamma \mathbf{s} \gamma^{-1} = \left(\gamma \sigma_1 \gamma^{-1}, \cdots, \gamma \sigma_{r-1} \gamma^{-1} \right),$$

on S_m^{r-1}. Denote these orbits by $\mathcal{O} = S_m^{r-1}/S_m$. There exists a map $\theta : \Delta \backslash S_m / \Delta \to \mathcal{O}$ defined for $\mathbf{s} = (\sigma_1, \cdots, \sigma_r)$ by

$$\theta(\Delta \mathbf{s} \Delta) = \{\gamma(\sigma_1 \sigma_r^{-1}, \sigma_2 \sigma_r^{-1}, \cdots, \sigma_{r-1}\sigma_r^{-1})\gamma^{-1} \mid \gamma \in S_m\}.$$

It is easy to see that θ is independent of the representative $\mathbf{s}$, and defines a bijective function from $\Delta \backslash S_m / \Delta$ to $\mathcal{O}$.

In Section 2, we show how to define a spanning set for the invariant tensors on $V(\mathbf{n}) \oplus V(\mathbf{n})^*$ parameterized by the set S_m^r. This is a simple consequence of Schur-Weyl duality (see Theorem 2). Then, in Section 3 we show how this spanning set projects onto the invariants in the symmetric tensors on $V(\mathbf{n}) \oplus V(\mathbf{n})^*$. After projecting, many equalities arise, and we show how a spanning set for the invariants is naturally parameterized by $\Delta \backslash S_m^r / \Delta$ and $\mathcal{O}$.

The right hand side of Equation 1.1 may be interpreted in certain graph enumeration problems, which we recall following [**KL01**]. Let $\mathcal{G} = \mathcal{G}(\mathcal{V}, \mathcal{E})$ be a simple connected graph, with vertex set $\mathcal{V}$ and edge set $\mathcal{E}$. Let $\beta(\mathcal{G}) = |\mathcal{E}| - |\mathcal{V}| + 1$, which is the number of independent cycles in $\mathcal{G}$ (the first Betti number). Let $N(v)$ denote the neighborhood[3] of a vertex $v \in \mathcal{V}$. A graph $\tilde{\mathcal{G}}$ is said to be a covering of $\mathcal{G}$ with projection $p : \tilde{\mathcal{G}} \to \mathcal{G}$ if there exists a surjection $p : \tilde{\mathcal{V}} \to \mathcal{V}$ such that $p|_{N(\tilde{v})} : N(\tilde{v}) \to N(v)$ is a bijection for any $v \in \mathcal{V}$ and $\tilde{v} \in p^{-1}(v)$. If p is n-to-one, we say $p : \tilde{\mathcal{G}} \to \mathcal{G}$ is an n-fold covering. In the image below, $\tilde{\mathcal{G}}$ is a 2-fold covering of $\mathcal{G}$. We see that the neighborhood of a black vertex of $\tilde{\mathcal{G}}$ maps injectively onto the neighborhood of the black vertex of $\mathcal{G}$.

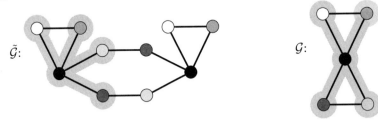

Two coverings $p_i : \tilde{\mathcal{G}}_i \to \mathcal{G}$, $i = 1, 2$ are said to be isomorphic if there exists a graph isomorphism $\Phi : \tilde{\mathcal{G}}_1 \to \tilde{\mathcal{G}}_2$ such that the following diagram commutes:

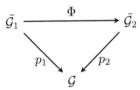

The right hand side of Equation 1.1 is equal to the number of isomorphism classes of m-fold coverings of $\mathcal{G}$ with $\beta(\mathcal{G}) = r - 1$ (see [**KL90**]).

In light of this graphical interpretation of Equation 1.1, one anticipates a bijective correspondence between finite graph coverings and measurements of quantum entanglement. Indeed, such a correspondence exists, which we illustrate in the next section.

1.3. The correspondence. If V is a complex vector space then we denote[4] the complex valued polynomial functions on V by $\mathcal{P}(V)$. Suppose that a compact Lie group, K, acts $\mathbb{C}$-linearly on V. The K-action on V gives rise to an action on $\mathcal{P}(V)$ by $k \cdot f(v) = f(k^{-1}v)$ for $k \in K$, $f \in \mathcal{P}(V)$ and $v \in V$. Both $\mathcal{P}(V)$ and $\mathcal{P}_\mathbb{R}(V)$ are complex vector spaces with a natural gradation by degree. As a graded representation, $\mathcal{P}_\mathbb{R}(V) \cong \mathcal{P}\left(V \oplus \overline{V}\right)$, where $\overline{V}$ denotes the complex vector space with the opposite complex structure (see [**MW02**]). Let V^* refer to the representation on the complex valued linear functionals on V defined by $(k \cdot \lambda)(v) = \lambda(k^{-1}v)$ for $v \in V$, $\lambda \in V^*$, and $k \in K$. As a representation of K, $\overline{V}$ is equivalent to V^*.

In what is to follow, we will complexify the compact group, K, to a complex reductive linear algebraic group. All representations of G will be assumed to be

---

[3]The neighborhood $N(v)$ of $v \in \mathcal{V}$ is the set of all vertices in $\mathcal{V}$ adjacent to v.

[4]Here we are viewing V as a *complex* space rather than a *real* space, as we do in defining $\mathcal{P}_\mathbb{R}(V)$.

regular. That is, the matrix coefficients are regular functions on the underlying affine variety G. An irreducible regular representation restricts to an irreducible complex representation of K. Furthermore, since K is Zariski dense in G, regular representations of G (and hence G-invariants) are determined on K. Note that $G = \mathrm{GL}(n)$ when $K = U(n)$.

We now specialize to $V = V(\mathbf{n}) = \mathbb{C}^{n_1} \otimes \cdots \otimes \mathbb{C}^{n_r}$, and set up notation for the coordinates in V and V^*. For positive integers k and n, let $\mathrm{Mat}_{n,k}$ denote the vector space of $n \times k$ complex matrices. Let $E^i_j \in \mathrm{Mat}_{n,k}$ denote the matrix with entry in row i and column j equal to 1 and all other entries 0. The group of $n \times n$ invertible matrices with complex number entries will be denoted by $\mathrm{GL}(n)$. This group acts on $\mathrm{Mat}_{n,k}$ by multiplication on the left. We identify $\mathbb{C}^n = \mathrm{Mat}_{n,1}$, which has a distinguished ordered basis consisting of $e_i = E^i_1 \in \mathrm{Mat}_{n,1}$ for $i = 1, \cdots, n$.

In the case of $G = \mathrm{GL}(n)$ we will identify $(\mathbb{C}^n)^*$ with the representation on $\mathrm{Mat}_{1,n}$ defined by the action $g \cdot v = vg^{-1}$ for $v \in \mathrm{Mat}_{1,n}$ and $g \in \mathrm{GL}(n)$. Set $e^i = E^i_i \in \mathrm{Mat}_{1,n}$ for $i = 1, \cdots, n$. Then, $(e^1, \cdots, e^n)$ is an ordered basis for $(\mathbb{C}^n)^*$, dual to $(e_1, \cdots, e_n)$.

Arbitrary tensors in $V(\mathbf{n})$ and $V(\mathbf{n})^*$ are of the form

$$\sum x^{i_1 \cdots i_r} e_{i_1} \otimes \cdots \otimes e_{i_r} \in V(\mathbf{n}),$$

and

$$\sum y_{i_1 \cdots i_r} e^{i_1} \otimes \cdots \otimes e^{i_r} \in V(\mathbf{n})^*,$$

where $x^{i_1 \cdots i_r}$ and $y_{i_1 \cdots i_r}$ are complex scalars. We may view the variables $x^{i_1 \cdots i_r}$ and $y_{i_1 \cdots i_r}$ as degree 1 polynomial functions in $\mathcal{P}_{\mathbb{R}}(V(\mathbf{n}))$, where $y_{i_1 \cdots i_r}$ are the complex conjugates of $x^{i_1 \cdots i_r}$.

Let $\mathcal{G}$ be a connected simple graph with $\beta(\mathcal{G}) = r - 1$. In [**KL01**] the isomorphism classes of m-fold covers of $\mathcal{G}$ are parameterized by the orbits in $S_m^{r-1} = S_m \times \cdots \times S_m$ ($r-1$ factors) under the conjugation action of S_m. Thus, one expects to form a basis[5] element of the space of degree $2m$ invariants from a choice, up to conjugation, of $r - 1$ permutations. Let

$$[\sigma_1, \cdots, \sigma_{r-1}] = \left\{ \tau(\sigma_1, \cdots, \sigma_{r-1})\tau^{-1} : \tau \in S_m \right\}$$

be such a choice. We present now an invariant associated with $[\sigma_1, \cdots, \sigma_{r-1}]$.

We define $f_{[\sigma_1, \cdots, \sigma_{r-1}]}$ as the sum over the indices

$$I_1 = (i_1^{(1)} i_2^{(1)} \cdots i_r^{(1)}), \cdots, I_m = (i_1^{(m)} i_2^{(m)} \cdots i_r^{(m)})$$

where $1 \leq i_k^{(j)} \leq n_k$ (with $j = 1, \cdots, m$) of

$$x^{I_1} \cdots x^{I_m} y_{i_1^{(\sigma_1(1))} \cdots i_{r-1}^{(\sigma_{r-1}(1))} i_r^{(1)}} \cdots y_{i_1^{(\sigma_1(m))} \cdots i_{r-1}^{(\sigma_{r-1}(m))} i_r^{(m)}}.$$

We simultaneously parameterize the degree $2m$ polynomial invariants and m-fold coverings of simple connected graphs in the following way. Given a double coset S_m^r, applying θ, one obtains an $(r-1)$-tuple of permutations $(\sigma_1, \cdots, \sigma_{r-1})$. Combinatorially, we can encode the S_m-orbit of $(\sigma_1, \cdots, \sigma_{r-1})$ under the simultaneous conjugation action by coloring each permutation. This action "forgets" the labels of the domain and range of each permutation. The resulting combinatorial data takes the form of an unlabeled directed graph with edges colored by $r - 1$ colors.

---

[5]In general, one obtains a spanning set for the invariants. However, if $n_i \geq m$ for all i then we have a basis for the degree $2m$ invariants.

We will now exhibit this process in the case $r = m = 3$. Consider

$$S_3(\tilde{\sigma}_1, \tilde{\sigma}_2, \tilde{\sigma}_3)S_3$$

where $\tilde{\sigma}_1 = (1\ 3\ 2), \tilde{\sigma}_2 = (2\ 3), \tilde{\sigma}_3 = (1\ 3)$. Then

$$\theta(\tilde{\sigma}_1, \tilde{\sigma}_2, \tilde{\sigma}_3) = (\tilde{\sigma}_1\tilde{\sigma}_3^{-1}, \tilde{\sigma}_2\tilde{\sigma}_3^{-1}) = (\sigma_1, \sigma_2) = ((1\ 2), (1\ 2\ 3)).$$

$((1\ 2), (1\ 2\ 3))$ $\tau((1\ 2), (1\ 2\ 3))\tau^{-1}$ 3-fold Cover of a "Figure 8"

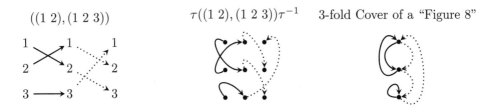

Using the invariant defined above, we have $f_{[\sigma_1, \sigma_2]}$ is the sum of terms of the form

$$x^{i_1^{(1)} i_2^{(1)} i_3^{(1)}} x^{i_1^{(2)} i_2^{(2)} i_3^{(2)}} x^{i_1^{(3)} i_2^{(3)} i_3^{(3)}} y_{i_1^{(\sigma_1(1))} i_2^{(\sigma_2(1))} i_3^{(1)}} y_{i_1^{(\sigma_1(2))} i_2^{(\sigma_2(2))} i_3^{(2)}} y_{i_1^{(\sigma_1(3))} i_2^{(\sigma_2(3))} i_3^{(3)}}$$

$$= x^{i_1^{(1)} i_2^{(1)} i_3^{(1)}} x^{i_1^{(2)} i_2^{(2)} i_3^{(2)}} x^{i_1^{(3)} i_2^{(3)} i_3^{(3)}} y_{i_1^{(2)} i_2^{(2)} i_3^{(1)}} y_{i_1^{(1)} i_2^{(3)} i_3^{(2)}} y_{i_1^{(3)} i_2^{(1)} i_3^{(3)}}$$

All possible diagrams for the $r = m = 3$ case are shown below:

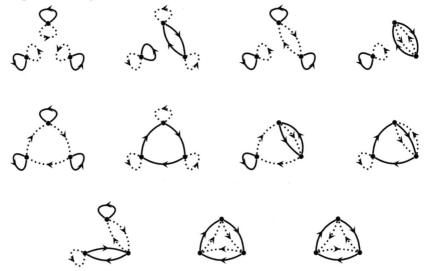

Each coloring of the directed graphs corresponds to an isomorphism class of m-fold covering of a connected simple graph G with $\beta(G) = r - 1$ (see [**KL01**]). We illustrate this correspondence for $m = 2$, and $r = 2, 3, 4$. In the following table, the simple graph is homotopic to a bouquet of loops (on the left) and the possible graph coverings are on the right. The colors and orientations determine the covering map. The corresponding $K(\mathbf{n})$-invariants are written out explicitly following the Einstein summation convention.

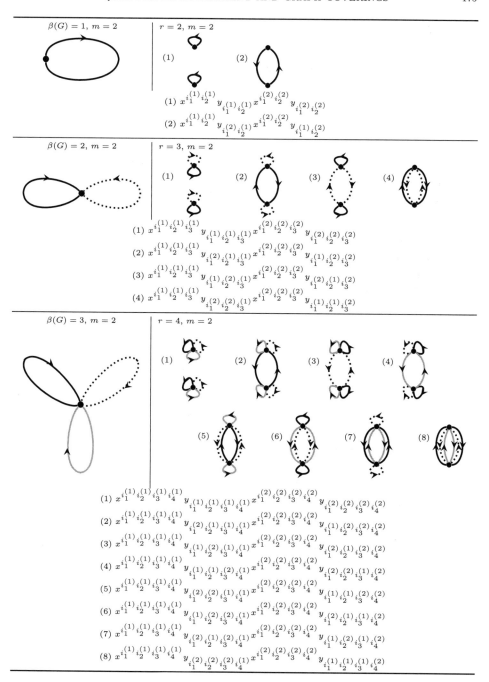

Finally, the fact that all invariants fall into this correspondence follows from

THEOREM 1.1. *For all $n_1, \cdots, n_r$ and $d = 2m$, we have*

$$Span\{f_{[\sigma_1, \cdots, \sigma_{r-1}]} : (\sigma_1, \cdots, \sigma_{r-1}) \in S_m^{r-1}\} = \mathcal{P}_{\mathbb{R}}{}^d(V(\mathbf{n}))^{K(\mathbf{n})}.$$

PROOF. Follows from Theorem 3.3 proved in Section 3.   □

Acknowledgments We would like to thank Jan Draisma for valuable comments in an initial draft of this manuscript, including the observation that Theorem 1.1 follows from standard arguments in classical invariant theory. We would also like to thank Nolan Wallach for many helpful comments.

2. Invariants in the tensor algebra

The group $\mathrm{GL}(n)$ has the structure of a reductive linear algebraic group over the field $\mathbb{C}$. This article concerns the regular representations of such groups, which are closed under the operations of direct sum, tensor product, and duality. That is, in general, if V_1 and V_2 are regular representations of a linear algebraic group G then $V_1 \oplus V_2$ is a regular representation of G defined by $g \cdot (v_1, v_2) = (gv_1, gv_2)$ and $V_1 \otimes V_2$ is a regular representation of G defined by $g \cdot (v_1 \otimes v_2) = (gv_1) \otimes (gv_2)$ (for $g \in G$ and $v_i \in V_i$), and extending by linearity. Also if V_1 (resp. V_2) is a regular representation of an algebraic group G_1 (resp. G_2) then $V_1 \otimes V_2$ also denotes the representation of $G_1 \times G_2$ defined by $(g_1, g_2) \cdot (v_1 \otimes v_2) = (g_1 v_1) \otimes (g_2 v_2)$. In the case where $G_1 = G_2 = G$ then both G and $G \times G$ act on $V_1 \otimes V_2$. The latter will be referred to as the "outer" action of $G \times G$ whose restriction to the diagonal subgroup, $\{(g, g) : g \in G\}$, is equivalent to the former action, referred to as the "inner" action of G. Throughout we will be careful to distinguish between these two actions, when there is ambiguity.

Let $\mathbf{n} = (n_1, \cdots, n_r)$ denote an r-tuple of positive integers. Let $V(\mathbf{n}) = \mathbb{C}^{n_1} \otimes \mathbb{C}^{n_2} \otimes \cdots \otimes \mathbb{C}^{n_r}$, which is an irreducible representation of the group $\mathrm{G}(\mathbf{n}) = \mathrm{GL}(n_1) \times \cdots \times \mathrm{GL}(n_r)$, under the outer action defined by

$$(g_1, \cdots, g_r) \cdot (v_1 \otimes \cdots \otimes v_r) = (g_1 v_1) \otimes \cdots \otimes (g_r v_r),$$

where for all i, $g_i \in \mathrm{GL}(n_i)$, $v_i \in \mathbb{C}^{n_i}$, and extending by linearity.

For a non-negative integer d, let

$$\mathcal{T}^d(\mathbf{n}) = \bigotimes\nolimits^d \left[V(\mathbf{n}) \oplus V(\mathbf{n})^* \right].$$

The group $\mathrm{G}(\mathbf{n})$ acts on $\mathcal{T}^d(\mathbf{n})$ by the inner action on the d-fold tensors on $V(\mathbf{n}) \oplus (V(\mathbf{n}))^*$. The goal of this section is to find a spanning set for the $\mathrm{G}(\mathbf{n})$-invariants, $\mathcal{T}^d(\mathbf{n})^{\mathrm{G}(\mathbf{n})}$. We shall see shortly that, upon examination of the action of the center of $\mathrm{G}(\mathbf{n})$, $\mathcal{T}^d(\mathbf{n})^{\mathrm{G}(\mathbf{n})} = \{0\}$ for d odd. Thus, we assume $d = 2m$ for a non-negative integer m.

If V is a vector space, we introduce the notation $V_0 = V$ and $V_1 = V^*$. We let $\mathbf{b} = (b_1, b_2, \cdots, b_d)$ denote a d-tuple of zeros and ones. Set $\mathbb{T}(\mathbf{n}, \mathbf{b}) = \bigotimes_{i=1}^d V(\mathbf{n})_{b_i}$. We have

$$\mathcal{T}^d(\mathbf{n}) = \bigoplus_{\mathbf{b}} \mathbb{T}(\mathbf{n}, \mathbf{b}).$$

where the sum is over all d-long $\{0, 1\}$-sequences, $\mathbf{b}$. This equality follows from the bi-linearity of the tensor product (ie: $(A \oplus B) \otimes (C \oplus D) = A \otimes C \oplus A \otimes D \oplus B \otimes C \oplus B \otimes D$). Thus, the problem of finding a basis for $\mathcal{T}^d(\mathbf{n})^{\mathrm{G}(\mathbf{n})}$ reduces to finding a basis for $\mathbb{T}(\mathbf{n}, \mathbf{b})^{\mathrm{G}(\mathbf{n})}$ for each $\mathbf{b}$. The tensor factors in $\mathbb{T}(\mathbf{n}, \mathbf{b})$ may be re-ordered by defining

$$T_n(\mathbf{b}) = \bigotimes_{i=1}^d (\mathbb{C}^n)_{b_i},$$

which is a representation of GL(n) wrt the inner action. Then, permuting the tensor factors so that those involving $\mathbb{C}^{n_i}$ are grouped together defines an isomorphism

$$\Phi_{\mathbf{n}}^{\mathbf{b}} : \mathbb{T}(\mathbf{n}, \mathbf{b}) \to T_{n_1}(\mathbf{b}) \otimes \cdots \otimes T_{n_r}(\mathbf{b})$$

of G($\mathbf{n}$)-representations. We will now obtain a basis for the G($\mathbf{n}$)-invariants of $\bigotimes_{i=1}^{r} T_{n_i}(\mathbf{b})$, for each $\mathbf{b}$, under the assumption that n_i are large with respect to d.

The center, denoted $\mathcal{Z}(\mathbf{n})$, of G($\mathbf{n}$) is $\{(x_1 I_{n_1}, \cdots, x_r I_{n_r}) | x_1, \cdots, x_r \in \mathbb{C}^\times\}$ where I_k is the $k \times k$ identity matrix ($k \in \mathbb{Z}^+$). Let $|\mathbf{b}|_1 = \sum_i b_i$ denote the number of 1's in $\mathbf{b}$, while $|\mathbf{b}|_0 = d - |\mathbf{b}|_1$ is the number of 0's in $\mathbf{b}$.

LEMMA 2.1. *For any $\mathbf{n}$ and $\mathbf{b}$, we have* $\mathbb{T}(\mathbf{n}, \mathbf{b})^{G(\mathbf{n})} = \{0\}$ *if* $|\mathbf{b}|_1 \neq |\mathbf{b}|_0$.

PROOF. The center of GL(n_i), $\{x_i I_{n_i} | x_i \in \mathbb{C}^\times\}$, acts on $T_{n_i}(\mathbf{b})$ by the scalar $x_i^{|\mathbf{b}|_0 - |\mathbf{b}|_1}$. Therefore $\mathcal{Z}(\mathbf{n})$ acts trivially on $\mathbb{T}(\mathbf{n}, \mathbf{b})$ exactly when $|\mathbf{b}|_1 = |\mathbf{b}|_0$. $\square$

As a consequence of Lemma 2.1, we will assume from this point on that $d = 2m$ is even, and that any $\mathbf{b}$ is to have m 1's and m 0's. The converse of Lemma 2.1 is also true which we address next.

Suppose that $\mathbf{b} = 0^m 1^m$ where 1^m (resp. 0^m) is a sequence of m 1's (resp. 0's). Then, for all n we have $T_n(\mathbf{b}) = (\bigotimes^m \mathbb{C}^n) \otimes (\bigotimes^m \mathbb{C}^n)^*$. For each permutation $\sigma \in S_m$, define:

$$t_n(\sigma) = \sum_{\substack{(j_1, \cdots, j_m) \\ 1 \le j_k \le n, \forall k}} (e_{j_1} \otimes \cdots \otimes e_{j_m}) \otimes (e^{j_{\sigma(1)}} \otimes \cdots \otimes e^{j_{\sigma(m)}})$$

(Recall that in the above sum, e_k, and e^k, are the ordered bases of $\mathbb{C}^n$ and $(\mathbb{C}^n)^*$ respectively defined previously.)

THEOREM (Schur-Weyl Duality). *Let n and m be positive integers. The tensor product $\bigotimes^m \mathbb{C}^n$ is a representation of GL(n) under the inner action, and also a representation of S_m as defined by permutation of the tensor factors. Each of these actions generates the full commuting associative algebra action. Thus, we obtain a surjective algebra homomorphism*

$$\mathbb{C}[S_m] \to End_{GL(n)}\left(\bigotimes^m \mathbb{C}^n\right),$$

which is an isomorphism if and only if $n \ge m$.

PROOF. See [**GW09**] Section 4.2.4 and Chapter 9. $\square$

PROPOSITION 2.2. *For all n, if $n \ge m$ then $\{t_n(\sigma) | \sigma \in S_m\}$ is a basis for $T_n(0^m 1^m)^{GL(n)}$; otherwise, outside of these inequalities, the above are a spanning set.*

PROOF. Observe that

$$T_n(\mathbf{b})^{GL(n)} = \left[\left(\bigotimes^m \mathbb{C}^n\right) \otimes \left(\bigotimes^m \mathbb{C}^n\right)^*\right]^{GL(n)} \cong End_{GL(n)}(\bigotimes^m \mathbb{C}^n)$$

The result follows from Schur-Weyl duality. $\square$

For an r-tuple of permutations $\mathbf{s} = (\sigma_1, \cdots, \sigma_r) \in S_m^r$, define $t(\mathbf{s}) = t_{n_1}(\sigma_1) \otimes \cdots \otimes t_{n_r}(\sigma_r)$. We now obtain

COROLLARY 2.3. *Given m and $\mathbf{n}$, if for all i with $1 \leq i \leq r$ we have $n_i \geq m$, then*

$$\{t(\mathbf{s}) | \mathbf{s} \in S_m^r\}$$

is a vector space basis for the $G(\mathbf{n})$-invariants in $T_{n_1}(\mathbf{b}) \otimes \cdots \otimes T_{n_r}(\mathbf{b})$ in the case when $\mathbf{b} = 0^m 1^m$. Otherwise, outside of these inequalities, the above are a spanning set.

PROOF. Given a finite collection of finite dimensional vector spaces $V_1, \cdots, V_r$, if B_i is a basis for V_i then $\{v_1 \otimes \cdots \otimes v_r | v_j \in B_j \text{ for all } 1 \leq j \leq r\}$ is a basis for $\bigotimes_{i=1}^r V_i$. Apply to the situation where $V_i = T_{n_i}(\mathbf{b})^{\mathrm{GL}(n_i)}$ and $B_i = \{t_\sigma^i | \sigma \in S_m\}$. Apply Proposition 2.2. $\qquad\square$

We now turn to the situation where $\mathbf{b}$ is not necessarily $0^m 1^m$. If M is an m-element subset of $\{1, \cdots, 2m\}$ define

$$\mathbf{b}_M = (b_1, \cdots, b_{2m}) \text{ where: } b_j = \begin{cases} 1, & j \in M; \\ 0, & j \notin M. \end{cases}$$

Let $\gamma_M \in S_{2m}$ denote a (indeed any[6]) permutation of $\{1, \cdots, 2m\}$ such that γ_M permutes the coordinates of $0^m 1^m$ to obtain b_M.

LEMMA 2.4. *For any m-element subset, M, of $\{1, \cdots, 2m\}$ there exists an isomorphism of $GL(n)$-representations,*

$$\Psi_n^M : T_n(0^m 1^m) \to T_n(\mathbf{b}_M)$$

PROOF. Permuting tensor factors does not change the isomorphism class of an inner tensor product. Define Ψ_n^M so as to permute the tensor factors using γ_M. $\qquad\square$

PROPOSITION 2.5. *Given m and $n \geq m$, we have that for any m-element subset M of $\{1, \cdots, 2m\}$, a basis for the $GL(n)$-invariants in $T_n(\mathbf{b}_M)$ is given by*

$$\left\{\Psi_n^M\left(t_n(\sigma)\right) | \sigma \in S_m\right\}.$$

Otherwise, outside of these inequalities, the above are a spanning set.

PROOF. Follows immediately from the isomorphism Ψ_n^M (Lemma 2.4) and the statement of Proposition 2.2. $\qquad\square$

For $\mathbf{s} = (\sigma_1, \cdots, \sigma_r) \in S_m^r$ define $t_{\mathbf{n}}^M(\mathbf{s}) = \Psi_{n_1}^M(t_{n_1}(\sigma_1)) \otimes \cdots \otimes \Psi_{n_r}^M(t_{n_r}(\sigma_r))$.

COROLLARY 2.6. *Given m and $\mathbf{n}$ such that for all i with $1 \leq i \leq r$ we have $n_i \geq m$, then for all m-element subsets, M, of $\{1, \cdots, 2m\}$ the set $\{t_{\mathbf{n}}^M(\mathbf{s}) | \mathbf{s} \in S_m^r\}$ is a basis for the $G(\mathbf{n})$-invariants in $\bigotimes_{i=1}^r T_{n_i}(\mathbf{b}_M)$. Otherwise, outside these inequalities, the above are a spanning set.*

PROOF. Apply the isomorphism $\Psi_{n_i}^M$ to each tensor factor in Corollary 2.3. $\qquad\square$

We note that when $\mathbf{b} = \mathbf{b}_M$, $t_{\mathbf{n}}^M(\mathbf{s})$ is in the range of $\Phi_{\mathbf{n}}^{\mathbf{b}}$, defined above Lemma 2.1. Finally, set $\phi_{\mathbf{n}}^M(\mathbf{s}) = (\Phi_{\mathbf{n}}^{\mathbf{b}_M})^{-1}\left(t_{\mathbf{n}}^M(\mathbf{s})\right)$. We obtain

---

[6]Note that there are $(m!)^2$ possible choices for γ_M

COROLLARY 2.7. *Given m and $\mathbf{n}$ such that for all i with $1 \leq i \leq r$ we have $n_i \geq m$, a basis for the $G(\mathbf{n})$-invariants in $\mathcal{T}^{2m}(\mathbf{n})$ is*

$$\left\{\phi_{\mathbf{n}}^{M}(\mathbf{s}) \middle| \mathbf{s} \in S_m^r \text{ and } M \subseteq \{1, \cdots, 2m\} \text{ with } |M| = m\right\}$$

Otherwise, outside of these inequalities, the above are a spanning set. Therefore,

$$\dim\left(\mathcal{T}^{2m}(\mathbf{n})\right)^{G(\mathbf{n})} \leq \binom{2m}{m}(m!)^r.$$

with equality holding exactly when $n_i \geq m$ for all $1 \leq i \leq r$.

3. Invariants in the symmetric algebra

For a complex vector space V, let $\bigotimes V = \bigoplus_{d \geq 0} \bigotimes^d V$ and $\mathcal{S}(V) = \bigoplus_{d=0}^{\infty} \mathcal{S}^d(V)$ denote the $\mathbb{N}$-graded tensor and symmetric algebras respectively. Recall that $\mathcal{S}(V)$ is defined as the quotient of the tensor algebra by the two sided ideal, $\langle x \otimes y - y \otimes x | x, y \in V\rangle$. The map $p : \bigotimes(V) \to \mathcal{S}(V)$ defined by

$$p(x_1 \otimes x_2 \otimes \cdots \otimes x_d) = x_1 \cdot x_2 \cdot \cdots \cdot x_d$$

on $\bigotimes^d V$ defines a surjective homomorphism of graded associative $\mathbb{C}$-algebras. Note, of course, that the product on $\bigotimes(V)$, denoted by $\otimes$, is non-commutative, while the product on $\mathcal{S}(V)$, denoted by $\cdot$, is commutative.

PROPOSITION 3.1. *Given m and $\mathbf{n}$ such that for all $1 \leq i \leq r$ we have $n_i \geq m$, the set*

$$\left\{p\left(\phi_{\mathbf{n}}^{M}(\mathbf{s})\right) \middle| \mathbf{s} \in S_m^r \text{ and } M \subseteq \{1, \cdots, 2m\} \text{ with } |M| = m\right\}$$

spans $\mathcal{S}^{2m}(V(\mathbf{n}) \oplus V(\mathbf{n})^*)^{G(\mathbf{n})}$.

PROOF. The group $G(\mathbf{n})$ is reductive, therefore the restriction of p to the $G(\mathbf{n})$-invariants maps onto the $G(\mathbf{n})$-invariants in $\mathcal{S}^{2m}(V(\mathbf{n}) \oplus V(\mathbf{n})^*)$. □

In light of Proposition 3.1, we now investigate the dependence of $p\left(\phi_{\mathbf{n}}^{M}(\mathbf{s})\right)$ on the subset M. In fact we have

PROPOSITION 3.2. *For all m and $\mathbf{n}$, if M is an m-element subset of $\{1, 2, \cdots, 2m\}$ then for all $\mathbf{s} \in S_m^r$ we have*

$$p\left(\phi_{\mathbf{n}}^{(0^m 1^m)}(\mathbf{s})\right) = p\left(\phi_{\mathbf{n}}^{M}(\mathbf{s})\right).$$

PROOF. In general, if V is a vector space and $v_1, \cdots, v_d \in V$ then we have $p(v_1 \otimes \cdots \otimes v_d) = p(v_{\sigma(1)} \otimes \cdots \otimes v_{\sigma(d)})$ for any permutation $\sigma \in S_d$. The map $(\Phi_{\mathbf{n}}^{\mathbf{b}_M})^{-1} \circ \left(\Psi_{n_1}^M \otimes \cdots \otimes \Psi_{n_r}^M\right) \circ \Phi_{\mathbf{n}}^{\mathbf{b}_M}$ defines a permutation of tensor factors according to γ_M. □

Given $\mathbf{s} = (\sigma_1, \cdots, \sigma_r) \in S_m^r$ we define $F_{\mathbf{s}} = p(\phi_{\mathbf{s}}^{(0^m 1^m)})$ which is equal to the product

(3.1)
$$\left[\left(e_{i_1^{(1)}} \otimes \cdots \otimes e_{i_r^{(1)}}\right) \cdot \left(e_1^{i^{(\sigma_1(1))}} \otimes \cdots \otimes e_{r-1}^{i^{(\sigma_{r-1}(1))}} \otimes e_r^{i^{(\sigma_r(1))}}\right)\right] \cdot$$
$$\vdots$$
$$\cdot \left[\left(e_{i_1^{(m)}} \otimes \cdots \otimes e_{i_r^{(m)}}\right) \cdot \left(e_1^{i^{(\sigma_1(m))}} \otimes \cdots \otimes e_{r-1}^{i^{(\sigma_{r-1}(m))}} \otimes e_r^{i^{(\sigma_r(m))}}\right)\right]$$

summed over all ordered m-tuples of r-tuples of indices $i_k^{(j)}$ with $1 \leq i_k^{(j)} \leq n_k$ where $1 \leq j \leq m$ and $1 \leq k \leq r$.

The product, $\cdot$, is commutative. A consequence of this fact is that the left (resp. right) factor in the rows may be permuted. That is we may replace $(\sigma_1, \cdots, \sigma_r)$ with $(\gamma \sigma_1, \cdots, \gamma \sigma_r)$ for any $\gamma \in S_m$. If $\gamma = \sigma_r^{-1}$, we may reduce to the assumption that $\sigma_r = 1$ (the identity permutation). This means that the product 3.1 is equal to:

$$(3.2) \quad \begin{aligned} &\left[\left(e_{i_1^{(1)}} \otimes \cdots \otimes e_{i_r^{(1)}} \right) \cdot \left(e^{i_1^{(\sigma_1(1))}} \otimes \cdots \otimes e^{i_{r-1}^{(\sigma_{r-1}(1))}} \otimes e^{i_r^{(1)}} \right) \right] \cdot \\ &\qquad\qquad\qquad\qquad \vdots \\ &\cdot \left[\left(e_{i_1^{(m)}} \otimes \cdots \otimes e_{i_r^{(m)}} \right) \cdot \left(e^{i_1^{(\sigma_1(m))}} \otimes \cdots \otimes e^{i_{r-1}^{(\sigma_{r-1}(m))}} \otimes e^{i_r^{(m)}} \right) \right] \end{aligned}$$

Note that *any one* of the permutations $\sigma_1, \cdots, \sigma_r$ could be assumed to be the identity. We have arbitrarily chosen $\sigma_r = 1$.

The domain of the permutations $\sigma_1, \cdots, \sigma_r$ is the set $\{1, \cdots, m\}$. Noting again that the product, $\cdot$, in the symmetric algebra is commutative, we see that the "rows" of the above expression may be permuted without changing the expression. Permuting the rows corresponds to a simultaneous permutation of the domain of each σ_k. That is, we may simultaneously conjugate $\sigma_1, \cdots, \sigma_r$ (since relabeling a permutation's domain corresponds to conjugating the permutation). We obtain that for any $\mathbf{s} = (\sigma_1, \cdots, \sigma_r) \in S_m^r$ and $\tau \in S_m$ we have $F_{(\sigma_1, \cdots, \sigma_r)} = F_{(\tau \sigma_1 \tau^{-1}, \cdots, \tau \sigma_r \tau^{-1})}$.

The commutative $\mathbb{C}$-algebra of polynomial functions on a vector space $\mathbb{V}$ is naturally isomorphic to $\mathcal{S}(\mathbb{V}^*)$. If $\mathbb{V} = V(\mathbf{n}) \oplus V(\mathbf{n})^*$ then $\mathbb{V}$ is self-dual. Therefore, we obtain a natural isomorphism between $\mathcal{S}(V(\mathbf{n}) \oplus V(\mathbf{n})^*)$ and $\mathcal{P}(V(\mathbf{n}) \oplus V(\mathbf{n})^*)$.

Next, we explicitly describe the value of each $F_{\mathbf{s}}$ on the vector space $V(\mathbf{n}) \oplus V(\mathbf{n})^*$. A general element of $V(\mathbf{n})$ is of the form

$$X = \sum_{\substack{1 \leq i_j \leq n_j \\ 1 \leq j \leq r}} x^{i_1 i_2 \cdots i_r} e_{i_1} \otimes \cdots \otimes e_{i_r},$$

where $x^{i_1 i_2 \cdots i_r} \in \mathbb{C}$ while a general element of $V(\mathbf{n})^*$ is of the form

$$Y = \sum_{\substack{1 \leq i_j \leq n_j \\ 1 \leq j \leq r}} y_{i_1 i_2 \cdots i_r} e^{i_1} \otimes \cdots \otimes e^{i_r}.$$

where $y_{i_1 i_2 \cdots i_r} \in \mathbb{C}$. The value of $\otimes_{j=1}^r e^{k_j} \in \mathcal{P}^1(V(\mathbf{n}))$ on $V(\mathbf{n}) \oplus V(\mathbf{n})^*$ is $x^{j_1 j_2 \cdots j_r}$, while the value of $\otimes_{j=1}^r e_{k_j} \in \mathcal{P}^1(V(\mathbf{n})^*)$ on $V(\mathbf{n}) \oplus V(\mathbf{n})^*$ is $y_{i_1 i_2 \cdots i_r}$.

Therefore, given $(\sigma_1, \cdots, \sigma_{r-1}) \in S_m^{r-1}$, let $f_{\mathbf{s}}$ denote the polynomial function on $V(\mathbf{n}) \oplus V(\mathbf{n})^*$ corresponding to $F_{\mathbf{s}}$ with $\mathbf{s} = (\sigma_1, \cdots, \sigma_{r-1}, \sigma_r)$ with $\sigma_r = 1$. That is, the value of $f_{\mathbf{s}}$ on (X, Y) is

$$\left[\left(x^{i_1^{(1)} \cdots i_r^{(1)}} \right) \left(y_{i_1^{(\sigma_1(1))} \cdots i_r^{(\sigma_r(1))}} \right) \right] \cdots \left[\left(x^{i_1^{(m)} \cdots i_r^{(m)}} \right) \left(y_{i_1^{(\sigma_1(m))} \cdots i_r^{(\sigma_r(m))}} \right) \right]$$

summed over all ordered m-tuples of r-tuples of indices $i_k^{(j)}$ with $1 \leq i_k^{(j)} \leq n_k$ where $1 \leq j \leq m$ and $1 \leq k \leq r$.

THEOREM 3.3. *Given m and r, let S_m^{r-1}/S_m denote the orbits of S_m on the set S_m^{r-1} under the action of simultaneous conjugation. Then let $t = |S_m^{r-1}/S_m|$,* *and choose distinct representatives from each S_m-orbit, $\mathbf{s}_1, \cdots \mathbf{s}_t$. For any $\mathbf{n}$ with $n_i \geq m$ for all $1 \leq i \leq r$, the set*

$$\mathcal{B} = \{f_{\mathbf{s}_1}, f_{\mathbf{s}_2}, \cdots, f_{\mathbf{s}_t}\}$$

is a basis for the $G(\mathbf{n})$-invariants in $\mathcal{S}^{2m}(V(\mathbf{n}) \oplus V(\mathbf{n})^*)$. Otherwise, outside of these inequalities, the above are a spanning set.

PROOF. We have seen that $\mathcal{B}$ spans. In [**HW09**] it is shown that $t = \widetilde{h}_{m,r} = \dim \mathcal{S}^{2m}(V(\mathbf{n}) \oplus V(\mathbf{n})^*)^{G(\mathbf{n})}$. Therefore, $\mathcal{B}$ is linearly independent. $\square$

References

[EPR35] A. Einstein, B. Podolsky, and N. Rosen, *Can quantum-mechanical description of physical reality be considered complete?*, Phys. Rev. **47** (1935), no. 777. Quantum information theory. ↑2

[Fey81] Richard P. Feynman, *Simulating physics with computers*, Internat. J. Theoret. Phys. **21** (1981/82), no. 6-7, 467–488. Physics of computation, Part II (Dedham, Mass., 1981). MR658311 ↑2

[GW09] Roe Goodman and Nolan R. Wallach, *Symmetry, representations, and invariants*, Graduate Texts in Mathematics, vol. 255, Springer, Dordrecht, 2009. MR2522486 ↑9

[HW09] Michael W. Hero and Jeb F. Willenbring, *Stable Hilbert series as related to the measurement of quantum entanglement*, Discrete Math **309** (2009), no. 23-24, 6508–6514. MR2558615 ↑2, 3, 13

[KL01] Jin Ho Kwak and Jaeun Lee, *Enumeration of graph coverings, surface branched coverings and related group theory*, Combinatorial & computational mathematics (Pohang, 2000), World Sci. Publ., River Edge, NJ, 2001, pp. 97–161. MR1868421 (2003b:05083) ↑2, 4, 5, 6

[KL90] _____, *Isomorphism classes of graph bundles*, Canad. J. Math. **42** (1990), no. 4, 747–761. ↑4

[MW02] David A. Meyer and Noland Wallach, *Invariants for multiple qubits: the case of 3 qubits*, Mathematics of quantum computation, Comput. Math. Ser., Chapman & Hall/CRC, Boca Raton, FL, 2002, pp. 77–97. MR2007943 (2004h:81034) ↑2, 3, 4

[Wal05] Nolan R. Wallach, *The Hilbert series of measures of entanglement for 4 qubits*, Acta Appl. Math. **86** (2005), no. 1-2, 203–220. MR2134319 (2006c:81020) ↑2, 3

UNIVERSITY OF WISCONSIN - MILWAUKEE, DEPARTMENT OF MATHEMATICAL SCIENCES, P.O. BOX 0413, MILWAUKEE WI 53201

Contemporary Mathematics
Volume **557**, 2011

The dual pair $(O_{p,q}, \widetilde{OSp}_{2,2})$ and Zuckerman translation

Dan Lu and Roger Howe

ABSTRACT. We study the joint action of the "super" dual pair $(O(p,q), \widetilde{OSp}_{2,2})$, consisting of the indefinite orthogonal group $O(p,q)$ and the "supergroup" $\widetilde{OSp}_{2,2}$, on the space of Schwartz-function valued differential forms, $\mathcal{S}(\mathbb{R}^n; \Lambda^*((\mathbb{R}^n)^*))$, where $n = p + q$. We show that the Casimir operators of $O(p,q)$ and $\widetilde{OSp}_{2,2}$ coincide. Except for the finitely many representations for which the Casimir operator vanishes, the analog of the local theta correspondence is shown to be valid. The main tool is the Zuckerman translation principle applied to the local theta correspondence for the dual pair $(O(p,q), \widetilde{SL}_2(\mathbb{R}))$ acting on $\mathcal{S}(\mathbb{R}^n)$.

1. Introduction

The notion of dual pair provides an effective tool for understanding classical invariant theory [Ho1] beyond what is described in Weyl's book [Wey], and, in a transcendental formulation [Ho4], has been used by many authors in the theory of automorphic forms [We1], [Ge1], [KR], [Ich].[S-P], [GJS], etc. As introduced in [Ho1], dual pairs embraced Lie superalgebras [Ka], and dual pairs involving superalgebras encompass a variety of useful results of multilinear algebra, such as the Hilbert Syzygy Theorem, the Hodge Decomposition, the Poincaré Lemma, and others. Dual pairs involving superalgebras are also relevant to physics [Ho2].

However, in analytic or transcendental settings, attention has been given predominantly to dual pairs of groups or Lie algebras [LTZ], [ØZ], [LZ], [PT], [LPTZ]; dual pairs involving a superalgebra have largely been avoided. The first author undertook investigation of the more general class of dual pairs by studying a basic example, the dual pair $(O(p,q), \widetilde{OSp}_{2,2})$, where $O(p,q)$ is the indefinite orthogonal group, the isometries of a bilinear form on $\mathbb{R}^n$ of signature (p,q) where $n = p + q$, and $\widetilde{OSp}_{2,2}$ is orthosymplectic Lie supergroup [Kap] of signature $(2,2)$. Here the Lie part of $\widetilde{OSp}_{2,2}$ is $\widetilde{SL}_2 \times SO_2$, where $\widetilde{SL}_2$ is the two-fold cover of SL_2. Part of the interest in this dual pair comes from its connection with Maxwell's equations in the case $(p,q) = (3,1)$. An application to a generalized version of Maxwell's equations is made in [LH].

It was also explained in [LH] that, at least for generic values of a key parameter (the eigenvalue of the Casimir operator, see §3.2), the main phenomenon of the

1991 *Mathematics Subject Classification.* Primary 22E46; Secondary 17B10 17B20.

Key words and phrases. Dual pair, Zuckerman's translation, oscillator-spin representation, Lie superalgebras, Dual correspondence.

DAN LU AND ROGER HOWE

theory for a dual pair of groups, namely the existence of a natural bijection between representations of the two members of the pair, persists. However, the existence of the bijection at a family of special values of the parameter was left open. The goal of this note is to explain how the Zuckerman translation functors allow the appropriate bijection to be confirmed for all but a finite number of values of the parameter. At most of the values in question, the Zuckerman functors show that the correspondence behaves essentially like the known correspondence [LTZ] for the non-super dual pair $(O(p,q), \widetilde{SL}_2)$. It seems possible that at the finite number of remaining values, the behavior is interestingly different.

The paper is organized as follows. In §2, we introduce the dual pair and describe its action on the space $\mathcal{S}(\mathbb{R}^n; \Lambda^*((\mathbb{R}^n)^*))$ of differential forms of Schwartz class on $\mathbb{R}^n$. In §3, we discuss representations of the supergroup $O\widetilde{Sp}_{2,2}$, and explain how its Casimir operator helps to organize its representation theory. In particular, we show that an irreducible representation of $O\widetilde{Sp}_{2,2}$ in which the Casimir operator does not act by zero, is induced irreducibly from a parabolic subgroup containing the Lie part $\widetilde{SL}_2 \times SO_2$. In §4, we discuss the relevant representations of $O(p,q)$, and show how the Zuckerman functors, together with the known structure [HT2] of the representations appearing in the correspondence for the dual pair $(O(p,q), \widetilde{SL}_2)$, gives a description of almost all of them. Finally, in §5, we combine the results of the earlier sections to discuss the correspondence for our pair. An interesting aspect of this situation is the relation of the Casimir operators of $O(p,q)$ and of $O\widetilde{Sp}_{2,2}$. For the pairs $(O(p,q), \widetilde{SL}_2)$, the Casimir operators are closely related to each other. However, for the pair $(O(p,q), O\widetilde{Sp}_{2,2})$, the Casimir operators simply coincide as operators on the space $\mathcal{S}(\mathbb{R}^n; \Lambda^*((\mathbb{R}^n)^*))$.

2. The oscillator-spin representation of $(O(p,q), O\widetilde{Sp}_{2,2})$

In this section we introduce the *oscillator-spin representation* of $(O(p,q), O\widetilde{Sp}_{2,2})$ acting on the space $\mathcal{S}(\mathbb{R}^n; \Lambda^*((\mathbb{R}^n)^*))$ of Schwartz class differential forms on $\mathbb{R}^n$ with $n = p + q$, and we prove an elegant relation between the Casimir operators of $O(p,q)$ and $O\widetilde{Sp}_{2,2}$.

2.1. The operators of $\mathfrak{so}(p,q)$ and $\mathfrak{osp}_{2,2}$. We have given a fairly detailed description of this action in [LH]. Here we just remind the reader of some essential facts from that account.

The space $\mathcal{S}(\mathbb{R}^n; \Lambda^*((\mathbb{R}^n)^*))$ can be identified with the tensor product of $\mathcal{S}(\mathbb{R}^n)$, the space of complex-valued Schwartz functions [SW] on $\mathbb{R}^n$, and the exterior algebra $\Lambda^*((\mathbb{R}^n)^*)$ by taking any element $f \otimes \omega$ in $\mathcal{S}(\mathbb{R}^n) \otimes \Lambda^*((\mathbb{R}^n)^*)$ to the $\Lambda^*((\mathbb{R}^n)^*)$-valued function $\alpha(f \otimes \omega)$ defined by

$$(2.1) \qquad \alpha(f \otimes \omega)(x) = f(x)\omega$$

We denote the basis elements of the copy of $(\mathbb{R}^n)^*$ that generates the exterior algebra $\Lambda^*((\mathbb{R}^n)^*)$ by dx_i (or below, also by dy_j). We interchangeably regard dx_i as an element of $\Lambda^*((\mathbb{R}^n)^*)$ or as an operator on $\Lambda^*((\mathbb{R}^n)^*)$ by multiplication (on the left). We let $\rfloor_i$ denote the "exterior differentiation" or "inner multiplication" dual to dx_i [Ho2], [La2]. The operators dx_i and $\rfloor_i$ satisfy the Canonical Anticommutation Relations (CAR):

$$(2.2) \qquad \{dx_i, \rfloor_j\} = \delta_{ij}, \qquad \{dx_i, dx_j\} = 0 = \{\rfloor_i, \rfloor_j\}.$$

Throughout the paper, we use the curly brackets $\{\ ,\ \}$ to indicate the anticommutator of two operators: $\{A, B\} = AB + BA$.

The group $GL_n(\mathbb{R})$ acts naturally on each of the factors $\mathcal{S}(\mathbb{R}^n)$ and $\Lambda^*((\mathbb{R}^n)^*))$, and so acts on $\mathcal{S}(\mathbb{R}^n; \Lambda^*((\mathbb{R}^n)^*))$ via the tensor product of these actions. Explicitly, we have for any $g \in GL_n(\mathbb{R})$,

$$(2.3) \quad \widetilde{\omega}(g)(f \otimes v)(x) = \left(|\det g|^{-\frac{1}{2}} f(g^{-1}x)\right) \otimes \left(|\det|^{\frac{1}{2}} g(\omega)\right) = f(g^{-1}x) \otimes g(\omega)$$

for $f \in \mathcal{S}(\mathbb{R}^n)$ and $\omega \in \Lambda^*((\mathbb{R}^n)^*)$.

We are interested in the restriction of this action to the orthogonal group $O(p,q)$, We let x_a for $1 \le a \le p$ and y_b for $1 \le b \le q$ be coordinates on $\mathbb{R}^n$, where $n = p + q$. Thus, a typical point in $\mathbb{R}^n$ is

$$(2.4) \qquad \vec{v} = \begin{bmatrix} x_1 \\ x_2 \\ \vdots \\ x_p \\ y_1 \\ y_2 \\ \vdots \\ y_q \end{bmatrix}$$

The *orthogonal group* $O(p,q)$ is the group of isometries of the indefinite quadratic form

$$(2.5) \qquad r_{p,q}^2 = \sum_{a=1}^{p} x_a^2 - \sum_{b=1}^{q} y_b^2,$$

If we differentiate the action of $O(p,q)$ on $\mathcal{S}(\mathbb{R}^n) \otimes \Lambda^*((\mathbb{R}^n)^*)$, we find that a basis for the action of Lie algebra $\mathfrak{o}_{p,q}$ consists of:

$$(2.6) \quad \begin{aligned} x_j \frac{\partial}{\partial x_k} - x_k \frac{\partial}{\partial x_j} + dx_j \rfloor_{x_k} - dx_k \rfloor_{x_j} & \qquad j, k = 1, \cdots, p \\ y_j \frac{\partial}{\partial y_k} - y_k \frac{\partial}{\partial y_j} + dy_j \rfloor_{y_k} - dy_k \rfloor_{y_j} & \qquad j, k = 1, \cdots, q \\ x_j \frac{\partial}{\partial y_k} + y_k \frac{\partial}{\partial x_j} + dx_j \rfloor_{y_k} + dy_k \rfloor_{x_j} & \qquad 1 \le j \le p, 1 \le k \le q \end{aligned}$$

A key point for this paper is that we can find operators commuting with the action of $O(p,q)$ that span a copy of the *Lie superalgebra* $\mathfrak{osp}_{2,2}$. The *even or Lie part* of this superalgebra is $\mathfrak{sl}_2 \oplus \mathfrak{so}_2$. This is spanned by

$$e^- = \frac{1}{2} \square_{p,q} = \frac{1}{2} \left(\frac{\partial^2}{\partial x_1^2} + \cdots + \frac{\partial^2}{\partial x_p^2} - \frac{\partial^2}{\partial y_1^2} - \cdots - \frac{\partial^2}{\partial y_q^2} \right)$$

$$e^+ = -\frac{1}{2} r_{p,q}^2 = -\frac{1}{2}(x_1^2 + \cdots + x_p^2 - y_1^2 - \cdots - y_q^2)$$

$$(2.7a) \qquad h = \sum_{i=1}^{p} x_i \frac{\partial}{\partial x_i} + \sum_{j=1}^{q} y_i \frac{\partial}{\partial y_i} + \frac{p+q}{2}$$

which are the generators of a Lie algebra isomorphic to $\mathfrak{sl}_2$; and

$$(2.7b) \qquad h' = \sum_{i=1}^{p} dx_i \rfloor_{x_i} + \sum_{j=1}^{q} dy_j \rfloor_{y_j} - \frac{p+q}{2}$$

which is the generator of a Lie algebra $\mathfrak{so}_2 \simeq \mathbb{R}$. The *odd or super part* is spanned by the four operators

$$d = \sum_{i=1}^{p} \frac{\partial}{\partial x_i} dx_i + \sum_{j=1}^{q} \frac{\partial}{\partial y_j} dy_j \qquad \delta = \sum_{i=1}^{p} x_i \rfloor_{x_i} + \sum_{j=1}^{q} y_j \rfloor_{y_j}$$

$$(2.7c) \qquad d^* = \sum_{i=1}^{p} \frac{\partial}{\partial x_i} \rfloor_{x_i} - \sum_{j=1}^{q} \frac{\partial}{\partial y_j} \rfloor_{y_j} \qquad \delta^* = \sum_{i=1}^{p} x_i dx_i - \sum_{j=1}^{q} y_j dy_i.$$

The operators (2.7 a) form a standard basis for $\mathfrak{sl}_2$; that is, they satisfy the commutation relations

$$(2.7d) \qquad [h, e^+] = 2e^+, \qquad [h, e^-] = -2e^-, \qquad [e^+, e^-] = h$$

The operators e^+ and e^- are formally self-adjoint with respect to the canonical inner product on $\mathcal{S}(\mathbb{R}^n)$, and the operator h is formally skew-adjoint. Thus, the three operators h, ie^+ and $-ie^-$ (where $i = \sqrt{-1}$) form a standard basis for a copy of $\mathfrak{sl}_2$ consisting of formally skew-adjoint operators. Although it will not play an explicit role in this paper, it is important that the operators of this Lie algebra exponentiate to define a unitary action of the metaplectic group $\widetilde{SL}_2$ on $\mathcal{S}(\mathbb{R}^n)$. See for example [HT]. It is well-known (again see [HT]) that the operator $\mathfrak{k} = e^+ - e^-$ generates a compact subgroup of $\widetilde{SL}_2$. The operator $\mathfrak{k}$ itself is diagonalizable on $\mathcal{S}(\mathbb{R}^n)$, with eigenvalues that are pure imaginary, of the form $i(m + \frac{n}{2})$, where m is an integer. Further, the operators

$$\tilde{h} = -i\mathfrak{k} = i(e^- - e^+)$$

$$\tilde{e}^+ = \mathfrak{n}^+ = \frac{1}{2}(h + i(e^+ + e^-))$$

$$(2.7e) \qquad \tilde{e}^- = \mathfrak{n}^- = \frac{1}{2}(h - i(e^- + e^+))$$

form a standard basis for a copy of $\mathfrak{sl}_2$ in the complex span of h and $e^\pm$, and the operator $\tilde{h}$ will be diagonalizable with half-integer eigenvalues. It is a basis such as this that makes the results described in §3 relevant to the dual pair $(O(p,q), \widetilde{OSp}_{2,2})$.

In summary, the Lie algebras $\mathfrak{sl}_2$ and $\mathfrak{so}_2$ generate groups $\widetilde{SL}_2$ and SO_2, and the product group $\widetilde{SL}_2 \times SO_2$ together with the operators of part c) generate the Lie supergroup $\widetilde{OSp}_{2,2}$. This action of $\widetilde{OSp}_{2,2}$ together with the action of $O(p,q)$ in (2.3) give the oscillator-spin representation $\tilde{\omega}_{p,q}$ of $(O(p,q), \widetilde{OSp}_{2,2})$.

2.2. Casimir operators. In this section we will prove a relationship between the Casimir operators of $\mathfrak{so}(p,q)$ and $\mathfrak{osp}_{2,2}$. (The Casimir operator of $\mathfrak{so}_n$ is standard [GW]. For the Casimir of $\mathfrak{osp}_{2,2}$, see Theorem 3.5. We will not need to know it until the end of the calculation below.) For simplicity, we will do this for the definite case, $q = 0$. In fact, the relation is an algebraic phenomenon, and this proof is valid in the general case.

In its action on $\mathcal{S}(\mathbb{R}^n; \Lambda^*((\mathbb{R}^n)^*))$, the Casimir operator of $\mathfrak{so}_n$ is given by

$$-\sum_{1 \leq j < k \leq n} (x_j D_k - x_k D_j + dx_j \rfloor_k - dx_k \rfloor_j)^2 = -\frac{1}{2} \sum_{1 \leq j \neq k \leq n} (x_j D_k - x_k D_j + dx_j \rfloor_k - dx_k \rfloor_j)^2$$

$$= -\frac{1}{2} \sum_{1 \leq j \neq k \leq n} (A_{jk} - B_{jk} + C_{jk} - D_{jk})^2.$$

Here $D_k = \frac{\partial}{\partial x_k}$. We want to relate this operator to the operators belonging to $\mathfrak{osp}_{2,2}$. To this end, we will evaluate the 16 sums $\sum_{1 \leq j < k \leq n} F_{jk} H_{jk}$, where F and H are one of A, B, C or D. These follow a common pattern, which we will illustrate with two of the sums, and then just state the results for the others. The terms A and B commute with the terms C and D, and we use this to reduce the number of calculations.

$$\sum_{1 \leq j \neq k \leq n} A_{jk}^2 = \sum_{1 \leq j \neq k \leq n} x_j D_k x_j D_k = \sum_{1 \leq j \neq k \leq n} x_j^2 D_k^2$$

$$= (\sum_{1 \leq j \leq n} x_j^2)(\sum_{1 \leq k \leq n} D_k^2) - \sum_{1 \leq j \leq n} x_j^2 D_j^2 = (-2e^+)(2e^-) - \sum_{1 \leq j \leq n} x_j^2 D_j^2.$$

$$\sum_{1 \leq j \neq k \leq n} A_{jk} B_{jk} = \sum_{1 \leq j \neq k \leq n} x_j D_k x_k D_j = \sum_{1 \leq j \neq k \leq n} x_j D_j D_k x_k$$

$$= (\sum_{1 \leq j \leq n} x_j D_j)(\sum_{1 \leq j \leq n} D_k x_k) - \sum_{1 \leq j \leq n} x_j D_j^2 x_j$$

$$= (h - \frac{n}{2})(h + \frac{n}{2}) - \sum_{1 \leq j \leq n} x_j D_j^2 x_j.$$

In similar fashion, we find that

$$\sum_{1 \leq j \neq k \leq n} B_{jk} A_{jk} = (h + \frac{n}{2})(h - \frac{n}{2}) - \sum_{1 \leq j \leq n} D_j x_j^2 D_j;$$

$$\sum_{1 \leq j \neq k \leq n} B_{jk}^2 = (2e^-)(-2e^+) - \sum_{1 \leq j \leq n} D_j^2 x_j^2.$$

$$\sum_{1 \leq j \neq k \leq n} A_{jk} C_{jk} = \sum_{1 \leq j \neq k \leq n} C_{jk} A_{jk} = \delta^* d^* - \sum_{1 \leq j \leq n} x_j dx_j D_j \rfloor_j$$

$$\sum_{1 \leq j \neq k \leq n} A_{jk} D_{jk} = \sum_{1 \leq j \neq k \leq n} D_{jk} A_{jk} = -\delta d + \sum_{1 \leq j \leq n} x_j \rfloor_j D_j dx_j.$$

$$\sum_{1 \leq j \neq k \leq n} B_{jk} C_{jk} = \sum_{1 \leq j \neq k \leq n} C_{jk} B_{jk} = d\delta - \sum_{1 \leq j \leq n} D_j dx_j x_j \rfloor_j$$

$$\sum_{1 \leq j \neq k \leq n} B_{jk} D_{jk} = \sum_{1 \leq j \neq k \leq n} D_{jk} B_{jk} = -d^* \delta^* + \sum_{1 \leq j \leq n} D_j \rfloor_j x_j dx_j.$$

$$\sum_{1 \leq j \neq k \leq n} C_{jk}^2 = 0; \qquad \sum_{1 \leq j \neq k \leq n} D_{jk}^2 = 0.$$

$$\sum_{1 \leq j \neq k \leq n} C_{jk} D_{jk} = (\frac{n}{2} - h')(h' + \frac{n}{2}) = \sum_{1 \leq j \neq k \leq n} D_{jk} C_{jk}.$$

In each of these equations, a sum $\sum_{1 \leq j < k \leq n} F_{jk} H_{jk}$ is expressed as a product of two operators from $\mathfrak{osp}_{2,2}$, plus a sum over a single index of a product of the basic generators. We will call the first type of term, the *main term*, and the second type of term, the *remainder term*. From the above formulas, we see that the sum of the main terms equals

$$
-4e^+e^- - (h - \frac{n}{2})(h + \frac{n}{2}) - (h + \frac{n}{2})(h - \frac{n}{2}) - 4e^-e^+
$$
$$
+ \ 0 + (h' - \frac{n}{2}))(h' + \frac{n}{2}) + (h' + \frac{n}{2})(h' - \frac{n}{2}) + 0 + 2(\delta^*d^* + \delta d - d\delta - d^*\delta^*)
$$
$$
= -4(e^+e^- + e^-e^+) - 2h^2 + \frac{n^2}{2} + 2h'^2 - \frac{n^2}{2} + 2([\delta^*, d^*] + [\delta, d])
$$
$$
= -2(h^2 + 2(e^+e^- + e^-e^+) - h'^2 + ([d, \delta] + [d^*, \delta^*])).
$$

Comparing with the formula of Theorem 3.5, we recognize the expression in parentheses as just the Casimir operator of $\mathfrak{osp}_{2,2}$.

It is not difficult to verify that the sum of the remainder terms is zero. Combined with our formula for the main terms, this establishes the following lovely fact. (We have actually only established this in the case of $O(n, 0)$, but the proof in general can be done in the same way.)

THEOREM 2.1. *The action on* $\mathcal{S}(\mathbb{R}^n; \Lambda^*((\mathbb{R}^n)^*))$ *of the Casimir operators of* $O(p, q)$ *and* $O\widetilde{S}p_{2,2}$ *are equal:*

$$
\Omega_{O(p,q)} = \Omega_{\mathfrak{osp}}.
$$

3. Irreducible representations of $O\widetilde{S}p_{2,2}$

In this section, we describe the irreducible representations of $O\widetilde{S}p_{2,2}$ satisfying appropriate technical conditions. The main method is to describe representations of $O\widetilde{S}p_{2,2}$ in terms of the well-understood representations of $\widetilde{SL}_2(\mathbb{R})$.

3.1. Admissible dual of $\widetilde{SL}_2$. Since the Lie part of $O\widetilde{S}p_{2,2}$ consists of $\widetilde{SL}_2 \times SO_2$, the classification of irreducible representation of $O\widetilde{S}p_{2,2}$ is closely related with the irreducible representation of $\widetilde{SL}_2$. In this part, we recall the well-known classification [HT], [Kn2] of irreducible representations of $\widetilde{SL}_2$. Let $\{h, e^+, e^-\}$ be defined as in (2.7a), which is a standard basis of $\mathfrak{sl}_2$. Recall that a representation of $\mathfrak{sl}_2$ is called *h-semisimple* if the representation space is spanned by eigenvectors for h. Similarly, a representation of $\mathfrak{sl}_2 \oplus \mathfrak{so}_2$ is called (h, h')-semisimple if the representation space is spanned by joint eigenvectors for h and h'. A representation of $\mathfrak{sl}_2$ is called *quasisimple* if the Casimir operator $\Omega_{\mathfrak{sl}_2}$ (see formula (3.2)) acts as a scalar. A representation of $\mathfrak{sl}_2 \oplus \mathfrak{so}_2$ is called quasisimple if both $\Omega_{\mathfrak{sl}_2}$ and h' act as scalars.

Following the notation in [HT], we have

THEOREM 3.1. ([HT]) *All h-admissible, h-semisimple quasisimple irreducible* $\mathfrak{sl}_2$-*modules are classified into the 4 classes:*
1) *Lowest weight modules* V_λ, $\lambda \notin \mathbb{Z}^-$;
2) *Highest weight modules* $\bar{V}_\lambda$, $\lambda \notin \mathbb{Z}^+$;
3) *Finite dimensional modules* $F_{-\lambda}, \lambda \in \mathbb{Z}^+$, *of dimension* $\lambda + 1$, *and with lowest weight* $-\lambda$ *(and highest weight* λ);
4) $U(\nu^+, \nu^-)$ $\nu^+, \nu^- \notin \mathbb{Z}$.

Furthermore, we have the admissible dual of SL_2,

THEOREM 3.2. ([HT]) *All irreducible admissible representations of SL_2 must be infinitesimally equivalent to one of the following:*
1) *Discrete series $V_\lambda, \lambda \in \mathbb{Z}^+; \overline{V}_\lambda, \lambda \in \mathbb{Z}^-$.*
2) *Finite dimensional representations F_{-m}, $m \in \mathbb{Z}^+$.*
3) *Principal series $S^{s,+} = U(\frac{s}{2}, \frac{s}{2})$, $s \notin 2\mathbb{Z}$, $S^{s,-} = U(\frac{s+1}{2}, \frac{s-1}{2})$, $s \notin 2\mathbb{Z}+1$*

REMARK 3.3. a) The basis of $\mathfrak{sl}_2$ that is appropriate for the infinitesimal analysis of representations of $\widetilde{SL}_2$ is the basis given in formulas (2.7 e). Since the element $i\tilde{h}$ of this basis generates a compact subgroup of $\widetilde{SL}_2$, we know that $\tilde{h}$ will be diagonalizable in the infinitesimal version of any representation of $\widetilde{SL}_2$.

b) As noted in §2.1, it is the group $\widetilde{SL}_2$, the two-fold cover of SL_2, that relates to the representation of $O\widetilde{Sp}_{2,2}$ in the duality correspondence $(O(p,q), O\widetilde{Sp}_{2,2})$. Hence, we are particularly interested in its representations. We say a representation of $\widetilde{SL}_2$ is *genuine* if it does not factor to SL_2. The irreducible genuine representations of $\widetilde{SL}_2$ are infinitesimally equivalent (in the action of the basis of formulas (2.7 e)) to the following:
1) Lowest weight modules, V_λ, $\lambda \in \mathbb{Z} + \frac{1}{2}$.
2) Highest weight modules, $\overline{V}_\lambda$, $\lambda \in \mathbb{Z} + \frac{1}{2}$.
3) Principal series $U(\frac{\nu}{2} + \frac{1}{4}, \frac{\nu}{2} + \frac{3}{4})$ or $U(\frac{\nu}{2} + \frac{3}{4}, \frac{\nu}{2} + \frac{1}{4})$, $\nu \notin \mathbb{Z} + \frac{1}{2}$.

3.2. The Casimir operator and structure of representations of $\mathfrak{osp}_{2,2}$.
In this section, we give a classification of irreducible representations of $\mathfrak{osp}_{2,2}$ subject to some technical restrictions. These are most of the $\mathfrak{osp}_{2,2}$-representations appearing in the the duality correspondence of $(O(p,q), O\widetilde{Sp}_{2,2})$. These results are summarized in [LH], and are treated in detail in [Lu]. Here we give some discussion of the key computations involved in the proofs that are especially germane to our main result.

We use the operators in (2.7) as the basis elements of Lie superalgebra $\mathfrak{osp}_{2,2}$. Then by simple computation, it is easy to get the following (anti)commutation relations.

For the Lie subalgebra $\mathfrak{sl}_2 \oplus \mathfrak{so}_2$:

(3.1a) $[h, e^+] = 2e^+$ $[h, e^-] = -2e^-$ $[e^+, e^-] = h$ $[h', x] = 0$

Here x is anything in $\mathfrak{sl}_2 \oplus \mathfrak{so}_2$.

Between the Lie algebra and the super part:

$$
\begin{array}{llll}
[e^+, d] = \delta^* & [e^+, \delta^*] = 0 & [e^+, d^*] = \delta & [e^+, \delta] = 0 \\
[e^-, d] = 0 & [e^-, \delta^*] = d & [e^-, d^*] = 0 & [e^-, \delta] = d^* \\
[h, d] = -d & [h, \delta^*] = \delta^* & [h, d^*] = -d^* & [h, \delta] = \delta \\
[h', d] = d & [h', \delta^*] = \delta^* & [h', d^*] = -d^* & [h', \delta] = -\delta
\end{array}
$$

(3.1b)

Anticommutation relations for the super part:

$$
d^2 = \{d, \delta^*\} = (\delta^*)^2 = 0 \qquad (d^*)^2 = \{d^*, \delta\} = \delta^2 = 0
$$

(3.1c) $\{d, d^*\} = 2e^-$ $\{d, \delta\} = h + h'$ $\{d^*, \delta^*\} = h - h'$ $\{\delta, \delta^*\} = -2e^+$

We can summarize the commutation relations above as follows:

THEOREM 3.4. $\mathfrak{osp}_{2,2}$ *is a classical Lie superalgebra. The pairs* $\{d, \delta^*\}$ *and* $\{d^*, \delta\}$ *each span a two dimensional irreducible module for* $\mathfrak{sl}_2$ *with eigenvalue* $+1$ *and* -1 *for* h' *respectively.*

The Casimir operator for $\mathfrak{sl}_2$,

$$(3.2) \qquad \Omega_{\mathfrak{sl}_2} = h^2 + 2(e^+ + e^-) = h^2 - 2h + 4e^+e^- = h^2 + 2h + 4e^-e^+,$$

which generates the center of the universal enveloping algebra of $\mathfrak{sl}_2$, plays a key role in dealing with the representations of $\mathfrak{sl}_2$. For understanding the representations of $\mathfrak{osp}_{2,2}$, it is likewise valuable to use an analogous central element of $\mathfrak{U}(\mathfrak{osp}_{2,2})$ [Lu1].

THEOREM 3.5. *Define*

$$\Omega_{\mathfrak{osp}} = \Omega_{\mathfrak{sl}_2} - h'^2 + ([d, \delta] + [d^*, \delta^*])$$

in $\mathfrak{U}(\mathfrak{osp}_{2,2})$. *Then* $\Omega_{\mathfrak{osp}}$ *is in the center of* $\mathfrak{U}(\mathfrak{osp}_{2,2})$, *in other words,* $\Omega_{\mathfrak{osp}}$ *commutes with every element in* $\mathfrak{U}(\mathfrak{osp}_{2,2})$. *Moreover,* $\Omega_{\mathfrak{osp}}$ *generates the center of* $\mathfrak{U}(\mathfrak{osp}_{2,2})$ *as an (associative) algebra, i.e.* $Z(\mathfrak{U}(\mathfrak{osp}_{2,2})) = \mathbb{C}[\Omega_{\mathfrak{osp}}]$.

PROOF. This is proved in [Lu1] (see also [LH]). We indicate here a typical part of the computations needed to show that $\Omega_{\mathfrak{osp}}$ is central. These are based on some general identities for elements A, B and C in an associative algebra. The first is the standard Jacobi identity for commutators, and the others are analogues for supercommutators.

$$[A, [B, C]] = [[A, B], C] + [B, [A, C]] = \{\{A, B\}, C\} - \{B, \{A, C\}\}$$

$$(3.3) \qquad [A, \{B, C\}] = \{[A, B], C\} + \{B, [A, C]\} = [\{A, B\}, C] - [B, \{A, C\}]$$

Using these identities and the supercommutation relations (3.1), we compute

$$[e^+, [d, \delta]] = [[e^+, d], \delta] + [d, [e^+, \delta]] = [\delta^*, \delta],$$

and

$$[e^+, [d^*, \delta^*]] = [[e^+, d^*], \delta^*] + [d^*, [e^+, \delta^*]] = [\delta, \delta^*].$$

Adding these equations, and using the commutation relation $[e^+, h'] = 0$, and the well-known fact that $\Omega_{\mathfrak{sl}_2}$ is central in $\mathfrak{U}(\mathfrak{sl}_2)$, we conclude that $\Omega_{\mathfrak{osp}}$ commutes with e^+.

We will also show that $\Omega_{\mathfrak{osp}}$ commutes with d. Making use of the identities (3.3) and the supercommutation relations (3.1), we find

$$[d, h^2] = [d, h]h + h[d, h] = dh + hd = \{d, h\}, \qquad \text{and similarly,} \qquad [d, h'^2] = -\{d, h'\};$$

$$[d, \{e^+, e^-\}] = \{[d, e^+], e^-\} + \{e^+, [d, e^-]\} = -\{\delta^*, e^-\},$$

$$[d, [d, \delta]] = \{\{d, d\}, \delta\} - \{d, \{d, \delta\}\} = -\{d, h + h'\},$$

$$[d, [d^*, \delta^*]] = \{\{d, d^*\}, \delta^*]\} + \{d^*, \{d, \delta^*\}\} = 2\{e^-, \delta^*\}.$$

Combining these equations, we find that $[d, \Omega_{\mathfrak{osp}}] = 0$, as desired. $\square$

To describe representations of $\mathfrak{osp}_{2,2}$, we relate them to representations of $\mathfrak{sl}_2 \oplus \mathfrak{so}_2$. We will be mainly (but not exclusively) interested in irreducible representations. Since the operator $\Omega_{\mathfrak{osp}}$ commutes with all of $\mathfrak{osp}_{2,2}$, it is reasonable to expect that $\Omega_{\mathfrak{osp}}$ would be represented by a scalar operator in an irreducible representation of $\mathfrak{osp}_{2,2}$, and given some mild technical assumptions, this will be so. Following the terminology of Harish-Chandra for Lie algebras, we will call a representation of $\mathfrak{osp}_{2,2}$ in which $\Omega_{\mathfrak{osp}}$ acts by a scalar, a *quasisimple* representation.

THE DUAL PAIR $(O_{p,q}, O\widetilde{Sp}_{2,2})$ AND ZUCKERMAN TRANSLATION

Denote by Z^- the subsuperalgebra of $\mathfrak{osp}_{2,2}$ spanned by d^* and δ, and by Z^+ the subsuperalgebra spanned by d and δ^*. Note that both $Z^{\pm}$ are normalized by $\mathfrak{sl}_2 \oplus \mathfrak{so}_2$. Denote by $\mathfrak{q}$ the subsuperalgebra spanned by $\mathfrak{sl}_2 \oplus \mathfrak{so}_2$ and by Z^-.

If we have a representation τ of $\mathfrak{sl}_2 \oplus \mathfrak{so}_2$, we can extend it to a representation of $\mathfrak{q}$, still denoted by τ, by letting d^* and δ act by the zero operator. We then form the induced representation

$$(3.4) \qquad \mathrm{Ind}_{\mathfrak{q}}^{\mathfrak{osp}_{2,2}}\tau = \mathrm{Ind}\ \tau$$

of $\mathfrak{osp}_{2,2}$, as the natural action by left multiplication on the space

$$\mathfrak{U}(\mathfrak{osp}_{2,2}) \otimes_{\mathfrak{q}} \tau \simeq \Lambda^*(Z^+) \otimes \tau \simeq \tau \oplus (Z^+ \otimes \tau) \oplus ([d, \delta^*]\tau) \simeq \tau \oplus (d \otimes \tau) \oplus (\delta^* \otimes \tau) \oplus (d\delta^* \otimes \tau).$$

Here we are abusing notation by letting τ stand for the space on which the representation τ is realized.

Given any representation of $\mathfrak{osp}_{2,2}$ on a vector space U, we can consider the subspace

$$(3.5) \qquad U_0 = \ker d^* \cap \ker \delta.$$

It is easy to check that U_0 is invariant under the action of $\mathfrak{sl}_2 \oplus \mathfrak{so}_2$. We will consider U_0 as a representation of $\mathfrak{sl}_2 \oplus \mathfrak{so}_2$ in the obvious way. Also, we can easily show that U_0 is non-zero if U is. Indeed, if $u \neq 0$ is a non-zero vector in U, then one of u, or $d^*(u)$, or $\delta(u)$, or $d^*\delta(u) = -\delta d^*(u)$ will be non-zero and will belong to U_0.

There is a natural $\mathfrak{osp}_{2,2}$-intertwining operator

$$(3.6) \qquad \alpha : \mathrm{Ind}\ U_0 \to U$$

defined in the obvious way:

$$\alpha : (u_1 + d \otimes u_2 + \delta^* \otimes u_3 + [d, \delta^*] \otimes u_4) = u_1 + d(u_2) + \delta^*(u_3) + [d, \delta^*](u_4).$$

If the space U is an irreducible module for $\mathfrak{osp}_{2,2}$, then α must clearly be surjective. Thus, every irreducible representation of $\mathfrak{osp}_{2,2}$ is a quotient of a representation induced from $\mathfrak{q}$. In fact, typically, the intertwining operator α is an isomorphism.

THEOREM 3.6. *a) Let U be a quasisimple $\mathfrak{osp}_{2,2}$ module. Suppose that the Casimir operator $\Omega_{\mathfrak{osp}}$ acts on U by a non-zero scalar. Then the intertwining operator α defined in equation (3.6) is an isomorphism.*

b) Suppose that V is a quasisimple $\mathfrak{sl}_2 \oplus \mathfrak{so}_2$ module in which the operator $\Omega_{\mathfrak{sl}_2} - h'^2 - 2h'$ acts by a non-zero scalar. Extend V to a $\mathfrak{q}$ module as explained above. Then

$$(\mathrm{Ind}\ V)_0 = V.$$

REMARK 3.7. The formula

$$[d, \delta] + [d^*, \delta^*] = -\{d, \delta\} + 2d\delta + \{d^*, \delta^*\} - 2\delta^*d^* = -(h + h') + (h - h') + 2(d\delta - \delta^*d^*)$$
$$= -2h' + 2(d\delta - \delta^*d^*)$$

allows us to write

$$(3.7a) \qquad \Omega_{\mathfrak{osp}} = \Omega_{\mathfrak{sl}_2} - h'^2 - 2h' + 2(d\delta - 2\delta^*d^*).$$

This in turn implies that

$$(3.7b) \qquad (\Omega_{\mathfrak{osp}})_{|U_0} = \Omega_{\mathfrak{sl}_2} - h'^2 - 2h'.$$

Thus, we can compute the value of $\Omega_{\mathfrak{osp}}$ on U in terms of the action of $\mathfrak{sl}_2 \oplus \mathfrak{so}_2$ on U_0.

PROOF. a) The kernel of α will be an $\mathfrak{osp}_{2,2}$ submodule of Ind U_0, so in particular, it will be an $\mathfrak{sl}_2 \oplus \mathfrak{so}_2$ submodule. The subspaces U_0, $Z^+ \otimes U_0$, and $d\delta^*(U_0)$ are the h' eigenspaces in Ind U_0, so ker α will be the sum of its intersection with these spaces. Also, $d\delta^*(U_0)$ is isomorphic to U_0 as an $\mathfrak{sl}_2$-module, and ker $\alpha \cap d\delta^*(U_0)$ will be an $\mathfrak{sl}_2$-submodule.

We compute

$$\begin{aligned}
d^*\delta d\delta^* &= d^*(\{d,\delta\} - d\delta)\delta^* = d^*(h + h')\delta^* - d^* d\delta\delta^* \\
&= 2d^*\delta^* + d^*\delta^*(h + h') - (\{d^*,d\} - dd^*)(\{\delta,\delta^*\} - \delta^*\delta) \\
&= 2(\{d^*,\delta^*\} - \delta^* d^*) + (\{d^*,\delta^*\} - \delta^* d^*)(h + h') - (2e^- - dd^*)(-2e^+ - \delta^*\delta) \\
&= 2(h - h') - 2\delta^* d^* + (h - h')(h + h') - \delta^* d^*(h + h') + 4e^- e^+ + 2dd^* e^+ \\
&\quad + 2e^-\delta^*\delta + dd^*\delta^*\delta \\
&= (h^2 + 2h + 4e^- e^+) - h'^2 - 2h' - 2\delta^* d^* - \delta^* d^*(h + h') + 2dd^* e^+ + 2e^-\delta^*\delta \\
&\quad + dd^*\delta^*\delta \\
&= (\Omega_{\mathfrak{sl}_2} - h'^2 - 2h') + (-2\delta^* d^* - \delta^* d^*(h + h') + 2dd^* e^+ + 2e^-\delta^*\delta + dd^*\delta^*\delta)
\end{aligned}$$

Now we observe that in the final expression, the five terms in parentheses on the right vanish on U_0, while equation (3.7b) tells us that the initial part agrees with $\Omega_{\mathfrak{osp}}$ on U_0. Therefore, when $\Omega_{\mathfrak{osp}} \neq 0$, the maps $d\delta^*$ and $d^*\delta$ are isomorphisms from U_0 to $d\delta^*(U_0)$ and back. In particular, since α is injective on U_0, $(d\delta^* \otimes U_0) \cap \ker \alpha = \{0\}$.

Now suppose that an element $d \otimes u_1 + \delta^* \otimes u_2$ in $Z^+ \otimes U_0$ belongs to ker α. For this to be non-zero, one of u_1 or u_2 must be non-zero. For convenience, suppose $u_2 \neq 0$. Then $d(d \otimes u_1 + \delta^* \otimes u_2) = d\delta^* \otimes u_2$ should also be in ker α. But we have just seen that this is not possible. We conclude that ker $\alpha = \{0\}$, as desired.

b) The above calculation shows that $d\delta^*(V) \cap \ker(d^*\delta) = \{0\}$, so a fortiori, $d\delta^*(V) \cap (\ker d^* \cap \ker \delta) = \{0\}$. We will show also that $(d(V) \oplus \delta^*(V)) \cap (\ker d \cap \ker \delta) = \{0\}$. This will establish part b).

Consider an element $y = dv + \delta^* w$ with v and w in V. We compute

$$d^*(y) = d^* dv + d^*\delta^* w = \{d, d^*\}v + \{d^*, \delta^*\}w = 2e^-(v) + (h - h')(w).$$

Similarly,

$$\delta(y) = (h + h')(v) - 2e^+(w).$$

Suppose that both of these are zero. Then also

$$0 = e^+ d^*(y) = 2e^+ e^-(v) + e^+(h - h')(w) = 2e^+ e^-(v) + (h - h' - 2)e^+(w),$$

and similarly,

$$0 = (h - h' - 2)\delta(y) = (h - h' - 2)(h + h')(v) - 2(h - h' - 2)e^+(w).$$

Adding the last equation to twice the preceding one gives

$$0 = 4e^+ e^-(w) + (h - h' - 2)(h + h')(v) = (4e^+ e^- + h^2 - h'^2 - 2h - 2h')(v) = (\Omega_{\mathfrak{sl}_2} - h'^2 - 2h')(v).$$

But $\Omega_{\mathfrak{sl}_2} - h'^2 - 2h'$ is assumed to be a non-zero scalar. Hence $v = 0$. A similar calculation now shows that $w = 0$ also. This is what we wanted to show. $\square$

REMARK 3.8. a) It is shown in [Lu1] that, when $\Omega_{\mathfrak{osp}}$ is the zero operator, that Ind U_0 breaks up into two irreducible representations. However, we will not need this finer analysis here.

b) The correspondence $V \to \operatorname{Ind} V$ may be considered as a functor from (the category of) $\mathfrak{sl}_2 \oplus \mathfrak{so}_2$-modules to (the category of) $\mathfrak{osp}_{2,2}$-modules. Evidently, it is exact. Similarly, $U \to U_0$ is a functor from $\mathfrak{osp}_{2,2}$-modules to $\mathfrak{sl}_2 \oplus \mathfrak{so}_2$. It is not obviously exact, but it is left exact.

We can restrict these functors to quasisimple modules. More precisely, we can consider the category of $\mathfrak{sl}_2 \oplus \mathfrak{so}_2$-modules which are generalized eigenspaces with non-zero eigenvalue for the operator $\Omega_{\mathfrak{sl}_2} - h'^2 - 2h'$, and the category of $\mathfrak{osp}_{2,2}$-modules which are generalized eigenspaces for the operator $\Omega_{\mathfrak{osp}}$. Then the theorem above implies that these functors are eigenvalue-preserving and mutually inverse. In particular, we have the following

COROLLARY 3.9. Any irreducible quasisimple representation of $\mathfrak{osp}_{2,2}$ with non-zero Casimir eigenvalue can be realized by inducing from $\mathfrak{q}$, from a unique irreducible quasisimple $\mathfrak{sl}_2 \oplus \mathfrak{so}_2$ on which $\Omega_{\mathfrak{sl}_2} - h'^2 - 2h'$ acts by the same eigenvalue.

3.3. Classification of $\mathfrak{osp}_{2,2}$ modules with non-zero Casimir eigenvalue. The results above combined with §3.1 allow us to to classify h-admissible, h, h'-semisimple representations for $\mathfrak{osp}_{2,2}$ that are irreducible and quasisimple, as long as $\Omega_{\mathfrak{osp}}$ is not represented by 0.

Since the Lie part is a direct sum of $\mathfrak{sl}_2$ and $\mathfrak{so}_2$, we classify the irreducible $\mathfrak{osp}_{2,2}$-modules according to their structures considered as $\mathfrak{sl}_2 \oplus \mathfrak{so}_2$-modules. From §3.1, we know the h-semisimple, h-admissible quasisimple irreducible representations of $\mathfrak{sl}_2$ are classified into four different classes. Since $\mathfrak{so}_2$ is an abelian Lie algebra, its irreducible representations can be identified by the eigenvalue of h', which is an element of $\mathbb{C}$. Also since $\mathfrak{so}_2$ commutes with $\mathfrak{sl}_2$, each h, h'-semisimple irreducible $\mathfrak{sl}_2 \oplus \mathfrak{so}_2$-module has only one eigenvalue for h'. We use $V_{\lambda,\mu}$, $\overline{V}_{\lambda,\mu}$, $F_{\lambda,\mu}$, $U(\nu^+, \nu^-, \mu)$ to denote the irreducible $\mathfrak{sl}_2 \oplus \mathfrak{so}_2$-modules of the same $\mathfrak{sl}_2$-types as in §3.1 with h'-eigenvalue μ. We will analyze h-admissible h, h'-semisimple quasisimple $\mathfrak{osp}_{2,2}$-modules which contain $\mathfrak{sl}_2 \oplus \mathfrak{so}_2$-modules from these four different classes.

Based on the results in [Lu1], we get the different classes of h, h'-semisimple, h-admissible, $\Omega_{\mathfrak{osp}}$-quasisimple irreducible $\mathfrak{osp}_{2,2}$-modules as follows.

1) Irreducible lowest weight modules $W(\lambda, \mu)(\lambda \notin \mathbb{Z}^-)$.

The condition that $\Omega_{\mathfrak{osp}} \neq 0$ is $\mu \neq \pm\lambda$.

W is composed of four $\mathfrak{sl}_2 \oplus \mathfrak{so}_2$-modules depicted below. The isomorphic types of $\mathfrak{sl}_2 \oplus \mathfrak{so}_2$-modules are described on the vertices of the diagram. The modules on the same row have the same h'-eigenvalues. The operators from the super part map these $\mathfrak{sl}_2 \oplus \mathfrak{so}_2$-modules back and forth. The space W_0 (see equation (3.5)) is the module $V_{\lambda+1,\mu-1}$ (not $V_{\lambda,\mu}$!).

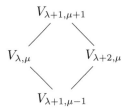

2) Irreducible highest weight modules $\overline{W}(\lambda,\mu)$ $(\lambda \notin \mathbb{Z}^+)$.

The condition that $\Omega_{\mathfrak{osp}} \neq 0$ is again $\mu \neq \pm\lambda$.

W is composed of four $\mathfrak{sl}_2 \oplus \mathfrak{so}_2$-modules depicted in the diagram below. Again, we point out that W_0 is the module $\overline{V}_{\lambda-1,\mu-1}$.

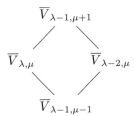

3) Irreducible finite dimensional modules $F_{\mathfrak{osp}}(\lambda,\mu)$ $(\lambda \in \mathbb{Z}^-)$.

Since these are both highest and lowest weight modules, the condition for non-vanishing of $\Omega_{\mathfrak{osp}} \neq 0$ is again that $\mu \neq \pm\lambda$.

a) $\lambda \neq -1$, $F_{\mathfrak{osp}}(\lambda,\mu)$ is composed of four finite dimensional $\mathfrak{sl}_2 \oplus \mathfrak{so}_2$-modules, with $F_{\mathfrak{osp}}(\lambda,\mu)_0 = F_{\lambda+1,\mu-1}$.

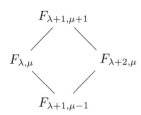

b) $\lambda = -1$,

$$F_{\mathfrak{osp}}(-1,\mu)|_{\mathfrak{sl}_2\oplus\mathfrak{so}_2} \cong F_{-1,\mu} \oplus F_{0,\mu-1} \oplus F_{0,\mu+1}$$

4) Irreducible modules $W(\nu^+,\nu^-,\mu)$, $\nu^+,\nu^- \notin \mathbb{Z}$, $\mu \in \mathbb{C}$

The condition of non-vanishing $\Omega_{\mathfrak{osp}}$ is that $\mu \neq \pm(\nu^+ + \nu^-)$.

W is composed of four $\mathfrak{sl}_2 \oplus \mathfrak{so}_2$-modules depicted below. The $\mathfrak{sl}_2 \oplus \mathfrak{so}_2$ submodule W_0 is $U(\nu^+, \nu^- + 1, \mu - 1)$.

$$
\begin{array}{ccc}
& U(\nu^+, \nu^- + 1, \mu + 1) & \\
\diagup & & \diagdown \\
U(\nu^+, \nu^-, \mu) & & U(\nu^+ + 1, \nu^- + 1, \mu) \\
\diagdown & & \diagup \\
& U(\nu^+, \nu^- + 1, \mu - 1) &
\end{array}
$$

THEOREM 3.10. *All h, h'-semisimple, h-admissible, quasisimple irreducible representations of $\mathfrak{osp}_{2,2}$ with $\Omega_{\mathrm{osp}} \neq 0$ are classified into four groups listed above:*
1) $W(\lambda, \mu), \lambda \in \mathbb{C}, \lambda \notin \mathbb{Z}^-, \mu \in \mathbb{C}$, and $\mu \neq \pm\lambda$;
2) $\overline{W}(\lambda, \mu), \lambda \notin \mathbb{Z}^+, \mu \in \mathbb{C}$ and $\mu \neq \pm\lambda$;
3) $F_{\mathrm{osp}}(\lambda, \mu), \lambda \in \mathbb{Z}^-$ and $\mu \neq \pm\lambda$;
4) $W(\nu^+, \nu^-, \mu), \nu^+, \nu^- \notin \mathbb{Z}, \mu \in \mathbb{C}$ and $\mu \neq \pm(\nu^+ + \nu^-)$.

We also obtain the eigenvalues of the Casimir operator in these irreducible $\mathfrak{osp}_{2,2}$-modules summarized in Theorem 3.10.

THEOREM 3.11. *The Casimir operator Ω_{osp} of $\mathfrak{osp}_{2,2}$ has the following eigenvalues in the irreducible representations of $\mathfrak{osp}_{2,2}$.*

1) $\Omega_{\mathrm{osp}}v = (\lambda^2 - \mu^2)v$ $v \in W(\lambda, \mu)$ $\lambda \notin \mathbb{Z}^-, \mu \in \mathbb{C}$ and $\mu \neq \pm\lambda$;

2) $\Omega_{\mathrm{osp}}v = (\lambda^2 - \mu^2)v$ $v \in \overline{W}(\lambda, \mu)$ $\lambda \notin \mathbb{Z}^+, \mu \in \mathbb{C}$ and $\mu \neq \pm\lambda$;

3) $\Omega_{\mathrm{osp}}v = (\lambda^2 - \mu^2)v$ $v \in F_{\mathrm{osp}}(\lambda, \mu)$ $\lambda \in \mathbb{Z}^-, \mu \in \mathbb{C}$ and $\mu \neq \pm\lambda$;

4) $\Omega_{\mathrm{osp}}v = ((\nu^+ + \nu^-)^2 - \mu^2)v$ $v \in W(\nu^+, \nu^-, \mu)$ $\nu^+, \nu^- \notin \mathbb{Z}, \mu \in \mathbb{C}$

$$\text{and } \mu \neq \pm(\nu^+ + \nu^-).$$

4. Some degenerate principal series of $O(p, q)$

In this section, we discuss a class of representations of $O(p, q)$ which are induced from a maximal parabolic subgroup, usually called degenerate principal series. In [HT2], one can find the structure of some of these degenerate principal series, the ones which are obtained from inducing a character on the Levi component of a maximal parabolic subgroup. These representations occur in the duality correspondence of $(O(p, q), \widetilde{SL}_2)$ [LTZ]. However, in the duality correspondence of $(O(p, q), O\widetilde{Sp}(2, 2))$, more representations will appear. To deal with these, we must study representations induced from a particular class of finite dimensional representations of a maximal parabolic subgroup. Except in a few exceptional cases, we are able to determine the structures of these new representations by using Zuckerman's translation functor. The exceptional cases are exactly those on which the action of the Casimir operator of $O(p, q)$ vanishes.

4.1. Zuckerman's translation functor. Let us review the Zuckerman's translation functor [Zu]. Let $\mathfrak{g}^{\mathbb{C}}$ be the complexification of the Lie algebra of a linear reductive group G. Let $\mathfrak{h}^{\mathbb{C}}$ be a Cartan subalgebra of $\mathfrak{g}^{\mathbb{C}}$ and W be the Weyl group [Kn2] of $\mathfrak{g}^{\mathbb{C}}$. We are interested in the case $G = O(p, q)$.

We say that a $(\mathfrak{g}, K)$ module [Kn2] M is a *Harish-chandra-module* if:

a) the center $Z(\mathfrak{g})$ of the universal enveloping algebra $\mathfrak{U}(\mathfrak{g})$ acts on M through a finite-dimensional quotient;

b) M is a direct sum of irreducible representations of K; and

c) each irreducible representation of K occurs with finite multiplicity.

Let $\triangle^+(\mathfrak{h}^{\mathbb{C}}, \mathfrak{g}^{\mathbb{C}})$ be a fixed choice of positive root system [Kn2] and M be an H-C module with infinitesimal character γ such that $\operatorname{Re}\gamma$ is dominant under $\triangle^+(\mathfrak{h}^{\mathbb{C}}, \mathfrak{g}^{\mathbb{C}})$. Suppose V_σ is a finite dimensional irreducible representation of $\mathfrak{g}$ with highest weight σ with respect to this same positive system of roots. The $(\mathfrak{g}, K)$-module $M \otimes V_\sigma$ is again a Harish-Chandra module, and has a summand with infinitesimal character $\gamma + \sigma$. Let $P^\gamma_{\gamma+\sigma}$ be the projection on this summand and let $\phi^\gamma_{\gamma+\sigma}$ be the functor $M \to P^\gamma_{\sigma+\gamma}(M \otimes V_\sigma)$. Let $V^*_{-\sigma}$ be the contragradient module to V_σ. Then $M \otimes V^*_{-\sigma}$ has a direct summand with infinitesimal character $\gamma - \sigma$. Let $P^\gamma_{\gamma-\sigma}$ be the projection on this summand and let $\psi^\gamma_{\gamma-\sigma}$ be the functor $M \to P^\gamma_{\gamma-\sigma}(M \otimes V^*_{-\sigma})$.

THEOREM 4.1. *Fix a Cartan subalgebra $\mathfrak{h}^{\mathbb{C}}$ of $\mathfrak{g}^{\mathbb{C}}$ and a positive system $\triangle^+(\mathfrak{h}^{\mathbb{C}}, \mathfrak{g}^{\mathbb{C}})$ such that $\operatorname{Re}\gamma$ is dominant. Let W_γ and $W_{\gamma+\sigma}$ be the stabilizer in the Weyl group of γ and $\gamma + \sigma$ respectively. If $W_\gamma = W_{\gamma+\sigma}$ then $\phi^\gamma_{\gamma+\sigma}, \psi^{\gamma+\sigma}_\gamma$ are exact functors in the category of H-C modules.*

PROOF. See [Zu].

$\square$

4.2. Degenerate principal series of $O(p,q)$ in the $(O(p,q), \widetilde{OSp}_{2,2})$ duality.
Let $O(p,q)$ be the orthogonal group acting on $\mathbb{R}^{p+q}$ with the basis $\{\epsilon_1, \ldots, \epsilon_n\}$ with $n = p + q$, as described in formulas (2.4) - (2.6). Let $\mathfrak{g}$ be its Lie algebra. We will assume that both p and q are positive. We let $O(p)$ denote the subgroup of $O(p,q)$ that leaves the vectors $\epsilon_j : p + 1 \leq j \leq n$ fixed. Similarly, $O(q)$ is the subgroup that leaves the vectors $\epsilon_j : 1 \leq j \leq p$, fixed. Then $O(p) \times O(q)$ is a maximal compact subgroup of $O(p,q)$. We will denote it by K, and $\mathfrak{k}$ is the Lie algebra of K.

The plane spanned by ϵ_1 and ϵ_n is a hyperbolic plane H, containing two null lines $L^\pm$, which are spanned by the vectors $\epsilon^\pm = \epsilon_1 \pm \epsilon_n$. Let A denote the subgroup of $O(p,q)$ that preserves the lines $L^\pm$ and that leaves fixed the vectors ϵ_j for $1 < j < n$. Note that $A \simeq \mathbb{R}^\times$. In fact, the mapping α defined by

$$(4.1) \qquad a(\epsilon^+) = \alpha(a)\epsilon^+, \qquad \text{for } a \in A,$$

defines an explicit isomorphism from A to $\mathbb{R}^\times$.

The principal series representations we will be dealing with come in families parametrized by characters of A. Note that A is not connected; it has two connected components. This complicates the explicit description of the characters of A. Our notation will not be quite standard. Here we indicate the standard conventions.

Let $\mathfrak{a} \simeq \mathbb{R}$ be the Lie algebra of A. Then $\exp : \mathfrak{a} \to A$ is an isomorphism from $\mathfrak{a}$ to A_0, the identity component of A. To define a character ν of A, one must specify the restriction of ν to A_0, and the value (namely ± 1) on the element $\alpha^{-1}(-1)$. We can write

$$(4.2a) \qquad \nu^0(\exp z) = e^{d\nu(z)}, \qquad \text{for } z \in \mathfrak{a},$$

where $d\nu$ is an appropriate element in $\mathfrak{a}^*$. Conversely, given an element $\lambda \in \mathfrak{a}^*$, we can define a character ν_λ^0 of A_0 by

(4.2b) $$\nu_\lambda^0(\exp z) = \exp(\lambda(z)).$$

We can extend ν_λ^0 to a character ν_λ of A by defining

(4.2c) $$\nu_\lambda(a) = \nu_\lambda^0(\lambda(|\alpha(a)|))$$

In order to specify a character ν, besides $d\nu$, we need to know the value of $\nu(\alpha^{-1}(-1))$, which will be ± 1. We will write

(4.2d) $$sgn(a) = \frac{\alpha(a)}{|\alpha(a)|}.$$

This is a character of A which is identically 1 on A_0, and takes the value -1 on $\alpha^{-1}(-1)$. The two possible characters whose restriction to A_0 is ν_λ^0 are then ν_λ and $\epsilon\nu_\lambda$. It is common in the literature [Kn2], [Wa] to focus on the value of characters on A_0, and to index them by their differentials. However, except in the calculations of infinitesimal characters below, we will use the global form of the characters of A, that describe the behavior on all of A, not just on A_0.

The subgroup $P \subset O(p,q)$ consisting of elements that preserve the null line $L^+ = \mathbb{R}\epsilon^+$ is a maximal parabolic subgroup. Let $O(p-1, q-1) = M$ denote the subgroup of $O(p,q)$ that leaves ϵ_1 and ϵ_n (equivalently, $\epsilon^\pm$) fixed. Let N^+ be the subgroup that acts as the identity on L^+ and on $(L^+)^\perp/L^+$, where $(L^+)^\perp$ is the orthogonal complement of L^+. Then $O(p-1, q-1) \cdot A$ is a Levi component of P, and N^+ is the unipotent radical of P, and

(4.3) $$P \simeq (O(p-1, q-1) \cdot A) \cdot N^+ = MAN^+$$

is the Langlands decomposition of P[Kn2].

Let $Q \subset P$ be a minimal parabolic subgroup of $O(p,q)$, with unipotent radical $(N^+)'$, Levi component M', containing maximal split torus A'. Denote the Lie algebras of these groups by $\mathfrak{q}$, $\mathfrak{n}^{+'}$ and $\mathfrak{a}'$. Then $\mathfrak{g} = \mathfrak{k} \oplus \mathfrak{a}' \oplus \mathfrak{n}^{+'}$ is an Iwasawa decomposition of $\mathfrak{g}$.

Define the character ρ_A of A by

(4.4a) $$\rho_A(a) = \det |(\text{Ad } a)_{|\mathfrak{n}^+}|^{\frac{1}{2}}$$

for $a \in A$. If z is in $\mathfrak{a}$, then we can write

(4.4b) $$\rho_A(\exp z) = e^{d\rho_A(z)},$$

where

(4.4c) $$2d\rho_A = \sum_{\mu \in \Gamma^+} (\dim\mu)\mu, \quad \text{where } \Gamma^+ = \{\text{roots of } (\mathfrak{g}, \mathfrak{a}) \text{ positive for } N^+\}.$$

Explicitly, in our case,

(4.4d) $$d\rho_A = (\frac{p-1+q-1}{2})d\alpha = (\frac{n}{2}-1)d\alpha, \quad \text{and} \quad \rho_A = |\alpha|^{\frac{n}{2}-1}.$$

Let σ be a representation of M, and ν a character of A. We call the induced representation $\pi(P, \sigma \otimes \nu) = \text{Ind}_P^G(\sigma \otimes \nu \otimes 1)$ a *generalized principal series representation or a degenerate principal series*. The representation space for $\text{Ind}_P^G(\sigma \otimes \nu \otimes 1)$ is

(4.5) $$\{f \in C^\infty(G, V) | f(mang) = \nu(a)\rho_A(a)\sigma(m)f(g)\}$$

with norm

$$\| f \|^2 = \int_K |f(k)|^2 \, dk.$$

G acts by

$$\pi(P, \sigma \otimes \nu)(g) f(x) = f(xg).$$

The inclusion of the factor $\rho_A(a)$ in equation (4.5) indicates that we are using the normalized induction [Kn2]. The normalization makes these degenerate principal series unitary when σ and ν are unitary. For us, σ will be finite dimensional, and will be unitary only when it is trivial or the determinant character.

In [HT2], Howe and Tan have analyzed the degenerate principal series induced from a character (i.e., when σ is trivial). In particular, they explicitly give all the composition factors for this class of degenerate principal series and the K-types in each composition factor. These are the $O(p,q)$-representations in the duality correspondence of $(O(p,q), \widetilde{SL_2})$ [LTZ]. This result is the other ingredient besides Zuckerman translation for describing the duality correspondence of $(O(p,q), O\widetilde{Sp}_{2,2})$.

In the oscillator-spin representation, we need to tensor the representation space $\mathcal{S}(\mathbb{R}^n)$ for $(O(p,q), \widetilde{SL_2})$ with the exterior algebra $\Lambda^*((\mathbb{R}^n)^*)$. Thus it is necessary for us to determine the structure of $\mathrm{Ind}_P^G(1 \otimes \nu \otimes 1) \otimes \Lambda^k((\mathbb{R}^n)^*)$ as $O(p,q)$-representation.

We have

$$\mathbb{R}^n = \mathbb{R}^{p+q} \simeq H \oplus H^\perp = (L^+ \oplus L^-) \oplus \mathbb{R}^{(p-1)+(q-1)} = (L^+ \oplus L^-) \oplus \mathbb{R}^{n-2}.$$

Also, $\mathbb{R}^n \simeq (\mathbb{R}^n)^*$ as $O(p,q)$-module. This means that

$$(4.6) \qquad \Lambda^*((\mathbb{R}^n)^*) \simeq \Lambda^*(L^+ \oplus L^-) \otimes \Lambda^*(\mathbb{R}^{n-2})$$

as modules for $O(p-1, q-1) \cdot A$. If we parse this into the homogeneous components $\Lambda^k((\mathbb{R}^n)^*)$, we get the following result.

LEMMA 4.2. *Let $P = MAN$ be defined as above, then we have the following decompositions of $\Lambda^k((\mathbb{R}^n)^*)$ as a representation for $M \cdot A$.*
i) when $2 \leqslant k \leqslant n-2$, $\quad \Lambda^k((\mathbb{R}^n)^*)|_{MA} \cong$
$\Lambda^k((\mathbb{R}^{n-2})^*) \oplus \Lambda^{k-1}((\mathbb{R}^{n-2})^*)) \otimes L^+ \oplus \Lambda^{k-1}((\mathbb{R}^{n-2})^*) \otimes L^- \oplus \Lambda^{k-2}((\mathbb{R}^{n-2})^*),$
ii) when $k = 1$, $\quad \Lambda^1((\mathbb{R}^n)^*)|_{MA} \cong$
$\Lambda^1((\mathbb{R}^{n-2})^*) \oplus \Lambda^0((\mathbb{R}^{n-2})^*) \otimes L^+ \oplus \Lambda^0((\mathbb{R}^{n-2})^*) \otimes L^-,$
iii) when $k = n-1$, $\quad \Lambda^{n-1}((\mathbb{R}^n)^*)|_{MA} \cong$
$\Lambda^{n-2}((\mathbb{R}^{n-2})^*) \otimes L^+ \oplus \Lambda^{n-2}((\mathbb{R}^{n-2})^*) \otimes L^- \oplus \Lambda^{n-3}((\mathbb{R}^{n-2})^*).$

REMARK 4.3. In applying Lemma 4.2, it is well to keep in mind that, as $O(p,q)$ modules, we have $\Lambda^k((\mathbb{R}^n)^*) \simeq \det \otimes \Lambda^{n-k}((\mathbb{R}^n)^*)$. Mutatis mutandis, similar remarks apply to the representations σ_k of $O(p-1, q-1)$ that appear in Theorem 4.3.

Taking into account that $L^+ \subset L^+ \oplus \mathbb{R}^{n-2} \subset \mathbb{R}^n$ is a composition series of $\mathbb{R}^n$ for the action of P, we see that Lemma 4.2 has the following consequence.

THEOREM 4.4. *Let* $\mathrm{Ind}_P^G(1 \otimes \nu \otimes 1)$ *be the induced representation defined above, and let* $O(p,q)$ *act on* $\Lambda^k((\mathbb{R}^n)^*)$, $\mathrm{Ind}_P^G(1 \otimes \nu \otimes 1) \otimes \Lambda^k((\mathbb{R}^n)^*)$ *is a representation of* $O(p,q)$.

1) When $2 \leqslant k \leqslant n-2$, *it has a natural chain*
$$0 \subset V_1 \subset V_2 \subset V_3 \subset V_4 = \mathrm{Ind}_P^G(1 \otimes \nu \otimes 1) \otimes \Lambda^k((\mathbb{R}^n)^*) \text{ with}$$

$$V_1 \cong \mathrm{Ind}_P^G(\sigma_{k-1} \otimes (\alpha\nu) \otimes 1) \qquad V_2/V_1 \cong \mathrm{Ind}_P^G(\sigma_k \otimes \nu \otimes 1)$$

$$V_3/V_2 \cong \mathrm{Ind}_P^G(\sigma_{k-2} \otimes \nu \otimes 1) \qquad V_4/V_3 \cong \mathrm{Ind}_P^G(\sigma_{k-1} \otimes (\alpha^{-1}\nu) \otimes 1)$$

Here σ_k *is the representation of* $O(p-1, q-1)$ *on* $\Lambda^k((\mathbb{R}^{n-2})^*)$, *and* α *is the character of equation (4.1).*

2) When $k = 1$, *it has a chain* $0 \subset V_1 \subset V_2 \subset V_3 = \mathrm{Ind}_P^G(1 \otimes \nu \otimes 1) \otimes \Lambda^1((\mathbb{R}^n)^*)$ *with*

$$V_1 \cong \mathrm{Ind}_P^G(1 \otimes (\alpha\nu) \otimes 1) \qquad V_2/V_1 \cong \mathrm{Ind}_P^G(\sigma_1 \otimes \nu \otimes 1)$$

$$V_3/V_2 \cong \mathrm{Ind}_P^G(1 \otimes (\alpha^{-1}\nu) \otimes 1)$$

3) When $k = n-1$, *it also has a chain* $0 \subset V_1 \subset V_2 \subset V_3 = \mathrm{Ind}_P^G(1 \otimes \nu \otimes 1) \otimes$ $\Lambda^{n-1}((\mathbb{R}^n)^*)$

$$V_1 \cong \mathrm{Ind}_P^G(\sigma_{n-2} \otimes (\alpha\nu) \otimes 1) \qquad V_2/V_1 \cong \mathrm{Ind}_P^G(\sigma_{n-3} \otimes \nu \otimes 1)$$

$$V_3/V_2 \cong \mathrm{Ind}_P^G(\sigma_{n-2} \otimes (\alpha^{-1}\nu) \otimes 1)$$

4) When $k = n$, $\mathrm{Ind}_P^G(1 \otimes \nu \otimes 1) \otimes \Lambda((\mathbb{R}^n)^*) \cong \mathrm{Ind}_P^G(1 \otimes \nu \otimes 1) \otimes \det$.

PROOF. From Lemma 4.2, we know that $\Lambda^k((\mathbb{R}^n)^*)|_{MA} \cong \Lambda^k((\mathbb{R}^{n-2})^*) \oplus$ $\Lambda^{k-1}((\mathbb{R}^{n-2})^*) \otimes L^+ \oplus \Lambda^{k-1}((\mathbb{R}^{n-2})^*) \otimes L^- \oplus \Lambda^{k-2}((\mathbb{R}^{n-2})^*)$ for $2 \leqslant k \leqslant n-2$. Thus $\Lambda^k((\mathbb{R}^n)^*)$ has a composition series of length 4 under the parabolic subgroup. By the theory of induced representations [Kn2, Ch.10], we get the conclusion for $2 \leqslant k \leqslant n-2$. Also from Lemma 4.2, we also have the result for the special cases of $k = 1$, $n-1$ and n. The theorem follows. $\qquad\square$

4.3. Structure of the relevant degenerate principal series.
We will now look further into the structure of the degenerate principal series identified in Theorem 4.4.

Let $P = MAN$ be the maximal parabolic subgroup of $O(p,q)$ above, $\mathfrak{t}^{\mathbb{C}}$ be a Cartan subalgebra of $\mathfrak{m}^{\mathbb{C}}$, then $\mathfrak{h}^{\mathbb{C}} = (\mathfrak{a} \oplus \mathfrak{t})^{\mathbb{C}}$ is a Cartan subalgebra of $\mathfrak{g}_{\mathbb{C}}$. We may if we wish assume that the split Cartan subalgebra $\mathfrak{a}'$ used to describe the Iwasawa decomposition of $O(p,q)$ (see formulas (4.4)) is contained in $\mathfrak{h}^{\mathbb{C}}$, but that is not needed for the discussion below.

Let v_1 be the element of $\mathfrak{a}$ such that $v_1(\epsilon^{\pm}) = \pm\epsilon^{\pm}$. In other words, if the character α is as in formula (4.1), we have $\alpha(\exp tv_1) = \exp t$. According to the well-known [GW] description of roots and weights for orthogonal groups, we can introduce coordinates on $\mathfrak{h}^{\mathbb{C}}$ and its dual as follows.

i) when $n = p+q = 2m$ is even, let $\mathfrak{h}^{\mathbb{C}} = \{v_1, v_2, \ldots, v_m\}$ be a basis of $\mathfrak{h}^{\mathbb{C}}$ and $\{e_i\}$ be a dual basis. Again with suitable choice of the v_j, we have a set of simple roots $\triangle = \{\alpha_1, \ldots, \alpha_m\}$ where $\alpha_i = e_i - e_{i+1}$ for $i \leqslant i \leqslant m-1$, $\alpha_m = e_{m-1} + e_m$. The associated set of positive roots is
$$\Phi^+ = \{e_i \pm e_j : 1 \leqslant i,j \leqslant m\}.$$
And its Weyl group

$W = \{$determinant one permutations and sign changes of $\{e_1, \ldots, e_m\}\}$.
(The determinant one condition holds in $SO(p,q)$; in $O(p,q)$, all permutations and sign changes are possible.)

ii) when $n = p + q = 2m + 1$ is odd, let $\mathfrak{h}^{\mathbb{C}} = \{v_1, v_2, \ldots, v_m\}$ be a basis of $\mathfrak{h}^{\mathbb{C}}$ and $\{e_i\}$ be a dual basis. Then, again if the v_j are suitably chosen, we have a set of simple roots $\triangle = \{\alpha_1, \ldots, \alpha_m\}$ where $\alpha_i = e_i - e_{i+1}$ for $1 \leqslant i \leqslant m-1$, $\alpha_m = e_m$. The associated set of positive roots is
$$\Phi^+ = \{e_i \pm e_j : 1 \leqslant 1 < j \leqslant m\} \cup \{e_i : 1 \leqslant i \leqslant m\}.$$
And its Weyl group
$$W = \{\text{permutations and sign changes of } \{e_1, \ldots, e_m\}\}.$$

Set $\eta = \frac{p+q}{2} - [\frac{p+q}{2}]$, where $[\frac{p+q}{2}]$ is the largest integer less than or equal to $\frac{p+q}{2}$. Then $\eta = 0$ if $p + q$ is even, $\eta = \frac{1}{2}$ if $p + q$ is odd.

Easy calculation [GW] shows that

$$(4.7) \quad \rho = \frac{1}{2} \sum_{\alpha \in \Phi^+} \alpha = (m - 1 + \eta)e_1 + (m - 2 + \eta)e_2 + \ldots + (\eta + 1)e_{m-1} + \eta e_m$$

In the discussion below of infinitesimal characters, we will let $d\nu$ stand for the element of $\mathfrak{a}^*$ that describes the character ν on the identity component of A. Thus we have

$$(4.8) \qquad\qquad \nu(\exp t v_1) = e^{t d\nu(v_1)}.$$

Also, we will abbreviate $d\nu(v_1) = d\nu$. Thus, below, $d\nu$ will be understood as a complex number.

THEOREM 4.5. *As the notation above, for $1 \leqslant k \leqslant \frac{p+q}{2} - 1$, $d\nu \in \mathbb{C}$, $\operatorname{Re} d\nu \geqslant 0$, let λ_k be the highest weight of $O(p,q)$ acting on $\Lambda^k((\mathbb{R}^{p+q})^*)$ and $u(d\nu)$ be the infinitesimal character of $\operatorname{Ind}_P^G(1 \otimes \nu \otimes 1)$. Then when $d\nu \neq \frac{p+q}{2} - k - 1$, $\phi_{u(d\nu)+\lambda_k}^{u(d\nu)}$ is exact and*
i) If $\operatorname{Re} d\nu \geqslant \frac{p+q}{2} - k - 1$, $\phi_{u(d\nu)+\lambda_k}^{u(d\nu)} \operatorname{Ind}_P^G(1 \otimes \nu \otimes 1) = \operatorname{Ind}_P^G(\sigma_{k-1} \otimes (\alpha\nu) \otimes 1)$.
ii) If $\operatorname{Re} d\nu < \frac{p+q}{2} - k - 1$, $\phi_{u(d\nu)+\lambda_k}^{u(d\nu)} \operatorname{Ind}_P^G(1 \otimes \nu \otimes 1) = \operatorname{Ind}_P^G(\sigma_k \otimes \nu \otimes 1)$.
In particular, when $d\nu \notin \mathbb{Z}$ for $p+q$ even or $d\nu \notin \mathbb{Z} + \frac{1}{2}$ for $p+q$ odd, $\operatorname{Ind}_P^G(\sigma_k \otimes d\nu \otimes 1)$ is irreducible.

PROOF. We use the above ordering. The infinitesimal character $u(d\nu)$ of $\operatorname{Ind}_P^G(\nu \otimes 1)$ is

e_1	e_2	e_3	$\ldots$	e_{m-1}	e_m
$d\nu$	$m'-2$	$m'-3$	$\ldots$	$1'$	$0'$

Here the numbers under e_i indicate the coefficient in front of e_i when we express the infinitesimal character as a linear combination of e_i's. Also, we use the notation that $j' = j + \eta$ for numbers $1 \leq j \leq m$. Thus above, $1' = 1 + \eta$, $m' = m + \eta$, etc. Note also that $m' = \frac{p+q}{2}$.

If $m' - i \leqslant \operatorname{Re} d\nu < m' - (i-1)$, then, after permutation by some element $w \in W$, the dominant version of the infinitesimal character is

	e_1	e_2	$\ldots$	e_{i-2}	e_{i-1}	e_i	$\ldots$	e_{m-1}	e_m
(4.9)	$m'-2$	$m'-3$	$\ldots$	$m'-(i-1)$	$d\nu$	$m'-i$	$\ldots$	$1'$	$0'$

If $m' - i < \operatorname{Re} d\nu < m' - (i-1)$, then this infinitesimal character is regular. If $d\nu = m' - i$, then it is singular and its stabilizer is $S_{(i-1,i)}$, the group (of order 2) that permutes $i-1$ and i.

We know that σ_k has the highest weight

e_1	e_2	e_3	$\cdots$	e_{k+1}	e_{k+2}	$\cdots$	e_{m-1}	e_m
0	1	1	$\cdots$	1	0	$\cdots$	0	0

thus $\operatorname{Ind}_P^G(\sigma_k \otimes \nu \otimes 1)$ has the infinitesimal character $\chi_{u(k,\nu)}$

e_1	e_2	e_3	$\cdots$	e_{k+1}	e_{k+2}	$\cdots$	e_{m-1}	e_m
$d\nu$	$m'-1$	$m'-2$	$\cdots$	$m'-k$	$m'-k-2$	$\cdots$	$1'$	$0'$

As above, the dominant version of this infinitesimal character depends on the value of $\operatorname{Re} d\nu$.

a) if $m' - i \leqslant \operatorname{Re} d\nu < m' - (i-1)$ for $i \leqslant k$, the dominant version is

e_1	e_2	$\cdots$	e_i	e_{i+1}	$\cdots$	e_{k+1}	e_{k+2}	$\cdots$	e_{m-1}	e_m
$m'-1$	$m'-2$	$\cdots$	$d\nu$	$m'-i$	$\cdots$	$m'-k$	$m'-k-2$	$\cdots$	$1'$	$0'$

(4.10a)

b) if $m' - k - 2 \leqslant \operatorname{Re} \nu < m' - k$, the dominant version is

e_1	e_2	$\cdots$	e_k	e_{k+1}	e_{k+2}	$\cdots$	e_{m-1}	e_m
$m'-1$	$m'-2$	$\cdots$	$m'-k$	$d\nu$	$m'-k-2$	$\cdots$	$1'$	$0'$

(4.10b)

c) if $m' - i \leqslant \operatorname{Re} d\nu < m' - (i-1)$ for $i \geqslant k+3$, the dominant version is

e_1	$\cdots$	e_k	e_{k+1}	$\cdots$	e_{i-2}	e_{i-1}	e_i	$\cdots$	e_{m-1}	e_m
$m'-1$	$\cdots$	$m'-k$	$m'-k-2$	$\cdots$	$m'-(i-1)$	$d\nu$	$m'-i$	$\cdots$	$1'$	$0'$

(4.10c)

It is clear that $\Lambda^k((\mathbb{R}^{p+q})^*)$ has highest weight $\lambda_k = e_1 + \cdots + e_k$. From the formulas above, it is easy to check $W_{\lambda_k + u(d\nu)} = W_{u(d\nu)}$ when $d\nu \neq \frac{p+q}{2} - k - 1 = m' - k - 1$. Theorem 4.1 then implies that $\phi_{u(d\nu)+\lambda_k}^{u(d\nu)}$ is exact. However, when $d\nu = \frac{p+q}{2} - k - 1$, we see that $u(d\nu)$ is singular and $\lambda_k + u(d\nu)$ is regular. Thus Theorem 4.1 does not apply for this value of $d\nu$.

Consider $\operatorname{Ind}_P^G(\nu \otimes 1) \otimes \Lambda^k((\mathbb{R}^{p+q})^*)$. We know that it has composition factors

$$\operatorname{Ind}_P^G(\sigma_k \otimes \nu \otimes 1) \qquad\qquad \operatorname{Ind}_P^G(\sigma_{k-1} \otimes (\alpha\nu) \otimes 1)$$
$$\operatorname{Ind}_P^G(\sigma_{k-1} \otimes (\alpha^{-1}\nu) \otimes 1) \qquad\qquad \operatorname{Ind}_P^G(\sigma_{k-2} \otimes \nu \otimes 1)$$

From formulas (4.10), we can see that when $d\nu \neq \frac{p+q}{2} - k \pm 1$, these four degenerate principal series have distinct infinitesimal characters. Indeed, we can summarize the results of those formulas as follows. We let C_ρ be the set of coefficients of ρ This is the set of integers (or half-integers, as appropriate) consecutive from $0'$ to $m'-1$. If we look at the coefficients of the infinitesimal character of $\operatorname{Ind}_P^G(\sigma_k \otimes \nu \otimes 1)$, we see that its coefficients consist of all the elements of C_ρ except for $m' - k - 1$, together with $d\nu$. We may assume that $d\nu \geq 0$. If $d\nu$ is not an element of C_ρ, then it is readily identifiable among the coefficients of $\operatorname{Ind}_P^G(\sigma_k \otimes \nu \otimes 1)$, regardless of where it is placed. Likewise, the absence of $m' - k - 1$ from the coefficients of the infinitesimal character of $\mathrm{Ind}_P^G(\sigma_k \otimes \nu \otimes 1)$ lets us identify k. Thus, if $d\nu$ does not belong to C_ρ, we can readily identify k and $d\nu$ from the infinitesimal character.

Similar considerations apply when $d\nu$ happens to be equal to an element of C_ρ, as long as it is not $m' - k - 1$. Then among the coefficients of the infinitesimal character, one element of C_ρ will be doubled up, and this will be $d\nu$. Also, the element $m' - k - 1$ will still be missing. Thus in this case also, we can readily identify $d\nu$ and k.

It is only when $d\nu = m' - k - 1$ that we cannot reconstruct $d\nu$ and k from the infinitesimal character of $\mathrm{Ind}_P^G(\sigma_k \otimes \nu \otimes 1)$. But in this case, the set of coefficients is exactly C_ρ, and the infinitesimal character is exactly ρ.

We can be slightly more precise. If $d\nu = m' - k - 1$, then $d(\alpha\nu) = d\nu + 1 = m' - k = m' - (k-1) - 1$, so that $\mathrm{Ind}_P^G(\sigma_k \otimes \nu \otimes 1)$ and $\mathrm{Ind}_P^G(\sigma_{k-1} \otimes (\alpha\nu) \otimes 1)$ have the same infinitesimal character, while the infinitesimal characters of $\mathrm{Ind}_P^G(\sigma_{k-1} \otimes (\alpha^{-1}\nu) \otimes 1)$ and of $\mathrm{Ind}_P^G(\sigma_{k-2} \otimes \nu \otimes 1)$ are distinct from this and from each other. On the other hand, if $d\nu = m - k + 1 = m' - (k-2) - 1$, then similar reasoning applies to show that the latter two components have infinitesimal character ρ, and the first two components have infinitesimal characters that are not ρ and distinct from each other.

Thus, as long as $d\nu \neq \frac{p+q}{2} - k \pm 1$, the tensor product $\mathrm{Ind}_P^G(\nu \otimes 1) \otimes \Lambda^k(\mathbb{R}^{p+q})$ is a direct sum of its component degenerate principal series and the Zuckerman translate $\phi_{u(d\nu)+\lambda_k}^{u(d\nu)} \mathrm{Ind}_P^G(1 \otimes \nu \otimes 1)$ is one of them. Referring again to the formulas (4.10) we see that when $\mathrm{Re}\, d\nu \geqslant \frac{p+q}{2} - k - 1$, $\lambda_k + u(d\nu)$ is the infinitesimal character of $\mathrm{Ind}_P^G(\sigma_{k-1} \otimes (\alpha\nu) \otimes 1)$, thus

$$\phi_{u(d\nu)+\lambda_k}^{u(d\nu)} \mathrm{Ind}_P^G(1 \otimes \nu \otimes 1) = \mathrm{Ind}_P^G(\sigma_{k-1} \otimes (\alpha\nu) \otimes 1)$$

Similarly when $\mathrm{Re}\, d\nu < \frac{p+q}{2} - k - 1$,

$$\phi_{u(d\nu)+\lambda_k}^{u(d\nu)} \mathrm{Ind}_P^G(1 \otimes \nu \otimes 1) = \mathrm{Ind}_P^G(\sigma_k \otimes \nu \otimes 1)$$

By [HT2], when $d\nu \notin \mathbb{Z}$ for $p + q$ even, or $d\nu \notin \mathbb{Z} + \frac{1}{2}$ for $p + q$ odd, $\mathrm{Ind}_P^G(1 \otimes \nu \otimes 1)$ is irreducible. From Zuckerman translation, we also get the irreducibility of $\mathrm{Ind}_P^G(\sigma_k \otimes \nu \otimes 1)$ in this situation. $\square$

It is worth noting that, in the above discussion, the cases when the hypothesis of the Zuckerman translation principle does not hold all have infinitesimal character equal to ρ.

PROPOSITION 4.6. For $0 \leqslant k \leqslant m - 1$, $k \in \mathbb{Z}$, $d\nu = \pm(\frac{p+q}{2} - k - 1)$, $\mathrm{Ind}_P^G(\sigma_k \otimes \nu \otimes 1)$ has the same infinitesimal character χ_ρ with

$$\rho = \frac{1}{2} \sum_{\alpha \in \Phi^+} \alpha = (m' - 1)e_1 + (m' - 2)e_2 + \ldots + (1')e_{m-1} + 0'e_m$$

as in (4.7). Furthermore, the eigenvalue of the Casimir operator $\Omega_{O(p,q)}$ is 0.

PROOF. We may as well take $d\nu \geq 0$. The infinitesimal characters for $\mathrm{Ind}_P^G(\sigma_k \otimes \nu \otimes 1)$ with $d\nu = \pm(\frac{p+q}{2} - k - 1)$ are already obtained in (4.10).

Concerning the eigenvalue of $\Omega_{O(p,q)}$, we recall that it is well known and easy to see that the trivial representation of $O(p,q)$ has also this infinitesimal character. For example, the representation induced from the trivial representation of P, which contains the trivial representation of $O(p,q)$, has infinitesimal character ρ. Thus the eigenvalue of the Casimir operator $\Omega_{O(p,q)}$ is 0. $\square$

Theorem 4.5 can be restated:

COROLLARY 4.7. For $0 \leqslant k \leqslant m - 1$, we have

$$\operatorname{Ind}_P^G(\sigma_k \otimes \nu \otimes 1) = \phi_{u(d\nu)+\lambda_k}^{u(d\nu)}(\operatorname{Ind}_P^G(1 \otimes \nu \otimes 1))$$

when $d\nu < \frac{p+q}{2} - k - 1$, and

$$\operatorname{Ind}_P^G(\sigma_k \otimes \nu \otimes 1) = \phi_{u(d\nu-1)+\lambda_{k+1}}^{u(d\nu-1)}(\operatorname{Ind}_P^G(\alpha^{-1}\nu \otimes 1))$$

when $d\nu > \frac{p+q}{2} - k - 1$. Moreover, in these cases, these functors establish a bijection of composition series of the relevant scalar principal series with $\operatorname{Ind}_P^G(\sigma_k \otimes \nu \otimes 1)$.

   When $d\nu = \frac{p+q}{2} - k - 1$, then the Casimir operator $\Omega_{O(p,q)}$ acts by zero on $\operatorname{Ind}_P^G(\sigma_k \otimes \nu \otimes 1)$.

   We now use Theorem 4.5 and Corollary 4.7 to transfer the known structure of the scalar degenerate principal series $\operatorname{Ind}_P^G(1 \otimes \nu \otimes 1)$ to the $\operatorname{Ind}_P^G(\sigma_k \otimes \nu \otimes 1)$. The explicit structure of the $\operatorname{Ind}_P^G(1 \otimes \nu \otimes 1)$ is described in [HT2]. The details vary according to the parities of p and q. We will discuss here the case when p and q are both even.

   There are two scalar principal series with the same infinitesimal character, namely $\operatorname{Ind}_P^G(1 \otimes \nu \otimes 1)$ and $\operatorname{Ind}_P^G(1 \otimes (sgn)\nu \otimes 1)$, which we denote below by $\pi(\nu \otimes 1)$ and $\pi((sgn)\nu \otimes 1)$. When $d\nu$ is integral, one of these representations is irreducible, and the other breaks up into several pieces: 4 when $d\nu$ is large ($|d\nu| \geqslant \frac{p+q}{2} - 1$), 3 when it is small ($0 \neq |d\nu| \leqslant \frac{p+q}{2} - 2$), and 2 when $d\nu = 0$.

   In [HT2], the constituents of $\pi(\nu \otimes 1)$ and $\pi((sgn)\nu \otimes 1)$ are described in terms of their K-types. As we have seen, the maximal compact subgroup of $O(p,q)$ is $K \simeq O(p,0) \times O(0,q) = O(p) \times O(q)$. The representations of K that appear in the restrictions of $\pi(\nu \otimes 1)$ and $\pi((sgn)\nu \otimes 1)$ to K are the tensor products of spherical harmonics $\mathcal{H}^\ell(\mathbb{R}^p) \otimes \mathcal{H}^m(\mathbb{R}^q)$. These K-types all appear with multiplicity one in the restrictions of $\pi(\nu \otimes 1)$ and $\pi((sgn)\nu \otimes 1)$. The ones with $\ell + m$ even appear in one of $\pi(\nu \otimes 1)$ or $\pi((sgn)\nu \otimes 1)$, and the ones with $\ell + m$ odd appear in the other. If these spaces are represented by points $\begin{bmatrix} \ell \\ m \end{bmatrix}$ in the positive quarter plane, then the K types that appear in the constituents of the reducible representation correspond to points in one of the regions cut off by certain straight lines, as illustrated in Figure 5.1. We will call these constituents

(4.11) $Y^{+p}(\nu), \quad Y^{+q}(\nu), \quad Y^c(\nu), \quad \text{and} \quad Y^f(\nu),$

as indicated in the Figure 4.1.

   It is easy enough to see from the Figure 4.1 that the constituents can be characterized by means of the K-types they contain, as follows:

        i) $Y^{+p}(\nu)$ contains $\mathcal{H}^\ell(\mathbb{R}^p) \otimes \mathcal{H}^m(\mathbb{R}^q)$ with ℓ arbitrarily large for any fixed m;

        or $\ell - m$ is bounded below, but not above.

        ii) $Y^{+q}(\nu)$ contains $\mathcal{H}^\ell(\mathbb{R}^p) \otimes \mathcal{H}^m(\mathbb{R}^q)$ with m arbitrarily large for fixed ℓ;

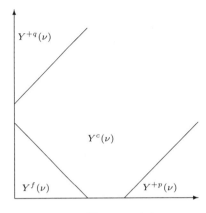

$$\text{Figure 4.1}$$

or $\ell - m$ is bounded above, but not below.

iii) $Y^c(\nu)$ contains $\mathcal{H}^\ell(\mathbb{R}^p) \otimes \mathcal{H}^m(\mathbb{R}^q)$ for which $|\ell - m|$ is bounded, but ℓ and m individually can become arbitrarily large.

iv) $Y^f(\nu)$ contains only finitely many K-types $\mathcal{H}^\ell(\mathbb{R}^p) \otimes \mathcal{H}^m(\mathbb{R}^q)$.

The K types of the $\operatorname{Ind}_P^G(\sigma_k \otimes \nu \otimes 1)$ are more complicated than those of $\operatorname{Ind}_P^G(1 \otimes \nu \otimes 1)$, but they exhibit qualitatively similar behavior. Let ${}^p\gamma_a$ for $1 \le a \le \frac{p}{2}$, be the fundamental weights for $O(p)$, and let ${}^q\gamma_b$ for $1 \le b \le \frac{q}{2}$ be the fundamental weights for $O(q)$. Let μ_p^ψ denote the irreducible representation of $O(p)$ with highest weight ψ. Note that

$$\mathcal{H}^\ell(\mathbb{R}^p) = \mu_p^{\ell \cdot ({}^p\gamma_1)}.$$

Some calculations with branching laws tell us that the K-types in $\operatorname{Ind}_P^G(\sigma_k \otimes \nu \otimes 1)$ have the form $\mu_p^\psi \otimes \mu_q^\xi$, where $\psi = \ell \cdot ({}^p\gamma_1) + {}^p\gamma_a$ for $\ell \ge 0$, and likewise, $\xi = m \cdot ({}^q\gamma_1) + {}^q\gamma_b$. There are some further restrictions on a and b, but these are not important for us here. The main point to observe is that ℓ and m are the only coefficients that are greater than 1, and that they can vary independently, and are arbitrary - in particular, arbitrarily large. The multiplicities of a particular representation of K can be one or two.

It is also well-known [Su] that

$$\Lambda^k((\mathbb{R}^p)^*) \otimes \mathcal{H}^\ell$$

is a sum of four constituents, with highest weights of the form $\ell' \cdot ({}^p\gamma_1) + {}^p\gamma_{k'}$, with $|\ell' - \ell| \le 1$ and $|k' - k| \le 1$. Likewise well-known is the decomposition

$$\Lambda^k((\mathbb{R}^{p+q})^*)_{|O(p) \times O(q)} \simeq \sum_{a+b=k} \Lambda^a((\mathbb{R}^p)^*) \otimes \Lambda^b((\mathbb{R}^q)^*),$$

(with the restriction that $a \le p$ and $b \le q$). These two facts together imply that the tensor products $\Lambda^k((\mathbb{R}^{p+q})^*) \otimes Y^{+p}(\nu)$, $\Lambda^k((\mathbb{R}^{p+q})^*) \otimes Y^{+q}(\nu)$, $\Lambda^k((\mathbb{R}^{p+q})^*) \otimes Y^c(\nu)$, and $\Lambda^k((\mathbb{R}^{p+q})^*) \otimes Y^f(\nu)$ will have K-types $\mu_p^\psi \otimes \mu_q^\xi$, where $\psi = \ell \cdot ({}^p\gamma_1) + {}^p\gamma_a$ for $\ell \ge 0$, and likewise, $\xi = m \cdot ({}^q\gamma_1) + {}^q\gamma_b$, where the coefficients ℓ and m obey the same qualitative behavior as specified above in the descriptions of the constituents $Y^{+p}(\nu)$, $Y^{+q}(\nu)$, $Y^c(\nu)$, and $Y^f(\nu)$.

Combining these results, we conclude that when $d\nu$ is large enough, the representation $\operatorname{Ind}_P^G(\sigma_k \otimes \nu \otimes 1)$ (or $\operatorname{Ind}_P^G(\sigma_k \otimes (sgn)\nu \otimes 1)$, as appropriate) consists of four constituents, namely

(4.12a)
$$Y^{+p}(\nu, k) = \phi_{u(d\nu-1)+\lambda_{k+1}}^{u(d\nu-1)}(Y^{+p}(\nu)).$$

(4.12b)
$$Y^{+q}(\nu, k) = \phi_{u(d\nu-1)+\lambda_{k+1}}^{u(d\nu-1)}(Y^{+q}(\nu)).$$

(4.12c)
$$Y^c(\nu, k) = \phi_{u(d\nu-1)+\lambda_{k+1}}^{u(d\nu-1)}(Y^c(\nu)).$$

(4.12d)
$$Y^f(\nu, k) = \phi_{u(d\nu-1)+\lambda_{k+1}}^{u(d\nu-1)}(Y^f(\nu)).$$

Again, these constituents are characterized by their K-types, which are $\mu_p^\psi \otimes \mu_q^\xi$, where $\psi = \ell \cdot ({}^p\gamma_1) + {}^p\gamma_a$ for $\ell \geq 0$, and likewise, $\xi = m \cdot ({}^q\gamma_1) + {}^q\gamma_b$, where the coefficients ℓ and m obey the same qualitative behavior as specified above in the descriptions of the constituents $Y^{+p}(\nu)$, $Y^{+q}(\nu)$, $Y^c(\nu)$, and $Y^f(\nu)$.

When $d\nu < \frac{p+q}{2} - 1$, the component $Y^f(\nu, k)$ will be missing; and also we may have to use $\phi_{u(d\nu)+\lambda_k}^{u(d\nu)}$ in place of $\phi_{u(d\nu-1)+\lambda_{k+1}}^{u(d\nu-1)}$. In order to avoid excessively complicated notation, we will just write

(4.12e)
$$Y^{+p}(\nu, k) = \phi(Y^{+p}(\nu)),$$

and similarly for the other constituents.

5. Duality Correspondence of $(O(p,q), \widetilde{OSP}_{2,2})$

In §3, we have given a classification of irreducible $\mathfrak{osp}_{2,2}$ representations with non-zero Casimir eigenvalue. In §4, we used the Zuckerman translator functor to determine the structure of a class of $O(p,q)$ degenerate principal series induced from a particular class of finite dimensional representations of a maximal parabolic subgroup. Combining these two results and the explicit duality correspondence of $(O(p,q), \widetilde{SL}_2)$, as described in [LTZ], in this part we show that the correspondence is bijective for the representations with nonzero eigenvalue for the Casimir operator of $O(p,q)$ (or of $\mathfrak{osp}_{2,2}$) in the oscillator-spin representation. There are actually only finitely many representations appearing in the oscillator-spin representation on which the Casimir operator vanishes. We conjecture that the duality correspondence will also be a bijection in these exceptional cases.

The Lie part of $\widetilde{OSp}_{2,2}$ is $\widetilde{SL}_2 \times SO_2$, hence the duality correspondence of $(O(p,q), \widetilde{OSp}_{2,2})$ is closely related to the duality correspondence of $(O(p,q), \widetilde{SL}_2)$ in the oscillator representation. We begin by recalling some basic facts about that correspondence.

For this discussion, we will use the abbreviation $\pi(\nu \otimes 1)$ for the degenerate principal series $\pi(P, 1 \otimes \nu \otimes 1) \simeq \operatorname{Ind}_P^G(1 \otimes \nu \otimes 1)$ of $O(p,q)$ defined in §4.2 . We denote by $\pi'(\nu \otimes 1)$ the normalized principal series [Kn2], [LTZ] representations of SL_2 (or $\widetilde{SL}_2$ if $p+q$ is odd). When $p+q$ is even, then if $d\nu \notin \mathbb{Z}$, $\pi'(\nu \otimes 1)$ is irreducible. However, if $d\nu \in \mathbb{Z}$, $\pi'(\nu \otimes 1)$ can have up to three composition factors.

Similarly, when $p + q$ is odd, if $d\nu \notin \mathbb{Z} + \frac{1}{2}$, $\pi'(\nu \otimes 1)$ is irreducible, and when $d\nu \in \mathbb{Z} + \frac{1}{2}$, $\pi'(\nu \otimes 1)$ has two composition factors [LTZ].

The following result from [LTZ] constructs an intertwining operator from the oscillator representation of $(O(p,q), \widetilde{SL_2})$ to the tensor product of degenerate principal series.

THEOREM 5.1. *Let $\omega_{p,q}$ be the oscillator representation of $O(p,q) \times \widetilde{SL_2}$ on the space $\mathcal{S}(\mathbb{R}^n)$ of Schwartz functions on $\mathbb{R}^n$, then for each $\nu \in \mathbb{C}$, there exists a nonzero intertwining operator*

$$\Pi_\nu : \omega_{p,q} \to \pi(sgn^{\frac{q-p}{2}} \nu \otimes 1) \otimes \pi'(\nu^{-1} \otimes 1)$$

Here sgn is the character of formula (4.2d).

This result is extended to $(O(p,q), \widetilde{OSp}_{2,2})$ in [LH], Proposition 2.5.1. (We note that conventions in [LH] are slightly different from [LTZ]. In particular, normalized induction is not used.)

When the representations $\pi(sgn^{\frac{q-p}{2}} \nu \otimes 1)$ and $\pi'(\nu^{-1} \otimes 1)$ are irreducible, Theorem 5.1 determines the $(O(p,q), \widetilde{SL_2})$ correspondence. It is well-known, and can be read off from the classification stated in §3.1, that the principal series of SL_2 are irreducible when $d\nu \notin \mathbb{Z}$, and the genuine principal series of $\widetilde{SL_2}$ are irreducible when $d\nu \notin \frac{1}{2} + \mathbb{Z}$. The corresponding irreducibility results for $O(p,q)$ are given in [HT2] for the scalar principal series, and they are extended to the representations $\text{Ind}_P^G(\sigma_k \otimes \nu \otimes 1)$ in Theorem 4.5. In particular, this takes care of all cases when $d\nu \notin \frac{p+q}{2} + \mathbb{Z}$. When $d\nu \in \frac{p+q}{2} + \mathbb{Z}$, and the principal series are reducible, the situation is more complicated. We now turn to these cases.

In [LTZ], the correspondence of $(O(p,q), \widetilde{SL_2})$ is also described explicitly when the induced representations are reducible. Let us denote by $R(G, \omega_{p,q})$ the set of irreducible admissible representations of G which could be realized as quotients of $\omega_{p,q}$.

THEOREM 5.2. *Let $\omega_{p,q}$ be the oscillator representation of $(O(p,q), \widetilde{SL_2})$ on $\mathcal{S}(\mathbb{R}^n)$, then for each $\pi \in R(O(p,q), \omega_{p,q})$, there is a unique $\pi' \in R(\widetilde{SL_2}, \omega_{p,q})$ such that the tensor product $\pi \otimes \pi'$ is in $R(O(p,q) \times \widetilde{SL_2}, \omega_{p,q})$. In this case, we write $\theta(\pi) = \pi'$ or $\theta(\pi') = \pi$.*

We remark that as stated, Theorem 5.2 is just a special case of the general local theta correspondence as established in [Ho4]. However, [LTZ] describes the correspondence explicitly for the pair $(O(p,q), \widetilde{SL_2})$, and it this description which we will want below. We will give more details about the [LTZ] description in our discussion of how to extend this result to $(O(p,q), \widetilde{OSp}_{2,2})$.

THEOREM 5.3. (**Duality correspondence of** $(O(p,q), \widetilde{OSp}_{2,2})$) *In the oscillator-spin representation $\widetilde{\omega}_{p,q}$ of $(O(p,q), \widetilde{OSp}_{2,2})$, when the eigenvalue of $\Omega_{O(p,q)}$ is nonzero, for each $\pi \in R(O(p,q), \widetilde{\omega}_{p,q})$, there is a unique $\pi' \in R(\widetilde{OSp}_{2,2}, \widetilde{\omega}_{p,q})$ such that $\pi \otimes \pi' \in R(O(p,q) \times \widetilde{OSp}_{2,2}, \widetilde{\omega}_{p,q})$. Again, we write $\theta(\pi) = \pi'$ or $\theta(\pi') = \pi$.*

PROOF. As we have remarked, when $d\nu \notin \frac{p+q}{2} + \mathbb{Z}$, both factors in the correspondence of Theorem 5.1, and its extension to $(O(p,q), \widetilde{OSP}_{2,2})$ given in [LH], are irreducible. This establishes the existence of the correspondence except when $d\nu \in \frac{p+q}{2} + \mathbb{Z}$, which will be the subject of the discussion below.

The details of the argument vary slightly according to the parities of p and q. We will sketch the proof in the case p, q even with $p, q \geqslant 2$. The other cases can be handled by similar arguments. Since Theorem 5.3 is not optimal in any case, we will ignore some exceptional cases, so we actually establish a slightly weaker result. The main point here is to show how Zuckerman translation can be used to reduce the super duality correspondence to the known correspondence in the scalar case.

Here are the specifics of the correspondence of Theorem 5.2 when p and q are even. For a given value $d\nu \in \mathbb{Z}$, there are two representations of SL_2, namely $\pi'(\nu \otimes 1)$ and $\pi'((sgn)\nu \otimes 1)$. To describe their structure, it is useful to make use of the operator $\mathfrak{k}$ of equation (2.7e), which is the infinitesimal generator of a maximal compact subgroup $T \simeq SO_2$ of SL_2. On one of the two representations, the operator $\mathfrak{k}$ will have eigenvalues $2m$, for $m \in \mathbb{Z}$, and on the other it will have eigenvalues $2m + 1$.

One of the two induced representations is irreducible, and the other breaks into 3 pieces (just 2 pieces when $d\nu = 0$). The pieces have $\mathfrak{k}$ eigenvalues congruent to $d\nu - 1$ modulo 2. One is finite dimensional, with eigenvalues of absolute value less than $|d\nu|$. In the notation of Theorem 3.2, this is the representation $F_{1-|d\nu|}$. The other two constituents have eigenvalues either all positive or all negative respectively, and of absolute value larger than $|d\nu|$. In Theorem 3.2, these are the lowest and highest weight representations $V_{|d\nu|+1}$ and $\overline{V}_{-(|d\nu|+1)}$.

If we think of the eigenvectors of $\mathfrak{k}$ as being located on a number line, each at the spot of its eigenvalue, then $F_{1-|d\nu|}$ occupies the middle portion of the line, located symmetrically around 0, and the representations $V_{|d\nu|+1}$ and $\overline{V}_{-(|d\nu|+1)}$ occupy the right and left ends of the line, respectively.

$$\overline{V}_{-(|d\nu|+1)} \qquad F_{1-|\nu|} \qquad V_{|d\nu|+1}$$

In terms of these constituent diagrams, [LTZ] describe the $(O(p,q), SL_2)$ duality correspondence as follows.

$$(5.1) \qquad V_{|d\nu|+1} \leftrightarrow Y^{+p}(\nu) \qquad F_{1-|d\nu|} \leftrightarrow Y^{f}(\nu), \quad \text{and} \quad \overline{V}_{-(|d\nu|+1)} \leftrightarrow Y^{+q}(\nu).$$

This is illustrated in Figure 5.1.

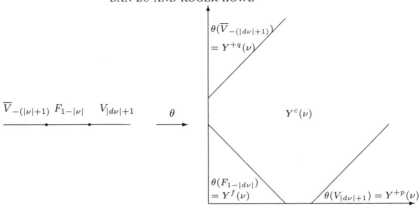

FIGURE 5.1

Note that the constituent $Y^c(\nu)$ of $\pi(\nu \otimes 1)$ represented by the central strip above the line cutting off the finite region in the corner of the quarter plane appears neither as a quotient nor a subrepresentation of $\pi(\nu \otimes 1)$, and does not appear in the duality correspondence. (However, when $Y^f(\nu)$ is not there (i.e., for $|d\nu| < \frac{p+q}{2} - 1$) then $Y^c(\nu)$ *is* a quotient (or subrepresentation), and *does* appear in the correspondence.)

Our strategy for describing the duality correspondence for $(O(p,q), \widetilde{OSp}_{2,2})$ is to combine what we know about the structure of the $O(p,q)$ degenerate principle series, obtained by using Zuckerman translation with the K-type knowledge coming from [Lu1], as described at the end of §4.

From (2.1), we know that the oscillator-spin representation $\widetilde{\omega}_{p,q}$ acts on the space $\mathcal{S}(\mathbb{R}^n) \otimes \Lambda^*((\mathbb{R}^n)^*)$. When we restrict $\widetilde{\omega}_{p,q}$ to $O(p,q) \times \widetilde{SL}_2 \times SO_2$, we find $\widetilde{SL}_2$ only acts on the function part and SO_2 only acts on the exterior part. Therefore, $\widetilde{\omega}_{p,q}|_{O(p,q)\times\widetilde{SL}_2\times SO_2}$ can be regarded as the extension from the oscillator representation $\omega_{p,q}$ of $(O(p,q), \widetilde{SL}_2)$ such that $O(p,q)$ acts both on the function part and exterior part.

Let τ be an irreducible representation of $\widetilde{OSp}_{2,2}$ that appears as a quotient of $\mathcal{S}(\mathbb{R}^n)\otimes\Lambda^*((\mathbb{R}^n)^*)$. Let τ_0 be the associated representation of $\widetilde{SL}_2 \times SO_2$ associated to τ as per formula (3.5). Then τ_0 considered as a representation of $\widetilde{SL}_2$ is a quotient of $\mathcal{S}(\mathbb{R}^n)$, and if the eigenvalue of h' on τ_0 is $k-1-m = k-1-\frac{p+q}{2}$, then τ_0 is a quotient of $\mathcal{S}(\mathbb{R}^n) \otimes \Lambda^{k-1}((\mathbb{R}^n)^*)$.

According to [Ho4], the maximal quotient of $\mathcal{S}(\mathbb{R}^n)$ that transforms according to $(\tau_0)_{|\widetilde{SL}_2}$ has the form $\hat{\tau}_0' \otimes \tau_0$, where $\hat{\tau}_0'$ is an admissible representation of $O(p,q)$, and $\hat{\tau}_0'$ has a unique irreducible quotient τ_0' that corresponds to $(\tau_0)_{|\widetilde{SL}_2}$.

Evidently, then, the maximal quotient of $\mathcal{S}(\mathbb{R}^n)\otimes\Lambda^{k-1}((\mathbb{R}^n)^*)$ that transforms according to τ_0 is $(\Lambda^{k-1}((\mathbb{R}^n)^*) \otimes \hat{\tau}_0') \otimes \tau_0$. Thus, if $\tau' \otimes \tau$ is an $O(p,q) \times \widetilde{OSp}_{2,2}$ quotient of $\mathcal{S}(\mathbb{R}^n) \otimes \Lambda^*((\mathbb{R}^n)^*)$, then τ' is a quotient of $\Lambda^{k-1}((\mathbb{R}^n)^*) \otimes \hat{\tau}_0'$. A preliminary implication of this is that the maximal quotient of $\mathcal{S}(\mathbb{R}^n) \otimes \Lambda^*((\mathbb{R}^n)^*)$ that transforms according to τ under $\widetilde{OSp}_{2,2}$ has the form $\hat{\tau}' \otimes \tau$, where $\hat{\tau}'$ is an admissible representation of $O(p,q)$. We want to show that $\hat{\tau}'$ has a unique irreducible quotient.

We know the infinitesimal character of $\hat{\tau}_0'$, so we know that its constituents belong to $\mathrm{Ind}_P^G(1 \otimes \nu \otimes 1)$ for a suitable character ν. To facilitate the argument, suppose that $\hat{\tau}_0'$ is actually a subrepresentation of $\mathrm{Ind}_P^G(1 \otimes \nu \otimes 1) = \pi(\nu \otimes 1)$. This means that $\hat{\tau}_0' \otimes \Lambda^\ell((\mathbb{R}^n)^*)$ is a subrepresentation of $\pi(\nu \otimes 1) \otimes \Lambda^\ell((\mathbb{R}^n)^*)$, which by Theorem 4.4 has the four constituents

$$\mathrm{Ind}_P^G(\sigma_{k-3} \otimes \nu \otimes 1), \qquad \mathrm{Ind}_P^G(\sigma_{k-2} \otimes (\alpha\nu) \otimes 1)$$
$$\mathrm{Ind}_P^G(\sigma_{k-2} \otimes (\alpha^{-1}\nu) \otimes 1), \qquad \mathrm{Ind}_P^G(\sigma_{k-1} \otimes \nu \otimes 1)$$

From the proof of Theorem 4.5, we know that only one of these constituents will have the correct infinitesimal character when the Casimir eigenvalue is nonzero. Hence we can conclude that the quotient map from $\Lambda^{k-1}((\mathbb{R}^n)^*) \otimes \hat{\tau}_0'$ to $\hat{\tau}'$ factors through one of these constituents.

The full representation τ of $\widetilde{OSp}_{2,2}$ consists of four representations of $\widetilde{SL}_2 \otimes SO_2$. We call them τ_0, τ_1^+, τ_1^-, and $\tau_2 = d\delta^* \tau_0$. Note that $\tau_1^+ \oplus \tau_1^- = \tau_0 \otimes Z^+$, in the notation of §3.

If we repeat this exercise with the $\widetilde{SL}_2 \times SO_2$ modules τ_1^+, τ_1^-, and τ_2, and compare the $O(p,q)$ constituents, we find that $\mathrm{Ind}_P^G(\sigma_{k-1} \otimes \nu \otimes 1)$ is the only constituent they all have in common.

From Theorem 4.5, we know that if $d\nu < m - k - 1$, then $\mathrm{Ind}_P^G(\sigma_{k-1} \otimes \nu \otimes 1) = \phi_{u(d\nu)+\lambda_{k-1}}^{u(d\nu)}(\mathrm{Ind}_P^G(1 \otimes \nu \otimes 1))$. On the other hand, if $d\nu > m-k-1$, then $\mathrm{Ind}_P^G(\sigma_{k-1} \otimes \nu \otimes 1) = \phi_{u(d\nu)+\lambda_k}^{u(d\nu)}(\mathrm{Ind}_P^G(1 \otimes \alpha^{-1}\nu \otimes 1))$. In either case, then, the maximal quotient of $\Lambda^{k-1}((\mathbb{R}^n)^*) \otimes \hat{\tau}_0'$ factors through the constituent of $\Lambda^{k-1}((\mathbb{R}^n)^*) \otimes \hat{\tau}_0'$ with the appropriate infinitesimal character. If $d\nu < m - k$, then this is the image of the Zuckerman functor , and the mapping is exact. It follows that, since $\hat{\tau}_0'$ has a unique irreducible quotient, so does $\phi_{u(d\nu)+\lambda_{k-1}}^{u(d\nu)}(\hat{\tau}_0')$, and therefore so does $\hat{\tau}'$.

If however, $d\nu > m - k - 1$, then $\mathrm{Ind}_P^G(\sigma_{k-1} \otimes \nu \otimes 1)$ is the Zuckerman quotient of $\Lambda^k((\mathbb{R}^n)^*) \otimes \mathrm{Ind}_P^G(1 \otimes \alpha^{-1}\nu \otimes 1)$, so again $\hat{\tau}'$ is a quotient of the exact functor $\phi_{u(d\nu-1)+\lambda_k}^{u(d\nu-1)}$ of the representation $(\hat{\tau}_1^-)'$. Since $(\hat{\tau}_1^-)'$ has a unique quotient, so will $\hat{\tau}'$. This concludes our sketch of the proof of Theorem 5.3. $\square$

REMARK 5.4. The argument sketched above shows that the duality correspondence for $(O(p,q), \widetilde{OSp}_{2,2})$ converts the induction functor $\tau_0 \to \mathrm{Ind}_q^{\mathrm{osp}} \tau_0$ of formula (3.4) into Zuckerman translation. In other words, the diagram of Figure 5.2 (next page) commutes.

In terms of the notation of §4, we have for $d\nu \in \mathbb{Z}^+$ and $d\nu \neq \frac{p+q}{2} - k - 1$

$$W(d\nu, k+1 - \frac{p+q}{2}) = \theta(Y^{+p}(\sigma_k, \nu)), \quad \overline{W}(-d\nu, k+1 - \frac{p+q}{2}) = \theta(Y^{+q}(\sigma_k, \nu)),$$

$$F_{\mathrm{osp}}(-d\nu, k+1 - \frac{p+q}{2}) = \theta(Y^f(\sigma_k, \nu))$$

In Theorem 5.3, we proved the duality correspondence except for the irreducible $\widetilde{OSp}_{2,2}$-representations whose eigenvalues of Casimir operator are zero. This is just a finite set of $\widetilde{OSp}_{2,2}$-representations. It is reasonable to believe, the duality correspondence also holds for these cases. Therefore, we make the following conjecture:

CONJECTURE 5.5. In the oscillator-spin representation of $(O(p,q), \widetilde{OSp}_{2,2})$, the joint spectrum $R(O(p,q) \times \widetilde{OSp}_{2,2}, \widetilde{\omega}_{p,q})$ is the graph of bijection between the spectra $R(O(p,q), \widetilde{\omega}_{p,q})$ and $R(\widetilde{OSp}_{2,2}, \widetilde{\omega}_{p,q})$ of the factors. In other words, for each $\pi \in R(O(p,q), \widetilde{\omega}_{p,q})$ there is a unique $\pi' \in R(\widetilde{OSp}_{2,2}, \widetilde{\omega}_{p,q})$ such that $\pi \otimes \pi' \in R(O(p,q) \times \widetilde{OSp}_{2,2}, \widetilde{\omega}_{p,q})$.

REMARK 5.6. The argument in Theorem 5.3 fails for $d\nu = \frac{p+q}{2} - k - 1$, since from $O(p,q)$-side, the Zuckerman's functor is not an isomorphism and we don't know its composition factors, from the $\widetilde{OSp}_{2,2}$-side, the irreducible module is only composed of two irreducible $\widetilde{SL}_2 \times \widetilde{SO}_2$-modules [Lu1], which makes the correspondence more complicated.

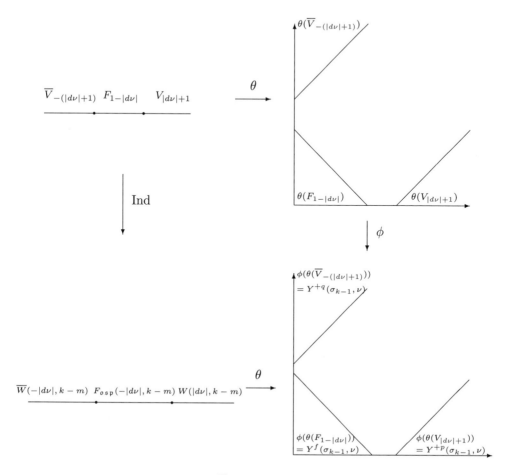

FIGURE 5.2

Acknowledgement: the authors are grateful to the referee for finding a lot of typos in a very short time.

References

[Co] L. Conlon, Differentiable manifolds: a first course. Birkhäuser Advanced Texts, Birkhäuser Boston, Inc., Boston, MA, 1993.

[Fo] G. Folland,. Harmonic analysis in phase space. Annals of Mathematics Studies **122**. Princeton University Press, Princeton, NJ, 1989.

[Ge1] S. Gelbart, Examples of dual reductive pairs, Automorphic forms, Representations and L-functions, Proc. Sympos. Pure Math XXXIII, Part 1 (1977), Amer. Math. Soc., Providence, R.I., 1979, 287–296.

[Ge2] S. Gelbart, Weil's representation and the spectrum of the metaplectic group. Lecture Notes in Mathematics **530**, Springer-Verlag, Berlin-New York, 1976.

[GJS] D. Ginzburg, D. Jiang, and D. Soudry, Poles of L-functions and theta liftings for orthogonal groups, J. Inst. Math. Jussieu **8** (2009), no. 4, 693–741.

[GW] R. Goodman and N. R. Wallach, Representations and invariants of the classical groups. Encyclopedia of Mathematics and its Applications **68**, Cambridge University Press, 1998.

[Ho1] R. Howe, Remarks on classical invariant theory, Trans. Amer. Math. Soc. **313** (1989), 539–570.

[Ho2] R. Howe, Dual pairs in physics: harmonic oscillators, photons, electrons, and singletons. Applications of group theory in physics and mathematical physics, Lectures in Appl. Math. **21**, Amer. Math. Soc., Providence, R.I., 1985, 179–207.

[Ho3] R. Howe, θ-series and invariant theory, Automorphic forms, representations and L-functions, Proc. Sympos. Pure Math XXXIII, Part 1 (1977), Amer. Math. Soc., Providence, R.I., 1979, 275 – 285.

[Ho4] R. Howe, Transcending classical invariant theory, J. Amer. Math. Soc. **2** (1989), 535–552.

[Ho5] R. Howe, On some results of Strichartz and of Rallis and Schiffman, J. Funct. Anal. **32** (1979), 297–303.

[Ho6] R. Howe, A century of Lie theory. American Mathematical Society centennial publications, Vol. II, Amer. Math. Soc., Providence, RI, 1992, 101 – 320.

[HT] R. Howe and E-C. Tan, Nonabelian harmonic analysis: Applications of SL(2,R). Universitext. Springer-Verlag, New York, 1992.

[HT2] R. Howe and E.C. Tan, Homogeneous functions on light cones: the infinitesimal structure of some degenerate principal series representations. Bull. Amer. Math. Soc. (N.S.) **28** (1993), 1 – 74.

[Ich] Ichino, Atsushi, On the regularized Siegel-Weil formula. (English summary) J. Reine Angew. Math. **539** (2001), 201–234.

[Ja] J. Jackson, Classical Electrodynamics, Wiley, New York, 1962.

[Ka] V. G. Kac, Lie superalgebras, Adv. in Math. **26** (1997), 8–96.

[Kap] I. Kaplansky, Z_2-graded algebras, Illinois J. Math. **35** (1991), 85–92.

[Kn1] A. Knapp, Lie groups beyond an introduction. Progress in Mathematics **140**, Birkhäuser Boston, Inc., Boston, MA, 1996.

[Kn2] A. Knapp, Representation theory of semisimple groups. An overview based on examples. Princeton Mathematical Series **36**. Princeton University Press, Princeton, NJ, 1986.

[KR] S. S. Kudla and S. Rallis, A regularized Siegel-Weil formula: the first term identity, Ann. Math. 140 (1994), 1–80.

[La1] S. Lang, $SL2(\mathbb{R})$. Reprint of the 1975 edition. Graduate Texts in Math. **105**, Springer-Verlag, New York, 1985.

[La2] S. Lang, Algebra. Revised third edition. Graduate Texts in Math. **211**. Springer-Verlag, New York, 2002.

[LZ] S.T. Lee, C. Zhu, Degenerate principal series and local theta correspondence, Trans. Amer. Math. Soc. **350** (1998), 5017 – 5046.

[LPTZ] J.S. Li, A. Paul, E.C. Tan, C. Zhu, The explicit duality correspondence of $(Sp(p,q), O^*(2n))$, J. Funct. Anal. **200** (2003), 71 – 100.

[LTZ] J.S. Li, E.C. Tan and C. Zhu, Tensor product of degenerate principal series and local theta correspondence, J. Funct. Anal. **186** (2001), 381–431.

[LH] D. Lu and R. Howe, The dual pair $(O(p,q), O\widetilde{Sp}_{2,2})$ and Maxwell's equations, in *Casimir Force, Casimir Operators and the Riemann Hypothesis*, M. Wakayama, G. van Dijk (eds.), de Gruyter, Berlin, 2010, 1 – 29.

[Lu1] D. Lu, Dual pairs, Lie superalgebra and Maxwell's equations. thesis, Yale Univ., 2009.

[Lu2] D. Lu, On the classification of the irreducible $\mathfrak{osp}_{2,2}$-representations. Preprint, Universität Leipzig, 2010, in preparation.

[ØZ] B. Ørsted and G. Zhang, L^2-versions of the Howe correspondence II, J. Math. Pures Appl. **74** (1995), 165 –183.

[PT] A. Paul and E.C. Tan, On the dual pairs $(O(p,q), SL(2,R))$, $(U(p,q), U(1,1))$ and $(Sp(p,q), O^*(4))$, Pac. J. Math. **187**, (1999) 349 - 378.

[S-P] R. Schulze-Pillot, Local theta correspondence and the lifting of Duke, Imamoglu and Ikeda, Osaka J. Math. **45** (2008), 965 – 971.

[Su] S. Sundaram, Tableaux in the representation theory of the classical Lie groups. Invariant theory and tableaux, IMA Vol. Math. Appl. **19**, Springer, New York, 1990, 191-225.

[SW] E. Stein and G. Weiss, Introduction to Fourier Analysis on Euclidean Space, Princeton Mathematical Series **32**, Princeton University Press, Princeton, N.J., 1971.

[Wa] N. R. Wallach, Real reductive groups. I, Pure and Applied Mathematics **132**. Academic Press, Inc., Boston, MA, 1988.

[We1] A. Weil, Sur certains groupes d'operateurs unitaires, Acta Math. **111** (1964), 143 – 211

[We2] A. Weil, Fonction zeta et Distributions, Seminaire Bourbaki 1966, 523-531.

[Wey] H. Weyl, The classical groups: their invariants and representations, Princeton University Press, Princeton, N.J., 1939.

[Zu] G. Zuckerman, Tensor products of finite and infinite dimensional representations of semisimple Lie groups. Ann. Math. (2) **106** (1977), 295 – 308.

DEPARTMENT OF MATHEMATICS, YALE UNIVERSITY, NEW HAVEN, 06511 USA
Current address: Department of Mathematics, Leipzig University, Leipzig, 04103 Germany
E-mail address: `dan.lu@math.uni-leipzig.de`

DEPARTMENT OF MATHEMATICS, YALE UNIVERSITY, NEW HAVEN, USA 06511
E-mail address: `howe@math.yale.edu`

Contemporary Mathematics
Volume **557**, 2011

On the algebraic set of singular elements
in a complex simple Lie algebra

Bertram Kostant and Nolan Wallach

ABSTRACT. Let G be a complex simple Lie group and let $\mathfrak{g} = \operatorname{Lie} G$. Let $S(\mathfrak{g})$ be the G-module of polynomial functions on $\mathfrak{g}$ and let $\operatorname{Sing}\mathfrak{g}$ be the closed algebraic cone of singular elements in $\mathfrak{g}$. Let $\mathcal{L} \subset S(\mathfrak{g})$ be the graded ideal defining $\operatorname{Sing}\mathfrak{g}$ and let $2r$ be the dimension of a G-orbit of a regular element in $\mathfrak{g}$. Then $\mathcal{L}^k = 0$ for any $k < r$. On the other hand, there exists a remarkable G-module $M \subset \mathcal{L}^r$ which already defines $\operatorname{Sing}\mathfrak{g}$. The main results of this paper are a determination of the structure of M.

0. Introduction

0.1. Let G be a complex simple Lie group and let $\mathfrak{g} = \operatorname{Lie} G$. Let $\ell = \operatorname{rank}\mathfrak{g}$. Then in superscript centralizer notation one has $\dim\mathfrak{g}^x \geq \ell$ for any $x \in \mathfrak{g}$. An element $x \in \mathfrak{g}$ is called regular (resp. singular) if $\dim\mathfrak{g}^x = \ell$ (resp. $> \ell$). Let $\operatorname{Reg}\mathfrak{g}$ be the set of all regular elements in $\mathfrak{g}$ and let $\operatorname{Sing}\mathfrak{g}$, its complement in $\mathfrak{g}$, be the set of all singular elements in $\mathfrak{g}$. Then one knows that $\operatorname{Reg}\mathfrak{g}$ is a nonempty Zariski open subset of $\mathfrak{g}$ and hence $\operatorname{Sing}\mathfrak{g}$ is a closed proper algebraic subset of $\mathfrak{g}$.

Let $S(\mathfrak{g})$ (resp. $\wedge\mathfrak{g}$) be the symmetric (resp. exterior) algebra over $\mathfrak{g}$. Both algebras are graded and are G-modules by extension of the adjoint representation. Let $\mathcal{B}$ be the natural extension of the Killing form to $S(\mathfrak{g})$ and $\wedge\mathfrak{g}$. (See Subsection 1.1 for further reference). The inner product it induces on u and v in either $S(\mathfrak{g})$ or $\wedge\mathfrak{g}$ is denoted by (u,v). The use of $\mathcal{B}$ permits an identification of $S(\mathfrak{g})$ with the algebra of polynomial functions on $\mathfrak{g}$. Since $\operatorname{Sing}\mathfrak{g}$ is clearly a cone the ideal, $\mathcal{L}$, of all $f \in S(\mathfrak{g})$ which vanish on $\operatorname{Sing}\mathfrak{g}$ is graded. Let $n = \dim\mathfrak{g}$ and let $r = (n - \ell)/2$. One knows that $n - \ell$ is even so that $r \in \mathbb{Z}_+$. It is easy to show that

$$\mathcal{L}^k = 0, \text{ for all } k < r. \tag{0.1}$$

The purpose of this paper is to define and study a rather remarkable G-submodule

$$M \subset \mathcal{L}^r \tag{0.2}$$

which in fact defines $\operatorname{Sing}\mathfrak{g}$. That is, if $x \in \mathfrak{g}$, then

$$x \in \operatorname{Sing}\mathfrak{g} \iff f(x) = 0, \ \forall f \in M \tag{0.3}$$

2010 *Mathematics Subject Classification.* Primary 20G05, 20G20, 22E60; Secondary, 22E10, 22E45.

Research partially supported by NSF grant DMS 0963035.

0.2. We will now give a definition of M. The use of $\mathcal{B}$ permits an identification of $\wedge \mathfrak{g}$ with the underlying space of the cochain complex defining the cohomology of $\mathfrak{g}$. The coboundary operator is denoted here by d (and δ in [Kz]) is a (super) derivation of degree 1 of $\wedge \mathfrak{g}$ so that $dx \in \wedge^2 \mathfrak{g}$ for any $x \in \mathfrak{g}$. Since $\wedge^{\text{even}} \mathfrak{g}$ is a commutative algebra there exists a homomorphism

$$\gamma : S(\mathfrak{g}) \to \wedge^{\text{even}} \mathfrak{g}$$

where for $x \in \mathfrak{g}$, $\gamma(x) = -dx$. One readily has that

$$S^k(\mathfrak{g}) \subset \operatorname{Ker} \gamma, \text{ for all } k > r. \tag{0.4}$$

Let $\gamma_r = \gamma | S^r(\mathfrak{g})$ so that

$$\gamma_r : S^r(\mathfrak{g}) \to \wedge^{2r} \mathfrak{g}. \tag{0.5}$$

If $x \in \mathfrak{g}$, one readily has

$$x^r \in \operatorname{Ker} \gamma_r \iff x \in \operatorname{Sing} \mathfrak{g}. \tag{0.6}$$

Let Γ be the transpose of γ_r so that one has a G-map

$$\Gamma : \wedge^{2r} \mathfrak{g} \to S^r(\mathfrak{g}). \tag{0.7}$$

By definition

$$M = \operatorname{Im} \Gamma. \tag{0.8}$$

0.3. Let $J = S(\mathfrak{g})^G$ so that (Chevalley) J is a polynomial ring $\mathbb{C}[p_1, \ldots, p_\ell]$ where the invariants p_j can be chosen to be homogeneous. In fact if m_j, $j = 1, \ldots, \ell$, are the exponents of $\mathfrak{g}$ we can take $\deg p_j = m_j + 1$. For any linearly independent $u_1, \ldots, u_\ell \in \mathfrak{g}$, let

$$\psi(u_1, \ldots, u_\ell) = \det \partial_{u_i} p_j \tag{0.9}$$

where, if $v \in \mathfrak{g}$, ∂_v is the operator of partial derivative by v in $S(\mathfrak{g})$. One has

$$\psi(u_1, \ldots, u_\ell) \in S^r(\mathfrak{g}) \tag{0.10}$$

since, as one knows, $\sum_{i=1}^\ell m_i = r$.

Let Σ_{2r} be the permutation group of $\{1, \ldots, 2r\}$ and let $\Pi_r \subset \Sigma_{2r}$ be a subset (of cardinality $(2r-1)(2r-3) \cdots 1$) with the property that all ν contained in Π_r are even, and such that, as sets of unordered pairs,

$$\{(\nu(1), \nu(2)), \ldots, (\nu(2r-1), \nu(2r)) \mid \nu \in \Pi_r\}$$

is the set of all partitions of $\{1, \ldots, 2r\}$ into a union of r subsets each of which has two elements. The following is one of our main theorems. Even more than explicitly determining $\psi(u_1, \ldots, u_\ell)$ one has

THEOREM 0.1. *Let $u_1, \ldots, u_\ell$ be any ℓ linearly independent elements in $\mathfrak{g}$ and let $w_1, \ldots, w_{2r}$ be a basis of the $\mathcal{B}$-orthogonal subspace to the span of the u_i. Then there exists some fixed $\kappa \in \mathbb{C}^\times$ such that, for all $x \in \mathfrak{g}$,*

$$\sum_{\nu \in \Pi_r} ([w_{\nu(1)}, w_{\nu(2)}], x) \cdots ([w_{\nu(2r-1)}, w_{\nu(2r)}], x) = \kappa \, \psi(u_1, \ldots, u_\ell)(x). \tag{0.11}$$

Moreover $\psi(u_1, \ldots, u_\ell) \in M$. In fact the left side of (0.11) is just $\Gamma(w_1 \wedge \cdots \wedge w_{2r})(x)$. In addition M is the span of $\psi(u_1, \ldots, u_\ell)$, over all $\{u_1, \ldots, u_\ell\}$, taken from the $\binom{n}{l}$ subsets of ℓ-elements in any given basis of $\mathfrak{g}$.

We now deal with the G-module structure of M. For any subspace $\mathfrak{s}$ of $\mathfrak{g}$, say of dimension k, let $[\mathfrak{s}] = \mathbb{C}v_1 \wedge \cdots \wedge v_k \subset \wedge^k \mathfrak{g}$ where the v_i are a basis of $\mathfrak{s}$. Let $\mathfrak{h}$ be a Cartan subalgebra of $\mathfrak{g}$ and let Δ be the set of roots for the pair $(\mathfrak{h}, \mathfrak{g})$. For any $\varphi \in \Delta$ let $e_\varphi \in \mathfrak{g}$ be a corresponding root vector. Let $\Delta_+ \subset \Delta$ be a choice of a set of positive roots and let $\mathfrak{b}$ be the Borel subalgebra spanned by $\mathfrak{h}$ and all e_φ for $\varphi \in \Delta_+$. For any subset $\Phi \subset \Delta$ let $\mathfrak{a}_\Phi \subset \mathfrak{g}$ be the span of e_φ for $\varphi \in \Phi$. Also let $\langle \Phi \rangle = \sum_{\varphi \in \Phi} \varphi$ so that

$$[\mathfrak{a}_\Phi] \text{ is an } \mathfrak{h}\text{-weight space for the } \mathfrak{h}\text{-weight } \langle \Phi \rangle. \tag{0.12}$$

A subset $\Phi \subset \Delta_+$ will be said to be an ideal in Δ_+ if $\mathfrak{a}_\Phi$ is an ideal of $\mathfrak{b}$. In such a case, if $\operatorname{card} \Phi = k$, then the span V_Φ of $G \cdot [\mathfrak{a}_\Phi]$ is an irreducible G-submodule of $\wedge^k \mathfrak{g}$ having $[\mathfrak{a}_\Phi]$ as highest weight space and $\langle \Phi \rangle$ as highest weight. Let $\mathcal{I}$ be the set of all ideals Φ in Δ_+ of cardinality ℓ. It is shown in [KW] that all ideals in $\mathfrak{b}$ of dimension ℓ are abelian and hence are of the form $\mathfrak{a}_\Phi$ for a unique $\Phi \in \mathcal{I}$. Specializing k in [K3] to ℓ one has that, by definition, $A_\ell \subset \wedge^\ell \mathfrak{g}$ is the span of $[\mathfrak{s}]$ over all abelian subalgebras $\mathfrak{s} \subset \mathfrak{g}$ of dimension ℓ. Using results in [K3] and that in [KW] above, one also has that A_ℓ is a multiplicity one G-module with the complete reduction

$$A_\ell = \oplus_{\Phi \in \mathcal{I}} V_\Phi \tag{0.13}$$

so that there are exactly $\operatorname{card} \mathcal{I}$ irreducible components. In addition it has been shown in [K3] that ℓ is the maximal eigenvalue of the ($\mathcal{B}$ normalized) Casimir operator, Cas, in $\wedge^\ell \mathfrak{g}$ and A_ℓ is the corresponding eigenspace. In the present paper the G-module structure of M is given in

THEOREM 0.2. *As G-modules one has an equivalence*

$$M \cong A_\ell \tag{0.14}$$

so that M is a multiplicity one module with $\operatorname{card} \mathcal{I}$ irreducible components. Moreover the components can be parameterized by $\mathcal{I}$ in such a way that the component corresponding to $\Phi \in \mathcal{I}$ has highest weight $\langle \Phi \rangle$. In addition Cas takes the value ℓ on each and every irreducible component of M.

1. Preliminaries

1.1. Let $\mathfrak{g}$ be a complex semisimple Lie algebra and let G be a Lie group such that $\mathfrak{g} = \operatorname{Lie} G$. Let $\mathfrak{h} \subset \mathfrak{g}$ be a Cartan subalgebra of $\mathfrak{g}$ and let ℓ be the rank of $\mathfrak{g}$ so that $\ell = \dim \mathfrak{h}$. Let Δ be the set of roots for the pair $(\mathfrak{h}, \mathfrak{g})$ and let $\Delta_+ \subset \Delta$ be a choice of a set of positive roots. Let $r = \operatorname{card} \Delta_+$ so that

$$n = \ell + 2r \tag{1.1}$$

where we let $n = \dim \mathfrak{g}$. Let $\mathcal{B}$ be Killing form (x, y) on $\mathfrak{g}$. For notational economy we identify $\mathfrak{g}$ with its dual $\mathfrak{g}^*$ using $\mathcal{B}$. The bilinear form $\mathcal{B}$ extends to an inner product (p, q), still denoted by $\mathcal{B}$, on the two graded algebras, the symmetric algebra $S(\mathfrak{g})$ of $\mathfrak{g}$ and the exterior algebra $\wedge \mathfrak{g}$ of $\mathfrak{g}$. Since we have Killing form identified as $\mathfrak{g}^*$ with $\mathfrak{g}$, the $\wedge \mathfrak{g}$ extension of $\mathcal{B}$ may be found in Section 2.2 in [Kz]. The $S(\mathfrak{g})$ extension may be found in Section 1, p. 335, in [K2]. If $x_i, y_j \in \mathfrak{g}$, $i = 1, \ldots, k$, $j =$

$1, \ldots, m$, then the product of x_i is orthogonal to the product of y_j in both $S(\mathfrak{g})$ and $\wedge \mathfrak{g}$ if $k \neq m$, whereas if $k = m$,

$$(x_1 \cdots x_k, y_1 \cdots y_k) = \sum_{\sigma \in \Sigma_k} (x_1, y_{\sigma(1)}) \cdots (x_k, y_{\sigma(k)}) \quad \text{in } S(\mathfrak{g})$$

$$(x_1 \wedge \cdots \wedge x_k, y_1 \wedge \cdots \wedge y_k) = \sum_{\sigma \in \Sigma_k} sg(\sigma)(x_1, y_{\sigma(1)}) \cdots (x_k, y_{\sigma(k)}) \quad \text{in } \wedge \mathfrak{g}.$$

(1.2)

Here Σ_k is the permutation group on $\{1, \ldots, k\}$ and sg abbreviates the signum character on Σ_k.

The identification of $\mathfrak{g}$ with its dual has the effect of identifying $S(\mathfrak{g})$ with the algebra of polynomial functions $f(y)$ on $\mathfrak{g}$. Thus if $x, y \in \mathfrak{g}$, then $x(y) = (x, y)$ and if $x_i \in \mathfrak{g}$, $i = 1, \ldots, k$, then

$$(x_1 \cdots x_k)(y) = \prod_{i=1}^{k} (x_i, y)$$

$$= (x_1 \cdots x_k, \frac{1}{k!} y^k).$$

(1.3)

The identification of $\mathfrak{g}$ with its dual also has the effect of identifying the (supercommutative) algebra $\wedge \mathfrak{g}$ with the underlying space of the standard cochain complex defining the cohomology of $\mathfrak{g}$. Let d be the (super) derivation of degree 1 of $\wedge \mathfrak{g}$, defined by putting

$$d = \frac{1}{2} \sum_{i=1}^{n} \varepsilon(w_i) \theta(z_i).$$

(1.4)

Here $\varepsilon(u)$, for any $u \in \wedge \mathfrak{g}$, is left exterior multiplication by u so that $\varepsilon(u) v = u \wedge v$ for any $v \in \wedge \mathfrak{g}$. Also $w_i, i = 1, \ldots, n$, is any basis of $\mathfrak{g}$ and $z_i \in \mathfrak{g}$, $i = 1, \ldots, n$, is the $\mathcal{B}$ dual basis. $\theta(x)$, for $x \in \mathfrak{g}$, is the derivation of $\wedge g$, of degree 0, defined so that $\theta(x) y = [x, y]$ for any $y \in \mathfrak{g}$. One readily notes that (1.4) is independent of the choice of the basis w_i. Thus if $x \in \mathfrak{g}$, then $dx \in \wedge^2 \mathfrak{g}$ is given by

$$dx = \frac{1}{2} \sum_{i=1}^{n} w_i \wedge [z_i, x].$$

(1.5)

Any element $\omega \in \wedge^2 \mathfrak{g}$ defines an alternating bilinear form on $\mathfrak{g}$. Its value $\omega(y, z)$ on $y, z \in \mathfrak{g}$ may be given in terms of $\mathcal{B}$ by

$$\omega(y, z) = (\omega, y \wedge z).$$

(1.6)

The rank of ω is necessarily even. In fact if $\text{rank} \, \omega = 2k$, then there exist $2k$ linearly independent elements $v_i \in \mathfrak{g}$, $i = 1, \ldots, 2k$, such that

$$\omega = v_1 \wedge v_2 + \cdots + v_{2k-1} \wedge v_{2k}.$$

(1.7)

The radical of ω, denoted by $\text{Rad} \, \omega$, is the space of all $y \in \mathfrak{g}$ such that $\omega(y, z) = 0$ for all $z \in \mathfrak{g}$. For $u \in \wedge \mathfrak{g}$, let $\iota(u)$ be the transpose of $\varepsilon(u)$ with respect to $\mathcal{B}$ on $\wedge \mathfrak{g}$. If $u = y \in \mathfrak{g}$, then one knows that $\iota(y)$ is the (super) derivation of degree minus 1 defined so that if $z \in \mathfrak{g}$, then $\iota(y) z = (y, z)$. (See p. 8 in [Kz]). From (1.6) one has

$$\text{Rad} \, \omega = \{y \in \mathfrak{g} \mid \iota(y) \omega = 0\}.$$

(1.8)

If $\mathfrak{s}$ is any subspace of $\mathfrak{g}$, let $\mathfrak{s}^{\perp}$ be the $\mathcal{B}$ orthogonal subspace to $\mathfrak{s}$. From (1.7) one then has that

$$\{v_i\}, \, i = 1, \ldots, 2k, \quad \text{is a basis of } (\text{Rad} \, \omega)^{\perp}.$$

(1.9)

If $\mathfrak{s} \subset \mathfrak{g}$ is any subspace, say of dimension m, let $[\mathfrak{s}] \in \wedge^m \mathfrak{g}$ be the $\mathbb{C}$ span of the decomposable element $u_1 \wedge \cdots \wedge u_m$ where $\{u_i,\ i = 1, \ldots, m\}$ is a basis of $\mathfrak{s}$. One notes that if $\omega \in \wedge^2 \mathfrak{g}$ is given as in (1.7), then

$$\omega^k = k!\ v_1 \wedge \cdots \wedge v_{2k} \tag{1.10}$$

so that

$$\omega^j \neq 0 \iff j \leq k \text{ and } \omega^k \in [\mathrm{Rad}\,\omega^\perp]. \tag{1.11}$$

$\{w_j,\ j = 1, \ldots, n\}$ be a $\mathcal{B}$ orthonormal basis of $\mathfrak{g}$. Put $\mu = w_1 \wedge \cdots \wedge w_n$ so that

$$(\mu, \mu) = 1 \tag{1.12}$$

so that μ is unique up to sign and $\wedge^n \mathfrak{g} = \mathbb{C}\mu$. For any $v \in \wedge \mathfrak{g}$ let $v^* = \iota(v)\mu$. We recall the more or less well known

PROPOSITION 1.1. *If $\mathfrak{s} \subset \mathfrak{g}$ is any subspace and $0 \neq u \in [\mathfrak{s}]$, then*

$$0 \neq u^* \in [\mathfrak{s}^\perp]. \tag{1.13}$$

Moreover if $s, t \in \wedge \mathfrak{g}$, one has

$$(s, t) = (s^*, t^*). \tag{1.14}$$

PROOF. Let $\{y_i,\ i = 1, \ldots, m\}$ be a basis of $\mathfrak{s}$ chosen so that $u = y_1 \wedge \cdots \wedge y_m$ and let $\{z_j,\ j = 1, \ldots, n - m\}$ be a basis of $\mathfrak{s}^\perp$. Then if $y'_k,\ k = 1, \ldots, m$, are chosen in $\mathfrak{g}$ such that $(y_i, y'_k) = \delta_{ik}$, it is immediate that the y'_k together with the z_j form a basis of $\mathfrak{g}$ so that for some $\lambda \in \mathbb{C}^\times$ one has

$$\lambda y'_1 \wedge \cdots \wedge y'_m \wedge z_1 \wedge \cdots \wedge z_{n-m} = \mu. \tag{1.15}$$

But since interior product is the transpose of exterior product one has

$$\iota(q)\,\iota(p) = \iota(p \wedge q) \tag{1.16}$$

for any $p, q \in \wedge^\mathfrak{g}$. Thus by (1.15) one has

$$u^* = \lambda z_1 \wedge \cdots \wedge z_{n-m}$$

establishing (1.13). To prove (1.14) it suffices by linearity to assume that both s and t are decomposable of some degree m. Thus we can assume $s = y_1 \wedge \cdots \wedge y_m$ and $t = z_1 \wedge \cdots \wedge z_m$ for $y_i, z_j \in \mathfrak{g}$. But now, as one knows, and readily establishes,

$$\varepsilon(y)\,\iota(z) + \iota(z)\varepsilon(y) = (y, z)\mathrm{Id}_\mathfrak{g} \tag{1.17}$$

for $y, z \in \mathfrak{g}$. Thus

$$\begin{aligned} (s^*, t^*) &= (\iota(s)\mu, \iota(t)\mu) \\ &= (\mu, \varepsilon(s)\iota(t)\mu). \end{aligned} \tag{1.18}$$

But then using (1.17) and the fact that $\varepsilon(y)\mu = 0$ for any $y \in \mathfrak{g}$, one has

$$(\mu, \varepsilon(s)\iota(t)\mu) = \sum_{j=0}^{m-1} (-1)^j (y_m, z_{m-j})(\mu, \varepsilon(y_1) \cdots \varepsilon(y_{m-1})\,\iota(z_m) \cdots \iota(\widehat{z_{m-j}}) \cdots \iota(z_1)\mu).$$

But then by induction and the expansion of the determinant defined by the last row one has

$$\begin{aligned} (\mu, \varepsilon(s)\iota(t)\mu) &= det\,(y_i, z_j)(\mu, \mu) \\ &= (s, t) \end{aligned}$$

proving (1.14). $\qquad \square$

1.2. The algebra $S(\mathfrak{g})$ is a G-module extending the adjoint representation. Let $J = S(\mathfrak{g})^G$ be the subalgebra of $\mathfrak{g}$-invariants. Let $H \subset S(\mathfrak{g})$ be the graded $\mathfrak{g}$-submodule of harmonic elements in $S(\mathfrak{g})$ (See §1.4 in [K2] for definitions). Then one knows

$$S(\mathfrak{g}) = J \otimes H. \tag{1.19}$$

See (1.4.3) in [K2].

Let r be as in (1.1). For the convenience of the reader we repeat a paragraph in §1.2 of [K4]. Let $\Sigma_{2r,2}$ be the subgroup of all $\sigma \in \Sigma_{2r}$ such that σ permutes the set of unordered pairs $\{(1,2),(3,4),\ldots,(2r-1,2r)\}$. It is clear that $\Sigma_{2r,2}$ has order $r!\,2^r$. Now let Π_r be a cross-section of the set of left cosets of $\Sigma_{2r,2}$ in Σ_{2r}. Thus one has a disjoint

$$\Sigma_{2r} = \bigcup_{\nu \in \Pi_r} \nu\,\Sigma_{2r,2}. \tag{1.20}$$

One notes that the cardinality of Π_r is $(2r-1)(2r-3)\cdots 1$ (the index of $\Sigma_{2r,2}$ in Σ_{2r}) and the correspondence

$$\nu \mapsto ((\nu(1),\nu(2)),(\nu(3),\nu(4)),\ldots,(\nu(2r-1),\nu(2r)) \tag{1.21}$$

sets up a bijection of Π_r with the set of all partitions of $(1,2,\ldots,2r)$ into a union of subsets, each of which has two elements. Furthermore, since the signum character restricted to $\Sigma_{2r,2}$ is nontrivial we may choose Π_r so that

$$sg(\nu) = 1$$

for all $\nu \in \Pi_r$.

In [K4] we defined a map $\Gamma : \wedge^{2r}\mathfrak{g} \to S(\mathfrak{g})$. (Its significance will become apparent later). Here, using Proposition 1.2 in [K4] we will give a simpler definition of Γ. By Proposition 1.2 in [K4] one has

PROPOSITION 1.2. *There exists a map*

$$\Gamma : \wedge^{2r}\mathfrak{g} \to S^r(\mathfrak{g}) \tag{1.21a}$$

such that for any $w_i \in \mathfrak{g}$, $i = 1,\ldots,2r$, one has

$$\Gamma(w_1 \wedge \cdots \wedge w_{2r}) = \sum_{\nu \in \Pi_r} [w_{\nu(1)}, w_{\nu(2)}] \cdots [w_{\nu(2r-1)}, w_{\nu(2r)}]. \tag{1.22}$$

As a polynomial function of degree r on $\mathfrak{g}$, one notes that

$$\Gamma(w_1 \wedge \cdots \wedge w_{2r})(x) = \sum_{\nu \in \Pi_r} ([w_{\nu(1)}, w_{\nu(2)}], x) \cdots ([w_{\nu(2r-1)}, w_{\nu(2r)}], x). \tag{1.23}$$

This is clear from (1.1.7) in [K4] and (1.3) here.

The algebra $\wedge\mathfrak{g}$ is a natural G-module by extension of the adjoint representation. It is clear that Γ is a G-map. Let $M \subset S^r(\mathfrak{g})$ be the image of Γ. The following is proved as Corollary 3.3 in [K4].

THEOREM 1.3. *One has $M \subset H^r$ so that M is a G-module of harmonic polynomials of degree r on $\mathfrak{g}$.*

Giving properties of M and determining its rather striking $\mathfrak{g}$-module structure is the main goal of this paper.

For any $y \in \mathfrak{g}$ one has the familiar supercommutation formula $\iota(y)d + d\iota(y) = \theta(y)$. See e.g., (92) in [K5]. Now let $x, y \in \mathfrak{g}$. Since $d\iota(y)(x) = 0$ one has $\iota(y)dx = [y, x]$. Thus, by (1.8), using superscript notation for centralizers one has

$$\operatorname{Rad} dx = \mathfrak{g}^x. \tag{1.24}$$

Clearly $[x, \mathfrak{g}]$ is the $\mathcal{B}$ orthogonal subspace in $\mathfrak{g}$ to $\mathfrak{g}^x$ so that

$$[x, \mathfrak{g}] = (\operatorname{Rad} dx)^\perp \tag{1.25}$$

for any $x \in \mathfrak{g}$.

For any $x \in \mathfrak{g}$ one knows $\dim \mathfrak{g}^x \geq \ell$. Recall that an element $x \in \mathfrak{g}$ is called regular if $\dim \mathfrak{g}^x = \ell$. The set $\operatorname{Reg} \mathfrak{g}$ of regular elements is nonempty and Zariski open. Its complement, $\operatorname{Sing} \mathfrak{g}$, is the Zariski closed set of singular elements. One notes, by (1.11), that

$$\operatorname{Sing} \mathfrak{g} = \{x \in \mathfrak{g} \mid (dx)^r = 0\}. \tag{1.26}$$

Now $\wedge^{\operatorname{even}}\mathfrak{g}$ is a commutative algebra and hence there exists a homomorphism

$$\gamma : S(\mathfrak{g}) \to \wedge^{\operatorname{even}}\mathfrak{g} \tag{1.27}$$

such that for $x \in \mathfrak{g}$,

$$\gamma(x) = -dx.$$

Let γ_r be the restriction of γ to $S^r(\mathfrak{g})$. The following result, established as Theorem 1.4 in [K4], asserts that Γ is the transpose of γ_r.

THEOREM 1.4. *Let $y_1, \ldots, y_r \in \mathfrak{g}$ and let $\zeta \in \wedge^{2r}(\mathfrak{g})$. Then*

$$(y_1 \cdots y_r, \Gamma(\zeta)) = (-1)^r (dy_1 \wedge \cdots \wedge dy_r, \zeta). \tag{1.28}$$

Now one knows that $S^r(\mathfrak{g})$ is (polarization) spanned by all powers x^r for $x \in \mathfrak{g}$. Using (1.3), (1.26) and Theorem 1.4 we recover Proposition 3.2 in [K4]. The key point is that M defines the variety $\operatorname{Sing} \mathfrak{g}$.

THEOREM 1.5. *Let $x \in \mathfrak{g}$ and $\zeta \in \wedge^{2r}\mathfrak{g}$. Then*

$$\Gamma(\zeta)(x) = \frac{(-1)^r}{r!}((dx)^r, \zeta). \tag{1.29}$$

In particular

$$f(x) = 0, \ \forall f \in M \iff x \in \operatorname{Sing}(\mathfrak{g}). \tag{1.30}$$

If $\mathfrak{a}$ is a Cartan subalgebra of $\mathfrak{g}$, then one knows that $\mathfrak{a} \cap \operatorname{Sing} \mathfrak{g}$ is a union of the root hyperplanes in $\mathfrak{a}$. Hence as a corollary of Theorem 1.5 one has

THEOREM 1.6. *Let $\mathfrak{a}$ be a Cartan subalgebra of $\mathfrak{g}$. Let $\Delta_+(\mathfrak{a})$ be a choice of positive roots for the pair $(\mathfrak{a}, \mathfrak{g})$. Then for any $f \in M$ one has*

$$f|\mathfrak{a} \in \mathbb{C} \prod_{\beta \in \Delta_+(\mathfrak{a})} \beta. \tag{1.31}$$

Going to the opposite extreme we recall that a nilpotent element e is called principal if it is regular. Let e be a principal nilpotent element. Then by Corollary 5.6 in [K1] there exists a unique nilpotent radical $\mathfrak{n}$ of a Borel subalgebra such that $e \in \mathfrak{n}$. Furthermore $\mathfrak{g}^e \cap [\mathfrak{n}, \mathfrak{n}]$ is a linear hyperplane in $\mathfrak{g}^e$ and $\mathfrak{g}^e \cap [\mathfrak{n}, \mathfrak{n}] = (\operatorname{Sing} \mathfrak{g}) \cap \mathfrak{g}^e$ by Theorem 5.3 and Theorem 6.7 in [K1]. Thus there exists a nonzero linear functional ξ on $\mathfrak{g}^e$ such that

$$\operatorname{Ker} \xi = (\operatorname{Sing} \mathfrak{g}) \cap \mathfrak{g}^e. \tag{1.32}$$

This establishes

THEOREM 1.7. *Let* $e \in \mathfrak{g}$ *be principal nilpotent. Let* $f \in M$. *Then using the notation of (1.32) one has*

$$f|\mathfrak{g}^e \in \mathbb{C}\,\xi^r. \tag{1.33}$$

Since Sing $\mathfrak{g}$ is clearly a cone it follows that the ideal $\mathcal{L}$ of $f \in S(g)$ which vanishes on Sing $\mathfrak{g}$ is graded. One of course has that $M \subset \mathcal{L}^r$. We now observe that r is the minimal value of k such that $\mathcal{L}^k \neq 0$

PROPOSITION 1.8. *Assume that* $0 \neq f \in \mathcal{L}^k$. *Then* $k \geq r$.

PROOF. Since $f \neq 0$ there clearly exists a Cartan subalgebra $\mathfrak{a}$ of $\mathfrak{g}$ such that $f|\mathfrak{a} \neq 0$. But then using the notation of Theorem 1.6 it follows from the prime decomposition that β divides $f|\mathfrak{a}$ for all $\beta \in \Delta_+(\mathfrak{a})$. Thus $k \geq r$. $\qquad\square$

2. The structure of M in terms of minors and as a G-module

2.1. For any $z \in \mathfrak{g}$ let ∂_z be the partial derivative of $S(\mathfrak{g})$ defined by z. Let $W(\mathfrak{g}) = S(\mathfrak{g}) \otimes \wedge \mathfrak{g}$ so that $W(\mathfrak{g})$ can be regarded as the supercommutative algebra of all differential forms on $\mathfrak{g}$ with polynomial coefficients. To avoid confusion with the already defined d, let d_W be the operator of exterior differentiation on $W(\mathfrak{g})$. That is, d_W is a derivation of degree 1 defined so that if $\{z_i, w_j\}$, $i,j = 1, \ldots, n$, are dual $\mathcal{B}$ bases of $\mathfrak{g}$, then

$$d_W(f \otimes u) = \sum_i^n \partial_{z_i} f \otimes \varepsilon(w_i)\,u \tag{2.1}$$

where $f \in S(\mathfrak{g})$ and $u \in \wedge \mathfrak{g}$. Of course d_W is independent of the choice of bases. In particular $d_W f$ is a differential form of degree 1 on $\mathfrak{g}$.

For any $x \in \mathfrak{g}$ one has a homomorphism

$$W(\mathfrak{g}) \to \wedge \mathfrak{g}, \quad \varphi \mapsto \varphi(x) \tag{2.2}$$

defined so that if $\varphi = f \otimes u$, using the notation of (2.1), then $\varphi(x) = f(x)u$. Next one notes that the G-module structures on $S(\mathfrak{g})$ and $\wedge \mathfrak{g}$ define, by tensor product, a G-module structure on $W(\mathfrak{g})$. Clearly d_W is a G map. If $a \in G$ and $\varphi \in W(\mathfrak{g})$, the action of a on φ will simply be denoted by $a \cdot \varphi$. If $x \in \mathfrak{g}$ one readily has

$$a \cdot (\varphi(x)) = a \cdot \varphi(a \cdot x). \tag{2.3}$$

One knows (Chevalley) that J is a polynomial ring $\mathbb{C}[p_1, \ldots, p_\ell]$ where the p_j are homogeneous polynomials. If $d_j = \deg p_j$, for $j = 1, \ldots, \ell$, and $m_j = d_j - 1$, then the m_j are exponents of $\mathfrak{g}$ so that

$$\sum_{j=1}^{\ell} m_j = r. \tag{2.4}$$

Moreover we can choose the p_j so that $\partial_y p_j \in H$ for any $y \in \mathfrak{g}$ (see Theorem 67 in [K5]). In fact, if H_{ad} is the primary component of H corresponding to the adjoint representation, then the multiplicity of the adjoint representation in H_{ad} is equal to ℓ and τ_j, $j = 1, \ldots, \ell$, is a basis of $\mathrm{Hom}_G(\mathfrak{g}, H_{\mathrm{ad}})$ where

$$\tau_j(y) = \partial_y p_j \tag{2.5}$$

for any $y \in \mathfrak{g}$. Again see Theorem 67 in [K5].

REMARK 2.2. Using the notation of (2.1) note that

$$\{w_{i_1} \wedge \cdots \wedge w_{i_\ell} \mid 1 \le i_1 < \cdots < i_\ell \le n\}$$

is a basis of $\wedge^\ell \mathfrak{g}$. Furthermore

$$\{z_{j_1} \wedge \cdots \wedge z_{j_\ell} \mid 1 \le j_1 < \cdots < j_\ell \le n\}$$

is the dual basis since clearly

$$(w_{i_1} \wedge \cdots \wedge w_{i_\ell}, z_{j_1} \wedge \cdots \wedge z_{j_\ell}) = \prod_{k=1}^{n} \delta_{i_k j_k}. \tag{2.6}$$

In addition if the w_i are a $\mathcal{B}$-orthonormal basis of $\mathfrak{g}$, then $w_i = z_i$, $i = 1, \ldots, n$, and hence (2.6) implies that $\{w_{i_1} \wedge \cdots \wedge w_{i_\ell} \mid 1 \le i_1 < \cdots < i_\ell \le n\}$ is a $\mathcal{B}$ orthonormal basis of $\wedge^\ell \mathfrak{g}$.

Now for any $y_i \in \mathfrak{g}$, $i = 1, \ldots, \ell$, let $\psi(y_1, \ldots, y_\ell) = \det \partial_{y_i} p_j$ so that

$$\psi(y_1, \ldots, y_\ell) \in S^r(\mathfrak{g}) \tag{2.7}$$

by (2.4). But now $d_W p_j$ is an invariant 1-form on $\mathfrak{g}$. If $x \in \mathfrak{g}$, then $d_W p_j(x) \in \wedge^1 \mathfrak{g}$. Explicitly, using the notation in (2.1), one has

$$d_W p_j(x) = \sum_{i=1}^{n} \partial_{z_i} p_j(x)\, w_i. \tag{2.8}$$

One notes that $\partial_{z_i} p_j$ is an $n \times \ell$ matrix of polynomial functions. There are $\binom{n}{\ell}$ $\ell \times \ell$ minors for this matrix. The determinants of these minors all lie in $S^r(\mathfrak{g})$ and appear in the following expansion.

PROPOSITION 2.1. *Let the notation be as in (2.1). Let $x \in \mathfrak{g}$. Then in $\wedge^\ell \mathfrak{g}$ one has*

$$d_W\, p_1(x) \wedge \cdots \wedge d_W\, p_\ell(x) = \sum_{1 \le i_1 < \cdots < i_\ell \le n} \psi(z_{i_1}, \ldots, z_{i_\ell})(x) w_{i_1} \wedge \cdots \wedge w_{i_\ell}. \tag{2.9}$$

PROOF. This is just standard exterior algebra calculus using (2.8).   □

THEOREM 2.2. *Let v_i, $i = 1, \ldots, n$, be a $\mathcal{B}$ orthonormal basis of $\mathfrak{g}$ chosen and ordered so that v_i, $i = 1, \ldots, \ell$, is a basis of $\mathfrak{h}$. Then there exists a scalar $\kappa \in \mathbb{C}^\times$ such that, for any $y \in \mathfrak{h}$,*

$$d_W\, p_1(y) \wedge \cdots \wedge d_W\, p_\ell(y) = \kappa\, \Big(\prod_{\varphi \in \Delta_+} \varphi(y) \Big)\, v_1 \wedge \cdots \wedge v_\ell. \tag{2.10}$$

PROOF. If $a \in G$, $x \in \mathfrak{g}$ and $j = 1, \ldots, \ell$, then since $d_W p_j$ is G-invariant one has

$$a \cdot d_W\, p_j(x) = d_W\, p_j(a \cdot x). \tag{2.11}$$

But this implies that

$$d_W\, p_j(x) \in \text{cent } \mathfrak{g}^x \tag{2.12}$$

since if we choose $a \in G^x$ in (2.11) it follows from (2.11) that $d_W p_j(x)$ commutes with $\mathfrak{g}^x$. But $x \in \mathfrak{g}^x$ so that $d_W p_j(x) \in \mathfrak{g}^x$. This establishes (2.12).

Now by Theorem 9, p. 382 in [K2] one has that if $x \in \mathfrak{g}$, then

$$\{d_W\, p_1(x), \ldots, d_W\, p_\ell(x)\} \text{ are linearly independent} \iff x \in \operatorname{Reg} \mathfrak{g}. \quad (2.12a)$$

Thus the left side of (2.10) vanishes if and only if $y \in \operatorname{Sing} \mathfrak{g} \cap \mathfrak{h}$. In particular, choosing the z_i in (2.9) so that $v_j = z_j$ for $j = 1, \ldots, \ell$, one has $\psi(v_1, \ldots, v_\ell)(y) = 0$ if y is singular by the expansion (2.9). One the other hand, if $y \in \mathfrak{h}$ is regular then, by (2.12), one must have that

$$\{d_W\, p_j(y),\ j = 1, \ldots, \ell\} \text{ is a basis of } \mathfrak{h}. \quad (2.13)$$

Thus if y is regular, the left side of (2.10) equals $\nu\, v_1 \wedge \cdots \wedge v_\ell$ for some $\nu \in \mathbb{C}^\times$. Comparing with the expansion (2.9) one must have $\nu = \psi(v_1, \ldots, v_\ell)(y)$. But then $\psi(v_1, \ldots, v_\ell)|\mathfrak{h}$ is a polynomial of degree r which vanishes on $y \in \mathfrak{h}$ if and only if $y \in \mathfrak{h}$ is singular. Thus

$$\psi(v_1, \ldots, v_\ell)|\mathfrak{h} = \kappa \prod_{\varphi \in \Delta_+} \varphi$$

for some nonzero constant κ. This proves (2.10). $\square$

2.2. For any root $\varphi \in \Delta$ let $e_\varphi \in \mathfrak{g}$ be a corresponding root vector. We will make choices so that

$$(e_\varphi, e_{-\varphi}) = 1. \quad (2.14)$$

For any $x \in \mathfrak{h}$, one then has

$$dx = \sum_{\varphi \in \Delta_+} \varphi(x)\, e_\varphi \wedge e_{-\varphi}. \quad (2.15)$$

See Proposition 37, p. 311 in [K5], noting (106), p. 302 and (142), p. 309 in [K5]. But then recalling (1.27) one has

$$\gamma_r(x^r) = r!(-1)^r \prod_{\varphi \in \Delta_+} \varphi(x)\, e_\varphi \wedge e_{-\varphi}. \quad (2.16)$$

But since $(e_\varphi \wedge e_{-\varphi}, e_\varphi \wedge e_{-\varphi}) = -1$, by (2.14), for any $\varphi \in \Delta_+$ one has that

$$\Big(\prod_{\varphi \in \Delta_+} e_\varphi \wedge e_{-\varphi},\ \prod_{\varphi \in \Delta_+} e_\varphi \wedge e_{-\varphi} \Big) = (-1)^r. \quad (2.17)$$

But then if $\{v_i \mid i = 1, \ldots, \ell\}$ is an orthonormal basis of $\mathfrak{h}$, one has

$$\Big(v_1 \wedge \cdots \wedge v_\ell \wedge \prod_{\varphi \in \Delta_+} e_\varphi \wedge e_{-\varphi},\ v_1 \wedge \cdots \wedge v_\ell \wedge \prod_{\varphi \in \Delta_+} e_\varphi \wedge e_{-\varphi} \Big) = (-1)^r. \quad (2.18)$$

But then we may choose an ordering of the v_i such that

$$\mu = i^r v_1 \wedge \cdots \wedge v_\ell \wedge \prod_{\varphi \in \Delta_+} e_\varphi \wedge e_{-\varphi} \quad (2.19)$$

so that

$$(v_1 \wedge \cdots \wedge v_\ell)^* = i^r \prod_{\varphi \in \Delta_+} e_\varphi \wedge e_{-\varphi}. \quad (2.20)$$

But then one has

THEOREM 2.3. *There exists $\kappa_o \in \mathbb{C}^\times$ such that for any $x \in \mathfrak{g}$,*

$$
\begin{aligned}
(d_W\, p_1(x) \wedge \cdots \wedge d_W\, p_\ell(x))^* &= \kappa_o\, \frac{(-dx)^r}{r!} \\
&= \kappa_o\, \gamma_r(\frac{x^r}{r!}).
\end{aligned}
\tag{2.21}
$$

PROOF. If $y \in \mathfrak{h}$ is regular, then (2.21), for $y = x$, follows from (2.16),(2.20) and Theorem 2.2. That is

$$
\begin{aligned}
(d_W\, p_1(y) \wedge \cdots \wedge d_W\, p_\ell(y))^* &= \kappa_o\, \frac{(-dy)^r}{r!} \\
&= \kappa_o\, \gamma_r(\frac{y^r}{r!}).
\end{aligned}
\tag{2.22}
$$

But now if $x \in \mathfrak{g}$ regular and semisimple there exist $a \in G$ and a regular $y \in \mathfrak{h}$ such that $a \cdot y = x$. But now since $*$ and γ_r are clearly G-maps one has (2.21) by applying the action of a to both sides of (2.22). However the set of regular semisimple elements in $\mathfrak{g}$ is dense (this nonempty set is Zariski open) one has (2.21) for all $x \in \mathfrak{g}$ by continuity. $\qquad\square$

Returning to our module M of harmonic polynomials on $\mathfrak{g}$ of degree r, it is obvious, by definition, that M is spanned by all $f \in S^r\mathfrak{g}$ of the form $f = \Gamma(w_1 \wedge \cdots \wedge w_{2r})$ where the $w_i \in \mathfrak{g}$ are linearly independent. Explicitly $\Gamma(w_1 \wedge \cdots \wedge w_{2r})$ is given by (1.22). We now show that $\Gamma(w_1 \wedge \cdots \wedge w_{2r})$ may also be given as the determinant of one of the $\ell \times \ell$ minors in the expansion (2.9).

THEOREM 2.4. *Let $w_k \in \mathfrak{g}$, $k = 1,\ldots,2r$, be linearly independent and let $\mathfrak{s} \subset \mathfrak{g}$ be the span of the w_k and let $u_i \in \mathfrak{g}, = 1,\ldots,\ell$, be a basis of $\mathfrak{s}^\perp$. Then there exists a constant $\kappa_1 \in \mathbb{C}^\times$ such that*

$$
\begin{aligned}
\Gamma(w_1 \wedge \cdots \wedge w_{2r}) &= \kappa_1\, \psi(u_1,\ldots,u_\ell) \\
&= \kappa_1 \det \partial_{u_i}\, p_j.
\end{aligned}
\tag{2.23}
$$

Furthermore M is the span of all $\ell \times \ell$ determinant minors $\psi(v_1,\ldots,v_\ell)$ where $v_i \in \mathfrak{g}$, $i = 1,\ldots,\ell$, are linearly independent.

PROOF. Clearly we may choose the two dual bases in (2.1) so that the given w_k are the first $2r$-elements of the w basis and the u_i are the last ℓ elements of the z basis. Thus there exists $\kappa_2 \in \mathbb{C}^\times$ such that

$$
(u_1 \wedge \cdots \wedge u_\ell)^* = \kappa_2\, w_1 \wedge \cdots \wedge w_{2r}.
\tag{2.24}
$$

Now let $x \in \mathfrak{g}$. Then by the expansion (2.9) one has

$$
(d_W\, p_1(x) \wedge \cdots \wedge d_W\, p_\ell(x), u_1 \wedge \cdots \wedge u_\ell) = \psi(u_1,\ldots,u_\ell)(x).
\tag{2.25}
$$

But then by (1.3), (1.14), (1.29) and (2.21) one has

$$
\begin{aligned}
\psi(u_1,\ldots,u_\ell)(x) &= ((d_W\, p_1(x) \wedge \cdots \wedge d_W\, p_\ell(x))^*, (u_1 \wedge \cdots \wedge u_\ell)^*) \\
&= \kappa_o \kappa_2\, (\gamma_r(\frac{x^r}{r!}), w_1 \wedge \cdots \wedge w_{2r}) \\
&= \kappa_1^{-1}\, \Gamma(w_1 \wedge \cdots \wedge w_{2r})(x)
\end{aligned}
\tag{2.26}
$$

where $\kappa_1^{-1} = \kappa_o\kappa_2$. The last statement in the theorem is obvious since clearly u_i, $i = 1, \ldots, \ell$, is an aritrary set of ℓ-independent elements in $\mathfrak{g}$. $\square$

2.3. Let $\{z_i, w_j\}$ be the arbitrary dual bases of $\mathfrak{g}$ as in (1.4). Then, independent of the choice of bases, the Casimir operator Cas on $\wedge^{\mathfrak{g}}$ is given by

$$\mathrm{Cas} = \sum_{i=1}^{n} \theta(z_i)\theta(w_i).$$

We recall special cases of some results in [K3]. Let $A_\ell \subset \wedge^\ell \mathfrak{g}$ be the span in $\wedge^\ell\mathfrak{g}$ of all $[\mathfrak{c}]$ where $\mathfrak{c} \in \mathfrak{g}$ is a commutative Lie subalgebra of dimension ℓ. Since the set of such subalgebras includes, for example, Cartan subalgebras it is obvious that $A_\ell \neq 0$. In fact note that

$$[\mathfrak{g}^y] \subset A_\ell \tag{2.27}$$

for any $y \in \mathrm{Reg}\,\mathfrak{g}$ since, as one knows, $\mathfrak{g}^y$ is abelian if y is regular. Clearly A_ℓ is a G submodule of $\wedge^\ell$. On the other hand, let m_ℓ be the maximal value of Cas on $\wedge^\ell$ and let M_ℓ be the corresponding Cas eigenspace. Again, clearly M_ℓ is a G-submodule of $\wedge^\ell\mathfrak{g}$. From the definition of M_ℓ it is obvious that $\mathrm{Hom}_G(M_\ell, \wedge^\ell\mathfrak{g}/M_\ell) = 0$. Since $\mathcal{B}|\wedge^\ell\mathfrak{g}$ is nonsingular it follows that

$$\mathcal{B}|M_\ell \text{ is nonsingular} \tag{2.28}$$

and hence M_ℓ is self-contragredient. Noting the $1/2$ in (2.1.7) of [K3] the following result is a special case of Theorem (5), p. 156 in [K3].

THEOREM 2.5. *One has*

$$A_\ell = M_\ell \tag{2.29}$$

and in addition

$$m_\ell = \ell. \tag{2.30}$$

For any ordered subset $\Phi \subset \Delta$, $\Phi = \{\varphi_1, \ldots, \varphi_k\}$, let $e_\Phi = e_{\varphi_1} \wedge \cdots \wedge e_{\varphi_k}$ and put $\langle\Phi\rangle = \sum_{\varphi\in\Phi}\varphi$ so that with respect to $\mathfrak{h}$,

$$e_\Phi \in \wedge^k\mathfrak{g} \text{ is a weight vector of weight } \langle\Phi\rangle. \tag{2.31}$$

Let $\mathfrak{b} \subset \mathfrak{g}$ be the Borel subalgebra of $\mathfrak{g}$ spanned by $\mathfrak{h}$ and $\{e_\varphi\}$, for $\varphi \in \Delta_+$, and put $\mathfrak{n} = [\mathfrak{b}, \mathfrak{b}]$. Any ideal $\mathfrak{a}$ of $\mathfrak{b}$ where $\mathfrak{a} \subset \mathfrak{n}$ is necessarily spanned by root vectors. We will say that Φ, as above, is an ideal of Δ_+ if $\Phi \subset \Delta_+$ and $\mathfrak{a}_\Phi = \sum_{i=1}^{k}\mathbb{C}e_{\varphi_i}$ is an ideal in $\mathfrak{b}$.

REMARK 2.6. One notes that if Φ is an ideal of Δ_+ and $V_\Phi \subset \wedge^k\mathfrak{g}$ is the G-module spanned by $G \cdot e_\Phi$, then V_Φ is irreducible having e_Φ as highest weight vector and $\langle\Phi\rangle$ as highest weight.

As already noted in [K3] (see bottom of p. 158) it is immediate that if $\mathfrak{a}$ is any abelian ideal in $\mathfrak{b}$, then $\mathfrak{a} \subset \mathfrak{n}$ so that $\mathfrak{a} = \mathfrak{a}_\Phi$ for an ideal $\Phi \subset \Delta_+$. Much more subtly it has been established in [KW] (see Lemma 12, p. 113 in [KW]) that any ideal $\mathfrak{a}$ of $\mathfrak{b}$ having dimension ℓ is in fact abelian. Let $\mathcal{I}$ be the (obviously finite) set of all ideals Φ in Δ_+ which have cardinality ℓ. If $\Phi_1, \Phi_2 \in \mathcal{I}$ are distinct, then $\langle\Phi_1\rangle \neq \langle\Phi_2\rangle$ by Theorem (7), p. 158 in [K3] so that V_{Φ_1} are inequivalent $\mathfrak{g}$ and G modules. Then Theorem (8), p. 159 in [K3] implies

THEOREM 2.7. M_ℓ is a multiplicity one G-module. In fact

$$M_\ell = \oplus_{\Phi \in \mathcal{I}} V_\Phi \qquad (2.32)$$

so the number of irreducible components in M_ℓ is the cardinality of $\mathcal{I}$.

REMARK 2.8. In the general case we do not have a formula for card $\mathcal{I}$ although computing this number in any given case does not seem to be too difficult. In the special case where $\mathfrak{g} \cong \text{Lie } Sl(n, \mathbb{C})$ one easily has a bijective correspondence of $\mathcal{I}$ with the set of all Young tableaux of size $n - 1$ so that in this case

$$\text{card } \mathcal{I} = p(n - 1) \qquad (2.33)$$

where p here is the classical partition function. Let

$$\tau : \wedge^\ell \mathfrak{g} \to \wedge^{2r} \mathfrak{g} \qquad (2.34)$$

be the G-ismorphism defined by putting $\tau(u) = u^*$ recalling that $u^* = \iota(u)\mu$. Let $M_{2r} = \tau(M_\ell)$.

THEOREM 2.9. τ is a $\mathcal{B}$-isomorphism so that $\mathcal{B}|M_{2r}$ is nonsingular. Furthermore ℓ is the maximal eigenvalue of Cas on $\wedge^{2r}\mathfrak{g}$ and M_{2r} is the corresponding eigenspace. As G modules one has

$$M_\ell \cong M_{2r} \qquad (2.35)$$

so that M_{2r} is a multiplicity 1 module where in fact

$$M_{2r} \cong \oplus_{\Phi \in \mathcal{I}} V_\Phi. \qquad (2.36)$$

We recall the V_Φ is an irreducible G-module with highest weight $\langle \Phi \rangle$. See (2.31).

PROOF. The first statement follows from Proposition 1.1. The remaining statements are immediate from Theorem 2.7 since τ is a G-isomorphism. $\qquad \square$

In light of equality $M_\ell = A_\ell$ (see (2.9)) Ranee Brylinski in her thesis (see [RB]) proved that M_ℓ is the span of $G \cdot [\mathfrak{h}]$. The thesis however has not been published. A stronger theorem (motivated by her result) appears in [KW]. The following result is just Corollary 2, p. 105 in [KW].

THEOREM 2.10. M_ℓ is the span of $G \cdot [\mathfrak{g}^x]$ for any $x \in \text{Reg } \mathfrak{g}$.

Now by (2.12) and (2.12a) one has

$$\mathbb{C} \, d_W \, p_1(x) \wedge \cdots \wedge d_W \, p_\ell(x) = [\mathfrak{g}^x] \qquad (2.37)$$

for any $x \in \text{Reg } \mathfrak{g}$. Using Theorem 2.3 we can now transfer Theorem 2.10 to M_{2r} where it will have consequences for the structure of the space of functions $M \subset H^r$.

THEOREM 2.11. M_{2r} is the span of $G \cdot (\gamma_r(\frac{x^r}{r!}))$ for any $x \in \text{Reg } \mathfrak{g}$.

PROOF. This is immediate from Theorem 2.3, Theorem 2.10, (2.37) and the fact that τ is a G-isomorphism. $\qquad \square$

Let N_{2r} be the $\mathcal{B}$ orthogonal subspace to M_{2r} in $\wedge^{2r}\mathfrak{g}$. By the first statement in Theorem 2.9 one has a $\mathcal{B}$ orthogonal G-module decomposition $\wedge^{2r}\mathfrak{g}$,

$$\wedge^{2r}\mathfrak{g} = N_{2r} \oplus M_{2r}. \qquad (2.38)$$

REMARK 2.12. Note that by Theorem 2.9 any eigenvalue of Cas in N_{2r} is less than ℓ.

We return now to our G-space M of homogeneous harmonic polynomials on $\mathfrak{g}$ of degree r which define Sing $\mathfrak{g}$. We recapitulate some of the properties of $M = \Gamma(\wedge^{2r}\mathfrak{g})$ already established in this paper. Let $w_k \in \mathfrak{g}$, $k = 1, \ldots, 2r$, be linearly independent and let $z_i \in \mathfrak{g}$, $i = 1, \ldots, \ell$, be linearly independent and $\mathcal{B}$ orthogonal to the w_k. Then for suitable generators p_j, $j = 1, \ldots, \ell$, of $J = S(\mathfrak{g})^G$, we have

(1) $\Gamma(w_i \wedge \cdots \wedge w_{2r})$ is explicitly given by (1.23)

(2) $\Gamma(w_i \wedge \cdots \wedge w_{2r})$ is given as (up to scalar multiplication) $\det \partial_{z_i} p_j$. See Th. 2.4.

(3) If $f \in M$, then $f|\mathfrak{a}$, where $\mathfrak{a}$ is any Cartan subalgebra or

$\qquad \mathfrak{a} = \mathfrak{g}^e$ for e principal nilpotent, is given in Theorems 1.6 and 1.7.

We now determine the G-module structure of M.

THEOREM 2.13. $N_{2r} = \mathrm{Ker}\,\Gamma$ and

$$\Gamma : M_{2r} \to M \tag{2.39}$$

is a G-isomorphism so that as G-modules

$$M \cong M_{2r} \cong M_\ell = A_\ell \tag{2.40}$$

where we recall $A_\ell \subset \wedge^\ell \mathfrak{g}$ has been defined in [K3] as the span of $[\mathfrak{s}]$ over all abelian subalgebras $\mathfrak{s} \subset \mathfrak{g}$ of dimension ℓ.

Furthermore we have defined $\mathcal{I}$ as the set of all ideals Φ in Δ_+ of cardinality ℓ, parameterizing with the notation $\mathfrak{a}_\Phi$, the set of all ideals $\mathfrak{a}$ of $\mathfrak{b}$ having dimension ℓ. See Remark 2.6.

Moreover M is a multiplicity one G-module with $\mathrm{card}\,\mathcal{I}$ irreducible components. In addition $\mathcal{I}$ parameterizes these components in the sense that the component corresponding to $\Phi \in \mathcal{I}$ is equivalent to V_Φ, using the notation of Remark 2.6, and hence has highest weight $\langle \Phi \rangle$. Finally Cas takes the value ℓ on each and every irreducible component of M.

PROOF. By (1.27) and (1.29) one has

$$(\Gamma(\zeta)(x) = (\zeta, \gamma_r(\frac{x^r}{r!})) \tag{2.41}$$

for any $x \in \mathfrak{g}$ and any $\zeta \in \wedge^{2r}\mathfrak{g}$. Of course $\gamma_r(\frac{x^r}{r!}) = 0$ for any $x \in \mathrm{Sing}\,\mathfrak{g}$ (see (2.12a) and Theorem 2.3). However M_{2r} is the span of $G \cdot \gamma_r(\frac{x^r}{r!})$ for any $x \in \mathrm{Reg}\,\mathfrak{g}$ by Theorem 2.11. Thus not only does (2.41) imply that $N_{2r} \subset \mathrm{Ker}\,\Gamma$ but $N_{2r} = \mathrm{Ker}\,\Gamma$ since if $\zeta \in M_{2r}$ and $x \in \mathrm{Reg}\,\mathfrak{g}$ there exists $a \in G$ such that if $y = a \cdot x$, then $\Gamma(\zeta)(y) \neq 0$ by Theorem 2.11 and the nonsingularity of $\mathcal{B}|M_{2r}$, as asserted in Theorem 2.9. Since Γ is a G-map one has the isomorphism (2.39). The remaining statements follow from Theorem 2.5 and Theorem 2.9. $\qquad \square$

References

[K1] B. Kostant, The Three Dimensional Sub-Group and the Betti Numbers of a Complex Simple Lie Group, *Amer. Jour. of Math.*, **81**(1959), 973–1032.

[K2] B. Kostant, Lie Group Representations on Polynomial Rings, *Amer. J. Math.*, **85**(1963), 327–404.

[K3] B. Kostant, Eigenvalues of a Laplacian and Commutative Lie Subalgebras, *Topology*, **13**(1965) 147–159.

[K4] B. Kostant, A Lie Algebra Generalization of the Amitsur-Levitski Theorem, *Adv. in Math.*, **40**(1981), No. 2, 155–175.

[K5] B. Kostant, Clifford Algebra Analogue of the Hopf-Koszul-Samelson Theorem, the ρ-Decomposition, $C(\mathfrak{g}) = \mathrm{End}\, V_\rho \otimes C(P)$, and the $\mathfrak{g}$-Module Structure of $\wedge \mathfrak{g}$, *Adv. in Math.*, **125**(1997), 275–350.

[KW] B. Kostant and N. Wallach, On a theorem of Ranee Brylinski, *Contemporary Mathematics*, **490**(2009), 105–142.

[Kz] J. L. Koszul, Homologie et cohomologie des algèbres de Lie, *Bull. Soc. Math. Fr.*, **78**(1950), 65–127.

[RB] R. Brylinski, *Abelian algebras and adjoint orbits*, Thesis MIT, 1981.

5 MERRILL ROAD, NEWTON CENTRE, MA 02459
E-mail address: kostant@math.mit.edu

DEPARTMENT OF MATHEMATICS, UNIVERSITY OF CALIFORNIA, SAN DIEGO, LA JOLLA, CA 92093
E-mail address: nwallach@ucsd.edu

Contemporary Mathematics
Volume **557**, 2011

An Explicit Embedding of Gravity and the Standard Model in E8

A. Garrett Lisi

ABSTRACT. The algebraic elements of gravitational and Standard Model gauge
fields acting on a generation of fermions may be represented using real matrices.
These elements match a subalgebra of spin(11,3) acting on a Majorana-Weyl
spinor, consistent with GraviGUT unification. This entire structure embeds
in the quaternionic real form of the largest exceptional Lie algebra, E8. These
embeddings are presented explicitly and their implications discussed.

1. Introduction

Since the inception of the Standard Model of particle physics in the 1970s,
theorists have speculated that the zoo of elementary particles might correspond to a
more unified mathematical structure. The most well established of these unification
attempts are the classical Grand Unified Theories (GUTs), involving the embedding
of the Standard Model Lie algebra, $su(3) \oplus su(2) \oplus u(1)$, in larger Lie algebras.[1]
Some evidence that this unification occurs in nature comes from the fact that the
coupling strengths of the strong $su(3)$ and electroweak $su(2) \oplus u(1)$ forces converge
to approximately the same value for short distances. Also, the fermions, which
live in representation spaces of the Standard Model Lie group, embed nicely in
representation spaces of the unified groups.

In the Georgi-Glashow Grand Unified Theory, the Standard Model Lie algebra
embeds in $su(5)$ and the fermions live in $\bar{5}$ and 10 representation spaces. Unfor-
tunately for this GUT, the new particles in $su(5)$ would allow protons to decay at
a rapid rate, which has been ruled out by experiment. In another Grand Unified
Theory, which has not yet been ruled out by proton decay, the Standard Model
Lie algebra embeds in $spin(10)$ and fermions live in a 16 spinor rep. This $spin(10)$
GUT contains the $su(5)$ GUT as a subalgebra and also contains a third GUT, the
Pati-Salam GUT, via

$$su(4) \oplus su(2) \oplus su(2) = spin(6) \oplus spin(4) \subset spin(10)$$

All of these Grand Unified Theories are good models for unifying the strong and
electroweak forces, but disregard gravity.

2010 *Mathematics Subject Classification.* Primary 20C35; Secondary 53Z05, 81T13, 83E99.

Geometrically, the gauge fields of the Standard Model are parts of the connection of a principal $SU(3) \otimes SU(2) \otimes U(1)$ bundle over four dimensional spacetime. Algebraically, this gauge connection is a 1-form valued in the Standard Model Lie algebra. In modern descriptions of gravity the gravitational field is also a connection – the *spin connection* 1-form, valued in the $spin(1,3)$ Lie algebra – which acts on Dirac fermions in 4-spinor reps. These spin and gauge connections correspond to the complete structure group, $SU(3) \otimes SU(2) \otimes U(1) \otimes Spin(1,3)$ (modulo discrete subgroups), of the associated fermion vector bundle. This structure group also acts on the *gravitational frame* – a 1-form field valued in the 4-vector of $spin(1,3)$ – and on a set of Higgs scalar fields.

Given the success of the $spin(10)$ Grand Unified Theory, it is natural to consider its further unification with gravity in a $spin(11,3)$ GraviGUT, with each generation of fermions in a 64-spinor rep.[2] In such a GraviGUT, the gravitational frame and Higgs fields naturally reside in the complement of $spin(1,3)$ and $spin(10)$ in $spin(11,3)$, as the *frame-Higgs* part of the *unified bosonic connection*.[3, 4] Geometrically, all elementary particle fields correspond to parts of the unified bosonic connection or the fermion fibers over spacetime. Algebraically, elementary particles are eigenstates (weight vectors) of the gravitational and Standard Model Lie algebras, and their eigenvalues (weights) are their charges with respect to the different forces.

In previous work it was proposed that all bosons *and* fermions may be unified as parts of a *superconnection* valued in the largest simple exceptional Lie algebra, E_8.[5] This proposal was based on the observation that the weights of gravitational and Standard Model bosons and fermions match roots of E_8. Such a description, using weight matching, is a valuable initial step, but does not fully account for the non-compact nature of the groups involved.

In this new work, a companion to [5], we provide an explicit match between algebraic elements of gravity and the Standard Model and some Lie algebra generators of $E_{8(-24)}$. This match is achieved via the intermediate step of embedding gravity and the Standard Model in the $spin(11,3)$ GraviGUT using explicit matrix representations. Throughout the exposition, terminology is purposefully kept as elementary, complete, and explicit as possible, in the interest of accessibility.

2. Algebra of Gravity and the Standard Model

The matrix representation building blocks of spin algebras are the traceless, Hermitian, $\mathbb{C}(2)$ Pauli matrices,

$$\sigma_1 = \begin{bmatrix} 0 & 1 \\ 1 & 0 \end{bmatrix} \qquad \sigma_2 = \begin{bmatrix} 0 & -i \\ i & 0 \end{bmatrix} \qquad \sigma_3 = \begin{bmatrix} 1 & 0 \\ 0 & -1 \end{bmatrix}$$

which, under the matrix product, satisfy

(2.1) $\sigma_I \sigma_J = \delta_{IJ} 1 + i \, \epsilon_{IJ}{}^K \sigma_K$

in which "1" stands for an identity matrix of appropriate size, and ϵ^{IJK} is the permutation symbol. These three $\mathbb{C}(2)$ matrices, σ_I, represent basis vector elements of the $Cl(3)$ Clifford algebra.

2.1. Gravity. Using the tensor product of Pauli matrices, the four basis vectors of the spacetime Clifford algebra, $Cl(1,3)$, are conventionally represented in

$\mathbb{C}(4)$ as

$$
(2.2) \quad
\begin{aligned}
\gamma_1 &= i\,\sigma_2 \otimes \sigma_1 \\
\gamma_2 &= i\,\sigma_2 \otimes \sigma_2 \\
\gamma_3 &= i\,\sigma_2 \otimes \sigma_3 \\
\gamma_4 &= \sigma_1 \otimes 1
\end{aligned}
\qquad
v^a \gamma_a =
\begin{bmatrix}
0 & 0 & v^3 + v^4 & v^1 - i\,v^2 \\
0 & 0 & v^1 + i\,v^2 & -v^3 + v^4 \\
-v^3 + v^4 & -v^1 + i\,v^2 & 0 & 0 \\
-v^1 - i\,v^2 & v^3 + v^4 & 0 & 0
\end{bmatrix}
$$

the chiral Weyl representation of the Dirac matrices. Taking the product of these, using (2.1) on both levels, the $\mathbb{C}(4)$ matrix representatives of the six independent $Cl^2(1,3) = spin(1,3)$ bivector basis elements

$$
\gamma_{ab} = \gamma_{[a}\gamma_{b]} = \tfrac{1}{2}(\gamma_a\gamma_b - \gamma_b\gamma_a) \qquad = \gamma_a\gamma_b \text{ for } a < b
$$

are

$$
\begin{aligned}
\gamma_{23} &= -i\,1\otimes\sigma_1 \\
\gamma_{13} &= i\,1\otimes\sigma_2 \\
\gamma_{12} &= -i\,1\otimes\sigma_3 \\
\gamma_{14} &= \sigma_3\otimes\sigma_1 \\
\gamma_{24} &= \sigma_3\otimes\sigma_2 \\
\gamma_{34} &= \sigma_3\otimes\sigma_3
\end{aligned}
$$

With these basis elements, the gravitational spin connection, written locally as $\underline{\omega} = \tfrac{1}{2}d\underline{x}^\mu \omega_\mu{}^{ab}\gamma_{ab}$, acts on fermions as $4^{\mathbb{C}}_S$ Dirac spinors,

$$
\underline{\omega}\,\psi =
\begin{bmatrix}
\underline{\omega}_L & 0 \\
0 & \underline{\omega}_R
\end{bmatrix}
\begin{bmatrix}
\psi_L \\
\psi_R
\end{bmatrix}
$$

in which the left (positive) chiral $2^{\mathbb{C}}_{S+}$ and right (negative) chiral $2^{\mathbb{C}}_{S-}$ parts of the Dirac spinor are acted on independently as

$$
(2.3) \quad
\underline{\omega}_L\psi_L =
\begin{bmatrix}
-i\,\underline{\omega}^{12} + \underline{\omega}^{34} & \underline{\omega}^{13} - i\,\underline{\omega}^{23} + \underline{\omega}^{14} - i\,\underline{\omega}^{24} \\
-\underline{\omega}^{13} - i\,\underline{\omega}^{23} + \underline{\omega}^{14} + i\,\underline{\omega}^{24} & i\,\underline{\omega}^{12} - \underline{\omega}^{34}
\end{bmatrix}
\begin{bmatrix}
\psi_L^\wedge \\
\psi_L^\vee
\end{bmatrix}
$$

and

$$
\underline{\omega}_R\psi_R =
\begin{bmatrix}
-i\,\underline{\omega}^{12} - \underline{\omega}^{34} & \underline{\omega}^{13} - i\,\underline{\omega}^{23} - \underline{\omega}^{14} + i\,\underline{\omega}^{24} \\
-\underline{\omega}^{13} - i\,\underline{\omega}^{23} - \underline{\omega}^{14} - i\,\underline{\omega}^{24} & i\,\underline{\omega}^{12} + \underline{\omega}^{34}
\end{bmatrix}
\begin{bmatrix}
\psi_R^\wedge \\
\psi_R^\vee
\end{bmatrix}
$$

with $\psi^\wedge$ and $\psi^\vee$ the "spin up" and "spin down" components of a fermion. Note that, with our choice of representation, $\underline{\omega}_R = -\underline{\omega}_L^\dagger$; and the traceless $\mathbb{C}(2)$ representation in (2.3) establishes the $spin(1,3) = sl(2,\mathbb{C})$ Lie algebra equivalence.

The $\mathbb{C}(4)$ chirality projection operators,

$$
(2.4) \quad
P_{L/R} = \tfrac{1}{2}(1 \pm i\,\gamma)
\qquad
P_L =
\begin{bmatrix}
1 & 0 \\
0 & 0
\end{bmatrix}
\qquad
P_R =
\begin{bmatrix}
0 & 0 \\
0 & 1
\end{bmatrix}
$$

are constructed using the $Cl^4(1,3)$ pseudoscalar,

$$
\gamma = \gamma_1\gamma_2\gamma_3\gamma_4 = -i\,\sigma_3 \otimes 1 =
\begin{bmatrix}
-i & 0 \\
0 & i
\end{bmatrix}
$$

and are needed to describe the action of Standard Model gauge fields on fermions.

2.2. The Standard Model. The three conventional basis $\mathbb{C}(2)$ matrix generators of the real $su(2)$ Lie algebra are $\frac{i}{2}\sigma_I$, with their Lie bracket satisfying

$$[\tfrac{i}{2}\sigma_I, \tfrac{i}{2}\sigma_J] = -\tfrac{1}{4}(\sigma_I\sigma_J - \sigma_J\sigma_I) = -\epsilon_{IJ}{}^K \tfrac{i}{2}\sigma_K$$

with real structure constants, $-\epsilon_{IJ}{}^K$. Curiously, in the Standard Model, the electroweak $su(2)$ acts only on the left chiral parts of leptons and quarks arranged in doublets, such as on $\begin{bmatrix} \nu_L \\ e_L \end{bmatrix}$ and $\begin{bmatrix} u_L \\ d_L \end{bmatrix}$. Acting on a doublet $8^{\mathbb{C}}$ column of two Dirac fermions, such as $\begin{bmatrix} \nu \\ e \end{bmatrix}$, the electroweak $su(2)_L$ generators are $\frac{i}{2}\sigma_I \otimes P_L$, employing the left chiral projection matrix (2.4).

The real $su(3)$ Lie algebra corresponding to the strong force of the Standard Model acts in the fundamental 3 representation on "colored" quarks. The eight $su(3)$ basis generators, $\frac{i}{2}\lambda_A$, are expressed in terms of the conventional traceless, Hermitian, $\mathbb{C}(3)$ Gell-Mann matrices, λ_A, written here in condensed form as

$$g^A\lambda_A = \begin{bmatrix} g^3 + \frac{1}{\sqrt{3}}g^8 & g^1 - ig^2 & g^4 - ig^5 \\ g^1 + ig^2 & -g^3 + \frac{1}{\sqrt{3}}g^8 & g^6 - ig^7 \\ g^4 + ig^5 & g^6 + ig^7 & \frac{-2}{\sqrt{3}}g^8 \end{bmatrix}$$

When considering how $su(3)$ acts on all fermions, it is convenient to define the $\mathbb{C}(4)$, "expanded" Gell-Mann matrices,

$$\Lambda_A = \begin{bmatrix} 0 & 0 \\ 0 & \lambda_A \end{bmatrix}$$

which act on a multiplet of leptons and red, green, and blue quarks in a $4^{\mathbb{C}}$ column,

$$[\begin{matrix} l & q^r & q^g & q^b \end{matrix}]$$

(written here in a row for convenience). These eight expanded Gell-Mann matrix generators span an $su(3)$ subalgebra of $su(4)$.

The electroweak $u(1)$ of the Standard Model acts on fermions according to their weak hypercharge, Y. With one complete generation of Standard Model fermions, including left and right chiral components, in a $16^{\mathbb{C}}$ column,

(2.5) $\qquad [\nu_L\ \nu_R\ e_L\ e_R\ u_L^r\ u_R^r\ d_L^r\ d_R^r\ u_L^g\ u_R^g\ d_L^g\ d_R^g\ u_L^b\ u_R^b\ d_L^b\ d_R^b]$

the electroweak $u(1)_Y$ generator acts on this as the diagonal matrix of their hypercharges,

(2.6) $\qquad \frac{i}{2}\,\mathrm{diag}(-1\ \ 0\ -1\ -2\ \ \frac{1}{3}\ \ \frac{4}{3}\ \ \frac{1}{3}\ -\frac{2}{3}\ \ \frac{1}{3}\ \ \frac{4}{3}\ \ \frac{1}{3}\ -\frac{2}{3}\ \ \frac{1}{3}\ \ \frac{4}{3}\ \ \frac{1}{3}\ -\frac{2}{3}\ )$

This $\mathbb{C}(16)$ generator (or a $\mathbb{C}(32)$ generator if spin components are included), with these hypercharges, happens to be the combination of an $su(2)_R$ generator and an $su(4)$ generator,

$$\tfrac{i}{2}(1 \otimes \sigma_3 \otimes P_R + \sqrt{\tfrac{2}{3}}\,\Lambda_{15} \otimes 1 \otimes 1)$$

in which

$$\Lambda_{15} = \sqrt{\tfrac{3}{2}}\,\mathrm{diag}(-1\ \ \tfrac{1}{3}\ \ \tfrac{1}{3}\ \ \tfrac{1}{3}\ )$$

is an "extended" Gell-Mann matrix, related to the $\frac{i}{2}\Lambda_{15}$ generator of $su(4)$. This $\frac{i}{2}\Lambda_{15}$ generator of $su(4)$ corresponds to a $u(1)_{B-L}$ subalgebra, with charges corresponding to baryon minus lepton number. This peculiar coincidence of hypercharges, (2.6), is the key to the successful embedding of the Standard Model Lie algebra in Grand Unified Theories.

The familiar electric charges of elementary particles are the components of the $u(1)_Q$ generator, a combination of the $u(1)_Y$ generator, (2.6), and an $su(2)_L$ generator,

$$\frac{i}{2}(1\otimes\sigma_3\otimes P_R + \sqrt{\frac{2}{3}}\Lambda_{15}\otimes1\otimes1) + \frac{i}{2}1\otimes\sigma_3\otimes P_L$$

$$= i\,\mathrm{diag}(0 \quad 0 \; -1 \; -1 \; \tfrac{2}{3} \; \tfrac{2}{3} \; -\tfrac{1}{3} \; -\tfrac{1}{3} \; \tfrac{2}{3} \; \tfrac{2}{3} \; -\tfrac{1}{3} \; -\tfrac{1}{3} \; \tfrac{2}{3} \; \tfrac{2}{3} \; -\tfrac{1}{3} \; -\tfrac{1}{3})$$

2.3. Together. Each generation of Standard Model fermions is a $32^{\mathbb{C}}$ column, with the field degrees of freedom of the first generation fermions labeled according to (2.5), with two spin components included for each. The six gravitational $spin(1,3)$, eight strong $su(3)$, three electroweak $su(2)_L$, and one hypercharge $u(1)_Y$ basis generators of the combined $su(3)\oplus su(2)_L\oplus u(1)_Y\oplus spin(1,3)$ Lie algebra of gravity and the Standard Model act on these as eighteen $\mathbb{C}(32)$ matrix generators:

(2.7)
$$
\begin{aligned}
T^G_{ab} &= \quad 1 \otimes 1 \otimes \gamma_{ab} \\
T^g_A &= \tfrac{i}{2}\Lambda_A \otimes 1 \otimes 1 \\
T^W_I &= \tfrac{i}{2} 1 \otimes \sigma_I \otimes P_L \\
T^Y &= \tfrac{i}{2} 1 \otimes \sigma_3 \otimes P_R + \tfrac{i}{2}\sqrt{\tfrac{2}{3}}\Lambda_{15}\otimes1\otimes1
\end{aligned}
$$

The spin and gauge parts of the connection, $(\frac{1}{2}\underline{\omega} + \underline{g} + \underline{W} + \underline{B})$, in terms of these basis generators, are $\underline{\omega} = \frac{1}{2}\omega^{ab}T^G_{ab}$ for the $spin(1,3)$ connection, $\underline{g} = g^A T^g_A$ for the $su(3)$ gluons, $\underline{W} = W^I T^W_I$ for $su(2)_L$, and $\underline{B} = B^Y T^Y$ for $u(1)_Y$.

Looking at these eighteen generators, as $\mathbb{C}(32)$ matrices, each has either real or imaginary components. And these act on a generation of Standard Model fermions in a $32^{\mathbb{C}}$ representation space. Nevertheless, they span a real Lie algebra, with real structure constants and a corresponding real Lie group. In order to represent these generators with real matrices, we use a standard trick to "realize" the complex structure. We convert every i in the generator matrices (2.7) to an $\mathbb{R}(2)$ matrix squaring to minus the identity,

$$i \mapsto \begin{bmatrix} 0 & -1 \\ 1 & 0 \end{bmatrix} = -i\sigma_2 \qquad f \mapsto \begin{bmatrix} f_r \\ f_i \end{bmatrix}$$

and convert every complex fermion component to a real $2^{\mathbb{R}}$, so that

$$if = \begin{bmatrix} 0 & -1 \\ 1 & 0 \end{bmatrix}\begin{bmatrix} f_r \\ f_i \end{bmatrix} = \begin{bmatrix} -f_i \\ f_r \end{bmatrix} = i(f_r + if_i) = -f_i + if_r$$

With this conversion, the basis generators (2.7) become $\mathbb{R}(64)$ matrices, corresponding to the same Lie algebra, and act on a generation of fermions in a $64^{\mathbb{R}}$ in the same way as the $\mathbb{C}(32)$ generators acted on a $32^{\mathbb{C}}$. To be complete and explicit, the eighteen basis generators (2.7) of the gravitational and Standard Model Lie algebra, represented as $\mathbb{R}(64)$ matrices, may be expressed numerically using the tensor products (at six levels) of $\mathbb{C}(2)$ Pauli matrices. These matrices, and a $64^{\mathbb{R}}$ of the first generation fermions, are:

$$
\begin{aligned}
T^G_{23} &= & i\ 1 \otimes 1 \otimes 1 \otimes 1 \otimes \sigma_1 \otimes \sigma_2 \\
T^G_{13} &= & i\ 1 \otimes 1 \otimes 1 \otimes 1 \otimes \sigma_2 \otimes 1 \\
T^G_{12} &= & i\ 1 \otimes 1 \otimes 1 \otimes 1 \otimes \sigma_3 \otimes \sigma_2 \\
T^G_{14} &= & 1 \otimes 1 \otimes 1 \otimes \sigma_3 \otimes \sigma_1 \otimes 1 \\
T^G_{24} &= & -\ 1 \otimes 1 \otimes 1 \otimes \sigma_3 \otimes \sigma_2 \otimes \sigma_2 \\
T^G_{34} &= & 1 \otimes 1 \otimes 1 \otimes \sigma_3 \otimes \sigma_3 \otimes 1 \\[2mm]
T^g_1 &= & -\tfrac{i}{4}\ \sigma_1 \otimes \sigma_1 \otimes 1 \otimes 1 \otimes 1 \otimes \sigma_2 \\
&& -\tfrac{i}{4}\ \sigma_2 \otimes \sigma_2 \otimes 1 \otimes 1 \otimes 1 \otimes \sigma_2 \\
T^g_2 &= & -\tfrac{i}{4}\ \sigma_1 \otimes \sigma_2 \otimes 1 \otimes 1 \otimes 1 \otimes 1 \\
&& +\tfrac{i}{4}\ \sigma_2 \otimes \sigma_1 \otimes 1 \otimes 1 \otimes 1 \otimes 1 \\
T^g_3 &= & -\tfrac{i}{4}\ \sigma_3 \otimes 1 \otimes 1 \otimes 1 \otimes 1 \otimes \sigma_2 \\
&& +\tfrac{i}{4}\ 1 \otimes \sigma_3 \otimes 1 \otimes 1 \otimes 1 \otimes \sigma_2 \\
T^g_4 &= & \tfrac{i}{4}\ \sigma_1 \otimes \sigma_3 \otimes 1 \otimes 1 \otimes 1 \otimes \sigma_2 \\
&& -\tfrac{i}{4}\ \sigma_1 \otimes 1 \otimes 1 \otimes 1 \otimes 1 \otimes \sigma_2 \\
T^g_5 &= & \tfrac{i}{4}\ \sigma_2 \otimes 1 \otimes 1 \otimes 1 \otimes 1 \otimes 1 \\
&& -\tfrac{i}{4}\ \sigma_2 \otimes \sigma_3 \otimes 1 \otimes 1 \otimes 1 \otimes 1 \\
T^g_6 &= & -\tfrac{i}{4}\ 1 \otimes \sigma_1 \otimes 1 \otimes 1 \otimes 1 \otimes \sigma_2 \\
&& +\tfrac{i}{4}\ \sigma_3 \otimes \sigma_1 \otimes 1 \otimes 1 \otimes 1 \otimes \sigma_2 \\
T^g_7 &= & -\tfrac{i}{4}\ \sigma_3 \otimes \sigma_2 \otimes 1 \otimes 1 \otimes 1 \otimes 1 \\
&& +\tfrac{i}{4}\ 1 \otimes \sigma_2 \otimes 1 \otimes 1 \otimes 1 \otimes 1 \\
T^g_8 &= -\tfrac{i}{4\sqrt3}\ \sigma_3 \otimes 1 \otimes 1 \otimes 1 \otimes 1 \otimes \sigma_2 \\
&& -\tfrac{i}{4\sqrt3}\ 1 \otimes \sigma_3 \otimes 1 \otimes 1 \otimes 1 \otimes \sigma_2 \\
&& +\tfrac{i}{2\sqrt3}\ \sigma_3 \otimes \sigma_3 \otimes 1 \otimes 1 \otimes 1 \otimes \sigma_2 \\[2mm]
T^W_1 &= & -\tfrac{i}{4}\ 1 \otimes 1 \otimes \sigma_1 \otimes 1 \otimes 1 \otimes \sigma_2 \\
&& -\tfrac{i}{4}\ 1 \otimes 1 \otimes \sigma_1 \otimes \sigma_3 \otimes 1 \otimes \sigma_2 \\
T^W_2 &= & \tfrac{i}{4}\ 1 \otimes 1 \otimes \sigma_2 \otimes 1 \otimes 1 \otimes 1 \\
&& \tfrac{i}{4}\ 1 \otimes 1 \otimes \sigma_2 \otimes \sigma_3 \otimes 1 \otimes 1 \\
T^W_3 &= & -\tfrac{i}{4}\ 1 \otimes 1 \otimes \sigma_3 \otimes \sigma_3 \otimes 1 \otimes \sigma_2 \\
&& -\tfrac{i}{4}\ 1 \otimes 1 \otimes \sigma_3 \otimes 1 \otimes 1 \otimes \sigma_2 \\[2mm]
T^Y &= & \tfrac{i}{4}\ 1 \otimes 1 \otimes \sigma_3 \otimes \sigma_3 \otimes 1 \otimes \sigma_2 \\
&& -\tfrac{i}{4}\ 1 \otimes 1 \otimes \sigma_3 \otimes 1 \otimes 1 \otimes \sigma_2 \\
&& +\tfrac{i}{6}\ \sigma_3 \otimes 1 \otimes 1 \otimes 1 \otimes 1 \otimes \sigma_2 \\
&& +\tfrac{i}{6}\ 1 \otimes \sigma_3 \otimes 1 \otimes 1 \otimes 1 \otimes \sigma_2 \\
&& +\tfrac{i}{6}\ \sigma_3 \otimes \sigma_3 \otimes 1 \otimes 1 \otimes 1 \otimes \sigma_2
\end{aligned}
$$

(2.8)

$$
\psi =
\begin{bmatrix}
\nu \\ e \\ u^r \\ d^r \\ u^g \\ d^g \\ u^b \\ d^b
\end{bmatrix}
\left\{
\begin{bmatrix}
e^{\wedge}_{Lr} \\
e^{\wedge}_{Li} \\
e^{\vee}_{Lr} \\
e^{\vee}_{Li} \\
e^{\wedge}_{Rr} \\
e^{\wedge}_{Ri} \\
e^{\vee}_{Rr} \\
e^{\vee}_{Ri}
\end{bmatrix}
\right.
$$

These eighteen $\mathbb{R}(64)$ matrices, representing gravitational and Standard Model generators acting on each other via their Lie bracket and on a generation of fermion components in a $64^{\mathbb{R}}$ column, constitute the conventional description of the gravitational and Standard Model algebra. Any successful unified description of gravity and the Standard Model must describe how these algebraic elements embed in a larger algebra.

3. Unification in a GraviGUT

The gravitational and Standard Model Lie algebra, $su(3) \oplus su(2)_L \oplus u(1)_Y \oplus spin(1,3)$, a subalgebra of $spin(11,3)$, acts on a generation of Standard Model fermions in a specific representation, (2.8), corresponding to the particle interactions of bosons and fermions. Remarkably, this $64^{\mathbb{R}}$ representation space of fermions is equal to a Majorana-Weyl spinor of $spin(11,3)$. To establish this equivalence, we build a matrix representation of $spin(11,3)$ from a representation of the $Cl(11,3)$ Clifford algebra. Specifically, we choose this *very nice* representation of fourteen $Cl^1(11,3)$ Clifford basis vectors as real $\mathbb{R}(128)$ matrices:

(3.1)

$$
\begin{aligned}
\Gamma_1 &= i\,\sigma_1 \otimes 1 \otimes 1 \otimes \sigma_2 \otimes \sigma_2 \otimes \sigma_1 \otimes \sigma_2 \\
\Gamma_2 &= -i\,\sigma_1 \otimes 1 \otimes 1 \otimes \sigma_2 \otimes \sigma_2 \otimes \sigma_2 \otimes 1 \\
\Gamma_3 &= i\,\sigma_1 \otimes 1 \otimes 1 \otimes \sigma_2 \otimes \sigma_2 \otimes \sigma_3 \otimes \sigma_2 \\
\Gamma_4 &= \sigma_1 \otimes 1 \otimes 1 \otimes \sigma_2 \otimes \sigma_1 \otimes 1 \otimes \sigma_2 \\
\Gamma_5 &= \sigma_1 \otimes 1 \otimes 1 \otimes \sigma_1 \otimes 1 \otimes 1 \otimes 1 \\
\Gamma_6 &= \sigma_1 \otimes 1 \otimes 1 \otimes \sigma_2 \otimes \sigma_3 \otimes 1 \otimes \sigma_2 \\
\Gamma_7 &= \sigma_1 \otimes 1 \otimes 1 \otimes \sigma_3 \otimes 1 \otimes 1 \otimes 1 \\
\Gamma_8 &= -\,\sigma_2 \otimes 1 \otimes 1 \otimes 1 \otimes 1 \otimes 1 \otimes \sigma_2 \\
\Gamma_9 &= \sigma_2 \otimes \sigma_3 \otimes \sigma_2 \otimes 1 \otimes \sigma_2 \otimes \sigma_2 \otimes \sigma_1 \\
\Gamma_{10} &= -\,\sigma_2 \otimes 1 \otimes \sigma_2 \otimes 1 \otimes \sigma_2 \otimes \sigma_2 \otimes \sigma_3 \\
\Gamma_{11} &= -\,\sigma_2 \otimes \sigma_2 \otimes 1 \otimes 1 \otimes \sigma_2 \otimes \sigma_2 \otimes \sigma_1 \\
\Gamma_{12} &= \sigma_2 \otimes \sigma_2 \otimes \sigma_3 \otimes 1 \otimes \sigma_2 \otimes \sigma_2 \otimes \sigma_3 \\
\Gamma_{13} &= \sigma_2 \otimes \sigma_1 \otimes \sigma_2 \otimes 1 \otimes \sigma_2 \otimes \sigma_2 \otimes \sigma_1 \\
\Gamma_{14} &= -\,\sigma_2 \otimes \sigma_2 \otimes \sigma_1 \otimes 1 \otimes \sigma_2 \otimes \sigma_2 \otimes \sigma_3
\end{aligned}
$$

These satisfy the Clifford algebra basis vector identity

$$\Gamma_i\Gamma_j = \tfrac{1}{2}(\Gamma_i\Gamma_j + \Gamma_j\Gamma_i) + \tfrac{1}{2}(\Gamma_i\Gamma_j - \Gamma_j\Gamma_i) = \Gamma_i \cdot \Gamma_j + \Gamma_i \times \Gamma_j = \eta_{ij}1 + \Gamma_{ij}$$

in which the Clifford product of these generators is equivalent to the product of the representative matrices. The η_{ij} is the Minkowski metric of signature $(11,3)$. The $\Gamma_{ij} = \Gamma_i\Gamma_j$, with $i < j$, are nintey-one independent bivector basis generators, spanning $Cl^2(11,3)$. As $\mathbb{R}(128)$ matrices, these bivectors are block diagonal,

(3.2)
$$\Gamma_{ij} = \begin{bmatrix} \Gamma_{ij}^+ & 0 \\ 0 & \Gamma_{ij}^- \end{bmatrix}$$

with Γ_{ij}^+ and Γ_{ij}^- the $\mathbb{R}(64)$ basis matrix generators spanning the positive and negative chiral representations of $spin(11,3)$. The $spin(11,3)$ Lie algebra generators may be represented by either chiral set of matricies, $\Gamma_{ij}^\pm$, or by the larger Γ_{ij}. The Γ_{ij}^+ matrices act on a real $64_{S+}^{\mathbb{R}}$ column matrix – a positive chiral Majorana-Weyl $spin(11,3)$ spinor.

The gravitational and Standard Model Lie algebra is a subalgebra of $spin(11,3)$, spanned by eighteen basis generators,

$$(3.3) \quad \begin{aligned}
&T^G_{23} = \Gamma^+_{23} && T^g_1 = \tfrac{1}{4}\Gamma^+_{9\,12} - \tfrac{1}{4}\Gamma^+_{10\,11} && T^W_1 = \tfrac{1}{4}\Gamma^+_{58} - \tfrac{1}{4}\Gamma^+_{67} \\
&T^G_{13} = \Gamma^+_{13} && T^g_2 = -\tfrac{1}{4}\Gamma^+_{9\,11} - \tfrac{1}{4}\Gamma^+_{10\,12} && T^W_2 = -\tfrac{1}{4}\Gamma^+_{57} - \tfrac{1}{4}\Gamma^+_{68} \\
&T^G_{12} = \Gamma^+_{12} && T^g_3 = -\tfrac{1}{4}\Gamma^+_{9\,10} + \tfrac{1}{4}\Gamma^+_{11\,12} && T^W_3 = -\tfrac{1}{4}\Gamma^+_{56} + \tfrac{1}{4}\Gamma^+_{78} \\
&T^G_{14} = \Gamma^+_{14} && T^g_4 = -\tfrac{1}{4}\Gamma^+_{9\,14} + \tfrac{1}{4}\Gamma^+_{10\,13} && \\
&T^G_{24} = \Gamma^+_{24} && T^g_5 = \tfrac{1}{4}\Gamma^+_{9\,13} + \tfrac{1}{4}\Gamma^+_{10\,14} && T^Y = \tfrac{1}{4}\Gamma^+_{56} + \tfrac{1}{4}\Gamma^+_{78} + \tfrac{1}{6}\Gamma^+_{9\,10} \\
&T^G_{34} = \Gamma^+_{34} && T^g_6 = \tfrac{1}{4}\Gamma^+_{11\,14} - \tfrac{1}{4}\Gamma^+_{12\,13} && \qquad + \tfrac{1}{6}\Gamma^+_{11\,12} + \tfrac{1}{6}\Gamma^+_{13\,14} \\
& && T^g_7 = -\tfrac{1}{4}\Gamma^+_{11\,13} - \tfrac{1}{4}\Gamma^+_{12\,14} && \\
& && T^g_8 = -\tfrac{1}{4\sqrt3}\Gamma^+_{9\,10} - \tfrac{1}{4\sqrt3}\Gamma^+_{11\,12} + \tfrac{1}{2\sqrt3}\Gamma^+_{13\,14} &&
\end{aligned}$$

Because of our *very nice* Clifford algebra representation, (3.1), these eighteen $\mathbb{R}(64)$ matrix generators, (3.3), are *numerically identical* to the conventional gravitational and Standard Model basis generator matrices, (2.8). With this identification, we also see directly that a generation of fermions, in a $64^{\mathbb{R}}$ of (2.8), is identified as a positive chiral Majorana-Weyl $spin(11,3)$ spinor, $64^{\mathbb{R}}_{S+}$, with identical components. This embedding of the gravitational and Standard Model algebra in a $spin(11,3)$ GraviGUT is in agreement with the recent work of Nesti and Percacci,[2] and with a unified GraviGUT action.[4] It is worth noting that the embedding of the gravitational and Standard Model Lie algebra in $spin(11,3)$ expressed in (3.3) in terms of bivector generators is valid regardless of the choice of matrix representation. Nevertheless, the main value of the present work is in describing this embedding directly and explicitly, as we have here, via a convenient choice of matrix representation. The Clifford basis vector matrices, (3.1), are essentially the square root of the conventional Standard Model generators.

This Clifford basis vector representation, (3.1), is useful not only for matching the Standard Model generators, but also for matching the $su(4) \oplus su(2)_L \oplus su(2)_R$ basis generators of the Pati-Salam GUT. For these, the seven standard $su(4)$ basis generators not in the strong $su(3)$ subalgebra, and the three $su(2)_R$ basis generators, are:

$$(3.4) \quad \begin{aligned}
&T^g_9 = \tfrac{1}{4}\Gamma^+_{11\,14} + \tfrac{1}{4}\Gamma^+_{12\,13} && T^{W'}_1 = \tfrac{1}{4}\Gamma^+_{58} + \tfrac{1}{4}\Gamma^+_{67} \\
&T^g_{10} = \tfrac{1}{4}\Gamma^+_{11\,13} - \tfrac{1}{4}\Gamma^+_{12\,14} && T^{W'}_2 = -\tfrac{1}{4}\Gamma^+_{57} + \tfrac{1}{4}\Gamma^+_{68} \\
&T^g_{11} = \tfrac{1}{4}\Gamma^+_{9\,14} + \tfrac{1}{4}\Gamma^+_{10\,13} && T^{W'}_3 = \tfrac{1}{4}\Gamma^+_{56} + \tfrac{1}{4}\Gamma^+_{78} \\
&T^g_{12} = \tfrac{1}{4}\Gamma^+_{9\,13} - \tfrac{1}{4}\Gamma^+_{10\,14} && \\
&T^g_{13} = \tfrac{1}{4}\Gamma^+_{9\,12} + \tfrac{1}{4}\Gamma^+_{10\,11} && \\
&T^g_{14} = \tfrac{1}{4}\Gamma^+_{9\,11} - \tfrac{1}{4}\Gamma^+_{10\,12} && \\
&T^g_{15} = \tfrac{1}{2\sqrt6}\Gamma^+_{9\,10} + \tfrac{1}{2\sqrt6}\Gamma^+_{11\,12} + \tfrac{1}{2\sqrt6}\Gamma^+_{13\,14} &&
\end{aligned}$$

In the Pati-Salam GUT, these twenty-one basis generators, spanning an $\mathbb{R}(64)$ matrix representation of $su(4) \oplus su(2)_L \oplus su(2)_R$, act on fermions as in (2.8). The twelve basis generators of the Standard Model in (3.3) are a subset of these twenty-one. Thus, the Standard Model Lie algebra embeds in $su(4) \oplus su(2)_L \oplus su(2)_R$, which is a subalgebra of $spin(10)$, which is a subalgebra of $spin(11,3)$. The six basis generators of gravitational $spin(1,3)$ in (3.3) are a subset of the ninety-one $spin(11,3)$ basis generators. The other forty basis generators in $spin(11,3)$, other than the forty-five of $spin(10)$ and six of $spin(1,3)$, may be used to describe a *frame-Higgs* multiplet.

3.1. Frame-Higgs and the Unified Bosonic Connection.

In modern formulations of gravity, the *frame*, locally $\underline{e} = \underline{dx}^\mu (e_\mu)^a \gamma_a$, in which γ_a are four $Cl^1(1,3)$ Clifford basis vectors, (2.2), describes a rest frame at every point of spacetime. The frame relates a set of four orthonormal basis vectors, $\vec{e}_a = (e_a^-)^\mu \vec{\partial}_\mu$, at each spacetime point to the γ_a spanning a local Minkowski spacetime, determining the spacetime metric, $g_{\mu\nu} = (e_\mu)^a \eta_{ab} (e_\nu)^b$. The *spin connection*, locally $\underline{\omega} = \frac{1}{2}\underline{\omega}^{ab}\gamma_{ab}$, valued in $spin(1,3) = Cl^2(1,3)$, acts on the frame as a 4-vector, and describes how it changes over spacetime. The covariant derivative of the frame is the gravitational torsion 2-form,

$$(3.5) \qquad \underline{\underline{T}} = \underline{D}\underline{e} = \underline{d}\underline{e} + \underline{\omega} \times \underline{e} = \underline{d}\underline{e} + \tfrac{1}{2}\underline{\omega}\underline{e} + \tfrac{1}{2}\underline{e}\underline{\omega} = (\underline{d}\underline{e}^a + \underline{\omega}^a{}_b \underline{e}^b)\gamma_a$$

which is usually zero, allowing the spin connection to be determined from the frame. In such expressions a wedge is always implied between differential forms, with underlines used to designate form grade, and we use the Clifford product – equivalent to the matrix product between representative matrices.

In the $spin(10)$ GUT there is usually a scalar Higgs multiplet that is a 10-vector, $\phi = \phi^P \Gamma_P$. Using the $Cl(10)$ Clifford product, the covariant derivative of such a Higgs is

$$\underline{D}\phi = \underline{d}\phi + 2\underline{A} \times \phi = \underline{d}\phi + \underline{A}\phi - \phi\underline{A} = (\underline{d}\phi^P + 2\underline{A}^P{}_Q \phi^Q)\Gamma_P$$

We can combine all bosonic fields – spin connection, frame, Higgs, and gauge fields – in the *unified bosonic connection* of $spin(11,3)$,

$$(3.6) \qquad \underline{H} = \tfrac{1}{2}\underline{\omega} + \tfrac{1}{4}\underline{e}\phi + \underline{A} = \tfrac{1}{4}\underline{\omega}^{ab}\Gamma_{ab} + \tfrac{1}{4}\underline{e}^a\phi^P\Gamma_{aP} + \tfrac{1}{2}\underline{A}^{PQ}\Gamma_{PQ}$$

in which we have substituted Γ_a's for γ_a's in the gravitational frame, $\underline{e}$, and spin connection, $\underline{\omega}$. In this way, the gravitational frame and a Higgs naturally emerge as the *frame-Higgs* part, $\underline{e}\phi$, of a $spin(11,3)$ connection. The curvature of this unified bosonic connection is

$$(3.7) \qquad \underline{\underline{F}} = \underline{d}\underline{H} + \underline{H}\underline{H} = \tfrac{1}{2}\left(\underline{\underline{R}} - \tfrac{1}{8}\underline{e}\underline{e}\phi^2\right) + \tfrac{1}{4}\left(\underline{\underline{T}}\phi - \underline{e}\underline{D}\phi\right) + \underline{\underline{F}}_A$$

in which $\underline{\underline{R}} = \underline{d}\underline{\omega} + \tfrac{1}{2}\underline{\omega}\underline{\omega}$ is the Riemann curvature 2-form, we recognize $\underline{\underline{T}}$ and $\underline{D}\phi$ as the torsion and covariant derivative of the Higgs, and $\underline{\underline{F}}_A = \underline{d}\underline{A} + \underline{A}\underline{A}$ is the curvature of the $spin(10)$ gauge field.

The dynamical symmetry breaking of $spin(11,3)$ to gravitational $spin(1,3)$ and $spin(10)$ is described in more detail in [**4**]. To summarize, the frame-Higgs, $\underline{e}\phi$, becomes nonzero over all of spacetime, reducing the structure group to that of gravity and the Standard Model, with the frame, $\underline{e}$, describing a de Sitter cosmology, and the Higgs, $\phi \simeq \phi_0$, having a nonzero vacuum expectation value. Gauge fields interact with ϕ_0, becoming massive, and this Higgs background also gives masses to fermions via the covariant Dirac derivative,

$$(3.8) \qquad \underline{D}\psi = \underline{d}\psi + \underline{H}\psi = (\underline{d} + \tfrac{1}{2}\underline{\omega} + \tfrac{1}{4}\underline{e}\phi + \underline{A})\psi$$

Interactions between elementary particles correspond to the geometric and algebraic structure expressed in the curvature (3.7) and covariant Dirac derivative (3.8) as they appear in an action.

3.2. GraviGUT Algebra. The structure of the $spin(11,3)$ GraviGUT Lie algebra, corresponding to interactions between bosons, is described explicitly by the Lie bracket between generators,

$$(3.9) \qquad [\Gamma_{ij}, \Gamma_{kl}] = 2\eta_{jk}\Gamma_{il} - 2\eta_{jl}\Gamma_{ik} + 2\eta_{il}\Gamma_{jk} - 2\eta_{ik}\Gamma_{jl}$$

Elements of $spin(11,3)$ also act as linear operators on the Majorana-Weyl spinor representation space, shuffling the components of fermions in a $64^{\mathbb{R}}_{S+}$, corresponding to interactions between bosons and fermions. If we write such a spinor as $\psi = \psi^\iota Q_\iota$, in terms of coefficients, ψ^ι, multiplying canonical basis column matrix *spinor generators*, Q_ι, then the multiplicative operation of $spin(11,3)$ generators on these can be used to define a generalized bracket,

$$(3.10) \qquad [\Gamma_{ij}, Q_\iota] = \Gamma_{ij}^+ Q_\iota = Q_\kappa (\Gamma_{ij}^+)^\kappa{}_\iota$$

with $(\Gamma_{ij}^+)^\kappa{}_\iota$ the numerical components of the Γ_{ij}^+ matrices. Together, the $spin(11,3)$ Lie algebra, (3.9), and its action on Majorana-Weyl spinors, (3.10), constitute the *GraviGUT algebra*. The word "algebra" is used here in a generalized sense, even though the bracket between two spinors may be undefined. If we define such a bracket trivially,

$$(3.11) \qquad [Q_\iota, Q_\kappa] = 0$$

and presume $[Q_\iota, \Gamma_{ij}] = -[\Gamma_{ij}, Q_\iota]$, then these brackets, (3.9-3.11), are the Lie brackets of a Lie algebra – completing the GraviGUT algebra – in which spinors are an ideal. Similarly, it is possible to construct the Standard Model algebra by using the Standard Model Lie algebra generators in (3.9) and the matrix representation, (2.8), in (3.10), with the $64^{\mathbb{R}}$ as an ideal. The Standard Model algebra is thus a subalgebra of the GraviGUT algebra.

4. Unification in E_8

Remarkably, the GraviGUT algebra, defined by (3.9) and (3.10), is a subalgebra of the quaternionic real form of the E_8 Lie algebra, $E_{8(-24)}$. This can be seen immediately by using an explicit construction of $E_{8(-24)}$ involving a $spin(12,4)$ subalgebra acting on a $128^{\mathbb{R}}_{S+}$.

To construct a convenient positive chiral real matrix representation of $spin(12,4)$ we first choose a set of representative $\mathbb{R}(256)$ matrices for the sixteen $Cl(12,4)$ Clifford basis vectors:

$$(4.1) \qquad \begin{aligned} \Gamma_i' &= & \sigma_1 \otimes \Gamma_i \\ \Gamma_{15}' &= & \sigma_1 \otimes \Gamma \\ \Gamma_{16}' &= -i\,\sigma_2 \otimes 1 \end{aligned}$$

in which the Γ_i are the fourteen $\mathbb{R}(128)$ matrices in (3.1), and

$$\Gamma = \Gamma_1\Gamma_2\Gamma_3\Gamma_4\Gamma_5\Gamma_6\Gamma_7\Gamma_8\Gamma_9\Gamma_{10}\Gamma_{11}\Gamma_{12}\Gamma_{13}\Gamma_{14} = \sigma_3 \otimes 1$$

From these Clifford basis vector matrices we can construct the representative $\mathbb{R}(256)$ matrices of the one-hundred-twenty bivector basis generators, Γ'_{xy}, of $spin(12, 4)$,

(4.2)

$$\Gamma'_{ij} = 1 \otimes \Gamma_{ij} = \begin{bmatrix} \Gamma'^+_{ij} & 0 \\ 0 & \Gamma'^-_{ij} \end{bmatrix} \qquad \Gamma'^+_{ij} = \Gamma_{ij} = \begin{bmatrix} \Gamma^+_{ij} & 0 \\ 0 & \Gamma^-_{ij} \end{bmatrix}$$

$$\Gamma'_{i\,15} = 1 \otimes \Gamma_i \Gamma \qquad \Gamma'^+_{i\,15} = \Gamma_i \Gamma = \begin{bmatrix} 0 & -\Gamma^+_i \\ \Gamma^-_i & 0 \end{bmatrix}$$

$$\Gamma'_{i\,16} = \sigma_3 \otimes \Gamma_i \qquad \Gamma'^+_{i\,16} = \Gamma_i = \begin{bmatrix} 0 & \Gamma^+_i \\ \Gamma^-_i & 0 \end{bmatrix}$$

$$\Gamma'_{15\,16} = \sigma_3 \otimes \Gamma \qquad \Gamma'^+_{15\,16} = \Gamma = \begin{bmatrix} 1 & 0 \\ 0 & -1 \end{bmatrix}$$

in which appear the $\mathbb{R}(128)$ Γ_{ij} and Γ_i matrices from (3.2) and (3.1). The positive chiral parts of these generators, Γ'^+_{xy}, span the $\mathbb{R}(128)$ positive chiral matrix representation of $spin(12, 4)$. They act on a real positive chiral spinor representation space, $\psi = \psi^\phi Q'_\phi \in 128^{\mathbb{R}}_{S+}$, spanned by the canonical set of unit column matrices, Q'_ϕ.

The quaternionic real form of the largest simple exceptional Lie algebra, $E_{8(-24)}$, is spanned by the one-hundred-twenty independent basis generators, Γ'_{xy}, of $spin(12, 4)$ and the one-hundred-twenty-eight Q'_ϕ, now interpreted as basis generators. The complete set of $E_{8(-24)}$ Lie brackets between these basis generators is:

(4.3)

$$[\Gamma'_{wx}, \Gamma'_{yz}] = 2\eta_{xy}\Gamma'_{wz} - 2\eta_{xz}\Gamma'_{wy} + 2\eta_{wz}\Gamma'_{xy} - 2\eta_{wy}\Gamma'_{xz}$$

$$[\Gamma'_{xy}, Q'_\phi] = -[Q'_\phi, \Gamma'_{xy}] = \Gamma'^+_{xy} Q'_\phi = Q'_\psi (\Gamma'^+_{xy})^\psi{}_\phi$$

$$[Q'_\phi, Q'_\psi] = -\Gamma'_{xy}(\Gamma'^{+xy})_{\psi\phi} = -\Gamma'_{xy}\eta^{xw}\eta^{yz}(\Gamma'^+_{wz})^\lambda{}_\phi g_{\lambda\psi}$$

in which η_{xy} is a Minkowski metric of signature $(12, 4)$ and $g_{\lambda\psi}$ is part of the $E_{8(-24)}$ Killing form.

The Killing form is a symmetric bilinear operator – a metric on the $E_{8(-24)}$ Lie algebra as a vector space. Its components between the one-hundred-twenty independent bivector basis elements are

$$g_{wx\,yz} = (\Gamma'_{wx}, \Gamma'_{yz}) = \eta_{xy}\eta_{wz} - \eta_{wy}\eta_{xz}$$

with signature $(48, 72)$ – the Killing form of $spin(12, 4)$. Between bivector and spinor generators the Killing form is zero, $g_{xy\,\phi} = (\Gamma'_{xy}, Q'_\phi) = 0$. The $\mathbb{R}(128)$ matrix of Killing form components between spinor generators, $g_{\lambda\psi} = (Q'_\lambda, Q'_\psi)$, appearing in (4.3), is the product of the odd signature Clifford vector matrices,

(4.4) $\qquad g = (\Gamma'_1 \Gamma'_2 \Gamma'_3 \Gamma'_{16})^+ = \Gamma'^+_{1\,16}\Gamma'^+_{2\,16}\Gamma'^+_{3\,16} = -\sigma_1 \otimes 1 \otimes 1 \otimes \sigma_2 \otimes \sigma_2 \otimes 1 \otimes 1$

having signature $(64, 64)$. Together, the signature of the $E_{8(-24)}$ Killing form is $(112, 136)$ – consistent with the $112 - 136 = -24$ label, and with the embedding of E_7 in $E_{8(-24)}$.[6]

The main observation of this paper is that the GraviGUT algebra, specified by (3.9) and (3.10), is a subalgebra of $E_{8(-24)}$, specified by (4.3). Specifically, the GraviGUT subalgebra of $E_{8(-24)}$ is spanned by the ninety-one Γ'_{ij} of $spin(11, 3) \subset$

$spin(12,4)$ and sixty-four Q'_ι of $64^{\mathbb{R}}_{S+} \subset 128^{\mathbb{R}}_{S+}$, with the subalgebra

$$(4.5) \qquad \begin{aligned} [\Gamma'_{ij}, \Gamma'_{kl}] &= 2\eta_{jk}\Gamma'_{il} - 2\eta_{jl}\Gamma'_{il} + 2\eta_{il}\Gamma'_{jk} - 2\eta_{ik}\Gamma'_{jl} \\ [\Gamma'_{ij}, Q'_\iota] &= \Gamma'^{+}_{ij}Q'_\iota = Q'_\kappa (\Gamma^{+}_{ij})^\kappa{}_\iota \end{aligned}$$

from (4.3) matching that of the $spin(11,3)$ GraviGUT, having used (4.2) to see that $(\Gamma'^{+}_{ij})^\kappa{}_\iota = (\Gamma^{+}_{ij})^\kappa{}_\iota$. In this completely direct and explicit way, we see that the $spin(11,3)$ GraviGUT, and hence the algebra of gravity and the Standard Model, is inside $E_{8(-24)}$. This embedding of the Standard Model algebra in $E_{8(-24)}$ includes not only the gravitational and Standard Model Lie algebra, but also its correct action on a generation of fermions – the structure of $E_{8(-24)}$ explains the existence and structure of the spinor, electroweak, and strong fermion multiplets of the Standard Model algebra.

If the complete structure of $E_{8(-24)}$ is interpreted directly, with each of two-hundred-forty-eight generators corresponding to a different kind of elementary particle, then it predicts the existence of many new particles in addition to those of the Standard Model. Of the two-hundred-forty-eight, one-hundred-twenty are basis generators of $spin(12,4)$ and one-hundred-twenty-eight of $128^{\mathbb{R}}_{S+}$. Of the $spin(12,4)$ generators, ninety-one are for $spin(11,3)$, consistent with a GraviGUT. That Gravi-GUT has six generators for gravitational $spin(1,3)$, forty-five for $spin(10)$, and forty for a frame-Higgs, including a Higgs 10 multiplet. The $spin(10)$ GUT includes the twelve g, W, and B bosons of the Standard Model, as well as predicting the existence of three new $W'^{\pm}$ and Z' particles related to $su(2)_R$, and a collection of thirty colored X bosons. The remaining generators of $spin(12,4)$, not in $spin(11,3)$, correspond to twenty more X bosons, one Peccei-Quinn w boson of $spin(1,1)$, and eight for more frame-Higgs' that includes two axions. Of the one-hundred-twenty-eight spinor generators, sixty-four are those of one generation of Standard Model fermions, while the other sixty-four are those of mirror fermions, with opposite charges – predicted to exist if the structure of $E_{8(-24)}$ is interpreted directly.

Given this explicit embedding of gravity and the Standard Model inside $E_{8(-24)}$, one might wonder how to interpret the paper "There is no 'Theory of Everything' inside E8."[7] In their work, Distler and Garibaldi prove that, using a direct decomposition of E_8, when one embeds gravity and the Standard Model in E_8, there are also mirror fermions. They then claim this prediction of mirror fermions (the existence of "non-chiral matter") makes E8 Theory unviable. However, since there is currently no good explanation for why any fermions have the masses they do, it is overly presumptuous to proclaim the failure of E_8 unification – since the detailed mechanism behind particle masses is unknown, and mirror fermions with large masses could exist in nature. Nevertheless, it was helpful of Distler and Garibaldi to emphasize the difficulty of describing the three generations of fermions, which remains an open problem.

Although it is possible to define a map, based on triality, between the generation of sixty-four Standard Model fermion generators in $E_{8(-24)}$, the sixty-four mirror fermion generators, and sixty-four non-Standard Model boson generators, these cannot be interpreted as three generations of fermions under a direct decomposition of $E_{8(-24)}$. The proposal that these three blocks of generators might correspond to the three generations of fermions, suggested in [5], remains a vague hint towards some more mysterious structure, and not a direct identification. The explanation for the existence of three generations of fermions, all with the same apparent algebraic structure, remains largely a mystery. Nevertheless, it is intriguing that the algebraic elements of all gravitational and Standard Model bosons acting correctly on a generation of fermions are found in E_8.

4.1. Superconnection. The fact that generators corresponding to bosons and fermions coexist in the $E_{8(-24)}$ Lie algebra suggests an interesting geometric construction. When one describes the geometry of a principal bundle over spacetime, the connection is a 1-form valued in the Lie algebra. However, fermion fields are not conventional 1-forms over spacetime, but are fields of anti-commuting numbers – called Grassmann numbers by physicists. These fields also arise in the BRST approach to gauge theory,[8] in which these anti-commuting fields are sometimes interpreted as 1-forms over the space of connections. As in BRST, it is useful to consider a unified field, called a *superconnection*, comprised of the formal sum of connection 1-forms and anti-commuting fields, both valued in the same algebra,

$$(4.6) \qquad\qquad \underline{A} = \underline{H} + \psi$$

Computing the curvature of this superconnection,

$$(4.7) \qquad\qquad \underline{\underline{F}} = \underline{d}\underline{A} + \tfrac{1}{2}[\underline{A}, \underline{A}] = \underline{\underline{F}} + \underline{D}\psi + \psi\psi$$

we find a gauge field curvature, a covariant derivative term, and a term quadratic in ψ. If the gauge field, $\underline{H}$, is the unified bosonic connection, (3.6), and the anti-commuting field, ψ, is a Standard Model fermion multiplet or the $E_{8(-24)}$ subspace corresponding to a fermion multiplet, then the covariant derivative term, $\underline{D}\psi$, is identical to the covariant Dirac derivative in (3.8). The superconnection, (4.6), constitutes a unified field comprised of all known types of elementary particles, and its curvature, (4.7), can be used to describe their dynamics.

5. Discussion and Conclusion

In this work we have described an explicit embedding of the Standard Model algebra, including gravity, in the $spin(11, 3)$ GraviGUT, and its subsequent embedding in $E_{8(-24)}$. If the geometry of our universe is fundamentally that of an $E_{8(-24)}$ principal bundle, this symmetry must be broken, with the frame-Higgs part of the connection attaining a vacuum expectation value, and the structure reducing to that of gravity and the Standard Model. If $E_{8(-24)}$ unification is interpreted directly, it predicts the existence of many new particles, including W', Z', and X bosons, a rich Higgs sector, and the existence of mirror fermions. The charges of these elementary particles – their eigenvalues with respect to a collection of mutually commuting force generators – are presented in the Elementary Particle Explorer[1], which displays various GUT and GraviGUT weight diagrams and embeddings. It is a distinct possibility that some of these new particles may be detected at the Large Hadron Collider.

Although the explicit embedding of gravity and the Standard Model in $E_{8(-24)}$ described here is incontrovertible, it raises many questions. Since a direct interpretation implies the existence of mirror fermions, which are not known to exist in nature, it will be necessary to understand how these particles obtain large masses or otherwise work out in the theory. Fortunately, the embedding in $E_{8(-24)}$ also predicts the existence of axions and other Higgs scalars, which interact with the

---

[1]http://deferentialgeometry.org/epe/

mirrors. These scalars could also help explain the existence of the three generations of fermions – possibly as scalar-fermion composites – and their masses. Another potential hint towards addressing the generation problem is the embedding of the Standard Model algebra in the split real form, $E_{8(8)}$, via the Pati-Salam GUT, which is important but has not been extensively discussed here. This issue of generations, as well as particle masses, the action for the theory, and ultimately its quantum description, remain open questions.

The explicit embedding of gravity and the Standard Model in the $spin(11,3)$ GraviGUT and in $E_{8(-24)}$ described here provides a solid base from which to develop these ideas further. The explicit real matrix representation of gravity and the Standard Model, (2.8), the compatible $Cl(11,3)$ matrix representation, (3.1), and the explicit description of the structure of $E_{8(-24)}$ in (4.3), may be especially useful to current and future researchers, whether considering unification in E_8 or other algebras. From the solid foundation of these explicit matrix representations and their embeddings, we might get a better perspective on many of the deep open questions remaining.

6. Acknowledgements

The author wishes to thank Andrew Golato for constructive feedback. This research was supported by a grant from The Foundational Questions Institute.

References

1. J. Baez and J. Huerta, "The Algebra of Grand Unified Theories," arxiv:0904.1556.
2. F. Nesti and R. Percacci, "Chirality in unified theories of gravity," arxiv:0909.4537.
3. A.G. Lisi, "Clifford bundle formulation of BF gravity generalized to the standard model," gr-qc/0511120.
4. A.G. Lisi, L. Smolin, and S. Speziale, "Unification of gravity, gauge fields, and Higgs bosons," arxiv:1004.4866.
5. A.G. Lisi, "An Exceptionally Simple Theory of Everything," arxiv:0711.0770.
6. J.F. Adams, *Lectures On Exceptional Lie Groups*, University of Chicago (1996).
7. J. Distler and S. Garibaldi, "There is no 'Theory of Everything' inside E8," arxiv:0905.2658.
8. J.W. van Holten, "Aspects of BRST Quantization," hep-th/0201124.

A. GARRETT LISI, 95 MINER PL, MAKAWAO, HI 96768
E-mail address: alisi@hawaii.edu

Contemporary Mathematics
Volume **557**, 2011

FROM GROUPS TO SYMMETRIC SPACES

G. LUSZTIG

This paper is based on a talk given at the conference "Representation theory of real reductive groups", Salt Lake City, July 2009.

We fix an algebraically closed field $\mathbf{k}$ of characteristic exponent p. (We assume, except in §18, that either $p = 1$ or $p \gg 0$.) We also fix a *symmetric space* that is a triple (G, θ, K) where G is a connected reductive algebraic group over $\mathbf{k}$, $\theta : G \to G$ is an involution and K is the identity component of the fixed point set of θ (K is a connected reductive algebraic group). We shall often write (G, K) instead of (G, θ, K). Let $\mathfrak{g} = \text{Lie } (G), \mathfrak{k} = \text{Lie } (K), \mathfrak{p} = \mathfrak{g}/\mathfrak{k}$. Note that K acts naturally on $\mathfrak{p}$ by the adjoint action.

If H is a connected reductive algebraic group over $\mathbf{k}$ then H gives rise to a symmetric space $(H \times H, H)$ where H is imbedded in $H \times H$ as the diagonal; here $\theta(a, b) = (b, a)$. (Such a symmetric space is said to be *diagonal*.)

In this paper we examine various properties/constructions which are known for groups (or for diagonal symmetric spaces) and we do some experiments to see to what extent they generalize to non-diagonal symmetric spaces.

CONTENTS

Supported in part by the National Science Foundation

1. Notation. Let $\mathcal{B}$ be the variety of Borel subgroups of G.

An element $x \in \mathfrak{p}$ is said to be *nilpotent* if the closure of the K-orbit of x in $\mathfrak{p}$ contains 0. Let $\mathcal{N}$ be the set of nilpotent elements in $\mathfrak{p}$ (a closed K-stable subvariety of $\mathfrak{p}$).

Let l be a prime number invertible in $\mathbf{k}$ and let $\bar{\mathbf{Q}}_l$ be an algebraic closure of $\mathbf{Q}_l$. Let $\bar{\ } : \bar{\mathbf{Q}}_l \to \bar{\mathbf{Q}}_l$ be the involution obtained by transporting complex conjugation under some field isomorphism $\bar{\mathbf{Q}}_l \to \mathbf{C}$.

We denote by $\mathbf{F}_q$ a finite field with q elements; $\bar{\mathbf{F}}_q$ denotes an algebraic closure of $\mathbf{F}_q$; $\psi : \mathbf{F}_q \to \bar{\mathbf{Q}}_l^*$ denotes a fixed nontrivial character.

For a finite set X, we denote by $|X|$ the cardinal of X.

In the remainder of this subsection we assume that $\mathbf{k} = \bar{\mathbf{F}}_q$. For an algebraic variety X over $\mathbf{k}$ let $\mathcal{D}(X)$ be the bounded derived category of constructible $\bar{\mathbf{Q}}_l$-sheaves on X. If $f : X \to \mathbf{k}$ is a morphism let $\mathcal{L}_f$ be the inverse image under f of the Artin-Schreier local system on $\mathbf{k}$ defined by f.

Let V be an n-dimensional vector space over $\mathbf{k}$ and let V^* be is its dual. Let $\langle , \rangle : V^* \times V \to \mathbf{k}$ be the canonical pairing. Then the local system $\mathcal{L}_{\langle , \rangle}$ on $V^* \times V$ is well defined. Consider the diagram $V^* \xleftarrow{a} V^* \times V \xrightarrow{b} V$ where a and b is the first and second projection. If $\mathcal{K} \in \mathcal{D}(V)$ we set $\mathfrak{F}(\mathcal{K}) = a_!(b^*(\mathcal{K}) \otimes \mathcal{L}_{\langle , \rangle})[n] \in \mathcal{D}(V^*)$ (Deligne-Fourier transform). If V is endowed with a nondegenerate symmetric bilinear form then we can use this to identify V and V^* and to regard $\mathfrak{F}(\mathcal{K})$ as an object of $\mathcal{D}(V)$.

2. Almost diagonal symmetric spaces. Let $\mathcal{Z}$ be the centre of G. Let $(\bar{G}, \bar{\theta}, \bar{K})$ be the symmetric space such that $\bar{G} = G/\mathcal{Z}$ and $\bar{\theta} : \bar{G} \to \bar{G}$ is the involution induced by θ. We say that $(\bar{G}, \bar{\theta}, \bar{K})$ is *almost diagonal* (resp. *quasi-split*) if for any involution $\bar{\theta}' : \bar{G} \to \bar{G}$ such that $\bar{\theta}, \bar{\theta}'$ induce the same involution of the Weyl group of $\bar{G}$ we have $\dim(\bar{K}) \geq \dim(\bar{K}')$ (resp. $\dim(\bar{K}) \leq \dim(\bar{K}')$); here $\bar{K}'$ is the identity component of the fixed point set of $\bar{\theta}'$. We say that (G, θ, K) is almost diagonal (resp. quasi-split) if $(\bar{G}, \bar{\theta}, \bar{K})$ is almost diagonal (resp. quasi-split).

For example, if (G, K) is diagonal then it is almost diagonal and quasi-split. In addition,

(a) (GL_{2n}, Sp_{2n}),

(b) $(SO_{2n}, SO_{2n-1}), (n \geq 2)$,

(c) (E_6, F_4)

are almost diagonal. Also, (G, G) is almost diagonal.

We say that (G, K) is of equal rank if G, K contain a common maximal torus.

3. A generalization of Schur's lemma. In this subsection we assume that $\mathbf{k} = \bar{\mathbf{F}}_q$ and that we are given an $\mathbf{F}_q$-rational structure on G compatible with θ. Then the finite groups $G(\mathbf{F}_q), K(\mathbf{F}_q)$ are well defined. For any irreducible representation ρ of $G(\mathbf{F}_q)$ over $\mathbf{C}$ let ρ_0 be the space of $K(\mathbf{F}_q)$-invariant vectors in ρ. If $(G, K) = (H \times H, H)$ is diagonal and the $\mathbf{F}_q$-structure on G comes from an $\mathbf{F}_q$-structure on H then $\dim \rho_0 \in \{0, 1\}$. Indeed we have $\rho = \rho' \boxtimes \rho''$ where ρ', ρ'' are irreducible representations of $H(\mathbf{F}_q)$ and the claim follows from Schur's lemma for $H(\mathbf{F}_q)$. If (G, K) is as in 2(a) then again $\dim \rho_0 \in \{0, 1\}$ (see [BKS]); if (G, K) is as in 2(b), 2(c) then $\dim \rho_0 \in \{0, 1\}$ (R. Lawther). For general (G, K), it is not true that $\dim \rho_0 \in \{0, 1\}$; but one can show that $\dim \rho_0 \leq c$ where c is a constant depending only on (G, K) not on q or ρ (see [L3]). Hence the algebra of double cosets $\mathbf{C}[K(\mathbf{F}_q) \backslash G(\mathbf{F}_q) / K(\mathbf{F}_q)]$ is a direct sum of matrix algebras of sizes $\leq c$.

4. Elliptic curves arising from a symmetric space. Let $\mathcal{O}, \mathcal{O}'$ be two K-orbits on $\mathcal{B}$ and let $gK \in G/K$. Following [Gi] we consider the subvariety $\mathcal{O} \cap g\mathcal{O}'$ of $\mathcal{B}$. This clearly depends only on the coset gK. (In the case where (G, K) is diagonal, the variety $\mathcal{O} \cap g\mathcal{O}'$ is a special case of a variety which appears in [L1, 2.5].) We will give an example (which I have found around 1990) where $\mathcal{O} \cap g\mathcal{O}'$ is an elliptic curve with finitely many points removed. Let V be a 3-dimensional **k**-vector space with a fixed nondegenerate symmetric bilinear form $(,)$. Let $SO(V)$ be the corresponding special orthogonal group. We take $(G, K) = (GL(V), SO(V))$. For any subspace V' of V we set $V'^{\perp} = \{x \in V; (x, V') = 0\}$. Let Q be the set of lines L in V such that $L \subset L^{\perp}$. Let Q' be the set of planes P in V such that $P^{\perp} \subset P$. We identify $\mathcal{B}$ with the set of pairs L, P where L is a line in V, P is a plane in V and $L \subset P$. There are four K-orbits on $\mathcal{B}$:

$\mathcal{O}_0 = \{L, P; L \subset P, L \in Q, P \in Q'\}$, $\mathcal{O}_1 = \{L, P; L \subset P, L \in Q, P \notin Q'\}$, $\mathcal{O}_2 = \{L, P; L \subset P, L \notin Q, P \in Q'\}$, $\mathcal{O}_3 = \{L, P; L \subset P, L \notin Q, P \notin Q'\}$.

The closure of $\mathcal{O}_1$ is $\bar{\mathcal{O}}_1 = \{L, P; L \subset P, L \in Q\}$. The closure of $\mathcal{O}_2$ is $\bar{\mathcal{O}}_2 = \{L, P; L \subset P, P \in Q'\}$. We can find $g \in GL(V)$ such that the quadrics Q, gQ in the projective plane are in general position that is $Q \cap gQ$ consists of four distinct points. Let $E = \bar{\mathcal{O}}_1 \cap g\bar{\mathcal{O}}_2 = \{L, P; L \subset P, L \in Q, P \in gQ'\}$. The non-singular surfaces $\{L, P; L \subset P, L \in Q\}$, $\{L, P; L \subset P, P \in gQ'\}$ in $\mathcal{B}$ intersect transversally hence E is a non-singular curve in $\mathcal{B}$. For any $L \in Q$ let F_L be the fibre of the first projection $E \to Q$. Now F_L may be identified with the set of tangents to the quadric gQ passing through L. It consists of two elements if L is not one of the four points in $Q \cap gQ$ and is a single element otherwise. Thus E is a double covering of the projective line Q branched at four points. Hence E is an elliptic curve. Let $E_{ij} = \bar{\mathcal{O}}_i \cap g\bar{\mathcal{O}}_j$ $(i, j \in [0, 3])$. We have

$$E = E_{00} \sqcup E_{02} \sqcup E_{10} \sqcup E_{12}.$$

We have

$$E_{00} = \{L, P; L \subset P, L \in Q \cap gQ, P \in Q' \cap gQ'\} = \emptyset.$$

(If L is one of the four elements of $Q \cap gQ$ and P is one of the four elements of $Q' \cap gQ'$ then $L \not\subset P$.) We have

$$E_{02} = \{L, P; L \subset P, L \in Q, L \notin gQ, P \in Q' \cap gQ'\}$$
$$= \{L, P; L \subset P, L \in Q, P \in Q' \cap gQ'\}.$$

Now for any of the four elements $P \in Q' \cap gQ'$ there is a unique $L \in Q$ such that $L \subset P$. Hence E_{02} consists of four points. Similarly, E_{10} consists of four points. It follows that $E_{12} = E - (E_{10} \sqcup E_{02})$ is an elliptic curve with eight points removed.

Note that if (G, K) is diagonal then the variety $\mathcal{O} \cap g\mathcal{O}'$ (with $gK \in G/K$) can never be an elliptic curve with finitely many points removed.

5. Dimension of a nilpotent orbit. If (G, K) is diagonal, any K-orbit in $\mathcal{N}$ has even dimension. The same holds if (G, K) is almost diagonal, but not in the general case.

Consider for example a **k**-vector space V of dimension $N \geq 4$ with a fixed nondegenerate symmetric bilinear form $(,)$ and let U be a codimension one subspace of V such that $(,)$ is nondegenerate when restricted to U. Let $SO(V)$, $SO(U)$ be the corresponding special orthogonal groups. We have an obvious imbedding $SO(U) \subset SO(V)$ which makes $(SO(V), SO(U))$ into a symmetric space. In this case $\mathfrak{p}$ may be identified with U with the obvious action of $SO(U)$. We have

$\mathcal{N} = \{x \in U; (x, x) = 0\}$. This is the union of two $SO(U)$-orbits $\{0\}$ and $\mathcal{N} - \{0\}$. Note that $\dim(\mathcal{N} - \{0\}) = N - 2$ is even precisely when N is even that is, precisely when $(SO(V), SO(U))$ is almost diagonal.

6. Some intersection cohomology sheaves. Let C be a K-orbit in $\mathcal{N}$. Let $IC(\bar{C}, \bar{\mathbf{Q}}_l) \in \mathcal{D}(\mathfrak{p})$ be the intersection cohomology complex of the closure $\bar{C}$ of C with coefficients in $\bar{\mathbf{Q}}_l$, extended by 0 on $\mathfrak{p} - \bar{C}$. Let $\mathcal{H}^i IC(\bar{C}, \bar{\mathbf{Q}}_l)$ be the i-th cohomology sheaf of $IC(\bar{C}, \bar{\mathbf{Q}}_l)$ and let $\mathcal{H}^i_x IC(\bar{C}, \bar{\mathbf{Q}}_l)$ be its stalk at $x \in \mathfrak{p}$.

Assume for example that $(G, K) = (SO(V), SO(U))$, with V, U, N as in §5. Let $C = \mathcal{N} - \{0\}$. Then $\bar{C} = \mathcal{N}$. Moreover:

(a) If N is even ≥ 4 then $IC(\bar{C}, \bar{\mathbf{Q}}_l) = \bar{\mathbf{Q}}_l$.

(b) If N is odd ≥ 3 then $\mathcal{H}^0 IC(\bar{C}, \bar{\mathbf{Q}}_l) = \bar{\mathbf{Q}}_l$, $\dim \mathcal{H}^{N-3}_x IC(\bar{C}, \bar{\mathbf{Q}}_l)$ is 1 if $x = 0$ and is 0 if $x \in \mathfrak{p} - \{0\}$; $\mathcal{H}^i IC(\bar{C}, \bar{\mathbf{Q}}_l) = 0$ if $i \neq 0, N - 3$.

Now let $(G, K) = (GL(V' \times V''), GL(V') \times GL(V''))$ where V', V'' are $\mathbf{k}$-vector spaces of dimension 2. In this case we may identify $\mathfrak{p}$ with $\operatorname{Hom}(V', V'') \times \operatorname{Hom}(V'', V')$ with the obvious action of K. Let $C_1 \subset \mathfrak{p}$ be the set of pairs A, B where $A \in \operatorname{Hom}(V', V'')$, $B \in \operatorname{Hom}(V'', V')$ are such that $\ker A = B(V'')$ is 1-dimensional and $\ker B = A(V')$ is one dimensional. Note that C_1 is a K-orbit in $\mathcal{N}$ of dimension 4. Moreover $\bar{C}_1$ is the set of pairs A, B where $A \in \operatorname{Hom}(V', V'')$, $B \in \operatorname{Hom}(V'', V')$ are such that A and B are singular and $AB = 0, BA = 0$. We have

(c) $\mathcal{H}^0 IC(\bar{C}_1, \bar{\mathbf{Q}}_l) = \bar{\mathbf{Q}}_l$, $\dim \mathcal{H}^2_x IC(\bar{C}_1, \bar{\mathbf{Q}}_l)$ is 2 if $x = 0$ and is 0 if $x \in \mathfrak{p} - \{0\}$; $\mathcal{H}^i IC(\bar{C}_1, \bar{\mathbf{Q}}_l) = 0$ if $i \neq 0, 2$.

7. Generalization of a theorem of Steinberg. In this subsection we assume that $\mathbf{k} = \bar{\mathbf{F}}_q$. Assume also that we are given an $\mathbf{F}_q$-rational structure on G compatible with θ. Then $\mathfrak{p}$ and $\mathcal{N}$ have a natural $\mathbf{F}_q$-structure. Steinberg [S] has shown that if (G, K) is diagonal, then the number of elements in $\mathcal{N}(\mathbf{F}_q)$ is an even power of q. The same holds if (G, K) is almost diagonal, but not in the general case.

For example, if (G, K) is as in 2(a) then $|\mathcal{N}(\mathbf{F}_q)| = q^{2n^2 - 2n}$. (Assume for simplicity that G is $\mathbf{F}_q$-split. The $K(\mathbf{F}_q)$-orbits in $\mathcal{N}(\mathbf{F}_q)$ are indexed by the partitions of n in such a way that the stabilizer in $K(\mathbf{F}_q)$ of an element in $\mathcal{N}(\mathbf{F}_q)$ has cardinal equal to the cardinal of the stabilizer in $GL_n(\mathbf{F}_q)$ of a unipotent element of the corresponding type (with q replaced by q^2), see [BKS]. Hence the desired equality follows from the corresponding equality in the diagonal case.)

If $V, U, (,), N$ are as in §5 with $V, U, ()$ defined over $\mathbf{F}_q$ and $(G, K) = (SO(V), SO(U))$ then $|\mathcal{N}(\mathbf{F}_q)|$ is equal to q^{N-2} if N is even and to $q^{N-2} \pm (q^{(N-1)/2} - q^{(N-3)/2})$ if N is odd. Note that $(SO(V), SO(U))$ is almost diagonal precisely when N is even.

8. $\mathfrak{F}$-thin, $\mathfrak{F}$-thick nilpotent orbits. In this subsection we assume that $\mathbf{k} = \bar{\mathbf{F}}_q$. Let C be a K-orbit in $\mathcal{N}$. Then the Deligne-Fourier transform $\mathfrak{F}(IC(\bar{C}, \bar{\mathbf{Q}}_l)) \in \mathcal{D}(\mathfrak{p})$ is up to shift a simple perverse sheaf on $\mathfrak{p}$. (We can choose a K-invariant nondegenerate symmetric bilinear form on $\mathfrak{p}$.)

We say that C is $\mathfrak{F}$-thin if the support of $\mathfrak{F}(IC(\bar{C}, \bar{\mathbf{Q}}_l))$ is contained in $\mathcal{N}$ (hence is the closure of a single K-orbit in $\mathcal{N}$, by the finiteness of the number of K-orbits in $\mathcal{N}$). We say that C is $\mathfrak{F}$-thick if the support of $\mathfrak{F}(IC(\bar{C}, \bar{\mathbf{Q}}_l))$ is equal to $\mathfrak{p}$ (hence $\mathfrak{F}(IC(\bar{C}, \bar{\mathbf{Q}}_l))$ restricted to some open dense subset of $\mathfrak{p}$ is up to shift an irreducible local system).

Note that the K-orbit 0 is always $\mathfrak{F}$-thick. The statement that

"any K-orbit in $\mathcal{N}$ is $\mathfrak{F}$-thick"
is true if (G,K) is diagonal. It is also true if (G,K) is as in 2(a) (see §15) and if (G,K) is as in 2(b) (in this last case this follows from the computations in §12). It is however false for general (G,K).

For example if $(G,K) = (SO(V), SO(U))$ with V, U, N as in §5, N odd, and if $C = \mathcal{N} - \{0\}$ then C is not $\mathfrak{F}$-thick but $\mathfrak{F}$-thin (this follows from the computations in §12).

If $(G,K) = (GL(V' \times V''), GL(V') \times GL(V''))$ and C_1 are as in §6 then C_1 is $\mathfrak{F}$-thick (this follows from the computations in §13) but no K-orbit in $\mathcal{N}$ other than 0 and C_1 is $\mathfrak{F}$-thick.

From these and other examples it appears that, $\mathfrak{F}$-thin K-orbits in $\mathcal{N}$ can exist only if (G,K) is of equal rank.

9. $\mathfrak{F}$-thin nilpotent orbits and affine canonical bases. In this subsection we assume that $(G,K) = (GL(V' \times V''), GL(V') \times GL(V''))$ where V', V'' are **k**-vector spaces of dimension N', N''. In this case we may identify $\mathfrak{p}$ with $E := \mathrm{Hom}(V', V'') \times \mathrm{Hom}(V'', V')$ with the obvious action of K.

Now $\mathcal{N}$ consists of all $(A, B) \in E$ such that $AB : V'' \to V''$ and $BA : V' \to V'$ are nilpotent. We describe the classification of K-orbits in $\mathcal{N}$. This is the same as the (known) classification of isomorphism classes of nilpotent representations of fixed degree of a cyclic quiver with 2 vertices. Let $\mathcal{P} = \{1, -1\} \times \mathbf{Z}_{>0}$. For $(r, m) \in \mathcal{P}$ we define $g_{r,m} \in \mathbf{N} \times \mathbf{N}$ by $g_{r,m} = (m/2, m/2)$ if m is even, $g_{r,m} = ((m+r)/2, (m-r)/2)$ if m is odd. Let $\tilde{\mathcal{P}} = \tilde{\mathcal{P}}_{N',N''}$ be the set of maps $\sigma : \mathcal{P} \to \mathbf{N}$ such that σ has finite support and $\sum_{(r,m)\in\mathcal{P}} \sigma(r,m) g_{r,m} = (N', N'')$. For $\sigma \in \tilde{\mathcal{P}}$ let $\mathcal{N}_\sigma$ be the set of all $(A, B) \in E$ with the following property: there exists direct sum decompositions $V' = \oplus_{i=1}^s V_i'$, $V'' = \oplus_{i=1}^s V_i''$ such that
$AV_i' \subset V_i'', BV_i'' \subset V_i', \dim(V_i' \oplus V_i'') = m_i$;
the linear map $T_i : V_i' \oplus V_i'' \to V_i' \oplus V_i''$, $(\xi', \xi'') \mapsto (B\xi'', A\xi')$ is nilpotent with a single Jordan block; set $r_i = 1$ if $T_i^{m_i-1} V' \neq 0$, $r_i = -1$ if $T_i^{m_i-1} V'' \neq 0$;
for any $(r, m) \in \mathcal{P}$, the number of $i \in [1, s]$ such that $(r_i, m_i) = (r, m)$ is equal to $\sigma(r, m)$.
Then $(\mathcal{N}_\sigma)_{\sigma\in\tilde{\mathcal{P}}}$ are precisely the K-orbits on $\mathcal{N}$.

Let $\tilde{\mathcal{P}}^0$ be the set of all $\sigma \in \tilde{\mathcal{P}}$ such that for any $m \geq 1$ we have either $\sigma(1, m) = 0$ or $\sigma(-1, m) = 0$.

In the remainder of this subsection we assume that $\mathbf{k} = \bar{\mathbf{F}}_q$. We show:

(a) *If $\sigma \in \tilde{\mathcal{P}}^0$ then the K-orbit $\mathcal{N}_\sigma$ is $\mathfrak{F}$-thin.*
The proof is based on the theory of (affine) canonical bases. From [L4, 5.9] we see that $\mathcal{K} = IC(\bar{\mathcal{N}}_\sigma, \bar{\mathbf{Q}}_l)$ is (up to shift) an element of the canonical basis attached to the cyclic quiver with 2 vertices. Using [L5, 10.2.3] we see that $\mathfrak{F}\mathcal{K}$ is (up to shift) also an element of the canonical basis attached to the same cyclic quiver with the opposite orientation. Using again [L4, 5.9] we see that $\mathfrak{F}\mathcal{K} = IC(\bar{\mathcal{N}}_{\sigma'}, \bar{\mathbf{Q}}_l)$ (up to shift) for a well defined $\sigma' \in \tilde{\mathcal{P}}$ (we use an isomorphism of the cyclic quiver with the one with opposed orientation). In particular the support of $\mathfrak{F}\mathcal{K}$ is contained in $\mathcal{N}$. This proves (a).

We will show elsewhere that, conversely, if $\sigma \in \tilde{\mathcal{P}}$ and $\mathcal{N}_\sigma$ is $\mathfrak{F}$-thin, then $\sigma \in \tilde{\mathcal{P}}^0$.

10. Character sheaves on $\mathfrak{p}$. In this subsection we assume that $\mathbf{k} = \bar{\mathbf{F}}_q$.
A simple perverse sheaf on $\mathfrak{p}$ is said to be *orbital* if it is K-equivariant and its support is the closure of a single K-orbit in $\mathcal{N}$. A simple perverse sheaf on $\mathfrak{p}$ is said

to be *anti-orbital* (or a *character sheaf*) if it is of the form $\mathfrak{F}(A)$ for some orbital simple perverse sheaf A on $\mathfrak{p}$. (These two definitions are identical to the definitions of an orbital or anti-orbital complex on $\mathfrak{g}$ given in [L2].) For example, if C is an $\mathfrak{F}$-thin nilpotent K-orbit in $\mathfrak{p}$ then $IC(\bar{C}, \bar{\mathbf{Q}}_l)$ is up to shift both an orbital and an antiorbital simple perverse sheaf.

11. Computation of a Fourier transform. We set $k = \mathbf{F}_q$. Let U be a k-vector space of dimension $M \geq 3$ with a fixed nondegenerate symmetric bilinear form $(,) : U \times U \to k$. When M is even we set $\delta = 1$ if $(,)$ is split over k and $\delta = -1$ otherwise. When M is odd we set $\delta = 0$. Define $f : U \to \bar{\mathbf{Q}}_l$ by $f(x) = 0$ if $(x, x) \neq 0$, $f(x) = 1$ if $(x, x) = 0, x \neq 0$, $f(x) = 1 + \delta q^{(M-2)/2}$ if $x = 0$.

Define $\hat{f} : U \to \bar{\mathbf{Q}}_l$ by $\hat{f}(x) = q^{-M/2} \sum_{y \in U} \psi((x, y)) f(y)$ for $x \in U$ (a Fourier transform). When M is odd we define a function $\zeta : \{x \in U; (x, x) \neq 0\} \to \{1, -1\}$ by $\zeta(x) = 1$ if $(,)$ is split on te subspace $\{x' \in U; (x, x') = 0\}$ and $\zeta(x) = -1$ otherwise. We show:

(a) For M even we have $\hat{f} = f$.

(b) For M odd we have $\hat{f}(x) = \zeta(x)$ if $(x, x) \neq 0$, $\hat{f}(x) = 0$ if $(x, x) = 0, x \neq 0$, $\hat{f}(x) = q^{(M-2)/2}$ if $x = 0$.

We have

$$\hat{f}(0) = q^{-M/2} |\{y \in U; (y, y) = 0\}| + \delta q^{-1}$$
$$= q^{-M/2}(q^{M-1} + \delta(q^{M/2} - q^{(M-2)/2})) + \delta q^{-1} = q^{(M-2)/2} + \delta.$$

Now assume that $x \in U - \{0\}, (x, x) = 0$. We can find $x' \in U$ such that $(x', x') = 0$, $(x, x') = 1$. Let $U' = \{z \in U; (z, x) = 0, (z, x') = 0\}$. Any $y \in U$ such that $(y, y) = 0$ can be written uniquely in the form $y = ax + bx' + z$ where $a, b \in k, z \in U'$ and $2ab + (z, z) = 0$. We have

(c) $$\hat{f}(x) = q^{-M/2} \sum_{a, b \in k, z \in U'; 2ab+(z,z)=0} \psi(b) + \delta q^{-1}.$$

If $b \neq 0$ and $z \in U'$ then a in the last sum is determined by $a = -(z, z)/(2b)$. Hence

$$\hat{f}(x) = q^{-M/2} \sum_{b \in k^*} |U'| \psi(b) + q^{-M/2} \sum_{a \in k, z \in U'; (z,z)=0} 1 + \delta q^{-1}$$
$$= -q^{(M-4)/2} + q^{-(M-2)/2}(q^{M-3} + \delta(q^{(M-2)/2} - q^{(M-4)/2})) + \delta q^{-1} = \delta.$$

Since $f \mapsto \hat{f}$ is an isometry, we have

$$\sum_{x \in U; (x,x) \neq 0} \hat{f}(x) \overline{\hat{f}(x)} = \sum_{x \in U} f(x) \overline{f(x)} - \sum_{x \in U; (x,x)=0} \hat{f}(x) \overline{\hat{f}(x)}$$
$$= |\{x \in U; (x, x) = 0, x \neq 0\}| + (1 + \delta q^{(M-2)/2})^2$$
$$- |\{x \in U; (x, x) = 0, x \neq 0\}| \delta^2 - (q^{(M-2)/2} + \delta)^2.$$

This is 0 if M is even hence in this case $\hat{f}(x) = 0$ for any $x \in U$ such that $(x, x) \neq 0$, proving (a).

In the remainder of this subsection we assume that M is odd. Since f is invariant under the $O(U) \times k^*$-action on V (k^* acts by homothety) the same holds for $\hat{f}$. Hence there exists $\lambda, \mu \in \bar{\mathbf{Q}}_l$ such that $\hat{f}(x) = \lambda$ if $x \in \zeta^{-1}(1)$, $\hat{f}(x) = \mu$ if $x \in \zeta^{-1}(-1)$. We fix a 2-dimensional subspace R of U such that $(,)$ is nondegenerate, split on R. Let $U' = \{x \in U; (x, R) = 0\}$. For any $c \in k^*$ we can find $x, x' \in R$ such that $(x, x) = c$, $(x', x') = 0$, $(x, x') = 1$. Any $y \in U$ such that $(y, y) = 0$ can be written uniquely in the form $y = ax + bx' + z$ where $a, b \in k$, $z \in U'$ and $a^2 c + 2ab + (z, z) = 0$. We have

$$\hat{f}(x) = \sum_{a,b\in k, z\in U'; a^2c+2ab+(z,z)=0} \psi(ac + b).$$

Thus $\hat{f}(x)$ does not depend on x but only on c (and U'). We denote it by $\phi(c)$. If $a \neq 0$ and $z \in U'$ then b in the last sum is determined uniquely by $b = -(a^2 c + (z, z))/(2a)$. Hence

$$\phi(c) = \sum_{a\in k^*, z\in U'} \psi(ac - (a^2c + (z, z))/(2a)) + \sum_{b\in k, z\in U'; (z,z)=0} \psi(b)$$

$$= \sum_{a\in k^*, z\in U'} \psi((ac - (z, z)a^{-1})/2).$$

We compute

$$\sum_{c\in k^*} \phi(c) = \sum_{z\in U'} \sum_{a\in k^*} (\psi(-(z, z)a^{-1})/2) \sum_{c\in k^*} \psi(ac/2))$$

$$= -\sum_{z\in U'} \sum_{a\in k^*} \psi(-(z, z)a^{-1})/2) =$$

$$- \sum_{z\in U'; (z,z)\neq 0} (-1) - \sum_{z\in U'; (z,z)=0} (q - 1) = (q^M - q^{M-1}) - q^{M-1}(q - 1) = 0.$$

Note also that $\phi(c)$ depends only on the image of c in k^*/k^{*2}. The last sequence of equalities shows that $\phi(c) = -\phi(c')$ if $c, c' \in k^*$, $c'/c \notin k^{*2}$. It follows that $\lambda = -\mu$.

By a property of Fourier transform we have $\sum_{x\in U} \hat{f}(x) = q^{M/2} f(0)$ hence $\sum_{x\in U; (x,x)\neq 0} \hat{f}(x) = q^{M/2} - q^{(M-2)/2}$ that is

$$\lambda|\zeta^{-1}(1)| + \mu|\zeta^{-1}(-1)| = |\zeta^{-1}(1)| - |\zeta^{-1}(-1)|.$$

Since $\lambda = -\mu$ and $|\zeta^{-1}(1)| - |\zeta^{-1}(-1)| \neq 0$ it follows that $\lambda = 1, \mu = -1$. This proves (b).

12. Another computation of a Fourier transform. We set $k = \mathbf{F}_q$. Let V', V'' be two k-vector spaces of dimension 2. Let $E = \mathrm{Hom}(V', V'') \times \mathrm{Hom}(V'', V')$. Let E° be the set of all $(A; B) \in E$ such that A and B are singular and $AB = 0$, $BA = 0$. (We write $(A; B)$ instead of (A, B) to avoid confusion with an inner product.) Define a nondegenerate symmetric bilinear form $(,) : E \times E \to k$ by $((A; B), (\tilde{A}; \tilde{B})) = \mathrm{tr}(A\tilde{B}, V'') + \mathrm{tr}(B\tilde{A}, V')$. Define $f : E \to \bar{\mathbf{Q}}_l$ by $f(A; B) = 1$ if $(A; B) \in E^\circ - \{0; 0\}$, $f(0; 0) = 1 + 2q$, $f(A; B) = 0$ if $(A; B) \in E - E^\circ$. Define $\hat{f} : E \to \bar{\mathbf{Q}}_l$ by

$$\hat{f}(\tilde{A}; \tilde{B}) = q^{-4} \sum_{(A;B)\in E} \psi(((A; B), (\tilde{A}; \tilde{B})))f(A; B).$$

(Fourier transform).

Let E_{rs} be the set of all $(A; B) \in E$ such that A and B are nonsingular and $AB : V'' \to V''$, $BA : V' \to V'$ are regular semisimple. Define $\zeta : E_{rs} \to \bar{\mathbf{Q}}_l$ by $\zeta(A : B) = 1$ if the two eigenvalues of AB (or BA) are in F_q^*; $\zeta(A : B) = -1$ if the two eigenvalues of AB (or BA) are not in F_q^*. We show that for any $(A; B) \in E_{rs}$ we have

(a)
$$\hat{f}(A; B) = q^{-2}\zeta(A; B).$$

For $(A : B) \in E$ let $f_1(A : B)$ be the number of $(L'; L'')$ where L' is a line in V', L'' is a line in V'' and $BV'' \subset L' \subset \ker A$, $AV' \subset L'' \subset \ker B$. This defines a function $f_1 : E \to \bar{\mathbf{Q}}_l$. Note that $f_1(A; B) = f(A; B)$ if $(A; B) \neq (0; 0)$ and $f_1(0; 0) = (q + 1)^2 = f(0; 0) + q^2$. Let $\hat{f}_1 : E \to \bar{\mathbf{Q}}_l$ be the Fourier transform of f_1. For $(\tilde{A}; \tilde{B}) \in E$ we have

$$\hat{f}_1(\tilde{A}; \tilde{B}) = q^{-4} \sum_{L', L''} \sum_{\substack{(A;B); \\ AV' \subset L'', AL'=0, \\ BV'' \subset L', BL''=0}} \psi(\mathrm{tr}(A\tilde{B}, V'') + \mathrm{tr}(B\tilde{A}, V'))$$

where L' runs through the lines in V' and L'' runs through the lines in V''. For fixed L', L'', the set of $(A; B)$ in the last sum is a k-vector space of dimenion 2 and $(A, B) \mapsto \mathrm{tr}(A\tilde{B}, V'') + \mathrm{tr}(B\tilde{A}, V')$ is a linear form on this vector space which is zero if $\tilde{A}L' \subset L''$, $\tilde{B}L'' \subset L'$ and is zero otherwise. Thus

$$\hat{f}_1(\tilde{A}; \tilde{B}) = q^{-2}|\{L', L''; \tilde{A}L' \subset L'', \tilde{B}L'' \subset L'\}|.$$

Assume now that $(\tilde{A}; \tilde{B}) \in E_{rs}$. The condition that $\tilde{A}L' \subset L''$, $\tilde{B}L'' \subset L'$ is equivalent to $\tilde{B}\tilde{A}L' = L'$, $\tilde{A}L' = L''$. Hence

$$\hat{f}_1(\tilde{A}; \tilde{B}) = q^{-2}|\{L'; \tilde{B}\tilde{A}L' = L'\}| = q^{-2}(\zeta(\tilde{A}; \tilde{B}) + 1).$$

We have
$$\hat{f}(\tilde{A}; \tilde{B}) = \hat{f}_1(\tilde{A}; \tilde{B}) - q^{-2} = q^{-2}\zeta(\tilde{A}; \tilde{B})$$

and (a) is proved.

13. Computation of a Deligne-Fourier transform.
The results in this subsection were found around 1990.

Let V be a $\mathbf{k}$-vector space of dimension $2n$ with a fixed nondegenerate symplectic form $\langle,\rangle : V \times V \to \mathbf{k}$. Let

$$E = \{T \in \mathrm{End}(V); \langle T(x), y\rangle = \langle x, T(y)\rangle \quad \forall x, y \in V\}.$$

The non-degenerate symmetric bilinear form $(,) : \mathrm{End}(V) \times \mathrm{End}(V) \to \mathbf{k}$ given by $T, T' \mapsto \mathrm{tr}(TT')$ remains nondegenerate when restricted to E. The symplectic group $Sp(V)$ acts naturally on E, preserving $(,)$. Let E' be the set of all $T \in E$ such that $T : V \to V$ is semisimple and any eigenspace of T is 2-dimensional. Note that E' is open dense in E. Let E_0 be the set of all $T \in E$ such that $T : V \to V$ is nilpotent. Note that E_0 is $Sp(V)$-stable and the set of $Sp(V)$-orbits on E_0 is in natural bijection with the set of partitions of n. (Any element of E_0 has Jordan blocks of sizes $n_1, n_1, n_2, n_2, \ldots, n_t, n_t$ where $n_1 \geq n_2 \geq \cdots \geq n_t$ is a partition of n.)

Let F be the set of all flags $V_* = (V_0 \subset V_1 \subset V_2 \ldots \subset V_{2n})$ in V such that $\dim V_i = i$, $V_{2n-i} = \{x \in V; \langle x, V_i \rangle = 0\}$ for all $i \in [0, 2n]$. For $T \in E, V_* \in F$ we write $T \dashv V_*$ instead of $TV_i \subset V_i$ for all i.

Let $V_* \in F$. Let $E^{V_*} = \{T \in E; T \dashv V_*\}$. Let $E_0^{V_*} = \{T \in E_0; T \dashv V_*\}$. We have:

(a) E^{V_*} is exactly the orthogonal of $E_0^{V_*}$ with respect to $(,) : E \times E \to \mathbf{k}$.

If $T \in E^{V_*}$, $T' \in E_0^{V_*}$ then $TV_i \subset V_i$, $T'V_i \subset V_{i-1}$ for all $i \geq 1$. Hence $T'TV_i \subset V_{i-1}$ for all i so that $\operatorname{tr}(T'T) = 0$ and $(T, T') = 0$. Thus E^{V_*} is contained in the orthogonal of $E_0^{V_*}$ with respect to $(,) : E \times E \to \mathbf{k}$. From the definitions we see that

$$\dim E^{V_*} + \dim E_0^{V_*} = n^2 + (n^2 - n) = 2n^2 - n = \dim E$$

and (a) follows.

Let $\tilde{E} = \{(T, V_*) \in E \times F; T \dashv V_*\}$. Define $\pi : \tilde{E} \to E$ by $(T, V_*) \mapsto T$. Let $\mathcal{K} = \pi_! \bar{\mathbf{Q}}_l \in \mathcal{D}(E)$. Let $\tilde{E}_0 = \{(T, V_*) \in E_0 \times F; T \dashv V_*\}$. Define $\pi_0 : \tilde{E}_0 \to E$ by $(T, V_*) \mapsto T$. Let $\mathcal{K}_0 = \pi_{0!} \bar{\mathbf{Q}}_l \in \mathcal{D}(E)$.

In the remainder of this subsection we assume that $\mathbf{k} = \bar{\mathbf{F}}_q$. We show:

(b)
$$\mathfrak{F}(\mathcal{K}_0) \cong \mathcal{K}[n].$$

Consider the diagram $E \xleftarrow{a} E \times E \xrightarrow{b} E$ where a (resp. b) is the first (resp. second) projection. By definition we have

$$\mathfrak{F}(\mathcal{K}_0) = a_!(b^*(\pi_{0!} \bar{\mathbf{Q}}_l) \otimes \mathcal{L}_{(,)})[2n^2 - n].$$

We have $b^*(\pi_{0!} \bar{\mathbf{Q}}_l) = \rho_!(\bar{\mathbf{Q}}_l)$ where $\rho : E \times \tilde{E}_0 \to E \times E$ is given by $(T; (T', V_*)) \mapsto (T; T')$ and

$$\mathfrak{F}(\mathcal{K}_0) = a_!(\rho_!(\mathcal{L}_f))[2n^2 - n] = \tilde{\rho}_! \mathcal{L}_f [2n^2 - n]$$

where $\tilde{\rho} = a\rho : E \times \tilde{E}_0 \to E$ and $f : E \times \tilde{E}_0 \to \mathbf{k}$ is given by $(T; (T', V_*)) \mapsto (T, T')$. We have a partition $E \times \tilde{E}_0 = Z' \sqcup Z''$ where $Z' = \{(T; (T', V_*)) \in E \times \tilde{E}_0; T \dashv V_*\}$, $Z'' = \{(T; (T', V_*)) \in E \times \tilde{E}_0; T \not\dashv V_*\}$. Let $\tilde{\rho}' : Z' \to E$, $\tilde{\rho}'' : Z'' \to E$ be the restrictions of $\tilde{\rho}$; let $f' : Z' \to \mathbf{k}$, $f'' : Z'' \to E \times E$ be the restrictions of f. We have a distinguished triangle

$$(\tilde{\rho}''_! \mathcal{L}_{f''}, \tilde{\rho}_! \mathcal{L}_f, \tilde{\rho}'_! \mathcal{L}_{f'})$$

in $\mathcal{D}(E)$. We show that $\tilde{\rho}''_! \mathcal{L}_{f''} = 0$. For each $T \in E$ the fibre Z''_T of $\tilde{\rho}''$ at T may be identified with $\{(T', V_*) \in \tilde{E}_0; T \not\dashv V_*\}$ and it is enough to show that $H_c^i(Z''_T; \mathcal{L}_{f''}) = 0$ for all $T \in E$ and all i. We can map Z''_T to $\{V_* \in F; T \not\dashv V_*\}$ by $(T', V_*) \mapsto V_*$. For each $T \in E$ and $V_* \in F$ such that $T \not\dashv V_*$ let Z''_{T,V_*} be the fibre of the last map at V_*. It is enough to show that $H_c^i(Z''_{T,V_*}; \mathcal{L}_{f''}) = 0$ for all $T \in E$, $V_* \in F$ such that $T \not\dashv V_*$ and all i. This follows from [L5, 8.1.13] since Z''_{T,V_*} is a $\mathbf{k}$-vector space and the restriction of f'' to this vector space (that is $T' \mapsto (T, T')$) is a nonzero linear form (see (a)).

Thus we have $\tilde{\rho}''_! \mathcal{L}_{f''} = 0$. From the distinguished triangle above it follows that $\tilde{\rho}_! \mathcal{L}_f = \tilde{\rho}'_! \mathcal{L}_{f'}$. From (a) we see that f' is identically zero. Hence $\mathcal{L}_{f'} = \bar{\mathbf{Q}}_l$. Moreover $\tilde{\rho}'$ factors as $Z' \xrightarrow{s} \tilde{E} \xrightarrow{\pi} E$ where $s(T; (T', V_*)) = (T, V_*)$. Thus $\tilde{\rho}'_! \mathcal{L}_f =$

$\pi_! s_! \bar{\mathbf{Q}}_l$. Note that s is a vector bundle; its fibre over (T, V_*) may be identified with $\{T' \in E_0; T' \dashv V_*\}$, a vector space of dimension $n^2 - n$. Hence $s_! \bar{\mathbf{Q}}_l = \bar{\mathbf{Q}}_l[-2n^2 + 2n]$ (we ignore the Tate twist). This proves (b).

Let $\mathcal{A}_0$ be the set of isomorphism classes of simple perverse sheaves on E which appear (possibly with a shift) as a direct summand of $\mathcal{K}_0$. Let $\mathcal{A}$ be the set of isomorphism classes of simple perverse sheaves on E which appear (possibly with a shift) as a direct summand of $\mathcal{K}$. From (b) we see that

(c) $\mathfrak{F}$ defines a bijection $\mathcal{A}_0 \xrightarrow{\sim} \mathcal{A}$.

Let $\mathcal{A}_0'$ be the set of isomorphism classes of $Sp(V)$-equivariant simple perverse sheaves on E with support contained in E_0. Clearly

(d) $\mathcal{A}_0 \subset \mathcal{A}_0'$.

Let $\pi' : \pi^{-1}(E') \to E'$ be the restriction of π. Let $\mathcal{A}'$ be the set of isomorphism classes of simple perverse sheaves on E which appear (possibly with a shift) as a direct summand of $\pi'_! \bar{\mathbf{Q}}_l$. Clearly we have a natural injective map

(e) $\mathcal{A}' \to \mathcal{A}$.

From (c),(d),(e) we see that

(f) $|\mathcal{A}'| \leq |\mathcal{A}| = |\mathcal{A}_0| \leq |\mathcal{A}_0'|$.

Now π' is a composition $\pi^{-1}(E') \xrightarrow{\tau} E'' \xrightarrow{\sigma} E'$ where E'' is the set of pairs (T, ω) where $T \in E'$ and ω is an indexing $E_{\omega(1)}, E_{\omega(2)}, \ldots, E_{\omega(n)}$ of the eigenspaces of T by $[1, n]$; σ is defined by $(T, \omega) \mapsto T$; τ is given by $(T, V_*) \mapsto (T, \omega)$ with ω defined as follows: for $i \in [1, n]$ we have $V_i = L_1 + L_2 + \cdots + L_i$ where $L_1, L_2, \ldots L_i$ are lines in $E_{\omega(1)}, E_{\omega(2)}, \ldots, E_{\omega(i)}$ respectively. Note that τ is a fibre bundle with fibres isomorphic to a product of n projective lines. Moreover σ is a finite principal covering with group S_n, the symmetric group in n letters. It follows that $\tau_!(\bar{\mathbf{Q}}_l)$ is a direct sum of shifts of $\bar{\mathbf{Q}}_l$. Hence the objects of $\mathcal{A}'$ are (up to shift) the same as the direct summands of $\sigma_! \bar{\mathbf{Q}}_l$. Now E'' is irreducible (since E' is so). It follows that the direct summands of $\sigma_! \bar{\mathbf{Q}}_l$ (up to shift and isomorphism) are in natural bijection with the irreducible representations of S_n (up to isomorphism). Thus, $|\mathcal{A}'| = p_n$ the number of partitions of n. Now the objects of $\mathcal{A}_0'$ are in bijection with the $Sp(V)$-orbits on E_0 (the stabilizer in $Sp(V)$ of an element in E_0 is connected). Hence $|\mathcal{A}_0'| = p_n$. Using now (f) we see that $|\mathcal{A}'| = |\mathcal{A}| = |\mathcal{A}_0| = |\mathcal{A}_0'| = p_n$. In particular we see that any object in $\mathcal{A}_0'$ is in $\mathcal{A}_0$, any object in $\mathcal{A}$ has support equal to E and $\mathfrak{F}$ defines a bijection between $\mathcal{A}_0'$ and $\mathcal{A}$. Thus,

(g) if A is an $Sp(V)$-equivariant simple perverse sheaf on E with support contained in E_0, then $\mathfrak{F}(A)$ has support equal to E.

14. Nilpotent K-orbits and conjugacy classes in a Weyl group. In this subsection we assume that $p = 1$. We show how the construction of [KL, 9.1] extends to the case of symmetric spaces.

We identify $\mathfrak{p}$ with $\{x \in \mathfrak{g}; \theta(x) = -x\}$ in the obvious way. Let $\mathcal{T}$ be the set of subspaces $\mathfrak{t}$ of $\mathfrak{p}$ such that for some torus T in G the Lie algebra of T equals $\mathfrak{t}$ and such that $\mathfrak{t}$ has maximum possible dimension. It is known that K acts transitively on $\mathcal{T}$ by conjugation. For $\mathfrak{t} \in \mathcal{T}$ let $\mathcal{W}$ be the group of components of the normalizer of $\mathfrak{t}$ in K. Let $\underline{\mathcal{W}}$ be the set of conjugay classes in the finite group $\mathcal{W}$; this is independent of the choice of $\mathfrak{t}$.

Let $\Phi = \mathbf{k}((\epsilon))$, $\phi = \mathbf{k}[[\epsilon]]$ where ϵ is an indeterminate. Let $\mathfrak{g}_\Phi = \Phi \otimes \mathfrak{p}$, $\mathfrak{p}_\Phi = \Phi \otimes \mathfrak{p}$, $\mathfrak{p}_\phi = \phi \otimes \mathfrak{p}$. Also, the groups $G(\Phi), K(\Phi)$ are well defined and the set $\mathcal{T}(\Phi)$ of Φ-points of $\mathcal{T}$ is well defined (it is a set of subspaces of $\mathfrak{p}_\Phi$).

The group $K(\Phi)$ acts naturally by conjugation on $\mathcal{T}(\Phi)$; as in [KL, §1, Lemma

2] we see that the set of $K(\Phi)$-orbits on $\mathcal{T}(\Phi)$ is naturally in $1-1$ correspondence with the set $\underline{\mathcal{W}}$. For $\gamma \in \underline{\mathcal{W}}$ let $\mathcal{O}_\gamma$ be the $K(\Phi)$-orbit on $\mathcal{T}(\Phi)$ corresponding to γ.

Let $\mathfrak{p}_{\Phi,rs}$ be the set of all $\xi \in \mathfrak{p}_\Phi$ for which there is a unique $\mathfrak{t}' \in \mathcal{T}(\Phi)$ such that $\xi \in \mathfrak{t}'$; we then set $\mathfrak{t}'_\xi = \mathfrak{t}'$. For $\gamma \in \underline{\mathcal{W}}$ let $\mathfrak{p}_{\Phi,rs,\gamma}$ be the set of all $\xi \in \mathfrak{p}_{\Phi,rs}$ such that $\mathfrak{t}'_\xi \in \mathcal{O}_\gamma$.

Let $x \in \mathcal{N}$. As in [KL, 9.1] there is a unique element $\gamma \in \underline{\mathcal{W}}$ such that the following holds: there exists a "Zariski open dense" subset V of $x + \epsilon \mathfrak{p}_\phi$ (a subset of $\mathfrak{p}_\phi$) such that $V \subset \mathfrak{p}_{\Phi,rs,\gamma}$. We set $\Psi(x) = \gamma$. Now Ψ is a map $\mathcal{N} \to \underline{\mathcal{W}}$. This map is constant on K-orbits hence it defines a map from the set of K-orbits on $\mathcal{N}$ to $\underline{\mathcal{W}}$ denoted again by Ψ. (In the case where $(G,K) = (H \times H, H)$ is diagonal, Ψ reduces to the map defined in [KL, 9.1].)

If (G,K) is diagonal then Ψ is expected to be injective (this is known to be true in almost all cases). If (G,K) is as in 2(a) then Ψ is bijective (see §15). If (G,K) is as in 2(b) then Ψ is bijective (see §16). For general (G,K), Ψ is neither injective nor surjective (see §17).

15. Example: (GL_{2n}, Sp_{2n}). In this subsection we assume that $(G,K) = (GL(V), Sp(V))$ with $\mathbf{k} = \mathbf{C}$ and we keep the notation of §13. Let Φ, ϕ be as in §14. Let $\bar{\Phi}$ be an algebraic closure of Φ. Define $v : \bar{\Phi} \to \mathbf{Q} \cup \{\infty\}$ by $v(0) = \infty$ and $v(a_0 v^f + \text{higher powers of } v) = f$ if $a_0 \in \mathbf{C}^*, f \in \mathbf{Q}$. Let $E_\Phi = \Phi \otimes E, E_\phi = \phi \otimes E$, $V_{\bar{\Phi}} = \bar{\Phi} \otimes V$. Let E'_Φ be the set of all $T \in E_\Phi$ such that T defines a semisimple endomorphism of $V_{\bar{\Phi}}$ whose eigenspaces are all 2-dimensional. For any conjugacy class σ in the symmetric group in n letters let $E'_{\Phi,\sigma}$ be the set of all $T \in E'_\Phi$ such that, if the eigenvalues of T are denoted by $\lambda_1 = \lambda_2, \lambda_3 = \lambda_4, \ldots, \lambda_{2n-1} = \lambda_{2n}$ (elements of $\mathbf{C}((\epsilon^{1/N}))$ for some $N \geq 1$) then the element γ of $\mathrm{Gal}(\mathbf{C}((\epsilon^{1/N}))/\Phi)$ which maps $\epsilon^{1/N}$ to $\exp(2\pi\sqrt{-1}/N)\epsilon^{1/N}$ permutes $\lambda_2, \lambda_4, \ldots, \lambda_{2n}$ according to a permutation in σ. Note that $E'_\Phi = \sqcup_\sigma E'_{\Phi,\sigma}$.

Let $x \in E_0$. Recall that $x : V \to V$ has Jordan blocks of sizes

$$n_1, n_1, n_2, n_2, \ldots, n_t, n_t$$

where $n_1 \geq n_2 \geq \cdots \geq n_t$ is a partition of n. The following statement is easily verified (it can be reduced to the case where $t = 1$).

(a) There exists $J \in E$ such that $x + \epsilon J : V_\Phi \to V_\Phi$ has eigenvalues $\exp(2\pi\sqrt{-1}s/n_i)a_i \epsilon^{1/n_i} \in \bar{\Phi}$
$(i = 1, 2, \ldots, 2t, s = 1, 2, \ldots, n_i)$ where $a_2, a_4, \ldots, a_{2t}$ are distinct in $\mathbf{C}$ and $a_1 = a_2, a_3 = a_4, \ldots, a_{2t-1} = a_{2t}$. In particular we have $x + \epsilon J \in E'_\Phi$.

Now let $X = X_0 + \epsilon X_1 + \epsilon^2 X_2 + \cdots \in E_\phi$ where $X_0, X_1, \ldots$ are in E. Let $\tilde{x} = x + \epsilon X \in E_\phi$. Let $\lambda_1, \lambda_2, \ldots, \lambda_{2n}$ be the eigenvalues of $\tilde{x}$ in $\bar{\Phi}$. We can assume that $\lambda_1 = \lambda_2, \lambda_3 = \lambda_4, \ldots, \lambda_{2n-1} = \lambda_{2n}$ and $v(\lambda_2) \leq v(\lambda_4) \leq \cdots \leq v(\lambda_{2n})$. Let $\mu_1, \mu_2, \ldots, \mu_{2n}$ be the eigenvalues of $x + \epsilon J$. We can assume that $\mu_1 = \mu_2, \mu_3 = \mu_4, \ldots, \mu_{2n-1} = \mu_{2n}$ and $v(\mu_2) \leq v(\mu_4) \leq \cdots \leq v(\mu_{2n})$. Applying [KL, 9.4] we have $v(\lambda_1\lambda_2\ldots\lambda_s) \geq v(\mu_1\mu_2\ldots\mu_s)$ for $s = 1, 2, \ldots, 2n$. Using this for $s = 2, 4, \ldots, 2n-2$ and adding the resulting inequalities we deduce

$$v(\lambda_2^{2n-2}\lambda_4^{2n-4}\ldots\lambda_{2n-2}^2) \geq v(\mu_2^{2n-2}\mu_4^{2n-4}\ldots\mu_{2n-2}^2)$$

hence

$$v(\lambda_2^{n-1}\lambda_4^{n-2}\ldots\lambda_{2n-2}^1) \geq v(\mu_2^{n-1}\mu_4^{n-2}\ldots\mu_{2n-2}^1).$$

For any $z \in E_\Phi$ we denote by $\Theta(z) \in \Phi$ the trace of the $(4n^2 - 4n)$-th exterior power of $\mathrm{ad}(z) : \mathrm{End}(V_\Phi) \to \mathrm{End}(V_\Phi)$. We have $\Theta(\tilde{x}) = \prod_{1 \le i < j \le n}(\lambda_{2i} - \lambda_{2j})^8$. Note that

$$v(\Theta(\tilde{x})) = 8 \sum_{i<j} v(\lambda_{2i} - \lambda_{2j}) \ge 8 \sum_{i<j} v(\lambda_{2i})$$
$$= 8v(\lambda_2^{n-1}\lambda_4^{n-2} \ldots \lambda_{2n-2}^1) \ge v(\mu_2^{n-1}\mu_4^{n-2} \ldots \mu_{2n-2}^1)$$
$$= 8 \sum_{i<j} v(\mu_{2i}) = 8 \sum_{i<j} v(\mu_{2i} - \mu_{2j}) = v\Theta(x + \epsilon J).$$

Moreover if $v\Theta(\tilde{x}) = v\Theta(x + \epsilon J)$ then $v(\lambda_{2i} - \lambda_{2j}) = v(\lambda_{2i})$ for all $i < j$ in $[1, n]$ and $v(\lambda_1\lambda_2 \ldots \lambda_s) = v(\mu_1\mu_2 \ldots \mu_s)$ for $s = 2, 4, \ldots, 2n - 2$ hence $v(\lambda_i^2) = v(\mu_i^2)$ that is, $v(\lambda_i) = v(\mu_i)$ for $i = 2, 4, \ldots, 2n - 2$.

Let $\mathcal{S}$ be the set of all $\tilde{x}$ in $x + \epsilon E_\phi$ such that $v(\Theta(\tilde{x})) = v(\Theta(x + \epsilon J))$. Then $\mathcal{S} \ne \emptyset$ (it contains $x + \epsilon J$) and is "Zariski open" in $x + \epsilon E_\phi$. Let $\tilde{x} \in \mathcal{S}$ and let $\lambda_1, \ldots, \lambda_{2n}$ be as above. We have

(b) $v(\lambda_1) = v(\lambda_2) = \cdots = v(\lambda_{2n_1}) = 1/n_1$,

$v(\lambda_{2n_1+1}) = v(\lambda_{2n_1+2}) = \cdots = v(\lambda_{2n_1+2n_2}) = 1/n_2$, etc.

except that when $n_t = 1$, $v(\lambda_{2n} - 1) = v(\lambda_{2n})$ is an integer (or ∞) not necessarily equal to $v(\mu_{2n})$. As in [KL, p.156] we see that

$$v(\lambda_i) = 1/m \implies \lambda_i \in \mathbf{C}[[\epsilon^{1/m}]].$$

(In [KL, p.156] lines $-7, -8$ one should replace "t not divisible by m" by "t not a divisor of m".) This and (b) implies that $\tilde{x} \in E'_{\Phi,\sigma}$ where σ contains a product of disjoint cycles of length $n_1, n_2, \ldots, n_t$. Thus we have $\mathcal{S} \subset E'_{\Phi,\sigma}$. We see that the map Ψ in §14 is bijective in this case.

16. Example: (SO_{2n}, SO_{2n-1}). In this subsection we assume that $\mathbf{k} = \mathbf{C}$ and that $(G, K) = (SO(V), SO(U))$ with $V, U, (,)$ as in §5 and with $\dim(U)$ odd. Let $\Phi, \phi, \bar{\Phi}, v$ be as in §15. Let $U_\Phi = \Phi \otimes U, U_\phi = \phi \otimes U, U_{\bar{\Phi}} = \bar{\Phi} \otimes U$. Let U'_Φ be the set of all $x \in U_\Phi$ such that $(x, x) \ne 0$. Let $U'_{\Phi,1}$ (resp. $U'_{\Phi,-1}$) be the set of all $x \in U'_\Phi$ such that $v((x, x)) \in 2\mathbf{Z}$ (resp. $v((x, x)) \in 2\mathbf{Z} + 1$). We have $U'_\Phi = U'_{\Phi,1} \sqcup U'_{\Phi,-1}$.

Let $U_0 = \{x \in U; (x, x) = 0\}$. Let $x \in U_0$. Let $X = u_0 + \epsilon u_1 + \epsilon^2 u_1 + \cdots \in U_\phi$ where $u_0, u_1, \ldots$ are in U. We have

$(x + \epsilon X, x + \epsilon X) = 2(x, u_0)\epsilon + (2(x, u_1) + (u_0, u_0))\epsilon^2 + a\epsilon^3$

with $a \in \phi$. If $x = 0$ let $\mathcal{S}$ be the set of all $x + \epsilon X$ (with X varying as above) such that $(u_0, u_0) \ne 0$.

If $x \in U_0 - \{0\}$ let $\mathcal{S}$ be the set of all $x + \epsilon X$ (with X varying as above) such that $(x, u_0) \ne 0$. In any case $\mathcal{S}$ is a nonempty "Zarisky open" subset of $x + \epsilon U_\phi$. If $x = 0$ we have $\mathcal{S} \subset U'_{\Phi,1}$. If $x \in U_0 - \{0\}$ we have $\mathcal{S} \subset U'_{\Phi,-1}$.

We see that the map Ψ in §14 is bijective in this case.

17. Example: $(GL_4, GL_2 \times GL_2)$. In this subsection we assume that $(G, K) = (GL(V' \times V''), GL(V') \times GL(V''))$ where V', V'' are $\mathbf{k}$-vector spaces of dimension 2 and $\mathbf{k} = \mathbf{C}$. We may identify $\mathfrak{p}$ with $\mathrm{Hom}(V', V'') \times \mathrm{Hom}(V'', V')$ with the obvious action of K. Let $\Phi, \phi, \bar{\Phi}, v$ be as in §15. Let $V'_\phi = \Phi \otimes V', V''_\phi = \Phi \otimes V'$. In our case $\mathfrak{p}_{\Phi,rs}$ consists of all $(A, B) \in \mathrm{Hom}(V'_\Phi, V''_\Phi) \times \mathrm{Hom}(V''_\Phi, V'_\Phi)$ such that $AB : V''_\Phi \to V''_\Phi$ is a regular semisimple automorphism (or equivalently, $BA : V'_\Phi \to V'_\Phi$ is a regular semisimple automorphism). The group $\mathcal{W}$ in §14 is in our case a dihedral group of order 8; hence $\underline{\mathcal{W}}$ has five elements, say $\gamma_1, \gamma_2, \gamma_3, \gamma_4, \gamma_5$. We describe the corresponding partition $\mathfrak{p}_{\Phi,rs} = \sqcup_{j \in [1,5]} \mathfrak{p}_{\Phi,rs,\gamma_j}$.

$\mathfrak{p}_{\Phi,rs,\gamma_j}$ consists of all $(A, B) \in \mathfrak{p}_{\Phi,rs}$ such that the eigenvalues λ, μ of AB in $\bar{\Phi}$ satisfy the following condition:

$j = 1$: $\lambda \in \Phi, \mu \in \Phi, v(\lambda) \in 2\mathbf{Z}, v(\mu) \in 2\mathbf{Z}$.

$j = 2$: $\lambda \in \Phi, \mu \in \Phi, v(\lambda) \in 2\mathbf{Z} + 1, v(\mu) \in 2\mathbf{Z} + 1$.

$j = 3$: $\lambda \in \Phi, \mu \in \Phi, v(\lambda) \in 2\mathbf{Z}, v(\mu) \in 2\mathbf{Z} + 1$.

$j = 4$: $v(\lambda) \in (1/2) + \mathbf{Z}, v(\mu) \in (1/2) + \mathbf{Z}$.

$j = 5$: $\lambda \notin \Phi, \mu \notin \Phi, v(\lambda) \in \mathbf{Z}, v(\mu) \in \mathbf{Z}$.

Let e_1, e_2 be a basis of V'; let e_3, e_4 be a basis of V''. The following elements form a set of representatives for the K-orbits on $\mathcal{N}$.

$N_1 : e_1 \mapsto e_4, e_2 \mapsto e_3, e_3 \mapsto 0, e_4 \mapsto e_2$;

$N_2 : e_1 \mapsto 0, e_2 \mapsto e_4, e_3 \mapsto e_2, e_4 \mapsto e_1$;

$N_3 : e_1 \mapsto e_4, e_2 \mapsto 0, e_3 \mapsto 0, e_4 \mapsto e_2$;

$N_4 : e_1 \mapsto 0, e_2 \mapsto e_4, e_3 \mapsto 0, e_4 \mapsto e_1$;

$N_5 : e_1 \mapsto e_4, e_2 \mapsto e_3, e_3 \mapsto 0, e_4 \mapsto 0$;

$N_6 : e_1 \mapsto 0, e_2 \mapsto 0, e_3 \mapsto e_2, e_4 \mapsto e_1$;

$N_7 : e_1 \mapsto e_4, e_2 \mapsto 0, e_3 \mapsto e_2, e_4 \mapsto 0$;

$N_8 : e_1 \mapsto e_4, e_2 \mapsto 0, e_3 \mapsto 0, e_4 \mapsto 0$;

$N_9 : e_1 \mapsto 0, e_2 \mapsto 0, e_3 \mapsto e_1, e_4 \mapsto 0$;

$N_1 0 = 0$.

For $i \in [1, 10]$ we consider $(A_i, B_i) = N_i + \epsilon \xi$ where $\xi \in \mathfrak{p}_\phi$ is given by

$e_1 \mapsto ae_3 + be_4, e_2 \mapsto ce_3 + de_4, e_3 \mapsto xe_1 + ye_2, e_4 \mapsto ze_1 + ue_2$

and $a, b, c, d, x, y, z, u \in \phi$. Define $a_0, b_0, c_0, d_0, x_0, y_0, z_0, u_0 \in \mathbf{C}$ to be the constant terms of a, b, c, d, x, y, z, u. The characteristic polynomial P_i of A_iB_i is given by

$P_1 = X^2 - O(\epsilon)X + \epsilon x_0 + O(\epsilon^2)$;

$P_2 = X^2 - O(\epsilon)X + \epsilon a_0 + O(\epsilon^2)$;

$P_3 = X^2 - (\epsilon(z_0 + d_0) + O(\epsilon^2))X + \epsilon^2 c_0 x_0 + O(\epsilon^3)$;

$P_4 = X^2 - (\epsilon(b_0 + u_0) + O(\epsilon^2))X + \epsilon^2 a_0 y_0 + O(\epsilon^3)$;

$P_5 = X^2 - (\epsilon(z_0 + y_0) + O(\epsilon^2))X + \epsilon^2(-x_0 u_0 + y_0 z_0) + O(\epsilon^3)$;

$P_6 = X^2 - (\epsilon(b_0 + c_0) + O(\epsilon^2))X + \epsilon^2(-a_0 d_0 + b_0 c_0) + O(\epsilon^3)$;

$P_7 = X^2 - (\epsilon(z_0 + c_0) + O(\epsilon^2))X + \epsilon^2 c_0 z_0 + O(\epsilon^3)$;

$P_8 = X^2 - (\epsilon z_0 + O(\epsilon^2))X + \epsilon^3 c_0(x_0 u_0 - y_0 z_0) + O(\epsilon^4)$;

$P_9 = X^2 - (\epsilon a_0 + O(\epsilon^2))X + \epsilon^3 u_0(a_0 d_0 - b_0 c_0) + O(\epsilon^4)$;

$P_{10} = X^2 - (\epsilon^2(a_0 x_0 + b_0 z_0 + c_0 y_0 + d_0 u_0) + O(\epsilon^3))X + \epsilon^4(a_0 d_0 - b_0 c_0)(x_0 u_0 - y_0 z_0) + O(\epsilon^5)$,

where $O(\epsilon^m)$ denotes an element of $\epsilon^m \phi$. We see that:

if $x_0 \neq 0$ then $N_1 + \epsilon \xi \in \mathfrak{p}_{\Phi,rs,\gamma_4}$;

if $a_0 \neq 0$ then $N_2 + \epsilon \xi \in \mathfrak{p}_{\Phi,rs,\gamma_4}$;

if $c_0 x_0((z_0 + d_0)^2 - 4c_0 x_0) \neq 0$, then $N_3 + \epsilon \xi \in \mathfrak{p}_{\Phi,rs,\gamma_2}$;

if $a_0 y_0((b_0 + u_0)^2 - 4a_0 y_0) \neq 0$, then $N_4 + \epsilon \xi \in \mathfrak{p}_{\Phi,rs,\gamma_2}$;

if $(x_0 u_0 - y_0 z_0)((z_0 - y_0)^2 + 4x_0 u_0) \neq 0$, then $N_5 + \epsilon \xi \in \mathfrak{p}_{\Phi,rs,\gamma_2}$;

if $(a_0 d_0 - b_0 c_0)((b_0 - c_0)^2 + 4a_0 d_0) \neq 0$, then $N_6 + \epsilon \xi \in \mathfrak{p}_{\Phi,rs,\gamma_2}$;

if $c_0 z_0(z_0 - c_0) \neq 0$ then $N_7 + \epsilon \xi \in \mathfrak{p}_{\Phi,rs,\gamma_2}$;

if $c_0 z_0(x_0 u_0 - y_0 z_0) \neq 0$ then $N_8 + \epsilon \xi \in \mathfrak{p}_{\Phi,rs,\gamma_3}$;

if $a_0 u_0(a_0 d_0 - b_0 c_0) \neq 0$ then $N_9 + \epsilon \xi \in \mathfrak{p}_{\Phi,rs,\gamma_3}$;

if $(a_0 d_0 - b_0 c_0)(x_0 u_0 - y_0 z_0) \neq 0$ and

$$(a_0x_0 + b_0z_0 + c_0y_0 + d_0u_0)^2 - 4(a_0d_0 - b_0c_0)(x_0u_0 - y_0z_0) \neq 0,$$
then $N_{10} + \epsilon\xi \in \mathfrak{p}_{\Phi,rs,\gamma_1}$.

Thus we have
$$\Psi(N_1) = \Psi(N_2) = \gamma_4,$$
$$\Psi(N_3) = \Psi(N_4) = \Psi(N_5) = \Psi(N_6) = \Psi(N_7) = \gamma_2,$$
$$\Psi(N_8) = \Psi(N_9) = \gamma_3,$$
$$\Psi(N_{10}) = \gamma_1.$$

18. Final comments. We want to define the notion of symmetric space without any assumption on p. Let G be a connected reductive group over $\mathbf{k}$ and let K be a closed connected reductive subgroup of G. For simplicity we assume that G, K contain a common maximal torus T. (A similar definition can be given without this assumption.) We have inclusions $R_K \subset R_G \subset X(T)$ where $X(T)$ is the character group of T and R_K (resp. R_G) is the set of roots of K (resp. G). We say that (G, K) is a symmetric space (of equal rank) if the inclusions $R_K \subset R_G \subset X(T)$ are the same as the corresponding inclusions for a symmetric space of equal rank in characteristic 0. Thus K does not necessarily come from an involution of G. For example if $p = 2$ we can take G of type E_8 and K a subgroup of type D_8 (such a subgroup exists but it is not the fixed point set of an involution of G). It would be interesting to see how many of the basic properties of symmetric spaces in characteristic 0 extend to this more general case (including the case $p = 2$).

REFERENCES

[BKS] E.Bannai, N.Kawanaka and S.Song, *The character table of the Hecke algebra* $H(GL_{2n}(F_q), Sp_{2n}(F_q))$, J.Algebra **129** (1990), 320-366.

[Gi] V.Ginzburg, *Admissible modules on a symmetric space*, Astérisque **173-174** (1989), 199-255.

[KL] D.Kazhdan and G.Lusztig, *Fixed point varieties on affine flag manifolds*, Isr.J.Math. **62** (1988), 129-168.

[L1] G.Lusztig, *Character sheaves I*, Adv.in Math. **56** (1985), 193-237.

[L2] G.Lusztig, *Fourier transforms on a semisimple Lie algebra over* F_q, Algebraic Groups Utrecht 1986, Lect. Notes in Math., vol. 1271, Springer Verlag, 1987, pp. 177-188.

[L3] G.Lusztig, $G(F_q)$-*invariants in irreducible* $G(F_{q^2})$-*modules*, Represent.Th. **4** (2000), 446-465.

[L4] G.Lusztig, *Affine quivers and canonical bases*, Publ.Math.I.H.E.S. **76** (1992), 111-163.

[L5] G.Lusztig, *Introduction to quantum groups*, Progr.in Math., vol. 110, Birkhauser Boston, 1993.

[S] R.Steinberg, *Endomorphisms of linear algebraic groups*, Mem.Amer.Math.Soc., vol. 80, 1968.

DEPARTMENT OF MATHEMATICS, M.I.T., CAMBRIDGE, MA 02139

Contemporary Mathematics
Volume **557**, 2011

STUDY OF ANTIORBITAL COMPLEXES

G. Lusztig

To Gregg Zuckerman on his 60th birthday

Introduction

0.1. Let U be an N-dimensional vector space over a finite field $\mathbf{F}_q$ and let U^* be the dual vector space. Let $\kappa : U \times U^* \to \mathbf{F}_q$ be the canonical pairing. We fix a prime number l such that $l \neq 0$ in $\mathbf{F}_q$ and a nontrivial character $\psi : \mathbf{F}_q \to \bar{\mathbf{Q}}_l^*$. (We denote by $\bar{\mathbf{Q}}_l$ an algebraic closure of the field of l-adic numbers.) If $f : U \to \bar{\mathbf{Q}}_l$ is a function, the Fourier transform $\hat{f} : U^* \to \bar{\mathbf{Q}}_l$ is defined by

$$\hat{f}(y) = q^{-N/2} \sum_{x \in U} \psi(\kappa(x, y)) f(x).$$

0.2. Assume that q is odd, N is even ≥ 4 and that U is endowed with a split nondegenerate symmetric bilinear form $(,) : U \times U \to \mathbf{F}_q$. This allows us to identify $U^* = U$. Let $SO(U)$ be the special orthogonal group of $(,)$. The following is the prototype of a problem that we are interested in.

(a) *Describe the space of all functions $f : U \to \bar{\mathbf{Q}}_l$ which are constant on each orbit of $SO(U)$ such that both f and $\hat{f}$ vanish on the complement of $\{x \in U ; (x, x) = 0\}$.*

It turns out that the space (a) consists of all scalar multiples of a single function namely:

$$f(x) = 0 \text{ if } (x, x) \neq 0, \quad f(x) = 1 \text{ if } (x, x) = 0, x \neq 0,$$
$$f(x) = 1 + q^{(N-2)/2} \text{ if } x = 0.$$

The proof is an easy exercise (the fact that $\hat{f} = f$ is shown in [L4,§11]; see also the last paragraph of 3.5).

Let $\mathbf{k}$ be an algebraic closure of $\mathbf{F}_q$. The function f can be interpreted as the characteristic function of the intersection cohomology complex of the quadric $(x, x) = 0$ in $\mathbf{k} \otimes U$. This quadric is singular at 0 and this accounts for the term $q^{(N-2)/2}$ in the formula for $f(0)$.

0.3. Let G be a connected reductive algebraic group over $\mathbf{k}$ with a given semisimple automorphism $\vartheta : G \to G$ with fixed point set G^ϑ. Let K be the identity component of G^ϑ. We assume that the characteristic of $\mathbf{k}$ is sufficiently large. Now ϑ induces a (semisimple) automorphism of the Lie algebra $\mathfrak{g}$ of G. For any $\zeta \in \mathbf{k}^*$ let $\mathfrak{g}_\zeta$ be the

Supported in part by the National Science Foundation

ζ-eigenspace of this automorphism. Note that K acts on $\mathfrak{g}_\zeta$ by the adjoint action. Let $\mathfrak{g}_\zeta^{nil}$ be the variety of elements in $\mathfrak{g}_\zeta$ which are nilpotent in $\mathfrak{g}$. We assume that we are given an $\mathbf{F}_q$-structure on G such that $\vartheta : G \to G$ is defined over $\mathbf{F}_q$, such that all eigenvalues of $\vartheta : \mathfrak{g} \to \mathfrak{g}$ are in $\mathbf{F}_q^*$ and such that there exists a Borel subgroup of G defined over $\mathbf{F}_q$ which is ϑ-stable. Then K, $\mathfrak{g}$ and $\mathfrak{g}_\zeta$ (for any ζ) inherit an $\mathbf{F}_q$-structure. Let $\kappa : \mathfrak{g} \times \mathfrak{g} \to \mathbf{k}$ be a fixed G-invariant nondegenerate symmetric bilinear form; we assume that it is also ϑ-invariant and defined over $\mathbf{F}_q$. It induces a nondegenerate bilinear pairing $\kappa_\zeta : \mathfrak{g}_\zeta \times \mathfrak{g}_{\zeta^{-1}} \to \mathbf{k}$ (for any ζ) which is again defined over $\mathbf{F}_q$. This allows us to identify $\mathfrak{g}_{\zeta^{-1}}$ with the dual space of $\mathfrak{g}_\zeta$ (compatibly with the $\mathbf{F}_q$-structures and K-actions). We fix $\zeta \in \mathbf{k}^*$. The following is a generalization of the problem 0.2(a).

(a) *Describe the space of all functions $f : \mathfrak{g}_\zeta(\mathbf{F}_q) \to \bar{\mathbf{Q}}_l$ (which are constant on each orbit of $K(\mathbf{F}_q)$) such that f vanishes on any non-nilpotent element of $\mathfrak{g}_\zeta(\mathbf{F}_q)$ and $\hat{f}$ vanishes on any non-nilpotent element of $\mathfrak{g}_{\zeta^{-1}}(\mathbf{F}_q)$. In particular, when is this space nonzero?*
(The example in 0.2 arises in the case where $G = SO_{2n+1}(\mathbf{k})$, $K = SO_{2n}(\mathbf{k})$, $\zeta = -1$.)

In the case where $\zeta = 1$ the solution of (a) can be found in [L1]. It turns out that in this case functions as in (a) exist very rarely and they are related to "cuspidal character sheaves". For example if $G = Sp_{2n}(\mathbf{k})$, $\vartheta = 1, \zeta = 1$, there is up to scalar at most one nonzero f as in (a); it exists if and only if $n = i(i+1)/2$ for some $i \in \mathbf{N}$. On the other hand, when ϑ is an inner automorphism and $\zeta \neq 1$, we will show (see 3.4(a)) that the space (a) is always nonzero.

0.4. Assume now that in 0.3 we take $G = GL(V)$ where V is a finite dimensional $\mathbf{k}$-vector space with a fixed $\mathbf{F}_q$-structure and a fixed grading $V = \oplus_{i \in \mathbf{Z}/m} V_i$ such that each V_i is defined over $\mathbf{F}_q$. Here m is a fixed integer ≥ 2 such that $m \neq 0$ in $\mathbf{k}$. Let $a \in \mathbf{k}^*$ be such that $a^m = 1$ and $1, a, \ldots, a^{m-1}$ are distinct. Define $g_0 \in G$ by $g_0(v) = a^i v$ for $v \in V_i, i \in \mathbf{Z}/m$. Define $\vartheta : G \to G$ as conjugation by g_0. Then $\mathfrak{g}_a$ can be viewed as the space of representations of a cyclic quiver with m vertices of fixed multidegree. In this case we will describe completely the space 0.3(a) by exhibiting an explicit basis for it. (See 2.14(a)). This basis is closely connected with the theory of canonical bases [L2] applied to affine SL_m.

0.5. As suggested by the example in 0.2, it is useful to study the problem 0.3(a) by passing to $\mathbf{k}$ and using geometric methods, namely using perverse sheaves instead of functions and using Deligne-Fourier (D-F) transform (see 0.7) instead of Fourier transform.

Let E be a finite dimensional $\mathbf{k}$-vector space and let $\check{E}$ be its dual space. Let K be a connected linear algebraic group with a given homomorphism of algebraic groups $K \to GL(E)$. Then K acts on E and (by duality) on $\check{E}$. A complex (of $\bar{\mathbf{Q}}_l$-sheaves) on E is said to be *orbital* if it is a simple K-equivariant perverse sheaf on E supported by the closure of a single K-orbit. A complex (of $\bar{\mathbf{Q}}_l$-sheaves) on $\check{E}$ is said to be *antiorbital* if it is isomorphic to the D-F transform of an orbital complex on E. (Then it is a simple K-equivariant perverse sheaf on $\check{E}$.) A complex (of $\bar{\mathbf{Q}}_l$-sheaves) on E is said to be *biorbital* if it is orbital for the K-action on E and its D-F transform is orbital for the K-action on $\check{E}$.

0.6. Assume now that $E = \mathfrak{g}_\zeta$, K are as in 0.3, $(\zeta \in \mathbf{k}^*)$. Some general results on K-orbits on $\mathfrak{g}_\zeta$ can be deduced from Vinberg's work [V]. In the case where $\zeta = 1$

an explicit description of the antiorbital complexes on $\mathfrak{g}_{\zeta^{-1}}$ is given in [L1].

In the case where $\mathfrak{g}_\zeta$ arises as in 0.4 from a cyclic quiver an explicit description of the antiorbital complexes on $\mathfrak{g}_{\zeta^{-1}}$ is given in Theorem 2.6. By "explicit" we mean that for each antiorbital complex we describe its support and the associated local system on an open dense smooth part of the support.

We also describe explicitly the collection of antiorbital complexes on $\check{E}$ in a case where E, K does not come from the setup in 0.3 (see 3.9). In this example, E is the coadjoint representation of a unipotent group K and there are infinitely many subvarieties of the adjoint representation $\check{E}$ which appear as supports of antiorbital complexes.

Assume again that $E = \mathfrak{g}_\zeta, K, \vartheta$ are as in 0.3 with $\zeta \in \mathbf{k}^*$, so that $\check{E} = \mathfrak{g}_{\zeta^{-1}}$. Let $\mathfrak{Q}_\zeta$ be the collection of simple K-equivariant perverse sheaves $\mathcal{K}$ on $\mathfrak{g}_\zeta$ such that $\mathrm{supp}(\mathcal{K}) \subset \mathfrak{g}_\zeta^{nil}$ and $\mathrm{supp}(\mathfrak{F}(\mathcal{K})) \subset \mathfrak{g}_{\zeta^{-1}}^{nil}$. Note that $\mathfrak{Q}_\zeta$ contains only finitely many objects up to isomorphism. The following is a geometric analogue of problem 0.3(a).

(a) *Describe explicitly the objects of $\mathfrak{Q}_\zeta$ (by describing the corresponding nilpotent K-orbits and associated local systems). In particular when is $\mathfrak{Q}_\zeta$ nonempty?*
In the case where $\zeta = 1$ the solution of this problem can be found in [L1]. It turns out that in this case $\mathfrak{Q}_\zeta$ is almost always empty. For example if $G = Sp_{2n}(\mathbf{k})$, $\vartheta = 1, \zeta = 1$, there is up to isomorphism at most one object of $\mathfrak{Q}_\zeta$; it exists if and only if $n = i(i+1)/2$ for some $i \in \mathbf{N}$. At the other extreme, in the case where ζ has large order in $\mathbf{k}^*$ (say $> \dim \mathfrak{g}$), $\mathfrak{g}_{\zeta^{-1}}$ and $\mathfrak{g}_\zeta$ consist of nilpotent elements and $\mathfrak{Q}_\zeta$ consists of all orbital complexes on $\mathfrak{g}_{\zeta^{-1}}$ (they are automatically antiorbital). If we assume only that $\zeta \neq 1$ (and that ϑ is an inner automorphism), we can show that $\mathfrak{Q}_\zeta \neq \emptyset$. (See 3.3.) In the case which arises as in 0.4 from a cyclic quiver, we determine $\mathfrak{Q}_\zeta$ explicitly. (See 2.13). In this case $\mathfrak{Q}_\zeta$ is exactly the collection of simple perverse sheaves on E defined by the theory of canonical bases for affine SL_m. (This gives a new characterization of that canonical basis.) In any case, the set (of isomorphism classes in) $\mathfrak{Q}_\zeta$ can be viewed as a basis of a finite dimensional $\mathbf{Q}(v)$-vector space $\mathbf{V}_\zeta$ and we have the operations of "induction" and "restriction" (see the Appendix) relating $\mathbf{V}_\zeta$ and the analogous vector space for a ϑ-stable Levi subgroup analogous to multiplication and comultiplication in a quantum group; we also have canonically $\mathbf{V}_\zeta = \mathbf{V}_{\zeta^{-1}}$ (compatibly with the bases) and this agrees with the operations of induction and restriction. Thus, $\mathfrak{Q}_\zeta$ behaves in many respects like a canonical basis. (This point of view has been already emphasized in [L5] which corresponds to the case where ζ has large order in $\mathbf{k}^*$.)

0.7. Notation. If X is an algebraic variety let $\mathcal{D}(X)$ be the bounded derived category of constructible $\bar{\mathbf{Q}}_l$-sheaves on X The objects of $\mathcal{D}(X)$ are said to be complexes. If $\mathcal{K} \in \mathcal{D}(X)$ we write $\mathcal{K}[?]$ instead of "$\mathcal{K}$ with some shift". We say "local system" instead of "$\bar{\mathbf{Q}}_l$-local system". If $\mathcal{L}$ is a local system on X and $x \in X$ we denote by $\mathcal{L}_x$ the stalk of $\mathcal{L}$ at x. If Y is a locally closed smooth irreducible subvariety of X and $\mathcal{L}$ is a local system on Y, then the intersection cohomology complex $IC(\bar{Y}, \mathcal{L})$ on the closure $\bar{Y}$ of Y is well defined. Extending this by 0 on $X - \bar{Y}$ we obtain a complex $IC(\bar{Y}, \mathcal{L})_X$ in $\mathcal{D}(X)$. Now ψ (see 0.1) defines an Artin-Schreier local system of rank 1 on $\mathbf{k}$; its inverse image under any morphism $\phi : X \to \mathbf{k}$ is a local system $\mathcal{L}^\phi$ of rank 1 on X.

Let $E, \check{E}$ be as in 0.5. Let $\langle , \rangle : E \times \check{E} \to \mathbf{k}$ be the obvious pairing. Let

$\mathfrak{F} : \mathcal{D}(E) \to \mathcal{D}(\check{E})$ be the Deligne-Fourier (D-F) transform

$$\mathcal{K} \mapsto \mathfrak{F}(\mathcal{K}) = \delta'_!(\delta^*(\mathcal{K}) \otimes \mathcal{L}^{\langle , \rangle})[\dim E];$$

here $\delta : E \times \check{E} \to E$, $\delta' : E \times \check{E} \to \check{E}$ are the two projections.

For $n \in \mathbf{N}$, $\mathfrak{S}_n$ denotes the symmetric group in n letters.

Contents

1. Cyclic quivers and orbital complexes.
2. Cyclic quivers and antiorbital complexes.
3. Further examples of antiorbital complexes.
Appendix. Induction, restriction.

1. Cyclic quivers and orbital complexes

1.1. In this and the next section we fix an integer $m \geq 1$ such that $m \neq 0$ in $\mathbf{k}$. We set $I = \mathbf{Z}/m$. We fix $\epsilon = \pm 1$ in I.

Let $\mathcal{C}$ be the category of finite dimensional $\mathbf{k}$-vector spaces V with a given I-grading that is, a direct sum decomposition $V = \oplus_{i \in I} V_i$ indexed by I (the morphisms are linear maps compatible with the grading). For $V \in \mathcal{C}$ we define $|V| = (|V|_i) \in \mathbf{N}^I$ by $|V|_i = \dim(V_i)$. Define $\mathbf{v} \in \mathcal{C}$ by $\mathbf{v}_i = \mathbf{k}$ for all $i \in I$. For $V \in \mathcal{C}$ we set

$$E_V^\epsilon = \{T \in \operatorname{End}(V); TV_i \subset V_{i+\epsilon} \quad \forall i\}.$$

Let $\mathcal{C}^\epsilon$ be the category whose objects are the pairs (V, T) where $V \in \mathcal{C}$ and $T \in E_V^\epsilon$ (the morphisms are linear maps compatible with the grading and which commute with the given endomorphisms). For $V \in \mathcal{C}$ let

$$G_V = \{A \in GL(V), AV_i = V_i \quad \forall i \in I\},$$

a closed subgroup of $GL(V)$ isomorphic to $\prod_{i \in I} GL(V_i)$. Note that G_V acts naturally by conjugation on E_V^ϵ. Clearly the G_V-orbits on E_V^ϵ are in natural bijection with the isomorphism classes of objects $(V', T) \in \mathcal{C}^\epsilon$ such that $V' \cong V$ in $\mathcal{C}$.

For any $T \in E_V^\epsilon$ and any $\lambda \in \mathbf{k}$ let $V_{T,\lambda}$ be the generalized λ-eigenspace of $T^m : V \to V$. Note that $V_{T,\lambda}$ is compatible with the grading of V hence is itself an object of $\mathcal{C}$. Moreover $V_{T,\lambda}$ is T-stable. We denote the restriction of T to $V_{T,\lambda}$ by $^\lambda T$. We have $(V_{T,\lambda}, {}^\lambda T) \in \mathcal{C}^\epsilon$ and $(V, T) = \oplus_{\lambda \in \mathbf{k}}(V_{T,\lambda}, {}^\lambda T)$ in $\mathcal{C}^\epsilon$. We set

$$\mathcal{S}_T = \{\lambda \in \mathbf{k}^*; V_{T,\lambda} \neq 0\}.$$

For any $\lambda \in \mathbf{k}$ let $E_V^{\epsilon,\lambda} = \{T \in E_V^\epsilon; V_{T,\lambda} = V\}$ (a closed G_V-stable subset of E_V^ϵ). We say that $E_V^{\epsilon,0}$ is the *nilpotent variety* in E_V^ϵ.

If $V \cong \mathbf{v}$ we set $E_V^{-\epsilon,*} = E_V^{-\epsilon} - E_V^{-\epsilon,0} = \cup_{\lambda \in \mathbf{k}^*} E_V^{-\epsilon,\lambda}$ (an open dense subset of $E_V^{-\epsilon}$); for $\lambda \in \mathbf{k}^*$ we define $\alpha_\lambda : E_V^{-\epsilon,*} \to \mathbf{k}^*$ by $\alpha_\lambda(T) = \lambda\lambda'$ where $T \in E_V^{-\epsilon,\lambda'}$.

We describe some objects of $\mathcal{C}^\epsilon$.

Let $\mathcal{A} = \{(a, b) \in \mathbf{Z} \times \mathbf{Z}; a \leq b\}$. Now $m\mathbf{Z}$ acts freely on $\mathcal{A}$ by $c : (a, b) \mapsto (a+c, b+c)$; let $\mathcal{A}_m$ be the space of orbits. We write $\overline{a, b}$ for the orbit of $(a, b) \in \mathcal{A}$.

Let $(a, b) \in \mathcal{A}$. Let $V = V_{a,b}$ be the $\mathbf{k}$-vector space with basis $e_a, e_{a+1}, \dots, e_b$ viewed as an object of $\mathcal{C}$ in which $e_j \in V_j$ for all $j = a, a+1, \dots, b$. (The index j

in V_j is viewed as an integer   mod m.) Define $T_1 \in E_V^1$ by $e_a \mapsto e_{a+1} \mapsto e_{a+2} \mapsto \ldots \mapsto e_b \mapsto 0$. Define $T_{-1} \in E_V^{-1}$ by $e_b \mapsto e_{b-1} \mapsto e_{b-2} \mapsto \ldots \mapsto e_a \mapsto 0$. We have $(V_{a,b}, T_\epsilon) \in \mathcal{C}^\epsilon$. Moreover the isomorphism class of $(V_{a,b}, T_\epsilon)$ in $\mathcal{C}^\epsilon$ depends only on the $m\mathbf{Z}$-orbit of (a, b). Hence we can use the notation (V_β, T_ϵ) for this isomorphism class where $\beta = \overline{a, b} \in \mathcal{A}_m$.

Let $\lambda \in \mathbf{k}^*$ and let $n \in \mathbf{Z}_{>0}$. Choose $\lambda' \in \mathbf{k}^*$ such that $\lambda'^m = \lambda$. Let $U = U(n)$ be the $\mathbf{k}$-vector space with basis $e_{a,i}$ ($a \in [1, n], i \in I$). For $i \in I$ let U_i be the subspace of U spanned by $e_{a,i}(a \in [1, n])$. These subspaces form an I-grading of U. Define $T_1 \in \operatorname{End}(U)$ by $e_{a,i} \mapsto e_{a,i+\epsilon}$ ($a \in [1, n], i \in I$). Define $T_2 \in \operatorname{End}(U)$ by $e_{a,i} \mapsto e_{a+1,i+\epsilon}$ ($a \in [1, n-1], i \in I$), $e_{n,i} \mapsto 0$ ($i \in I$). We have $T_1, T_2 \in E_U^\epsilon$, $T_1 T_2 = T_2 T_1$. We have $T_1^m = 1$ hence T_1 is semisimple. Clearly, T_2 is nilpotent hence $T_\epsilon(\lambda) := \lambda' T_1 + T_2 \in \operatorname{End}(U)$ is such that $T_\epsilon(\lambda)^m - \lambda : U \to U$ is nilpotent. We have $(U(n), T_\epsilon(\lambda)) \in \mathcal{C}^\epsilon$. We show that the isomorphism class of $(U, T_\epsilon(\lambda))$ is independent of the choice of λ'. Let $\lambda'' \in \mathbf{k}^*$ be such that $\lambda''^m = \lambda$. We must show that there exists $R \in G_U$ such that $R(\lambda' T_1 + T_2) = (\lambda'' T_1 + T_2)R$. Let $z = \lambda''/\lambda'$. Define $R \in G_U$ by $e_{a,i} \mapsto z^{-a} z^{\epsilon i} e_{a,i}$ ($a \in [1, n], i \in I$). (Here $z^{\epsilon i}$ makes sense since $z^m = 1$.) We have $RT_1 = zT_1$, $RT_2 = T_2 R$ and our claim follows.

Now (V_β, T_ϵ) and $(U(n), T_\epsilon(\lambda))$ are indecomposable objects of $\mathcal{C}^\epsilon$; it is easy to see that, conversely, any indecomposable object of $\mathcal{C}^\epsilon$ is isomorphic to (V_β, T_ϵ) (for a well defined $\beta \in \mathcal{A}_m$) or to $(U(n), T_\epsilon(\lambda))$ (for a well defined $n \geq 1$, $\lambda \in \mathbf{k}^*$).

1.2. Define $\mathbf{1} \in \mathbf{N}^I$ by $\mathbf{1}_i = 1$ for all $i \in I$. Let $\mathcal{R}$ be the set of maps $\rho : \mathbf{Z}_{>0} \to \mathbf{N}$ such that $\rho(n) = 0$ for all but finitely many n. For $\rho \in \mathcal{R}$ we set $\underline{\rho} = \sum_{n \in \mathbf{Z}_{>0}} \rho(n)n \in \mathbf{N}$. For any $t \in \mathbf{N}$ let $\mathcal{R}_t = \{\rho \in \mathcal{R}; \underline{\rho} = t\}$.

Let $\mathcal{P}$ be the set of maps $\sigma : \mathcal{A}_m \to \mathbf{N}$ such that $\sigma(\beta) = 0$ for all but finitely many $\beta \in \mathcal{A}_m$. Let

$$\mathcal{P}^{ap} = \{\sigma \in \mathcal{P}; \prod_{\overline{a,b} \in \mathcal{A}_m; b-a=n} \sigma(\overline{a,b}) = 0 \quad \forall n \in \mathbf{N}\}.$$

(This is a finite product.) The elements of $\mathcal{P}^{ap}$ are said to be *aperiodic*. For $\sigma \in \mathcal{P}$ we set $|\sigma| = \sum_{\beta \in \mathcal{A}_m} \sigma(\beta)|V_\beta| \in \mathbf{N}^I$. For any $\nu \in \mathbf{N}^I$ we set $\mathcal{P}_\nu = \{\sigma \in \mathcal{P}; |\sigma| = \nu\}$. Let $\mathcal{P}_\nu^{ap} = \{\sigma \in \mathcal{P}^{ap}; |\sigma| = \nu\}$.

Let $\tilde{Z}$ be the set of all collections $\theta = (\theta_\lambda)_{\lambda \in \mathbf{k}}$ where $\theta_\lambda \in \mathcal{R}$ for $\lambda \in \mathbf{k}^*$, $\theta_0 \in \mathcal{P}$ are such that $\theta_\lambda = 0$ for all but finitely many λ. For $\theta \in \tilde{Z}$ we set $|\theta| = \sum_{\lambda \in \mathbf{k}^*} \underline{\theta_\lambda} \mathbf{1} + |\theta_0| \in \mathbf{N}^I$. For $\nu \in \mathbf{N}^I$ let $\tilde{Z}_\nu = \{\theta \in \tilde{Z}; |\theta| = \nu\}$.

For any $\rho \in \mathcal{R}$ let Π_ρ be the corresponding irreducible representation of $\mathfrak{S}_{\underline{\rho}}$ over $\bar{\mathbf{Q}}_l$. (In particular if $\rho(1) = N$ and $\rho(n) = 0$ for $n > 1$ then Π_ρ is the unit representation of $\mathfrak{S}_N$.) Let $N_\rho = \dim \Pi_\rho$.

1.3. Let $V \in \mathcal{C}$. From the results in 1.1 we see that the G_V-orbits on E_V^ϵ are naturally indexed by the set $\tilde{Z}_{|V|}$: the G_V-orbit $E_{V,\theta}^\epsilon$ corresponding to $\theta \in \tilde{Z}_{|V|}$ consists of all $T \in E_V^\epsilon$ such that

$$(V, T) \cong \oplus_{\lambda \in \mathbf{k}^*, n \geq 1}(U(n), T_\epsilon(\lambda))^{\oplus \theta_\lambda(n)} \oplus \oplus_{\beta \in \mathcal{A}_m}(V_\beta, T_\epsilon)^{\oplus \theta_0(\beta)}$$

in $\mathcal{C}^\epsilon$. For any $\theta \in \tilde{Z}_{|V|}$, the complex

(a) $$IC(\overline{E_{V,\theta}^\epsilon}, \bar{\mathbf{Q}}_l)_{E_V^\epsilon}$$

is well defined. It is up (to shift) an orbital complex on E_V^ϵ with its natural G_V-action. Since the only G_V-equivariant irreducible local system on $E_{V,\theta}^\epsilon$ is $\bar{\mathbf{Q}}_l$ we see that any orbital complex on E_V^ϵ is (up to shift) of the form (a) for a well defined θ.

Let $\sigma \in \mathcal{P}_{|V|}$. Define $\bar\sigma \in \tilde{Z}_{|V|}$ by $\bar\sigma_0 = \sigma$, $\bar\sigma_\lambda = 0$ for $\lambda \in \mathbf{k}^*$. Note that the G_V-orbits contained in the nilpotent variety $E_V^{\epsilon,0}$ are exactly the subsets $E_{V,\bar\sigma}^\epsilon$ for various $\sigma \in \mathcal{P}_{|V|}$.

1.4. Let $V \in \mathcal{C}$ be such that $V \cong \mathbf{v}^{\oplus s}$ for some $s \in \mathbf{N}$. We fix $\lambda \in \mathbf{k}^*$. Let $D = \{S \in E_V^\epsilon; S^m = \lambda\}$. For $S \in D$ let $\dot{S} = \{N \in E_V^{\epsilon,0}; NS = NS\}$. Clearly, the map

$$\{(S,N); S \in D, N \in \dot{S}\} \xrightarrow{g} E_V^{\epsilon,\lambda}, \quad (S,N) \mapsto S+N$$

is a bijection. Let $\mathfrak{N} = \{f \in \mathrm{End}(V_0); f \text{ nilpotent}\}$. For any $S \in D$ we define a bijection $\mathfrak{N} \xrightarrow{\sim} \dot{S}$ by $\kappa \mapsto N = N_{S,\kappa}$ where $N(S^h v) = S^{h+1}\kappa(v)$ for $v \in V_0$, $h \in [0, m-1]$; we then have also $N(S^h v) = S^{h+1}\kappa(v)$ for $v \in V_0$, $h \in \mathbf{N}$. The inverse bijection is $N \mapsto \kappa$ where $\kappa(v_0) = S^{-1}N(v_0)$ for $v_0 \in V_0$. We see that we have a bijection $u : D \times \mathfrak{N} \xrightarrow{\sim} E_V^{\epsilon,\lambda}$ given by $(S,\kappa) \mapsto S + N_{S,\kappa}$. Now let $\mathbf{S}$ be the set of all (T,f) where $f = (0 = V^0 \subset V^1 \subset V^2 \subset \ldots \subset V^s = V)$ is a sequence of subobjects of V such that $V^j/V^{j-1} \cong \mathbf{v}$ for $j \in [1,s]$ and $T \in E_V^\epsilon$ is such that $T(V^j) \subset V^j$ for $j \in [1,s]$ and the element of $E_{V^j/V^{j-1}}^\epsilon$ induced by T is in $E_{V^j/V^{j-1}}^{\epsilon,\lambda}$. This is a smooth irreducible variety. Let $\mathbf{S}_0$ be the set of all (κ, f_0) where $f_0 = (0 = \mathcal{V}^0 \subset \mathcal{V}^1 \subset \mathcal{V}^2 \subset \ldots \subset \mathcal{V}^s = V_0)$ is a sequence of subspaces of V_0 such that $\dim(\mathcal{V}^j/\mathcal{V}^{j-1}) = 1$ for $j \in [1,s]$ and $\kappa \in \mathfrak{N}$ is such that $\kappa(\mathcal{V}^j) \subset \mathcal{V}^j$ for $j \in [1,s]$. We define an isomorphism $u' : D \times \mathbf{S}_0 \xrightarrow{\sim} \mathbf{S}$ by $(S, (\kappa, f_0)) \mapsto (T, f)$ where $T = S + N_{S,\kappa}$ and $f = (V^j)$ with $V_h^j = S^h(\mathcal{V}^j)$ for $h = 0, 1, \ldots, m-1$. We have a commutative diagram

$$
\begin{array}{ccccc}
\mathbf{S} & \xleftarrow{\ u'\ } & D \times \mathbf{S}_0 & \xrightarrow{\ a'\ } & \mathbf{S}_0 \\
{\scriptstyle g_1}\downarrow & & {\scriptstyle g_2}\downarrow & & {\scriptstyle g_3}\downarrow \\
E_V^{\epsilon,\lambda} & \xleftarrow{\ u\ } & D \times \mathfrak{N} & \xrightarrow{\ a\ } & \mathfrak{N}
\end{array}
$$

where $g_1(T,f) = T$, $g_2(S, (\kappa, f_0)) = (S, \kappa)$, $g_3(\kappa, f_0) = \kappa$ and a, a' are the obvious projections. It is well known that $g_{3!}\bar{\mathbf{Q}}_l \cong \oplus_{\rho \in \mathcal{R}_s} P_\rho^{\oplus N_\rho}[d]$ where P_ρ are simple mutually nonisomorphic perverse sheaves on $\mathfrak{N}$ and $d \in \mathbf{Z}$. Using the fact that the right square in the diagram above is cartesian and a is smooth with connected fibres we deduce that $g_{2!}\bar{\mathbf{Q}}_l \cong \oplus_{\rho \in \mathcal{R}_s} P_\rho'^{\oplus N_\rho}[d']$ where P_ρ' are simple mutually nonisomorphic perverse sheaves on $S \times \mathfrak{N}$ and $d' \in \mathbf{Z}$. Using the fact that u, u' are isomorphisms we deduce that $g_{1!}\bar{\mathbf{Q}}_l \cong \oplus_{\rho \in \mathcal{R}_s} P_\rho''^{\oplus N_\rho}[d']$ where P_ρ'' are simple mutually nonisomorphic perverse sheaves on $E_V^{\epsilon,\lambda}$ and d' is as above.

1.5. Assume that $V \in \mathcal{C}$, $U^1, U^2, \ldots, U^s \in \mathcal{C}$ are such that $U^1 \oplus U^2 \oplus \ldots \oplus U^s \cong V$. We have a diagram

$$E_{U^1}^\epsilon \times E_{U^2}^\epsilon \times \ldots \times E_{U^s}^\epsilon \xleftarrow{p_1} E' \xrightarrow{p_2} E'' \xrightarrow{p_3} E_V^\epsilon$$

with the following notation.

E'' is the set of all pairs (T,f) where f is a sequence $(0 = V^0 \subset V^1 \subset V^2 \subset \ldots \subset V^s = V)$ of subobjects of V such that $V^j/V^{j-1} \cong U^j$ for $j \in [1,s]$ and

$T \in E_V^\epsilon$ is such that $TV^j \subset V^j$ for all $j \in [1, s]$. E' is the set of all triples (T, f, ϕ) where $(T, f) \in E''$ (with $f = (V^j)$) and $\phi = (\phi_j)$ is a collection of isomorphisms $\phi_j : V^j/V^{j-1} \xrightarrow{\sim} U^j$ for $j \in [1, s]$. We have $p_1(T, f, \phi) = (T_j)$ where $T_j \in E_{U^j}^\epsilon$ is obtained by transporting the element of $E_{V^j/V^{j-1}}^\epsilon$ induced by T via ϕ_j for $j \in [1, s]$. We have $p_2(T, f, \phi) = (T, f)$, $p_3(T, f) = T$. Note that p_3 is a proper morphism.

Let $\mathcal{K}_j$ be a G_{U^j}-equivariant perverse sheaf (up to shift) on $E_{U^j}^\epsilon$ ($j \in [1, s]$). Then $\boxtimes_j \mathcal{K}_j$ is a $\prod_j G_{U^j}$-equivariant perverse sheaf (up to shift) on $\prod_j E_{U^j}^\epsilon$. Since p_1 is a smooth morphism with connected fibres, $\mathcal{K}' := p_1^*(\boxtimes_j \mathcal{K}_j)$ is a $\prod_j G_{U^j}$-equivariant perverse sheaf (up to shift) on E'. Since p_2 is a principal $\prod_j G_{U^j}$-bundle there is a well defined perverse sheaf (up to shift) $\mathcal{K}''$ on E'' such that $p_2^* \mathcal{K}'' = p_1^* \mathcal{K}'$. We set $\text{Ind}(\boxtimes_j \mathcal{K}_j) = p_{3!} \mathcal{K}'' \in \mathcal{D}(E_V^\epsilon)$.

1.6. We preserve the setup of 1.5. We fix $\lambda \in \mathbf{k}^*$. Assume that $U^j \cong \mathbf{v}$ for $j \in [1, s]$. For $j \in [1, s]$ let $\mathcal{K}_j = IC(E_{U^j}^{\epsilon, \lambda}, \bar{\mathbf{Q}}_l)_{E_{U^j}^\epsilon}$. Let $\mathbf{S}$ be the set of all $(T, f) \in E''$ such that the following holds: $f = (V^j)$ is such that for $j \in [1, s]$, $V^j/V^{j-1} \cong U^j$ and the element of $E_{V^j/V^{j-1}}^\epsilon$ induced by T is in $E_{V^j/V^{j-1}}^{\epsilon, \lambda}$. This is a closed smooth irreducible subset of E''. From the definitions we have $\mathcal{K}'' = IC(\mathbf{S}, \bar{\mathbf{Q}}_l)_{E''}$. We have $p_3(\mathbf{S}) \subset E_V^{\epsilon, \lambda}$. From 1.4 we see that

$$\text{Ind}(\boxtimes_j \mathcal{K}_j) \cong \oplus_{\rho \in \mathcal{R}_s} P_\rho''^{\oplus N_\rho}[d']$$

where P_ρ'' are simple mutually nonisomorphic perverse sheaves on E_V^ϵ with support on $E_V^{\epsilon, \lambda}$ and $d' \in \mathbf{Z}$. The perverse sheaves P_ρ'' are G_V-equivariant. Since the number of G_V-orbits in $E_V^{\epsilon, \lambda}$ is equal to $\sharp(\mathcal{R}_s)$ we see that the P_ρ'' are precisely the orbital complexes on E_V^ϵ with support contained in $E_V^{\epsilon, \lambda}$.

1.7. We preserve the setup of 1.5. Assume that $s = t + 1$ where $t \in \mathbf{N}$. Assume that $U^j \cong \mathbf{v}$ for $j \in [1, t]$ and $\sigma \in \mathcal{P}^{ap}$ is such that $|U^s| = |\sigma|$. For $j \in [1, t]$ let $\mathcal{K}_j = IC(\{0\}, \bar{\mathbf{Q}}_l)_{E_{U^j}^\epsilon}$. Let $\mathcal{K}_s = IC(\overline{E_{U^s, \bar{\sigma}}^\epsilon}, \bar{\mathbf{Q}}_l)_{E_{U^s}^\epsilon}$. Let S'' be the set of all $(T, f) \in E''$ such that the following holds: $f = (V^j)$ is such that $V^j/V^{j-1} \cong U^j$ for $j \in [1, s]$ and the element of $E_{V^j/V^{j-1}}^\epsilon$ induced by T is 0 if $j \in [1, t]$ and is in $E_{V^s/V^{s-1}, \bar{\sigma}}^\epsilon$ if $j = s$. This is a locally closed smooth irreducible subset of E''. From the definitions we have $\mathcal{K}'' = IC(\bar{S}'', \bar{\mathbf{Q}}_l)_{E''}$. We have $p_3(S'') \subset E_{V,0}^\epsilon$. Since $E_{V,0}^\epsilon$ is closed in E_V^ϵ we see that $p_3(\bar{S}'') \subset E_V^{\epsilon, 0}$. Thus $\text{supp}(\text{Ind}(\boxtimes_j \mathcal{K}_j)) \subset E_V^{\epsilon, 0}$. Moreover, since p_3 is proper, $\text{Ind}(\boxtimes_j \mathcal{K}_j)$ is isomorphic to $\oplus_{e \in [1, M]} P_e[d_e]$ where P_e are simple G_V-equivariant perverse sheaves on E_V^ϵ with $\text{supp}(P_e) \subset E_V^{\epsilon, 0}$ and $d_e \in \mathbf{Z}$ (we use the decomposition theorem). Let $\mathcal{Y}_{t, \sigma}$ be the set of isomorphism classes of simple perverse sheaves on E_V^ϵ that are isomorphic to P_e for some $e \in [1, M]$.

1.8. We preserve the setup of 1.5. Assume $s = t + 1$, $t \in \mathbf{N}$. Let $\theta \in \tilde{Z}_{|V|}$. Assume that for $j \in [1, t]$ we have $|U^j| = \theta_{\lambda_j} \mathbf{1}$ where $\lambda_1, \lambda_2, \ldots, \lambda_t$ are distinct elements of $\mathbf{k}^*$ and $|U^s| = |\theta_{\lambda_s}|$ (we set $\lambda_s = 0$). Note that $\theta_\lambda = 0$ for any $\lambda \notin \{\lambda_1, \ldots, \lambda_s\}$. For $j \in [1, s]$ define $\theta^j \in \tilde{Z}_{|U^j|}$ by $\theta_{\lambda_j}^j = \theta_{\lambda_j}$, $\theta_\lambda^j = 0$ if $\lambda \neq \lambda^j$. Let $S_j = E_{U^j, \theta^j}^\epsilon$, a G_{U^j}-orbit in $E_{U^j}^\epsilon$; let $\mathcal{K}_j = IC(\bar{S}_j, \bar{\mathbf{Q}}_l)_{E_{U^j}^\epsilon}$. Let S'' be the set of all $(T, f) \in E''$ such that the following holds: $f = (V^j)$ is such that $V^j/V^{j-1} \cong U^j$ for $j \in [1, s]$ and $T \in E_V^\epsilon$ is such that for $j \in [1, s]$, the element of $E_{V^j/V^{j-1}}^\epsilon$ induced by T belongs to $E_{V^j/V^{j-1}, \theta^j}^\epsilon$. We have $\mathcal{K}'' = IC(\bar{S}'', \bar{\mathbf{Q}}_l)_{E''}$. Let S be the set of all $T \in E_V^\epsilon$ such that for $j \in [1, s]$ we have $V_{T,\lambda_j} \cong U^j$ and the element of $E_{V_{T,\lambda_j}}^\epsilon$ induced by T belongs to $E_{V_{T,\lambda_j},\theta^j}^\epsilon$. Note that $S = E_{V,\theta}^\epsilon$. For any $T \in \bar{S}$ (closure of S) the fibre $p_3^{-1}(T)$ consists of exactly one element, namely (T, f) where $f = (V^j)$ is given by $V^1 = V_{T,\lambda_1}$, $V^2 = V_{T,\lambda_1} \oplus V_{T,\lambda_2}$, etc. More precisely, p_3 defines an isomorphism $\bar{S}'' \xrightarrow{\sim} \bar{S}$. We see that $\mathrm{Ind}(\boxtimes_j \mathcal{K}_j) = IC(\overline{E_{V,\theta}^\epsilon}, \bar{\mathbf{Q}}_l)_{E_V^\epsilon}$.

2. CYCLIC QUIVERS AND ANTIORBITAL COMPLEXES

2.1. We preserve the setup in 1.1. If $V \in \mathcal{C}$, we have a perfect bilinear pairing $E_V^\epsilon \times E_V^{-\epsilon} \to \mathbf{k}$ given by $(T, T') = \mathrm{tr}(TT', V)$; it is compatible with the G_V-action. We use it to identify $E_V^{-\epsilon}$ with the dual of E_V^ϵ. Hence the D-F transform $\mathfrak{F} : \mathcal{D}(E_V^\epsilon) \to \mathcal{D}(E_V^{-\epsilon})$ is well defined. In particular the notion of antiorbital complex on $E_V^{-\epsilon}$ is well defined; it is a complex of the form $\mathfrak{F}(\mathcal{K})$ where $\mathcal{K}$ is an orbital complex on E_V^ϵ (with respect to the G_V-action).

2.2. Define $h : (\mathbf{k}^*)^m \to \mathbf{k}$ and $h' : (\mathbf{k}^*)^m \to \mathbf{k}^*$ by

$$h(x_1, x_2, \ldots, x_m) = x_1 + x_2 + \cdots + x_m, \quad h'(x_1, x_2, \ldots, x_m) = x_1 x_2 \ldots x_m.$$

Then the local system $\mathcal{L}^h$ on $(\mathbf{k}^*)^m$ is defined (see 0.7). According to Deligne [D, Thm.7.8, p.221], the complex $\mathfrak{K}^m := h_!'\mathcal{L}^h[m-1] \in \mathcal{D}(\mathbf{k}^*)$ (a sheaf theory version of a family of generalized Kloosterman sums) is a local system of rank m on $\mathbf{k}^*$. (The rank 1 local system $\mathfrak{K}^1$ is of Artin-Schreier type. The rank 2 local system $\mathfrak{K}^2$ is implicit in Weil's paper [W].)

Assume now that $V \in \mathcal{C}$, $V \cong \mathbf{v}$. Let $\lambda \in \mathbf{k}^*$. Let $\mathcal{K} = IC(E_V^{\epsilon,\lambda}, \bar{\mathbf{Q}}_l)_{E_V^\epsilon}$. Note that $\alpha_\lambda^*(\mathfrak{K}^m)$ is a local system of rank m on $E_V^{-\epsilon,*}$ (α_λ as in 1.1). Let $\mathcal{K}' = IC(E_V^{-\epsilon}, \alpha_\lambda^*(\mathfrak{K}^m))_{E_V^{-\epsilon}}$. We show:

(a) $\mathcal{K}'[m] \cong \mathfrak{F}(\mathcal{K})[m-1]$.

Since $\mathfrak{F}(\mathcal{K})[m-1]$ is a simple perverse sheaf on $E_V^{-\epsilon}$ it is enough to show that

$$\mathfrak{F}(\mathcal{K})[m-1]|_{E_V^{-\epsilon,*}} \cong \mathcal{K}'[m]|_{E_V^{-\epsilon,*}}$$

in $\mathcal{D}(E_V^{-\epsilon,*})$ or equivalently that

$$\delta_{1!}'(\delta_1^* \bar{\mathbf{Q}}_l \otimes \mathcal{L}^{(,)})[m][m-1] = \alpha_\lambda^*(\mathfrak{K}^m)[m]$$

where $\delta_1 : E_V^{\epsilon,\lambda} \times E_V^{-\epsilon,*} \to E_V^{\epsilon,\lambda}$, $\delta_1' : E_V^{\epsilon,\lambda} \times E_V^{-\epsilon,*} \to E_V^{-\epsilon,*}$ are the projections and the restriction of $(,)$ to $E_V^{\epsilon,\lambda} \times E_V^{-\epsilon,*}$ is denoted again by $(,)$. Thus it is enough to show that $\delta_{1!}'\mathcal{L}^{(,)} = \alpha_\lambda^* h_!'\mathcal{L}^h$. We can assume that $V = \mathbf{v}$. Then $E_V^{\epsilon,\lambda}$ can be identified with $h'^{-1}(\lambda)$, $E_V^{-\epsilon,*}$ can be identified with $(\mathbf{k}^*)^m$, $(,)$ can be identified with the map

$$h'' : h'^{-1}(\lambda) \times (\mathbf{k}^*)^m \to \mathbf{k}, \quad ((x_1, \ldots, x_m), (y_1, \ldots, y_m)) \mapsto x_1 y_1 + \cdots + x_m y_m,$$

δ_1' can be identified with $\delta'' : h'^{-1}(\lambda) \times (\mathbf{k}^*)^m \to (\mathbf{k}^*)^m$ (second projection) and α_λ can be identified with

$$\alpha' : (\mathbf{k}^*)^m \to \mathbf{k}^*, \quad (z_1, \ldots, z_m) \mapsto \lambda z_1 \ldots z_m.$$

Define $j : h'^{-1}(\lambda) \times (\mathbf{k}^*)^m \to (\mathbf{k}^*)^m$ by

$$j((x_1, \ldots, x_m), (y_1, \ldots, y_m)) = (x_1 y_1, \ldots, x_m y_m).$$

It is enough to show that $\alpha'^* h_!' h^* = \delta_1'' h''^*$. Since $h'' = hj$ it is enough to show that $\alpha'^* h_!' h^* = \delta_1'' j^* h^*$, or that $\alpha'^* h_!' = \delta_1'' j^*$. This follows from the fact that the diagram consisting of α', h', δ'', j is cartesian.

2.3. Let $V \in \mathcal{C}$. Let $\sigma \in \mathcal{P}^{ap}, t \in \mathbf{N}$ be such that $|V| = |\sigma| + t\mathbf{1}$. Let $E_{V,t,\sigma}^{-\epsilon}$ be the set of all $T \in E_V^{-\epsilon}$ such that $\sharp(\mathcal{S}_T) = t$, $V_{T,\lambda} \cong \mathbf{v}$ for $\lambda \in \mathcal{S}_T$ and $^0T \in E_{V_{T,0},\bar{\sigma}}^{-\epsilon}$ (^0T as in 1.1). Note that $E_{V,t,\sigma}^{-\epsilon}$ is a locally closed, smooth, irreducible, G_V-stable subvariety of $E_V^{-\epsilon}$. We define a finite principal covering $\zeta : \tilde{E}_{V,t,\sigma}^{-\epsilon} \to E_{V,t,\sigma}^{-\epsilon}$ as follows. By definition, $\tilde{E}_{V,t,\sigma}^{-\epsilon}$ is the set of all pairs (T, ω) where $T \in E_{V,t,\sigma}^{-\epsilon}$ and ω is a total order on $\mathcal{S}(T)$. The group of this covering is $\mathfrak{S}_t$ (it acts freely in an obvious way on $\tilde{E}_{V,t,\sigma}^{-\epsilon}$. The map ζ is $(T, \omega) \mapsto T$.

2.4. Let $\hat{\mathcal{P}} = \mathcal{P}^{ap} \times \mathcal{R}$. For $\nu \in \mathbf{N}^I$ let $\hat{\mathcal{P}}_\nu = \{(\sigma, \rho) \in \hat{\mathcal{P}}; |\sigma| + \rho\mathbf{1} = \nu\}$. We define a map $\hat{\mathcal{P}}_\nu \to \mathcal{P}_\nu$ by $(\sigma, \rho) \mapsto \tilde{\sigma}$ where $\tilde{\sigma}(\beta) = \sigma(\beta) + \rho(b-a)$ for any $\beta = \overline{a,b} \in \mathcal{A}_m$. Note that:

(a) *this map is a bijection.*

The inverse map is $\tilde{\sigma} \mapsto (\sigma, \rho)$ where $\rho \in \mathcal{R}$ is defined by
$$\rho(n) = \min_{\beta=\overline{a,b}\in\mathcal{A}_m; b-a=n} \tilde{\sigma}(\beta)$$
and $\sigma \in \mathcal{P}^{ap}$ is defined by $\sigma(\beta) = \tilde{\sigma}(\beta) - \rho(b-a)$ for any $\beta = \overline{a,b} \in \mathcal{A}_m$.

Let Z be the set of all collections $\pi = (\pi^\lambda)_{\lambda \in \mathbf{k}}$ where $\pi^\lambda \in \mathcal{R}$ for $\lambda \in \mathbf{k}^*$, $\pi^0 \in \hat{\mathcal{P}}$ are such that $\pi^\lambda = 0$ for all but finitely many $\lambda \in \mathbf{k}^*$. For $\nu \in \mathbf{N}^I$ let $Z_\nu = \{\pi \in Z; \sum_{\lambda \in \mathbf{k}^*} \pi^\lambda \mathbf{1} + |\pi^0| = \nu\}$. For $\pi \in Z$ and $\lambda \in \mathbf{k}$ we define $\pi_\lambda \in \mathcal{R}$ by $\pi_\lambda = \pi^\lambda$ if $\lambda \in \mathbf{k}^*$, $\pi_0 = \rho$ if $\lambda = 0$ and $\pi^0 = (\sigma, \rho)$.

Assume that we are given a collection of bijections

(b) $\Phi_{n,\lambda} : \mathcal{R}_n \xrightarrow{\sim} \mathcal{R}_n$ ($\lambda \in \mathbf{k}^*, n \in \mathbf{N}$), $\Psi_{\nu'} : \mathcal{P}_{\nu'} \xrightarrow{\sim} \hat{\mathcal{P}}_{\nu'}$ ($\nu' \in \mathbf{N}^I$).

Such a collection exists by (a). We define a bijection $\Phi : \tilde{Z} \xrightarrow{\sim} Z$ by $(\theta_\lambda) \mapsto (\pi_\lambda)$ where $\pi_\lambda = \Phi_{n,\lambda}(\theta_\lambda)$ for $\lambda \in \mathbf{k}^*$ (here $n = \underline{\theta_\lambda}$) and $\pi_0 = \Psi_{\nu'}(\theta_0)$ (here $\nu' = |\theta_0|$). This restricts for any $\nu \in \mathbf{N}^I$ to a bijection

(c) $\Phi_{(\nu)} : \tilde{Z}_\nu \xrightarrow{\sim} Z_\nu$.

2.5. Let $V \in \mathcal{C}$ and let $\nu = |V|$. Let $\pi \in Z_\nu$. Let $z = \sum_{\lambda \in \mathbf{k}^*} \pi_\lambda$. We have $\pi^0 = (\sigma, \pi_0) \in \hat{\mathcal{P}}$ where $\sigma \in \mathcal{P}^{ap}$ and $|V| = |\sigma| + z\mathbf{1}$. We associate to π a finite unramified covering $\xi : {}'E_{V,z,\sigma}^{-\epsilon} \to E_{V,z,\sigma}^{-\epsilon}$ as follows. By definition, ${}'E_{V,z,\sigma}^{-\epsilon}$ is the set of all pairs (T, g) where $T \in E_{V,z,\sigma}^{-\epsilon}$ and $g : \mathcal{S}(T) \to \mathbf{k}$ is a map such that $\sharp(g^{-1}(\lambda')) = \pi_{\lambda'}$ for $l' \in \mathbf{k}$. We define ξ by $(T, g) \mapsto T$. Note that for $T \in E_{V,z,\sigma}^{-\epsilon}$ we have
$$\sharp(\xi^{-1}(T)) = z!(\prod_{\lambda' \in \mathbf{k}} \pi_{\lambda'}!)^{-1}.$$

Let $\mathbf{L}$ be the local system of rank $m^{z-\pi_0}$ on ${}'E_{V,z,\sigma}^{-\epsilon}$ whose stalk at (T, g) is $\boxtimes_{\lambda \in \mathcal{S}_T, g(\lambda) \neq 0} \mathfrak{R}_{\lambda g(\lambda)}^m$. We define a finite principal covering $\zeta : {}''E_{V,z,\sigma}^{-\epsilon} \to {}'E_{V,z,\sigma}^{-\epsilon}$ as follows. By definition, ${}''E_{V,z,\sigma}^{-\epsilon}$ is the set of all triples (T, g, ω) where $(T, g) \in {}'E_{V,z,\sigma}^{-\epsilon}$ and ω is a collection of total orders on each of the sets $g^{-1}(\lambda')$ ($\lambda' \in \mathbf{k}$). The group of this covering is $\mathcal{G} := \prod_{\lambda' \in \mathbf{k}} \mathfrak{S}_{\pi_{\lambda'}}$. (It acts freely in an obvious way on ${}''E_{V,z,\sigma}^{-\epsilon}$.) Let $\tilde{\mathcal{E}}$ be the local system on ${}'E_{V,z,\sigma}^{-\epsilon}$ associated to the principal covering ζ and the irreducible representation $\boxtimes_{\lambda' \in \mathbf{k}} \Pi_{\pi_{\lambda'}}$ of $\mathcal{G}$. (The rank of $\tilde{\mathcal{E}}$ is $\prod_{\lambda' \in \mathbf{k}} N_{\pi_{\lambda'}}$.) Let $\mathcal{E}_\pi = \xi_!(\mathbf{L} \otimes \tilde{\mathcal{E}})$, a local system on $E_{V,z,\sigma}^{-\epsilon}$ of rank
$$z!(\prod_{\lambda' \in \mathbf{k}} \pi_{\lambda'}!)^{-1} m^{z-\pi_0} \prod_{\lambda' \in \mathbf{k}} N_{\pi_{\lambda'}}.$$

Hence the complex

(a) $$IC(\overline{E_{V,z,\sigma}^{-\epsilon}}, \mathcal{E}_\pi)_{E_V^{-\epsilon}}$$

is defined.

In the following theorem we assume that the characteristic of $\mathbf{k}$ is sufficiently large. (The reason is explained in 2.9.) It is likely that this assumption is unnecessary.

Theorem 2.6. *Let* $V \in \mathcal{C}$. *Let* $\nu = |V|$.

(a) *For any* $\pi \in Z_\nu$, *the local system* $\mathcal{E}_\pi$ *on* $E_{V,t,\sigma}^{-\epsilon}$ *(notation of 2.5) is irreducible.*

(b) *There exists a bijection* $\tilde{Z}_\nu \xrightarrow{\sim} Z_\nu$ *as in 2.4(c) such that if* $\theta \in \tilde{Z}_\nu$ *and* $\pi \in Z_\nu$ *is the corresponding element under this bijection, then*

$$\mathfrak{F}(IC(\overline{E_{V,\theta}^\epsilon}, \bar{\mathbf{Q}}_l)_{E_V^\epsilon}[?]) \cong IC(\overline{E_{V,t,\sigma}^{-\epsilon}}, \mathcal{E}_\pi)_{E_V^{-\epsilon}}[?].$$

The proof will occupy much of the remainder of this section. We see that the complexes 2.5(a) for various $\pi \in Z_\nu$ are (up to shift) exactly the antiorbital complexes on $E_V^{-\epsilon}$.

2.7. We preserve the setup of 1.5. Note that for $j \in [1,s]$, $\mathfrak{F}(\mathcal{K}_j)$ is a G_{U^j}-equivariant perverse sheaf (up to shift) on $E_{U^j}^{-\epsilon}$. Hence $\mathrm{Ind}(\boxtimes_j(\mathfrak{F}(\mathcal{K}_j)) \in \mathcal{D}(E_V^{-\epsilon})$ is defined. We have:

(a) $$\mathfrak{F}(\mathrm{Ind}(\boxtimes_j \mathcal{K}_j)) \cong \mathrm{Ind}(\boxtimes_j(\mathfrak{F}(\mathcal{K}_j))[?] \text{ in } \mathcal{D}(E_V^{-\epsilon}).$$

This is proved in [L2, 5.4] assuming that $m \geq 2$ and $s = 2$; the proof for the case when $m \geq 2$, $s \neq 2$ is similar to that for $m \geq 2, s = 2$ or can be deduced from the case $s = 2$ by repetition. The proof for $m = 1$ is identical to that for $m \geq 2$. (Alternatively, (a) can be deduced from Theorem A.2 in the Appendix.)

2.8. We preserve the setup of 1.5 but we replace ϵ by $-\epsilon$. We fix $\lambda \in \mathbf{k}^*$. Assume that $U^j \cong \mathbf{v}$ for $j \in [1,s]$. Let $\alpha_{\lambda,j} : E_{U^j}^{-\epsilon,*} \to \mathbf{k}^*$ be the map α_λ of 1.1 with V replaced by U^j. Let $\mathcal{K}_j = IC(E_{U^j}^{-\epsilon}, \alpha_{\lambda,j}^* \mathfrak{K}^m)$. Let S'' be the set of all $(T, f) \in E''$ such that the following holds: $f = (V^j)$ is such that for $j \in [1,s]$, $V^j/V^{j-1} \cong U^j$ and the element of $E_{V^j/V^{j-1}}^\epsilon$ induced by T is in $E_{V^j/V^{j-1}}^{-\epsilon,*}$. This is an open dense smooth irreducible subset of E''. Define $\alpha : S'' \to (\mathbf{k}^*)^s$ by $\alpha(T, f) = (a_1\lambda, a_2\lambda, \ldots, a_s\lambda)$ where $T^m = a_j$ on V^j/V^{j-1}. Let $\mathcal{L}'' = \alpha^*((\mathfrak{K}^m)^{\boxtimes s})$, a local system on S''. We have $\mathcal{K}'' = IC(E'', \mathcal{L}'')$. Let S be the set of all $T \in E_V^{-\epsilon}$ such that $\sharp(\mathcal{S}_T) = s$ and $V_{T,\lambda} \cong \mathbf{v}$ for $\lambda \in \mathcal{S}_T$. (Thus, $S = E_{V,s,0}^{-\epsilon}$, see 2.3.) This is an open dense subset of $E_V^{-\epsilon}$. Also $p_3^{-1}(S)$ is an open dense (hence irreducible) subset of S''; let $p_3' : p_3^{-1}(S) \to S$ be the restriction of p_3. We have an isomorphism $u : \tilde{E}_{V,s,0}^{-\epsilon} \xrightarrow{\sim} p_3^{-1}(S)$ (notation of 2.3) given by $(T, \omega) \mapsto (T, f)$ where $f = (V^j)$ is defined by $V^1 = V_{T,\lambda_1}$, $V^2 = V_{T,\lambda_1} \oplus V_{T,\lambda_2}$, etc. Here $\lambda_1, \lambda_2, \ldots, \lambda_s$ are the elements of $\mathcal{S}(T)$ arranged in the order given by ω. Under the isomorphism u, the map p_3' becomes the map ζ in 2.3 hence is a finite principal covering whose group is $\mathfrak{S}_s$. Thus $p_{3!}' \bar{\mathbf{Q}}_l \cong \oplus_{\rho \in \mathcal{R}_s}(\mathcal{L}^\rho)^{\oplus N_\rho}$ as local systems on S where $\mathcal{L}^\rho$ is the local system on S associated to p_3' and the irreducible representation Π_ρ of $\mathfrak{S}_s$. Note that the local systems $\mathcal{L}^\rho, \rho \in \mathcal{R}_s$, are irreducible and mutually nonisomorphic (by the irreducibility of $p_3^{-1}(S)$). Define $\bar{\alpha} : S \to (\mathbf{k}^*)^s/\mathfrak{S}_s$ by $\bar{\alpha}(T) = (a_1\lambda, a_2\lambda, \dots, a_s\lambda)$ (unordered) where $a_j \in \mathbf{k}^*$ are the scalars such that $V_{T,a_j} \neq 0$. Let $\bar{\mathbf{L}}$ be the local system on $(\mathbf{k}^*)^s/\mathfrak{S}_s$ whose inverse image under the obvious map $(\mathbf{k}^*)^s \to (\mathbf{k}^*)^s/\mathfrak{S}_s$ is $(\mathfrak{K}^m)^{\boxtimes s}$. Let $\mathbf{L} = \bar{\alpha}^*(\bar{\mathbf{L}})$, a local system on S. We have $\mathcal{L}''|_{p_3^{-1}(S)} = p_3'^*\mathbf{L}$. Hence

$$\mathrm{Ind}(\boxtimes_j \mathcal{K}_j)|_S = p_{3!}'(p_3'^*\mathbf{L}) = \mathbf{L} \otimes p_{3!}'\bar{\mathbf{Q}}_l = \oplus_{\rho \in \mathcal{R}_s} \mathbf{L} \otimes (\mathcal{L}^\rho)^{\oplus N_\rho}.$$

By 2.2, for $j \in [1, s]$, our $\mathcal{K}_j$ is (up to shift) the D-F transform of $\mathcal{K}_j$ in 1.6. Using 2.7 we deduce that our $\mathrm{Ind}(\boxtimes_j \mathcal{K}_j)$ is (up to shift) the D-F transform of $\mathrm{Ind}(\boxtimes_j \mathcal{K}_j)$ of 1.6. Hence

$$\mathrm{Ind}(\boxtimes_j \mathcal{K}_j) \cong \oplus_{\rho \in \mathcal{R}_s} \mathfrak{F}(P_\rho'')^{\oplus N_\rho}[d]$$

with P_ρ'' as in 1.6 and $d \in \mathbf{Z}$. Note that P_ρ'' are simple mutually nonisomorphic perverse sheaves on $E_V^{-\epsilon}$. Restricting to S we obtain

(a) $$\oplus_{\rho \in \mathcal{R}_s} \mathbf{L} \otimes (\mathcal{L}^\rho)^{\oplus N_\rho} \cong \oplus_{\rho \in \mathcal{R}_s} \mathfrak{F}(P_\rho'')|_S^{\oplus N_\rho}[d]$$

in $\mathcal{D}(S)$. It follows that for each ρ, $\mathfrak{F}(P_\rho'')|_S[d]$ is a local system on S (possibly zero). Let $\mathcal{R}_s'$ be the set of all $\rho \in \mathcal{R}_s$ such that $\mathfrak{F}(P_\rho'')|_S[d] \neq 0$. Since S is open dense in $E_V^{-\epsilon}$, $\mathfrak{F}(P_\rho'')|_S[d]$ ($\rho \in \mathcal{R}_s'$) are irreducible mutually nonisomorphic local system on S. Thus in (a), the right hand side is a direct sum of $\sum_{\rho \in \mathcal{R}_s'} N_\rho$ irreducible local systems while the left hand side is a direct sum of $\sum_{\rho \in \mathcal{R}_s} N_\rho$ nonzero local systems. This forces $\mathcal{R}_s = \mathcal{R}_s'$ and that each $\mathbf{L} \otimes \mathcal{L}^\rho$ is irreducible. If $\rho \neq \rho'$ then $\mathbf{L} \otimes \mathcal{L}^\rho \not\cong \mathbf{L} \otimes \mathcal{L}^{\rho'}$; otherwise the number of nonisomorphic irreducible local systems which appear in the left hand side of (a) would be $< \sharp(\mathcal{R}_s)$ while the analogous number for the right hand side of (a) would be equal to $\sharp(\mathcal{R}_s)$. We see that there is a unique permutation $\rho \mapsto \rho'$ of $\mathcal{R}_s$ such that $\mathfrak{F}(P_\rho'')|_S[d] \cong \mathbf{L} \otimes \mathcal{L}^{\rho'}$ for any $\rho \in \mathcal{R}_s$. It follows that $\mathfrak{F}(P_\rho'') \cong IC(E_V^{-\epsilon}, \mathbf{L} \otimes \mathcal{L}^{\rho'})_{E_V^{-\epsilon}}[?]$. Thus the D-F transforms of orbital complexes on E_V^ϵ with support contained in $E_V^{\epsilon,\lambda}$ are exactly the complexes of the form $IC(E_V^{-\epsilon}, \mathbf{L} \otimes \mathcal{L}^{\rho'})_{E_V^{-\epsilon}}[?]$ for various $\rho \in \mathcal{R}_s$. The bijection $\mathcal{R}_s \to \mathcal{R}_s$, $\rho \mapsto \rho'$ is denoted by $\Phi_{s,\lambda}$.

2.9. Let $V \in \mathcal{C}$. Combining two results [L2, 10.14], [L3, 5.9] in the theory of canonical bases, we see that

(a) *the collection of simple perverse sheaves* $IC(\overline{E_{V,\bar{\sigma}}^\epsilon}, \bar{\mathbf{Q}}_l)_{E_V^\epsilon}[?]$, $(\sigma \in \mathcal{P}_{|V|}^{ap})$ *is mapped bijectively by* $\mathfrak{F}$ *onto the collection of simple perverse sheaves*
$IC(\overline{E_{V,\bar{\sigma}}^{-\epsilon}}, \bar{\mathbf{Q}}_l)_{E_V^{-\epsilon}}[?]$, $(\sigma \in \mathcal{P}_{|V|}^{ap})$.

(Both collections index the canonical basis in degree $|V|$ of a certain algebra associated to affine SL_2. Note that in [L3] the arguments are in characteristic 0 and they imply what we need only in sufficiently large characteristic.)

It follows that there exists a bijection $\sigma \mapsto \sigma^*$, $\mathcal{P}_{|V|}^{ap} \xrightarrow{\sim} \mathcal{P}_{|V|}^{ap}$ such that

(b) $$\mathfrak{F}(IC(\overline{E_{V,\bar{\sigma}}^\epsilon}, \bar{\mathbf{Q}}_l)_{E_V^\epsilon}[?]) \cong IC(\overline{E_{V,\bar{\sigma}^*}^{-\epsilon}}, \bar{\mathbf{Q}}_l)_{E_V^{-\epsilon}}[?]$$

for any $\sigma \in \mathcal{P}_{|V|}^{ap}$.

2.10. Let $V \in \mathcal{C}$. Let $\mathcal{K} = IC(\{0\}, \bar{\mathbf{Q}}_l)_{E_V^\epsilon}$, $\mathcal{K}' = \bar{\mathbf{Q}}_l = IC(E_V^{-\epsilon}, \bar{\mathbf{Q}}_l)$. From the definitions we have $\mathfrak{F}(\mathcal{K}) = \mathcal{K}'[n]$ where $n = \dim E_V^{-\epsilon}$.

2.11. We preserve the setup of 1.5 but we replace ϵ by $-\epsilon$. Assume that $s = t + 1$ where $t \in \mathbf{N}$. Assume that U^j, σ are as in 1.7. Let $\sigma^* \in \mathcal{P}^{ap}_{|U^s|}$ be as in 2.9. For $j \in [1, t]$ let $\mathcal{K}_j = \bar{\mathbf{Q}}_l \in \mathcal{D}(E^{-\epsilon}_{U^j})$. Let $\mathcal{K}_s = IC(\overline{E^{-\epsilon}_{U^s, \bar{\sigma}^*}}, \bar{\mathbf{Q}}_l)_{E^{-\epsilon}_{U^s}}$, ($\bar{\sigma}^*$ as in 1.3). Let S'' be the set of all $(T, f) \in E''$ such that the following holds: $f = (V^j)$ is such that $V^j/V^{j-1} \cong U^j$ for $j \in [1, s]$ and the element of $E^{-\epsilon}_{V/V^{s-1}}$ induced by T is in $E^{-\epsilon}_{V/V^{s-1}, \bar{\sigma}^*}$. This is a locally closed smooth irreducible subset of E''. In our case the complex $\mathcal{K}''$ of 1.5 is $\mathcal{K}'' = IC(\bar{S}'', \bar{\mathbf{Q}}_l)_{E''}$. Let $S = E^{-\epsilon}_{V, t, \sigma^*}$, see 2.3. Now $p_3^{-1}(S)$ is an open dense (hence irreducible) subset of S''; let $p_3' : p_3^{-1}(S) \to S$ be the restriction of p_3. We have an isomorphism $u : \tilde{E}^{-\epsilon}_{V, t, \sigma^*} \xrightarrow{\sim} p_3^{-1}(S)$ (notation of 2.3) given by $(T, \omega) \mapsto (T, f)$ where $f = (V^j)$ is defined by $V^1 = V_{T, \lambda_1}$, $V^2 = V_{T, \lambda_1} \oplus V_{T, \lambda_2}$, etc. Here $\lambda_1, \lambda_2, \ldots, \lambda_t$ are the elements of $\mathcal{S}(T)$ arranged in the order given by ω. Under the isomorphism u, the map p_3' becomes the map ζ in 2.3 hence is a finite principal covering whose group is $\mathfrak{S}_t$. Thus $p'_{3!}\bar{\mathbf{Q}}_l \cong \oplus_{\rho \in \mathcal{R}_t}(\mathcal{L}^\rho)^{\oplus N_\rho}$ as local systems on S where $\mathcal{L}^\rho$ is the local system on S associated to p_3' and the irreducible representation Π_ρ of $\mathfrak{S}_t$. Note that the local systems $\mathcal{L}^\rho, \rho \in \mathcal{R}_t$, are irreducible and mutually nonisomorphic (by the irreducibility of $p_3^{-1}(S)$). Since p_3 is proper, we have $p_3(\bar{S}'') = p_3(\overline{p_3^{-1}(S)}) = \overline{p_3(p_3^{-1}(S))} = \bar{S}$ where $^-$ denotes closure. Hence $\operatorname{supp}(p_{3!}\mathcal{K}'') \subset \bar{S}$.

By 2.9(b), 2.10, for $j \in [1, s]$, our $\mathcal{K}_j$ is (up to shift) the D-F transform of $\mathcal{K}_j$ in 1.7. Using 2.7 we deduce that our $\operatorname{Ind}(\boxtimes_j \mathcal{K}_j)$ is (up to shift) the D-F transform of $\operatorname{Ind}(\boxtimes_j \mathcal{K}_j)$ of 1.7. Hence $\operatorname{Ind}(\boxtimes_j \mathcal{K}_j) = p_{3!}\mathcal{K}''$ is isomorphic to $\oplus_{e \in [1, M]}\mathfrak{F}(P_e)[d'_e]$ where P_e are as in 1.7 and $d'_e \in \mathbf{Z}$. Note that $\mathfrak{F}(P_e)$ are simple perverse sheaves on $E^{-\epsilon}_V$ with support contained in $\bar{S}$. Restricting to S we obtain

$$\oplus_{\rho \in \mathcal{R}_t}(\mathcal{L}^\rho)^{\oplus N_\rho} \cong \oplus_{e \in [1, M]}\mathfrak{F}(P_e)|_S[d'_e]$$

in $\mathcal{D}(S)$. It follows that for any e, $\mathfrak{F}(P_e)|_S[d'_e]$ is a local system on S (possibly zero). Since $\operatorname{supp}(\mathfrak{F}(P_e)) \subset \bar{S}$ we see that $\mathfrak{F}(P_e)|_S[d'_e]$ is either an irreducible local system on S or is 0. We deduce that $\mathfrak{F}(P_e)|_S$ is either 0 or is isomorphic to $\mathcal{L}^\rho[-d'_e]$ for a well defined $\rho \in \mathcal{R}_t$. Hence either $\operatorname{supp}(\mathfrak{F}(P_e)) \subset \bar{S} - S$ or $\mathfrak{F}(P_e) \cong IC(\bar{S}, \mathcal{L}^\rho)_{E^{-\epsilon}_V}[?]$. Conversely, we see that for any $\rho \in \mathcal{R}_t$ we have $IC(\bar{S}, \mathcal{L}^\rho)_{E^{-\epsilon}_V}[?] \cong \mathfrak{F}(P_e)$ for some e. Let $\mathcal{X}_{t, \sigma^*}$ be the set of isomorphism classes of simple perverse sheaves on $E^{-\epsilon}_V$ that are isomorphic to $\mathfrak{F}(P_e)$ for some $e \in [1, M]$. Let $\mathcal{X}'_{t, \sigma^*}$ be the set of isomorphism classes of simple perverse sheaves on $E^{-\epsilon}_V$ that are of the form $IC(\bar{S}, \mathcal{L}^\rho)_{E^{-\epsilon}_V}[?]$ for some $\rho \in \mathcal{R}_t$. We see that

(a) $\sharp(\mathcal{R}_t) = \sharp(\mathcal{X}'_{t, \sigma^*}) \leq \sharp(\mathcal{X}_{t, \sigma^*})$.

Let $\mathcal{Y}'_{t, \sigma}$ be the subset of $\mathcal{Y}_{t, \sigma}$ (see 1.7) which corresponds to the subset $\mathcal{X}'_{t, \sigma^*}$ under the bijection

(b) $\mathcal{Y}_{t, \sigma} \xrightarrow{\sim} \mathcal{X}_{t, \sigma^*}$.

induced by $\mathfrak{F}$. Then $\mathfrak{F}$ defines a bijection

(c) $\mathcal{Y}'_{t, \sigma} \xrightarrow{\sim} \mathcal{X}'_{t, \sigma^*}$.

We show:

(d) if $t, \tilde{t} \in \mathbf{N}$ and $\sigma, \tilde{\sigma} \in \mathcal{P}^{ap}$ are such that $|V| = |\sigma| + t\mathbf{1} = |\tilde{\sigma}| + \tilde{t}\mathbf{1}$ and $(t, \sigma) \neq (\tilde{t}, \tilde{\sigma})$, then $\mathcal{X}'_{t, \sigma^*} \cap \mathcal{X}'_{\tilde{t}, \tilde{\sigma}^*} = \emptyset$.

It is enough to show that $E^{-\epsilon}_{V, t, \sigma^*} \cap E^{-\epsilon}_{V, \tilde{t}, \tilde{\sigma}^*} = \emptyset$. This is clear from the definition.

Applying the inverse of $\mathfrak{F}$ we obtain:

(e) in the setup of (d), we have $\mathcal{Y}'_{t,\sigma} \cap \mathcal{Y}'_{t,\tilde{\sigma}} = \emptyset$.

Let $\nu = |V|$. Let X be the set of all $(t,\tilde{\sigma}) \in \mathbf{N} \times \mathcal{P}^{ap}$ such that $|\tilde{\sigma}| + t1 = \nu$. We have

$$\sharp(\mathcal{P}_\nu) \geq \sharp(\cup_{(t,\tilde{\sigma})\in X}\mathcal{Y}_{t,\tilde{\sigma}}) \geq \sharp(\cup_{(t,\tilde{\sigma})\in X}\mathcal{Y}'_{t,\tilde{\sigma}}) = \sum_{(t,\tilde{\sigma})\in X} \sharp(\mathcal{Y}'_{t,\tilde{\sigma}})$$

(f)
$$= \sum_{(t,\tilde{\sigma})\in X} \sharp(\mathcal{X}'_{t,\tilde{\sigma}*}) = \sum_{(t,\tilde{\sigma})\in X} \sharp(\mathcal{R}_t) = \sharp(\hat{\mathcal{P}}_\nu) = \sharp(\mathcal{P}_\nu).$$

The first $\geq$ follows from the fact that each $\mathcal{Y}_{t,\sigma}$ consists of simple perverse sheaves of the form $IC(\overline{E^\epsilon_{V,\tilde{\sigma}}}, \bar{\mathbf{Q}}_l)_{E^\epsilon_V}[?]$ with $\sigma \in \mathcal{P}_\nu$. The second $\geq$ follows from the inclusion $\mathcal{Y}'_{t,\tilde{\sigma}} \subset \mathcal{Y}_{t,\tilde{\sigma}}$. The first $=$ follows from (e). The second $=$ follows from (c). The third $=$ follows from (b). The fourth $=$ follows from definitions. The fifth $=$ follows from 2.4(a). It follows that each inequality in (f) is an equality.

In particular we see that any simple perverse sheaf of the form $IC(\overline{E^\epsilon_{V,\overline{\sigma_1}}}, \bar{\mathbf{Q}}_l)_{E^\epsilon_V}[?]$ with $\sigma_1 \in \mathcal{P}_\nu$ belongs to $\mathcal{Y}'_{t,\sigma}$ for a unique $(t,\sigma) \in X$. Hence any simple perverse sheaf of the form $\mathfrak{F}(IC(\overline{E^\epsilon_{V,\overline{\sigma_1}}}, \bar{\mathbf{Q}}_l)_{E^\epsilon_V}[?])$ with $\sigma_1 \in \mathcal{P}_\nu$ belongs to $\mathcal{X}'_{t,\sigma*}$ for a unique $(t,\sigma) \in X$. In particular it is of the form $IC(\overline{E^{-\epsilon}_{V,t,\sigma*}}, \mathcal{L}^\rho)_{E^{-\epsilon}_V}[?]$ for a unique $(\sigma^*,\rho) \in \hat{\mathcal{P}}_{|V|}$, $t = \underline{\rho}$.

Since $\sharp(\mathcal{P}_\nu) = \sharp(\hat{\mathcal{P}}_\nu)$, see 2.4, we see that there exists a unique bijection $\Psi_\nu : \mathcal{P}_\nu \xrightarrow{\sim} \hat{\mathcal{P}}_\nu$, $\sigma' \mapsto \hat{\sigma}'$, such that

(g)
$$\mathfrak{F}(IC(\overline{E^\epsilon_{V,\tilde{\sigma}'}}, \bar{\mathbf{Q}}_l)_{E^\epsilon_V}[?]) \cong IC(\overline{E^{-\epsilon}_{V,t,\sigma*}}, \mathcal{L}^\rho)_{E^{-\epsilon}_V}[?]$$

for any $\sigma' \in \mathcal{P}_\nu$; here $\hat{\sigma}' = (\sigma^*,\rho)$, $t = \underline{\rho}$.

2.12. We preserve the setup of 1.5 but we replace ϵ by $-\epsilon$. Assume that $s = t+1$, $t \in \mathbf{N}$. Let $\lambda_1, \lambda_2, \ldots, \lambda_s$, θ, U^j be as in 1.8. Let $\nu = |V|$. Let $\Phi_{(\nu)} : \tilde{Z}_\nu \xrightarrow{\sim} Z_\nu$ be the bijection associated in 2.4 to the collection of bijections $\Phi_{n,\lambda}$ (as in 2.8) and $\Psi_{\nu'}$ (as in 2.11). Let $\pi = \Phi_{(\nu)}(\theta) \in Z_\nu$. For $j \in [1,t]$ let $d_j = \pi_{\lambda_j}$. Let $S_j = E^{-\epsilon}_{U^j,d_j,0}$, see 2.3. Let $\mathbf{L}_j$ be the local system on S_j defined as $\mathbf{L}$ in 2.8 (with V,s,λ replaced by U^j, d_j, λ_j). Let $\mathcal{L}^{\pi\lambda_j}$ be the local system on S_j defined as $\mathcal{L}^{\rho'}$ in 2.8 (with V,s,λ,ρ replaced by $U^j, d_j, \lambda_j, \theta_{\lambda_j}$). Let $\mathcal{K}_j = IC(E^{-\epsilon}_{U^j}, \mathbf{L}_j \otimes \mathcal{L}^{\pi\lambda_j})_{E^{-\epsilon}_{U^j}}$. By 2.8, this $\mathcal{K}_j$ is (up to shift) the D-F transform of the $\mathcal{K}_j$ in 1.8. We write $\pi^{\lambda_s} = \pi^0$ as $(\sigma,\rho) \in \mathcal{P}^{ap} \times \mathcal{R}$. Let $d_s = \underline{\rho}$. Let $S_s = E^{-\epsilon}_{U^s,d_s,\sigma}$. Let $\mathcal{L}^{\pi_0}$ be the local system on S_s defined as $\mathcal{L}^\rho$ in 2.11 (with V,t,ρ replaced by U^s, d_s, π_0). Let $\mathcal{K}_s = IC(\overline{E^{-\epsilon}_{U^s,d_s,\sigma}}, \mathcal{L}^{\pi_0})_{E^{-\epsilon}_{U^s}}$. By 2.11, this $\mathcal{K}_s$ is (up to shift) the D-F transform of the $\mathcal{K}_s$ in 1.8.

Let S'' be the set of all $(T,f) \in E''$ such that the following holds: $f = (V^j)$ is such that $V^j/V^{j-1} \cong U^j$ for $j \in [1,s]$ and T is such that the element T_j of $E^{-\epsilon}_{V^j/V^{j-1}}$ induced by T is in $E^{-\epsilon}_{V^j/V^{j-1},d_j,0}$ (if $j \in [1,t]$) and in $E^{-\epsilon}_{V/V^s,d_s,\sigma}$ (if $j = s$). This is a locally closed smooth irreducible subset of E''. We define a finite principal covering $\zeta'' : \tilde{S}'' \to S''$ as follows. By definition, $\tilde{S}''$ is the set of all triples (T,f,ω) where $(T,f) \in S''$ and $\omega = (\omega_j)$ is a collection of total orders ω_j on $\mathcal{S}(T_j)$ $(j = 1, \ldots, s)$. The group of this covering is $\mathcal{G} = \prod_{j\in[1,s]} \mathfrak{S}_{d_j}$ (it acts freely in an obvious way on $\tilde{E}^{-\epsilon}_{V,t,\sigma}$). The map ζ'' is $(T,f,\omega) \mapsto (T,f)$. Let $\tilde{\mathcal{E}}$ be the local system on S''

associated to the principal covering ζ'' and the irreducible representation $\boxtimes_{\lambda' \in \mathbf{k}} \Pi_{\pi_{\lambda'}}$ of $\mathcal{G}$. Let $\mathbf{L}''$ be the local system on S'' whose stalk at (T, f) is $\boxtimes_{j \in [1,t]} \boxtimes_{\lambda \in S_{T_j}} \mathcal{R}^m_{\lambda \lambda_j}$. In our case the complex $\mathcal{K}''$ of 1.5 is $\mathcal{K}'' = IC(\bar{S}'', \mathbf{L}'' \otimes \tilde{\mathcal{E}})_{E_V^{-\epsilon}}$. From 2.7 we see that our $\mathrm{Ind}(\boxtimes_j \mathcal{K}_j) = p_{3!} \mathcal{K}''$ is (up to shift) the D-F transform of $\mathrm{Ind}(\boxtimes_j \mathcal{K}_j) = IC(\overline{E^\epsilon_{V,\theta}}, \bar{\mathbf{Q}}_l)_{E_V^\epsilon}$ of 1.8.

Let $z = \sum_{j \in [1,s]} \pi_{\lambda_j}$. Let $S = E^{-\epsilon}_{V,z,\sigma}$, see 2.3. Now $p_3^{-1}(S)$ is an open dense subset of S''; let $p_3' : p_3^{-1}(S) \to S$ be the restriction of p_3. We have an isomorphism $u : {}'E^{-\epsilon}_{V,z,\sigma} \xrightarrow{\sim} p_3^{-1}(S)$ (see 2.5) given by $(T, g) \mapsto (T, f)$ where $f = (V^j)$ is defined by

$$V^1 = \oplus_{\lambda \in g^{-1}(\lambda_1)} V_{T,\lambda}, \quad V^2 = \oplus_{\lambda \in g^{-1}(\lambda_1) \cup g^{-1}(\lambda_2)} V_{T,\lambda}, \text{ etc.}$$

Under the isomorphism u, the map p_3' becomes the map ξ of 2.5. Since p_3 is proper, we have $p_3(\bar{S}'') = p_3(\overline{p_3^{-1}(S)}) = \overline{p_3(p_3^{-1}(S))} = \bar{S}$ where $\bar{}$ denotes closure. Hence $\mathrm{supp}(p_{3!} \mathcal{K}'') \subset \bar{S}$ so that $p_{3!} \mathcal{K}''$ is completely determined by its restriction to S. Since $p_{3!} \mathcal{K}''[?]$ is the D-F transform of a simple perverse sheaf, it is itself a simple perverse sheaf. We have

$$(p_{3!} \mathcal{K}'')|_S = p_{3!}'((\mathbf{L}'' \otimes \tilde{\mathcal{E}})_{p_3^{-1}(S)}) = \xi_!(\mathbf{L} \otimes \tilde{\mathcal{E}}) = \mathcal{E}_\pi$$

(notation of 2.5). We see that $p_{3!} \mathcal{K}'' = IC(\overline{E^{-\epsilon}_{V,z,\sigma}}, \mathcal{E}_\pi)_{E_V^{-\epsilon}}$. Thus $IC(\overline{E^{-\epsilon}_{V,z,\sigma}}, \mathcal{E}_\pi)_{E_V^{-\epsilon}}$ is (up to shift) the D-F transform of $IC(\overline{E^\epsilon_{V,\theta}}, \bar{\mathbf{Q}}_l)_{E_V^\epsilon}$. This forces $\mathcal{E}_\pi$ to be an irreducible local system, proving 2.6(a); 2.6(b) follows as well.

2.13. We state a converse of 2.9(a). Let $V \in \mathcal{C}$. Let $\nu = |V|$. Let $\mathcal{K}$ be a G_V-equivariant simple perverse sheaf on E_V^ϵ.

(a) *The following three conditions on $\mathcal{K}$ are equivalent:*

(i) $\mathrm{supp}(\mathcal{K})$ is contained in the nilpotent variety and $\mathrm{supp}(\mathfrak{F}(\mathcal{K}))$ is contained in the nilpotent variety;

(ii) $\mathcal{K}$ is biorbital.

(iii) $\mathcal{K} \cong IC(\overline{E^\epsilon_{V,\bar{\sigma}}}, \bar{\mathbf{Q}}_l)_{E_V^\epsilon}[?]$ for some $\sigma \in \mathcal{P}^{ap}_\nu$.

(This gives a new characterization of the perverse sheaves which constitute the canonical basis [L2] associated to a cyclic quiver.)

Let S_1 (resp. S_2 or S_3) be the set of $\mathcal{K}$ (up to isomorphism) as in (i) (resp. as in (ii) or (iii)). From 2.9(a) we see that $S_3 \subset S_1$. From 2.11 we see that

$$\sharp(S_1) = \sharp((\sigma^*, \rho) \in \hat{\mathcal{P}}_\nu; \underline{\rho} = 0\} = \sharp(\mathcal{P}^{ap}_\nu) = \sharp(S_3).$$

It follows that $S_1 = S_3$.

Clearly, $S_1 \subset S_2$. Assume now that $\mathcal{K} \in S_2$. If $\mathrm{supp}(\mathcal{K})$ is the closure of a non-nilpotent orbit then from 2.12 we see that the support of $\mathfrak{F}(\mathcal{K})$ is a closure of a subvariety of the form $E^{-\epsilon}_{V,z,\sigma}$ where $z > 0$, $\sigma \in \mathcal{P}^{ap}$; in particular, the support of $\mathfrak{F}(\mathcal{K})$ is not the closure of a single orbit so that $\mathfrak{F}(\mathcal{K})$ is not orbital, a contradiction. Thus, $\mathrm{supp}(\mathcal{K})$ is the closure of a nilpotent orbit. The same argument shows that if $\mathrm{supp}(\mathfrak{F}(\mathcal{K}))$ is the closure of a non-nilpotent orbit then $\mathfrak{F}(\mathfrak{F}(\mathcal{K}))$ is not orbital hence $\mathcal{K}$ is not orbital, a contradiction. Thus, $\mathrm{supp}(\mathfrak{F}(\mathcal{K}))$ is the closure of a nilpotent orbit. We see that $\mathcal{K} \in S_1$. Thus $S_2 \subset S_1$ hence $S_1 = S_2$. This proves (a).

2.14. Let $V \in \mathcal{C}$. Let $\nu = |V|$. Assume that we are given an $\mathbf{F}_q$-structure on each V_i. Then E_V^ϵ inherits an $\mathbf{F}_q$-structure and for each $\sigma \in \mathcal{P}_\nu$, the subset $E^\epsilon_{V,\bar{\sigma}}$ of E_V^ϵ is defined over $\mathbf{F}_q$. Let $U = E_V^\epsilon(\mathbf{F}_q)$. Then the dual space U^* can be identified with

$E_V^{-\epsilon}(\mathbf{F}_q)$. For $\sigma \in \mathcal{P}_\nu$ let $U_\sigma = E_{V,\bar\sigma}^{\epsilon}(\mathbf{F}_q)$, $U_\sigma^* = E_{V,\bar\sigma}^{-\epsilon}(\mathbf{F}_q)$; we define $f_\sigma : U \to \bar{\mathbf{Q}}_l$, $f_\sigma' : U^* \to \bar{\mathbf{Q}}_l$ as follows. If $x \in U \cap \overline{E_{V,\bar\sigma}^{\epsilon}}$ (resp. $x' \in U^* \cap \overline{E_{V,\bar\sigma}^{-\epsilon}}$) we define $f_\sigma(x)$ (resp. $f_\sigma'(x')$) as the alternating sum of the traces of the Frobenius map on the stalks of the cohomology sheaves of $IC(\overline{E_{V,\bar\sigma}^{\epsilon}}, \bar{\mathbf{Q}}_l)$ (resp. $IC(\overline{E_{V,\bar\sigma}^{-\epsilon}}, \bar{\mathbf{Q}}_l)$) at x (resp. x'); if $x \in U - \overline{E_{V,\bar\sigma}^{\epsilon}}$ (resp. $x' \in U^* - \overline{E_{V,\bar\sigma}^{-\epsilon}}$) we set $f_\sigma(x) = 0$ (resp. $f_\sigma'(x') = 0$). Note that $(f_\sigma)_{\sigma \in \mathcal{P}_\nu}$ (resp. $(f_\sigma')_{\sigma \in \mathcal{P}_\nu}$) is a $\bar{\mathbf{Q}}_l$-basis of the vector space $\mathcal{F}$ (resp. $\mathcal{F}'$) of functions $U \to \bar{\mathbf{Q}}_l$ (resp. $U^* \to \bar{\mathbf{Q}}_l$) which are constant on the orbits of $G_V(\mathbf{F}_q)$ and vanish on non-nilpotent elements.

For any $(\sigma, \rho) \in \hat{\mathcal{P}}_\nu$ we define a function $f_{\sigma,\rho}'' : U^* \to \bar{\mathbf{Q}}_l$ as follows. Let $t = \rho$. If $x \in U^* \cap \overline{E_{V,t,\sigma}^{-\epsilon}}$ we define $f_{\sigma,\rho}''(x)$ as the alternating sum of the traces of the Frobenius map on the stalks of the cohomology sheaves of $IC(\overline{E_{V,t,\sigma}^{-\epsilon}}, \mathcal{L}^\rho)$ at x; if $x \in U^* - \overline{E_{V,t,\sigma}^{-\epsilon}}$, we set $f_{\sigma,\rho}''(x) = 0$. We have a partition $E_{V,t,\sigma}^{-\epsilon}(\mathbf{F}_q) = \sqcup_g E_{V,t,\sigma}^{-\epsilon}(\mathbf{F}_q)_g$ (g runs through the conjugacy classes in $\mathfrak{S}_t$) where $E_{V,t,\sigma}^{-\epsilon}(\mathbf{F}_q)_g$ is the set of all $T \in E_{V,t,\sigma}^{-\epsilon}(\mathbf{F}_q)$ such that the action of Frobenius on $\mathcal{S}_T$ is by a permutation of type g. Clearly each $E_{V,t,\sigma}^{-\epsilon}(\mathbf{F}_q)_g$ is nonempty. From the definitions for any $T \in E_{V,t,\sigma}^{-\epsilon}(\mathbf{F}_q)$ we have $f_{\sigma,\rho}''(T) = \text{tr}(g, \Pi_\rho)$ where g is defined by $T \in E_{V,t,\sigma}^{-\epsilon}(\mathbf{F}_q)_g$.

From 2.11(g) we see that there exists a bijection $\mathcal{P}_\nu \xrightarrow{\sim} \hat{\mathcal{P}}_\nu$, $\sigma' \mapsto \hat\sigma'$, such that for any $\sigma' \in \mathcal{P}_\nu$ we have $\hat{f}_{\sigma'} = b_{\sigma'} f_{\sigma^*,\rho}''$ where $(\sigma^*, \rho) = \hat\sigma'$ and $b_{\sigma'} \in \bar{\mathbf{Q}}_l^*$. Moreover from 2.9(b) we see that we have $\sigma' \in \mathcal{P}_\nu^{ap}$ if and only if $(\sigma^*, \rho) = \hat\sigma'$ satisfies $\rho = 0$.

Let $f \in \mathcal{F}$ be such that $\hat{f} \in \mathcal{F}'$. We can write uniquely $f = \sum_{\sigma \in \mathcal{P}_\nu} a_\sigma f_\sigma$ where $a_\sigma \in \bar{\mathbf{Q}}_l$. Assume that $a_\sigma \neq 0$ for some $\sigma \in \mathcal{P}_\nu - \mathcal{P}_\nu^{ap}$. Applying Fourier transform we obtain $\hat{f} = \sum_{(\sigma,\rho)\in\hat{\mathcal{P}}_\nu} a_{\sigma,\rho}' f_{\sigma,\rho}''$ where $a_{\sigma,\rho}' \in \bar{\mathbf{Q}}_l$ and $a_{\sigma_0,\rho_0}' \in \bar{\mathbf{Q}}_l^*$ for some $(\sigma_0, \rho_0) \in \hat{\mathcal{P}}_\nu$ with $\rho_0 > 0$. We can assume in addition that $\dim E_{V,\rho_0,\sigma_0}^{-\epsilon} \geq \dim E_{V,\rho,\sigma}^{-\epsilon}$ for any $(\sigma, \rho) \in \hat{\mathcal{P}}_\nu$ such that $a_{\sigma,\rho}' \neq 0$, $\rho > 0$. Let $t = \rho_0$. Then for any $(\sigma, \rho) \in \hat{\mathcal{P}}_\nu$ such that $a_{\sigma,\rho}' \neq 0$ we have $\overline{E_{V,\rho,\sigma}^{-\epsilon}} \cap E_{V,\rho_0,\sigma_0}^{-\epsilon} = \emptyset$ unless $\rho = t$, $\sigma = \sigma_0$. Since $\hat{f} \in \mathcal{F}'$ we have $\hat{f}|_{E_{V,\rho_0,\sigma_0}^{-\epsilon}} = 0$. Hence

$$\sum_{(\sigma,\rho)\in\hat{\mathcal{P}}_\nu} a_{\sigma,\rho}' f_{\sigma,\rho}''|_{E_{V,t,\sigma_0}^{-\epsilon}} = 0$$

and

$$\sum_{\rho\in\mathcal{R}_t} a_{\sigma_0,\rho}' f_{\sigma_0,\rho}''|_{E_{V,t,\sigma_0}^{-\epsilon}} = 0,$$
$$\sum_{\rho\in\mathcal{R}_t} a_{\sigma_0,\rho}' \text{tr}(g, \Pi_\rho) = 0$$

for any conjugacy class g in $\mathfrak{S}_t$. It follows that $a_{\sigma_0,\rho}' = 0$ for any $\rho \in \mathcal{R}_t$, contradicting $a_{\sigma_0,\rho_0}' = 0$. This contradiction shows that $a_\sigma = 0$ for any $s \in \mathcal{P}_\nu - \mathcal{P}_\nu^{ap}$. Thus f is a linear combination of the functions $f_\sigma, (\sigma \in \mathcal{P}_\nu^{ap})$. Conversely, if $\sigma \in \mathcal{P}_\nu^{ap}$ then from 2.9(b) we see that $\hat{f}_\sigma = c f_{\sigma^*}'$ where $\sigma^* \in \mathcal{P}_\nu^{ap}$ and $c \in \bar{\mathbf{Q}}_l^*$. Thus we have $\hat{f}_s \in \mathcal{F}'$. Thus any linear combination f of the functions $f_\sigma, (\sigma \in \mathcal{P}_\nu^{ap})$ satisfies $f \in \mathcal{F}$, $\hat{f} \in \mathcal{F}'$ and we have the following result.

(a) *The functions* $f_\sigma, (\sigma \in \mathcal{P}_\nu^{ap})$ *form a basis of the vector space of all functions* $f \in \mathcal{F}$ *such that* $\hat{f} \in \mathcal{F}'$.

2.15. The results in 2.11 suggest a way to organize the nilpotent K-orbits in $\mathfrak{g}_\zeta$ (in the context of 0.3) with $\zeta \neq 1$. (Here we assume for simplicity that all nilpotent elements in $\mathfrak{g}_\zeta$ have connected isotropy group in K but a similar picture should hold in general.) Namely, each nilpotent K-orbit should be attached to a ϑ-stable Levi subgroup L of G up to K-conjugacy (such that ϑ acts on L_{der}, the derived group of L, as an inner automorphism) and a biorbital complex $\mathcal{K}$ on the ζ-part of the Lie algebra of L_{der}. Moreover, the nilpotent K-orbits corresponding to a given $(L, \mathcal{K})$ should be such that the D-F transforms of the corresponding orbital complexes have the same support and they should be indexed by something similar to the irreducible representations of a Weyl group. Thus something like the "generalized Springer correspondence" should hold even though the small and semismall maps in the usual theory are missing in general.

3. FURTHER EXAMPLES OF ANTIORBITAL COMPLEXES

3.1. Let $G, \mathfrak{g}, \vartheta, K, \kappa$ be as in 0.3. We assume that the characteristic of $\mathbf{k}$ is sufficiently large. For $\zeta \in \mathbf{k}^*$ let $\mathfrak{g}_\zeta, \kappa_\zeta$ be as in 0.3. For any subspace V of $\mathfrak{g}$ we set $V^\perp = \{x \in \mathfrak{g}; \kappa(x, V) = 0\}$.

We say that $\vartheta : G \to G$ is *inner* if there exists a semisimple element $g_0 \in G$ such that $\vartheta(g) = g_0 g g_0^{-1}$ for all $g \in G$.

For any parabolic subgroup P of G we denote by U_P the unipotent radical of P and by $\underline{P}, \underline{U}_P$ the Lie algebras of P, U_P. Assume now that $\vartheta(P) = P$. Then $\underline{P} = \oplus_{\zeta \in \mathbf{k}^*} \underline{P}_\zeta, \underline{U}_P = \oplus_{\zeta \in \mathbf{k}^*} \underline{U}_{P,\zeta}$ where $\underline{P}_\zeta = \underline{P} \cap \mathfrak{g}_\zeta, \underline{U}_{P,\zeta} = \underline{U}_P \cap \mathfrak{g}_\zeta$.

Let $\zeta \in \mathbf{k}^*$. Assume that ϑ is inner. Let X be a K-orbit in the variety of ϑ-stable parabolic subgroups of G (or equivalently a connected component of that variety). Note that X is a projective variety. Let $\tilde{X}_\zeta = \{(x, P); P \in X, x \in \underline{P}_\zeta\}$, $\tilde{X}'_\zeta = \{(x, P); P \in X, x \in \underline{U}_{P,\zeta}\}$. Define $\pi_\zeta : \tilde{X}_\zeta \to \mathfrak{g}_\zeta, \pi'_\zeta : \tilde{X}'_\zeta \to \mathfrak{g}_\zeta$ by $(x, P) \mapsto x$. We show:

(a) $$\mathfrak{F}(\pi_{\zeta!}(\bar{\mathbf{Q}}_l)) = \pi'_{\zeta^{-1}!}(\bar{\mathbf{Q}}_l))[?].$$

Here we view $\mathfrak{F}$ as a functor $\mathcal{D}(\mathfrak{g}_\zeta) \to \mathcal{D}(\mathfrak{g}_{\zeta^{-1}})$. We have a commutative diagram

$$
\begin{array}{ccccc}
\tilde{X}_\zeta & \xleftarrow{\;b\;} & \Xi & \xrightarrow{\;c\;} & \bar{\Xi} \\
\pi_\zeta \downarrow & & r \downarrow & & d \downarrow \\
\mathfrak{g}_\zeta & \xleftarrow{\;s\;} & \mathfrak{g}_\zeta \times \mathfrak{g}_{\zeta^{-1}} & \xrightarrow{\;t\;} & \mathfrak{g}_{\zeta^{-1}}
\end{array}
$$

where

$\Xi = \{(x, x', P); P \in X, x \in \underline{P}_\zeta, x' \in \mathfrak{g}_{\zeta^{-1}}\}, \bar{\Xi} = \{(x', P); x' \in \mathfrak{g}_{\zeta^{-1}}, P \in X\}$,
s, t are the obvious projections and b, c, r, d are the obvious maps. Let $\tilde{t} = tr : \Xi \to \mathfrak{g}_{\zeta^{-1}}$. Let $\kappa' : \Xi \to \mathbf{k}$ be $(x, x', P) \mapsto \kappa_\zeta(x, x')$. We have

$$\mathfrak{F}(\pi_{\zeta!}(\bar{\mathbf{Q}}_l)) = t_!(r_!(\bar{\mathbf{Q}}_l) \otimes \mathcal{L}^{\kappa_\zeta})[?] = \tilde{t}_!(\mathcal{L}^{\kappa'})[?] = d_! c_!(\mathcal{L}^{\kappa'})[?].$$

We have a partition $\Xi = \Xi_0 \cup \Xi_1$ where Ξ_0 is defined by the condition that $x' \in \underline{U}_P$. Let $c_1 : \Xi_1 \to \bar{\Xi}$ be the restriction of c. We show that

(b) $$c_{1!}(\mathcal{L}^{\kappa'}) = 0.$$

The fibre of c_1 at (x', P) is $\underline{P}_\zeta$. The restriction of κ' to this fibre is the linear map $x \mapsto \kappa(x, x')$. It is enough to show that this linear map is not identically zero. (Assume that $\kappa(x, x') = 0$ for any $x \in \underline{P}_\zeta$ that is, $x' \in (\underline{P}_\zeta)^\perp$. Since $x' \in \mathfrak{g}_{\zeta^{-1}}$

we have automatically $x' \in (\underline{P}_{\zeta'})^\perp$ for $\zeta' \neq \zeta$ hence $x' \in \underline{P}^\perp$. Hence $x' \in \underline{U}_P$ a contradiction.) This proves (b).

From (b) we deduce

$$d_! c_! (\mathcal{L}^{\kappa'}) = d_! c_! j_! j^* \mathcal{L}^{\kappa'} = t_{0!}(\mathcal{L}^{\kappa'}|_{\Xi_0})$$

where $j : \Xi_0 \to \Xi$ is the inclusion and $t_0 : \Xi_0 \to \mathfrak{g}_\zeta$ is $(x, x', P) \mapsto x'$. Now t_0 is a composition $\Xi_0 \to \tilde{X}'_{\zeta-1} \xrightarrow{\pi'_{\zeta-1}} \mathfrak{g}_{\zeta-1}$ (the first map, $(x, x', P) \mapsto (x', P)$, is an affine bundle whose fibre at (x', P) is isomorphic to $\underline{P}_\zeta$, a vector space of constant dimension as P runs through X which is a K-orbit.) If $(x, x', P) \in \Xi_0$ then $\kappa'(x, x', P) = \kappa(x, x')$. This is zero since $x \in \underline{P}_\zeta$ and $x' \in \underline{U}_P$. Thus $t_{0!}(\mathcal{L}^{\kappa'}|_{\Xi_0}) = t_{0!}(\bar{\mathbf{Q}}_l) = \pi'_{\zeta-1!}(\bar{\mathbf{Q}}_l)[?]$ and (a) follows.

3.2. We preserve the setup of 3.1 and assume that ϑ is inner. Let $\zeta \in \mathbf{k}^*$. A ϑ-stable parabolic subgroup P of G is said to be ζ-*tight* if $\underline{P}_\zeta = \underline{U}_{P,\zeta}$. In this case, P is also ζ^{-1}-tight; indeed, we have

$$\underline{U}_{P,\zeta^{-1}} = \underline{P}^\perp \cap \mathfrak{g}_{\zeta-1} = \{x \in \mathfrak{g}_{\zeta-1}; \kappa(x, \underline{P}_\zeta) = 0\}$$
$$= \{x \in \mathfrak{g}_{\zeta-1}; \kappa(x, \underline{U}_{P,\zeta}) = 0\} = \underline{U}_P^\perp \cap \mathfrak{g}_{\zeta-1} = \underline{P}_{\zeta-1}.$$

Now let

(a) X be a K-orbit on the variety of ϑ-stable parabolic subgroups of G such that some (or equivalently, any) $P \in X$ is ζ-tight.
In this case we have $\tilde{X}_\zeta = \tilde{X}'_\zeta$, $\tilde{X}_{\zeta-1} = \tilde{X}'_{\zeta-1}$ hence $\pi_{\zeta!}(\bar{\mathbf{Q}}_l) = \pi'_{\zeta!}(\bar{\mathbf{Q}}_l)$, $\pi_{\zeta-1!}(\bar{\mathbf{Q}}_l) = \pi'_{\zeta-1!}(\bar{\mathbf{Q}}_l)$. Combining this with 3.1(a) we obtain

(b) $$\mathfrak{F}(\pi_{\zeta!}(\bar{\mathbf{Q}}_l)) = \mathfrak{F}(\pi'_{\zeta!}(\bar{\mathbf{Q}}_l)) = \pi'_{\zeta-1!}(\bar{\mathbf{Q}}_l)[?] = \pi_{\zeta-1!}(\bar{\mathbf{Q}}_l)[?].$$

Let $\mathcal{B}$ be the variety of Borel subgroups of G. We show:

(c) *if* $B \in \mathcal{B}$, $\vartheta(B) = B$ *and* $\zeta \in \mathbf{k}^* - \{1\}$ *then* B *is* ζ-tight.
Let $\mathfrak{n} = \underline{U}_B$. Now ϑ acts naturally on $\underline{B}, \mathfrak{n}, \underline{B}/\mathfrak{n}$ and we have an obvious direct sum decomposition $\underline{B}/\mathfrak{n} = \oplus_{\zeta' \in \mathbf{k}^*}(\underline{B}/\mathfrak{n})_{\zeta'}$; moreover, $(\underline{B}/\mathfrak{n})_{\zeta'}$ is the image of $\underline{B}_{\zeta'}$ under $\underline{B} \to \underline{B}/\mathfrak{n}$. We can find a semisimple element $g_0 \in G$ such that $\vartheta = \mathrm{Ad}(g_0)$. Since $\vartheta(B) = B$ we have $g_0 \in B$. Hence the centralizer of g_0 in B contains a maximal torus of B. Hence $\underline{B}_1$ contains a Cartan subalgebra $\mathfrak{h}$ of $\underline{B}$. The image of $\mathfrak{h}$ under $\underline{B} \to \underline{B}/\mathfrak{n}$ is on the one hand equal to $\underline{B}/\mathfrak{n}$ and on the other hand is contained in $(\underline{B}/\mathfrak{n})_1$. Thus $\underline{B}/\mathfrak{n} = (\underline{B}/\mathfrak{n})_1$. It follows that $(\underline{B}/\mathfrak{n})_{\zeta'} = 0$ for any $\zeta' \neq 1$. In particular $(\underline{B}/\mathfrak{n})_\zeta = 0$. Hence the image of $\underline{B}_\zeta$ under $\underline{B} \to \underline{B}/\mathfrak{n}$ is 0. In other words, $\underline{B}_\zeta \subset \mathfrak{n}$ and (c) follows.

From (c) we see that any K-orbit X on $\mathcal{B}^\vartheta$ (the variety of ϑ-stable Borel subgroups of G) is as in (a) hence (b) is applicable to it.

3.3. We preserve the setup of 3.1 and assume that ϑ is inner. Let $\zeta \in \mathbf{k}^* - \{1\}$. We define a collection $\mathfrak{Q}'_\zeta$ of simple perverse sheaves on $\mathfrak{g}_\zeta$ as follows. A simple perverse sheaf A on $\mathfrak{g}_\zeta$ is said to be in $\mathfrak{Q}'_\zeta$ if there exists a K-orbit X on $\mathcal{B}^\vartheta$ such that some shift of A is a direct summand of $\pi_{\zeta!}\bar{\mathbf{Q}}_l$ where $\pi_\zeta : \tilde{X}_\zeta \to \mathfrak{g}_\zeta$ is defined in terms of X as in 3.1. Note that any object A of $\mathfrak{Q}'_\zeta$ is K-equivariant and $\mathrm{supp}(A)$ is contained in $\mathfrak{g}_\zeta^{nil}$, the variety of nilpotent elements of $\mathfrak{g}_\zeta$ (we use 3.2(b)).

Note that if X is as in 3.2(a) and $\pi_\zeta : \tilde{X}_\zeta \to \mathfrak{g}_\zeta$ is defined in terms of X as in 3.1 then by the decomposition theorem, $\pi_{\zeta!}\bar{\mathbf{Q}}_l \cong \oplus_h A_h[d_h]$ where A_h are simple perverse sheaves on $\mathfrak{g}_\zeta$ and $d_h \in \mathbf{Z}$. We show that

(a) $A_h \in \mathfrak{Q}'_\zeta$ for any h.

Let X' be the variety of all $B \in \mathcal{B}^\vartheta$ such that B is contained in some (necessarily unique) $P \in X$. Define $j : X' \to X$ by $B \mapsto P$. Note that X' is a union of (finitely many) K-orbits $X'_1, \ldots, X'_s$ on $\mathcal{B}^\vartheta$. Let $Y_i = \{(x, B); B \in X'_i; x \in \underline{B}_\zeta\}$; define $\rho_i : Y_i \to \mathfrak{g}_\zeta$ by $(x, B) \mapsto x$. Let $Y = \{(x, B); B \in X'; x \in \underline{B}_\zeta\}$; define $\rho : Y \to \mathfrak{g}_\zeta$ by $(x, B) \mapsto x$. We have $Y = \sqcup_i Y_i$ hence $\rho_! \bar{\mathbf{Q}}_l = \oplus_i \rho_{i!} \bar{\mathbf{Q}}_l$. Define $\sigma : Y \to \tilde{X}_\zeta$ by $\sigma(x, B) = (x, j(B))$. For $(x, P) \in \tilde{X}_\zeta$ we can identify $\sigma^{-1}(x, P)$ with $\{B \in \mathcal{B}^\vartheta; B \subset P\}$ (if $B \in \mathcal{B}^\vartheta$, $B \subset P$ we have automatically $x \in \underline{B}_\zeta$; indeed we have $x \in \underline{P}_\zeta = \underline{U}_{P,\zeta} \subset \underline{U}_{B,\zeta} \subset \underline{B}_\zeta$). We see that σ is a locally trivial fibration whose fibres are finite unions of flag manifolds hence $\sigma_! \bar{\mathbf{Q}}_l \cong \oplus_j \bar{\mathbf{Q}}_l[2c_j]$ where c_j are integers. We have $\rho = \pi_\zeta \sigma$ hence $\rho_! \bar{\mathbf{Q}}_l = \oplus_j \pi_{\zeta!}\bar{\mathbf{Q}}_l[2c_j]$. Hence $A_h[?]$ is a direct summand of $\rho_! \bar{\mathbf{Q}}_l$. Hence $A_h[?]$ is a direct summand of $\rho_{i!}\bar{\mathbf{Q}}_l$ for some i. Thus (a) holds.

Let $A \in \mathfrak{Q}'_\zeta$. It is known [V,§2, Prop.2] that $\mathfrak{g}_\zeta^{nil}$ is a union of finitely many K-orbits. It follows that $A = IC(\bar{\mathcal{O}}, \mathcal{E})_{\mathfrak{g}_\zeta}[?]$ where $\mathcal{O}$ is a K-orbit in $\mathfrak{g}_\zeta^{nil}$, $\bar{\mathcal{O}}$ is the closure of $\mathcal{O}$ and $\mathcal{E}$ is an irreducible K-equivariant local system on $\mathcal{O}$. Thus A is an orbital complex on $\mathfrak{g}_\zeta$ with unipotent support.

From 3.2(b) we see that $\mathfrak{F}$ defines a bijection from $\mathfrak{Q}'_\zeta$ (up to isomorphism) to $\mathfrak{Q}'_{\zeta^{-1}}$ (up to isomorphism). We see that

(b) *any object of $\mathfrak{Q}'_\zeta$ is biorbital. Moreover $\mathfrak{Q}'_\zeta \subset \mathfrak{Q}_\zeta$ (see 0.6).*
We show:

(c) *$\mathfrak{Q}'_\zeta$ is nonempty. Hence $\mathfrak{Q}_\zeta$ is nonempty. More precisely, there exists a K-orbit $\mathcal{O}$ in $\mathfrak{g}_\zeta^{nil}$ such that $IC(\bar{\mathcal{O}}, \bar{\mathbf{Q}}_l)_{\mathfrak{g}_\zeta}[?] \in \mathfrak{Q}'_\zeta$ where $\bar{\mathbf{Q}}_l$ is viewed as a local system on $\mathcal{O}$.*
It is well known that $\mathcal{B}^\vartheta \neq \emptyset$. Let X be a K-orbit on $\mathcal{B}^\vartheta$. Define $\pi_\zeta : \tilde{X}_\zeta \to \mathfrak{g}_\zeta$ as in 3.1 in terms of X. Note that $\tilde{X}_\zeta$ is vector bundle over X (whose fibre over $B \in X$ is $\underline{B}_\zeta$); in particular $\tilde{X}_\zeta$ is smooth irreducible. Hence $\pi_\zeta(\tilde{X}_\zeta)$ is an irreducible subvariety of $\mathfrak{g}_\zeta^{nil}$ (we use 3.2(b)). There is a unique K-orbit $\mathcal{O}$ in $\mathfrak{g}_\zeta^{nil}$ such that $\mathcal{O}$ is open in $\pi_\zeta(\tilde{X}_\zeta)$. Let $n = \dim \pi_\zeta^{-1}(x)$ for any $x \in \mathcal{O}$ (note that $\pi_\zeta^{-1}(x) \neq \emptyset$ for $x \in \mathcal{O}$). For $x \in \mathcal{O}$ let S_x be the set of irreducible components of dimension n of $\pi_\zeta^{-1}(x)$. We have a finite covering $\tau : S \to \mathcal{O}$ whose fibre at $x \in \mathcal{O}$ is S_x. Note that the 2δ-th cohomology sheaf $\mathcal{F}$ of $(\pi_{\zeta!}\bar{\mathbf{Q}}_l)|_{\mathcal{O}}$ may be identified with $\tau_! \bar{\mathbf{Q}}_l$ (we ignore Tate twists). Hence it contains $\bar{\mathbf{Q}}_l$ as a direct summand. Since $(\pi_{\zeta!}\bar{\mathbf{Q}}_l)|_{\mathcal{O}}$ is a direct sum of shifts of irreducible local systems on $\mathcal{O}$ it follows that some shift of $\bar{\mathbf{Q}}_l$ is a direct summand of $(\pi_{\zeta!}\bar{\mathbf{Q}}_l)|_{\mathcal{O}}$. By the decomposition theorem, $\pi_{\zeta!}\bar{\mathbf{Q}}_l$ is a direct sum of shifts of irreducible perverse sheaves with support contained in the closure $\bar{\mathcal{O}}$ of $\mathcal{O}$. Hence some shift of $IC(\bar{\mathcal{O}}, \bar{\mathbf{Q}}_l)_{\mathfrak{g}_\zeta}$ is a direct summand of $\pi_{\zeta!}\bar{\mathbf{Q}}_l$. This proves (c).

From (b), (c) we deduce:

(d) *there exists a K-orbit $\mathcal{O}$ in $\mathfrak{g}_\zeta^{nil}$ such that $IC(\bar{\mathcal{O}}, \bar{\mathbf{Q}}_l)_{\mathfrak{g}_\zeta}[?]$ belongs to $\mathfrak{Q}_\zeta$. Here $\bar{\mathbf{Q}}_l$ is viewed as a local system on $\mathcal{O}$.*
Let $\mathbf{V}'_\zeta$ be the $\mathbf{Q}(v)$-vector space with basis given by the isomorphism classes of objects in $\mathfrak{Q}'_\zeta$. (v is an indeterminate). For any K-orbit X on $\mathcal{B}^\vartheta$ we set $[X] =$

$\sum_{A,j} n_{A,j} v^j \in \mathbf{V}'_\zeta$ (sum over $A \in \mathfrak{Q}'_\zeta$ up to isomorphism and $j \in \mathbf{Z}$) where $\pi_{\zeta!} \bar{\mathbf{Q}}_l \cong \oplus_{A,j} A[j]$ ($\pi_\zeta : \tilde{X}_\zeta \to \mathfrak{g}_\zeta$ is defined in terms of X as in 3.1). We conjecture that:

(e) *the elements $[X]$ (for various K-orbits X on $\mathcal{B}^\vartheta$) generate the vector space* $\mathbf{V}'_\zeta$.

This is known to be true in the case arising from a cyclic quiver (see [L2]) and also in the case where ζ has large order in $\mathbf{k}^*$ (see [L5]).

3.4. We preserve the setup of 0.3 and assume that ϑ is inner. Let $\zeta \in \mathbf{k}^* - \{1\}$. The following result is analogous to 3.3(d).

(a) *There exists a nonzero function $f : \mathfrak{g}_\zeta(\mathbf{F}_q) \to \bar{\mathbf{Q}}_l$ (which is constant on each orbit of $K(\mathbf{F}_q)$) such that f vanishes on any non-nilpotent element of $\mathfrak{g}_\zeta(\mathbf{F}_q)$ and $\hat{f} : \mathfrak{g}_{\zeta^{-1}}(\mathbf{F}_q) \to \bar{\mathbf{Q}}_l$ vanishes on any non-nilpotent element of $\mathfrak{g}_{\zeta^{-1}}(\mathbf{F}_q)$.*
From the assumptions in 0.3, $\mathcal{B}^\vartheta$ has at least one $\mathbf{F}_q$-rational point. Hence it has some irreducible component X which is defined over $\mathbf{F}_q$. We have $X(\mathbf{F}_q) \neq \emptyset$. The morphism $\pi_\zeta : \tilde{X}_\zeta \to \mathfrak{g}_\zeta$ in 3.1 restricts to a map $t : \tilde{X}_\zeta(\mathbf{F}_q) \to \mathfrak{g}_\zeta(\mathbf{F}_q)$. Define $f_\zeta : \mathfrak{g}_\zeta(\mathbf{F}_q) \to \bar{\mathbf{Q}}_l$ by $f_\zeta(x) = \sharp(t^{-1}(x))$. We have $f_\zeta(0) = \sharp(X(\mathbf{F}_q)) \neq 0$. Thus $f_\zeta \neq 0$. From 3.2(b) we deduce $\hat{f}_\zeta = c f_{\zeta^{-1}}$ where $c \in \bar{\mathbf{Q}}_l^*$ and that f_ζ vanishes on any non-nilpotent element of $\mathfrak{g}_\zeta(\mathbf{F}_q)$; similarly, $f_{\zeta^{-1}}$ vanishes on any non-nilpotent element of $\mathfrak{g}_{\zeta^{-1}}(\mathbf{F}_q)$. Hence $\hat{f}_\zeta$ vanishes on any non-nilpotent element of $\mathfrak{g}_{\zeta^{-1}}(\mathbf{F}_q)$. Clearly, f_ζ is constant on each orbit of $K(\mathbf{F}_q)$. This proves (a).

3.5. Assume that $2 \neq 0$ in $\mathbf{k}$. Let V be a $\mathbf{k}$-vector space of finite dimension $N = 2n \geq 4$ with a given nondegenerate symmetric bilinear form $(,) : V \times V \to \mathbf{k}$. Let K be the corresponding special orthogonal group acting on V in an obvious way. For any $\lambda \in \mathbf{k}$ let $Q_\lambda = \{x \in V; (x,x)/2 = \lambda\}$. Note that if $\lambda \neq 0$, then Q_λ is a single K-orbit in V; moreover $Q_0 - \{0\}$ is a single K-orbit in V. Also the isotropy group in K of any point in V is connected. For $\lambda \in \mathbf{k}^*$ let $\mathcal{K}_\lambda = IC(Q_\lambda, \bar{\mathbf{Q}}_l)_V \in \mathcal{D}(V)$. We set $\mathcal{K}_0 = IC(Q_0, \bar{\mathbf{Q}}_l)_V \in \mathcal{D}(V)$ where $\bar{\mathbf{Q}}_l$ is viewed as a local system on $Q_0 - \{0\}$. We set $\mathcal{K}'_0 = IC(\{0\}, \bar{\mathbf{Q}}_l)_V \in \mathcal{D}(V)$. Let $V_* = V - Q_0$, an open subset of V. For $\lambda \in \mathbf{k}^*$ we define $\alpha_\lambda : V_* \to \mathbf{k}^*$ by $\alpha_\lambda(x) = \lambda(x,x)/2$. We identify V with its dual via $(,)$. Hence $\mathfrak{F} : \mathcal{D}(V) \to \mathcal{D}(V)$ is well defined. The following result describes the antiorbital complexes on V.

(i) $\mathfrak{F}(\mathcal{K}_\lambda) = IC(V, \alpha_\lambda^* \mathfrak{K}^2)[?]$ for any $\lambda \in \mathbf{k}^*$;

(ii) $\mathfrak{F}(\mathcal{K}_0) = \mathcal{K}_0[?]$;

(iii) $\mathfrak{F}(\mathcal{K}'_0) = \bar{\mathbf{Q}}_l[?]$.

Now (iii) is obvious and (ii) is proved in [L4]. We prove (i). It is enough to check this at the level of functions on the set of rational points of V over a finite field. We set $k = \mathbf{F}_q$. Let U be a k-vector space of dimension $N = 2n \geq 4$ with a fixed nondegenerate symmetric bilinear form $(,) : U \times U \to k$ which is split over k. Let $\lambda \in k^*$. Define $f : U \to \bar{\mathbf{Q}}_l$ by $f(x) = 1$ if $(x,x)/2 = \lambda$, $f(x) = 0$ if $(x,x)/2 \neq \lambda$. By 0.1, $\hat{f} : U \to \bar{\mathbf{Q}}_l$ is given by $\hat{f}(x) = q^{-n} \sum_{y \in U; (y,y)=2\lambda} \psi(x,y)$ for $x \in U$. We compute $\hat{f}(x)$ assuming that $(x,x)/2 = \lambda' \neq 0$. We can find a 2-dimensional subspace P of U such that $x \in P$ and $(,)$ is nondegenerate, split on P. Let $P' = \{z \in U; (z,P) = 0\}$. Note that $(,)$ is nondegenerate split on P'. We have

$$\hat{f}(x) = q^{-n} \sum_{p \in P, p' \in P'; (p,p)+(p',p')=2\lambda} \psi(x,p)$$

$$= q^{-n} \sum_{p \in P} \sharp(p' \in P'; (p',p') = 2\lambda - (p,p))\psi(x,p)$$

$$= q^{-n}\Big(\sum_{p \in P; (p,p)=2\lambda} ((q^n - 1)(q^{n-1} + 1) + 1)\psi(x,p)$$

$$+ \sum_{p \in P; (p,p) \neq 2\lambda} q^{n-1}(q^n - 1)\psi(x,p)\Big)$$

$$= \sum_{p \in P; (p,p)=2\lambda} \psi(x,p) + q^{-1}(q^n - 1) \sum_{p \in P} \psi(x,p).$$

The last sum is 0 since $p \mapsto \psi(x,p)$ is a non-trivial character $P \to \bar{\mathbf{Q}}_l^*$. Thus $\hat{f}(x) = \sum_{p \in P; (p,p)=2\lambda} \psi(x,p)$. We pick a basis e_1, e_2 of P such that $(e_1, e_2) = 1$, $(e_i, e_i) = 0$ for $i = 1, 2$. We set $x = x_1 e_1 + x_2 e_2$ where $x_i \in k$, $x_1 x_2 = \lambda'$. We set $p = p_1 e_1 + p_2 e_2$ where $p_i \in k$, $p_1 p_2 = \lambda$. We have

$$\hat{f}(x) = \sum_{p_1, p_2 \in k^*; p_1 p_2 = \lambda} \psi(x_1 p_1 + x_2 p_2) = \sum_{p_1', p_2' \in k^*; p_1' p_2' = \lambda\lambda'} \psi(p_1' + p_2').$$

This identity (and the analogous identities where q is replaced by a power of q) implies (i).

We now give an alternative proof of (ii) at the level of functions on U. For any n-dimensional isotropic subspace L of U we consider the function $f_L : U \to \bar{\mathbf{Q}}_l$ which takes the constant value 1 on L and is 0 on $U - L$. It is clear that $\hat{f}_L = f_L$. Let $f = c^{-1} \sum_L f_L$ where L runs over all n-dimensional isotropic subspaces of U and $c = 2(q+1)(q^2 + 1)\ldots(q^{n-2} + 1)$. Clearly, $f(x) = 0$ if $(x,x) \neq 0$, $f(x) = 1$ if $(x,x) = 0$, $x \neq 0$, $f(x) = 1 + q^{n-1}$ if $x = 0$. We have $\hat{f} = f$. This gives the required identity.

Let $\mathcal{K}$ be a K-equivariant simple perverse sheaf on V. From (i),(ii),(iii) above we see that the following three conditions on $\mathcal{K}$ are equivalent:

(I) both $\operatorname{supp}(\mathcal{K})$ and $\operatorname{supp}(\mathfrak{F}(\mathcal{K}))$ are contained in $\{x \in V; (x,x) = 0\}$;

(II) $\mathcal{K}$ is biorbital;

(III) $\mathcal{K} \cong \mathcal{K}_0[?]$.

3.6. Let V be a $\mathbf{k}$-vector space of dimension $2n + 2 \geq 6$ with a fixed nondegenerate symplectic form $\langle,\rangle : V \times V \to \mathbf{k}$ and with a fixed grading $V = V_0 \oplus V_1$ such that $\langle V_0, V_1 \rangle = 0$ and such that $\dim V_0 = 2, \dim V_1 = 2n$. Let $\mathfrak{sp}(V) = \{T \in \operatorname{End}(V); \langle T(x), y \rangle + \langle x, T(y) \rangle = 0 \quad \forall x, y \in V\}$. Let

$$E = \{T \in \mathfrak{sp}(V); TV_0 \subset V_1, TV_1 \subset V_0\}.$$

Note that $E = \mathfrak{sp}_{-1}$ where $\mathfrak{sp}_{\pm 1}$ are the ± 1 eigenspaces of an involution of $\mathfrak{sp}(V)$ (induced by an involution of $Sp(V)$ whose fixed point set is $K = Sp(V_0) \times Sp(V_1)$) which acts naturally on E. Hence the notion of antiorbital complex on E is well defined (a special case of 0.6). (In our case the function κ_ζ is the restriction to $E \times E$ of the symmetric bilinear form $T, T' \mapsto \operatorname{tr}(TT')$ on $\operatorname{End}(V)$.) The variety of nilpotent elements in E decomposes into a union of three K-orbits $\{0\}, \mathcal{O}, \mathcal{O}'$ represented by $0, N, N'$ where $(V, N) \cong (V_{1,3}, T_1)^{\oplus 2} \oplus (V_{1,1}, T_1)^{\oplus(2n-4)}$, $(V, N') \cong (V_{0,1}, T_1) \oplus (V_{1,0}, T_1) \oplus (V_{1,1}, T_1)^{\oplus(2n-2)}$ (as objects of $\mathcal{C}^1$ that is without a symplectic form). Note that the isotropy groups of $0, N, N'$ in K are connected. Let $\mathcal{K}_0 = IC(\{0\}, \bar{\mathbf{Q}}_l)_E[?]$, $\mathcal{K} = IC(\bar{\mathcal{O}}, \bar{\mathbf{Q}}_l)_E[?]$, $\mathcal{K} = IC(\bar{\mathcal{O}}', \bar{\mathbf{Q}}_l)_E[?]$ be the corresponding orbital complexes on E. We have the following result.

(a) $\mathfrak{F}(\mathcal{K}_0) = IC(E, \bar{\mathbf{Q}}_l)[?]$, $\mathfrak{F}(\mathcal{K}) \cong \mathcal{K}$, $\mathfrak{F}(\mathcal{K}') \cong \mathcal{K}'$. *In particular, $\mathcal{K}$ and $\mathcal{K}'$ are biorbital.*

The first equality in (a) is obvious. Now let X be the set of all (T, W) where $T \in E$ and W is an n-dimensional isotropic subspace of V_1 such that $T(V_0) \subset W \subset \ker(T)$. Let X' be the set of all (T, W') where $T \in E$ and W' is a line of V_0 such that $T(V_1) \subset W' \subset \ker(T)$. Note that X, X' are smooth varieties and the obvious projections $\rho : X \to E$, $\rho' : X' \to E$ are proper maps. If $T \in \rho(X)$ then clearly $T^3 = 0$ so that T is nilpotent. Similarly if $T \in \rho'(X')$ then $T^3 = 0$ so that T is nilpotent; actually in this case we have $T^2 = 0$. (It is enough to show that $T^2 V_0 = 0$ or that $T^2 x = 0, T^2 x' = 0$ where x, x' is a basis of V_0 such that $\langle x, x' \rangle = 1$. Clearly, $Tx = 0$, $T^2 x' = cx$ where $c \in \mathbf{k}$. We have $c = \langle cx, x' \rangle = \langle T^2 x', x' \rangle = -\langle Tx', Tx' \rangle = 0$. Thus $c = 0$ and $T^2 x' = 0$.) Note that $\rho^{-1}(N)$ can be identified with the variety of all n-dimensional isotropic subspaces W of V_1 such that W contains $N(V_0)$ (a 2-dimensional isotropic subspace of V_1) and is contained in $\ker(N) \cap V_1$ (a codimension 2 subspace of V_1 equal to the perpendicular of $N(V_0)$). Thus ρ restricts to a map $\rho^{-1}(\mathcal{O}) \to \mathcal{O}$ which is a (locally trivial) fibre bundle whose fibre is isomorphic to the space of Lagrangian subspaces of a $(2n-4)$ dimensional symplectic vector space. Since $\mathcal{O}$ is open in the nilpotent variety of E we see (using the decomposition theorem) that $\rho_! \bar{\mathbf{Q}}_l$ is isomorphic to a direct sum of complexes of the form $\mathcal{K}[?]$ (at least one) and of some complexes of the form $\mathcal{K}'[?]$ or $\mathcal{K}_0[?]$.

Next we note that $\rho'^{-1}(N')$ is a single point of X' namely (N', W') where $W' = N'(V_1)$; more precisely, the restriction of ρ' from $\rho'^{-1}(\mathcal{O}')$ to $\mathcal{O}'$ is an isomorphism. Since $\rho'(X') \subset \mathcal{O}' \cup \{0\}$ and $\mathcal{O}'$ is open in $\mathcal{O}' \cup \{0\}$ we see (using the decomposition theorem) that $\rho'_! \bar{\mathbf{Q}}_l$ is isomorphic to $\mathcal{K}'[?]$ direct sum with some complexes of the form $\mathcal{K}_0[?]$.

Let W be an n-dimensional isotropic subspace of V_1 and let W' be a line in V_0. Let

$\mathfrak{p} = \{T \in \mathfrak{sp}(V); TW \subset W\}$, $\mathfrak{p}' = \{T \in \mathfrak{sp}(V); TW' \subset W'\}$,

$\mathfrak{n} = \{T \in \mathfrak{sp}(V); TV \subset W^\perp, TW^\perp \subset W, TW = 0\}$, $\mathfrak{n}' = \{T \in \mathfrak{sp}(V); TV \subset W'^\perp, TW'^\perp \subset W', TW' = 0\}$.

Here $W^\perp, W'^\perp$ denote the perpendicular to W, W' with respect to $\langle, \rangle$. Note that $\mathfrak{p}, \mathfrak{p}'$ are parabolic subalgebras of $\mathfrak{sp}(V)$ with nil-radicals $\mathfrak{n}, \mathfrak{n}'$. Moreover we have $\mathfrak{p} = \mathfrak{p}_1 \oplus \mathfrak{p}_{-1}, \mathfrak{p}' = \mathfrak{p}'_1 \oplus \mathfrak{p}'_{-1}$, $\mathfrak{n} = \mathfrak{n}_1 \oplus \mathfrak{n}_{-1}$, $\mathfrak{n}' = \mathfrak{n}'_1 \oplus \mathfrak{n}'_{-1}$ where $()_i = () \cap \mathfrak{sp}_i$. We show that $\mathfrak{p}_{-1} = \mathfrak{n}_{-1}, \mathfrak{p}'_{-1} = \mathfrak{n}'_{-1}$ (that is, the parabolic subgroups of $Sp(V)$ corresponding to $\mathfrak{p}, \mathfrak{p}'$ are (-1)tight). Let $T \in \mathfrak{p}_{-1}$. We have $TW \subset V_0 \cap W = 0$ so that $TW = 0$. It follows that $TV \subset W^\perp$. We have $W^\perp = V_0 \oplus W$ hence $TW^\perp = TV_0 + TW = TV_0 \subset W^\perp \cap V_1 = W$. We see that $\mathfrak{p}_{-1} = \mathfrak{n}_{-1}$. A similar proof shows that $\mathfrak{p}'_{-1} = \mathfrak{n}'_{-1}$. Now $\rho_! \bar{\mathbf{Q}}_l = \widetilde{\mathrm{Ind}}_{\mathfrak{p}_{-1}}^{\mathfrak{sp}_{-1}} \bar{\mathbf{Q}}_l$, $\rho'_! \bar{\mathbf{Q}}_l = \widetilde{\mathrm{Ind}}_{\mathfrak{p}'_{-1}}^{\mathfrak{sp}_{-1}} \bar{\mathbf{Q}}_l$ (see A.1). Here $\bar{\mathbf{Q}}_l$ is viewed as a complex on $\mathfrak{p}_{-1}/\mathfrak{n}_{-1} = 0$ or $\mathfrak{p}'_{-1}/\mathfrak{n}'_{-1} = 0$. From Theorem A.2 (or from 3.2(b)) we see that $\mathfrak{F}(\rho_! \bar{\mathbf{Q}}_l) = \rho_! \bar{\mathbf{Q}}_l[?]$, $\mathfrak{F}(\rho'_! \bar{\mathbf{Q}}_l) = \rho'_! \bar{\mathbf{Q}}_l[?]$.

It follows that $\mathfrak{F}(\rho'_! \bar{\mathbf{Q}}_l)$ is isomorphic to $\mathcal{K}'[?]$ direct sum with some complexes of the form $\mathcal{K}_0[?]$; it is also isomorphic to $\mathfrak{F}(\mathcal{K}')[?]$ direct sum with some complexes of the form $\mathfrak{F}(\mathcal{K}_0)[?]$. But $\mathrm{supp}\mathfrak{F}(\mathcal{K}_0) = E$ showing that $\mathfrak{F}(\mathcal{K}_0)[?]$ cannot be a direct summand of $\mathfrak{F}(\rho_!'\bar{\mathbf{Q}}_l)$. It follows that $\rho_!'\bar{\mathbf{Q}}_l \cong \mathcal{K}'[?] \cong \mathfrak{F}(\mathcal{K}')[?]$. Hence $\mathfrak{F}(\mathcal{K}') \cong \mathcal{K}'$.

We also see that $\mathfrak{F}(\rho_!\bar{\mathbf{Q}}_l)$ is isomorphic to a direct sum of complexes $\mathcal{K}[?]$ (at least one) and complexes of the form $\mathcal{K}'[?]$ or $\mathcal{K}_0[?]$; it is also isomorphic to a direct sum of complexes $\mathfrak{F}(\mathcal{K})[?]$ (at least one) and complexes of the form $\mathfrak{F}(\mathcal{K}')[?] = \mathcal{K}'[?]$ or $\mathfrak{F}(\mathcal{K}_0)[?]$. Again $\mathfrak{F}(\mathcal{K}_0)[?]$ cannot be a direct summand of $\mathfrak{F}(\rho_!\bar{\mathbf{Q}}_l)$. It follows that $\mathfrak{F}(\mathcal{K})$ is isomorphic to $\mathcal{K}$ or to $\mathcal{K}'$. If $\mathfrak{F}(\mathcal{K}) \cong \mathcal{K}'$ then $\mathcal{K} \cong \mathfrak{F}(\mathcal{K}')$ hence $\mathcal{K} \cong \mathcal{K}'$ which is not the case. Hence we have $\mathfrak{F}(\mathcal{K}) \cong \mathcal{K}$. This proves (a).

3.7. Let V be a **k**-vector space of dimension $2n$ with a fixed nondegenerate symplectic form $\langle,\rangle : V \times V \to \mathbf{k}$. Let

$$E = \{T \in \mathrm{End}(V); \langle T(x), y\rangle = \langle x, T(y)\rangle \quad \forall x, y \in V\}.$$

Note that E can be viewed as the (-1) eigenspace of an involution of $\mathrm{End}(V)$ (induced by an involution of $GL(V)$ whose fixed point set is the symplectic group $Sp(V)$ which acts naturally on E). Hence the notion of antiorbital complex on E is well defined (a special case of 0.6). (In our case the function κ_ζ is the restriction to $E \times E$ of the symmetric bilinear form $T, T' \mapsto \mathrm{tr}(TT')$ on $\mathrm{End}(V)$.) Let E_0 be the set of all $T \in E$ such that $T : V \to V$ is semisimple and any eigenspace of T is 2-dimensional. Note that E_0 is open dense in E. Using methods similar to those in §2 we see that any antiorbital complex on E is of the form $IC(E_0, \mathcal{L})[?]$ for a suitable local system $\mathcal{L}$ on E_0 (compare with [L4, §13].) It follows that, if $n > 0$, there are no biorbital complexes on E.

3.8. We preserve the setup of 3.1. Let $\zeta \in \mathbf{k}^*$. For any K-orbit $\mathcal{O}$ in $\mathfrak{g}_\zeta^{nil}$ let $\mathcal{O}^! = \{(x, y) \in \mathfrak{g}_\zeta \times \mathfrak{g}_{\zeta^{-1}}; x \in \mathcal{O}, [x, y] = 0\}$ where $[,]$ is the bracket in $\mathfrak{g}$. As in [L5, 22.2], we identify $\mathcal{O}^!$ with the conormal bundle of $\mathcal{O}$ in $\mathfrak{g}_\zeta$. Hence it is smooth, irreducible of dimension $\dim \mathfrak{g}_\zeta$. Hence $A := \{(x, y) \in \mathfrak{g}_\zeta^{nil} \times \mathfrak{g}_{\zeta^{-1}}; [x, y] = 0\}$ is a (closed) subvariety of $\mathfrak{g}_\zeta \times \mathfrak{g}_{\zeta^{-1}}$ of pure dimension $\dim \mathfrak{g}_\zeta$; its irreducible components are $\overline{\mathcal{O}^!}$ (closure of $\mathcal{O}^!$) for various $\mathcal{O}$ as above. Similarly for any K-orbit $\mathcal{V}$ in $\mathfrak{g}_{\zeta^{-1}}^{nil}$ let $\mathcal{V}^! = \{(x, y) \in \mathfrak{g}_\zeta \times \mathfrak{g}_{\zeta^{-1}}; y \in \mathcal{V}, [x, y] = 0\}$. Then $\mathcal{V}^!$ is smooth, irreducible of dimension $\dim \mathfrak{g}_{\zeta^{-1}} = \dim \mathfrak{g}_\zeta$. Hence $A' := \{(x, y) \in \mathfrak{g}_\zeta \times \mathfrak{g}_{\zeta^{-1}}^{nil}; [x, y] = 0\}$ is a (closed) subvariety of $\mathfrak{g}_\zeta \times \mathfrak{g}_{\zeta^{-1}}$ of pure dimension $\dim \mathfrak{g}_\zeta$; its irreducible components are $\overline{\mathcal{V}^!}$ (closure of $\mathcal{V}^!$) for various $\mathcal{V}$ as above. Let

$$\Lambda = A \cap A' = \{(x, y) \in \mathfrak{g}_\zeta^{nil} \times \mathfrak{g}_{\zeta^{-1}}^{nil}; [x, y] = 0\},$$

a (closed) subvariety of $\mathfrak{g}_\zeta \times \mathfrak{g}_{\zeta^{-1}}$ of dimension $\leq \dim \mathfrak{g}_\zeta$. The irreducible components of Λ of dimension $\dim \mathfrak{g}_\zeta$ are of the form $\overline{\mathcal{O}^!}$ (where $\mathcal{O}$ runs through a subset H_ζ of the set of K-orbits on $\mathfrak{g}_\zeta^{nil}$). They are also of the form $\overline{\mathcal{V}^!}$ (where $\mathcal{V}$ runs through a subset $H_{\zeta^{-1}}$ of the set of K-orbits on $\mathfrak{g}_{\zeta^{-1}}^{nil}$). Hence there is a unique bijection $\iota : H_\zeta \to H_{\zeta^{-1}}$ such that $\overline{\mathcal{O}^!} = \overline{\iota(\mathcal{O})^!}$ for any $\iota \in H_\zeta$.

Now let $\mathcal{K} \in \mathfrak{Q}_\zeta$ and assume that $\mathrm{supp}\mathcal{K} = \bar{\mathcal{O}}$ (closure of $\mathcal{O}$ as above), $\mathrm{supp}\mathfrak{F}(\mathcal{K}) = \bar{\mathcal{V}}$ (closure of $\mathcal{V}$ as above). Let S (resp. S') be the singular support of $\mathcal{K}$ (resp. $\mathfrak{F}(\mathcal{K})$); they are subvarieties of $\mathfrak{g}_\zeta \times \mathfrak{g}_{\zeta^{-1}}$. Note that $S = S'$ and $\mathcal{O}^! \subset S \subset \cup_{\mathcal{O}' \subset \bar{\mathcal{O}}} \mathcal{O}'^! \subset A$, $\mathcal{V}^! \subset S' \subset \cup_{\mathcal{V}' \subset \bar{\mathcal{V}}} \mathcal{V}'^! \subset A'$ where $\mathcal{O}'$ runs over the K-orbits in $\bar{\mathcal{O}}$ and $\mathcal{V}'$ runs over the K-orbits in $\bar{\mathcal{V}}$. It follows that $\mathcal{O}^! \subset A'$. Since $\mathcal{O}^! \subset A$ we see that $\mathcal{O}^! \subset A \cap A' = \Lambda$. Similarly, $\mathcal{V}^! \subset \Lambda$. We see that:

(a) *if* $\mathcal{K} \in \mathfrak{Q}_\zeta$ *then* $\mathrm{supp}(\mathcal{K}) = \bar{\mathcal{O}}$ *where* $\mathcal{O} \in H_\zeta$. *In particular if* $x \in \mathcal{O}$ *and* $y \in \mathfrak{g}_{\zeta^{-1}}$ *satisfies* $[x, y] = 0$ *then* y *is nilpotent.*

In the case where $\mathfrak{g}_\zeta$ arises as in 0.4 from a cyclic quiver, the variety Λ is the same as that defined in [L2, §12]. In this case Λ is of pure dimension. Also in this case the bijection ι can be viewed as a bijection $\mathcal{P}_\nu^{ap} \xrightarrow{\sim} \mathcal{P}_\nu^{ap}$ (notation of 1.2). It would be interesting to describe this bijection explicitly. (It is analogous to the involution of $\mathcal{R}_t$ which takes a partition to the conjugate partition.)

If ζ has large order in $\mathbf{k}^*$ then Λ is again of pure dimension (see [L5, 22.2]).

3.9. Let G (resp. $\mathfrak{g}$) be the set of all 4×4 matrices $a = (a_{ij})$ with $a_{ij} \in \mathbf{k}$ for $i, j \in [1, 4]$, $a_{ij} = 0$ for all $i > j$ and $a_{ii} = 1$ (resp. $a_{ii} = 0$) for all i. Note that G is naturally a unipotent algebraic group with Lie algebra $\mathfrak{g}$. Let $\mathfrak{h}$ be the set all 4×4 matrices $b = (b_{ij})$ with $b_{ij} \in \mathbf{k}$ for $i, j \in [1, 4]$, $b_{ij} = 0$ for all $i \le j$. We identify $\mathfrak{g}$ with the dual space of $\mathfrak{h}$ via the nondegenerate bilinear pairing $(a, b) = \sum_{i<j} a_{ij} b_{ji} \in \mathbf{k}$. Now G acts on $\mathfrak{g}$ by conjugation and this induces a G-action on $\mathfrak{h}$ (by duality). Hence the orbital complexes on $\mathfrak{h}$ are well defined and the antiorbital complexes on $\mathfrak{g}$ are well defined.

Consider the partition $\mathfrak{h} = U_1 \cup U_2 \cup U_3 \cup U_4 \cup U_5$ where
$U_1 = \{b; b_{31} = b_{41} = b_{42} = 0\}$; $U_2 = \{b; b_{31} = b_{41} = 0, b_{42} \neq 0\}$;
$U_3 = \{b; b_{41} = b_{42} = 0, b_{31} \neq 0\}$; $U_4 = \{b; b_{41} \neq 0\}$;
$U_5 = \{b; b_{41} = 0, b_{31} \neq 0, b_{42} \neq 0\}$.
For $i = 1, 2, 3, 4, 5$, U_i is a union of G_n-orbits of fixed dimension: $0, 2, 2, 4, 2$.
Let
$V_1 = \mathfrak{g}$, $V_2 = \{a \in \mathfrak{g}; a_{23} = a_{34} = 0\}$, $V_3 = \{a \in \mathfrak{g}; a_{12} = 0, a_{23} = 0\}$,
$V_4 = \{a \in \mathfrak{g}; a_{12} = a_{34} = 0\}$, $V_4' = \{a \in V_4; a_{23} \neq 0\}$,
$V_{5;x,y} = \{a \in \mathfrak{g}; a_{23} = 0, xa_{12} - ya_{34} = 0\}$ $(x \in \mathbf{k}^*, y \in \mathbf{k}^*)$.
The class of antiorbital complexes on $\mathfrak{g}$ consists of:
(1) $\mathcal{L}^f[6]$ where $f : V_1 \to \mathbf{k}$ is $a \mapsto xa_{12} + ya_{23} + za_{34}$ $(x, y, z \in \mathbf{k})$;
(2) $\mathcal{L}_\mathfrak{g}^f[4]$ where $f : V_2 \to \mathbf{k}$ is $a \mapsto xa_{12} + ya_{24}$, $(x \in \mathbf{k}, y \in \mathbf{k}^*)$;
(3) $\mathcal{L}_\mathfrak{g}^f[4]$ where $f : V_3 \to \mathbf{k}$ is $a \mapsto xa_{13} + ya_{34}$, $(x \in \mathbf{k}^*, y \in \mathbf{k})$;
(4) $IC(V_4, \mathcal{L}^f)_\mathfrak{g}[4]$ where $f : V_4' \to \mathbf{k}$ is $a \mapsto xa_{23} + y(a_{14} - \frac{a_{24}a_{13}}{a_{23}})$, $(x \in \mathbf{k}, y \in \mathbf{k}^*)$;
(5) $\mathcal{L}_\mathfrak{g}^f[4]$ where $f : V_{5,c,d} \to \mathbf{k}$ is $a \mapsto xa_{13} + ya_{24} + zx^{-1}a_{34}$, $(z \in \mathbf{k})$.
Note that the complexes $(1), (2), (3), (4), (5)$ are obtained by applying $\mathfrak{F}$ to the orbital complexes on $\mathfrak{h}$ with support contained in U_1, U_2, U_3, U_4, U_5 respectively.

APPENDIX. INDUCTION, RESTRICTION

A.1. Let $G, \mathfrak{g}, \vartheta, K, \kappa$ be as in 0.3. We assume that the characteristic of $\mathbf{k}$ is sufficiently large. For $\zeta \in \mathbf{k}^*$ let $\mathfrak{g}_\zeta, \kappa_\zeta$ be as in 0.3. We consider the datum
$(P, U, l, \mathfrak{p}, \mathfrak{n}, \mathfrak{l})$
where P is a parabolic subgroup of G such that $\vartheta(P) = P$, U is the unipotent radical of P, $L = P/U$ and $\mathfrak{p}, \mathfrak{n}, \mathfrak{l}$ denote the Lie algebras of P, U, L. Note that $\mathfrak{p} = \oplus_{\zeta \in \mathbf{k}^*} \mathfrak{p}_\zeta$, $\mathfrak{n} = \oplus_{\zeta \in \mathbf{k}^*} \mathfrak{n}_\zeta$ where $\mathfrak{p}_\zeta = \mathfrak{p} \cap \mathfrak{g}_\zeta$, $\mathfrak{n}_\zeta = \mathfrak{n} \cap \mathfrak{g}_\zeta$. Moreover ϑ induces a semisimple automorphism of L and of $\mathfrak{l}$ denoted again by ϑ and we have a decomposition $\mathfrak{l} = \oplus_{\zeta \in \mathbf{k}^*} \mathfrak{l}_\zeta$ where $\mathfrak{l}_\zeta$ is the ζ-eigenspace of $\vartheta : \mathfrak{l} \to \mathfrak{l}$ (it is also equal to $\pi(\mathfrak{p}_z)$ where $\pi : \mathfrak{p} \to \mathfrak{l}$ is the obvious map). Let L_K be the identity component of the fixed point set of $\vartheta : L \to L$. It acts naturally on $\mathfrak{l}_\zeta$ for any $\zeta \in \mathbf{k}^*$.

Let $\zeta \in \mathbf{k}^*$. Let $\pi_\zeta : \mathfrak{p}_\zeta \to \mathfrak{l}_\zeta$ be the restriction of π. Let $U_K = U \cap K$ (a connected unipotent group), $P_K = (P \cap K)^0$. Let $E_\zeta' = K \times_{U_K} \mathfrak{p}_\zeta, E_\zeta'' = K \times_{P_K} \mathfrak{p}_\zeta$.

we have a diagram

$$\mathfrak{l}_\zeta \xleftarrow{p_1^\zeta} E'_\zeta \xrightarrow{p_2^\zeta} E''_\zeta \xrightarrow{p_3^\zeta} \mathfrak{g}_\zeta$$

where $p_i = p_i\zeta$ are given by $p_1(g,x) = \pi_\zeta(x)$, $p_2(g,x) = (g,x)$, $p_3(g,x) = \mathrm{Ad}(g)(x)$. Note that p_1 is smooth with connected fibres, p_2 is a principal bundle with group $P_K/U_K = L_K$ and p_3 is a proper map. Let A be a semisimple L_K-equivariant complex on $\mathfrak{l}_\zeta$. Then p_1^*A is a semisimple L_K-equivariant complex on E'_ζ. Hence $p_1^*A = p_2^*A'$ for a well defined semisimple complex A' on E''_ζ. We set

$$\mathrm{Ind}_{\mathfrak{p}_\zeta}^{\mathfrak{g}_\zeta}(A) = p_{3!}A' \in \mathcal{D}(\mathfrak{g}_\zeta),$$

$$\widetilde{\mathrm{Ind}}_{\mathfrak{p}_\zeta}^{\mathfrak{g}_\zeta}(A) = \mathrm{Ind}_{\mathfrak{p}_\zeta}^{\mathfrak{g}_\zeta}(A)[\dim \mathfrak{n}_0 + \dim \mathfrak{n}_\zeta].$$

The following result is of the same type as [L1, 7(a)], [L2, Theorem 5.4], [L5, Cor.10.5], [H, Theorem 4.1]. The proof we give is almost a word by word repetition of that of Theorem 5.4 in [L2].

Theorem A.2. *We preserve the setup of A.1. We have*

$$\mathfrak{F}(\widetilde{\mathrm{Ind}}_{\mathfrak{p}_\zeta}^{\mathfrak{g}_\zeta}(A)) \cong \widetilde{\mathrm{Ind}}_{\mathfrak{p}_{\zeta^{-1}}}^{\mathfrak{g}_{\zeta^{-1}}}(\mathfrak{F}(A)).$$

We consider the commutative diagram

$$
\begin{array}{ccccccc}
X_a & \xleftarrow{u_{ba}} & X_b & \xrightarrow{u_{bc}} & X_c & \xrightarrow{u_{cd}} & X_d \\
 & & \uparrow u_{eb} & & \uparrow u_{fc} & & \uparrow u_{gd} \\
 & & X_e & \xrightarrow{u_{ef}} & X_f & \xrightarrow{u_{fg}} & X_g \\
\uparrow u_{ja} & & \uparrow u_{he} & & \uparrow u_{if} & & \\
 & & X_h & \xrightarrow{u_{hi}} & X_i & & \\
 & & \downarrow u_{hk} & & \downarrow u_{il} & & \downarrow u_{gp} \\
X_j & \xleftarrow{u_{kj}} & X_k & \xrightarrow{u_{kl}} & X_l & & \\
\downarrow u_{jm} & & \downarrow u_{kn} & & \downarrow u_{lo} & & \\
X_m & \xleftarrow{u_{mn}} & X_n & \xrightarrow{u_{no}} & X_o & \xrightarrow{u_{op}} & X_p
\end{array}
$$

in which the notation is as follows.

$X_a = \mathfrak{l}_\zeta$, $X_c = E''_\zeta$, $X_b = E'_\zeta$, $X_d = \mathfrak{g}_\zeta$.

$X_m = \mathfrak{l}_{\zeta^{-1}}$, $X_o = E''_{\zeta^{-1}}$, $X_n = E'_{\zeta^{-1}}$, $X_p = \mathfrak{g}_\zeta$.

$X_g = \mathfrak{g}_\zeta \times \mathfrak{g}_{\zeta^{-1}}$, $X_f = K \times_{P_K} (\mathfrak{p}_\zeta \times \mathfrak{g}_{\zeta^{-1}})$, $X_e = K \times_{U_K} (\mathfrak{p}_\zeta \times \mathfrak{g}_{\zeta^{-1}})$.

$X_i = K \times_{P_K} (\mathfrak{p}_\zeta \times \mathfrak{p}_{\zeta^{-1}})$, $X_h = K \times_{U_K} (\mathfrak{p}_\zeta \times \mathfrak{p}_{\zeta^{-1}})$.

$X_j = \mathfrak{l}_\zeta \times \mathfrak{l}_{\zeta^{-1}}$, $X_k = K \times_{U_K} (\mathfrak{l}_\zeta \times \mathfrak{p}_{\zeta^{-1}})$, $X_l = K \times_{P_K} (\mathfrak{l}_\zeta \times \mathfrak{p}_{\zeta^{-1}})$.

$u_{ba} = p_1^\zeta$, $u_{bc} = p_2^\zeta$, $u_{cd} = p_3^\zeta$, $u_{mn} = p_1^{\zeta^{-1}}$, $u_{no} = p_2^{\zeta^{-1}}$, $u_{op} = p_3^{\zeta^{-1}}$.

u_{fg} is $(g,x,x') \mapsto (\mathrm{Ad}(g)x, \mathrm{Ad}(g)x')$, u_{kj} is $(g,x,x') \mapsto (x, \pi_{\zeta^{-1}}(x'))$.

$u_{ef}, u_{hi}, u_{kl}, u_{eb}, u_{fc}, u_{gd}, u_{ja}, u_{he}, u_{if}, u_{hk}, u_{il}, u_{gp}, u_{kn}, u_{lo}, u_{jm}$ are the obvious maps.

Note that $\kappa : \mathfrak{g} \times \mathfrak{g} \to \mathbf{k}$ restricted to $\mathfrak{p} \times \mathfrak{p}$ induces a function $\mathfrak{l} \times \mathfrak{l} \to \mathbf{k}$ which is analogous to κ. Hence we can define local systems of rank 1 on $X_j, X_k, X_h, X_l, X_i, X_e, X_f, X_g$ (equal to $\mathcal{L}^{\kappa_\zeta}$ in the case of X_g) which correspond to each other under inverse image by

$$u_{fg}, u_{ef}, u_{he}, u_{if}, u_{hi}, u_{kj}, u_{hk}, u_{kl}, u_{il}.$$

We shall denote each of these local systems by $\mathcal{L}$.

Let $L_a = A \in \mathcal{D}(X_a)$. Let $L_b = u_{ba}^* L_a \in \mathcal{D}(X_b)$; let $L_c \in \mathcal{D}(X_c)$ be the unique semisimple complex such that $u_{bc}^* L_c = L_b$ and let $L_d = u_{cd!} L_c \in \mathcal{D}(X_d)$. Let $L_g = u_{gd}^* L_d \in \mathcal{D}(X_g)$ and let $L_p = u_{gp!}(L_g \otimes \mathcal{L}) \in \mathcal{D}(X_p)$. By definition we have $\mathfrak{F}(\mathrm{Ind}_{\mathfrak{p}_\zeta}^{\mathfrak{g}_\zeta}(A)) = L_p[D]$ where $D = \dim \mathfrak{g}_\zeta$. Now let $L_j = u_{ja}^* L_a \in \mathcal{D}(X_j)$ and $L_m = u_{jm!}(L_j \otimes \mathcal{L}) \in \mathcal{D}(X_m)$. This is a semisimple, L_K−equivariant complex (it is $\mathfrak{F}(A)[-\dim \mathfrak{l}_\zeta]$). Let $L_n = u_{nm}^* L_m \in \mathcal{D}(X_n)$ and let $L_o \in \mathcal{D}(X_o)$ be the unique semisimple complex such that $u_{no}^* L_o = L_n$. Let $L_p' = u_{op!} L_o \in \mathcal{D}(X_p)$. By definition we have $\mathrm{Ind}_{\mathfrak{p}_{\zeta-1}}^{\mathfrak{g}_{\zeta-1}}(\mathfrak{F}(A)) = L_p'[\dim \mathfrak{l}_\zeta]$. Hence it suffices to prove that

(a) $$L_p \cong L_p'[\dim \mathfrak{l}_\zeta - \dim \mathfrak{g}_\zeta + \dim \mathfrak{n}_{\zeta-1} - \dim \mathfrak{n}_\zeta].$$

Let $L_e = u_{eb}^* L_b \in \mathcal{D}(X_e)$, $L_f = u_{fc}^* L_c \in \mathcal{D}(X_f)$. Then L_f is a semisimple complex (since L_c is semisimple and u_{fc} is smooth with connected fibres) and $L_e = u_{ef}^* L_f$. Moreover, $u_{fg!} L_f = L_g$ (since the diagram $u_{fc}, u_{cd}, u_{fg}, u_{gd}$ is cartesian). Hence we may go from L_a to L_p by the chain $L_e = (u_{ba} u_{eb})^* L_a$, $L_e = u_{ef}^* L_f$ (L_f semisimple), $L_p = (u_{gp} u_{fg})_! (L_f \otimes \mathcal{L})$. Similarly, we may go from L_a to L_p' by the chain $L_k = (u_{ja} u_{kj})^* L_a \in \mathcal{D}(X_k)$, $L_k = u_{kl}^* L_l$ ($L_l \in \mathcal{D}(X_l)$ semisimple), $L_p' = (u_{op} u_{lo})_! (L_l \otimes \mathcal{L})$.

Let $L_h = u_{hk}^* L_k \in \mathcal{D}(X_h)$, $L_i = u_{il}^* L_l \in \mathcal{D}(X_i)$. Note that u_{il} is a vector bundle with fibres of dimension $\dim \mathfrak{n}_\zeta$. It follows that L_i is semisimple (recall that L_l is semisimple) and that $u_{il!} L_i = L_l[-2 \dim \mathfrak{n}_\zeta]$. Hence we have $(u_{op} u_{lo} u_{il})_! (L_i \otimes \mathcal{L}) = L_p'[-2 \dim \mathfrak{n}_\zeta]$. We have $\dim \mathfrak{n}_{\zeta-1} - \dim \mathfrak{n}_\zeta = \mathfrak{g}_\zeta - \mathfrak{l}_\zeta - 2 \dim \mathfrak{n}_\zeta$. Hence it is enough to prove that

(a1) $$L_p \cong L_p'[-2 \dim \mathfrak{n}_\zeta].$$

We have $u_{hi}^* L_i = L_h$ and it follows that we may go from L_a to L_p' by the chain $L_h = (u_{ja} u_{kj} u_{hk})^* L_a$, $L_h = u_{hi}^* L_i$ ($L_i \in \mathcal{D}(X_i)$ semisimple), $L_p'[-2 \dim \mathfrak{n}_\zeta] = (u_{op} u_{lo} u_{il})_! (L_i \otimes \mathcal{L})$.

Since L_c is semisimple and $u_{fc} u_{if}$ a vector bundle we see that $u_{if}^* L_f = (u_{fc} u_{if})^* L_c$ is semisimple. Now both $u_{if}^* L_f$ and L_i are semisimple and they have the same inverse image L_h under u_{hi} (a smooth morphism with connected fibres). It follows that $u_{if}^* L_f \cong L_i$. Since $u_{op} u_{lo} u_{il} = u_{gp} u_{fg} u_{if}$ we see that $L_p'[-2 \dim \mathfrak{n}_\zeta] = (u_{gp} u_{fg} u_{if})_! (u_{if}^* (L_f \otimes \mathcal{L}))$. We now see that (a1) would be a consequence of the following statement:

$L_f \otimes \mathcal{L}$ and $u_{if!} u_{if}^* (L_f \otimes \mathcal{L})$ have the same image under $(u_{gp} u_{fg})_!$.

An equivalent statement is:

(b) if u' denotes the inclusion of $X_f - X_i$ into X_f (as an open subset), then $(u_{gp} u_{fg} u')_! u'^* (L_f \otimes \mathcal{L}) = 0$.

(We use the distinguished triangle associated with the partition $X_f = X_i \cup (X_f - X_i)$.) We now consider the commutative diagram

$$
\begin{array}{ccccccc}
X_b & \xrightarrow{u_{bc}} & X_c & \xleftarrow{u_{fc}} & X_f & \xleftarrow{u'} & X_f - X_i \\
\downarrow{\scriptstyle u_b} & & \downarrow{\scriptstyle u_c} & & \downarrow{\scriptstyle u_f} & & \downarrow{\scriptstyle u''} \\
Y_b & \xrightarrow{w_{bc}} & Y_c & \xleftarrow{w_{fc}} & Y_f & \xleftarrow{w'} & Y_f - X_l \\
\downarrow{\scriptstyle v'} & & & & \downarrow{\scriptstyle w''} & & \\
X_a & & & & X_p & &
\end{array}
$$

where the notation is as follows.

$Y_b = (K/U_K) \times \mathfrak{l}_\zeta$, $Y_c = K \times_{P_K} \mathfrak{l}_\zeta$, $Y_f = K \times_{P_K} (\mathfrak{l}_\zeta \times \mathfrak{g}_{\zeta^{-1}})$.

$u_b, u_c, u_f, w_{bc}, w_{fc}, w', v'$ are the obvious maps, w'' is $(g, x, x') \mapsto \mathrm{Ad}(g)x'$.

Let $M_b = v'^* L_a \in \mathcal{D}(Y_b)$; this is a L_K-equivariant semisimple complex hence there is a well defined semisimple complex $M_c \in \mathcal{D}(Y_c)$ such that $w_{bc}^* M_c = M_b$. Let $M_f = w_{fc}^* M_c \in \mathcal{D}(Y_f)$. It is clear that $u_b^* M_b = L_b$, $u_c^* M_c = L_c$, $u_f^* M_f = L_f$ (note that u_b, u_c, u_f are vector bundles). Hence we have $u'^* L_f = u'^* u_f^* M_f = u''^* w'^* M_f$. The statement (b) can now be rewritten in terms of M_f instead of L_f:

$(u_{gp} u_{fg} u')_!(u''^* w'^* M_f \otimes \mathcal{L}) = 0$

or equivalently (using $u_{gp} u_{fg} u' = w'' w' u''$):

$(w'' w' u'')_!(u''^* w'^* M_f \otimes \mathcal{L}) = 0$.

This would be a consequence of the following statement:

$u''_!(u''^* w'^* M_f \otimes \mathcal{L}) = 0$.

We have $u''_!(u''^* w'^* M_f \otimes \mathcal{L}) = w'^* M_f \otimes (u''_! \mathcal{L})$, hence it suffices to prove that $u''_! \mathcal{L} = 0$ in $\mathcal{D}(Y_f - X_l)$.

Let us fix a point $(g, x, x') \in Y_f - X_l$ and let Γ be the fibre of w' over this point. Let $\tilde{\kappa} : \Gamma \to \mathbf{k}$ be the function $(g, x, x') \mapsto \kappa(x, x')$. By base change, it is enough to prove that the cohomology with compact support of Γ with coefficients in $\mathcal{L}^{\tilde{\kappa}}|_{\Gamma}$ is zero. By a known property of Artin-Schreier local systems, it is enough to verify the following statement: one can identify Γ with $\mathbf{k}^N$ for some N so that $\tilde{\kappa}$ is given by a non-constant affine linear form on $\mathbf{k}^N$. We can find a ϑ-stable Levi subgroup $\tilde{L}$ of P. Let $\tilde{\mathfrak{l}}$ be the Lie algebra of $\tilde{L}$. Let $\tilde{\mathfrak{l}}_\zeta = \tilde{\mathfrak{l}} \cap \mathfrak{g}_\zeta$. We have $\mathfrak{p}_\zeta = \tilde{\mathfrak{l}}_\zeta \oplus \mathfrak{n}_\zeta$. We can identify $\mathfrak{l}_\zeta = \tilde{\mathfrak{l}}_\zeta$ in an obvious way; hence we can view x as an element of $\tilde{\mathfrak{l}}_\zeta$. We have an isomorphism $\mathfrak{n}_\zeta \xrightarrow{\sim} \Gamma$ given by $z \mapsto (g, x + z, x')$. The function $\tilde{\kappa}$ on Γ can be identified with the function $z \mapsto \kappa(x + z, x') = \kappa(z, x') + \kappa(x, x')$ on $\mathfrak{n}_\zeta$. It is enough to show that the linear function $z \mapsto \kappa(z, x')$ is not identically zero. Assume that it is identically zero that is $\kappa(\mathfrak{n}_\zeta, x') = 0$. If $\zeta' \neq \zeta$ we have $\kappa(\mathfrak{g}_{\zeta'}, \mathfrak{g}_{\zeta^{-1}}) = 0$. Hence $\kappa(\mathfrak{n}_{\zeta'}, x') = 0$. We see that $\kappa(\mathfrak{n}, x') = 0$ hence $x' \in \mathfrak{p}'$ and $x' \in \mathfrak{p}_{\zeta^{-1}}$. This contradicts $x' \in \mathfrak{g}_{\zeta^{-1}} - \mathfrak{p}_{\zeta^{-1}}$. The theorem is proved.

A.3. In the setup of A.1 let us assume that A is an orbital complex (on $\mathfrak{l}_\zeta$). We show:

(a) $\tilde{\mathrm{Ind}}_{\mathfrak{p}_\zeta}^{\mathfrak{g}_\zeta}(A)$ *is a direct sum of finitely many orbital complexes (with shifts) on* $\mathfrak{g}_\zeta$.

Let A', p_3 be as in A.1. In our case A' can be viewed as a mixed complex, pure of weight 0 (up to shift); hence by the decomposition theorem, $p_{3!} A'$ is a direct sum of shifts of simple perverse sheaves. It is then enough to show that $S := \mathrm{supp}(p_{3!} A)$

is contained in the union of finitely many K-orbits on $\mathfrak{g}_\zeta$. Since the intersection of any G-orbit on $\mathfrak{g}$ with $\mathfrak{g}_\zeta$ is a union of finitely many K-orbits, it is enough to show that S is contained in the union of finitely many G-orbits on $\mathfrak{g}$. Let $S_0 = \mathrm{supp}(A) \subset \mathfrak{l}_\zeta$. From the definitions, S is contained in the union of K-orbits in $\mathfrak{g}_\zeta$ that meet $\pi_\zeta^{-1}(S_0)$. Hence S is contained in the union of G-orbits in $\mathfrak{g}$ that meet $\pi^{-1}(S_0)$. Hence it is enough to show that $\pi^{-1}(S_0)$ is contained in a finite union of G-orbits in $\mathfrak{g}$. It is also enough to show that the set $\mathcal{S}$ of semisimple parts of the various elements in $\pi^{-1}(S_0)$ is contained in a single G-orbit in $\mathfrak{g}$. Now the set $\mathcal{S}_0$ of semisimple parts of the various elements in S_0 is contained in a single L_K-orbit in $\mathfrak{l}_\zeta$ hence in a single L-orbit in $\mathcal{O}_0$ in $\mathfrak{l}$. Since $\mathcal{S} \subset \pi^{-1}(\mathcal{S}_0)$ it is enough to note that the set of semisimple elements in $\pi^{-1}(\mathcal{S}_0)$ is a single P-orbit on $\mathfrak{p}$. This completes the proof of (a).

Next we assume, in the setup of A.1, that A is an antiorbital complex (on $\mathfrak{l}_\zeta$). We show:

(b) $\tilde{\mathrm{Ind}}_{\mathfrak{p}_\zeta}^{\mathfrak{g}_\zeta}(A)$ *is a direct sum of finitely many antiorbital complexes (with shifts) on* $\mathfrak{g}_\zeta$.

By assumption we have $A = \mathfrak{F}(A_1)$ where A_1 is an orbital complex on $\mathfrak{l}_{\zeta^{-1}}$. Using A.2 with A replaced by A_1 and ζ by ζ^{-1} we see that

$$\tilde{\mathrm{Ind}}_{\mathfrak{p}_\zeta}^{\mathfrak{g}_\zeta}(A) = \mathfrak{F}(\tilde{\mathrm{Ind}}_{\mathfrak{p}_{\zeta^{-1}}}^{\mathfrak{g}_{\zeta^{-1}}}(A_1)).$$

Using (a) with A replaced by A_1 and ζ by ζ^{-1} we see that

$$\tilde{\mathrm{Ind}}_{\mathfrak{p}_{\zeta^{-1}}}^{\mathfrak{g}_{\zeta^{-1}}}(A_1) \cong \oplus_i P_i[d_i]$$

where P_i are orbital complexes on $\mathfrak{g}_{\zeta^{-1}}$ and $d_i \in \mathbf{Z}$. It follows that

$$\tilde{\mathrm{Ind}}_{\mathfrak{p}_\zeta}^{\mathfrak{g}_\zeta}(A) \cong \oplus_i \mathfrak{F}(P_i)[d_i']$$

where $d_i' \in \mathbf{Z}$ and (b) is proved.

Finally we assume, in the setup of A.1, that A is a biorbital complex (on $\mathfrak{l}_\zeta$). We show:

(c) $\tilde{\mathrm{Ind}}_{\mathfrak{p}_\zeta}^{\mathfrak{g}_\zeta}(A)$ *is a direct sum of finitely many biorbital complexes (with shifts) on* $\mathfrak{g}_\zeta$.

Using (a) we see that $\tilde{\mathrm{Ind}}_{\mathfrak{p}_\zeta}^{\mathfrak{g}_\zeta}(A) \cong \oplus_i P_i[d_i]$ where P_i are orbital complexes on $\mathfrak{g}_\zeta$ and $d_i \in \mathbf{Z}$. Using (b) we see that $\oplus_i P_i[d_i] \cong \oplus_j Q_j[e_j]$ where Q_j are antiorbital complexes on $\mathfrak{g}_\zeta$ and $e_j \in \mathbf{Z}$. It follows that the isomorphism classes in $\{P_i\}$ coincide with the isomorphism classes in $\{Q_j\}$. Hence each P_i is biorbital. This proves (c).

A.4. Let $\zeta \in \mathbf{k}^*$. For $B \in \mathcal{D}(\mathfrak{g}_\zeta)$ we define (in the setup of A.1) $\mathrm{Res}_{\mathfrak{p}_\zeta}^{\mathfrak{g}_\zeta}(B) = \pi_{\zeta!}(B|_{\mathfrak{p}_\zeta}) \in \mathcal{D}(\mathfrak{l}_\zeta)$, $\tilde{\mathrm{Res}}_{\mathfrak{p}_\zeta}^{\mathfrak{g}_\zeta}(B) = \mathrm{Res}_{\mathfrak{p}_\zeta}^{\mathfrak{g}_\zeta}(B)[\dim \mathfrak{n}_\zeta - \dim \mathfrak{n}_1]$.

Proposition A.5. *We have* $\mathfrak{F}(\tilde{\mathrm{Res}}_{\mathfrak{p}_\zeta}^{\mathfrak{g}_\zeta}(B)) \cong \tilde{\mathrm{Res}}_{\mathfrak{p}_{\zeta^{-1}}}^{\mathfrak{g}_{\zeta^{-1}}}(\mathfrak{F}(B))$ *in* $\mathcal{D}(\mathfrak{l}_{\zeta^{-1}})$.

The proof is exactly the same as that in [L5, 10.4] if we use Lemma A.6 below instead of [L5, 10.3] and we replace $\mathfrak{g}_n, \mathfrak{p}_n, \mathfrak{l}_n, \mathfrak{n}_n, \mathfrak{g}_{-n}, \mathfrak{p}_{-n}, \mathfrak{l}_{-n}, \mathfrak{n}_{-n}, \mathfrak{n}_0, \tau, \mathcal{L}_\tau$ by $\mathfrak{g}_\zeta, \mathfrak{p}_\zeta, \mathfrak{l}_\zeta, \mathfrak{n}_\zeta, \mathfrak{g}_{\zeta^{-1}}, \mathfrak{p}_{\zeta^{-1}}, \mathfrak{l}_{\zeta^{-1}}, \mathfrak{n}_{\zeta^{-1}}, \mathfrak{n}_1, \kappa_\zeta, \mathcal{L}^{\kappa_\zeta}$.

Lemma A.6. *Consider the diagram*

$$
\begin{array}{ccc}
(\mathfrak{g}_\zeta \times \mathfrak{p}_{\zeta^{-1}}) - (\mathfrak{p}_\zeta \times \mathfrak{p}_{\zeta^{-1}}) & \longrightarrow & \mathfrak{g}_\zeta \times \mathfrak{p}_{\zeta^{-1}} \\
\chi' \downarrow & & \chi \downarrow \\
(\mathfrak{g}_\zeta \times \mathfrak{l}_{\zeta^{-1}}) - (\mathfrak{p}_\zeta \times \mathfrak{l}_{\zeta^{-1}}) & \longrightarrow & \mathfrak{g}_\zeta \times \mathfrak{l}_{\zeta^{-1}}
\end{array}
$$

where the horizontal maps are the obvious inclusions, χ is the obvious projection and χ' makes the diagram commutative. We have $\chi'_!(\mathcal{L}^{\kappa_\zeta}) = 0$. (Here $\mathcal{L}^{\kappa_\zeta}$ is the local system on $(\mathfrak{g}_\zeta \times \mathfrak{p}_{\zeta^{-1}}) - (\mathfrak{p}_\zeta \times \mathfrak{p}_{\zeta^{-1}})$ defined by restricting the analogous local system on $\mathfrak{g}_\zeta \times \mathfrak{p}_{\zeta^{-1}}$, see 0.3.)

We choose a ϑ-stable Levi subgroup $\tilde{L}$ of P. Let $\tilde{\mathfrak{l}}$ be the Lie algebra of $\tilde{L}$. We have $\tilde{\mathfrak{l}} = \oplus_{\zeta' \in \mathbf{k}^*} \tilde{\mathfrak{l}}_{\zeta'}$ where $\tilde{\mathfrak{l}}_{\zeta'} = \tilde{\mathfrak{l}} \cap \mathfrak{g}_{\zeta'}$. We identify $\tilde{\mathfrak{l}}$ with $\mathfrak{l}$ via the obvious map $\mathfrak{p} \to \mathfrak{l}$. Let $(x, y) \in \mathfrak{g}_\zeta \times \mathfrak{l}_{\zeta^{-1}}$ be such that $x \notin \mathfrak{p}_\zeta$. We have

(a) $\chi'^{-1}(x, y) = \{(x, y + z); z \in \mathfrak{n}_{\zeta^{-1}}\}$.

It is enough to show that the cohomology with compact support of the variety (a) with coefficients in $\mathcal{L}^{\kappa_\zeta}$ is zero. We identify this variety with $\mathfrak{n}_{\zeta^{-1}}$ via the coordinate z. The restriction of κ_ζ to this variety is of the form $z \mapsto \kappa(x, z) + c$ where c is a constant. It suffices to show that the linear function $z \mapsto \kappa(x, z)$ on $\mathfrak{n}_{\zeta^{-1}}$ is not identically zero. If it were identically zero, then $\kappa(x, \mathfrak{n}) = 0$ hence $x \in \mathfrak{p}$ and $x \in \mathfrak{p}_\zeta$ contradicting our assumptions. The lemma is proved.

A.7. Assume now that B is an orbital complex (on $\mathfrak{g}_\zeta$). We show:

(a) *any composition factor of any perverse cohomology sheaf of $\tilde{\mathrm{Res}}_{\mathfrak{p}_\zeta}^{\mathfrak{g}_\zeta}(B)$ is an orbital complex on $\mathfrak{l}_\zeta$.*

Since each perverse cohomology sheaf of $D := \tilde{\mathrm{Res}}_{\mathfrak{p}_\zeta}^{\mathfrak{g}_\zeta}(B)$ is L_K-equivariant, it is enough to show that $\mathrm{supp}(D)$ is contained in the union of finitely many L_K-orbits on $\mathfrak{l}_\zeta$. Since the intersection of any L-orbit on $\mathfrak{l}$ with $\mathfrak{l}_\zeta$ is a union of finitely many L_K-orbits, it is enough to show that $\mathrm{supp}(D)$ is contained in the union of finitely many L-orbits in $\mathfrak{l}$. Clearly, $\mathrm{supp}(D)$ is contained in $\pi(\mathrm{supp}(B) \cap \mathfrak{p})$ hence it is enough to show that $\pi(\mathrm{supp}(B) \cap \mathfrak{p})$ is contained in the union of finitely many L-orbits in $\mathfrak{l}$. Since $\mathrm{supp}(B)$ is the union of finitely many K-orbits, it is contained in the union of finitely many G-orbits on $\mathfrak{g}$, and it is enough to note that for any G-orbit $\mathcal{O}$ in $\mathfrak{g}$, the set $\pi(\mathcal{O} \cap \mathfrak{p})$ is the union of finitely many L-orbits in $\mathfrak{l}$. This completes the proof of (a).

Next we assume that B is an antiorbital complex (on $\mathfrak{g}_\zeta$). We show:

(b) *any composition factor of any perverse cohomology sheaf of $\tilde{\mathrm{Res}}_{\mathfrak{p}_\zeta}^{\mathfrak{g}_\zeta}(B)$ is an antiorbital complex on $\mathfrak{l}_\zeta$.*

Let R be a composition factor of a perverse cohomology sheaf of $\tilde{\mathrm{Res}}_{\mathfrak{p}_\zeta}^{\mathfrak{g}_\zeta}(B)$. By assumption we have $B = \mathfrak{F}(B_1)$ where B_1 is an orbital complex on $\mathfrak{g}_{\zeta^{-1}}$. Using A.5 with B replaced by B_1 and ζ by ζ^{-1} we see that $\tilde{\mathrm{Res}}_{\mathfrak{p}_\zeta}^{\mathfrak{g}_\zeta}(B) = \mathfrak{F}(\tilde{\mathrm{Res}}_{\mathfrak{p}_{\zeta^{-1}}}^{\mathfrak{g}_{\zeta^{-1}}}(B_1))$. Let R_1 be a simple perverse sheaf on $\mathfrak{l}_{\zeta^{-1}}$ such that $R = \mathfrak{F}(R_1)$. Then R_1 is a composition factor of a perverse cohomology sheaf of $\tilde{\mathrm{Res}}_{\mathfrak{p}_{\zeta^{-1}}}^{\mathfrak{g}_{\zeta^{-1}}}(B_1)$. Using (a) with B replaced by B_1 and ζ by ζ^{-1} we see that R_1 is orbital. Hence R is antiorbital and (b) is proved.

Finally we assume that B is a biorbital complex (on $\mathfrak{g}_\zeta$). Combining (a),(b) we obtain:

(c) *any composition factor of any perverse cohomology sheaf of $\tilde{\mathrm{Res}}_{\mathfrak{p}_\zeta}^{\mathfrak{g}_\zeta}(B)$ is a biorbital complex on $\mathfrak{l}_\zeta$.*

References

[D] P.Deligne, *Cohomologie étale (SGA $4\frac{1}{2}$)*, Lect.Notes in Math., vol. 569, Springer Verlag, 1977.

[H]   A.Henderson, *Fourier transform, parabolic induction, and nilpotent orbits*, Transfor. Groups
      6 (2001), 353-370.

[L1] G.Lusztig, *Fourier transform on semisimple Lie algebras over* $\mathbf{F}_q$, Algebraic Groups Utrecht
      1986, Lect.Notes in Math., vol. 1271, Springer Verlag, 1987, pp. 177-188.

[L2] G.Lusztig, *Quivers, perverse sheaves and quantized enveloping algebras*, Jour. Amer. Math.
      Soc. **4** (1991), 365-421.

[L3] G.Lusztig, *Affine quivers and canonical bases*, Publ. Math. IHES **76** (1992), 111-163.

[L4] G.Lusztig, *From groups to symmetric spaces*, arxiv:0908.4414.

[L5] G.Lusztig, *Study of perverse sheaves arising from graded Lie algebras*, Adv.Math. **112** (1995),
      147-217.

[V]   E.B.Vinberg, *The Weyl group of a graded Lie algebra*, Izvestiya Akad.Nauk SSSR **40** (1976),
      488-526.

[W]   A.Weil, *On some exponential sums*, Proc.Nat.Acad.Sci. **34** (1948), 204-207.

DEPARTMENT OF MATHEMATICS, M.I.T., CAMBRIDGE, MA 02139

Contemporary Mathematics
Volume **557**, 2011

Adelization of Automorphic Distributions and Mirabolic Eisenstein Series

Stephen D. Miller and Wilfried Schmid

Dedicated to Gregg Zuckerman on his 60th birthday

ABSTRACT. Automorphic representations can be studied in terms of the embeddings of abstract models of representations into spaces of functions on Lie groups that are invariant under discrete subgroups. In this paper we describe an adelic framework to describe them for the group $GL(n, \mathbb{R})$, and provide a detailed analysis of the automorphic distributions associated to the mirabolic Eisenstein series. We give an explicit functional equation for some distributional pairings involving this mirabolic Eisenstein distribution, and the action of intertwining operators.

1. Introduction

Ever since the Poisson integral formula, the principal of recovering an eigenfunction from its "boundary values" (which are in general distributions) has been a useful tool in analysis. For automorphic forms, which are eigenfunctions of a ring of invariant differential operators, the boundary values can alternatively be described in terms of embeddings of models of representations into spaces of functions, embeddings which share the invariance of the automorphic forms. These automorphic distributions then control an entire automorphic representation in terms of a single object.

In previous papers we have applied automorphic distributions to studying summation formulas and the analytic continuation of L-functions [18–20], mainly for the full level congruence subgroup $GL(n, \mathbb{Z}) \subset GL(n, \mathbb{R})$. In this paper we present automorphic distributions in an adelic setting, in order to use them for general congruence subgroups. We also provide a thorough treatment of the automorphic distributions for a special but prominent type of Eisenstein series, the mirabolic Eisenstein series for the congruence subgroup $\Gamma_0(N) \subset GL(n, \mathbb{Z})$. We derive a

1991 *Mathematics Subject Classification.* 11M36, 22E45, 11F55, 11F66 .

Key words and phrases. Eisenstein series, automorphic distributions, adelization, invariant pairings, intertwining operators.

Partially supported by NSF grant DMS-0901594 and an Alfred P. Sloan Foundation Fellowship.

Partially supported by DARPA grant HR0011-04-1-0031 and NSF grant DMS-0500922.

precise form of their Fourier expansions, which also gives the analytic continuation of this mirabolic series, and prove an intertwining relation that is analogous to a functional equation. We also show that these properties extend to a relevant automorphic pairing established in [22] that involves these mirabolic Eisenstein distributions. In our forthcoming paper [23] this pairing will be calculated as the exterior square L-function times a precise ratio of Gamma factors, thereby giving a new construction of this L-function that leads to a stronger analytic continuation than previously known, as well as a functional equation.

The notion of adelic automorphic distribution is designed so that the action of the p-adic groups $GL(n, \mathbb{Q}_p)$ matches its usual action on adelic automorphic forms. This has the advantage of being able to quote certain calculations, such as local integrals, that have already been performed in related problems. One could also attempt stronger generalizations, which more generally treat the boundary values on a finite number of p-adic groups simultaneously with those on the real group, or which extend to number fields and different groups.

Sections 2, 3, and 4 contain, respectively, some properties of cuspidal automorphic distributions, mirabolic Eisenstein distributions, and the pairings of automorphic distributions. These topics are then reconsidered in section 5 using adelic terminology, which re-expresses them in a different notation that is useful in many applications. We also include an appendix recalling the known description of the generic unitary dual of $GL(n, \mathbb{R})$, as well as Langlands' recipe for defining the Gamma factors of the tensor product, symmetric square, and exterior square L-functions. Both are useful in analytic number theory, where one inputs the structure of a functional equation, and uses constraints on the shifts in the Gamma factors to obtain estimates.

It is a pleasure to dedicate this paper to Gregg Zuckerman on his 60th birthday, as his early work on Whittaker functions is essential to clarity with which we now understand the generic unitary dual. The first author in particular extends his appreciation to Zuckerman for his friendliness and helpfulness as a colleague at an early stage in his career. We also wish to thank Bill Casselman, Erez Lapid, and Freydoon Shahidi for helpful discussions, and the referee for a careful reading of the paper.

2. Automorphic Distributions

In this section we recall the notion of automorphic distribution. We let G denote the group of real points of a reductive matrix group defined over $\mathbb{Q}$, and $\Gamma \subset G$ an arithmetic subgroup. The particular examples that will matter to us are $G = GL(n, \mathbb{R})$, and a rational conjugate of a congruence subgroup[1] $\Gamma \subset GL(n, \mathbb{Z})$. We let $Z_G =$ denote the center of G, and fix a unitary central character

$$\omega \, : \, Z_G \, \longrightarrow \, \{ z \in \mathbb{C}^* \mid |z| = 1 \} \, . \tag{2.1}$$

Then G acts unitarily, by right translation, on the Hilbert space

$$L^2_\omega(\Gamma \backslash G) \; = $$

$$\{ f \in L^2_{\mathrm{loc}}(\Gamma \backslash G) \mid \int_{\Gamma \backslash G / Z_G} |f|^2 \, dg < \infty \ \text{and} \ f(gz) = \omega(z) f(g), \ z \in Z_G \} \, . \tag{2.2}$$

---

[1]The principal congruence subgroup $\Gamma(m) \subset GL(n, \mathbb{Z})$ is the kernel of the reduction map from $GL(n, \mathbb{Z})$ to $GL(n, \mathbb{Z}/m\mathbb{Z})$. A congruence subgroup is one which contains $\Gamma(m)$ for some m. For $n > 2$, they are precisely the finite index subgroups.

Automorphic distributions are associated to classical[2] automorphic representations, i.e., to G-invariant unitary embeddings

$$j : V \hookrightarrow L^2_\omega(\Gamma \backslash G) \qquad (2.3)$$

of an irreducible unitary representation (π, V) of G. The space of C^∞ vectors $V^\infty \subset V$ is dense in V, and carries a canonical Frechét topology. The linear map

$$\tau = \tau_j : V^\infty \longrightarrow \mathbb{C}, \qquad \tau(v) = j(v)(e), \qquad (2.4)$$

is well defined and Γ-invariant because j maps V^∞ to $C^\infty(\Gamma \backslash G)$. It is also continuous with respect to the topology of V^∞, and thus may be regarded as a Γ-invariant *distribution vector* for the dual unitary representation (π', V'),

$$\tau \in \left((V')^{-\infty}\right)^\Gamma. \qquad (2.5)$$

This is the *automorphic distribution* corresponding to the automorphic representation (2.3). The former determines the latter completely: for $v \in V^\infty$ and $g \in G$,

$$j(v)(g) = j(\pi(g)v)(e) = \langle \tau, \pi(g)v \rangle = \langle \pi'(g^{-1})\tau, v \rangle, \qquad (2.6)$$

so one can reconstruct the functions $j(v)$, $v \in V^\infty$, in terms of τ; because of the density of V^∞ in V, τ determines $j(v) \in L^2_\omega(\Gamma \backslash G)$ for all vectors $v \in V$.

In the following, we shall also consider automorphic distributions that do not correspond to irreducible summands of $L^2_\omega(\Gamma \backslash G)$, as in (2.3). These are Γ-invariant distribution vectors for admissible representations of finite length which need not be unitary, in particular the distribution analogues of Eisenstein series.

Most traditional approaches to automorphic forms work with finite dimensional K-invariant spaces of *automorphic functions*, meaning collections of functions $\{j(v)\}$ with v ranging over a basis of a finite dimensional, K-invariant subspace of V; here $K \subset G$ denotes a maximal compact subgroup. Finite dimensional, K-invariant subspaces necessarily consist of C^∞ vectors, so these automorphic functions are smooth. When (π, V) happens to be a spherical representation, it is natural to consider the single automorphic function $j(v_0)$ determined by the – unique, up to scaling – K-fixed vector $v_0 \in V$, $v_0 \neq 0$. In that case $j(v_0)$ can be interpreted as a Γ-invariant function on the symmetric space G/K. For non-spherical representations, typically no such canonical choice exists, and making a definite choice may in fact be delicate. In the theory of integral representations of L-functions, for example, a wrong choice may result in an integral being identically zero instead of the L-function one is interested in, or it may result in an archimedean integral that is more difficult to compute, possibly even not computable at all [2, §2.6]. By working directly with the automorphic distribution τ, our approach avoids these issues; in particular it does not matter whether (π, V) is spherical or not.

Results of Casselman [5] and Casselman-Wallach [6, 31] imply that $(V')^{-\infty}$ can be realized as a closed subspace of the space of distribution vectors for a not-necessarily-unitary principal series representation,

$$(V')^{-\infty} \hookrightarrow V^{-\infty}_{\lambda,\delta}; \qquad (2.7)$$

the subscripts λ, δ refer to the parameters of the principal series and will be explained shortly. Thus

$$\tau \in \left(V^{-\infty}_{\lambda,\delta}\right)^\Gamma \qquad (2.8)$$

---

[2]As distinguished from adelic automorphic representations.

becomes a Γ-invariant distribution vector for a principal series representation[3] with parameters (λ, δ). The embedding (2.7) is equivalent to the representation V being a quotient of the dual principal series representation $V_{-\lambda, \delta}$.

In describing the principal series, we specialize the choice of G to keep the discussion concrete,

$$G = GL(n, \mathbb{R}).\qquad(2.9)$$

Its two subgroups

$$B = \left\{ \begin{pmatrix} b_1 & 0 & \dots & 0 \\ * & b_2 & \dots & 0 \\ \vdots & \vdots & \ddots & \vdots \\ * & * & \dots & b_n \end{pmatrix} \;\middle|\; b_j \in \mathbb{R}^*, \ 1 \le j \le n \right\},$$

$$(2.10)$$

$$N = \left\{ \begin{pmatrix} 1 & * & \dots & * \\ 0 & 1 & \dots & * \\ \vdots & \vdots & \ddots & \vdots \\ 0 & 0 & \dots & 1 \end{pmatrix} \right\}$$

are, respectively, maximal solvable and maximal unipotent. The quotient

$$X = G/B\qquad(2.11)$$

is compact, and is called the *flag variety* of G. Since N acts freely on its orbit through the identity coset in $X = G/B$ and has the same dimension as X, one can identify N with an dense open subset of the flag variety,

$$N \ \simeq \ N \cdot eB \ \hookrightarrow \ X.\qquad(2.12)$$

This is the *open Schubert cell* in X.

The principal series is parameterized by pairs $(\lambda, \delta) \in \mathbb{C}^n \times (\mathbb{Z}/2\mathbb{Z})^n$. For any such pair, we define the character

$$\chi_{\lambda, \delta} : B \longrightarrow \mathbb{C}^*,$$

$$(2.13)$$

$$\chi_{\lambda, \delta} \begin{pmatrix} b_1 & 0 & \dots & 0 \\ * & b_2 & \dots & 0 \\ \vdots & \vdots & \ddots & \vdots \\ * & * & \dots & b_n \end{pmatrix} = \prod_{j=1}^{n} \left((\operatorname{sgn} b_j)^{\delta_j} |b_j|^{\lambda_j} \right).$$

The parametrization also involves the quantity

$$\rho = \left(\tfrac{n-1}{2}, \tfrac{n-3}{2}, \dots, \tfrac{1-n}{2} \right) \in \mathbb{C}^n.\qquad(2.14)$$

Each pair (λ, δ) determines a G-equivariant C^∞ line bundle $\mathcal{L}_{\lambda, \delta} \to X$, on whose fiber at the identity coset the isotropy group B acts via $\chi_{\lambda, \delta}$. By pullback from $X = G/B$ to G, the space of C^∞ sections becomes naturally isomorphic to a space of C^∞ functions on G,

$$C^\infty(X, \mathcal{L}_{\lambda, \delta}) \simeq \{ f \in C^\infty(G) \mid f(gb) = \chi_{\lambda, \delta}(b^{-1}) f(g) \text{ for } g \in G, \ b \in B \}.\quad(2.15)$$

This isomorphism relates the translation action of G on sections of $\mathcal{L}_{\lambda, \delta}$ to left translation of functions. By definition,

$$V_{\lambda, \delta}^\infty = C^\infty(X, \mathcal{L}_{\lambda - \rho, \delta})\qquad(2.16)$$

---

[3]This convention differs slightly from our earlier papers [18, 19], where we had switched the role of (π, V) and (π', V') at this stage for notational convenience. However, that switch causes a notational inconsistency for our adelic automorphic distributions in section 5 that we have elected to avoid.

is the space of C^∞ vectors of the principal series representation $V_{\lambda,\delta}$; the shift by ρ serves the purpose of making the labeling compatible with Harish-Chandra's parametrization of infinitesimal characters. Analogously

$$
\begin{aligned}
V_{\lambda,\delta}^{-\infty} &= C^{-\infty}(X, \mathcal{L}_{\lambda-\rho,\delta}) \\
&\simeq \{f \in C^{-\infty}(G) \mid f(gb) = \chi_{\lambda-\rho,\delta}(b^{-1})f(g) \text{ for } g \in G, b \in B\}
\end{aligned}
\tag{2.17}
$$

is the space of distribution vectors. The isomorphism in the second line is entirely analogous to (2.15).

The group N, which we had identified with the open Schubert cell, intersects B only in the identity. Thus, when the equivariant line bundle $\mathcal{L}_{\lambda-\rho,\delta} \to X$ is restricted to the open Schubert cell, it becomes canonically trivial, and distribution sections of the restricted line bundle become scalar-valued distributions,

$$
C^{-\infty}(N, \mathcal{L}_{\lambda-\rho,\delta}) = C^{-\infty}(N). \tag{2.18}
$$

This identification is N-invariant, of course. In particular any automorphic distribution

$$
\tau \in (V_{\lambda,\delta}^{-\infty})^{\Gamma} = C^{-\infty}(X, \mathcal{L}_{\lambda-\rho,\delta})^{\Gamma} \tag{2.19}
$$

restricts to a $\Gamma \cap N$-invariant distribution on the open Schubert cell:

$$
\tau \in C^{-\infty}(\Gamma \cap N \backslash N). \tag{2.20}
$$

Two comments are in order. Ordinarily, a distribution on a manifold is not completely determined by its restriction to a dense open subset. Since the Γ-translates of the open Schubert cell cover X, any automorphic distribution *is* determined by its restriction to N. The containment (2.20) should be interpreted in this sense. Secondly, when one views τ this way, the invariance under $\Gamma \cap N$ is directly visible. The invariance under any $\gamma \in \Gamma$ that does not lie in N can be described in terms of an appropriate factor of automorphy.

The abelianization $N/[N,N]$ – i.e., the quotient of N by the derived subgroup $[N,N]$ – is isomorphic to the additive group $\mathbb{R}^{n-1}$. Concretely, let

$$
n(x) = \begin{pmatrix}
1 & x_1 & 0 & \cdots & 0 \\
0 & 1 & x_2 & \cdots & 0 \\
\vdots & 0 & 1 & \ddots & \vdots \\
\vdots & \vdots & \vdots & \ddots & x_{n-1} \\
0 & 0 & 0 & \cdots & 1
\end{pmatrix}
\quad (\, x = (x_1, x_2, \ldots, x_{n-1}) \in \mathbb{R}^{n-1} \,); \tag{2.21}
$$

then $\mathbb{R}^{n-1} \simeq N/[N,N]$ via

$$
\mathbb{R}^{n-1} \ni x \longmapsto \text{image of } n(x) \in N/[N,N]. \tag{2.22}
$$

A congruence subgroup $\Gamma \subset G$ intersects N in a cocompact subgroup of N, and similarly $[N,N]$ in a cocompact subgroup of itself. This allows us to define

$$
\tau_{\text{abelian}} = \frac{1}{\text{covol}(\Gamma \cap [N,N])} \int_{(\Gamma \cap [N,N]) \backslash [N,N]} \ell(n)\tau \, dn, \tag{2.23}
$$

the sum of the *abelian Fourier component* of the automorphic distribution τ, as in (2.19–20); $\ell(n)$ denotes left translation by n. Equivalently

$$
\tau = \tau_{\text{abelian}} + \cdots, \tag{2.24}
$$

where $\cdots$ refers to the sum of Fourier components of τ on which $[N,N]$ acts non-trivially. By construction, $\tau_{\text{abelian}} \in V_{\lambda,\delta}^{-\infty}$, and the restriction of τ_{abelian} to N lies in $C^{-\infty}(([N,N] \cdot (\Gamma \cap N))\backslash N)$.

The quotient $([N, N] \cdot (\Gamma \cap N))\backslash N$ is compact, connected, abelian, hence a torus. Like any distribution on a torus, τ_{abelian} can be expressed as an infinite linear combination of characters. We may write

$$\tau_{\text{abelian}}(n(x)) \;=\; \sum\nolimits_{k\in\mathbb{Q}^{n-1}} c_k \, e(k_1 \, x_1 + k_2 \, x_2 + \cdots + k_{n-1} \, x_{n-1}) \qquad (2.25)$$

in which the coefficients c_k are tacitly assumed to vanish unless k lies inside $M^{-1}\mathbb{Z}$, for some appropriate integer M (which takes into account the size of the torus). Here, as from now on, we use the notational convention

$$e(z) \;=_{\text{def}}\; e^{2\pi i z} \,. \qquad (2.26)$$

In the case that Γ equals the full level congruence group $GL(n, \mathbb{Z})$, $\Gamma \cap N = N(\mathbb{Z})$ and k lies in $\mathbb{Z}^{n-1}$, because the isomorphism (2.22) induces $([N, N] \cdot (\Gamma \cap N))\backslash N \simeq \mathbb{Z}^{n-1}\backslash\mathbb{R}^{n-1}$.

Recall the notion of a cuspidal automorphic representation: an automorphic representation in the same sense as (2.3), such that

$$\int_{(\Gamma\cap N)\backslash N} j(v)(ng) \, dn \;=\; 0 \quad \text{for every } v \in V^{\infty}, \, g \in G, \qquad (2.27)$$

whenever $N \subset G$ is the unipotent radical of a proper parabolic subgroup, defined over $\mathbb{Q}$. We call an automorphic distribution $\tau \in (V^{-\infty})^{\Gamma}$ cuspidal if the corresponding automorphic representation has that property; this is equivalent to

$$\int_{N/(\Gamma\cap N)} \ell(n)\tau \, dn \;=\; 0 \qquad (2.28)$$

for every N as in (2.27) [21, Lemma 2.16]. In our particular setting of $GL(n)$ the cuspidality of τ implies

$$k \in \mathbb{Q}^{n-1} \,, \; k_j = 0 \text{ for at least one } j, \, 1 \le j \le n-1 \;\implies\; c_k = 0 \,, \qquad (2.29)$$

as can be seen by averaging the u-translates of τ over $U_{j,n-j}(\mathbb{Z})\backslash U_{j,n-j}$, the quotient of the unipotent radical of the $(j, n-j)$ parabolic modulo its group of integral points. However, the cuspidality of τ cannot be characterized solely in terms of the vanishing of certain Fourier coefficients at each cusp; it also involves conditions "at infinity" – see, for example, [18, §5].

The Casselman embedding (2.7) does not necessarily determine the parameters (λ, δ) uniquely. For example, when $V_{\lambda,\delta}$ is an irreducible principal series representation, (λ, δ) is determined only up to the action of the Weyl group. The abelian Fourier coefficients c_k, $k \in \mathbb{Q}^{n-1}$, do depend on the choice of Casselman embedding. When τ is cuspidal, one can introduce its *renormalized Fourier coefficients*

$$a_{(k_1,k_2,\ldots,k_{n-1})} \;=\; \prod_{j=1}^{n-1} \left((\operatorname{sgn} k_j)^{\delta_1+\delta_2+\cdots+\delta_j} \, |k_j|^{\lambda_1+\lambda_2+\cdots+\lambda_j} \right) c_{(k_1,k_2,\ldots,k_{n-1})} \,, \qquad (2.30)$$

which have canonical meaning. The L-functions of τ can be most naturally expressed in terms of the a_k. For k coprime to a finite set of primes depending on τ, the a_k are actually the eigenvalues of certain Hecke operators T_k acting on the automorphic representation, provided the Hecke action preserves the automorphic representation. This applies to all k when $\Gamma = GL(n, \mathbb{Z})$, demonstrating that the a_k are independent of the particular Casselman embedding. This independence can also be shown directly, without reference to Hecke operators – meaning that this independence holds for congruence subgroups Γ as well. We shall see this from a different point of view later in section 5, in terms of adelic Whittaker functions.

The terms in (2.25) have a canonical extension from the big Schubert cell N to G/B (i.e., the opposite of the restriction in (2.18-2.20)); see [7], where this issue is considered and resolved in greater generality. Let us consider the canonical extension of the additive character $n(x) \mapsto e(x_1 + \cdots + x_{n-1})$ in (2.25), which we will call the "Whittaker distribution" $w_{\lambda,\delta} \in V_{\lambda,\delta}^{-\infty}$ to emphasize its dependence on the principal series parameters. Its restriction to the big Bruhat cell $NB \subset G$ is determined by the transformation formula

$$
w_{\lambda,\delta}\left(\begin{pmatrix} 1 & x_1 & \star & \star & \star \\ & 1 & x_2 & \star & \star \\ & & \ddots & \star & \star \\ & & & 1 & x_{n-1} \\ & & & & 1 \end{pmatrix} \begin{pmatrix} b_1 & & & \\ \star & b_2 & & \\ \star & \star & \ddots & \\ \star & \star & \star & b_{n-1} \\ \star & \star & \star & \star & b_n \end{pmatrix} \right)
$$
$$
= \ e(x_1 + \cdots + x_{n-1}) \prod_{j=1}^{n} |b_j|^{(n+1)/2 - j - \lambda_j} \operatorname{sgn}(b_j)^{\delta_j} . \tag{2.31}
$$

For $k = (k_1, k_2, \ldots, k_n)$, let $D(k) = \operatorname{diag}(k_1 \cdots k_{n-1}, k_2 \cdots k_{n-1}, \ldots, k_{n-1}, 1)$ denote the diagonal matrix with diagonal entries $k_1 \cdots k_{n-1}, k_2 \cdots k_{n-1}, \ldots, k_{n-1}, 1$. If each $k_j \neq 0$ (as is automatically true for the indices corresponding to a cuspidal τ) conjugation by $D(k)$ transforms the character $n(x) \mapsto e(x_1 + \cdots + x_{n-1})$ into the character $n(x) \mapsto e(k_1 x_1 + \cdots + k_{n-1} x_{n-1})$. The canonical extension of the latter is therefore given by

$$
w_{\lambda,\delta}\left(D(k) g D(k)^{-1} \right) =
$$
$$
= \ w_{\lambda,\delta}\left(D(k) g \right) \prod_{j=1}^{n-1} \left(\prod_{i=j}^{n-1} |k_i|^{-(n+1)/2 + j + \lambda_j} \prod_{i=1}^{j} k_i^{\delta_j} \right) . \tag{2.32}
$$

In view of (2.25) and (2.30), the canonical extension of τ_{abelian} to G can be written as

$$
\tau_{\text{abelian}}(g) \ = \ \sum_{k \in \mathbb{Q}^{n-1}} \frac{a_k}{\left| \prod_{j=1}^{n-1} k_j^{j(n-j)/2} \right|} \, w_{\lambda,\delta}(D(k) g) . \tag{2.33}
$$

One then also has the following equality between distributions on G:

$$
\frac{1}{\operatorname{covol}(\Gamma \cap N)} \int_{\Gamma \cap N \backslash N} \tau(u g) \, e(-k_1 \, u_{1,2} - \cdots - k_{n-1} \, u_{n-1,n}) \, du
$$
$$
= \ \frac{a_k}{\left| \prod_{j=1}^{n-1} k_j^{j(n-j)/2} \right|} \, w_{\lambda,\delta}(D(k) g) , \tag{2.34}
$$

where $u_{i,j}$ denote the entries of $u \in N$ and $\operatorname{covol}(\Gamma \cap N)$ denotes the volume of the quotient $\Gamma \cap N \backslash N$ under the Haar measure du, normalized so that $\operatorname{covol}(N(\mathbb{Z})) = 1$.

A number of relations involving automorphic distributions, such as the functional equations of their L-functions, involve not only a particular automorphic distribution – or equivalently, the corresponding automorphic representation – but also its contragredient. The map

$$
g \ \mapsto \ \widetilde{g}, \ \ \widetilde{g} = w_{\text{long}}(g^t)^{-1} w_{\text{long}}^{-1}, \ \ \text{with} \ \ w_{\text{long}} = \begin{pmatrix} & & 1 \\ & \cdot^{\cdot^{\cdot}} & \\ 1 & & \end{pmatrix} , \tag{2.35}
$$

defines an outer automorphism of $G = GL(n, \mathbb{R})$, which preserves the subgroups $GL(n, \mathbb{Z})$, B and N. One easily checks that

$$
\widetilde{\tau}(g) \ =_{\text{def}} \tau(\widetilde{g}) \ \in \ (V_{\lambda,\delta}^{-\infty})^{\widetilde{\Gamma}} , \ \ \text{with} \ \ \widetilde{\Gamma} = \{ \widetilde{\gamma} \, | \, \gamma \in \Gamma \} , \tag{2.36}
$$

the contragredient of τ, has abelian Fourier coefficients

$$\widetilde{c}_{(k_1, k_2, \ldots, k_{n-1})} \;=\; c_{(-k_{n-1}, -k_{n-2}, \ldots, -k_1)} \tag{2.37}$$

and principal series parameters

$$\widetilde{\lambda} \;=\; (-\lambda_n, -\lambda_{n-1}, \ldots, -\lambda_1), \quad \widetilde{\delta} \;=\; (\delta_n, \delta_{n-1}, \ldots, \delta_1). \tag{2.38}$$

3. Mirabolic Eisenstein series for $GL(n)$

The Epstein zeta functions on $GL(n, \mathbb{R})$, which are sums of powers of the norms of lattice vectors in $\mathbb{R}^n$, were an early example of higher rank Eisenstein series. They have a functional equation and analytic continuation coming from Poisson summation, in complete analogy with the Riemann zeta function. Langlands, and later Jacquet and Shalika [10], studied mirabolic Eisenstein series, which are an adelic generalization involving homogenous functions other than the norm. They play a crucial role in the functional equation and analytic continuation of a number of integral representations of L-functions, e.g. [3, 4, 8, 25]. In this section we describe their distributional counterparts. Proposition 3.16 gives the analytic continuation and an explicit formula for their Fourier coefficients in terms of L-functions and arithmetic sums. These have direct applications elsewhere, most recently to string theory where they describe fine details of graviton scattering amplitudes (see, for example, [9, 24]). A functional equation is given in proposition 3.48. The analytic properties later transfer to the pairings in section 4. They are understood most easily in classical terminology; in section 5 we shall convert them into adelic expressions whose analytic properties rest on what is proven here. It is possible to recover the results here from [10], using sophisticated machinery of Casselman and Wallach. However, the translation between the two is somewhat lengthy and unenlightening, and so we have chosen to rederive them from basic principles instead, highlighting the role of degenerate principal series and intertwining operators.

Mirabolic Eisenstein series are induced from one dimensional representations of the so-called mirabolic subgroup of $GL(n)$, colloquially dubbed the "miraculous parabolic"[4]. In fact, the functional equation involves not just one, but two different mirabolic subgroups and Eisenstein series. The mirabolic subgroups and the "opposites" of their unipotent radicals are

$$P = \left\{ \begin{pmatrix} a & 0 \cdots 0 \\ * & \\ \vdots & C \\ * & \end{pmatrix} \;\middle|\; C \in GL(n-1, \mathbb{R}), \; a \in \mathbb{R}^* \right\},$$

$$\widetilde{P} = \left\{ \begin{pmatrix} & & 0 \\ C & & \vdots \\ & & 0 \\ * \cdots * & a \end{pmatrix} \;\middle|\; C \in GL(n-1, \mathbb{R}), \; a \in \mathbb{R}^* \right\}, \tag{3.1}$$

$$U = \left\{ \begin{pmatrix} 1 & * & \cdots & * \\ & 1 & & 0 \\ 0 & & \ddots & \\ & & & 1 \end{pmatrix} \right\}, \quad \widetilde{U} = \left\{ \begin{pmatrix} 1 & & & * \\ & \ddots & 0 & \vdots \\ 0 & & 1 & * \\ & & & 1 \end{pmatrix} \right\};$$

---

[4]The terminology in the literature is not entirely consistent: some reserve the term "mirabolic" for the stabilizer of a line in $\mathbb{R}^n$, e.g. $\widetilde{P}$, but not P.

note that the outer automorphism (2.35) relates P to $\widetilde{P}$ and U to $\widetilde{U}$. In analogy to the flag variety $X = G/B$,

$$Y \;=\; G/P \quad \text{and} \quad \widetilde{Y} \;=\; G/\widetilde{P} \tag{3.2}$$

are *generalized flag varieties*. The former can be naturally identified with the projective space of hyperplanes in $\mathbb{R}^n$, the latter with the projective space of lines. Since $U \cap P = \widetilde{U} \cap \widetilde{P} = \{e\}$, we can identify U and $\widetilde{U}$ with the open Schubert cells in these two spaces,

$$U \;\simeq\; U \cdot eP \;\hookrightarrow\; Y\,, \qquad \widetilde{U} \;\simeq\; \widetilde{U} \cdot e\widetilde{P} \;\hookrightarrow\; \widetilde{Y}\,. \tag{3.3}$$

This is again entirely analogous to (2.12).

For $\nu \in \mathbb{C}$ and $\varepsilon \in \mathbb{Z}/2\mathbb{Z}$, we define

$$\chi_{\nu,\varepsilon} : P \to \mathbb{C}^*\,, \quad \chi_{\nu,\varepsilon}\begin{pmatrix} a\; 0 \ldots 0 \\ * \\ \vdots \quad B \\ * \end{pmatrix} \;=\; |a|^{\frac{(n-1)\nu}{n}}(\operatorname{sgn} a)^{\varepsilon}|\det B|^{-\frac{\nu}{n}}\,,$$

$$\tag{3.4}$$

$$\widetilde{\chi}_{\nu,\varepsilon} : \widetilde{P} \to \mathbb{C}^*\,, \quad \widetilde{\chi}_{\nu,\varepsilon}\begin{pmatrix} B \quad \begin{matrix} 0 \\ \vdots \\ 0 \end{matrix} \\ * \cdots * \; a \end{pmatrix} \;=\; |\det B|^{\frac{\nu}{n}}|a|^{-\frac{(n-1)\nu}{n}}(\operatorname{sgn} a)^{\varepsilon}\,.$$

We study these two characters without any loss of generality, because they account for all characters of P and $\widetilde{P}$, up to tensoring by central characters. Taking these other choices amounts to multiplying our eventual Eisenstein distributions by $\operatorname{sgn}(\det g)$, and has no analytic impact. The quantity

$$\rho_{\mathrm{mir}} \;=\; \frac{n}{2} \tag{3.5}$$

plays the role of ρ in the present context.

There exist unique G-equivariant C^∞ line bundles $\mathcal{L}_{\nu,\varepsilon} \to Y$, $\widetilde{\mathcal{L}}_{\nu,\varepsilon} \to \widetilde{Y}$, on whose fibers at the identity cosets the isotropy groups act by, respectively, $\chi_{\nu,\varepsilon}$ and $\widetilde{\chi}_{\nu,\varepsilon}$. The group G acts via left translation on

$$\begin{aligned} W^\infty_{\nu,\varepsilon} \;&=\; C^\infty(Y, \mathcal{L}_{\nu-\rho_{\mathrm{mir}},\varepsilon}) \\ &\simeq\; \{f \in C^\infty(G) \mid f(gp) = \chi_{\nu-\rho_{\mathrm{mir}},\varepsilon}(p^{-1})f(g) \text{ for } g \in G, p \in P\}\,, \\ \widetilde{W}^\infty_{\nu,\varepsilon} \;&=\; C^\infty(\widetilde{Y}, \widetilde{\mathcal{L}}_{\nu-\rho_{\mathrm{mir}},\varepsilon}) \\ &\simeq\; \{f \in C^\infty(G) \mid f(g\tilde{p}) = \widetilde{\chi}_{\nu-\rho_{\mathrm{mir}},\varepsilon}(\tilde{p}^{-1})f(g) \text{ for } g \in G, \tilde{p} \in \widetilde{P}\}\,. \end{aligned} \tag{3.6}$$

In particular, functions $f \in W^\infty_{\nu,\varepsilon}$ and $\tilde{f} \in \widetilde{W}^\infty_{\nu,\varepsilon}$ obey the respective transformation laws

$$\begin{aligned} f\left(g\left(\begin{smallmatrix} a \\ * & B \end{smallmatrix}\right)\right) \;&=\; |a|^{n/2-\nu}(\operatorname{sgn} a)^{\varepsilon}f(g) \quad \text{and} \\ \tilde{f}\left(g\left(\begin{smallmatrix} B \\ * & a \end{smallmatrix}\right)\right) \;&=\; |a|^{\nu-n/2}(\operatorname{sgn} a)^{\varepsilon}\tilde{f}(g)\,, \quad \text{provided } |a||\det B| = 1\,. \end{aligned} \tag{3.7}$$

These are the spaces of C^∞ vectors for degenerate principal series representations $W_{\nu,\varepsilon}$, $\widetilde{W}_{\nu,\varepsilon}$.

As in the case of the principal series, the line bundle $\mathcal{L}_{\nu-\rho_{\mathrm{mir}},\varepsilon}$ is equivariantly trivial over the open Schubert cell $U \subset Y$. Since $\delta_e \in C^{-\infty}(U)$, the Dirac delta function at $e \in U$, evidently has compact support in U, we may regard it as a distribution section of $\mathcal{L}_{\nu-\rho_{\mathrm{mir}},\varepsilon}$, or in other words, as a vector in $W_{\nu,\varepsilon}^{-\infty}$. This makes

$$\delta_\infty \ =_{\mathrm{def}} \ \ell(w_{\mathrm{long}})\delta_e \ \in \ W_{\nu,\varepsilon}^{-\infty} \tag{3.8}$$

well defined. By construction, δ_∞ is supported at $w_{\mathrm{long}}P \in Y$, the unique fixed point of U, also known as the closed Schubert cell in Y. Similarly there exists a delta function $\delta_{\widetilde{\infty}} \in \widetilde{W}_{\nu,\varepsilon}^{-\infty}$ supported on the closed Schubert cell $w_{\mathrm{long}}\widetilde{P} \in \widetilde{Y}$.

Mirabolic Eisenstein series are globally induced from a character of P or $\widetilde{P}$. As for their analytic properties, it suffices to study them for the congruence subgroups

$$\Gamma_0(N) \ = \ \left\{ \gamma \in GL(n,\mathbb{Z}) \mid \gamma \equiv \begin{pmatrix} \star & \cdots & \star & \star \\ \vdots & \ddots & \vdots & \vdots \\ \star & \cdots & \star & \star \\ 0 & \cdots & 0 & \star \end{pmatrix} \pmod{N} \right\} \tag{3.9}$$

or

$$\widetilde{\Gamma}_0(N) \ = \ \left\{ \gamma \in GL(n,\mathbb{Z}) \mid \gamma \equiv \begin{pmatrix} \star & \star & \cdots & \star \\ 0 & \star & & \star \\ \vdots & \vdots & \ddots & \vdots \\ 0 & \star & \cdots & \star \end{pmatrix} \pmod{N} \right\}, \tag{3.10}$$

by means of a reduction we will discuss in section 5. Of course $\Gamma_0(N)$ and $\widetilde{\Gamma}_0(N)$ are related by the outer automorphism (2.35). Any Dirichlet character ψ modulo N lifts to characters α of $\Gamma_0(N)$ and $\widetilde{\alpha}$ of $\widetilde{\Gamma}_0(N)$ defined through the formulas

$$\alpha(\gamma) \ = \ \psi(\gamma_{nn})^{-1} \quad \text{and} \quad \widetilde{\alpha}(\gamma) \ = \ \psi(\gamma_{11}), \qquad \gamma = (\gamma_{ij}). \tag{3.11}$$

The reason for the inverse is to ensure $\widetilde{\alpha}(\widetilde{\gamma}) = \alpha(\gamma)$, a property used below in (3.14). These characters are respectively trivial on the subgroups $\Gamma_1(N) \subset \Gamma_0(N)$ and $\widetilde{\Gamma}_1(N) \subset \widetilde{\Gamma}_0(N)$, which are defined by the congruence $\gamma_{nn} \equiv 1 \pmod{N}$ in the former case, and $\gamma_{11} \equiv 1 \pmod{N}$ in the latter case.

We let $\Gamma = \Gamma_0(N)$ and $\Gamma_\infty = \Gamma \cap w_{\mathrm{long}}Pw_{\mathrm{long}}$ denote its isotropy subgroup at $w_{\mathrm{long}}P \in Y$. Because $-e \in \Gamma_\infty$, we insist that $\psi(-1) = (-1)^\varepsilon$ so that Γ_∞ acts trivially on δ_e. (Otherwise the Eisenstein series we presently define would be identically zero.) With this choice of parity parameter define

$$E_{\nu,\psi} \ = \ L(\nu + \tfrac{n}{2}, \psi) \sum_{\gamma \in \Gamma/\Gamma_\infty} \alpha(\gamma)\, \ell(\gamma)\delta_\infty \ \in \ W_{\nu,\varepsilon}^{-\infty}. \tag{3.12}$$

For $\operatorname{Re}\nu > \rho_{\mathrm{mir}} = n/2$ this sum converges in the strong distribution topology. In the region $\{\operatorname{Re}\nu > \rho_{\mathrm{mir}}\}$, the resulting distribution vector depends holomorphically on ν and satisfies the condition $\ell(\gamma)E_{\nu,\psi} = \alpha(\gamma)^{-1}E_{\nu,\psi}$ for all $\gamma \in \Gamma$. Entirely analogously, with $\widetilde{\Gamma} = \widetilde{\Gamma}_0(N)$,

$$\widetilde{E}_{\nu,\psi} \ = \ L(\nu + \tfrac{n}{2}, \psi) \sum_{\gamma \in \widetilde{\Gamma}/\widetilde{\Gamma}_{\widetilde{\infty}}} \widetilde{\alpha}(\gamma)\, \ell(\gamma)\delta_{\widetilde{\infty}} \ \in \ \widetilde{W}_{\nu,\varepsilon}^{-\infty} \tag{3.13}$$

converges and depends holomorphically on ν in $\{\operatorname{Re}\nu > \rho_{\mathrm{mir}} = n/2\}$. The two Eisenstein series are related by the involution (2.35):

$$E_{\nu,\psi}(g) \ = \ \widetilde{E}_{\nu,\psi}(\widetilde{g}). \tag{3.14}$$

The following proposition gives a simpler formula for these Eisenstein distributions when restricted to the open, dense Bruhat cells $U, w_{\mathrm{long}}U \subset Y$, and $\widetilde{U}, w_{\mathrm{long}}\widetilde{U} \subset \widetilde{Y}$, respectively. Since both Eisenstein series are invariant under a congruence group, and the translates of any of these cells by that invariance group cover Y and $\widetilde{Y}$, respectively, restriction to either determines them completely. The statement involves the finite Fourier transform

$$\widehat{\psi}(m) \ = \ \sum_{a \,(\mathrm{mod}\, N)} \psi(a)\, e(\tfrac{am}{N}) \tag{3.15}$$

of a Dirichlet character of modulus N. (Note that $\psi(a) = 0$ when a is not relatively prime to N.)

3.16. PROPOSITION. (Analytic continuation and Fourier expansion of mirabolic Eisenstein distributions.) *Let $\mathrm{Re}\,\nu > n/2$. The restriction of the distribution $E_{\nu,\psi}$ to U as well as the restriction of the distribution $\widetilde{E}_{\nu,\psi}$ to $\widetilde{U}$ are determined by the common formulas*

$$E_{\nu,\psi}\begin{pmatrix} 1 & -u_{n-1} & \cdots & -u_1 \\ & 1 & & 0 \\ 0 & & \ddots & \\ & & & 1 \end{pmatrix} = \widetilde{E}_{\nu,\psi}\begin{pmatrix} 1 & & & u_1 \\ & \ddots & 0 & \vdots \\ 0 & & 1 & u_{n-1} \\ & & & 1 \end{pmatrix}$$

$$= \sum_{\substack{v \in \mathbb{Z}^n,\ v_1 > 0 \\ N | v_1, \ldots, v_{n-1}}} \psi(v_n)\, v_1^{-\nu-n/2}\, \delta_{v_n/v_1}(u_1) \cdots \delta_{v_2/v_1}(u_{n-1})$$

$$= N^{-\nu-n/2} \sum_{\substack{v \in \mathbb{Z}^n,\ v_1 > 0}} v_1^{-\nu-n/2+1} e(v_1 v_n u_1) \delta_{v_{n-1}/v_1}(u_2) \cdots \delta_{v_2/v_1}(u_{n-1})\, \widehat{\psi}(-v_n)$$

$$= \sum_{r \in \mathbb{Z}^{n-1}} a_r\, e(r_1 u_1 + \cdots + r_{n-1} u_{n-1}),$$

where

$$a_r = N^{-\nu-n/2} \sum_{\substack{d > 0 \\ d \,|\, r_1, \ldots, r_{n-1}}} d^{-\nu+n/2-1}\, \widehat{\psi}(-r_1/d).$$

Their restrictions to $w_{long}U$ and $w_{long}\widetilde{U}$ are determined by the common formula

$$\left(\ell(w_{long})\, E_{\nu,\psi}\right)\begin{pmatrix} 1 & u_{n-1} & \cdots & u_1 \\ & 1 & & 0 \\ 0 & & \ddots & \\ & & & 1 \end{pmatrix} = \left(\ell(w_{long})\, \widetilde{E}_{\nu,\psi}\right)\begin{pmatrix} 1 & & & u_1 \\ & \ddots & 0 & \vdots \\ 0 & & 1 & u_{n-1} \\ & & & 1 \end{pmatrix}$$

$$= \sum_{\substack{v \in \mathbb{Z}^n,\ v_n > 0 \\ N | v_1, \ldots, v_{n-1}}} \psi(v_n)\, v_n^{-\nu-n/2}\, \delta_{v_1/v_n}(u_1) \cdots \delta_{v_{n-1}/v_n}(u_{n-1})$$

$$= \sum_{r \in \mathbb{Z}^{n-1}} c_r\, e\left(\frac{r_1 u_1 + \cdots + r_{n-1} u_{n-1}}{N}\right),$$

where

$$c_r = \frac{1}{N^{n-1}} \sum_{\substack{d > 0 \\ d \,|\, r_1, \ldots, r_{n-1}}} \psi(d)\, d^{-\nu+n/2-1}.$$

These sums, and hence also both $E_{\nu,\psi}$ and $\widetilde{E}_{\nu,\psi}$, can be holomorphically continued to $\mathbb{C} - \{n/2\}$. They are entire if ψ is nontrivial, and have a simple pole at $\nu = n/2$ otherwise.

Proof: Because of the relation (3.14) and the visible transformation properties of the asserted formulas, the formulas for $E_{\nu,\psi}$ and $\widetilde{E}_{\nu,\psi}$ are equivalent. We shall thus work with $\widetilde{E}_{\nu,\psi}$, first deriving the formulas as sums of δ-functions, then the alternative expressions in terms of Fourier series, and finally deduce the meromorphic continuation from these.

We begin with the second set of formulas, for the restriction to $w_{long}\widetilde{U}$. Letting Γ instead stand for $w_{long}\widetilde{\Gamma}_0(N)w_{long}$, the expression for $\widetilde{E}_{\nu,\psi} \in \widetilde{W}_{\nu,\varepsilon}^{-\infty}$ in (3.13) may be rewritten as

$$\ell(w_{\text{long}})\widetilde{E}_{\nu,\psi} \;=\; L(\nu + \tfrac{n}{2}, \psi) \sum_{\gamma \in \Gamma/\Gamma \cap \widetilde{P}} \widetilde{\alpha}(w_{\text{long}}\,\gamma\, w_{\text{long}})\, \ell(\gamma)\, \delta_{\tilde{e}}\,. \qquad (3.17)$$

The last column of a matrix is unchanged, up to sign, after right multiplication by an element of $\Gamma \cap \widetilde{P}$. Moreover, every n-tuple of relatively prime integers occurs as the last column of some matrix in $GL(n, \mathbb{Z})$. Its subgroup Γ is defined by the congruence that all entries except for the final one in its last column are divisible by N. Therefore, the cosets $\Gamma/\Gamma \cap \widetilde{P}$ are in bijective correspondence with the set

$$\{\text{vectors } v = (v_1, \ldots, v_n) \in \mathbb{Z}^n \text{ with } GCD(v) = 1 \text{ and } N | v_1, \ldots, v_{n-1}\}/\{\pm 1\}.$$

Given $v \in \mathbb{Z}^n$ whose entries are relatively prime and satisfy the above divisibility condition, we let γ_v denote a coset representative in $\Gamma/\Gamma \cap \widetilde{P}$.

When (3.17) is restricted to $\widetilde{U}$, some of the terms in the sum on the right hand side vanish because the γ-translate of $\delta_{\tilde{e}}$ does not lie in the big cell. The nonvanishing terms are precisely those for which $\gamma \in \Gamma \subset G$ projects into the big cell $\widetilde{U} \subset \widetilde{Y} = G/\widetilde{P}$. A matrix whose final column is the vector v projects to the big cell $\widetilde{U}$ if and only if its last entry is nonzero; in this situation, applied to γ_v, we have the explicit matrix decomposition

$$\gamma_v \;=\; \begin{pmatrix} I & u \\ & 1 \end{pmatrix} \begin{pmatrix} A & \\ \star & v_n \end{pmatrix}, \qquad (3.18)$$

where $u = \frac{1}{v_n}(v_1, \ldots, v_{n-1}) \in \mathbb{R}^{n-1}$ and A is a matrix with determinant $\pm 1/v_n$. Therefore the range of summation in (3.17) is in bijection with

$$\{v = (v_1, \ldots, v_n) \in \mathbb{Z}^n \text{ with } GCD(v) = 1, \, N \mid v_1, \ldots, v_{n-1}, \text{ and } v_n > 0\}. \quad (3.19)$$

The decomposition (3.18) allows us to compute the following action of γ_v^{-1} on the delta function $\delta_{\left(\frac{v_1}{v_n}, \ldots, \frac{v_{n-1}}{v_n}\right)}$ on $\widetilde{U}$:

$$
\begin{aligned}
\ell(\gamma_v^{-1})\, \delta_{\left(\frac{v_1}{v_n}, \ldots, \frac{v_{n-1}}{v_n}\right)} &= \ell\left(\gamma_v^{-1} \begin{pmatrix} 1 & & & v_1/v_n \\ & \ddots & 0 & \vdots \\ 0 & & 1 & v_{n-1}/v_n \\ & & & 1 \end{pmatrix}\right) \delta_{\tilde{e}} \\
&= \ell\left(\begin{pmatrix} A & \\ \star & v_n \end{pmatrix}^{-1}\right) \delta_{\tilde{e}} \qquad (\det A = \pm 1/v_n) \\
&= (\operatorname{sgn} v_n)^{\varepsilon}\, |v_n|^{\nu + n/2}\, \delta_{\tilde{e}}\,.
\end{aligned}
\qquad (3.20)
$$

In this last equation, the transformation rule (3.7) has provided a factor of $(\operatorname{sgn} v_n)^{\varepsilon}$ $|v_n|^{\nu - n/2}$, while the δ-function identity $\delta_{\tilde{e}}(\frac{Au}{v_n}) = |v_n|^n \delta_{\tilde{e}}(u)$ is responsible for the rest of the exponent. Using $\widetilde{\alpha}(w_{\text{long}}\gamma w_{\text{long}}) = \psi(v_n)$, the summand for γ_v in (3.17) can be written as

$$\widetilde{\alpha}(w_{\text{long}}\, \gamma_v\, w_{\text{long}})\, \ell(\gamma_v)\, \delta_{\tilde{e}} \;=\; \psi(v_n)\, (\operatorname{sgn} v_n)^{\varepsilon}\, |v_n|^{-\nu - n/2}\, \delta_{\left(\frac{v_1}{v_n}, \ldots, \frac{v_{n-1}}{v_n}\right)}. \quad (3.21)$$

Summing this expression over the coset representatives from (3.19) gives, in terms of the coordinates $(u_1, \ldots, u_{n-1})$ on $\widetilde{U}$ in the second set of statements in the proposition, an expression similar to the one claimed there for $\ell(w_{\text{long}})\widetilde{E}_{\nu,\psi}$. They differ only in that the latter has no condition on $GCD(v)$. However, the first set consists of scalar multiples, by positive integers relatively prime to N, of the second set, and multiplication by the pre-factor $L(\nu + \tfrac{n}{2}, \psi)$ in (3.17) – unused until now – accounts for the discrepancy. (Note that terms for which $(v_n, N) > 1$ vanish.)

At this point, we have established the δ-function formula for the restriction of $\ell(w_{\text{long}})\widetilde{E}_{\nu,\psi}$ to $\widetilde{U}$, and therefore also the one for the restriction of $\ell(w_{\text{long}})E_{\nu,\psi}$ to U, to which it is equivalent. Had we instead considered the series $\widetilde{E}_{\nu,\psi}$ instead of its w_{long}-translate, the last column of γ would have entries $(v_n, \ldots, v_1)$, the reverse of the situation we encountered above. The identical reasoning produces the same formula, but with v_j replaced by v_{n+1-j} in the summand – exactly the first claim of the proposition.

Next we turn to the assertions about the Fourier expansions, starting first with the common expression for the w_{long} translates. It is periodic in each u_i with period N, so the coefficient c_r is computed by the integral

$$\frac{1}{N^{n-1}} \int_{(N\mathbb{Z}\backslash\mathbb{R})^{n-1}} \sum_{\substack{v \in \mathbb{Z}^n,\, v_n > 0 \\ N | v_1, \ldots, v_{n-1}}} \psi(v_n)\, v_n^{-\nu-n/2}\, e\left(-\frac{\sum_{i=1}^{n-1} r_i u_i}{N}\right) \times$$

$$\times\, \delta_{v_1/v_n}(u_1) \cdots \delta_{v_{n-1}/v_n}(u_{n-1})\, du_1 \cdots du_{n-1}$$

$$= \frac{1}{N^{n-1}} \sum_{v_n > 0} \sum_{\substack{v_1, \ldots, v_{n-1} \in \mathbb{Z}/Nv_n\mathbb{Z} \\ N | v_1, \ldots, v_{n-1}}} \psi(v_n)\, v_n^{-\nu-n/2}\, e\left(-\frac{\sum_{i=1}^{n-1} r_i v_i}{Nv_n}\right) \qquad (3.22)$$

$$= \frac{1}{N^{n-1}} \sum_{d > 0} \sum_{v_1, \ldots, v_{n-1} \in \mathbb{Z}/d\mathbb{Z}} \psi(d)\, d^{-\nu-n/2}\, e\left(\frac{\sum_{i=1}^{n-1} r_i v_i}{d}\right).$$

The sum over any fixed v_j, for $1 \le j \le n-1$, equals d if $d | r_j$, and zero otherwise. Therefore c_r is given by the formula stated in the proposition. The formula for a_r is computed by the same procedure. The hybrid formula for the restriction $E_{\nu,\psi}$ or $\widetilde{E}_{\nu,\psi}$ which involves a Fourier series in u_1, and δ-functions in the other variables, is proven by taking a Fourier integral only in the variable u_1, and leaving the other u_j alone.

Finally we come to the analytic continuation, which is equivalent for each of the expressions involved. We therefore consider the last formula in the statement of the proposition. The coefficient c_r equals a finite sum which is entire in ν, unless $r = (0, 0, \ldots, 0)$. In this exceptional case $c_0 = N^{1-n} L(\nu-n/2+1, \psi)$, which is entire for all nontrivial characters ψ, and has a simple pole at $\nu = n/2$ when ψ is trivial. This establishes the asserted meromorphic continuation of the restriction of the Eisenstein series $E_{\nu,\psi}$ to the open Schubert cell $w_{\text{long}}U$. Since $E_{\nu,\psi}$ is automorphic under $\Gamma_0(N)$, and the $\Gamma_0(N)$-translates of $w_{\text{long}}U$ cover $Y = G/P$, the continuation is valid on all of Y. Likewise, the identical meromorphic continuation applies to $\widetilde{E}_{\nu,\psi}$ because of (3.14). $\qquad\square$

We have now shown the analytic continuation of the mirabolic Eisenstein distributions. We next turn to their functional equations. The two degenerate principal series representations (3.6) are related by the *standard intertwining operator*

$$I_\nu\; :\; W^\infty_{-\nu,\varepsilon}\; \longrightarrow\; \widetilde{W}^\infty_{\nu,\varepsilon}, \qquad (3.23)$$

defined in terms of the realization by C^∞ functions by the integral

$$(I_\nu f)(g)\; =\; \int_U f(g\, w_{\text{long}}\, u)\, du\,; \qquad (3.24)$$

recall the definition of w_{long} in (2.35). It is well known that the integral converges absolutely[5] for $\operatorname{Re} \nu > n/2 - 1$, and we shall also see this directly. Two properties of I_ν are crucial for our purposes:

a) I_ν has a meromorphic continuation to all $\nu \in \mathbb{C}$, and

b) it extends continuously to a linear operator $I_\nu : W^{-\infty}_{-\nu,\varepsilon} \to \widetilde{W}^{-\infty}_{\nu,\varepsilon}$;

(3.25)

see [12] for the former, and [6] for the latter.

We now give an explicit formula for the action of I_ν in terms of the restriction of C^∞ functions to the open Schubert cells $U \subset G/P$, $\widetilde{U} \subset G/\widetilde{P}$, for ν in the range of convergence – i.e., for $\operatorname{Re} \nu > n/2 - 1$.

3.26. PROPOSITION. *Let* $f \in W^\infty_{-\nu,\varepsilon}$ *, and regard* f *as a function on* $U \cong \mathbb{R}^{n-1}$ *via its restriction to* U *and the identification*

$$\mathbb{R}^{n-1} \ni x \mapsto u(x) =_{\text{def}} \begin{pmatrix} 1 & x_{n-1} & \cdots & & x_1 \\ & 1 & & 0 & \\ & & \ddots & & \\ 0 & & & & 1 \end{pmatrix} \in U.$$

Similarly, regard $I_\nu f \in \widetilde{W}^\infty_{\nu,\varepsilon}$ *as a function on* $\widetilde{U} \cong \mathbb{R}^{n-1}$ *via the identification[6]*

$$\mathbb{R}^{n-1} \ni y \mapsto \widetilde{u}(y) =_{\text{def}} \begin{pmatrix} 1 & & & -y_1 \\ & \ddots & 0 & \vdots \\ 0 & & 1 & -y_{n-1} \\ & & & 1 \end{pmatrix} \in \widetilde{U}.$$

Then, for $\operatorname{Re} \nu > n/2 - 1$ *,* $(I_\nu f)(\widetilde{u}(y))$ *is given by the integral*

$$\int_{z \in \mathbb{R}^{n-1}} f(u(z)) \left| \sum_{j=2}^{n-1} y_j z_{n+1-j} - y_1 - z_1 \right|^{\nu - n/2} \operatorname{sgn}\left(\sum_{j=2}^{n-1} y_j z_{n+1-j} - y_1 - z_1\right)^\varepsilon dz.$$

Proof: By construction, the intertwining operator I_ν is invariant under left translation by any $g \in G$. To establish the assertion of the proposition, it therefore suffices to establish the integral expression for $y = 0$, and then to check that it is compatible with translation from $\widetilde{u}(0) = e$ to $\widetilde{u}(y)$.

First the compatibility with translation. On the one hand, $(I_\nu f)(\widetilde{u}(y)) = (\ell(\widetilde{u}(-y))(I_\nu f))(e) = (I_\nu \ell(\widetilde{u}(-y))f)(\widetilde{u}(0))$; on the other,

$$\int_{z \in \mathbb{R}^{n-1}} (\ell(\widetilde{u}(-y))f)(u(z)) \, |z_1|^{\nu - n/2} \operatorname{sgn}(-z_1)^\varepsilon dz =$$
$$= \int_{z \in \mathbb{R}^{n-1}} f(\widetilde{u}(y) \cdot u(z)) \, |z_1|^{\nu - n/2} \operatorname{sgn}(-z_1)^\varepsilon dz.$$

(3.27)

Since

$$\widetilde{u}(y) \cdot u(z) = \begin{pmatrix} 1 & z_{n-1} & z_{n-2} & \cdots & z_2 & \widetilde{z}_1 \\ & 1 & & 0 & \cdots & 0 & 0 \\ & & \ddots & & & \vdots \\ 0 & & & & 1 & 0 \\ & & & & & 1 \end{pmatrix} \begin{pmatrix} 1 & 0 & 0 & \cdots & 0 & 0 \\ & 1 & 0 & \cdots & 0 & -y_2 \\ & & \ddots & \ddots & & \vdots \\ 0 & & & & 1 & -y_{n-1} \\ & & & & & 1 \end{pmatrix},$$

(3.28)

with $\widetilde{z}_1 = z_1 - y_1 + \sum_{2 \le j \le n-1} z_j y_{n+1-j}$, the transformation law (3.7) implies that the integral (3.27) coincides with the integral in the proposition.

---

[5]For the sake of notational simplicity we are dropping the subscript ε for I_ν , since the action of the intertwining operator affects only ν, not ε.

[6]The minus signs are necessary to make (2.35) consistent with (3.14).

At this point, it suffices to treat the case $y = 0$. According to the definition of the intertwining operator,

$$
\begin{aligned}
(I_\nu f)\,(\widetilde{u}(0)) \;&=\; \int_{z\in\mathbb{R}^{n-1}} f(w_{\text{long}}\, u(z))\, dz \\[4pt]
&=\; \int_{z\in\mathbb{R}^{n-1}} f(u(\tfrac{1}{z_1}, \tfrac{-z_{n-1}}{z_1}, \ldots, \tfrac{-z_3}{z_1}, \tfrac{-z_2}{z_1}))\, |\,z_1\,|^{-\nu-n/2}\, \operatorname{sgn}(-z_1)^\varepsilon\, dz \qquad (3.29) \\[4pt]
&=\; \int_{z\in\mathbb{R}^{n-1}} f(u(z))\, |\,z_1\,|^{\nu-n/2}\, \operatorname{sgn}(-z_1)^\varepsilon\, dz\,;
\end{aligned}
$$

at the second step, we have used the transformation law (3.7) and the matrix identity

$$
w_{\text{long}}\, u(z) \;=\;
\begin{pmatrix}
1 & -z_2/z_1 & \cdots & -z_{n-1}/z_1 & 1/z_1 \\
 & 1 & & 0 & 0 \\
 & & \ddots & \ddots & \vdots \\
0 & & & 1 & 0 \\
 & & & & 1
\end{pmatrix}
\begin{pmatrix}
-1/z_1 & 0 & \cdots & 0 & 0 \\
0 & 0 & \cdots & 1 & 0 \\
\vdots & & \ddots & & \vdots \\
0 & 1 & 0 & \cdots & 0 \\
1 & z_{n-1} & z_{n-2} & \cdots & z_1
\end{pmatrix},
\qquad (3.30)
$$

and at the third step, the change of variables

$$
(z_1, z_2, \ldots, z_{n-1}) \;\mapsto\; \left(\tfrac{1}{z_1}, \tfrac{-z_{n-1}}{z_1}, \ldots, \tfrac{-z_3}{z_1}, \tfrac{-z_2}{z_1}\right). \qquad (3.31)
$$

The identity (3.29) completes the proof of the proposition. $\qquad\square$

The identity (3.30) and the transformation law (3.7) directly imply a simple estimate: along the line $\{x_2 = x_3 = \cdots = x_{n-1} = 0\}$, any $f \in W^\infty_{-\nu,\varepsilon}$ satisfies the bound $|f(u(x))| = O(\|x\|^{-\operatorname{Re}\nu - n/2})$ as $\|x\| \to \infty$; the implied constant depends on a bound for $\ell(w_{\text{long}})f$ on a neighborhood of the origin. We consider $SO(n-1)$ as a subgroup of $GL(n)$ by embedding it into the bottom right corner. Then $SO(n-1)$ acts transitively, by conjugation, on the set of lines in $\mathbb{R}^{n-1} \cong U$. By compactness, the translates $\ell(w_{\text{long}}m)f$, for $m \in SO(n-1)$, are uniformly bounded on bounded subsets of $\mathbb{R}^{n-1} \cong U$. Since $f \in W^\infty_{-\nu,\varepsilon}$ is invariant under right translation by elements of $SO(n-1)$, the estimate we gave holds not on just a single line, but globally on U:

$$
f \in W^\infty_{-\nu,\varepsilon} \;\implies\; \|f(u(x))\| = O(\|x\|^{-\operatorname{Re}\nu - n/2}) \text{ as } \|x\| \to \infty. \qquad (3.32)
$$

This bound and its derivation are valid for all $\nu \in \mathbb{C}$. When $\operatorname{Re}\nu > n/2 - 1$, it implies the convergence of the integral (3.29), both near the origin and at infinity. Since I_ν is G-invariant, we have established that the integral (3.24) does converge for $\operatorname{Re}\nu > n/2 - 1$ and any $g \in G$, as was mentioned earlier.

In complete analogy to $I_\nu : W^\infty_{-\nu,\varepsilon} \to \widetilde{W}^\infty_{\nu,\varepsilon}$ in (3.23–24), one can define the operator $\widetilde{I}_\nu : \widetilde{W}^\infty_{-\nu,\varepsilon} \to W^\infty_{\nu,\varepsilon}$; this involves integrating over $\widetilde{U}$ instead of U. Then $I_\nu, \widetilde{I}_\nu$ are dual to each other, in the sense that

$$
\int_{\widetilde{U}} I_\nu f_1(\widetilde{u})\, \widetilde{f}_2(\widetilde{u})\, d\widetilde{u} \;=\; \int_U f_1(u)\, \widetilde{I}_\nu \widetilde{f}_2(u)\, du\,,
$$
$$
\text{for all } f_1 \in W^\infty_{-\nu,\varepsilon} \text{ and } \widetilde{f}_2 \in \widetilde{W}^\infty_{-\nu,\varepsilon}\,; \qquad (3.33)
$$

the integrals on the two sides implement the natural G-equivariant pairings between $\widetilde{W}^\infty_{\nu,\varepsilon}$ and $\widetilde{W}^\infty_{-\nu,\varepsilon}$, respectively $W^\infty_{-\nu,\varepsilon}$ and $W^\infty_{\nu,\varepsilon}$. For $\operatorname{Re}\nu > n/2 - 1$, i.e., when the integrals defining I_ν and $\widetilde{I}_\nu$ converge, the identity follows from the explicit formula for I_ν in proposition 3.26 and the analogous formula for $\widetilde{I}_\nu$. Meromorphic continuation implies the identity for other values of ν.

Since I_ν extends continuously to $I_\nu : W_{-\nu,\varepsilon}^{-\infty} \to \widetilde{W}_{\nu,\varepsilon}^{-\infty}$, the identity (3.33) implies a concrete description of the effect of I_ν on distribution vectors,

$$\int_{\widetilde{U}} I_\nu \tau(\widetilde{u})\, \widetilde{f}(\widetilde{u})\, d\widetilde{u} \;=\; \int_U \tau(u)\, \widetilde{I}_\nu \widetilde{f}(u)\, du\,,$$

$$\text{for all } \tau \in W_{-\nu,\varepsilon}^{-\infty},\ \widetilde{f} \in \widetilde{W}_{-\nu,\varepsilon}^{\infty}\,. \tag{3.34}$$

Unlike in (3.33), the integrals in this identity have merely symbolic meaning: the pairings $\widetilde{W}_{\nu,\varepsilon}^{-\infty} \times \widetilde{W}_{-\nu,\varepsilon}^{\infty} \to \mathbb{C}$ and $W_{-\nu,\varepsilon}^{-\infty} \times W_{\nu,\varepsilon}^{\infty} \to \mathbb{C}$ involve "integration" over $\widetilde{Y} = G/\widetilde{P}$ and $Y = G/P$, not only over the dense open cells $\widetilde{U} \subset \widetilde{Y}$, $U \subset Y$. The integrals as written do extend naturally to $\widetilde{Y}$ and Y.

Let $E_{1,n} \in \mathfrak{gl}(n,\mathbb{R})$ denote the matrix with the entry 1 in the $(1,n)$-slot, and zero entries otherwise. If $f \in W_{-\nu,\varepsilon}^{\infty}$ and Re $\nu > 1 - n/2$, the estimate (3.32) shows that the integrals

$$J_\nu f(g) \;=_{\text{def}}\; \int_{\mathbb{R}} f\left(g \exp(t\, E_{1,n})\right) dt \qquad (f \in W_{-\nu}^{\infty},\ g \in G) \tag{3.35}$$

converge. For other values of ν, $\nu \notin 1 - n/2 - \mathbb{Z}_{\geq 0}$, the integrals still make sense by meromorphic continuation (the unspecified integer in $\mathbb{Z}_{\geq 0}$ in fact has the same parity as ε at any singularity). This can be seen by translating the point $\lim_{t\to\infty} \exp(tE_{1,n})P \in Y$ to the origin.

3.36. LEMMA. *Suppose $I_\nu : W_{-\nu,\varepsilon}^{\infty} \to \widetilde{W}_{\nu,\varepsilon}^{\infty}$ has no pole at ν, $W_{-\nu,\varepsilon}^{\infty}$ and $\widetilde{W}_{\nu,\varepsilon}^{\infty}$ are irreducible, and $\nu \notin 1 - n/2 - \mathbb{Z}_{\geq 0}$. Then for any $f \in W_{-\nu,\varepsilon}^{\infty}$, the integrals $J_\nu f(u)$ vanish for all $u \in U$ if and only if $I_\nu f \in \widetilde{W}_\nu^{\infty}$, viewed as C^∞ section of the line bundle $\widetilde{\mathcal{L}}_{\nu - \rho_{mir},\varepsilon} \to \widetilde{Y}$, vanishes on the entire complement of $\widetilde{U}$ in $\widetilde{Y}$.*

Both representations are generically irreducible, and I_ν depends meromorphically on ν, so the hypotheses are satisfied outside a discrete set of values of the parameter ν. The automorphism (2.35) preserves the one parameter group $t \mapsto \exp(t\, E_{1,n})$. Since this automorphism switches the roles of I_ν and $\widetilde{I}_\nu$, $W_{\nu,\varepsilon}^{\infty}$ and $\widetilde{W}_{\nu,\varepsilon}^{\infty}$, etc., the lemma applies analogously to $\widetilde{I}_\nu$.

The explicit formula for $I_\nu f$ – for $f \in C_c^\infty(U)$, so that convergence is not an issue – shows that I_ν cannot vanish. Because of the other hypotheses of the lemma, I_ν must then be one-to-one and have dense image. But the image is necessarily closed [6], hence in the situation of the lemma,

$$I_\nu : W_{-\nu,\varepsilon}^{\infty} \longrightarrow \widetilde{W}_{\nu,\varepsilon}^{\infty} \text{ is a topological isomorphism.} \tag{3.37}$$

Proof of Lemma 3.36: The $J_\nu f(u)$ depend meromorphically on ν, provided $f \in W_{-\nu,\varepsilon}^{\infty}$ varies meromorphically with ν. Evaluation of $I_\nu f$ at any particular point is also a meromorphic function of ν. Thus, without loss of generality, we may suppose

$$\text{Re } \nu \gg 0. \tag{3.38}$$

We shall relate I_ν and J_ν to the $GL(n-1)$-analogue of I_ν. This requires a temporary change in notation: in this proof we write $W_{n,\nu}^{\infty}$, $I_{n,\nu}$, etc., to signify the dependence on n (we omit the subscript ε since it is fixed and does not play an essential role).

We define

$$R_{n,\nu} : \widetilde{W}_{n,\nu}^{\infty} \longrightarrow \widetilde{W}_{n-1,\,\nu-1/2}^{\infty},$$

$$(R_{n,\nu}\widetilde{f})(g_1) = |\det g_1|^{\frac{n/2-\nu}{n(n-1)}} \widetilde{f}\left(\begin{pmatrix} 0 & & & \\ \vdots & & 1_{(n-1)\times(n-1)} & \\ 0 & & & \\ 1 & 0 & \cdots & 0 \end{pmatrix} \begin{pmatrix} 1 & 0 & \cdots & 0 \\ 0 & & & \\ \vdots & & g_1 & \\ 0 & & & \end{pmatrix} \right); \qquad (3.39)$$

the fractional power of $|\det g_1|$ is necessary to relate the transformation law (3.6) for $\widetilde{f} \in \widetilde{W}_{n,\nu}^{\infty}$ to that for $R_{n,\nu}\widetilde{f} \in \widetilde{W}_{n-1,\,\nu-1/2}^{\infty}$. The first matrix factor in the argument of $\widetilde{f}$ makes this restriction operator $GL(n-1)$-invariant relative to the tautological action on $\widetilde{W}_{n-1,\,\nu-1/2}^{\infty}$ and the action on $\widetilde{W}_{n,\nu}^{\infty}$ via the embedding $GL(n-1) \hookrightarrow GL(n)$ into the top left corner. This top left copy of $GL(n-1)$ acts transitively on the complement of $\widetilde{U}$ in $\widetilde{Y}$, hence

$$R_{n,\nu}\widetilde{f} \equiv 0 \quad \Longleftrightarrow \quad \widetilde{f} \text{ vanishes on the complement of } \widetilde{U} \text{ in } \widetilde{Y}. \qquad (3.40)$$

Next we define

$$A_{n,\nu} : W_{n,\nu}^{\infty} \longrightarrow W_{n-1,\,\nu+1/2}^{\infty},$$

$$(A_{n,\nu}f)(g_1) = |\det g_1|^{\frac{n/2+\nu}{n(n-1)}} J_{n,\nu}f\begin{pmatrix} & & 0 \\ g_1 & & \vdots \\ & & 0 \\ 0 & \cdots & 0 & 1 \end{pmatrix}. \qquad (3.41)$$

In this case, the power of $|\det g_1|$ reflects not only the discrepancy between the transformation laws (3.6) for n and $n-1$, but also the commutation of the appropriate factor across $\exp(tE_{1,n})$ in the defining relation (3.35) for J_ν. It is clear from the definition that $A_{n,\nu}$ relates the tautological action of $GL(n-1)$ on $W_{n-1,\,\nu+1/2}^{\infty}$ to that on $W_{\nu,\varepsilon}^{\infty}$ via the embedding $GL(n-1) \hookrightarrow GL(n)$ into the top left corner. We claim:

$$A_{n,\nu}f \equiv 0 \quad \Longleftrightarrow \quad J_{n,\nu}f(u) = 0 \text{ for all } u \in U. \qquad (3.42)$$

Indeed, since U is dense in G/P, f vanishes identically if and only if f vanishes on U. We use the analogous assertion about $A_\nu f$, coupled with the following observation: let U_1 denote the intersection of U with the image of $GL(n-1) \hookrightarrow GL(n)$; then $U_1 \cdot \{\exp(tE_{1,n})\} = U$.

The intertwining operators $I_{n,\nu}$, $I_{n-1,\,\nu-1/2}$ and the operators we have just defined constitute the four edges of a commutative diagram,

$$
\begin{array}{ccc}
W_{n,-\nu}^{\infty} & \xrightarrow{\quad I_{n,\nu} \quad} & \widetilde{W}_{n,\nu}^{\infty} \\
{\scriptstyle A_{n,-\nu}}\downarrow & & \downarrow{\scriptstyle R_{n,\nu}} \\
W_{n-1,\,-\nu+1/2}^{\infty} & \xrightarrow{\quad I_{n-1,\,\nu-1/2} \quad} & \widetilde{W}_{n-1,\,\nu-1/2}^{\infty}.
\end{array} \qquad (3.43)
$$

The commutativity is a consequence of two matrix identities. The first,

$$\begin{pmatrix} 0 & \cdots & 0 & 1 \\ & & 1 & 0 \\ \vdots & \cdot^{\cdot^{\cdot}} & & \vdots \\ 1 & & & 0 \end{pmatrix} \begin{pmatrix} 1 & x_{n-1} & \cdots & x_1 \\ 0 & 1 & \cdots & 0 \\ \vdots & & \ddots & \\ 0 & \cdots & & 1 \end{pmatrix} = $$

$$= \begin{pmatrix} 1 & 0 & \cdots & 0 \\ 0 & & & 1 \\ \vdots & & \cdot^{\cdot^{\cdot}} & \\ 0 & 1 & & \end{pmatrix} \begin{pmatrix} 1 & 0 & \cdots & \cdots & 0 \\ 0 & 1 & x_{n-1} & \cdots & x_2 \\ \vdots & & \ddots & & \\ 0 & \cdots & \cdots & & 1 \end{pmatrix} \begin{pmatrix} 0 & \cdots & 0 & 1 \\ 1 & 0 & \cdots & 0 & x_1 \\ & \cdot^{\cdot^{\cdot}} & & \\ 0 & \cdots & 1 & 0 \end{pmatrix}, \qquad (3.44)$$

implies a factorization of $I_{n,\nu}$ as the composition of $I_{n-1,\,\nu-1/2}$ with a certain intermediate operator, which involves an integration over the one parameter group

$\{\exp(tE_{2,n})\}$ instead of $\{\exp(tE_{1,n})\}$, as in the case of J_ν. The second,

$$\begin{pmatrix} 0 & & \\ \vdots & 1_{(n-1)\times(n-1)} & \\ 0 & & \\ 1 & 0 & \cdots \end{pmatrix} \begin{pmatrix} 1 & 0 & \cdots & 0 \\ 0 & & & \\ \vdots & & g_1 & \\ 0 & & & \end{pmatrix} \begin{pmatrix} 0 & \cdots & 0 & 1 \\ 1 & 0 \cdots & 0 & x_1 \\ & & \ddots & \\ 0 & \cdots & 1 & 0 \end{pmatrix} =$$

$$= \begin{pmatrix} & & & 0 \\ & g_1 & & \vdots \\ & & & 0 \\ 0 & \cdots & 0 & 1 \end{pmatrix} \begin{pmatrix} 1 & 0 & \cdots & 0 & x_1 \\ 0 & 1 & 0 & \cdots \\ \vdots & & \ddots & \\ 0 & & \cdots & 1 \end{pmatrix}, \tag{3.45}$$

relates this intermediate operator to $J_{n,\nu}$.

Under the hypotheses of the lemma $I_{n,\nu}$ is an isomorphism – recall (3.37). One can show that under the same hypotheses $I_{n-1,\,\nu-1/2}$ is also an isomorphism. Alternatively one can use the meromorphic dependence on ν to disregard the discrete set on which $I_{n-1,\,\nu-1/2}$ might fail to be an isomorphism. In any case, when both $I_{n-1,\,\nu-1/2}$ and $I_{n-1,\,\nu-1/2}$ are isomorphisms, (3.40), (3.42), and the commutativity of the diagram (3.43) imply the assertion of the lemma. $\qquad\square$

The functional equation of the mirabolic Eisenstein series relates $E_{-\nu,\psi}$ to $\widetilde{E}_{\nu,\psi^{-1}}$ via the intertwining operator $I_\nu : W_{-\nu,\varepsilon}^{-\infty} \to \widetilde{W}_{\nu,\varepsilon}^{-\infty}$. For the statement, we follow the notational convention

$$G_\delta(s) = \int_{\mathbb{R}} e(x)\,(\mathrm{sgn}(x))^\delta\,|x|^{s-1}\,dx = \begin{cases} 2(2\pi)^{-s}\,\Gamma(s)\,\cos\frac{\pi s}{2} & \text{if } \delta = 0 \\ 2(2\pi)^{-s}\,\Gamma(s)\,\sin\frac{\pi s}{2} & \text{if } \delta = 1 \end{cases} \tag{3.46}$$

[17], which we shall also use later in this paper. Note that the integral converges, conditionally only, for $0 < \mathrm{Re}\,s < 1$, but the expression on the right provides a meromorphic continuation to the entire s-plane. The two cases on the right hand side of (3.46) can be written uniformly using Γ-function identities as

$$G_\delta(s) = i^\delta\,\frac{\Gamma_{\mathbb{R}}(s+\delta)}{\Gamma_{\mathbb{R}}(1-s+\delta)}\,, \quad \text{with } \Gamma_{\mathbb{R}}(s) = \pi^{-s/2}\Gamma(\tfrac{s}{2}) \text{ and } \delta \in \{0,1\}. \tag{3.47}$$

We also need some notation pertaining to the finite harmonic analysis of Dirichlet characters. Let $\tau_\psi = \widehat{\psi}(1) = \sum_{b\,(\mathrm{mod}\,N)} \psi(b)e(\frac{b}{N})$ denote the Gauss sum for ψ, a Dirichlet character of modulus N (cf. (3.15)). We let $\widehat{(\mathbb{Z}/N\mathbb{Z})}^*$ denote the group of characters of $\mathbb{Z}/N\mathbb{Z}^*$ and $\phi(N)$, the Euler ϕ-function, its order.

3.48. PROPOSITION (Functional Equation).

$$I_\nu E_{-\nu,\psi} =$$
$$(-1)^\varepsilon N^{2\nu-\frac{\nu}{n}-\frac{1}{2}} G_\varepsilon(\nu - \tfrac{n}{2}+1)\,\frac{1}{\phi(N)} \sum_{\substack{a\,(\mathrm{mod}\,N) \\ \xi\in\widehat{(\mathbb{Z}/N\mathbb{Z})}^*}} \widehat{\psi}(a)\xi(a)^{-1}\,\ell(w_{long})\,\ell\left(\begin{smallmatrix} & N \\ I_{n-1} & \end{smallmatrix}\right)\,\widetilde{E}_{\nu,\xi}\,.$$

Consequently, if ψ is a primitive Dirichlet character of modulus N, then

$$I_\nu E_{-\nu,\psi} = (-1)^\varepsilon \tau_\psi N^{2\nu-\frac{\nu}{n}-\frac{1}{2}} G_\varepsilon(\nu - \tfrac{n}{2} + 1)\,\ell(w_{long})\,\ell\left(\begin{smallmatrix} & N \\ I_{n-1} & \end{smallmatrix}\right)\,\widetilde{E}_{\nu,\psi^{-1}}\,.$$

In particular

$$I_\nu E_{-\nu,\mathbb{1}} = G_0(\nu - \tfrac{n}{2} + 1)\,\widetilde{E}_{\nu,\mathbb{1}}\,,$$

where $\mathbb{1}$ is the trivial Dirichlet character of conductor $N = 1$.

PROOF. Since both sides of the equation depend meromorphically on ν, we may assume that the hypotheses of lemma 3.36 hold, both at ν and $-\nu$. We shall also require

$$\operatorname{Re} \nu \gg n/2, \qquad (3.49)$$

so that the integral defining $\widetilde{I}_\nu$ converges. Because of (3.34), the proposition is equivalent to the equality

$$\frac{1}{(-1)^\varepsilon N^{2\nu - \frac{\nu}{n} - \frac{1}{2}} G_\varepsilon(\nu - \frac{n}{2} + 1)} \int_U E_{-\nu,\psi}(u) \, \widetilde{I}_\nu \widetilde{f}(u) \, du \;\; =$$

$$\frac{1}{\phi(N)} \sum_{\substack{a \ (\mathrm{mod}\, N) \\ \xi \in (\widehat{\mathbb{Z}/N\mathbb{Z}})^*}} \widehat{\psi}(a) \xi(a)^{-1} \int_{\widetilde{U}} \ell\left(\begin{smallmatrix} I_{n-1} & \\ & N \end{smallmatrix}\right) \ell(w_{long}) \, \widetilde{E}_{\nu,\xi}(\widetilde{u}) \, \widetilde{f}(\widetilde{u}) \, d\widetilde{u}, \quad (3.50)$$

for all $\widetilde{f} \in \widetilde{W}^\infty_{-\nu}$. Both $E_{-\nu,\psi}$ and $\widetilde{E}_{\nu,\xi}$ are invariant under congruence subgroups of $GL(n,\mathbb{Z})$, and $\widetilde{I}_\nu : W^\infty_{-\nu,\varepsilon} \simeq \widetilde{W}^\infty_{\nu,\varepsilon}$ by (3.37). It therefore suffices to establish this equality when $\widetilde{I}_\nu \widetilde{f}$ has – necessarily compact – support in the open cell $U \subset Y$,

$$\operatorname{supp}\left(\widetilde{I}_\nu \widetilde{f}\right) \text{ is compact in } U. \qquad (3.51)$$

We shall make one other assumption, namely

$$\int_{\mathbb{R}} \widetilde{I}_\nu \widetilde{f}(u(x_1, x_2, \ldots x_{n-1})) \, dx_1 = 0, \quad \text{for all } x_2, \ldots, x_{n-1} \in \mathbb{R}. \qquad (3.52)$$

Indeed, if (3.50) were to hold subject to the condition (3.52), the restriction to $\widetilde{U}$ of the difference between $I_\nu E_{-\nu,\psi}$ and the formula we have asserted it is equal to could be expressed as a Fourier series

$$\sum_{r_2, \ldots, r_{n-1} \in \mathbb{Z}} a_{r_2, \ldots, r_{n-1}} \, e(r_2 y_2 + \cdots + r_{n-1} y_{n-1}), \qquad (3.53)$$

without dependence on y_1. But no such expression can be the restriction to $\widetilde{U}$ of a distribution vector invariant under a congruence subgroup Γ: any generic $\gamma \in \Gamma$ will transform the expression (3.53) to a distribution that does depend non-trivially on y_1. This justifies the additional hypothesis (3.52).

In effect, the integrals (3.52) coincide with the integrals $J_{-\nu}\left(\widetilde{I}_\nu \widetilde{f}\right)(u)$, as in (3.35), for $u \in U$. Consequently lemma 3.36 implies the vanishing of $I_{-\nu} \circ \widetilde{I}_\nu f$ on the complement of $\widetilde{U}$. But our hypotheses ensure that $I_{-\nu} \circ \widetilde{I}_\nu$ is a multiple of the identity, so

$$\widetilde{f} \text{ vanishes on the complement of } \widetilde{U} \text{ in } \widetilde{Y}. \qquad (3.54)$$

Having compact support in U, $\widetilde{I}_\nu \widetilde{f}$ surely vanishes on the complement of U in Y. Thus, applying the lemma in reverse, we find

$$\int_{\mathbb{R}} \widetilde{f}(\widetilde{u}(y_1, y_2, \ldots y_{n-1})) \, dy_1 = 0, \quad \text{for all } y_2, \ldots, y_{n-1} \in \mathbb{R}. \qquad (3.55)$$

We shall also need the estimate

$$\left| P\left(\tfrac{\partial}{\partial y_1}, \ldots, \tfrac{\partial}{\partial y_{n-1}}\right) \widetilde{f}(\widetilde{u}(y)) \right| = O(\|y\|^{-\operatorname{Re} \nu - n/2}) \quad \text{as } \|y\| \to \infty, \qquad (3.56)$$

for all constant coefficient differential operators $P\left(\tfrac{\partial}{\partial y_1}, \ldots, \tfrac{\partial}{\partial y_{n-1}}\right)$. It follows from (3.32), combined with the fact that the elements of the Lie algebra $\widetilde{\mathfrak{u}}$ of $\widetilde{U}$ act on

$\widetilde{W}^{\infty}_{-\nu}$ by constant coefficient vector fields on $\widetilde{U} \cong \mathbb{R}^{n-1}$. In view of (3.49), (3.56) implies the decay of $\widetilde{f}(\widetilde{u}(y))$ and all its derivatives.

We compute the integral on the right hand side of (3.50) using the last restriction formula in proposition 3.16:

$$\int_{\widetilde{U}} \ell\left(\begin{smallmatrix} I_{n-1} \\ & N \end{smallmatrix}\right) \ell(w_{\text{long}}) \widetilde{E}_{\nu,\xi}(\widetilde{u}) \, \widetilde{f}(\widetilde{u}) \, d\widetilde{u} \;\; =$$

$$= \; \frac{N^{(1/2-\nu/n)(n-1)}}{N^{n-1}} \sum_{\substack{r \in \mathbb{Z}^{n-1} \\ r_1 \neq 0 \\ d | GCD(r)}} \xi(d) \, d^{-\nu+n/2-1} \int_{\mathbb{R}^{n-1}} \widetilde{f}(\widetilde{u}(y)) \, e(r \cdot y) \, dy \, ; \quad (3.57)$$

here we have used the fact that $\left(\begin{smallmatrix} I_{n-1} \\ & N \end{smallmatrix}\right)^{-1} \widetilde{u}(y) = \widetilde{u}(Ny) \left(\begin{smallmatrix} I_{n-1} \\ & N \end{smallmatrix}\right)^{-1}$, and the transformation law (3.6) to pull out the power of N in the numerator. The terms corresponding to $r_1 = 0$ have been dropped because of (3.55). The sum in (3.57) is absolutely convergent because of the derivative bound (3.56).

Let us now consider the finite sum over a and ξ to its left in (3.50). By orthogonality of characters

$$\frac{1}{\phi(N)} \sum_{\substack{a \,(\text{mod } N) \\ \xi \in \widehat{(\mathbb{Z}/N\mathbb{Z})^*}}} \widehat{\psi}(a) \, \xi(a)^{-1} \xi(d) \;\; = \;\; \begin{cases} 0, & (d, N) > 1 \\ \widehat{\psi}(d), & (d, N) = 1 \, . \end{cases} \quad (3.58)$$

Therefore the right hand side of (3.50) is equal to

$$N^{(1-n)(1/2+\nu/n)} \sum_{\substack{r \in \mathbb{Z}^{n-1} \\ r_1 \neq 0 \\ d | GCD(r)}} \widehat{\psi}(d) \, d^{-\nu+n/2-1} \int_{\mathbb{R}^{n-1}} \widetilde{f}(\widetilde{u}(y)) \, e(r \cdot y) \, dy \, . \quad (3.59)$$

The compact support of $\widetilde{I}_\nu \widetilde{f}$ and (3.52) imply the analogous expression for the integral on the other side of (3.50), but using the hybrid formula for the restriction of $E_{-\nu,\psi}$ to U in proposition 3.16:

$$\int_U E_{-\nu,\psi}(u) \, \widetilde{I}_\nu \widetilde{f}(u) \, du \;\; = \;\; N^{\nu-n/2} \sum_{\substack{v \in \mathbb{Z}^n \\ v_1 > 0 \\ v_n \neq 0}} \widehat{\psi}(v_n) \, v_1^{\nu-n/2+1} \; \times$$

$$\times \int_{\mathbb{R}^{n-1}} \widetilde{I}_\nu \widetilde{f}(u(x)) \, e(v_1 v_n x_1) \delta_{v_{n-1}/v_1}(x_2) \cdots \delta_{v_2/v_1}(x_{n-1}) dx \, . \quad (3.60)$$

It is important to note that this sum converges absolutely. Indeed,

$$\sum_{v_2, \dots, v_{n-1} \in \mathbb{Z}} \left| \int_{\mathbb{R}} \phi(x_1, \tfrac{v_{n-1}}{v_1}, \dots, \tfrac{v_2}{v_1}) \, e(v_1 v_n x_1) \, dx_1 \right| \; \leq$$

$$\leq \; C \, v_1^{n-2} \sup_{x_2, \dots, x_{n-1} \in \mathbb{R}} \left| \int_{\mathbb{R}} \phi(x_1, x_2, \dots, x_{n-1}) \, e(v_1 v_n x_1) \, dx_1 \right| , \quad (3.61)$$

for any $\phi \in C_c^\infty(U)$ such as $\phi = \widetilde{I}_\nu \widetilde{f}$, with C depending only on the diameter of the support of ϕ; the supremum on the right decays faster than any negative power of $|v_1 v_n|$.

In view of (3.59) and (3.60), a notation change reduces (3.50) to the following assertion: under the hypotheses (3.49) and (3.51–52),

$$(-1)^\varepsilon G_\varepsilon(\nu - \tfrac{n}{2} + 1) \sum_{\substack{r \in \mathbb{Z}^{n-1} \\ r_1 \neq 0 \\ d>0}} \widehat{\psi}(d)\, d^{n/2-\nu-1} \int_{\mathbb{R}^{n-1}} \widehat{f}(\widetilde{u}(y))\, e\Big(\sum_j d\, r_j\, y_j\Big) dy$$

$$= \sum_{\substack{d>0 \\ k \neq 0}} \widehat{\psi}(d)\, k^{\nu-n/2+1} \sum_{r_2,\dots,r_{n-1} \in \mathbb{Z}} \int_{\mathbb{R}} \widetilde{I_\nu}\widehat{f}(x_1, \tfrac{r_2}{k}, \dots, \tfrac{r_{n-1}}{k})\, e(dkx_1)\, dx_1 \,. \tag{3.62}$$

The explicit formula for I_ν in proposition 3.26 – or more accurately, the analogous formula for $\widetilde{I_\nu}$ – implies

$$\int_{\mathbb{R}} \widetilde{I_\nu}\widehat{f}(x_1, \tfrac{r_2}{k}, \dots, \tfrac{r_{n-1}}{k})\, e(dkx_1)\, dx_1 =$$

$$= \int_{\mathbb{R}} \int_{\mathbb{R}^{n-1}} \widehat{f}(\widetilde{u}(z))\, e(dkx_1)\, |\sum_{j\geq 2} \tfrac{r_j z_{n+1-j}}{k} - z_1 - x_1|^{\nu-n/2} \times$$

$$\times\ \operatorname{sgn}(\sum_{j\geq 2} \tfrac{r_j z_{n+1-j}}{k} - z_1 - x_1)^\varepsilon\, dz\, dx_1$$

$$= \int_{\mathbb{R}} \int_{\mathbb{R}^{n-1}} \widehat{f}(\widetilde{u}(z))\, e(dkx_1 + d\sum_{j\geq 2} r_j z_{n+1-j} - dkz_1) \times \tag{3.63}$$

$$\times\ |-x_1|^{\nu-n/2}\, \operatorname{sgn}(-x_1)^\varepsilon\, dz\, dx_1$$

$$= \int_{\mathbb{R}} |x_1|^{\nu-n/2}\, \operatorname{sgn}(-x_1)^\varepsilon e(dkx_1) \times$$

$$\times \int_{\mathbb{R}^{n-1}} \widehat{f}(\widetilde{u}(z))\, e\Big(-dkz_1 + d\sum_{j\geq 2} r_j z_j\Big) dz \,.$$

The change of variables $x_1 \mapsto x_1 - z_1 + d^{-1} \sum r_j z_{n+1-j}$ at the second step depends on interchanging the order of the two integrals. The z-integral is an ordinary, convergent integral, whereas the x_1-integral is that of a distribution against a C^∞ function. It can be turned into an ordinary, convergent integral by repeated integration by parts near $x_1 = \infty$ to bring down the real part of the exponent $\nu - n/2$. Away from infinity the x_1-integral already is an ordinary convergent integral since $\operatorname{Re}\nu \gg 0$; the two phenomena must be separated by a suitable cutoff function. Our paper [17] describes these techniques in detail. They apply equally to the evaluation of the integral

$$\int_{\mathbb{R}} |x_1|^{\nu-n/2}\, \operatorname{sgn}(-x_1)^\varepsilon\, e(dkx_1)\, dx_1 = (-1)^\varepsilon\, |dk|^{n/2-\nu-1}\, G_\varepsilon(\nu - \tfrac{n}{2} + 1), \tag{3.64}$$

reducing it to (3.46) in the convergent range. Identifying k with r_1 and summing over $d > 0$ and $r \in \mathbb{Z}^{n-1}$, $r_1 \neq 0$, gives the identity (3.62), and hence completes the proof. $\qquad\square$

The parameter ν is natural from the representation theoretic point of view. In applications to functional equations, we set

$$\nu = ns - \rho_{\mathrm{mir}} = n(s - 1/2), \tag{3.65}$$

which has the effect of translating the symmetry $\nu \mapsto -\nu$ into $s \mapsto 1 - s$.

4. Pairing of Distributions

In this section we discuss some pairings of automorphic distributions that were constructed in [22], and how the analytic continuation and functional equations of Eisenstein distributions carry over to these pairings. In some cases the pairings can be computed as a product of shifts of the functions G_δ defined in (3.46), times certain L-functions. This gives a new construction of these L-functions, and a new method to directly study their analytic properties. In particular the results here are used crucially in our forthcoming paper [23] to give new results about the analytic continuation that were not available by the two existing methods, the Rankin-Selberg and Langlands-Shahidi methods.

We begin with a discussion of the distributional pairings in [22], though not in the same degree of generality as in that paper. We consider the semidirect product $G \cdot U$ of a real linear group G with a unipotent group U. We suppose that $G \cdot U$ acts on flag varieties or generalized flag varieties Y_j of real linear groups G_j, $1 \leq j \leq r$, in each case either by an inclusion $G \cdot U \hookrightarrow G_j$, or via $G \hookrightarrow G_j$ composed with the quotient map $G \cdot U \to G$. Then $G \cdot U$ acts on the product $Y_1 \times \cdots \times Y_r$. We suppose further that

$$G \cdot U \text{ has an open orbit } \mathcal{O} \subset Y_1 \times \cdots \times Y_r \text{, and at points of } \mathcal{O}$$
$$\text{the isotropy subgroup of } G \cdot U \text{ coincides with } Z_G = \text{center of } G, \tag{4.1}$$

so that $\mathcal{O} \simeq (G \cdot U)/Z_G$, and that

$$\text{the conjugation action of } G \text{ on } U \text{ preserves Haar measure on } U. \tag{4.2}$$

We let $\Gamma \subset G$, $\Gamma_U \subset U$, $\Gamma_j \subset G_j$ denote arithmetically defined subgroups such that $\Gamma \cdot \Gamma_U \hookrightarrow \Gamma_1 \times \cdots \times \Gamma_r$.

Our theorem also involves automorphic distributions $\tau_j \in C^{-\infty}(Y_j, \mathcal{L}_j)^{\Gamma_j}$, in other words, Γ_j-invariant distribution sections of G_j-equivariant C^∞ line bundles $\mathcal{L}_j \to Y_j$, $1 \leq j \leq r$. The exterior tensor product

$$\mathcal{L}_1 \boxtimes \cdots \boxtimes \mathcal{L}_r \longrightarrow Y_1 \times \cdots \times Y_r \tag{4.3}$$

restricts to a $G \cdot U$-equivariant line bundle over $\mathcal{O} \simeq (G \cdot U)/Z_G$. If

$$\text{the isotropy group } Z_G \text{ acts trivially on the fiber}$$
$$\text{of } \mathcal{L}_1 \boxtimes \cdots \boxtimes \mathcal{L}_r \text{ at points of } \mathcal{O}, \tag{4.4}$$

as we shall assume from now on, the restriction of the line bundle (4.3) to the open orbit $\mathcal{O}$ is canonically trivial. We can then regard

$$\tau = \text{ restriction of } \tau_1 \boxtimes \cdots \boxtimes \tau_r \text{ to } \mathcal{O} \tag{4.5}$$

as a scalar valued distribution on $(G \cdot U)/Z_G$ – a $\Gamma \cdot \Gamma_U$-invariant distribution, since the τ_j are Γ_j-invariant:

$$\tau \in C^{-\infty}((\Gamma \cdot \Gamma_U)\backslash(G \cdot U)/Z_G). \tag{4.6}$$

As the final ingredient, we fix a character

$$\chi : U \to \{ z \in \mathbb{C}^* \mid |z| = 1 \} \text{ such that } \chi(gug^{-1}) = \chi(u)$$
$$\text{for all } g \in G, \ u \in U, \text{ and } \chi(\gamma) = 1 \text{ for all } \gamma \in \Gamma_U. \tag{4.7}$$

Since $\Gamma_U \backslash U$ is compact,

$$\left\{ g \mapsto \int_{\Gamma_U \backslash U} \chi(u)\, \tau(ug)\, du \right\} \in C^{-\infty}(\Gamma \backslash G / Z_G) \tag{4.8}$$

is a well defined distribution on G/Z_G – a Γ-invariant scalar valued distribution because of (4.6–7). Finally, we require that

$$\text{at least one of the } \tau_i \text{ is cuspidal.} \tag{4.9}$$

4.10. THEOREM. [22, Theorem 2.29]. *Under the hypotheses just stated, for every test function* $\phi \in C_c^\infty(G)$, *the function*

$$g \; \mapsto \; F_{\tau,\chi,\phi}(g) \;=\; \int_{h \in G} \int_{\Gamma_U \backslash U} \chi(u)\, \tau(ugh)\, \phi(h)\, du\, dh$$

is a well defined C^∞ *function on* G/Z_G, *invariant on the left under* Γ. *This function is integrable over* $\Gamma \backslash G/Z_G$, *and the resulting integral*

$$P(\tau_1, \dots, \tau_r) \;=\; \int_{\Gamma \backslash G/Z_G} \int_{h \in G} \int_{\Gamma_U \backslash U} \chi(u)\, \tau(ugh)\, \phi(h)\, du\, dh\, dg$$

does not depend on the choice of ϕ, *provided* ϕ *is normalized by the condition* $\int_G \phi(g)\, dg = 1$. *The* r-*linear map* $(\tau_1, \dots, \tau_r) \mapsto F_{\tau,\chi,\phi} \in L^1(\Gamma \backslash G/Z_G)$ *is continuous, relative to the strong distribution topology, in each of its arguments, and relative to the* L^1 *norm on the image. If any one of the* τ_j *depends holomorphically on a complex parameter* s, *then so does* $P(\tau_1, \dots, \tau_r)$.

At first glance, the hypothesis (4.1) does not seem to include the hypothesis (2.4b) in [22]. However, since Z_G acts trivially on the orbit $\mathcal{O}$, the hypothesis (2.4b) does hold if we replace G by its derived group. Thus, instead of integrating over $\Gamma \backslash G/Z_G$, we could integrate over $(\Gamma \cap [G,G]) \backslash [G,G]/Z_G$. The hypotheses (4.1–2) are therefore sufficient to apply the results of [22].

We shall now describe two interesting cases of this pairing that both involve a similar setup of flag varieties and the mirabolic Eisenstein series as a factor. Because we shall work with more than one group and flag variety, we use subscripts: G_k will denote $GL(k, \mathbb{R})$ and $X_k = G_k/B_k$ its flag variety; cf. (2.10–11). The Eisenstein distributions $E_{\nu,\psi}$ from (3.12) are $\Gamma_1(N)$-invariant sections of the line bundle $\mathcal{L}_{\nu - \rho_{\mathrm{mir}}, \varepsilon}$ over the generalized flag variety $Y_n \cong \mathbb{RP}^{n-1}$. In addition to these series and representations $W_{\nu,\varepsilon}$ and $\widetilde{W}_{\nu,\varepsilon}$, we also consider their products with the character $\mathrm{sgn}(\det)^\eta$, $\eta \in \mathbb{Z}/2\mathbb{Z}$ (see the remark above (3.5)). Our two particular pairings depend crucially on the following geometric fact:

$$\begin{array}{c} G_n \text{ acts on } X_n \times X_n \times Y_n \text{ with a dense open orbit; the action on} \\ \text{this open orbit is free modulo the center, which acts trivially.} \end{array} \tag{4.11}$$

Indeed, the diagonal action of G_n on $X_n \times X_n$ has a dense open orbit. At any point in the open orbit, the isotropy subgroup consists of the intersection of two opposite Borel subgroups – equivalently, a G_n-conjugate of the diagonal subgroup. That group has a dense open orbit in Y_n, and only Z_n = center of G_n acts trivially.

In the first example, which represents the Rankin-Selberg L-function for automorphic distributions τ_1, τ_2 on $GL(n, \mathbb{R})$, the integer $r = 3$, $U = \{e\}$, $Y_1 = Y_2 = X_n$, and $Y_3 = \mathbb{RP}^{n-1}$. We require both τ_1 and τ_2 to be cuspidal, but impose no such condition on τ_3, which is taken to be the mirabolic Eisenstein distribution.

The second example, which represents the exterior square L-function of a cuspidal automorphic distribution τ on $GL(2n, \mathbb{R})$, involves a nontrivial unipotent group, has $r = 2$, and only a single cusp form $\tau_1 = \tau$ (τ_2 is the mirabolic Eisenstein distribution). The decomposition $\mathbb{R}^{2n} = \mathbb{R}^n \oplus \mathbb{R}^n$ induces embeddings

$$G_n \times G_n \;\hookrightarrow\; G_{2n}, \qquad X_n \times X_n \;\hookrightarrow\; X_{2n}. \tag{4.12}$$

The translates of $X_n \times X_n$ under the abelian subgroup

$$U = \left\{ \begin{pmatrix} I_n & A \\ 0_n & I_n \end{pmatrix} \ \bigg| \ A \in M_{n \times n}(\mathbb{R}) \right\} \subset G_{2n} \qquad (4.13)$$

sweep out an open subset of X_{2n}; moreover the various U-translates are disjoint, so that

$$U \times X_n \times X_n \hookrightarrow X_{2n}. \qquad (4.14)$$

Let $\tau \in C^{-\infty}(X_{2n}, \mathcal{L}_{\lambda-\rho,\delta})^{\Gamma}$ be a cuspidal automorphic distribution as in (2.19), and du be the Haar measure on U identified with the standard Lebesgue measure on $M_{n \times n}(\mathbb{R})$. The group of integral matrices $U(\mathbb{Z})$ lies in the kernel of the character

$$\theta : U \longrightarrow \mathbb{C}^*, \qquad \theta\left(\begin{smallmatrix} I_n & A \\ 0_n & I_n \end{smallmatrix} \right) = e(\operatorname{tr} A), \qquad (4.15)$$

and because $\Gamma \cap U(\mathbb{Z})$ has finite index in $U(\mathbb{Z})$, the integral

$$S_\theta \tau =_{\mathrm{def}}$$
$$\frac{1}{\operatorname{covol}(\Gamma \cap U(\mathbb{Z}))} \int_{\Gamma \cap U(\mathbb{Z}) \backslash U} \theta(u)\, \ell(u) \tau \, du \ \in \ C^{-\infty}(X_{2n}, \mathcal{L}_{\lambda-\rho,\delta}) \qquad (4.16)$$

is well defined, even if Γ is replaced by a finite index subgroup. It restricts to a distribution section of $\mathcal{L}_{\lambda-\rho,\delta}$ over the image of the open embedding (4.14). As such, it is smooth in the first variable, since $\ell(u) S_\theta \tau = \theta(u)^{-1} S_\theta \tau$ for $u \in U$. We can therefore evaluate this distribution section at $e \in U$, and define

$$S\tau = S_\theta \tau|_{X_n \times X_n} \ \in \ C^{-\infty}(X_n \times X_n, \mathcal{L}_{\lambda-\rho,\delta}|_{X_n \times X_n})^{\Gamma_n}. \qquad (4.17)$$

Here Γ_n is a congruence subgroup of $G_n(\mathbb{Z})$ whose diagonal embedding into $G_n \times G_n \subset G_{2n}$ leaves τ invariant under the left action, and preserves $\Gamma \cap U(\mathbb{Z})$ by conjugation. The superscript signifies invariance under the diagonal action of Γ_n on $X_n \times X_n$. This invariance is a consequence of the fact that conjugation by the diagonal embedding of any $\gamma \in \Gamma_n$ also preserves the character θ as well as U, without changing the measure.

We restrict the product of the G_n-equivariant line bundles $\mathcal{L}_{\lambda-\rho,\delta}|_{X_n \times X_n}$ and $\mathcal{L}_{\nu-\rho_{\mathrm{mir}},\varepsilon} \to Y_n$ to the open orbit and pull it back to G_n/Z_n ($Z_n = Z_{G_n} =$ center of G_n), resulting in a G_n-equivariant line bundle $\mathcal{L} \to G_n/Z_n$; $S\tau \cdot E_{\nu,\psi}$ is then a Γ'-invariant distribution section of $\mathcal{L}$ for

$$\Gamma' = \Gamma_n \cap \Gamma_1(N). \qquad (4.18)$$

The center Z_n acts on the fibers of $\mathcal{L}$ by the restriction to Z_n of the character $\chi_{\lambda-\rho,\delta} \cdot \chi_{\nu-\rho_{\mathrm{mir}},\varepsilon} \cdot \operatorname{sgn}(\det)^\eta$, where $\eta \in \mathbb{Z}/2\mathbb{Z}$; recall (2.13) and (3.4), and note that $\chi_{\lambda-\rho,\delta}$ takes values on Z_n via its diagonal embedding into $Z_{2n} \subset G_{2n}$. We shall assume that Z_n lies in the kernel of $\chi_{\lambda-\rho,\delta} \cdot \chi_{\nu-\rho_{\mathrm{mir}},\varepsilon} \cdot \operatorname{sgn}(\det)^\eta$ – equivalently,

$$\lambda_1 + \lambda_2 + \cdots + \lambda_{2n} = 0, \qquad \delta_1 + \delta_2 + \cdots + \delta_{2n} \equiv \varepsilon + n\eta \pmod 2. \qquad (4.19)$$

The first of these conditions involves no essential loss of generality, since twisting an automorphic representation by a central character does not affect the automorphy. The character $\chi_{\nu-\rho_{\mathrm{mir}},0}$ takes the value 1 on Z_n regardless of the choice of ν, hence (4.19) makes $\mathcal{L} \to G_n/Z_n$ a G_n-equivariantly trivial line bundle. In this situation, $S\tau \cdot E_{\nu,\psi}$ becomes a Γ'-invariant scalar valued distribution on G_n/Z_n,

$$S\tau \cdot E_{\nu,\psi} \in C^{-\infty}(G_n/Z_n)^{\Gamma'}. \qquad (4.20)$$

Theorem 4.10 applies to this specific setting and states

4.21. COROLLARY ([22]). *Under the hypotheses just stated, for every test function* $\phi \in C_c^\infty(G_n)$

$$P(\tau, E_{\nu,\psi}) = \int_{\Gamma' \backslash G_n / Z_n} \int_{h \in G_n} (S\tau \cdot E_{\nu,\psi})(gh)\, \phi(h)\, dh\, dg$$

does not depend on the choice of ϕ, *provided* ϕ *is normalized by the condition* $\int_{G_n} \phi(g)\, dg = 1$. *The function* $\nu \mapsto P(\tau, E_{\nu,\psi})$ *is holomorphic for* $\nu \in \mathbb{C} - \{n/2\}$, *with at most a simple pole at* $\nu = n/2$.

To make (4.20) concrete, we identify $X_{2n} \cong G_{2n}/B_{2n}$, $Y_n \cong G_n/P_n$ as before. We regard τ and $E_{\nu,\psi}$ as scalar distributions on G_{2n} and G_n respectively, with τ left invariant under $\Gamma \subset G_{2n}(\mathbb{Z})$, transforming according to $\chi_{\lambda-\rho,\delta}$ on the right under B_{2n}, and $E_{\nu,\psi}$ left invariant under $\Gamma_1(N) \subset G_n(\mathbb{Z})$, transforming according to $\chi_{\nu-\rho_{\mathrm{mir}},\varepsilon}$ on the right under P_n. The averaging process (4.16) makes sense also on this level. When we choose f_1, f_2, $f_3 \in G_n$ so that $(f_1 B_n, f_2 B_n, f_3 P_n)$ lies in the open orbit, we obtain an explicit description of $S\tau \cdot E_{\nu,\psi}$,

$$S\tau \cdot E_{\nu,\psi}(g) =$$
$$\frac{1}{\mathrm{covol}(\Gamma \cap U(\mathbb{Z}))} \int_{\Gamma \cap U(\mathbb{Z}) \backslash U} \theta(u)\, (\ell(u)\, \tau) \begin{pmatrix} gf_1 & 0_n \\ 0_n & gf_2 \end{pmatrix} E_{\nu,\psi}(gf_3)\, du. \tag{4.22}$$

We note that the f_j are determined up to *simultaneous left translation* by some $f_0 \in G_n$ and *individual right translation* by factors in B_n, respectively P_n. Translating the f_j by f_0 on the left has the effect of translating $S\tau \cdot E_{\nu,\psi}$ by f_0^{-1} on the right; it does *not* change the value of $P(\tau, E_{\nu,\psi})$ because the ambiguity can be absorbed by ϕ. Translating any one of the f_j on the right by an element of the respective isotropy group affects both $S\tau \cdot E_{\nu,\psi}$ and $P(\tau, E_{\nu,\psi})$ by a multiplicative factor – a non-zero factor depending on (λ, δ) in the case of f_1 or f_2, and the factor $\chi_{\nu-\rho_{\mathrm{mir}},\varepsilon}(p^{-1})$ when f_3 is replaced by $f_3 p$, $p \in P_n$.

One can eliminate the potential dependence on ν in this factor by requiring $f_3 \in U_n$; cf. (3.1). Specifically, in the following, we choose

$$f_1 = I_n, \quad f_2 = \begin{pmatrix} 0 & \cdots & 0 & 1 \\ \vdots & & \iddots & \\ 0 & 1 & & \\ 1 & 0 & \cdots & 0 \end{pmatrix}, \quad \text{and} \quad f_3 = \begin{pmatrix} 1 & 1 & \cdots & 1 \\ 0 & & & \\ \vdots & & I_{n-1} & \\ 0 & & & \end{pmatrix}, \tag{4.23}$$

which do determine a point $(f_1 B_n, f_2 B_n, f_3 P_n) \in X_n \times X_n \times Y_n$ lying in the open orbit. Note that $f_3 \in U_n$ and $f_2 = w_{\mathrm{long}}$, in the notation of (2.35).

The pairing $P(\tau, E_{\nu,\psi})$ inherits a functional equation from that of $E_{\nu,\psi}$, which involves the contragredient automorphic distribution $\widetilde{\tau}$ defined in (2.36–38). The argument we give below for it works *mutatis mutandis* to provide an analogous statement for the Rankin-Selberg pairing as well.

4.24. PROPOSITION.

$$P(\tau, E_{-\nu,\psi}) =$$

$$(-1)^{\varepsilon+\delta_{n+1}+\cdots+\delta_{2n}} N^{2\nu-\frac{\nu}{n}-\frac{1}{2}} \prod_{j=1}^{n} G_{\delta_{n+j}+\delta_{n+1-j}+\eta}(\lambda_{n+j} + \lambda_{n+1-j} + \tfrac{\nu}{n} + \tfrac{1}{2}) \times$$

$$\times \frac{1}{\phi(N)} \sum_{\substack{a \pmod N \\ \xi \in (\widehat{\mathbb{Z}/N\mathbb{Z}})^*}} \widehat{\psi}(a)\,\xi(a)^{-1}\, P\left(\ell\left(\begin{pmatrix} -w_{long} & \\ & w_{long} \end{pmatrix} \begin{pmatrix} N & & \\ & I_{n-1} & \\ & & N \\ & & & I_{n-1} \end{pmatrix} \right) \widetilde{\tau}, E_{\nu,\xi} \right).$$

The pairings on the right hand side are integrations over the quotient $\Gamma^* \backslash G_n / Z_n$, where

$$\Gamma^* = w_{long} \begin{pmatrix} N & \\ & I_{n-1} \end{pmatrix} \widetilde{\Gamma}' \begin{pmatrix} N & \\ & I_{n-1} \end{pmatrix}^{-1} w_{long} \qquad (\widetilde{\Gamma}' = \{\widetilde{\gamma} \mid \gamma \in \Gamma'\}) \qquad (4.25)$$

is the subgroup that $S\widetilde{\tau} \cdot E_{\nu,\xi}$ is naturally invariant under (cf. (4.18)). In the special case that τ is invariant under $GL(2n,\mathbb{Z})$, $N = 1$, $\psi = \mathbb{1}$ is the trivial Dirichlet character, and $\varepsilon \equiv \eta \equiv 0 \pmod 2$, the relation simplifies to

$$P(\tau, E_{\nu,\mathbb{1}}) =$$
$$(-1)^{\delta_1+\cdots+\delta_n} \prod_{j=1}^{n} G_{\delta_{n+j}+\delta_{n+1-j}}(\lambda_{n+j} + \lambda_{n+1-j} - \tfrac{\nu}{n} + \tfrac{1}{2})\, P(\widetilde{\tau}, E_{-\nu,\mathbb{1}}). \qquad (4.26)$$

A similar formula using the second displayed line in proposition 3.48 of course also gives a simplified functional equation when ψ is primitive, though we will not need to use this formula in what follows.

Proof: In analogy to $S\tau \cdot E_{\nu,\psi}$ in (4.22), one can define a product $S\tau \cdot \widetilde{\rho}$ of $S\tau$ and any distribution section $\widetilde{\rho}$ of $\mathcal{L}_{\nu-\rho_{\mathrm{mir}},\varepsilon} \to \widetilde{Y}$ as

$$S\tau \cdot \widetilde{\rho}(g) =$$
$$\frac{1}{\mathrm{covol}(\Gamma \cap U(\mathbb{Z}))} \int_{\Gamma \cap U(\mathbb{Z}) \backslash U} \theta(u)\, (\ell(u)\,\tau) \begin{pmatrix} g\widetilde{f_2} & 0_n \\ 0_n & g\widetilde{f_1} \end{pmatrix} \widetilde{\rho}(g\widetilde{f_3})\, du. \qquad (4.27)$$

Here we have applied the outer automorphism (2.35) to the base points $f_1 B_n$, $f_2 B_n$, $f_3 P_n$, and also switched the order of the two factors X_n. This choice of base points is in effect *only* when we multiply $S\tau$, or $S\widetilde{\tau}$, by a section of $\widetilde{\mathcal{L}}_{\nu-\rho_{\mathrm{mir}},\varepsilon} \to \widetilde{Y}$ such as $\widetilde{E}_{\nu,\xi}$ or $I_\nu E_{-\nu,\psi}$, rather than by $E_{-\nu,\psi}$; it is used internally in this proof, but not elsewhere in the paper.

Though corollary 4.21 as stated does not apply to (4.27) when $\widetilde{\rho} = \widetilde{E}_{\nu,\xi}$ or $I_\nu E_{-\nu,\psi}$, its conclusions apply so long as Γ' is appropriately modified to take into account the invariance group of $\widetilde{\rho}$. This can be seen either as a consequence of the general statement theorem 4.10, or alternatively deduced directly from corollary 4.21 using the outer automorphism (2.35). Let $\phi \in C_c^\infty(G_n)$ have $\int_{G_n} \phi(h)dh = 1$. The proof of the proposition involves computing the integral

$$\mathcal{I} = \int_{\Gamma' \backslash G_n / Z_n} \int_{G_n} (S\tau \cdot I_\nu E_{-\nu,\psi})(gh)\, \phi(h)\, dh\, dg \qquad (4.28)$$

in two different ways. The first involves inserting the formula for $I_\nu E_{-\nu,\psi}$ from proposition 3.48, obtaining

$$\mathcal{I} = (-1)^\varepsilon N^{2\nu-\frac{\nu}{n}-\frac{1}{2}} G_\varepsilon(\nu - \tfrac{n}{2} + 1) \frac{1}{\phi_{\mathrm{Euler}}(N)} \times$$

$$\sum_{\substack{a \,(\mathrm{mod}\, N) \\ \xi \in (\widehat{\mathbb{Z}/N\mathbb{Z}})^*}} \widehat{\psi}(a)\,\xi(a)^{-1} \int_{\Gamma'\backslash G_n/Z_n} \int_{G_n} \left(S\tau \cdot \ell\left(\begin{pmatrix} I_{n-1} & \\ & N \end{pmatrix} w_{long} \right) \widetilde{E}_{\nu,\xi} \right)(gh)\, \phi(h)\, dh\, dg$$

(4.29)

(we have denoted the Euler ϕ-function as ϕ_{Euler} here in order to avoid confusing it with the smooth function ϕ in the integrand). The integral can be written as

$$\frac{1}{\mathrm{covol}(\Gamma \cap U(\mathbb{Z}))} \int_{\Gamma'\backslash G_n/Z_n} \int_{G_n} \int_{\Gamma \cap U(\mathbb{Z})\backslash U} \theta(u) \times$$

$$\times \left(\ell(u)\,\tau\right)\begin{pmatrix} gh\widetilde{f_2} & 0_n \\ 0_n & gh\widetilde{f_1} \end{pmatrix} \widetilde{E}_{\nu,\xi}\left(w_{long} \begin{pmatrix} I_{n-1} & \\ & N \end{pmatrix}^{-1} gh\widetilde{f_3}\right) du\,\phi(h)\, dh\, dg\,.$$

(4.30)

We now change variables $g \mapsto \widetilde{g}$, $h \mapsto \widetilde{h}$, and then apply identities (2.36) and (3.14), after which we must replace Γ' by $\widetilde{\Gamma'}$: the integral becomes

$$\frac{1}{\mathrm{covol}(\Gamma \cap U(\mathbb{Z}))} \int_{\widetilde{\Gamma'}\backslash G_n/Z_n} \int_{G_n} \int_{\Gamma \cap U(\mathbb{Z})\backslash U} \theta(u) \times$$

$$\times \widetilde{\tau}\left(\widetilde{u}^{-1} \begin{pmatrix} ghf_1 & 0_n \\ 0_n & ghf_2 \end{pmatrix}\right) E_{\nu,\xi}\left(w_{long} \begin{pmatrix} N & \\ & I_{n-1} \end{pmatrix} ghf_3\right) du\,\phi(\widetilde{h})\, dh\, dg\,.$$

(4.31)

The above expression is unchanged if both instances of Γ are replaced by any finite index subgroup, in particular the principal congruence subgroup $\Gamma(m) = \{\gamma \in G_{2n}(\mathbb{Z}) | \gamma \equiv I_{2n} \,(\mathrm{mod}\, m)\}$ for some m (and hence any positive multiple of it). The change of variables $u \mapsto \widetilde{u}^{-1} = w_{long} u^t w_{long}$ preserves $\Gamma(m)$, $U(\mathbb{Z})$, U, the character θ, and the Haar measure du; it allows us to rewrite (4.31) as

$$\frac{1}{\mathrm{covol}(\Gamma(m) \cap U(\mathbb{Z}))} \int_{\widetilde{\Gamma'}\backslash G_n/Z_n} \int_{G_n} \int_{\Gamma(m) \cap U(\mathbb{Z})\backslash U} \theta(u) \times$$

$$\times \widetilde{\tau}\left(u\begin{pmatrix} ghf_1 & 0_n \\ 0_n & ghf_2 \end{pmatrix}\right) E_{\nu,\xi}\left(w_{long} \begin{pmatrix} N & \\ & I_{n-1} \end{pmatrix} ghf_3\right) du\,\phi(\widetilde{h})\, dh\, dg\,.$$

(4.32)

We may freely replace $\phi(h)$ with $\phi(\widetilde{h})$ because corollary 4.21 guarantees that this substitution of smoothing function does not affect the overall value. Since

$$\widetilde{\tau}\left(\begin{pmatrix} I_n & A \\ 0_n & I_n \end{pmatrix}\begin{pmatrix} g_1 & 0_n \\ 0_n & g_2 \end{pmatrix}\right) = \widetilde{\tau}\left(\begin{pmatrix} -I_n & 0_n \\ 0_n & I_n \end{pmatrix}\begin{pmatrix} I_n & A \\ 0_n & I_n \end{pmatrix}^{-1}\begin{pmatrix} g_1 & 0_n \\ 0_n & g_2 \end{pmatrix}\begin{pmatrix} -I_n & 0_n \\ 0_n & I_n \end{pmatrix}\right)$$

$$= (-1)^{\delta_{n+1}+\cdots+\delta_{2n}}\,\widetilde{\tau}\left(\begin{pmatrix} -I_n & 0_n \\ 0_n & I_n \end{pmatrix}\begin{pmatrix} I_n & A \\ 0_n & I_n \end{pmatrix}^{-1}\begin{pmatrix} g_1 & 0_n \\ 0_n & g_2 \end{pmatrix}\right),$$

(4.33)

we may replace the u in the argument of $\widetilde{\tau}$ by u^{-1}, so long as we left translate it by $\begin{pmatrix} -I_n & \\ & I_n \end{pmatrix}$ and multiply the overall expression by $(-1)^{\delta_{n+1}+\cdots+\delta_{2n}}$. Replacing g by $g \mapsto \begin{pmatrix} N & \\ & I_{n-1} \end{pmatrix}^{-1} w_{long} g$ converts $\widetilde{\Gamma'}$ into Γ^*, and nearly converts (4.32) into

$$(-1)^{\delta_{n+1}+\cdots+\delta_{2n}} P\left(\ell\left(\begin{pmatrix} -w_{long} & \\ & w_{long} \end{pmatrix} \begin{pmatrix} N & & \\ & I_{n-1} & \\ & & N \\ & & & I_{n-1} \end{pmatrix}\right) \widetilde{\tau}, E_{\nu,\xi}\right);$$

(4.34)

the only difference is that the u-integration is changed by the presence of these two matrices that left-translate $\widetilde{\tau}$. The compensating change of variables in u that undoes this conjugation preserves the character θ, but alters $\Gamma(m)$ because some nondiagonal entries are multiplied or divided by N. Were m replaced by mN in

(4.32) this conjugate would still be a subgroup of Γ, and hence its normalized u-integration would have the same value. We conclude that

$$
\mathcal{I} = (-1)^{\varepsilon + \delta_{n+1} + \cdots + \delta_{2n}}\, N^{2\nu - \frac{\nu}{n} - \frac{1}{2}}\, G_\varepsilon(\nu - \tfrac{n}{2} + 1)\, \tfrac{1}{\phi(N)} \times
$$

$$
\sum_{\substack{a\,(\mathrm{mod}\,N) \\ \xi \in (\widehat{\mathbb{Z}/N\mathbb{Z}})^*}} \widehat{\psi}(a)\, \xi(a)^{-1}\, P\left(\ell\left(\begin{pmatrix} & -w_{long} \\ w_{long} & \end{pmatrix} \begin{pmatrix} N & & \\ & I_{n-1} & \\ & & N \\ & & & I_{n-1} \end{pmatrix} \right) \widetilde{\tau}, E_{\nu,\xi} \right). \quad (4.35)
$$

The proof of the proposition now reduces to demonstrating that

$$
\mathcal{I} = \frac{G_\varepsilon(\nu - \tfrac{n}{2} + 1)}{\prod_{j=1}^n G_{\delta_{n+j} + \delta_{n+1-j} + \eta}(\lambda_{n+j} + \lambda_{n+1-j} + \tfrac{\nu}{n} + \tfrac{1}{2})}\, P(\tau, E_{-\nu,\psi}). \quad (4.36)
$$

By combining (4.27) and (4.28), $\mathcal{I}$ can be written as

$$
\mathcal{I} = \frac{1}{\mathrm{covol}(\Gamma \cap U(\mathbb{Z}))} \int_{\Gamma' \backslash G_n / Z_n} \int_{G_n} \int_{\Gamma \cap U(\mathbb{Z}) \backslash U} \theta(u) \times
$$

$$
\times\, (\ell(u)\,\tau) \begin{pmatrix} gh\widetilde{f_2} & 0_n \\ 0_n & gh\widetilde{f_1} \end{pmatrix} I_\nu E_{-\nu,\psi}(gh\widetilde{f_3})\, du\, \phi(h)\, dh\, dg. \quad (4.37)
$$

Right translating h by w_{long} converts $h\widetilde{f_1} = h$ to $hw_{long} = hf_2$, and $h\widetilde{f_2} = hw_{long}$ to $h = hf_1$. It also changes $\phi(h)$ to $\phi(hw_{long})$; however, this change can be undone by replacing $\phi(g)$ with $\phi(gw_{long})$, as both functions have the same total integral over G_n. Hence $\mathcal{I}$ can be expressed as

$$
\mathcal{I} = \frac{1}{\mathrm{covol}(\Gamma \cap U(\mathbb{Z}))} \int_{\Gamma' \backslash G_n / Z_n} \int_{G_n} \int_{\Gamma \cap U(\mathbb{Z}) \backslash U} \theta(u) \times
$$

$$
\times\, (\ell(u)\,\tau) \begin{pmatrix} ghf_1 & 0_n \\ 0_n & ghf_2 \end{pmatrix} I_\nu E_{-\nu,\psi}(ghw\widetilde{f_3})\, du\, \phi(h)\, dh\, dg. \quad (4.38)
$$

We shall now use the definition (3.24) of the intertwining operator I_ν. Since this involves an integral over the non-compact manifold U_n, it might seem that the formula cannot be applied to the distribution $E_{-\nu,\psi}$. However, the self-adjointness property (3.33) justifies the calculations we are about to present. In effect, the calculations with $E_{-\nu,\psi}$ reflect legitimate operations on the dual side. This is completely analogous to applying the calculus of differential operators to distributions as if they were functions. The duality depends on interpreting ϕ as a C^∞ section of a line bundle over $X_n \times X_n \times Y_n$, the mirror image of viewing the distribution section $S\tau \cdot E_{-\nu,\psi}$ as a scalar distribution[7] on G_n. In effect, we interpret the h-integration as the pairing of a distribution section of one line bundle against a smooth section of the dual line bundle, tensored with the line bundle of differential forms of top degree, by integration over the compact manifold $X_n \times X_n \times Y_n$. In a slightly different setting, this process is carried out in the proof of lemma 3.9 in [22]. What matters is that G_n acts on $X_n \times X_n \times Y_n$ with an open orbit. In any case, applying the definition (3.24) of I_ν, the notation $u(x)$ in proposition 3.26, and

---

[7]Strictly speaking, we should work with a smoothing function $\phi \in C_c^\infty(G_n/Z_n)$ instead of $\phi \in C_c^\infty(G_n)$, but this makes little difference for the rest of the argument.

the definition (2.35) of the automorphism $g \mapsto \widetilde{g}$, we find

$$I_\nu E_{-\nu,\psi}(g h w_{\text{long}} \widetilde{f_3})$$

$$= \int_{\mathbb{R}^{n-1}} E_{-\nu,\psi}(g h (f_3^{-1})^t u(x))\, dx$$

$$= \int_{\mathbb{R}^{n-1}} E_{-\nu,\psi}\left(g\, h\, u\left(\tfrac{x}{1-\sum x_j}\right)\right) \left|1 - \textstyle\sum_j x_j\right|^{-\nu - n/2} \operatorname{sgn}(1 - \textstyle\sum_j x_j)^\varepsilon dx \qquad (4.39)$$

$$= \int_{\mathbb{R}^{n-1}} E_{-\nu,\psi}\left(g\, h\, u(x)\right) \left|1 + \textstyle\sum_j x_j\right|^{\nu - n/2} \operatorname{sgn}(1 + \textstyle\sum_j x_j)^\varepsilon dx \,;$$

the equality at the second step follows from the transformation law (3.7) and the matrix identity

$$\begin{pmatrix} 1 & 0 & \cdots & 0 \\ -1 & 1 & & \\ \vdots & & \ddots & \\ -1 & & & 1 \end{pmatrix} \begin{pmatrix} 1 & x_{n-1} & \cdots & x_1 \\ 0 & 1 & & \\ \vdots & & \ddots & \\ 0 & & & 1 \end{pmatrix} = $$

$$= \begin{pmatrix} 1 & \frac{x_{n-1}}{1-\sum x_j} & \cdots & \frac{x_1}{1-\sum x_j} \\ 0 & 1 & & \\ \vdots & & \ddots & \\ 0 & & & 1 \end{pmatrix} \begin{pmatrix} \frac{1}{1-\sum x_j} & 0 & \cdots & 0 \\ -1 & & & \\ \vdots & & * & \\ -1 & & & \end{pmatrix}, \qquad (4.40)$$

and the third step in (4.39) from the change of coordinates $x_j \mapsto x_j(1 + \sum_j x_j)^{-1}$. To ensure convergence of the integral – or rather, of the corresponding integral on the dual side – we suppose $\operatorname{Re} \nu > n/2 - 1$.

We now combine (4.38) with (4.39). The resulting expression for $\mathcal{I}$ involves four integrals: the integrals over $\mathbb{R}^{n-1}$ and $(\Gamma \cap U(\mathbb{Z})) \backslash U$ on the inside – in either order, since they are independent – then the h-integral, and finally the integral over $G_n(\mathbb{Z}) \backslash G_n / Z_n$ on the outside. We claim that we can interchange the order of integration, to put the integration over $\mathbb{R}^{n-1}$ on the outside[8]: we can use partitions of unity to make the integrands for all the integrals have compact support. Then, using the definition of operations on distributions using the duality between distributions and smooth functions, the expression is converted into one for which Fubini's theorem applies. In terms of our specific choice of flags (4.23), this means

$$\mathcal{I} = \frac{1}{\operatorname{covol}(\Gamma \cap U(\mathbb{Z}))} \int_{\mathbb{R}^{n-1}} \int_{\Gamma' \backslash G_n / Z_n} \int_{G_n} \int_{\Gamma \cap U(\mathbb{Z}) \backslash U} \theta(u) \times$$

$$\times (\ell(u)\,\tau) \begin{pmatrix} gh & 0_n \\ 0_n & gh w_{\text{long}} \end{pmatrix} E_{-\nu,\psi}\left(g\, h\, u(x)\right) \left|1 + \textstyle\sum_j x_j\right|^{\nu - n/2} \qquad (4.41)$$

$$\times \operatorname{sgn}(1 + \textstyle\sum_j x_j)^\varepsilon \phi(h)\, du\, dh\, dg\, dx\,.$$

Neglecting a set of measure zero, we may integrate over $(\mathbb{R}^*)^{n-1}$ instead of $\mathbb{R}^{n-1}$. For $x \in (\mathbb{R}^*)^{n-1}$, $u(x)$ is conjugate to f_3 under the diagonal Cartan subgroup of G_n,

$$u(x) = a_x^{-1} f_3 \, a_x\,, \quad \text{with } a_x = \begin{pmatrix} 1 & & & \\ & x_{n-1} & & 0 \\ & & \ddots & \\ 0 & & & x_2 \\ & & & & x_1 \end{pmatrix}. \qquad (4.42)$$

---

[8]The integration over $U(\mathbb{Z}) \backslash U$ must remain on the inside; it is necessary to make sense of $S\widetilde{\tau}$ as a distribution section over $X_n \times X_n$.

We now change variables to replace h by ha_x. The identity

$$E_{-\nu,\psi}(g\,h\,a_x\,u(x)) \;=\; \left|\prod_{j=1}^{n-1} x_j\right|^{-\frac{\nu}{n}-\frac{1}{2}} \operatorname{sgn}\left(\prod_{j=1}^{n-1} x_j\right)^{\eta} E_{-\nu,\psi}(g\,h\,f_3)\,. \qquad (4.43)$$

follows from the transformation law (3.4) because the representation $W_{\nu,\delta}$ has been tensored by $\operatorname{sgn}(\det(\cdot))^{\eta}$ (see the comments between (4.17) and (4.19)). Similarly the identity

$$w_{\text{long}}^{-1}\,a_x\,w_{\text{long}} \;=\; \begin{pmatrix} x_1 & & & \\ & x_2 & & 0 \\ & & \ddots & \\ 0 & & & x_{n-1} \\ & & & & 1 \end{pmatrix} \qquad (4.44)$$

and the transformation law (2.17) imply

$$(\ell(u)\,\tau)\begin{pmatrix} gha_x & 0_n \\ 0_n & gha_x w_{\text{long}} \end{pmatrix} =$$

$$= \left(\prod_{j=1}^{n-1} |x_j|^{-\lambda_{n+j}-\lambda_{n+1-j}} (\operatorname{sgn} x_j)^{\delta_{n+j}+\delta_{n+1-j}}\right) (\ell(u)\,\tau)\begin{pmatrix} gh & 0_n \\ 0_n & ghw_{\text{long}} \end{pmatrix}. \qquad (4.45)$$

Therefore these characters of the x_j may be moved to the outermost integral in (4.41). The only remaining instance of x in the inner three integrations is in the argument of the test function, $\phi(ha_x)$. By the same reasoning as before, $h \mapsto \phi(ha_x)$ has total integral one, just like ϕ. Since these inner three integrations define the pairing $P(\tau, E_{-\nu,\psi})$, they depend only on this total integral, and hence their value is unchanged if a_x is removed from the argument of ϕ. The x-integral in (4.41) splits off to give

$$\mathcal{I} \;=\; \mathcal{H} \times P(\tau, E_{-\nu,\psi})\,, \qquad (4.46)$$

with

$$\mathcal{H} \;=\; \int_{\mathbb{R}^{n-1}} \left|1 + \sum_{j=1}^{n-1} x_j\right|^{\nu-n/2} \operatorname{sgn}\left(1 + \sum_{j=1}^{n-1} x_j\right)^{\varepsilon} \times$$

$$\times \left(\prod_{j=1}^{n-1} |x_j|^{-\lambda_{n+j}-\lambda_{n+1-j}-\nu/n-1/2}(\operatorname{sgn} x_j)^{\delta_{n+j}+\delta_{n+1-j}+\eta}\right) dx\,. \qquad (4.47)$$

This integral can be explicitly evaluated: according to lemma 4.50 below,

$$\mathcal{H} \;=\; (-1)^{\delta_2+\cdots+\delta_{2n-1}+(n-1)\eta} \times$$

$$\times \frac{G_{\varepsilon}(\nu-\frac{n}{2}+1)\prod_{j=1}^{n-1} G_{\delta_{n+j}+\delta_{n+1-j}+\eta}(-\lambda_{n+j}-\lambda_{n+1-j}-\frac{\nu}{n}+\frac{1}{2})}{G_{\varepsilon+\delta_2+\cdots+\delta_{2n-1}+(n-1)\eta}(\nu-\frac{n}{2}+1-\lambda_2+\cdots-\lambda_{2n-1}-\frac{n-1}{n}\nu+\frac{n-1}{2})}\,. \qquad (4.48)$$

At this point, the hypothesis (4.19) and the identity

$$G_{\delta}(s)\,G_{\delta}(1-s) \;=\; (-1)^{\delta} \qquad (4.49)$$

(which follows directly from (3.47)), establish (4.36) and hence the proposition. $\square$

4.50. LEMMA. *For $t \in \mathbb{R}^n$, $t_n \neq 0$, the integral*

$$\int_{\mathbb{R}^{n-1}} \left| t_n - \sum_{j=1}^{n-1} t_j \right|^{\beta_0 - 1} \mathrm{sgn}\left(t_n - \sum_{j=1}^{n-1} t_j \right)^{\eta_0} \times$$

$$\times \prod_{j=1}^{n-1} \left(|t_j|^{\beta_j - 1} \mathrm{sgn}(t_j)^{\eta_j} \right) dt_1 \cdots dt_{n-1} ,$$

converges absolutely when the real parts of $1 - \beta_0 - \beta_1 - \cdots - \beta_{n-1}$ and of the β_j are all positive. As a function of the β_j it extends meromorphically to all of $\mathbb{C}^n$, and equals

$$\frac{G_{\eta_0}(\beta_0) G_{\eta_1}(\beta_1) \cdots G_{\eta_{n-1}}(\beta_{n-1})}{G_{\eta_0 + \eta_1 + \cdots + \eta_{n-1}}(\beta_0 + \beta_1 + \cdots + \beta_{n-1})} |t_n|^{\beta_0 + \beta_1 + \cdots + \beta_{n-1} - 1} (\mathrm{sgn}\, t_n)^{\eta_0 + \eta_1 + \cdots + \eta_{n-1}} .$$

Proof: First we show that absolute convergence implies the formula we want to prove. We let $I(t_n)$ denote the value of the integral. Changing variables appropriately one finds

$$I(t_n) = |t_n|^{\beta_0 + \beta_1 + \cdots + \beta_{n-1} - 1} \mathrm{sgn}(t_n)^{\eta_0 + \eta_1 + \cdots + \eta_{n-1}} I(1) . \tag{4.51}$$

Recall the defining formula (3.46). Integration of the right hand side of the equality (4.51) against the function $e(t_n)$ results in the expression

$$G_{\eta_0 + \eta_1 + \cdots + \eta_{n-1}}\left(\sum_{j=0}^{n-1} \beta_j \right) I(1) , \tag{4.52}$$

whereas multiplication of the actual integral with $e(t_n)$, subsequent integration with respect to t_n, interchanging the order of integration, and the change of variables $t_j \mapsto t_j$ for $1 \leq j \leq n-1$, $t_n \mapsto \sum t_j$, result in the integral

$$\int_{\mathbb{R}^n} e(t_1 + \cdots + t_n) |t_n|^{\beta_0 - 1} \mathrm{sgn}(t_n)^{\eta_0} \prod_{j=1}^{n-1} \left(|t_j|^{\beta_j - 1} \mathrm{sgn}(t_j)^{\eta_j} \right) dt_1 \cdots dt_n . \tag{4.53}$$

Strictly speaking these integrals converge only conditionally, in the range Re $\beta_j \in (0,1)$. They can be turned into convergent integrals by a partition of unity argument and repeated integration by parts; for details see [17]. The integral (4.53) splits into a product of integrals of the type (3.46). The explicit formula for this integral, equated to the expression (4.52), gives the formula we want for $I(1)$, and hence for $I(t_n)$. Absolute convergence of $I(t_n)$ in the range Re $\beta_j > 0$, Re $(\sum \beta_j) < 1$ can be established by induction on n. For $n = 2$, the assertion follows from direct inspection. For the induction step, one integrates out one variable first and uses the uses the induction hypothesis, coupled with the explicit formula for the remaining integral in $n - 2$ variables. $\qquad \square$

5. Adelization of Automorphic Distributions

The definition of automorphic distribution in section 2 used classical language, as it is better suited for describing the necessary analysis of distributions on Lie groups. However, modern automorphic forms heavily uses the language of adeles to simplify and organize calculations, especially for general congruence subgroups Γ. In this section, we extend the notions there to the adeles by illustrating two different methods. In the first, we use strong approximation to derive an adelization of cuspidal automorphic distributions, analogous to the usual procedure of adelizing automorphic forms; in the second, we construct adelic Eisenstein distributions directly. Both constructions can be adapted to either case, and rely on the analysis in earlier sections at their core: it should be emphasized that the role of the adeles here is nothing more significant than a bookkeeping mechanism. However, there are deeper generalizations of this adelization which simultaneously take into account embeddings of several components of an automorphic representation. Such distributions are more complicated, and are useful for extending our theory to nonarchimedean places and number fields. The section concludes with the adelic analog of the pairing of the previous section.

For the sake of clarity, we have chosen to give an explicit, detailed discussion of this adelization for the linear algebraic group $GL(n)$ over $\mathbb{Q}$; this suffices for the application in [23]. However, the method generalizes to adelic automorphic representations for arbitrary connected, reductive linear algebraic groups defined over arbitrary number fields. We will make comments about the general case after describing the specifics for $GL(n)$ over $\mathbb{Q}$.

We for the most part use standard notation: $\mathbb{A}$ refers to the adeles of $\mathbb{Q}$, and $\mathbb{A}_f$ denotes the finite adeles, i.e., the restricted direct product of all $\mathbb{Q}_p$ with respect to $\mathbb{Z}_p$, $p < \infty$. If H denotes a group defined over $\mathbb{Z}$ such as $G = GL(n)$ or the unit upper triangular matrices N, we use the notation $H(R)$ to represent its R-points for the rings $R = \mathbb{Z}, \mathbb{Q}, \mathbb{Q}_p, \mathbb{R}, \mathbb{A}$, and $\mathbb{A}_f$. The maximal compact subgroup $\prod_{p<\infty} G(\mathbb{Z}_p)$ of $G(\mathbb{A}_f)$ will be denoted by K_f. We often stress membership in one of these groups with an appropriate subscript; for example, the general adele $g_\mathbb{A} \in G(\mathbb{A})$ can be decomposed as the product $g_\mathbb{A} = g_\infty \times g_2 \times g_3 \times g_5 \times \cdots$, or more concisely as $g_\infty \times g_f$, where the finite part $g_f \in G(\mathbb{A}_f)$ is the remaining product over the primes. The group $G(\mathbb{Q})$ sits inside each $G(\mathbb{Q}_p)$, and so at the same time embeds diagonally into $G(\mathbb{A})$. In order to avoid confusion here we shall use $G_\mathbb{Q}$ to denote this diagonally-embedded image; likewise, we let $H_\mathbb{Q} \subset G_\mathbb{Q}$ denote the diagonally embedded image of the rational points of an algebraic subgroup $H \subset G$ defined over $\mathbb{Z}$. Thus strong approximation, for example, asserts that $G(\mathbb{A}) = G_\mathbb{Q} G(\mathbb{R}) K_f$.

Suppose now that $\pi = \otimes_{p\leq\infty} \pi_p$ is an irreducible, cuspidal adelic automorphic representation of $G(\mathbb{A})$, with representation space $U \subset L^2_\omega(G_\mathbb{Q}\backslash G(\mathbb{A}))$ under the right action of $G(\mathbb{A})$. Here ω denotes a character of the center $Z(\mathbb{A})$, which we may assume is a finite order character after twisting π by a character of the determinant. Each function $\phi_\mathbb{A} \in U$ restricts to a function $\phi_\mathbb{R}$ on $G(\mathbb{R}) \subset G(\mathbb{A})$. Since the representation π acts continuously, $\phi_\mathbb{A}$ is stabilized by a congruence subgroup K of K_f. At the same time it is invariant on the left under $G_\mathbb{Q}$; since the K_f factor commutes across the $G(\mathbb{R})$ factor, we conclude that $\phi_\mathbb{R}$ is left-invariant under a congruence subgroup Γ of $G(\mathbb{Z})$. The same holds true (with different K and Γ) if we restrict $\phi_\mathbb{A}$ to a different section of $G(\mathbb{R})$ inside $G(\mathbb{A})$, for example one of the form $G(\mathbb{R}) \times \{g_f\}$: this is simply the restriction to $G(\mathbb{R})$ of $\pi(g_f)\phi_\mathbb{A}$. By strong approximation and the left invariance of $\phi_\mathbb{A}$ under $G_\mathbb{Q}$, this is tantamount to left translating $\phi_\mathbb{R} = \phi_\mathbb{A}|_{G(\mathbb{R})}$ by a rational, real matrix whose inverse approximates g_f. Thus adelic automorphic forms are functions from $G(\mathbb{A}_f)$ to smooth automorphic forms on $G(\mathbb{R})$. We shall use this vantage point as a template for adelizing automorphic distributions.

We now assume, as we may, that $\phi_\mathbb{A}$ corresponds to a nonzero pure tensor for $\pi = \otimes_{p\leq\infty} \pi_p$ that is furthermore a smooth vector for π_∞. Right translation by $G(\mathbb{R})$ commutes with the above correspondence, so $\phi_\mathbb{R}$ sits inside a classical automorphic representation equivalent to π_∞. It is therefore the image of an embedding of the form (2.3). By connecting these two constructions, an automorphic distribution τ now defines an embedding J of (π_∞, V_∞) into a subspace U_∞ of U: the closure of the subspace spanned by right $G(\mathbb{R})$-translates of $\phi_\mathbb{A}$.

Again as in section 2, τ is a distribution vector for π'_∞, and hence may be viewed as a distribution on $G(\mathbb{R})$ once a principal series embedding $\pi'_\infty \hookrightarrow V_{\lambda,\delta}$ has been chosen (cf. (2.7)). In what follows we fix such an embedding. The above procedure of course associates a distribution in $C^{-\infty}(G(\mathbb{R}))$ to any right translate of $\phi_\mathbb{A}$ by $g_f \in G(\mathbb{A}_f)$, a distribution which is left invariant under a discrete group that depends on g_f. Assembling these together, we form a map from $G(\mathbb{A}_f)$ to $C^{-\infty}(G(\mathbb{R}))$ which we call an "adelic automorphic distribution" for the automorphic representation π. More concretely, $\tau_\mathbb{A}(g_\mathbb{A}) = \tau_\mathbb{A}(g_\infty \times g_f)$ is defined to be the automorphic distribution in the variable g_∞ which describes the embedding of (π_∞, V_∞) into the space {restrictions of functions in $\pi(g_f)U_\infty$ to $G(\mathbb{R})$}.

The fixed principal series embedding for π'_∞ naturally exhibits π_∞ as the quotient of the dual principal series $V_{-\lambda,\delta}$. In particular, we may regard the pairing between $\tau(g_\infty \times g_f)$ and smooth vectors $v(g_\infty)$ in V_∞ as integration in g_∞ over a flag variety. We shall use the following notation generalizing (2.6):

$$
\begin{aligned}
J(v)(h_\infty \times h_f) &= \langle \tau_\mathbb{A}(g_\infty \times h_f), \pi_\infty(h_\infty)v(g_\infty) \rangle \\
&= \langle \tau_\mathbb{A}(h_\infty g_\infty \times h_f), v(g_\infty) \rangle,
\end{aligned}
\tag{5.1}
$$

where g_∞ is again the variable of integration in the pairing.

By convention $\tau_\mathbb{A}$ behaves like a function under diffeomorphisms and is dual to smooth, compactly supported measures in the g_∞ variable. Right translation of $\tau_\mathbb{A}$ by $G(\mathbb{A}_f)$ corresponds to right translation of functions in U. The group $G(\mathbb{A})$ also acts on $\tau_\mathbb{A}$ by left translation,

$$
(\ell(h_\mathbb{A})\tau_\mathbb{A})(g_\mathbb{A}) = \tau_\mathbb{A}(h_\mathbb{A}^{-1}g_\mathbb{A}).
\tag{5.2}
$$

This action on $\tau_\mathbb{A}$, restricted to $G(\mathbb{R})$, is consistent with (2.6) and (5.1), but note however that its restriction to $G(\mathbb{A}_f)$ acts on the left (as opposed to on the right, as it does for functions in U). Because the purpose of (5.2) is merely notational, this discrepancy will be harmless. Conjugates of the congruence subgroup $K \subset K_f$ that stabilizes $\phi_\mathbb{A}$ also stabilize $\tau_\mathbb{A}$:

$$
(\ell(k)\tau_\mathbb{A})(g_\infty \times g_f) = \tau_\mathbb{A}(g_\infty \times g_f) \quad \text{for each } k \in g_f K g_f^{-1}.
\tag{5.3}
$$

We claim that $G_\mathbb{Q}$ acts trivially on $\tau_\mathbb{A}$ under ℓ, i.e.,

$$
\tau_\mathbb{A}(\gamma g_\mathbb{A}) = \tau_\mathbb{A}(g_\mathbb{A}) \quad \text{for each } \gamma \in G_\mathbb{Q}.
\tag{5.4}
$$

Indeed, writing γ as $\gamma_\infty \times \gamma_f$, this amounts to checking that

$$
\langle \tau_\mathbb{A}(\gamma_\infty g_\infty \times \gamma_f g_f), v(g_\infty) \rangle = \langle \tau_\mathbb{A}(g_\infty \times g_f), v(g_\infty) \rangle,
\tag{5.5}
$$

or equivalently,

$$
J(v)(\gamma_\infty \times \gamma_f g_f) = J(v)(g_f)
\tag{5.6}
$$

for arbitrary $g_f \in \mathbb{A}_f$ and smooth vectors $v \in V_\infty$. The left hand side, $J(v)(\gamma g_f)$, equals the right hand side because the function $J(v) \in U$ is automorphic under $G_\mathbb{Q}$.

Let us now briefly indicate how this adelization works for a general connected, reductive linear algebraic group defined over a number field F and its adele ring $\mathbb{A} = \mathbb{A}_F$ (we refer to [1] as a general reference for the definition, and facts quoted below). Let $\phi_\mathbb{A}$ again denote a smooth vector for an automorphic representation

$\pi = \otimes_v \pi_v$ of $G(\mathbb{A})$, where v runs over all places of F. The function $\phi_{\mathbb{A}}$ on $G(\mathbb{A})$ is left invariant under the diagonally embedded G_F, and is right invariant under a congruence subgroup K of K_f, the product of maximal compact subgroups of $G(F_v)$ over all nonarchimedean places v of F. Though strong approximation fails in this setting (even for $G = GL(n)$ when the class number of F is greater than 1), the restriction of $\phi_{\mathbb{A}}$ to G_{∞}, the product of $G(F_v)$ over all archimedean places v, is left invariant under

$$\Gamma \ = \ \{\, \gamma \in G(F) \mid \gamma_f \in K \,\} , \tag{5.7}$$

regarded as a subgroup of G_{∞}. Since Γ is an arithmetic subgroup of G_{∞}, the quotient $Z_{\infty}\Gamma \backslash G_{\infty}$ has finite volume, where Z is the maximal F-split torus of the center of G, and Z_{∞} denotes the product of $Z(F_v)$ over all archimedean places v. Automorphic representations are assumed to transform according to a character of the adelic points $Z(\mathbb{A})$ of Z. Thus, as before, the restriction of a vector in the adelic automorphic representation gives rise to a classical automorphic representation of the real group G_{∞}, and hence an automorphic distribution on G_{∞} (this uses the fact that the Casselman-Wallach embedding theorem holds for arbitrary real reductive groups). Right translation by $G(\mathbb{A}_f)$ then allows us to construct an adelic automorphic distribution $\tau_{\mathbb{A}}$ following the same procedure as before.

We now return to some features of the earlier discussion about $G = GL(n)$ over $F = \mathbb{Q}$, starting with a description of the adelic version of the Whittaker distribution $w_{\lambda,\delta}$ from (2.31). Let ψ_+ denote the standard choice of additive character on $\mathbb{Q}\backslash\mathbb{A}$: the unique such character whose archimedean component maps $x \mapsto e^{2\pi i x}$. (What we say below needs to be modified slightly if a different nontrivial character of $\mathbb{Q}\backslash\mathbb{A}$ is chosen instead.) There is a standard group homomorphism c defined on the group of unipotent upper triangular matrices N, given by summing the entries just above the diagonal:

$$c \ : \ (n_{ij}) \ \mapsto \ n_{1,2} + n_{2,3} + \cdots . \tag{5.8}$$

The composition $\psi_+ \circ c$ is a nondegenerate character of $N_{\mathbb{Q}}\backslash N(\mathbb{A})$, and is used to define global Whittaker integrals on the automorphic representation π:

$$W_{\phi_{\mathbb{A}}}(g) \ = \ \int_{N_{\mathbb{Q}}\backslash N(\mathbb{A})} \phi_{\mathbb{A}}(ng)\, \psi_+(c(n))^{-1}\, dn , \quad \phi_{\mathbb{A}} \in U . \tag{5.9}$$

Here, as usual, dn denotes Haar measure on $N(\mathbb{A})$, normalized to give the quotient $N_{\mathbb{Q}}\backslash N(\mathbb{A})$ volume equal to 1. Likewise, we define an analogous adelic Whittaker integral for π using $\tau_{\mathbb{A}}$:

$$w(g) \ = \ \int_{N_{\mathbb{Q}}\backslash N(\mathbb{A})} \tau_{\mathbb{A}}(ng)\, \psi_+(c(n))^{-1}\, dn , \tag{5.10}$$

or more succinctly

$$w \ = \ \int_{N(\mathbb{A})/N_{\mathbb{Q}}} \ell(n)\tau_{\mathbb{A}}\, \psi_+(c(n))\, dn . \tag{5.11}$$

Like $\tau_{\mathbb{A}}$, $w(g) = w(g_{\infty} \times g_f)$ should be thought of as a function of $g_f \in G(\mathbb{A}_f)$ with values in $C^{-\infty}(G(\mathbb{R}))$. Indeed, for any fixed $g_f \in G(\mathbb{A}_f)$, (5.3) shows that $\tau_{\mathbb{A}}$ is stabilized by a finite index subgroup of $K_f \cap N(\mathbb{A}_f)$; strong approximation then shows this integration is therefore actually over a finite cover of the compact quotient $N(\mathbb{Z})\backslash N(\mathbb{R})$. Hence it reduces to (2.34) and gives a valid distribution in the g_{∞} variable. If v is a smooth vector for V_{∞} and $\phi_{\mathbb{A}} = J(v)$, then it is easily seen that the distribution w embeds v to (5.9). This is because the pairing between distributions and vectors here involves integration on the right, whereas the above integrations take place on the left.

When $\phi_{\mathbb{A}}$ is a pure tensor for $\pi = \otimes_{p \leq \infty} \pi_p$, the integral (5.9) factors into a product of local Whittaker functions:

$$W_{\phi_{\mathbb{A}}}(g_\infty \times g_f) \;=\; W_\infty(g_\infty) W_f(g_f) \;, \quad W_f(g_f) \;=\; \prod_{p<\infty} W_p(g_p). \qquad (5.12)$$

Here the W_p lie in the Whittaker model $\mathcal{W}_p$ for π_p, and are constrained to be the standard spherical Whittaker function (i.e., $W_p|_{G(\mathbb{Z}_p)} \equiv 1$) for almost all primes p. Importantly, by varying the pure tensor $\phi_{\mathbb{A}}$, the W_p can be chosen arbitrarily in $\mathcal{W}_p$ for any given finite set of primes. Were we to instead start with such a modified choice of $\phi_{\mathbb{A}} \in U$ and construct $\tau_{\mathbb{A}}$ from it as above, its adelic Whittaker integral (5.10) would have a similar factorization:

$$w(g_\infty \times g_f) \;=\; w_\infty(g_\infty) W_f(g_f). \qquad (5.13)$$

The distribution $w_\infty \in C^{-\infty}(G(\mathbb{R}))$ coincides with a nonzero multiple of the distribution $w_{\lambda,\delta}$ from (2.31), where (λ, δ) are the principal series parameters for the Casselman embedding of π'_∞. The paper [7] provides a rather complete study of the connection between the archimedean Whittaker distributions w_∞ and Whittaker functions for general Lie groups. The remaining product over primes is itself naturally related to the coefficient in (2.33).

We have therefore shown the following fact, which is useful in constructing adelic automorphic distributions with prescribed behavior at finite places.

5.14. PROPOSITION. *Let $\pi = \otimes \pi_p$ be a cuspidal automorphic representation of $GL(n)/\mathbb{Q}$, and S any finite set of primes. For each $p \in S$ chose a function W_p in the Whittaker model for π_p, and set W_p equal to the standard spherical vector for each prime $p \notin S$. Then there exists a pure tensor $\phi_{\mathbb{A}}$ for π whose corresponding adelic automorphic distribution $\tau_{\mathbb{A}}$ satisfies (5.13).*

A famous theorem independently proven by Piatetski-Shapiro and Shalika [26, 27] states that a smooth vector $\phi_{\mathbb{A}} \in U$ can be reconstructed as the sum of left translates of its global Whittaker function (5.9) by coset representatives $\mathcal{C}$ for $N_{\mathbb{Q}}^{(n-1)} \backslash GL(n-1)_{\mathbb{Q}}$, where $N^{(n-1)} = \{(n-1) \times (n-1)$ unit upper triangular matrices$\}$. The analogous formula

$$\tau_{\mathbb{A}}(g) \;=\; \sum_{\gamma \in \mathcal{C}} w\left(\left({}^\gamma \; {}_1 \right) g \right) \qquad (5.15)$$

holds for $\tau_{\mathbb{A}}$, as a consequence of the above relationships between embeddings of smooth vectors $v \in V_\infty$. It can also be proven using Fourier analysis on the nilpotent group $N(\mathbb{A})$, following along the lines of the original argument in [26, 27]. In particular, integrating (5.15) over $N'_{\mathbb{Q}} \backslash N'(\mathbb{A})$, where $N' = [N, N]$ is the derived subgroup of N, gives the following formula for the adelization of τ_{abelian}:

$$\tau_{\text{abelian}}(g) \;=\; \sum_{k \in \mathbb{Q}^{n-1}} w(D(k)g), \qquad (5.16)$$

where $D(k) \in G_{\mathbb{Q}}$ is the matrix defined just after (2.31). It is evident that $W_f(D(k)g)$ from (5.13) must equal the ratio multiplying $w_{\lambda,\delta}(D(k)g) = w_\infty(D(k)g)$ in (2.33). This observation also demonstrates that the normalized coefficients a_k are independent of the chosen Casselman embedding.

Next we turn to the adelic version of the mirabolic Eisenstein series distributions that were defined and analytically continued in section 3. Though these can be constructed as a special case of the adelic automorphic distributions just described, it is more useful to construct them directly, and then verify that they match the earlier construction. Jacquet and Shalika studied adelic mirabolic Eisenstein series as part of their integral representations of the Rankin-Selberg L-functions on $GL(n) \times GL(n)$ [10] and the exterior square L-functions on $GL(2n)$ [11]. As we commented earlier, it is also possible to derive the results here from theirs, using sophisticated machinery of Casselman and Wallach.

Our adelic construction involves modifying the archimedean data in the Jacquet-Shalika construction in order to mimic the δ-function that is averaged in (3.12), but leaving the nonarchimedean data intact. We begin by recalling the Schwartz-Bruhat space of $\mathbb{Q}_p^n$, which is the usual Schwartz space in n real variables when $p = \infty$, and is the space of locally constant, compactly supported functions when $p < \infty$. The latter are precisely the finite linear combinations of characteristic functions of sets of the form $v + p^N \mathbb{Z}_p^n$, where $v \in \mathbb{Q}_p^n$ and $N \in \mathbb{Z}$; for v fixed one need only consider N large, because of overlap among these sets. The global adelic Schwartz-Bruhat space consists of all finite linear combinations of functions which are global products $\Phi(g) = \prod_{p \leq \infty} \Phi_p(g_p)$ of Schwartz-Bruhat functions Φ_p on $\mathbb{Q}_p^n$, in which all but a finite number of functions Φ_p are constrained to be the "standard unramified choice" of the characteristic function of $\mathbb{Z}_p^n$.

The adelic Eisenstein series distributions are designed to have central character ω^{-1}, the inverse of the central character of $\tau_{\mathbb{A}}$; this is done in anticipation of the pairing between these objects at the end of the section. Strong approximation for $\mathbb{A}^*$ equates the double cosets $\mathbb{Q}^* \backslash \mathbb{A}^* / \mathbb{R}^*_{>0}$ to the inverse limit of all $(\mathbb{Z}/N\mathbb{Z})^*$, $N \in \mathbb{N}$. Therefore any Dirichlet character ψ, in particular the one in (3.12), has an adelization to a global character $\psi_{\mathbb{A}} = \prod_{p \leq \infty} \psi_p$ of $\mathbb{A}^*$ that is trivial on $\mathbb{Q}^*$.[9] We assume for the rest of the paper that

$$\psi_{\mathbb{A}} = \chi^n \omega^{-1}, \tag{5.17}$$

where χ is also a finite order character of $\mathbb{Q}^* \backslash \mathbb{A}^*$ of parity $\eta \in \mathbb{Z}/2\mathbb{Z}$, consistent with (4.19).

Set P' equal to the $(n-1,1)$ standard parabolic subgroup of G, so that $P' = w_{\mathrm{long}} P w_{\mathrm{long}}$ (cf. (3.1)). Jacquet-Shalika form their Eisenstein series as averages of the function

$$I(g, s) = \chi(\det g)^{-1} |\det g|^s \int_{\mathbb{A}^*} \Phi(e_n t g) |t|^{ns} \psi_{\mathbb{A}}(t)^{-1} d^*t, \tag{5.18}$$

where $e_n = (0, 0, \ldots, 0, 1)$ is the n-dimensional elementary basis row vector. Our construction of the Eisenstein distribution differs in that we modify the archimedean component Φ_∞ of each summand of Φ to be the δ-function of a nonzero point in $\mathbb{R}^n$. To emphasize this distinction, we sometimes refer to Jacquet-Shalika's choice as $\Phi_{\mathrm{JS},\infty}$ and ours as $\Phi_{\mathrm{D},\infty}$. When $\Phi(g) = \prod_{p \leq \infty} \Phi_p(g_p)$ is a pure tensor, the integral (5.18) splits as a product of local integrals over $\mathbb{Q}_p$, $p \leq \infty$, so that $I(g, s) = I_\infty(g_\infty, s) I_f(g_f, s)$, $I_f(g_f, s)$ being the product over all $p < \infty$. The computation of $I_f(g_f, s)$ is unchanged from the setting of Jacquet-Shalika, but the

[9]Please note this identification between Dirichlet and global characters is inverse to the one used by Jacquet-Shalika.

archimedean integral

$$I_\infty(g_\infty, s) \;=\; \operatorname{sgn}(\det g_\infty)^\eta \,|\det g_\infty|^s \int_{\mathbb{R}^*} \Phi_\infty(e_n t g_\infty) \,|t|^{ns} \operatorname{sgn}(t)^\varepsilon \, d^*t \qquad (5.19)$$

differs in that it defines a distribution on G instead of a smooth function when $\Phi_\infty = \Phi_{\mathrm{D},\infty}$. The local integrals obey the transformation law

$$I_p\left(\left(\begin{smallmatrix} B & * \\ & a \end{smallmatrix}\right) g_p\right) \;=\; \psi_p(a)\,\chi_p(a)^{-1}\,|a|^{-(n-1)s}\,\chi_p(\det B)^{-1}\,|\det B|^s\, I_p(g_p)\,, \qquad (5.20)$$

as can be seen by the change of variables $t \mapsto t/a$ in the integral.

We shall now describe how the respective local integrals $I_{\mathrm{JS},\infty}$ and $I_{\mathrm{D},\infty}$ are related by right smoothing. If ϕ is any smooth, compactly supported function on $G(\mathbb{R})$, we may choose

$$\Phi_{\mathrm{JS},\infty}(v) \;=\; \int_{G(\mathbb{R})} \Phi_{\mathrm{D},\infty}(vh)\,\phi(h)\,\operatorname{sgn}(\det h)^\eta\,|\det h|^s\, dh\,, \qquad v \in \mathbb{R}^n, \qquad (5.21)$$

since the integral defines a smooth function of compact support in v. The respective local integrals (5.19) of $\Phi_{\mathrm{JS},\infty}$ and $\Phi_{\mathrm{D},\infty}$ are related by

$$I_{\mathrm{JS},\infty}(g_\infty, s) \;=\; \int_{G(\mathbb{R})} I_{\mathrm{D},\infty}(g_\infty h, s)\,\phi(h)\, dh\,. \qquad (5.22)$$

It follows that right convolution of our distributional $I(g,s)$ over $G(\mathbb{R})$ results in an instance of Jacquet-Shalika's (5.18).

We now consider the computation of $I(g,s)$ for a particular type of pure tensor Φ, namely when Φ_∞ is the δ-function supported at $e_1 = (1, 0, \ldots, 0)$ and Φ_p is the characteristic function of $e_n + p^{N_p}\mathbb{Z}_p^n$, where $N = \prod p^{N_p}$ is the factorization of a positive integer N. Then $I_\infty(g,s)$ is supported on $P'(\mathbb{R})w_{\mathrm{long}} = w_{\mathrm{long}}P(\mathbb{R})$ by construction, and is in fact a constant multiple of the distribution

$$\delta_\infty \;\in\; W_{\nu,\varepsilon}^{-\infty} \otimes \operatorname{sgn}(\det)^\eta \qquad (5.23)$$

defined in (3.8), with $\nu = n(s - 1/2)$ (cf. (3.65)). The local integral for $p < \infty$ is computed as

$$I_p(g_p, s) \;=\; \chi_p(\det g_p)^{-1}|\det g_p|^s \int_{\substack{t \in \mathbb{Q}_p^* \\ tv \in e_n + p^{N_p}\mathbb{Z}_p^n}} |t|^{ns}\,\psi_p(t)^{-1}\, d^*t\,, \qquad (5.24)$$

where v is the bottom row of g. The transformation law (5.20) reduces the computation to $g_p \in GL(n, \mathbb{Z}_p)$, a set of coset representatives for the subgroup of upper triangular matrices, so that in particular we may assume $v \in \mathbb{Z}_p^n$, $p \nmid v$. In the case that $N_p = 0$, the set in the second constraint is simply $\mathbb{Z}_p^n$, and the integration is over $0 < |t| \le 1$. The integral is then a Tate integral for $L(s, \psi)$: it represents $(1 - \psi(p)p^{-ns})^{-1}$ if $\psi_{\mathbb{A}}$ is unramified at p, and zero otherwise. If $N_p \ge 1$, the second constraint reads $tv_j \equiv 0 \pmod{p^{N_p}}$ for $j < n$, while $tv_n \equiv 1 \pmod{p^{N_p}}$. This forces $v_n \in \mathbb{Z}_p - p\mathbb{Z}_p$, and the range of integration to $t \in v_n^{-1} + p^{N_p}\mathbb{Z}_p$. The integral vanishes if the ramification degree of ψ_p exceeds N_p, and equals a constant times $\psi(v_n)$ otherwise.

In particular, if $\gamma \in GL(n, \mathbb{Z})$, then $I(\gamma g_\infty, s)$ is the product of a constant, $\delta_\infty(g_\infty)$, $L(ns, \phi)$, and the characteristic function of $\Gamma_0(N)$. The sum of this over all cosets for $P'(\mathbb{Z})\backslash G(\mathbb{Z})$ is precisely the Eisenstein series in (3.12), up to a constant multiple. The coset space $P'(\mathbb{Z})\backslash G(\mathbb{Z})$ is in bijective correspondence with $P'(\mathbb{Q})\backslash G(\mathbb{Q})$ via the inclusion map $G(\mathbb{Z}) \hookrightarrow G(\mathbb{Q})$, because of the fact that every invertible rational matrix can be decomposed as an upper triangular rational matrix times an invertible integral one. We conclude that with this particular choice of local data,

$$E_{\mathbb{A}}(g_{\mathbb{A}}, s) \;=\; \sum_{\gamma \in P_{\mathbb{Q}}' \backslash G_{\mathbb{Q}}} I(\gamma g_{\mathbb{A}}, s) \qquad (5.25)$$

is a constant multiple of (3.12) when $g_{\mathbb{A}} \in G(\mathbb{R})$, and in particular converges in the strong distributional topology for Re $s > 1$. Strong approximation reduces the evaluation of the general $g_{\mathbb{A}} \in G(\mathbb{A})$ to this case, so the sum makes sense in general for Re $s > 1$ and defines an adelic automorphic distribution: a map from $G(\mathbb{A}_f)$ to automorphic distributions in $C^{-\infty}(G(\mathbb{R}))$. Because of (5.22), the right smoothing of $E_{\mathbb{A}}(g_{\mathbb{A}}, s)$ over $G(\mathbb{R})$ results in a smooth Eisenstein series on $G(\mathbb{Q}) \backslash G(\mathbb{A})$ considered by Jacquet-Shalika. Thus $E_{\mathbb{A}}$ is also an automorphic distribution in the earlier sense of a distribution which embeds into smooth automorphic forms.

The general choice of local data involves broader choices in two respects: Φ_∞ may be a δ-function supported at another nonzero point, and Φ_p may be the characteristic function of $v + p^N \mathbb{Z}_p^n$, N large. Right translating $E_{\mathbb{A}}(g_{\mathbb{A}}, s)$ by some $h \in GL(n, \mathbb{A})$ has the effect of replacing $\Phi(v)$ by $\Phi(vh)$. Since $GL(n)$ acts with two orbits on n-dimensional vectors, this means the general δ-function for Φ_∞ can be reduced to the case above, and that the characteristic functions for Φ_p can be reduced to the situation that $v = 0$ or $v = e_n$. Since $e_n + \mathbb{Z}_p^n = \mathbb{Z}_p^n$, the sets $e_n + p^N \mathbb{Z}_p^n$ for $N \geq 0$ we considered above indeed cover all possibilities. Thus the analytic properties of the general instance of (5.25) for linear combinations of such pure tensors Φ reduce to those we have just considered. In particular they have a meromorphic continuation to $s \in \mathbb{C} - \{1\}$, with at most a simple pole at $s = 1$ that occurs only when ψ is trivial.

Finally, we conclude by writing the general form of the automorphic pairing in terms of adeles, generalizing (4.10). We need to slightly adapt the notation there to the adelic setting. Let U denote the algebraic group

$$U \;=\; \left\{ \begin{pmatrix} I_n & A \\ 0_n & I_n \end{pmatrix} \;\middle|\; A \in M_{n \times n} \right\} \;\subset\; GL(2n) \qquad (5.26)$$

whose real points were previously denoted by U in (4.13). The character θ from (4.15) has a natural adelic extension,

$$\theta : U(\mathbb{A}) \longrightarrow \mathbb{C}^*, \qquad \theta \left(\begin{smallmatrix} I_n & A \\ 0_n & I_n \end{smallmatrix} \right) = \psi_+(\operatorname{tr} A), \qquad (5.27)$$

where ψ_+ is the additive character defined just above (5.8). Let du denote the Haar measure on $U(\mathbb{A})$ which gives the quotient $U_{\mathbb{Q}} \backslash U(\mathbb{A})$ volume 1.

With f_1, f_2, and f_3 still standing for flag representatives in $G(\mathbb{R})$ and $\psi \in C_c^\infty(G(\mathbb{R}))$ having total integral 1, the general adelic pairing is defined as

$$P(\tau_{\mathbb{A}}, E_{\mathbb{A}}(s)) \;=$$

$$= \int_{Z(\mathbb{A})G_{\mathbb{Q}} \backslash G(\mathbb{A})} \int_{G(\mathbb{R})} \left[\int_{U_{\mathbb{Q}} \backslash U(\mathbb{A})} \tau_{\mathbb{A}} \left(u \begin{pmatrix} ghf_1 & \\ & ghf_2 \end{pmatrix} \right) \overline{\theta(u)} du \right] \times \qquad (5.28)$$

$$\times \, E_{\mathbb{A}}(ghf_3, s) \, \psi(h) \, dh \, dg \,.$$

Several comments are in order to explain why the above makes sense. Firstly, for the same reason as in (5.10), the bracketed inner integration is over a finite cover of the compact quotient $U(\mathbb{Z}) \backslash U(\mathbb{R})$, and so defines a map from $G(\mathbb{A}_f)$ to distributions in $G(\mathbb{R})$ that corresponds to (4.16). This map is left invariant under the diagonal rational subgroup $G_{\mathbb{Q}}$ because of (5.4), and because conjugation through u changes neither $\theta(u)$ nor the measure du. It is also invariant under $Z(\mathbb{A})$ because (4.19) ensures that the central characters of $\tau_{\mathbb{A}}$ and $E_{\mathbb{A}}$ are inverse to each other. The invariance under both $G_{\mathbb{Q}}$ and $Z(\mathbb{A})$ is not affected by the second integration, which only involves h on the right. The second integration simultaneously smooths both the bracketed expression and $E_\nu(ghf_3)$ over $G(\mathbb{R})$: it gives a map from $G(\mathbb{A}_f)$ to smooth automorphic functions on $G(\mathbb{R})$. According to corollary 4.21 these restrictions to $G(\mathbb{R})$ are each integrable over their fundamental domain. Because of (5.3) and strong approximation, the last integration takes place on a finite cover of $Z(\mathbb{R})G(\mathbb{Z})\backslash G(\mathbb{R})$ – again by the same reasoning used for the bracketed inner integration in (5.28), and for (5.10) before it. Corollary 4.21 shows that the last integral is independent of the choice of ψ, assuming its normalization $\int_{G(\mathbb{R})} \psi(g)dg = 1$.

The above pairing inherits the meromorphic continuation to $s \in \mathbb{C} - \{1\}$ that its classical counterpart possesses (corollary 4.21), as well as a functional equation from (4.24):

$$P(\tau_{\mathbb{A}}, E_{\mathbb{A}}(1-s)) =$$
$$N^{2ns-s-n} \prod_{j=1}^{n} G_{\delta_{n+j}+\delta_{n+1-j}+\eta}(s + \lambda_{n+j} + \lambda_{n+1-j}) \, P(\tau'_{\mathbb{A}}, E'_{\mathbb{A}}(s)), \qquad (5.29)$$

where $\tau'_{\mathbb{A}}$ and $E'_{\mathbb{A}}$ correspond to the translated contragredient cusp form $\widetilde{\tau}$ and sum of the remaining Eisenstein data, respectively, from the right hand side in proposition 4.24. This formula simplifies when both π_p and the Eisenstein data Φ_p are unramified at all $p < \infty$ (which put us in the situation that $N = 1$). If Φ_∞ is the delta function at $e_1 \in \mathbb{R}^n$, then $E'_{\mathbb{A}}(s) = (-1)^{\delta_1 + \cdots + \delta_n} E_{\mathbb{A}}(s)$ and $\tau'_{\mathbb{A}} = \widetilde{\tau}_{\mathbb{A}}$ (cf. (4.26)).

Appendix A. Archimedean components of automorphic representations on $GL(n, \mathbb{R})$

Recall from section 2 that we study automorphic distributions in terms of the embedding (2.7) of π'_∞ into principal series representations $V_{\lambda,\delta}$. These embeddings are not unique. For full principal series representations, the parameters (λ, δ) are determined only up to simultaneous permutation of the λ_j and δ_j. In general, there is a smaller choice of embedding parameters. On the other hand, the Gamma factors predicted by Langlands also depend on the nature of the archimedean component of the automorphic representation in question. We use this connection between multiple embeddings and Gamma factors to exclude unwanted poles of L-functions.

In this appendix we collect the relevant results about embeddings into principal series and Langlands Gamma factors. All of these are well known to experts, but do not appear in the literature – at least not in convenient form.

A.1. The Generic unitary dual of $GL(n, \mathbb{R})$ and embeddings into the principal series. The possible real representations of $GL(n, \mathbb{R})$ that can occur as the archimedean component π_∞ of a cuspidal automorphic representation π are extremely limited by a number of local and global constraints. The latter are extremely subtle, and hence a complete classification seems hopeless at present. In this subsection we will instead describe the representations that satisfy perhaps the most well known local constraints for π_∞, namely those that are unitary and generic (i.e., have a Whittaker model).

The unitary dual for $GL(n,\mathbb{R})$ was first described by Vogan [30], and later by Tadić [28] using different methods. Tadić describes the unitary dual as certain parabolically induced representations from an explicit set $\mathcal{B}$ of representations of $GL(n',\mathbb{R})$, $n' \leq n$. He also proves that permuting the order of the induction data yields the same irreducible representation of $GL(n,\mathbb{R})$. His set $\mathcal{B}$ is defined in terms of not only induced representations of square integrable (modulo the center) representations of $GL(1,\mathbb{R})$ and $GL(2,\mathbb{R})$, but also certain irreducible quotients. These quotients, however, are not "large" in the sense of [29], and hence neither are any representations induced from them. It is a result of Casselman, Zuckerman, and Kostant (see [14]) that all generic representations of $GL(n,\mathbb{R})$ are large, and conversely that all large representations are generic.

Hence Tadić's list gives a description of the generic unitary dual, once these quotients are removed from $\mathcal{B}$. We now summarize this description, after making further simplifications using transitivity of induction. Let $n = n_1 + \cdots + n_r$ be a partition of n, and let $P \subset G = GL(n,\mathbb{R})$ be the standard parabolic subgroup of block upper triangular matrices corresponding to this partition. The Levi subgroup M of P is isomorphic to $GL(n_1,\mathbb{R}) \times \cdots \times GL(n_r,\mathbb{R})$. Let σ_i denote an irreducible, square integrable (modulo the center) representation of $GL(n_i,\mathbb{R})$. This forces n_i to equal 1 or 2, and σ_i to be one of the following possibilities:

(1) If $n = 1$, σ_i is either the trivial representation of $GL(1,\mathbb{R}) \simeq \mathbb{R}^*$, or else the sign character $\mathrm{sgn}(x)$.

(2) If $n = 2$, σ_i is a discrete series representation D_k (indexed to correspond to holomorphic forms of weight k, $k \geq 2$).

These representations are self dual. For each $1 \leq i \leq r$ and $s_i \in \mathbb{C}$, the twist $\sigma_i[s_i] = \sigma_i \otimes |\det(\cdot)|^{s_i}$ defines a representation of $GL(n_i,\mathbb{R})$. The tensor product of these twists defines a representation of M which extends to P by allowing the unipotent radical of P to act trivially. Let $I(P;\sigma_1[s_1],\ldots,\sigma_r[s_r])$ denote the representation of G parabolically induced from this representation of P, where the induction is normalized to carry unitary representations to unitary representations. In order to be consistent with the conventions of [16], the group action in this induced representation operates on the right, on functions which transform under P on the left.

We now give the constraints on the parameters s_i that govern precisely when $I(P;\sigma_1[s_1],\ldots,\sigma_r[s_r])$ is irreducible, generic, and unitary according to the results of Casselman, Kostant, Tadić, Vogan, and Zuckerman mentioned above. We assume that this representation is normalized to have a unitary central character, as we of course may by tensoring with a character of the determinant.

- **Unitarity constraint:** the multisets $\{\sigma_i[s_i]\}$ and $\{\sigma_i[-\bar{s}_i]\}$ must be equal, i.e., these lists are equal up to permutation (recall the σ_i are self dual). This is because the representation dual to $I(P;\sigma_1[s_1],\ldots,\sigma_r[s_r])$ is $I(P;\sigma_1[-s_1],\ldots,\sigma_r[-s_r])$.
- **Unitary dual estimate:** $|\mathrm{Re}\, s_i| < 1/2$. In the case of the principal series, this is commonly called the "trivial bound".
- **Permutation of order:** for any permutation $\tau \in S_r$, the induced representations $I(P;\sigma_1[s_1],\ldots,\sigma_r[s_r])$ and $I(P^\tau;\sigma_{\tau(1)}[s_{\tau(1)}],\ldots,\sigma_{\tau(r)}[s_{\tau(r)}])$ are equal, where P^τ is the standard parabolic whose Levi component is $GL(n_{\tau(1)},\mathbb{R}) \times \cdots \times GL(n_{\tau(r)},\mathbb{R})$.

The principal series representations $V_{\lambda,\delta}$ in (2.16) are induced representations, but induced from a lower triangular Borel subgroup (2.10). Our convention is well-suited for studying automorphic distributions, but induction from an upper triangular Borel subgroup is the more common convention in the literature on Langlands' classification of representations of real reductive groups [16] (e.g., his prediction of Γ-factors for automorphic L-functions). Using the Weyl group element w_{long} from (2.35) and the inverse map between the two, it is straightforward to show that $V_{\lambda,\delta}$ is equivalent to $I(B_+; \text{sgn}^{\delta_n}[\lambda_n], \ldots, \text{sgn}^{\delta_1}[\lambda_1])$, where B_+ is the upper triangular Borel subgroup of $GL(n, \mathbb{R})$. More generally, induction on the right from a lower triangular parabolic involves reversing the order of the inducing data, though the order is irrelevant for the representations in Tadić's classification of the unitary dual anyhow.

Embeddings into principal series are of course tautological for $n = 1$, where all irreducible representations are one dimensional. When $n = 2$, the discrete series representation D_k is a subrepresentation of the principal series representation $V_{\lambda,\delta}$ with parameters $\lambda = (-\frac{k-1}{2}, \frac{k-1}{2})$ and $\delta = (k, 0)$. This embedding is not unique: actually $D_k \otimes \text{sgn} \simeq D_k$, so $\delta = (k+1, 1)$ is an equally valid parameter. An irreducible principal series representation $V_{(\lambda_1,\lambda_2),(\delta_1,\delta_2)}$ embeds not only into itself, but also into $V_{(\lambda_2,\lambda_1),(\delta_2,\delta_1)}$. However, D_k is not a subrepresentation, but instead a quotient, of the representation $V_{(\frac{k-1}{2}, -\frac{k-1}{2}),(0,k)}$. If $\rho_1 \hookrightarrow \rho_2$, then $\rho_1[s] \hookrightarrow \rho_2[s]$. The twist $V_{\lambda,\delta}[s]$ is the principal series representation $V_{\lambda+(s,s,\ldots,s),\delta}$, so $D_k[s]$ embeds both into $V_{(s-\frac{k-1}{2}, s+\frac{k-1}{2}),(k,0)}$ and also $V_{(s-\frac{k-1}{2}, s+\frac{k-1}{2}),(k+1,1)}$. The description above shows that these are the only types of unitary generic representations of $GL(2, \mathbb{R})$.

Next we move to $GL(n, \mathbb{R})$ and consider a unitary, generic representation $\pi_\infty = I(P; \sigma_1[s_1], \ldots, \sigma_r[s_r])$ as above. Embeddings for $\pi_\infty' = I(P; \sigma_1[-s_1], \ldots, \sigma_r[-s_r])$ may be deduced from the previous paragraph, using the principle of transitivity of induction as follows. Let k_i denote the weight of the discrete series in block i (provided $n_i = 2$, of course). Now define vectors $\lambda \in \mathbb{C}^n$ and $\delta \in (\mathbb{Z}/2\mathbb{Z})^n$ in the following manner. If the integer $1 \leq j \leq n$ is contained in the i-th block n_i of the partition $n = (n_1, \ldots, n_r)$, set λ_j to be

$$\lambda_j = \begin{cases} -s_i, & n_i = 1; \\ -s_i - \frac{k_i-1}{2}, & n_i = 2 \text{ and } j = n_1 + \ldots + n_{i-1} + 1; \\ -s_i + \frac{k_i-1}{2}, & n_i = 2 \text{ and } j = n_1 + \ldots + n_{i-1} + 2. \end{cases} \tag{A.1}$$

Similarly, set

$$\delta_j \equiv \begin{cases} \varepsilon, & n_i = 1 \text{ and } \sigma_i = \text{sgn}(\cdot)^\varepsilon; \\ k_i, & n_i = 2 \text{ and } j = n_1 + \ldots + n_{i-1} + 1; \\ 0, & n_i = 2 \text{ and } j = n_1 + \ldots + n_{i-1} + 2. \end{cases} \tag{A.2}$$

One may alternatively replace k_i and 0 in the last two cases by $k_i + 1$ and 1, respectively. In other words, λ and δ are formed by concatenating the corresponding vectors which describe the embedding parameters for the $\sigma_i[-s_i]$, $1 \leq i \leq r$. By transitivity of induction, $\pi_\infty' = I(P; \sigma_1[-s_1], \ldots, \sigma_r[-s_r])$ is a subrepresentation of $V_{\lambda,\delta}$.

A.2. Langlands' Γ-factors. The Γ-factors which accompany an automorphic L-function $L(s, \pi, \rho)$ in its functional equation are conjectured to always be products, with shifts, of the functions

$$\Gamma_{\mathbb{R}}(s) = \pi^{-s/2} \Gamma(s/2) \quad \text{and} \quad \Gamma_{\mathbb{C}}(s) = 2(2\pi)^{-s} \Gamma(s) = \Gamma_{\mathbb{R}}(s) \Gamma_{\mathbb{R}}(s+1). \tag{A.3}$$

Langlands [16] gives a procedure to compute this archimedean factor $L_\infty(s, \pi, \rho)$ in terms of his description of π_∞ as a subquotient of an induced representation, along with a calculation involving the L-group representation ρ and the Weil group. When dealing with the group $GL(n)$, however, it is much more convenient to avoid the Weil group, and instead describe these Γ-factors in terms of the (freely permuted) induction data. We give a description of this for some notable examples, following the description in [13].

It is convenient to use Langlands' *isobaric* notation [15] for induced representations

$$\pi_\infty \;=\; I(P; \sigma_1[s_1], \ldots, \sigma_r[s_r]) \;=\; \sigma_1[s_1] \boxplus \cdots \boxplus \sigma_r[s_r], \qquad (A.4)$$

in which the operation $\boxplus$ on the right hand side should be thought of as a formal, abelian addition. Recall that the classification in section A.1 shows that every generic unitary representation of $GL(n)$ is an isobaric sum of the form (A.4), independent of the order. We use these formal sums here only as a bookkeeping device used to define Γ-factors; they do not always correspond to irreducible, archimedean components of cuspidal automorphic representations. This formal addition satisfies the following two properties. First, two isobaric sums Π_1, Π_2 may themselves be concatenated into a longer isobaric sum $\Pi_1 \boxplus \Pi_2$. Second, an isobaric sum can be twisted by the rule $(\Pi_1 \boxplus \Pi_2)[s] = \Pi_1[s] \boxplus \Pi_2[s]$.

We shall explain how to define $L(s, \Pi)$ for such a formal sum Π of twists of the σ_i, and how ρ transforms Π into another such formal sum $\rho(\Pi)$ for some examples of representations ρ of $GL(n, \mathbb{C}) = {}^L GL(n)^0$. Then $L_\infty(s, \pi, \rho)$ is defined as $L(s, \rho(\Pi))$, where Π is an isobaric sum for π_∞. We start with the definition of $L(s, \Pi)$ when Π is one of the basic building blocks σ_i, the self-dual, square integrable representations from section A.1:

$$L(s, triv) \;=\; \Gamma_\mathbb{R}(s) \;, \qquad L(s, \mathrm{sgn}) \;=\; \Gamma_\mathbb{R}(s+1) \;, \qquad (A.5)$$

$$\text{and} \qquad L(s, D_k) \;=\; \Gamma_\mathbb{C}\!\left(s + \tfrac{k-1}{2}\right). \qquad (A.6)$$

Next are rules for isobaric sums and twists:

$$L(s, \Pi[s']) = L(s + s', \Pi) \quad \text{and} \quad L(s, \Pi_1 \boxplus \Pi_2) = L(s, \Pi_1)\, L(s, \Pi_2). \quad (A.7)$$

Therefore $L(s, \Pi)$, for a general isobaric sum $\Pi = \sigma_1[s_1] \boxplus \cdots \boxplus \sigma_r[s_r]$, is given by

$$L(s, \Pi) \;=\; \prod_{i=1}^{r} L(s + s_i, \sigma_i) \,, \qquad (A.8)$$

and is explicitly determined by the definitions (A.5-A.6).

Let now $\Pi = \Pi_1 \boxplus \Pi_2 \boxplus \cdots \boxplus \Pi_r$ be an isobaric representation of $GL(n, \mathbb{R})$, and $\Pi' = \Pi'_1 \boxplus \Pi'_2 \boxplus \cdots \boxplus \Pi'_{r'}$ be an isobaric representation of $GL(m, \mathbb{R})$. The isobaric sum for the Rankin-Selberg tensor product representation $\Pi \times \Pi'$ of $GL(nm, \mathbb{R})$ is given by

$$\Pi \times \Pi' \;=\; \boxplus_{j=1}^{r} \boxplus_{k=1}^{r'} (\Pi_j \times \Pi'_k) \,, \qquad (A.9)$$

where now the meaning of $\Pi_j \times \Pi'_k$ must be explained. It is in general *not* the usual tensor product of two representations (more on this below). One has the relations

$$\Pi[s] \times \Pi'[s'] \;=\; (\Pi \times \Pi')[s + s'] \qquad (A.10)$$

and

$$\Pi \times \Pi' \;=\; \Pi' \times \Pi \,, \qquad (A.11)$$

which along with (A.9) may be regarded as formal rules for the calculation of tensor product on isobaric representations. They boil the general calculation down to the examples of $\sigma \times \sigma'$, where $\sigma, \sigma' \in \{triv, \text{sgn}, D_k \mid k \geq 2\}$. First, if σ or σ' is one of the representations $triv$ or sgn, then the Rankin-Selberg product corresponds to the usual tensor product. The only other case is when σ and σ' are both discrete series representations of $GL(2, \mathbb{R})$. In this situation one has $D_k \times D_\ell = D_{k+\ell-1} \boxplus D_{|k-\ell|+1}$. In summary $\sigma \times \sigma'$ is given by the following table:

$\sigma \setminus \sigma'$	$triv$	sgn	D_k
$triv$	$triv$	sgn	D_k
sgn	sgn	$triv$	D_k
D_ℓ	D_ℓ	D_ℓ	$D_{k+\ell-1} \boxplus D_{|k-\ell|+1}$

If $k = \ell$ there is no representation D_1, yet we use the convention (A.6) to write $L(s, D_1) = \Gamma_{\mathbb{C}}(s)$. In light of (A.3), it is equivalent to regard D_1 as $triv \boxplus$ sgn.

We now come to the exterior square representation Ext^2 that maps $GL(n) \to GL(\frac{n(n-1)}{2})$. It satisfies the following formal rules:

$$Ext^2 \left(\boxplus_{j=1}^r \Pi_j \right) = \left(\boxplus_{j=1}^r Ext^2 \Pi_j \right) \boxplus \left(\boxplus_{1 \leq j < k \leq r} (\Pi_j \times \Pi_k) \right) \tag{A.12}$$

and

$$Ext^2 (\Pi[s]) = \left(Ext^2 \Pi \right)[2s]. \tag{A.13}$$

Similarly to the above situation of tensor products, it is completely determined by the following table:

σ	$triv$	sgn	D_k
$Ext^2 \sigma$	$\emptyset$	$\emptyset$	sgn^k

The notation $\emptyset$ here indicates not to include a corresponding term in the formal sum; equivalently, $L(s, \emptyset) = 1$.

As a final example, consider the symmetric square representation Sym^2 that maps $GL(n) \to GL(\frac{n(n+1)}{2})$. It satisfies both rules (A.12) and (A.13), with the substitution of Sym^2 for Ext^2, and is completely determined by the table

σ	$triv$	sgn	D_k
$Sym^2 \sigma$	$triv$	$triv$	$D_{2k-1} \boxplus \text{sgn}^{k+1}$

To illustrate, we will conclude by explicitly calculating $L_\infty(s, \pi, Ext^2 \otimes \chi)$ when π is a cuspidal automorphic representation of $GL(n)$ over $\mathbb{Q}$, and χ is a Dirichlet character. We write π_∞ as the isobaric sum

$$\Pi = \left(\boxplus_{i=1}^{r_1} \text{sgn}^{\varepsilon_i}[s_i] \right) \boxplus \left(\boxplus_{j=1}^{r_2} D_{k_j}[s_{r_1+j}] \right), \tag{A.14}$$

as this is its most general form according to the description in section A.1. The rules (A.12-A.13) show that

$$Ext^2 \Pi = \Pi_1 \boxplus \Pi_2 \boxplus \Pi_3 \boxplus \Pi_4 \boxplus \Pi_5, \tag{A.15}$$

where

$$\Pi_1 = \boxplus_{i=1}^{r_1} (Ext^2 \text{sgn}^{\varepsilon_i})[2s_i] = \boxplus_{i=1}^{r_1} \emptyset[2s_i] = \emptyset,$$

$$\Pi_2 = \boxplus_{j=1}^{r_2} (Ext^2 D_{k_j})[2s_{r_1+j}] = \boxplus_{j=1}^{r_2} \text{sgn}^{k_j}[2s_{r_1+j}],$$

$$\Pi_3 = \boxplus_{\substack{i \leq r_1 \\ j \leq r_2}} (\text{sgn}^{\varepsilon_i} \times D_{k_j})[s_i + s_{r_1+j}] = \boxplus_{\substack{i \leq r_1 \\ j \leq r_2}} D_{k_j}[s_i + s_{r_1+j}],$$

$$\Pi_4 = \boxplus_{1 \leq i < k \leq r_1} (\text{sgn}^{\varepsilon_i} \times \text{sgn}^{\varepsilon_k})[s_i + s_k] = \boxplus_{1 \leq i < k \leq r_1} \text{sgn}^{\varepsilon_i + \varepsilon_k}[s_i + s_k],$$

$$\tag{A.16}$$

and

$$\begin{aligned}
\Pi_5 &= \boxplus_{1 \le j < \ell \le r_2} (D_{k_j} \times D_{k_\ell})[s_{r_1+j} + s_{r_1+\ell}] \\
&= \boxplus_{1 \le j < \ell \le r_2} \left(D_{k_j+k_\ell-1}[s_{r_1+j} + s_{r_1+\ell}] \boxplus D_{|k_j-k_\ell|+1}[s_{r_1+j} + s_{r_1+\ell}] \right).
\end{aligned}$$

$$\text{(A.17)}$$

If we choose ε_{ik} and $\varepsilon'_j \in \{0,1\}$ to be congruent to $\varepsilon_i + \varepsilon_k$ and k_j modulo 2, respectively, then

$$L(s, \Pi_1) = 1,$$

$$L(s, \Pi_2) = \prod_{j=1}^{r_2} \Gamma_{\mathbb{R}}(s + 2s_{r_1+j} + \varepsilon'_j),$$

$$L(s, \Pi_3) = \prod_{\substack{i \le r_1 \\ j \le r_2}} \Gamma_{\mathbb{C}}(s + s_i + s_{r_1+j} + \tfrac{k_j-1}{2}),$$

$$L(s, \Pi_4) = \prod_{1 \le i < k \le r_1} \Gamma_{\mathbb{R}}(s + s_i + s_k + \varepsilon_{ik}), \quad \text{and}$$

$$L(s, \Pi_5) =$$

$$\prod_{1 \le j < \ell \le r_2} \Gamma_{\mathbb{C}}(s + s_{r_1+j} + s_{r_1+\ell} + \tfrac{k_j+k_\ell-2}{2}) \Gamma_{\mathbb{C}}(s + s_{r_1+j} + s_{r_1+\ell} + \tfrac{|k_j-k_\ell|}{2}).$$

$$\text{(A.18)}$$

Consequently, $L_\infty(s, \pi, Ext^2) = L(s, Ext^2\Pi) = L(s, \Pi_2)L(s, \Pi_3)L(s, \Pi_4)L(s, \Pi_5)$ is the product of these factors.

The archimedean component χ_∞ of the character χ is sgn^η, where η is the parity parameter of χ defined by $\chi(-1) = (-1)^\eta$. The isobaric decomposition of

$$Ext^2\pi_\infty \otimes \chi_\infty = \boxplus_{j=1}^5 (\Pi_j \otimes \text{sgn}^\eta) \qquad \text{(A.19)}$$

may be computed using the tensoring rules above. These imply that Π_1, Π_3, and Π_5 are unchanged by tensoring with χ_∞, and that Π_2 and Π_4 change by adding η to their exponents of sgn. The result is that

$$L_\infty(s, \pi, Ext^2 \otimes \chi) = L(s, \Pi_1)L(s, \Pi_2 \otimes \text{sgn}^\eta)L(s, \Pi_3)L(s, \Pi_4 \otimes \text{sgn}^\eta)L(s, \Pi_5),$$

$$\text{(A.20)}$$

where

$$L(s, \Pi_2 \otimes \text{sgn}^\eta) = \prod_{j=1}^{r_2} \Gamma_{\mathbb{R}}(s + 2s_{r_1+j} + \varepsilon'_{j\eta}),$$

$$L(s, \Pi_4 \otimes \text{sgn}^\eta) = \prod_{1 \le i < k \le r_1} \Gamma_{\mathbb{R}}(s + s_i + s_k + \varepsilon_{ik\eta}),$$

$$\text{(A.21)}$$

and $\varepsilon'_{j\eta}$ and $\varepsilon_{ik\eta} \in \{0,1\}$ are congruent to $\varepsilon'_j + \eta \equiv k_j + \eta$ and $\varepsilon_{ik} + \eta \equiv \varepsilon_i + \varepsilon_k + \eta$ (mod 2), respectively.

References

[1] A. Borel and H. Jacquet, *Automorphic forms and automorphic representations*, Automorphic forms, representations and *L*-functions (Proc. Sympos. Pure Math., Oregon State Univ., Corvallis, Ore., 1977), Part 1, Proc. Sympos. Pure Math., XXXIII, Amer. Math. Soc., Providence, R.I., 1979, pp. 189–207. With a supplement "On the notion of an automorphic representation" by R. P. Langlands. MR546598 (81m:10055)

[2] Daniel Bump, *The Rankin-Selberg method: a survey*, Number theory, trace formulas and discrete groups (Oslo, 1987), 1989, pp. 49–109.

[3]   Daniel Bump and Solomon Friedberg, *The exterior square automorphic L-functions on* GL(n), Festschrift in honor of I. I. Piatetski-Shapiro on the occasion of his sixtieth birthday, Part II (Ramat Aviv, 1989), 1990, pp. 47–65.

[4]   Daniel Bump and David Ginzburg, *Symmetric square L-functions on* GL(r), Ann. of Math. (2) **136** (1992), no. 1, 137–205.

[5]   W. Casselman, *Jacquet modules for real reductive groups*, Proceedings of the International Congress of Mathematicians (Helsinki, 1978), 1980, pp. 557–563.

[6]   ――――, *Canonical extensions of Harish-Chandra modules to representations of G*, Canad. J. Math. **41** (1989), no. 3, 385–438.

[7]   William Casselman, Henryk Hecht, and Dragan Miličić, *Bruhat filtrations and Whittaker vectors for real groups*, The mathematical legacy of Harish-Chandra (Baltimore, MD, 1998), 2000, pp. 151–190.

[8]   David Ginzburg, *On standard L-functions for E_6 and E_7*, J. Reine Angew. Math. **465** (1995), 101–131, DOI 10.1515/crll.1995.465.101. MR1344132 (96m:11040)

[9]   Michael Green, Stephen D. Miller, Jorge Russo, and Pierre Vanhove, *Eisenstein series for higher rank groups and string theory amplitudes* (2010). http://arxiv.org/abs/1004.0163.

[10] H. Jacquet and J. A. Shalika, *On Euler products and the classification of automorphic representations. I*, Amer. J. Math. **103** (1981), no. 3, 499–558, DOI 10.2307/2374103. MR618323 (82m:10050a)

[11] Hervé Jacquet and Joseph Shalika, *Exterior square L-functions*, Automorphic forms, Shimura varieties, and L-functions, Vol. II (Ann Arbor, MI, 1988), 1990, pp. 143–226.

[12] Anthony W. Knapp, *Representation theory of semisimple groups*, Princeton Landmarks in Mathematics, Princeton University Press, Princeton, NJ, 2001. An overview based on examples; Reprint of the 1986 original.

[13] ――――, *Local Langlands correspondence: the Archimedean case*, Motives (Seattle, WA, 1991), Proc. Sympos. Pure Math., vol. 55, Amer. Math. Soc., Providence, RI, 1994, pp. 393–410. MR1265560 (95d:11066)

[14] Bertram Kostant, *On Whittaker vectors and representation theory*, Invent. Math. **48** (1978), no. 2, 101–184, DOI 10.1007/BF01390249. MR507800 (80b:22020)

[15] Robert P. Langlands, *Automorphic representations, Shimura varieties, and motives. Ein Märchen*, Automorphic forms, representations and L-functions (Proc. Sympos. Pure Math., Oregon State Univ., Corvallis, Ore., 1977), Part 2, 1979, pp. 205–246.

[16] ――――, *On the classification of irreducible representations of real algebraic groups*, Representation theory and harmonic analysis on semisimple Lie groups, 1989, pp. 101–170.

[17] Stephen D. Miller and Wilfried Schmid, *Distributions and analytic continuation of Dirichlet series*, J. Funct. Anal. **214** (2004), no. 1, 155–220.

[18] ――――, *Automorphic distributions, L-functions, and Voronoi summation for* GL(3), Ann. of Math. (2) **164** (2006), no. 2, 423–488. MR2247965

[19] ――――, *The Rankin-Selberg method for automorphic distributions*, Representation theory and automorphic forms, Progr. Math., vol. 255, Birkhäuser Boston, Boston, MA, 2008, pp. 111–150.

[20] ――――, *A general Voronoi summation formula for* GL(n, ℤ), Geometric analysis: Present and Future, Advanced Lectures in Math series, International Press. to appear.

[21] ――――, *On the rapid decay of cuspidal automorphic forms* (2010). preprint.

[22] ――――, *Pairings of automorphic distributions* (2010). preprint.

[23] ――――, *The archimedean theory of the Exterior Square L-functions over* ℚ (2010). preprint.

[24] Niels A. Obers and Boris Pioline, *Eisenstein series in string theory*, Classical Quantum Gravity **17** (2000), no. 5, 1215–1224, DOI 10.1088/0264-9381/17/5/330. Strings '99 (Potsdam). MR1764377 (2001g:81218)

[25] S. J. Patterson and I. I. Piatetski-Shapiro, *The symmetric-square L-function attached to a cuspidal automorphic representation of* GL₃, Math. Ann. **283** (1989), no. 4, 551–572, DOI 10.1007/BF01442854. MR990589 (90d:11070)

[26] I. I. Piatetski-Shapiro, *Multiplicity one theorems*, Automorphic forms, representations and L-functions (Proc. Sympos. Pure Math., Oregon State Univ., Corvallis, Ore., 1977), Part 1, 1979, pp. 209–212.

[27] J. A. Shalika, *The multiplicity one theorem for* GLₙ, Ann. of Math. (2) **100** (1974), 171–193.

[28] Marko Tadić, $\widehat{\mathrm{GL}}(n, \mathbb{C})$ *and* $\widehat{\mathrm{GL}}(n, \mathbb{R})$, Automorphic forms and L-functions II. Local aspects, Contemp. Math., vol. 489, Amer. Math. Soc., Providence, RI, 2009, pp. 285–313. MR2537046

[29] David A. Vogan Jr., *Gel'fand-Kirillov dimension for Harish-Chandra modules*, Invent. Math. **48** (1978), no. 1, 75–98. MR0506503 (58 #22205)

[30] _____, *The unitary dual of* GL(n) *over an Archimedean field*, Invent. Math. **83** (1986), no. 3, 449–505.

[31] Nolan R. Wallach, *Asymptotic expansions of generalized matrix entries of representations of real reductive groups*, Lie group representations, I (College Park, Md., 1982/1983), 1983, pp. 287–369.

DEPARTMENT OF MATHEMATICS, HILL CENTER – BUSCH CAMPUS, RUTGERS, THE STATE UNIVERSITY OF NEW JERSEY, 110 FRELINGHUYSEN RD, PISCATAWAY, NJ 08854-8019
E-mail address: miller@math.rutgers.edu

DEPARTMENT OF MATHEMATICS, HARVARD UNIVERSITY, CAMBRIDGE, MA 02138
E-mail address: schmid@math.harvard.edu

Contemporary Mathematics
Volume **557**, 2011

CATEGORIES OF INTEGRABLE $sl(\infty)$-, $o(\infty)$-, $sp(\infty)$-MODULES

IVAN PENKOV AND VERA SERGANOVA

SUMMARY. We investigate several categories of integrable $sl(\infty)$-, $o(\infty)$-, $sp(\infty)$-modules. In particular, we prove that the category of integrable $sl(\infty)$-, $o(\infty)$-, $sp(\infty)$-modules with finite-dimensional weight spaces is semisimple. The most interesting category we study is the category $\widetilde{\mathrm{Tens}}_\mathfrak{g}$ for $\mathfrak{g} = sl(\infty)$-, $o(\infty)$-, $sp(\infty)$. Its objects M are defined as integrable $\mathfrak{g}$-modules of finite Loewy length such that the algebraic dual M^* is also integrable and of finite Loewy length.

We prove that the simple objects of $\widetilde{\mathrm{Tens}}_\mathfrak{g}$ are precisely the simple tensor modules, i.e. the simple subquotients of the tensor algebra of the direct sum of the natural and conatural representations. We also study injectives in $\widetilde{\mathrm{Tens}}_\mathfrak{g}$ and compute the Ext^1's between simple modules.

Finally, we characterize a certain subcategory $\mathrm{Tens}_\mathfrak{g}$ of $\widetilde{\mathrm{Tens}}_\mathfrak{g}$ as the unique minimal abelian full subcategory of the category of integrable modules which contains a non-trivial module and is closed under tensor product and algebraic dualization.

1. INTRODUCTION

The category of finite-dimensional representations of a Lie algebra is endowed with a natural contravariant involution

$$(1) \qquad\qquad M \rightsquigarrow M^*,$$

where * indicates dual space. For categories of infinite-dimensional modules (1) is never an involution as $M \not\simeq M^{**}$. This is why one usually looks for a "restricted dual" or a "continuous dual" which might still yield a contravariant involution on a given category of infinite-dimensional modules. In this paper, we study two categories of infinite-dimensional modules of certain infinite-dimensional Lie algebras and show, in particular, that there exists an interesting category $\widetilde{\mathrm{Tens}}_\mathfrak{g}$ of infinite-dimensional representations on which the functor (1) of algebraic dualization is well-defined and preserves the property of a module to be of finite Loewy length.

More precisely, we study representations of locally finite Lie algebras, i.e. of direct limits of finite-dimensional Lie algebras. There are three well-known classical simple locally finite Lie algebras $sl(\infty), o(\infty), sp(\infty)$, each of them being defined by an obvious direct limit. None of these Lie algebras admits non-trivial finite-dimensional representations, and instead one studies integrable representations (the

Date: June 3, 2010.

2010 *Mathematics Subject Classification*. Primary 17B65, 17B10.

Key words and phrases. Locally finite Lie algebra, tensor representation, integrable module, Loewy length, semisimplicity, injective hull, injective resolution.

We thank Gregg Zuckerman for his supportive interest and constructive criticism of this project. We thank also A. Baranov for pointing out the connection of Proposition 4.3 to his work. Both authors acknowledge partial support through DFG Grants PE 980/2-1 and PE 980/3-1, and the second author acknowledges partial support through NSF grant 0901554.

definition see in section 2 below). However, the category of integrable $\mathfrak{g}$-modules for $\mathfrak{g} = sl(\infty)$-, $o(\infty)$-, $sp(\infty)$ is vast (and "wild" in the technical sense), so it is reasonable to look for interesting subcategories.

One subcategory we study is the category of integrable weight modules with finite-dimensional weight spaces, and this is obviously an analog of the category of finite-dimensional representations of a classical finite-dimensional Lie algebra. It is less obvious that for $\mathfrak{g} = sl(\infty)$ this category contains some rather interesting simple modules, which are not highest weight modules. The first main result of this paper is the proof of the semisimplicity of this category: an extension of Hermann Weyl's semisimplicity theorem to the classical Lie algebras $sl(\infty), o(\infty), sp(\infty)$.

The above category is clearly not the only reasonable generalization of the category of finite-dimensional representations, as for instance it does not contain the adjoint representation. Indeed, note that the adjoint representation has an infinite-dimensional weight space, the Cartan subalgebra itself. On the other hand, the adjoint representation is naturally a simple tensor module as defined in [PS]. More generally, we define the category $\widetilde{\text{Tens}}_{\mathfrak{g}}$ for $\mathfrak{g} \cong sl(\infty)$, $o(\infty)$, $sp(\infty)$ simply as the largest category of integrable $\mathfrak{g}$-modules which is closed under algebraic dualization and such that every object has finite Loewy length. This category is a (non-rigid) tensor category with respect to the usual tensor product.

The second main contribution of the present paper is the study of the category $\widetilde{\text{Tens}}_{\mathfrak{g}}$. In particular, we study injectives in $\widetilde{\text{Tens}}_{\mathfrak{g}}$ and compute the Ext^1's between simple modules. We also give an alternative characterization of $\widetilde{\text{Tens}}_{\mathfrak{g}}$ by proving that an integrable $\mathfrak{g}$-module is an object of $\widetilde{\text{Tens}}_{\mathfrak{g}}$ if and only if it admits only finitely many non-isomorphic simple subquotients each of which is a submodule of a suitable finite tensor product of natural and conatural modules.

Finally, we describe a certain subcategory $\text{Tens}_{\mathfrak{g}}$ of $\widetilde{\text{Tens}}_{\mathfrak{g}}$ as the unique minimal abelian full subcategory of the category of integrable modules which contains a non-trivial module and is closed under tensor product and algebraic dualization.

2. Basic definitions

The ground field is $\mathbb{C}$ and $\otimes$ stands for $\otimes_{\mathbb{C}}$. If $\mathcal{C}$ is a category, $C \in \mathcal{C}$ indicates that C is an object of $\mathcal{C}$. If P is a set, we denote by 2^P the power set of P. We recall that the cardinal numbers $\beth_n$ are defined inductively: $\beth_0 = \text{card}\,\mathbb{Z}$, $\beth_1 = \text{card}\,2^{\mathbb{Z}}$, $\beth_n = \text{card}\,2^{P_{n-1}}$, where P_{n-1} is a set of cardinality $\beth_{n-1}$.

In this paper $\mathfrak{g}$ stands for a *locally semisimple* (complex) Lie algebra. By definition, $\mathfrak{g} = \bigcup_{i \in \mathbb{Z}_{>0}} \mathfrak{g}_i$ where

$$(2) \qquad\qquad \mathfrak{g}_1 \subset \mathfrak{g}_2 \subset \mathfrak{g}_3 \subset \cdots$$

is a sequence of inclusions of semisimple finite-dimensional Lie algebras. We call the sequence (2) an *exhaustion* of $\mathfrak{g}$, and we will assume that it is fixed. A locally semisimple Lie algebra is *locally simple* if it admits an exhaustion (2) so that all $\mathfrak{g}_i$ are simple. It is clear that a locally simple Lie algebra is simple. If no restrictions on $\mathfrak{g}$ are clearly stated, in what follows $\mathfrak{g}$ is assumed to be an arbitrary locally semisimple Lie algebra.

A locally simple algebra $\mathfrak{g}$ is *diagonal* if an exhaustion (2) can be chosen so that all $\mathfrak{g}_i$ are classical simple Lie algebras and the natural representation V_i of $\mathfrak{g}_i$, when restricted to $\mathfrak{g}_{i-1}$, has the form $k_i V_{i-1} \oplus l_i V_{i-1}^* \oplus \mathbb{C}^{s_i}$ for some k_i, l_i and $s_i \in \mathbb{Z}_{\geq 0}$. Here V_{i-1} stands for the natural representation of $\mathfrak{g}_{i-1}$, $\mathbb{C}^{s_i}$ stands for the trivial module of dimension s_i, and $k_i V_{-1}$ (respectively, $l_i V_{-1}^*$) denotes the direct sum of k_i (respectively, l_i) copies of V_{-1} (respectively, V_{-1}^*).

The three classical simple Lie algebras $sl(\infty), o(\infty)$ and $sp(\infty)$ (defined respectively as $sl(\infty) = \cup_i sl(i)$, $o(\infty) = \cup_i o(i)$, $sp(\infty) := \cup_i sp(2i)$ via the natural inclusions $sl(i) \subset sl(i+1)$) etc.) are clearly diagonal. Moreover, $sl(\infty), o(\infty), sp(\infty)$ are (up to isomorphism) the only finitary locally simple Lie algebras $\mathfrak{g}$; *finitary* means by definition that $\mathfrak{g}$ admits a faithful countable-dimensional $\mathfrak{g}$-module with a basis in which each element $g \in \mathfrak{g}$ acts through a finite matrix, [Ba1], [Ba3]. More generally, there exists also a classification of locally simple diagonal Lie algebras up to isomorphism, [BZh]. We do not use this classification in the present paper and present only the simplest example of a diagonal Lie algebra not isomorphic to $sl(\infty), o(\infty)$ or $sp(\infty)$. This is the Lie algebra $sl(2^\infty)$ defined as the direct limit $\varinjlim sl(2^i)$ under the inclusions

$$sl(2^i) \to sl(2^{i+1}), A \to \begin{pmatrix} A & 0 \\ 0 & A \end{pmatrix}.$$

A $\mathfrak{g}$-module M is *integrable* if $\dim \operatorname{span}\{m, g \cdot m, g^2 \cdot m, \cdots\} < \infty$ for any $m \in M$ and $g \in \mathfrak{g}$. Since $\mathfrak{g}$ is locally semisimple, this is equivalent to the condition that, when restricted to any semisimple finite-dimensional subalgebra $\mathfrak{f}$ of $\mathfrak{g}$, M is isomorphic to a (not necessarily countable) direct sum of finite-dimensional $\mathfrak{f}$-modules. We denote by $\operatorname{Int}_\mathfrak{g}$ the category of integrable $\mathfrak{g}$-modules; $\operatorname{Int}_\mathfrak{g}$ is a full subcategory of the category of $\mathfrak{g}$-modules $\mathfrak{g}$-mod.

Any countable-dimensional $\mathfrak{g}$-module $M \in \operatorname{Int}_\mathfrak{g}$ can be exhausted by finite dimensional $\mathfrak{g}_i$-modules M_i, i. e. there exists a chain of finite-dimensional $\mathfrak{g}_i$-submodules $M_1 \subset M_2 \subset \ldots$ such that $M = \varinjlim M_i$. We call M *locally simple* if all M_i can be chosen to be simple modules. It is clear that a locally simple module is simple. Note also that if M is locally simple then any two exhaustions $\{M_i\}$ and $\{M_i'\}$ coincide from some point on: that follows from the fact that $M_i \cap M_i' \neq 0$ for some i and hence $M_j = M_j' = M_j \cap M_j'$ for any $j \geq i$. We say that a locally simple $\mathfrak{g}$-module $M = \varinjlim M_i$ is a *highest weight module* if there is a chain of nested Borel subalgebras $\mathfrak{b}_i$ of $\mathfrak{g}_i$ such that the $\mathfrak{b}_i$-highest weight space of M_i is mapped into the $\mathfrak{b}_{i+1}$-highest weight space of M_{i+1} under the inclusion $M_i \subset M_{i+1}$. The direct limit of highest weight spaces is then the $\mathfrak{b}$-*highest weight space of M*, where $\mathfrak{b} = \varinjlim \mathfrak{b}_i$.

By

$$\Gamma_\mathfrak{g} : \mathfrak{g} - \operatorname{mod} \rightsquigarrow \operatorname{Int}_\mathfrak{g},$$

$$M \mapsto \Gamma_\mathfrak{g}(M) := \{m \in M, \dim \operatorname{span}\{m, g \cdot m, g \cdot m^2, \cdots\} < \infty \quad \forall g \in \mathfrak{g}\}$$

we denote the *functor of $\mathfrak{g}$-integrable vectors*. It is an exercise to check that $\Gamma_\mathfrak{g}(M)$ is indeed a well-defined $\mathfrak{g}$-submodule of M; the fact that $\Gamma_\mathfrak{g}(M)$ is integrable is obvious. Furthermore, $\Gamma_\mathfrak{g}$ is a left-exact functor.

If $\mathfrak{g}$ is a diagonal (locally simple) Lie algebra, then one can define a *natural module* V of $\mathfrak{g}$. Indeed, the reader will verify that one can choose a subexhaustion of (2) such that the natural $\mathfrak{g}_i$-module V_i is a $\mathfrak{g}_i$-submodule of V_{i+1} for any i. Therefore, fixing arbitrary injective homomorphisms $V_i \to V_{i+1}$ of $\mathfrak{g}_i$-modules, we obtain a direct system and we set $V := \varinjlim V_i$. Note that V depends on the choice of the homomorphisms $V_i \to V_{i+1}$. In the special case when $\mathfrak{g} \cong sl(\infty)$, $o(\infty)$, $sp(\infty)$, the homomorphisms $V_i \to V_{i+1}$ are unique up to proportionality, and one can prove that as a result V is unique up to isomorphism, i.e. in particular does not depend on the fixed exhaustion of $\mathfrak{g}$. In these latter cases we speak about *the natural representation*.

By choosing injective homomorphisms of $\mathfrak{g}_i$-modules $V_i^* \to V_{i+1}^*$, we obtain a direct system defining a *conatural representation* of $\mathfrak{g}$. We denote such a representation by V_*. For $\mathfrak{g} \cong sl(\infty)$, $o(\infty)$, $sp(\infty)$ V_* is unique up to isomorphism. In fact, $V \simeq V_*$ for $\mathfrak{g} \cong o(\infty)$, $sp(\infty)$.

3. Injective modules in $\mathrm{Int}_\mathfrak{g}$ and semisimplicity of the category $\mathrm{Int}_{\mathfrak{g},\mathfrak{h}}^{\mathrm{fin}}$

Proposition 3.1. $\mathrm{Ext}_\mathfrak{g}^1(X, M^*) = 0$ *for any* $X, M \in \mathrm{Int}_\mathfrak{g}$.

Proof. We use that
$$\mathrm{Ext}_\mathfrak{g}^1(X, M^*) = \mathrm{Ext}_\mathfrak{g}^1(\mathbb{C}, \mathrm{Hom}_\mathbb{C}(X, M^*)) \simeq H^1(\mathfrak{g}, \mathrm{Hom}_\mathbb{C}(X, M^*)) = H^1(\mathfrak{g}, (X \otimes M)^*),$$
see for instance [W]. Therefore, it suffices to show that $H^1(\mathfrak{g}, R^*) = 0$ for any integrable $\mathfrak{g}$-module R.

Consider the standard complex for the cohomology of $\mathfrak{g}$ with coefficients in R^*:

$$(3) \qquad\qquad 0 \to R^* \to (\mathfrak{g} \otimes R)^* \to (\Lambda^2(\mathfrak{g}) \otimes R)^* \to \ldots$$

It is dual to the standard homology complex

$$0 \leftarrow R \leftarrow \mathfrak{g} \otimes R \leftarrow \Lambda^2(\mathfrak{g}) \otimes R \leftarrow \ldots,$$

which is the direct limit of complexes

$$0 \leftarrow R \leftarrow \mathfrak{g}_i \otimes R \leftarrow \Lambda^2(\mathfrak{g}_i) \otimes R \leftarrow \ldots.$$

Since $H_1(\mathfrak{g}_i, R) = 0$ for each i, we get $H_1(\mathfrak{g}, R) = 0$. Therefore, the dual complex (3) has trivial first cohomology, i.e. $H^1(\mathfrak{g}, R^*) = 0$. $\qquad\square$

Proposition 3.2. *For any* $M \in \mathrm{Int}_\mathfrak{g}$, $\Gamma_\mathfrak{g}(M^*)$ *is an injective object of* $\mathrm{Int}_\mathfrak{g}$.

Proof. Let $X \in \mathrm{Int}_\mathfrak{g}$. The exact sequence of $\mathfrak{g}$-modules
$$0 \to \Gamma_\mathfrak{g}(M^*) \to M^* \to M^*/\Gamma_\mathfrak{g}(M^*) \to 0$$
induces an exact sequence of vector spaces

$$0 \to \mathrm{Hom}_\mathfrak{g}(X, \Gamma_\mathfrak{g}(M^*)) \xrightarrow{\varphi} \mathrm{Hom}_\mathfrak{g}(X, M^*) \to \mathrm{Hom}_\mathfrak{g}(X, M^*/\Gamma_\mathfrak{g}(M^*)) \to$$

$$\to \mathrm{Ext}_\mathfrak{g}^1(X, \Gamma_\mathfrak{g}(M^*)) \xrightarrow{\psi} \mathrm{Ext}_\mathfrak{g}^1(X, M^*) = 0.$$

Since $\mathrm{Hom}_\mathfrak{g}(X, M^*/\Gamma_\mathfrak{g}(M^*)) = 0$ (this follows from the facts that a quotient of an integrable $\mathfrak{g}$-module is again an integrable $\mathfrak{g}$-module and that $\mathrm{Int}_\mathfrak{g}$ is closed with respect to extensions) we conclude that ψ is an isomorphism, i.e. that
$$\mathrm{Ext}_\mathfrak{g}^1(X, \Gamma_\mathfrak{g}(M^*)) = 0.$$

$\qquad\square$

Corollary 3.3. $\mathrm{Int}_\mathfrak{g}$ *has enough injectives.*

Proof. Let $M \in \mathrm{Int}_\mathfrak{g}$. Then $M \subset M^{**}$ and it is easy to check (using the local semisimplicity of $\mathfrak{g}$) that the projection $M^{**} \to \Gamma_\mathfrak{g}(M^*)^*$ induces an injection $M \to \Gamma_\mathfrak{g}(M^*)^*$. This in turn induces an injection $M \to \Gamma_\mathfrak{g}(\Gamma_\mathfrak{g}(M^*)^*)$, and $\Gamma_\mathfrak{g}(\Gamma_\mathfrak{g}(M^*)^*)$ is an injective object of $\mathrm{Int}_\mathfrak{g}$ by Proposition 3.2. $\qquad\square$

Note that there is a simpler proof of Corollary 3.3 not referring to Proposition 3.2. Indeed, it is enough to notice that the functor $\Gamma_{\mathfrak{g}} : \mathfrak{g}\text{-mod}\rightsquigarrow \text{Int}_{\mathfrak{g}}$ is right adjoint to the inclusion functor $\text{Int}_{\mathfrak{g}} \subset \mathfrak{g}\text{-mod}$. Then the equality

$$\text{Hom}_{\mathfrak{g}}(M, J_M) = \text{Hom}_{\mathfrak{g}}(M, \Gamma_{\mathfrak{g}}(J_M))$$

allows us to conclude that, if $i : M \to J_M$ is an injective homomorphism of $M \in \text{Int}_{\mathfrak{g}}$ into an injective $\mathfrak{g}$-module, then $\Gamma_{\mathfrak{g}}(J_M)$ is an injective object of $\text{Int}_{\mathfrak{g}}$ and i factors through the inclusion $\Gamma_{\mathfrak{g}}(J_M) \subset J_M$. In particular, this argument allows to reduce the existence of injective hulls in $\text{Int}_{\mathfrak{g}}$ to the well-known existence of injective hulls in $\mathfrak{g}$-mod.

With this in mind, we can view Propositions 3.1 and 3.2 as yielding an explicit construction of an injective module $\Gamma_{\mathfrak{g}}(M^*)$ associated to any $M \in \text{Int}_{\mathfrak{g}}$.

In the rest of this section we assume that $\mathfrak{g}$ admits a splitting Cartan subalgebra $\mathfrak{h} \subset \mathfrak{g}$, i.e. an abelian subalgebra $\mathfrak{h} \subset \mathfrak{g}$ such that $\mathfrak{g}$ decomposes as

$$\mathfrak{h} \oplus \bigoplus_{0 \neq \alpha \in \mathfrak{h}^*} \mathfrak{g}^{\alpha},$$

where

$$\mathfrak{g}^{\alpha} = \{g \in \mathfrak{g} | [h, g] = \alpha(h)g \text{ for any } h \in \mathfrak{h}\}.$$

It is well-known that in this case $\mathfrak{g}$ is isomorphic to a direct sum of copies of $sl(\infty), o(\infty), sp(\infty)$ and finite-dimensional simple Lie algebras, see [PStr].

We define the category $\text{Int}_{\mathfrak{g},\mathfrak{h}}^{\text{wt}}$ as the full subcategory of $\text{Int}_{\mathfrak{g}}$ which consists of *weight modules* M, i.e. objects $M \in \text{Int}_{\mathfrak{g}}$ which admit a decomposition

$$(4) \qquad M = \bigoplus_{\alpha \in \mathfrak{h}^*} M^{\alpha},$$

where

$$M^{\alpha} = \{m \in M | h \cdot m = \alpha(h)m \text{ for any } h \in \mathfrak{h}\}.$$

Note that (4) is automatically a decomposition of $\mathfrak{h}$-modules. It is also clear that there is a left exact functor

$$\Gamma_{\mathfrak{h}}^{\text{wt}} : \text{Int}_{\mathfrak{g}} \rightsquigarrow \text{Int}_{\mathfrak{g},\mathfrak{h}}^{\text{wt}}, \quad M \mapsto \bigoplus_{\alpha \in \mathfrak{h}^*} M^{\alpha}.$$

By $\Gamma_{\mathfrak{g},\mathfrak{h}}^{\text{wt}}$ we denote the composition

$$\Gamma_{\mathfrak{h}}^{\text{wt}} \circ \Gamma_{\mathfrak{g}} : \mathfrak{g}\text{-mod} \rightsquigarrow \text{Int}_{\mathfrak{g},\mathfrak{h}}^{\text{wt}}.$$

Lemma 3.4. *If X is an injective object of $\text{Int}_{\mathfrak{g}}$, then $\Gamma_{\mathfrak{h}}^{\text{wt}}(X)$ is an injective object of $\text{Int}_{\mathfrak{g},\mathfrak{h}}^{\text{wt}}$.*

Proof. It suffices to note that $\Gamma_{\mathfrak{h}}^{\text{wt}}$ is right adjoint to the inclusion functor $\text{Int}_{\mathfrak{g},\mathfrak{h}}^{\text{wt}} \subset \text{Int}_{\mathfrak{g}}$. $\square$

Example 3.5. Let $\mathfrak{g} = sl(\infty)$ and $M = V \otimes V_*$. Consider the $\mathfrak{g}$-module M^*. Let's think of $M^* = (V \otimes V_*)^*$ as the space of all infinite matrices $B = (b_{ij}), i, j \in \mathbb{Z}_{>0}$, and of M as the space of finitary infinite matrices $A = (a_{ij}), i, j \in \mathbb{Z}_{>0}$, where $B(A) = \sum_{i,j} b_{ij} a_{ji}$. Then $\mathfrak{g}$ is identified with the subspace $F \subset (V \otimes V_*)^*$ of finitary matrices with trace zero, and the $\mathfrak{g}$-module structure on M^* is given by $A \cdot B = [A, B]$. We fix the Cartan subalgebra $\mathfrak{h}$ to be the algebra of finitary diagonal matrices, and we claim that $\Gamma_{\mathfrak{h}}^{\text{wt}}(M^*) = F + D$ where D is the subspace of diagonal matrices. Indeed, clearly D equals the $\mathfrak{h}$-weight space $(M^*)^0$ of weight 0.

Furthermore, any weight space of non-zero weight is the span of an elementary non-diagonal matrix, hence $\Gamma_{\mathfrak{h}}^{\mathrm{wt}}(M^*) = F + D$. Note also that we have a non-splitting exact sequence of $\mathfrak{g}$-modules

$$0 \to \mathfrak{g} \to \Gamma_{\mathfrak{h}}^{\mathrm{wt}}(M^*) \to T \to 0,$$

where $T = D/D \cap F$ is a trivial $\mathfrak{g}$-module of dimension $\beth_1$.

Corollary 3.6. *For any $M \in \mathrm{Int}_{\mathfrak{g}}$, $\Gamma_{\mathfrak{g},\mathfrak{h}}^{\mathrm{wt}}(M^*)$ is an injective object of $\mathrm{Int}_{\mathfrak{g},\mathfrak{h}}^{\mathrm{wt}}$.*

Define now $\mathrm{Int}_{\mathfrak{g},\mathfrak{h}}^{\mathrm{fin}}$ as the full subcategory of $\mathrm{Int}_{\mathfrak{g},\mathfrak{h}}^{\mathrm{wt}}$ consisting of $\mathfrak{h}$-weight modules $M = \bigoplus_{\alpha \in \mathfrak{h}^*} M^\alpha$ such that $\dim M^\alpha < \infty$ for any $\alpha \in \mathfrak{h}^*$.

Theorem 3.7. *The category $\mathrm{Int}_{\mathfrak{g},\mathfrak{h}}^{\mathrm{fin}}$ is semisimple.*

Proof. Let $M \in \mathrm{Int}_{\mathfrak{g},\mathfrak{h}}^{\mathrm{fin}}$ be simple. There is an $\mathfrak{h}$-module isomorphism

$$M = \oplus_{\alpha \in \mathfrak{h}^*} M^\alpha.$$

Therefore, $M^* = \prod_{\alpha \in \mathfrak{h}^*} (M^\alpha)^*$. A non-difficult computation shows that $\Gamma_{\mathfrak{h}}^{\mathrm{wt}}(M^*)$ is isomorphic to $\oplus_{\alpha \in \mathfrak{h}^*}(M_\alpha)^*$. Moreover, using the fact that $\dim M^\alpha < \infty$ for all α, it is easy to check that $M_* := \oplus_{\alpha \in \mathfrak{h}^*}(M_\alpha)^*$ is a simple integrable $\mathfrak{g}$-module. Hence $M_* = \Gamma_{\mathfrak{g},\mathfrak{h}}^{\mathrm{wt}}(M^*)$. Applying $\Gamma_{\mathfrak{g},\mathfrak{h}}^{\mathrm{wt}}$ again, we see that

$$\Gamma_{\mathfrak{g},\mathfrak{h}}^{\mathrm{wt}}(\Gamma_{\mathfrak{g},\mathfrak{h}}^{\mathrm{wt}}(M^*)^*) = M.$$

Therefore, M is injective in $\mathrm{Int}_{\mathfrak{g},\mathfrak{h}}^{\mathrm{wt}}$, and thus also in $\mathrm{Int}_{\mathfrak{g},\mathfrak{h}}^{\mathrm{fin}}$, by Corollary 3.6. $\square$

Example 3.8.

a) Let $\mathfrak{g} = sl(\infty)$. One checks immediately that all tensor powers $V^{\otimes k}$, V being the natural module, are objects of $\mathrm{Int}_{\mathfrak{g},\mathfrak{h}}^{\mathrm{fin}}$. The same applies to the tensor powers of the conatural module V_*. However, the category $\mathrm{Int}_{\mathfrak{g},\mathfrak{h}}^{\mathrm{fin}}$ contains also more interesting modules as the following one: $M = \varinjlim S^i(V_i)$, V_i being the natural representation of $sl(i)$. The module M has 1-dimensional weight spaces, but is not a highest weight module, see [DP1, Example 3]. Note also that the adjoint representation is not an object of $\mathrm{Int}_{\mathfrak{g},\mathfrak{h}}^{\mathrm{fin}}$.

b) Let $\mathfrak{g} = o(\infty)$ and let $\mathfrak{g}$ be exhausted by $\mathfrak{g}_i = o(2i)$, $i \geq 3$. Denote by S_i^1 and S_i^2 the two non-isomorphic spinor $\mathfrak{g}_i$-modules. Then S_i^1 and S_i^2 are both isomorphic to $S_{i-1}^1 \oplus S_{i-1}^2$ as $\mathfrak{g}_{i-1}$-modules. Therefore, there is an injective homomorphism of $\mathfrak{g}_{i-1}$-modules $\varphi_{i-1}^{ks} : S_{i-1}^k \to S_i^s$ for $k, s \in \{1, 2\}$, and moreover φ_{i-1}^{ks} is unique up to proportionality. Any sequence $\{t_i\}_{i \geq 3}$ of elements in $\{1, 2\}$ defines a direct system

$$S_3^{t_3} \xrightarrow{\varphi_3^{t_3,t_4}} S_4^{t_4} \xrightarrow{\varphi_4^{t_4,t_5}} S_5^{t_5} \xrightarrow{\varphi_5^{t_5,t_6}} \dots,$$

and hence a simple $\mathfrak{g}$-module $S(\{t_i\})$. Using the fact that $S(\{t_i\})$ is locally simple, it is easy to see that $S(\{t_i\}) = S(\{t_i'\})$ if and only if the "tails" of the sequence $\{t_i\}$ and $\{t_i'\}$ coincide, i.e. $t_i = t_i'$ for large enough i.

The modules $S(\{t_i\})$ are weight modules with 1-dimensional spaces for any Cartan subalgebra $\mathfrak{h}$ of the form $\mathfrak{h} = \cup_i \mathfrak{h}_i$ where $\mathfrak{h}_3 \subset \mathfrak{h}_4 \subset \dots$ are nested Cartan subalgebras of $\mathfrak{g}_3 = o(6) \subset \mathfrak{g}_4 = o(8) \subset \dots$. In particular, $S(\{t_i\}) \in \mathrm{Int}_{\mathfrak{g},\mathfrak{h}}^{\mathrm{fin}}$.

4. ON THE INTEGRABILITY OF M^* FOR $M \in \mathrm{Int}_{\mathfrak{g}}$

Lemma 4.1. *Let* $M \in \mathrm{Int}_{\mathfrak{g}}$. *Then* $M^* \in \mathrm{Int}_{\mathfrak{g}}$ *if and only if for any* $i > 0$ $\mathrm{Hom}_{\mathfrak{g}_i}(N, M) \neq 0$ *only for finitely many non-isomorphic simple* $\mathfrak{g}_i$-*modules* N.

Proof. Fix i. Let Λ_i be the set of integral dominant weights of $\mathfrak{g}_i$ (for some fixed Borel subalgebra $\mathfrak{b}_i$ of $\mathfrak{g}_i$ with fixed Cartan subalgebra $\mathfrak{h}_i \subset \mathfrak{b}_i$) and V_λ^i be the simple $\mathfrak{g}_i$-module with highest weight λ. Denote by $\Lambda_i(M)$ the set of all $\lambda \in \Lambda_i$ such that $\mathrm{Hom}_{\mathfrak{g}_i}(V_\lambda^i, M) \neq 0$. Since M is a semisimple $\mathfrak{g}_i$-module, we can write M as

$$M = \oplus_{\lambda \in \Lambda_i(M)} M^\lambda \otimes V_\lambda^i,$$

where $M^\lambda := \mathrm{Hom}_{\mathfrak{g}_i}(V_\lambda^i, M)$ is a trivial $\mathfrak{g}_i$-module. We have

$$M^* = \prod_{\lambda \in \Lambda_i(M)} (V_\lambda^i)^* \otimes (M^\lambda)^*.$$

Suppose that $\Lambda_i(M)$ is finite. Then for any fixed $g \in \mathfrak{g}_i$ there is a polynomial $p_\lambda(z)$ such that $p_\lambda(g) \cdot (V_\lambda^i)^* = 0$. Set $p(z) := \prod_{\lambda \in \Lambda_i(M)} p_\lambda(z)$. Then $p(g) \cdot M^* = 0$. Hence g acts integrably on M^*, i.e. M^* is integrable over $\mathfrak{g}_i$.

Now let $\Lambda_i(M)$ be infinite. Let v_λ be a non-zero vector of weight $-\lambda$ in $(V_\lambda^i)^* \otimes (M^\lambda)^*$. One can choose h in the Cartan subalgebra of $\mathfrak{g}_i$ such that $\lambda(h) \neq \mu(h)$ for any $\mu \neq \lambda \in \Lambda_i(M)$. Let $v := \prod_{\lambda \in \Lambda_i(M)}(v_\lambda) \in \prod_{\lambda \in \Lambda_i(M)}(V_\lambda^i)^* \otimes (M^\lambda)^*$. Then $\dim(\mathbb{C}[h] \cdot v) = \infty$, and M^* is not $\mathfrak{g}_i$-integrable. $\square$

Corollary 4.2. *Let* $M, M' \in \mathrm{Int}_{\mathfrak{g}}$. *If* $M^*, (M')^* \in \mathrm{Int}_{\mathfrak{g}}$, *then* $(M \otimes M')^* \in \mathrm{Int}_{\mathfrak{g}}$ *and* $M^{**} \in \mathrm{Int}_{\mathfrak{g}}$.

Proposition 4.3. *Let* $\mathfrak{g}$ *be a locally simple Lie algebra. There exists a non-trivial module* $M \in \mathrm{Int}_{\mathfrak{g}}$ *such that* M^* *is integrable if and only if* $\mathfrak{g}$ *is diagonal.*

Proof. First of all, if $\mathfrak{g}$ is diagonal, then any natural module $V = \varinjlim V_n$ satisfies the finiteness condition of Lemma 4.1, hence V^* is integrable.

Before we prove the other direction, note that, by passing to a subexhaustion, we can always assume that $\mathfrak{g}$ is exhausted by classical simple Lie algebras $\mathfrak{g}_i$ of the same type (A, B, C or D). Let now $M \in \mathrm{Int}_{\mathfrak{g}}$ be non-trivial and M^* be integrable. We will show that $\mathfrak{g}$ is diagonal. Since M satisfies the finiteness condition of Lemma 4.1, $\mathrm{End}_{\mathbb{C}} M$ and its submodules satisfy this condition too. The adjoint module $\mathfrak{g}$ is a submodule of $\mathrm{End}_{\mathbb{C}} M$, hence this implies that for each i the number of $\mathfrak{g}_i$-isotypic components in $\mathfrak{g}_{i+k}$ is uniformly bounded for all $k > 0$. Since the adjoint module of $\mathfrak{g}_i$ is isomorphic to $(V_i \otimes V_i^*)/\mathbb{C}$ in the type A case, to $S^2(V_i)$ in type C, and to $\Lambda^2(V_i)$ in types B or D, one can easily check that for each i the number of $\mathfrak{g}_i$-isotypic components in V_{i+k} is also uniformly bounded by for all $k > 0$. Our goal is to show that for all sufficiently large i, V_{i+1} restricted to $\mathfrak{g}_i$ is isomorphic to a direct sum of copies of V_i, V_i^* and $\mathbb{C}$.

Let us start with the type A case. Pick an $sl(2)$-subalgebra in $\mathfrak{g}_n$ for some n. The set of $sl(2)$-weights in V is finite. Thus we can let $k \in \mathbb{Z}_{>0}$ be the maximal weight in this set and fix i such that k is a weight of V_i. Then $sl(2) \subset \mathfrak{g}_i$. Furthermore, we have an isomorphism of $\mathfrak{g}_i$-modules

$$V_{i+1} = T_{\lambda_1}(V_i) \oplus \cdots \oplus T_{\lambda_s}(V_i),$$

where each λ_j is a Young diagram and $T_{\lambda_j}(V_i)$ is the image of the corresponding Young projector in the appropriate tensor power of V_i. Since V_{i+1} does not have any weight greater than k, each diagram λ_j has only one column. Indeed, otherwise we can put a vector of weight k in each box of the first row and put other weight vectors in all other boxes of λ_j so that the total sum of all weights of vectors is greater than k, which contradicts the fact that k is the maximal weight. Next we claim that the length of this column equals $0, 1$, $\dim V_i$, or $\dim V_i - 1$. Indeed, if we put in the boxes of λ_i linearly independent vectors of maximal possible sum of weights, the total sum is not greater than k only in these four cases. Hence each simple $\mathfrak{g}_i$-constituent of V_{i+1} is isomorphic to V_i, V_i^* or $\mathbb{C}$ (the numbers 0 and $\dim V_i$ correspond both to the trivial 1-dimensional $\mathfrak{g}_i$-module).

If each $\mathfrak{g}_i$ is of type B or C, D, let $\mathfrak{s}_i \subset \mathfrak{g}_i$ be a maximal root subalgebra of type A. Notice that by the previous argument the restriction of V_{i+1} on $\mathfrak{s}_i$ is a sum of natural, conatural and trivial modules. That is only possible if the restriction of V_{i+1} to $\mathfrak{g}_i$ is a sum of natural and trivial modules. $\square$

Proposition 4.3 follows also from Corollary 3.9 in [Ba2].

Example 4.4.

a) Let $\mathfrak{g} = sl(\infty)$, and let $M = \varinjlim S^i(V_i)$ be as in Example 3.8, a). Then $\mathrm{Hom}_{\mathfrak{g}_i}(S^k(V_i), S^j(V_j)) \neq 0$ for all i, $k \leq j$. Hence $\mathrm{Hom}_{\mathfrak{g}_i}(S^k(V_i), M) \neq 0$ for all $k > 0$, and by Lemma 4.1 M^* is not an object of $\mathrm{Int}_{\mathfrak{g}}$.

b) Consider the case $\mathfrak{g} = o(\infty)$ and let $S(\{t_i\})$ be the $\mathfrak{g}$-module defined in Example 3.8, b). Then if N is a simple $\mathfrak{g}_i$-module, $\mathrm{Hom}_{\mathfrak{g}_i}(N, S(\{t_i\})) \neq 0$ iff $N \simeq S_i^1$ or $N \simeq S_i^2$. Hence $S(\{t_i\})^* \in \mathrm{Int}_{\mathfrak{g}}$ by Lemma 4.1. Moreover, $S(\{t_i\})^*$ is injective in $\mathrm{Int}_{\mathfrak{g}}$ by Proposition 3.2.

c) Let $\mathfrak{g} = sl(\infty)$ and let M be as in Example 3.5. Then $\mathrm{Hom}_{\mathfrak{g}_i}(N, M) \neq 0$ if N is isomorphic to one of the following simple $\mathfrak{g}_i$-modules: trivial, natural, conatural, adjoint. Therefore, M^* is $\mathfrak{g}$-integrable and injective in $\mathrm{Int}_{\mathfrak{g}}$. Furthermore, $M^* \cong \mathbb{C} \oplus \mathfrak{g}^*$.

5. On the Loewy length of $\Gamma_{\mathfrak{g}}(M^*)$ for $M \in \mathrm{Int}_{\mathfrak{g}}$

Recall that the *socle*, $\mathrm{soc}(M)$, of a $\mathfrak{g}$-module M is the largest semisimple submodule of M. The *socle filtration* of M is the filtration of $\mathfrak{g}$-modules

$$0 \subset \mathrm{soc}(M) \subset \mathrm{soc}^1(M) \subset \cdots \subset \mathrm{soc}^i(M) \subset \dots,$$

where $\mathrm{soc}^i(M) = p_i^{-1}(\mathrm{soc}(M/\mathrm{soc}^{i-1}(M)))$ and $p_i : M \to M/\mathrm{soc}^{i-1}(M)$ is the natural projection. We say that the socle filtration of M is *exhaustive* if $M = \varinjlim(\mathrm{soc}^i(M))$. We say that M has *finite Loewy length* if the socle filtration of M is finite and exhaustive. The *Loewy length* of M equals $k + 1$ where $k = \min\{r \mid \mathrm{soc}^r(M) = M\}$.

Proposition 5.1. *Let $M \in \mathrm{Int}_{\mathfrak{g}}$ be a simple $\mathfrak{g}$-module such that $\Gamma_{\mathfrak{g}}(M^*)$ has finite Loewy length. Then there exist $n \in \mathbb{Z}_{>0}$ and a direct system M_i of simple finite-dimensional $\mathfrak{g}_i$-modules such that $M = \varinjlim M_i$ and $\dim \mathrm{Hom}_{\mathfrak{g}_i}(M_i, M_j) = 1$ for all $j > i > n$.*

We first prove several lemmas.

Lemma 5.2. *Let $Q = \varinjlim Q_i \in \mathrm{Int}_{\mathfrak{g}}$, where Q_i are finite-dimensional, not necessarily simple, $\mathfrak{g}_i$-modules. Assume that for all sufficiently large i there are simple*

$\mathfrak{g}_i$-submodules $X_i \subset Q_i$ such that $\dim \operatorname{Hom}_{\mathfrak{g}_i}(X_i, X_{i+1}) > 2$. Then there exists a locally simple module $X = \varinjlim X_i \in \operatorname{Int}_{\mathfrak{g}}$ and a non-trivial extension of $\mathfrak{g}$-modules

$$0 \to Q \to Z \to X \to 0.$$

Proof. Fix a sequence of injective homomorphisms of $\mathfrak{g}_i$-modules $f_i : X_i \to X_{i+1}$ and set $X = \varinjlim X_i$. Let $Z_i := X_i \oplus Q_i$ and consider the injective homomorphisms of $\mathfrak{g}_i$-modules

$$a_i : Z_i \to Z_{i+1}, \ a_i((x,q)) := (f_i(x), t_i(x) + e_i(q)),$$

where t_i are some injective homomorphisms $X_i \to Q_{i+1}$, $e_i : Q_i \to Q_{i+1}$ are the given inclusions, and $q \in Q_i$, $x \in X_i$. Put $Z := \varinjlim Z_i$.

Then, clearly, Q is a submodule of Z and the quotient Z/Q is isomorphic to X. Thus we have constructed an extension of X by Q. This extension splits if and only if for all sufficiently large i there exist non-zero homomorphisms $p_i : X_i \to Q_i$ such that $t_i = p_{i+1} \circ f_i - e_i \circ p_i$, see the following diagram:

$$
\begin{array}{ccc}
X_{i+1} & \overset{p_{i+1}}{\to} & Q_{i+1} \\
\uparrow f_i & t_i \nearrow & \uparrow e_i \\
X_i & \overset{p_i}{\to} & Q_i.
\end{array}
$$

Assume that for any choice of $\{t_i\}$ such a splitting exists. If $n_i := \dim \operatorname{Hom}_{\mathfrak{g}_i}(X_i, Q_i)$, this assumption implies

$$\dim \operatorname{Hom}_{\mathfrak{g}_i}(X_i, Q_{i+1}) \le n_i + n_{i+1}.$$

On the other hand, $\dim \operatorname{Hom}_{\mathfrak{g}_i}(X_i, Q_{i+1}) \ge k_i n_{i+1}$ where $k_i := \dim \operatorname{Hom}_{\mathfrak{g}_i}(X_i, X_{i+1})$. Since $k_i > 2$, we have $n_{i+1} < n_i$. As $n_i > 0$ for all i, we obtain a contradiction. $\square$

Corollary 5.3. *Let $Q \in \operatorname{Int}_{\mathfrak{g}}$ be a simple $\mathfrak{g}$-module satisfying the assumption of Lemma 5.2. Then Q admits no non-zero homomorphism into an injective object of $\operatorname{Int}_{\mathfrak{g}}$ of finite Loewy length.*

Proof. For any $m > 0$ we will now construct an integrable module $Z^{(m)} \supset Q$ whose socle equals Q and whose Loewy length is greater than m. For $m = 1$ this was done in Lemma 5.2. Proceeding by induction, we set

$$Z_i^{(m)} := X_i \oplus Z_i^{(m-1)} = X_i \oplus (X_i \oplus Z_i^{(m-2)})$$

and define $a_i^{(m)} : Z_i^{(m)} \to Z_{i+1}^{(m)}$ by

$$a_i^{(m)}(x, x', z) = (f_i(x), r_i^{(m-1)}(x) + f_i(x'), t_i^{(m-2)}(x') + q_i^{(m-2)}(z)),$$

where now $\{t_i^{(m-2)}\}$ is a set of non-zero homomorphisms $t_i^{(m-2)} : X_i \to Z_{i+1}^{(m-2)}$ and $\{r_i^{(m-1)}\}$ is a set of non-zero homomorphisms $r_i^{(m-1)} : X_i \to X_{i+1}$. As in the proof of Lemma 5.2 one can choose $\{t_i^{(m-2)}\}$ and $\{r_i^{(m-1)}\}$ so that $Z^{(m)}$ is a non-split extension of X by $Z^{(m-1)}$, and $Z^{(m)}/Z^{(m-2)}$ is a non-split self-extension of X. Therefore, the Loewy length of $Z^{(m)}$ is greater than m. The statement follows. $\square$

Lemma 5.4. *Let $Q = \varinjlim Q_i \in \operatorname{Int}_{\mathfrak{g}}$ be a simple $\mathfrak{g}$-module which admits a non-zero homomorphism into an injective object of $\operatorname{Int}_{\mathfrak{g}}$ of finite Loewy length. Then there exist $n \in \mathbb{Z}_{>0}$ and a direct system of simple $\mathfrak{g}_i$-submodules S_i of Q such that $Q = \varinjlim S_i$ and $\dim \operatorname{Hom}_{\mathfrak{g}_i}(S_i, S_j) = 1$ for all $j > i > n$.*

Proof. Decompose each Q_i into a direct sum of isotypic components, $Q_i = Q_i^1 \oplus \cdots \oplus Q_i^{l(i)}$. We define a directed graph Γ as follows. The set of vertices $V(\Gamma)$ is by definition $\{Q_i^j\}$, and $V(\Gamma) = \cup_{i>0} V(\Gamma)_i$, where $V(\Gamma)_i = \{Q_i^1, \ldots, Q_i^{l(i)}\}$. There is an edge $A \to B$ in Γ if $A \in V(\Gamma_i)$, $B \in V(\Gamma_{i+1})$ and $\mathrm{Hom}_{\mathfrak{g}_i}(A, B) \neq 0$.

Let $\Gamma_{>i}$ be the full subgraph of Γ whose set of vertices equals $\cup_{k>i} V(\Gamma)_k$. For any vertex A of Γ we denote by $V(A)$ the set of vertices B such that there is a directed path from A to B. Let $\Gamma(A)$ be the full subgraph of Γ whose set of vertices equals $V(A)$, and $\Gamma(A)_{>i}$ be the full subgraph of $\Gamma(A)$ whose set of vertices equals $\cup_{k>i}(V(\Gamma)_k \cap V(A))$. Note that the simplicity of Q implies that $\Gamma_{>i}$ and $\Gamma(A)_{>i}$ are connected (as undirected graphs). In particular, if $\Gamma(A)$ is a tree, then $\Gamma(A)$ is just a string.

We will now prove that there exists a vertex A such that $\Gamma(A)$ is a tree. Indeed, assume the contrary. This implies that one can find an infinite sequence of vertices $A_1 \in V(\Gamma)_{i_1}, A_2 \in V(\Gamma)_{i_2}, \ldots$ such that the number of paths from A_n to A_{n+1} is greater than 2 for all n. Then $Q = \varinjlim Q_{i_k}$. In addition, one can easily see that Q satisfies the assumption of Lemma 5.2 and hence Q admits no non-zero homomorphism into an injective object of $\mathrm{Int}_{\mathfrak{g}}$ of finite Loewy length. Contradiction.

Fix now $A \in V(\Gamma)_i$ such that $\Gamma(A)$ is a tree. Then, as we mentioned above, $V(\Gamma)$ is necessarily a string $A_i = \{A \to A_{i+1} \to A_{i+2} \ldots\}$. Let S_j be a simple submodule of A_j, $j \geq i$. By Lemma 5.2 there exists n, such that $\dim \mathrm{Hom}_{\mathfrak{g}_j}(S_j, S_k) = 1$ for any $k > j \geq n$. Fix $s \in S_n$ and set $S_j = U(\mathfrak{g}_j) \cdot s$ for all $j \geq n$. Then S_j are simple and $Q = \varinjlim S_j$ satisfies the condition in the lemma. $\square$

Lemma 5.5. *Let $Q = \varinjlim S_i \in \mathrm{Int}_{\mathfrak{g}}$, where S_i are simple $\mathfrak{g}_i$-modules such that, for some n, $\dim \mathrm{Hom}_{\mathfrak{g}_i}(S_i, S_j) = 1$ for all $j > i > n$. Then Q^* has a unique simple submodule Q_*, and $Q_* \in \mathrm{Int}_{\mathfrak{g}}$.*

Proof. The condition on Q implies that $\dim \mathrm{Hom}_{\mathfrak{g}_i}(S_i, Q) = 1$ for all sufficiently large i. Therefore, $\dim \mathrm{Hom}_{\mathfrak{g}_i}(S_i^*, Q^*) = 1$ for all sufficiently large i. Note also that $Q_* = \varinjlim S_i^*$ is uniquely defined (as $\dim \mathrm{Hom}_{\mathfrak{g}_i}(S_i, S_{i+1}) = 1$) and is a simple integrable submodule of Q^*. Let S be some simple submodule of Q^*. Since $Q^* = \varprojlim S_i^*$ and $\mathrm{Hom}_{\mathfrak{g}}(S, Q^*) \neq 0$, we have $\mathrm{Hom}_{\mathfrak{g}_i}(S, S_i^*) \neq 0$ for some i. Therefore, $S_i^* \subset S$ as the multiplicity of S_i^* in Q^* is 1. This implies $S = Q_*$. $\square$

We are now ready to prove Proposition 5.1.
Proof of Proposition 5.1. Fix $0 \neq m \in M$ and put $M_i := U(\mathfrak{g}_i) \cdot m$. Then, by the simplicity of M, we have $M = \varinjlim M_i$. Since $\Gamma_{\mathfrak{g}}(M^*)$ has finite Loewy length, M^* has a simple submodule Q. By Lemma 5.4, Q satisfies the assumption of Lemma 5.5. The composition of the canonical injection $M \to (M^*)^*$ and the dual map $(M^*)^* \to Q^*$ defines an injective homomorphism $M \to Q^*$. By Lemma 5.5 $M \simeq Q_*$ and, since Q_* also satisfies the assumption of Lemma 5.5, we conclude that the claim of Proposition 5.1 holds for M. $\square$

The following statement is a direct consequence of Proposition 5.1.

Corollary 5.6. *Let $M \in \mathrm{Int}_{\mathfrak{g}}$ be a simple $\mathfrak{g}$-module such that $\Gamma_{\mathfrak{g}}(M^*)$ has finite Loewy length. Then for any sufficiently large i there exists a simple $\mathfrak{g}_i$-module N such that $\dim \mathrm{Hom}_{\mathfrak{g}_i}(N, M) = 1$.*

The next corollary is a direct consequence of Lemma 5.5 and Proposition 5.1.

Corollary 5.7. *Let $M \in \mathrm{Int}_{\mathfrak{g}}$ be a simple $\mathfrak{g}$-module such that $\Gamma_{\mathfrak{g}}(M^*)$ has finite Loewy length. Then M^* has a unique simple submodule M_*, and $M_* \in \mathrm{Int}_{\mathfrak{g}}$.*

Theorem 5.8. *Let $\mathfrak{g}$ be a locally simple Lie algebra which has a non-trivial module $M \in \mathrm{Int}_{\mathfrak{g}}$ such that M^* is integrable and has finite Loewy length, then $\mathfrak{g}$ is isomorphic to $sl(\infty)$, $o(\infty)$ or $sp(\infty)$.*

Proof. By Proposition 4.3 we know that $\mathfrak{g}$ is diagonal. Assume that $\mathfrak{g}$ is not finitary and there exists M satisfying the conditions of the theorem. Also assume that in the restriction of V_i to $\mathfrak{g}_{i-1}$ there is no costandard module (for types B, C and D it is automatic). Let $\mathfrak{g} = \varinjlim \mathfrak{g}_i$. Fix n and let $\varphi_k : \mathfrak{g}_n \to \mathfrak{g}_{n+k}$ denote the inclusion defined by our fixed exhaustion of $\mathfrak{g}$. Since $\mathfrak{g}$ is diagonal, there exists a root subalgebra $\mathfrak{l}_k \subset \mathfrak{g}_{n+k}$ such that $\mathfrak{l}_k \simeq \mathfrak{g}_n \oplus \cdots \oplus \mathfrak{g}_n$ and $\varphi_k(\mathfrak{g}_n)$ is the diagonal subalgebra in $\mathfrak{l}_k$. Let a_k be the number of simple direct summands in $\mathfrak{l}_k$. Since $\mathfrak{g}$ is not finitary, $a_k \to \infty$.

Note next that our condition on M implies that M admits a simple subquotient whose dual is integrable and of finite Loewy length. Therefore, without loss of generality, we may assume that M is simple. Then, by Corollary 5.6 $M = \varinjlim M_i$ is a direct limit of simple modules and, by possibly increasing n, we have $\dim \mathrm{Hom}_{\mathfrak{g}_n}(M_n, M_{n+k}) = 1$ for all k. Choose a set of Borel subalgebras $\mathfrak{b}_i \subset \mathfrak{g}_i$ such that $\varphi_k(\mathfrak{b}_n) \subset \mathfrak{b}_{n+k}$. Let h be the highest coroot of $\mathfrak{g}_n$ and let λ be the highest weight of some simple $\mathfrak{l}_k$-constituent L of M_{n+k}. Since M^* is integrable, Lemma 4.1 implies that $\lambda(\varphi_k(h))$ is bounded by some number t. If $h_1, \ldots, h_{a_k}$ are the images of $\varphi_k(h)$ in the simple direct summands of $\mathfrak{l}_k$ under the natural projections, we have $\lambda(h_j) \neq 0$ for at most t direct summands. Therefore, L isomorphic to an outer tensor product of at most t non-trivial simple $\mathfrak{g}_n$-modules. Since M_{n+k} is invariant under permutation of direct summands of $\mathfrak{l}_k$, we have at least $a_k - t$ simple constituents of M_{n+k} obtained from L by permutation of the simple direct summands of $\mathfrak{l}_k$. Note that all these simple constituents are isomorphic as $\varphi_k(\mathfrak{g}_n)$-modules. Thus the multiplicity of any simple $\varphi_{n+k}(\mathfrak{g}_n)$-module in M_{n+k} is at least $a_k - t$. Since $a_k \to \infty$, this contradicts Proposition 5.1.

The case when the restriction of V_n to $\mathfrak{g}_{n-1}$ contains a costandard simple constituent can be handled by a similar argument which we leave to the reader. $\square$

6. The category $\widetilde{\mathrm{Tens}}_{\mathfrak{g}}$ for $\mathfrak{g} \simeq sl(\infty), o(\infty), sp(\infty)$

Define $\widetilde{\mathrm{Tens}}_{\mathfrak{g}}$ as the largest full subcategory of $\mathrm{Int}_{\mathfrak{g}}$ which is closed under algebraic dualization and such that every object in it has finite Loewy length.

It is clear that $\widetilde{\mathrm{Tens}}_{\mathfrak{g}}$ is closed with respect to finite direct sums, however $\widetilde{\mathrm{Tens}}_{\mathfrak{g}}$ is not closed with respect to arbitrary direct sums (see Corollary 6.17 below). Note also that, if $\mathfrak{g}$ is finite-dimensional and semisimple, the objects of $\widetilde{\mathrm{Tens}}_{\mathfrak{g}}$ are integrable modules which have finitely many isotypic components.

It follows from Theorem 5.8 that if $\mathfrak{g}$ is locally simple and $\widetilde{\mathrm{Tens}}_{\mathfrak{g}}$ contains a non-trivial module, then $\mathfrak{g}$ is finitary.

In the rest of this section we assume that $\mathfrak{g} \simeq sl(\infty)$, $o(\infty)$ or $sp(\infty)$. Set $T^{p,q} := V^{\otimes p} \otimes (V_*)^{\otimes q}$, where V and V_* are respectively the natural and conatural $\mathfrak{g}$-modules ($V_* \simeq V$ when $\mathfrak{g} \simeq o(\infty), sp(\infty)$). The modules $T^{p,q}$ have been studied in [PS]; in particular, $T^{p,q}$ has finite length and is semisimple only if $pq = 0$ for $\mathfrak{g} = sl(\infty)$, and if $p + q \leq 1$ for $\mathfrak{g} = o(\infty), sp(\infty)$. Moreover, the Loewy length of

$T^{p,q}$ equals $\min\{p,q\}+1$ for $\mathfrak{g} = sl(\infty)$ and $[\frac{p+q}{2}]+1$ for $\mathfrak{g} = o(\infty), sp(\infty)$. A simple module M is called a *simple tensor module* if it is a submodule (or, equivalently, a subquotient) of $T^{p,q}$ for some p, q.

It is well known that there is a choice of nested Borel subalgebras $\mathfrak{b}_i \subset \mathfrak{g}_i$ such that all simple tensor modules are $\mathfrak{b}$-highest weight modules for $\mathfrak{b} = \varinjlim \mathfrak{b}_i$, see [PS]. (Moreover, the positive roots of any such $\mathfrak{b}$ are not generated by the simple roots of $\mathfrak{b}$. However, in the present paper we will make no further reference to this fact.)

Denote by Θ the set of all highest weights of simple tensor modules. If $\lambda \in \Theta$, by V_λ we denote the simple tensor module with highest weight λ, and, as in section 4, by V_λ^i we denote the simple $\mathfrak{g}_i$-highest weight module with highest weight λ (here λ is considered as a weight of $\mathfrak{g}_i$). It is clear that every $\lambda \in \Theta$ can be written in the form $\lambda = \sum a_i \gamma_i$ for some finite set $\gamma_1, ..., \gamma_s$ of linearly independent weights of V and some $a_i \in \mathbb{Z}$ (see [PS] for an explicit description of Θ). We put $|\lambda| := \sum |a_i|$. It is not hard to see that for any k the set of all $\mu \in \Theta$ with $|\mu| \leq k$ is finite. Moreover, all simple subquotients of $T^{p,q}$ are isomorphic to V_μ with $|\mu| \leq p+q$, and it follows from [PS] that if V_λ is a submodule in $T^{p,q}$ then $|\lambda| = p+q$.

Note that $(T^{p,q})^*, (T^{p,q})^{**}$, etc., are integrable modules. Indeed, it is easy to see (cf. [PS]) that for any fixed λ and any fixed $i > 0$ the non-vanishing of $\mathrm{Hom}_{\mathfrak{g}_i}(N, V_\lambda)$ for a simple $\mathfrak{g}_i$-module N implies $N \simeq V_\mu^i$ for $|\mu| \leq |\lambda|$. Hence the condition of Lemma 4.1 is satisfied for $T^{p,q}$ for fixed p, q. This shows that $(T^{p,q})^* \in \mathrm{Int}_{\mathfrak{g}}$. By Corollary 4.2, $(T^{p,q})^{**} \in \mathrm{Int}_{\mathfrak{g}}$, etc..

Lemma 6.1. *Fix $p, q \in \mathbb{Z}_{\geq 0}$.*

a) $(T^{p,q})^*$ has finite Loewy length, and all simple subquotients of $(T^{p,q})^*$ are tensor modules of the form V_λ for $|\lambda| \leq p+q$.

b) The direct product $\prod_{f \in \mathcal{F}} T_f^{p,q}$ of any family $\{T_f^{p,q}\}_{f \in \mathcal{F}}$ of copies of $T^{p,q}$ has finite Loewy length, and all simple subquotients of $\prod_{f \in \mathcal{F}} T_f^{p,q}$ are tensor modules of the form V_λ for $|\lambda| \leq p+q$.

Proof. First we prove b) using induction in $p+q$. The case $p+q = 0$ is trivial. If $p+q > 0$, without loss of generality we can assume that $p > 0$ (if $p = 0$ and $q > 0$ we replace V by V_* in the argument below). There is a canonical injective homomorphism $U \to \prod_{f \in \mathcal{F}} T_f^{p,q}$, where $U := V \otimes \prod_{f \in \mathcal{F}} T_f^{p-1,q}$, so we can consider U as a submodule of $\prod_{f \in \mathcal{F}} T_f^{p,q}$. By the induction assumption b) holds for $\prod_{f \in \mathcal{F}} T_f^{p-1,q}$. Since $T^{r,s}$ has finite length for all r, s, [PS], this implies that U has finite Loewy length and all simple subquotients of U are simple tensor modules of the form V_λ for $|\lambda| \leq p+q$. The quotient $(\prod_{f \in \mathcal{F}} T_f^{p,q})/U$ is isomorphic to a submodule of
$$R := \prod_{f \in \mathcal{F}} (V' \otimes T_f^{p-1,q}),$$
where V' is a copy of the vector space V with trivial $\mathfrak{g}$-module structure. Since $R \simeq \prod_{f \in \mathcal{F}} (\bigoplus_{i \in \mathbb{Z}} T_{f,i}^{p-1,q})$, by the induction assumption b) holds for R. Therefore, b) holds for $\prod_{f \in \mathcal{F}} T_f^{p,q}$.

a) To prove that $(T^{p,q})^*$ has finite Loewy length, we consider $U' := V_* \otimes (T^{p-1,q})^*$ as a submodule of $(T^{p,q})^*$. By the induction assumption, U' has finite Loewy length. The quotient $(T^{p,q})^*/U'$ is a submodule of $R' = \prod_{i \in \mathbb{Z}} (T_i^{p-1,q})^*$. The latter $\mathfrak{g}$-module has finite Loewy length by the induction assumption and b). The statement about the simple subquotients of $(T^{p,q})^*$ follows by an induction argument similar to the one in the proof of b). This proves a) for $(T^{p,q})^*$. $\qquad\square$

Example 6.2.

a) We start with the simplest example. Let $\mathfrak{g} = sl(\infty), o(\infty), sp(\infty)$ and $M = V^* = (T^{1,0})^*$. Then $M \in \widetilde{\text{Tens}}_{\mathfrak{g}}$ by Lemma 6.1. Furthermore, M is an injective object of $\text{Int}_{\mathfrak{g}}$ by Proposition 3.2. It is easy to see that $\text{soc}(M) = V_*$ and that $M/\text{soc}(M) = V^*/V_*$ is a trivial module of cardinality $\beth_1$. Since $\text{soc}(M)$ is simple, M is an injective hull of V_*.

b) Let $\mathfrak{g}$ be as in a) but let now $M = V^{**} = (T^{1,0})^{**}$. The exact sequence

$$0 \to V_* \to V^* \to V^*/V_* \to 0$$

yields an exact sequence

(5) $$0 \to (V^*/V_*)^* \to M \to (V_*)^* \to 0.$$

Since $(V^*/V_*)^*$ is a trivial $\mathfrak{g}$-module (cf. a)), it is injective, and hence (5) splits. This yields an isomorphism $M = V^{**} = (V_*)^* \oplus T$, T being a trivial $\mathfrak{g}$-module of cardinality $\beth_2$.

c) Here is a more interesting example. We consider the $\mathfrak{g}$-module M^* where $\mathfrak{g} = \text{sl}(\infty)$ and $M = V \otimes V_* = T^{1,1}$ as in Example 3.5. Recall the notation introduced in Example 3.5. In addition, let Sc be the one-dimensional space of scalar matrices, and F_r (respectively F_c) denote respectively the spaces of matrices with finitely many non-zero rows (resp., columns) (F has codimension 1 in $F_r \cap F_c$). It is important to notice that $\mathfrak{g} \cdot M^* \subset F_r + F_c$.

We first show that $\text{soc}(M^*) = Sc \oplus F = \mathbb{C} \oplus \mathfrak{g}$. It is obvious that $Sc \oplus F \subset \text{soc}(M^*)$ and that Sc is the largest trivial $\mathfrak{g}$-submodule of M^*. To see that $Sc \oplus F = \text{soc}(M^*)$, let X be any non-trivial simple submodule of $\text{soc}(M^*)$. Consider $0 \neq x \in X$. Then $\mathfrak{g} \cdot x \subset F_r + F_c$. Furthermore, it is easy to check that for any $0 \neq y \in F_r + F_c$, there exists $A \in \mathfrak{g}$ such that $A \cdot y \in F$ and $A \cdot y \neq 0$. Hence $X = F$, and we have shown that $\text{soc}(M^*) = Sc \oplus F$.

We now compute $\text{soc}^1(M^*)$. We claim that $F_r + F_c \subset \text{soc}^1(M^*)$. Since $BA \in F$ for $B \in F_r$, $A \in F$, the action of $\mathfrak{g}$ on F_r/F is simply left multiplication. Using this it is not difficult to establish an isomorphism of $\mathfrak{g}$-modules $F_r/F \simeq \bigoplus_{q \in Q} V_q$, where Q is a family of copies of V of cardinality $\beth_1$. Similarly, $F_c/F \simeq \bigoplus_{q \in Q} (V_*)_q$. (It is convenient to think here of V_* as the space of all row vectors each of which have finitely many non-zero entries.) This implies $F_r + F_c \subset \text{soc}^1(M^*)$.

Furthermore, $M^*/(F_r + F_c)$ is a trivial $\mathfrak{g}$-module as $\mathfrak{g} \cdot M^* \subset F_r + F_c$. Therefore, in order to compute $\text{soc}^1(M^*)$ we need to find all $z \in M^*$ such that $\mathfrak{g} \cdot z \subset Sc + F$. A direct computation shows that $\mathfrak{g} \cdot z \in Sc + F$ if and only $z \in J$, where J denotes the set of matrices each row and each column of which have finitely many non-zero elements. (In fact, $\mathfrak{g} \cdot J \subset F$). Thus $\text{soc}^1(M^*) = F_r + F_c + J$, and we obtain the socle filtration of M^*:

$$0 \subset Sc \oplus F \subset F_r + F_c + J \subset M^*.$$

In particular, the Loewy length of M^* equals 3, the irreducible subquotients of M^* up to isomorphism are $\mathbb{C}, V, V_*, \mathfrak{g}$, and all of them occur with multiplicity $\beth_1$, except $\mathfrak{g}$ which occurs with multiplicity 1.

Note that M^* is decomposable and is isomorphic to $\mathbb{C} \oplus \mathfrak{g}^*$. As the socle of $\mathfrak{g}^*$ is simple (being isomorphic to $\mathfrak{g}$), $\mathfrak{g}^*$ is indecomposable. Moreover $\mathfrak{g}^*$ is an injective hull of $F = \mathfrak{g}$.

d) We now give an example illustrating statement b) of Lemma 6.1. Let $\mathfrak{g} = sl(\infty), o(\infty), sp(\infty)$ and $M = \prod_{f \in \mathcal{F}} V_f$, $\{V_f\}_{f \in \mathcal{F}}$ being an infinite family of copies of the natural module V. Set $M^{\mathrm{fin}} = \{\psi : \mathcal{F} \to V \,|\, \dim(\psi(\mathcal{F})) < \infty\}$. Then M^{fin} is a $\mathfrak{g}$-submodule of M, and $\mathfrak{g} \cdot M \subset M^{\mathrm{fin}}$. Hence M/M^{fin} is a trivial $\mathfrak{g}$-module. Moreover, $M^{\mathrm{fin}} \simeq \bigoplus_{g \in 2^{\mathcal{F}}} V_g$, where $2^{\mathcal{F}}$ is the set of subsets of $\mathcal{F}$. Indeed,

$$M^{\mathrm{fin}} = \varinjlim(\prod_{f \in \mathcal{F}} (V^i)_f) = \varinjlim((\prod_{f \in \mathcal{F}} \mathbb{C}_f) \otimes V^i) \cong \varinjlim \bigoplus_{g \in 2^{\mathcal{F}}} (\mathbb{C}_g \otimes V^i) =$$

$$\varinjlim(\bigotimes_{g \in 2^{\mathcal{F}}} (V^i)_g) = \bigoplus_{g \in 2^{\mathcal{F}}} V_g.$$

This yields an exact sequence

(6) $$0 \to \bigoplus_{g \in 2^{\mathcal{F}}} V_g \to M \to T \to 0,$$

T being trivial module of dimension card $2^{\mathcal{F}}$. Since M has no non-zero trivial submodules, (6) is in fact the socle filtration of M. Consequently the Loewy length of M equals 2.

Corollary 6.3. *Let $M \in \mathrm{Int}_{\mathfrak{g}}$ have finite Loewy length and all simple subquotients of M be isomorphic to V_λ where $|\lambda|$ is less or equal than a fixed $k \in \mathbb{Z}_{>0}$. Then*

a) for any family $\{M_f\}_{f \in \mathcal{F}}$ of copies of M, the $\mathfrak{g}$-module $\prod_{f \in \mathcal{F}} M_f$ has finite Loewy length and all simple subquotients of $\prod_{f \in \mathcal{F}} M_f$ are isomorphic to V_λ with $|\lambda| \leq k$;

b) M^* has finite Loewy length and all simple subquotients of M^* are isomorphic to V_λ with $|\lambda| \leq k$;

c) $M \in \widetilde{\mathrm{Tens}}_{\mathfrak{g}}$.

Proof. a) The socle filtration of M induces a finite filtration on $\prod_{f \in \mathcal{F}} M_f$

$$0 \subset \prod_{f \in \mathcal{F}} \mathrm{soc}(M_f) \subset \cdots \subset \prod_{f \in \mathcal{F}} \mathrm{soc}^i(M_f) \subset \cdots \subset \prod_{f \in \mathcal{F}} M_f.$$

Furthermore,

(7) $$\mathrm{soc}^i(M)/\mathrm{soc}^{i-1}(M) \simeq \bigoplus_{|\lambda| \leq k} \bigoplus_{g \in \mathcal{F}_\lambda} (V_\lambda)_g$$

for some families $\{(V_\lambda)_g\}_{g \in \mathcal{F}_\lambda}$ of copies of V_λ. Hence

$$\prod_{f \in \mathcal{F}} (\mathrm{soc}^i(M_f)/\mathrm{soc}^{i-1}(M_f)) \simeq \bigoplus_{|\lambda| \leq k} \prod_{f \in \mathcal{F}} (\bigoplus_{g \in \mathcal{F}_\lambda} (V_\lambda)_g)_f.$$

Note that for each λ

$$\prod_{f \in \mathcal{F}} (\bigoplus_{g \in \mathcal{F}_\lambda} (V_\lambda)_g)_f \subset \prod_{(f,g) \in \mathcal{F} \times \mathcal{F}_\lambda} (V_\lambda)_{(f,g)}.$$

By Lemma 6.1 b), the $\mathfrak{g}$-module $\prod_{(f,g) \in \mathcal{F} \times \mathcal{F}_\lambda} (V_\lambda)_{(f,g)}$ has finite Loewy length and all its simple subquotients are isomorphic ot V_μ with $|\mu| \leq |\lambda| \leq k$. The same holds for $\prod_{f \in \mathcal{F}} (\mathrm{soc}^i(M_f)/\mathrm{soc}^{i-1}(M_f))$. Therefore, a) holds.

b) Since all V_λ with $|\lambda| \leq k$ satisfy the conditions of Lemma 4.1, M satisfies the condition of Lemma 4.1 and therefore $M^* \in \mathrm{Int}_{\mathfrak{g}}$.

The socle filtration of M induces a finite filtration on M^*

$$\cdots \subset \mathrm{soc}^i(M)^* \subset \mathrm{soc}^{i-1}(M)^* \subset \cdots .$$

Using (7) we get

$$\mathrm{soc}^{i-1}(M)^*/\mathrm{soc}^i(M)^* \simeq \bigoplus_{|\lambda| \leq k} \prod_{g \in \mathcal{F}_\lambda} (V_\lambda^*)_g.$$

By Lemma 6.1 b) V_λ^* has finite Loewy length and its simple subquotients are isomorphic to V_μ with $|\mu| \leq |\lambda|$, hence by a) the same holds for $\prod_{g \in \mathcal{F}_\lambda} (V_\lambda^*)_g$. This implies that b) holds.

c) Note that if M satisfies the assumptions of the corollary, then M^* and all higher duals, M^{**} etc., satisfy the assumptions of the corollary. Hence $M \in \widetilde{\mathrm{Tens}_{\mathfrak{g}}}$.
□

Remarkably, there is following abstract characterization of simple tensor modules.

Theorem 6.4. *If $M \in \mathrm{Int}_{\mathfrak{g}}$ is simple and $\Gamma_{\mathfrak{g}}(M^*)$ has finite Loewy length, then M is a simple tensor module.*

Proof. By Proposition 5.1, $M = \varinjlim M_i$ for some $n \in \mathbb{Z}_+$ and simple nested $\mathfrak{g}_i$-submodules $M_i \subset M$ with $\dim \mathrm{Hom}_{\mathfrak{g}_i}(M_i, M) = 1$ for all $i \geq n$. If $\mathfrak{g} = sl(\infty)$, it is useful to consider M as a $gl(\infty)$-module by extending the $sl(i)$-module structure on M_i to a $gl(i)$-module structure in a way compatible with the injections $M_i \to M_{i+1}$. It is easy to see that the condition $\dim \mathrm{Hom}_{\mathfrak{g}_i}(M_i, M) = 1$ for all $i \geq n$ ensures the existence of such an extension. Note, furthermore, that $\dim \mathrm{Hom}_{gl(i)}(M_i, M) = 1$. This allows us to assume that $\mathfrak{g} = gl(\infty)$ and $\mathfrak{g}_i = gl(i)$.

Let now $\mathfrak{c}$ denote the derived subalgebra of the centralizer of $\mathfrak{g}_n$ in $\mathfrak{g}$. Then obviously $\mathfrak{c}$ is a simple finitary Lie algebra whose action on M induces a trivial action on M_n. Hence, as a $\mathfrak{c}$-module, M is isomorphic to a quotient of $U(\mathfrak{g}) \otimes_{U(\mathfrak{c} \oplus \mathfrak{g}_n)} M_n$, or equivalently to a quotient of $S^{\cdot}(\mathfrak{g}/(\mathfrak{c} \oplus \mathfrak{g}_n)) \otimes M_n$. Note that $\mathfrak{g}/(\mathfrak{c} \oplus \mathfrak{g}_n)$, considered as a $\mathfrak{c}$-module has finite length and that its simple subquotients are natural, conatural, and possibly 1-dimensional trivial $\mathfrak{c}$-modules. This implies that every simple $\mathfrak{c}$-subquotient of M is a simple tensor $\mathfrak{c}$-module. In addition, for $i \geq n$, the number of non-zero marks of the highest weight of any simple $\mathfrak{g}_i$-submodule of M is not greater than n plus the multiplicity of the non-trivial simple constituents of the $\mathfrak{g}_n$-module $\mathfrak{g}/(\mathfrak{c} \oplus \mathfrak{g}_n)$. In particular, if λ_i denotes the highest weight of M_i then λ_i has at most $3n$ non-zero marks.

Consider first the case when $\mathfrak{g} = gl(\infty)$. Then every weight λ_i can be written uniquely in the form

$$a_1^i \varepsilon_1 + \cdots + a_k^i \varepsilon_k + b_1^i \varepsilon_{n-k} + \cdots + b_k^i \varepsilon_n$$

for some fixed k, $a_1^i \geq a_2^i \geq \cdots \geq a_k^i \geq 0$ and $0 \geq b_1^i \geq \cdots \geq b_k^i$. We claim that for sufficiently large i the weight stabilizes, i.e. $a_j^i = a_j^{i+1} = \cdots = a_j^p = \ldots$ and $b_j^i = b_j^{i+1} = \cdots = b_j^p = \ldots$ for all j, $1 \leq j \leq k$. Indeed, assume the contrary. Let j be the smallest index such that the sequence $\{a_j^i\}$ does not stabilize. By the branching rule for $gl(m) \subset gl(m+1)$ (see for instance [GW]) the sequence $\{a_j^i\}$ is non-decreasing. Hence there is p such that $a_j^{p+1} > a_j^p$. Set $\mu = \lambda_p + \varepsilon_j$. Then the multiplicity of M_{p-1} in V_μ^p is not zero and the multiplicity of V_μ^p in M_{p+1} is not zero. Since $V_\mu^p \neq M_p$, this shows that the multiplicity of M_{p-1} in M_{p+1} is at least 2. Contradiction. Similarly the sequence $\{b_j^i\}$ stabilizes. As it is easy to see, this is sufficient to conclude that $M \simeq V_\lambda$ for some $\lambda \in \Theta$.

Let $\mathfrak{g} = o(\infty)$ or $sp(\infty)$. In the first case we assume that $\mathfrak{g}_i = o(2i+1)$. Then $\lambda_i = a_1^i \varepsilon_1 + \cdots + a_k^i \varepsilon_k$ for some fixed k and $a_1^i \geq a_2^i \geq \cdots \geq a_k^i \geq 0$. The sequence $\{a_j^i\}$ is non-decreasing for every fixed j as follows from the branching laws for the respective pairs $o(2m+1) \subset o(2m+3)$ and $sp(2n) \subset sp(2m+2)$, see [GW]. Then by repeating the argument in the previous paragraph we prove that $\{a_j^i\}$ stabilizes, and consequently $M \simeq V_\lambda$ for some $\lambda \in \Theta$. $\qquad\square$

Corollary 6.3 and Theorem 6.4 show that a simple module $M \in \mathrm{Int}_\mathfrak{g}$ is an object of $\widetilde{\mathrm{Tens}}_\mathfrak{g}$ if and only if $\Gamma_\mathfrak{g}(M^*)$ has finite Loewy length. Below we will use this fact to give an equivalent definition of $\widetilde{\mathrm{Tens}}_\mathfrak{g}$ (Corollary 6.13). Furthermore, it is easy to check (see also [PS]) that for sufficiently large i the simple $\mathfrak{g}_i$-module V_λ^i occurs in Y with multiplicity 1, and all other simple $\mathfrak{g}_i$-constituents have infinite multiplicity and are isomorphic to V_μ^i with $|\mu| < |\lambda|$. In what follows we call this unique $\mathfrak{g}_i$-constituent the *canonical $\mathfrak{g}_i$-constituent of V_λ*. Note also that by Corollary 5.7 for each simple object M of $\widetilde{\mathrm{Tens}}_\mathfrak{g}$, M_* is a well-defined simple object in $\widetilde{\mathrm{Tens}}_\mathfrak{g}$. Hence M_* is well defined also for any semisimple object M of $\widetilde{\mathrm{Tens}}_\mathfrak{g}$: if $M = \bigoplus_{\lambda \in \Theta} M^\lambda \otimes V_\lambda$

(M^λ being trivial $\mathfrak{g}$-modules), then $M_* = \bigoplus_{\lambda \in \Theta} M^\lambda \otimes (V_\lambda)_*$. It is clear that $M_* \cong M$ for $\mathfrak{g} \cong o(\infty)$, $sp(\infty)$.

Corollary 6.5. *The simple objects of $\widetilde{\mathrm{Tens}}_\mathfrak{g}$ are precisely the simple tensor modules.*

Lemma 6.6. *Let $M \cong V_\lambda$ be a simple tensor module. Then $\mathrm{soc}((M_*)^*) \simeq M$. If V_μ is a subquotient of $(M_*)^*$ and $\mu \neq \lambda$, then $|\mu| < |\lambda|$.*

Proof. The first statement follows from Corollary 5.7.

The second statement follows immediately from the fact that $\mathrm{Hom}_{\mathfrak{g}_i}(V_\mu^i, (M_*)^*) \neq 0$ implies $|\mu| < |\lambda|$. $\qquad\square$

Corollary 6.7. *a) For any simple $M \in \widetilde{\mathrm{Tens}}_\mathfrak{g}$, $(M_*)^*$ is an injective hull of M in $\mathrm{Int}_\mathfrak{g}$ (and hence also in $\widetilde{\mathrm{Tens}}_\mathfrak{g}$).*

b) Any indecomposable injective object in $\widetilde{\mathrm{Tens}}_\mathfrak{g}$ is isomorphic to M^* for some simple module $M \in \widetilde{\mathrm{Tens}}_\mathfrak{g}$. In particular, any indecomposable injective module is isomorphic to a direct summand of $(T^{p,q})^*$ for some p, q.

c) For any $M \in \widetilde{\mathrm{Tens}}_\mathfrak{g}$, any injective hull I_M of M in $\mathrm{Int}_\mathfrak{g}$ is an object of $\widetilde{\mathrm{Tens}}_\mathfrak{g}$.

Proof. a) Follows directly from Proposition 3.2 and Lemma 6.6.

b) To derive b) from a) it suffices to note that an injective module in $\widetilde{\mathrm{Tens}}_\mathfrak{g}$ is indecomposable if and only if it has simple socle.

c) follows from the fact that I_M is isomorphic to a submodule of $\Gamma_{\mathfrak{g}}(M^{**})$, see Corollary 3.3. □

In what follows we set $I_\lambda := ((V_\lambda)_*)^*$.

Corollary 6.8. $\mathrm{End}_{\mathfrak{g}}(I_\lambda) = \mathbb{C}$.

Proof. If $\varphi \in \mathrm{End}_{\mathfrak{g}}(I_\lambda)$, then $\varphi|_{V_\lambda} = c\,\mathrm{Id}$ for $c \in \mathbb{C}$. Therefore, $V_\lambda \subset \mathrm{Ker}(\varphi - c\,\mathrm{Id})$. Furthermore, any non-zero $\mathfrak{g}$-submodule of I_λ contains $\mathrm{soc}(I_\lambda) = V_\lambda$, hence $V_\lambda \subset \mathrm{Im}(\varphi - c\,\mathrm{Id})$. This implies $\varphi - c\,\mathrm{Id} = 0$, as otherwise V_λ would be isomorphic to a subquotient of I_λ/V_λ contrary to Lemma 6.6. □

Lemma 6.9. *Let $X, Y, Z, M \in \widetilde{\mathrm{Tens}}_{\mathfrak{g}}$. Assume furthermore that Y is simple, $Y = \mathrm{soc}(M)$, and there exists an exact sequence*

$$0 \to X \to Z \xrightarrow{p} Y \to 0.$$

Then there exists $\tilde{M} \in \mathrm{Int}_{\mathfrak{g}}$ such that $Z \subset \tilde{M}$ and $\tilde{M}/X \simeq M$.

Proof. Let Y_i be the canonical $\mathfrak{g}_i$-constituent of Y. Then $Y = \varinjlim Y_i$. Set $Z_i := p^{-1}(Y_i)$ and $Q_i := Z_i \cap X$. Then $Z_i = Y_i \oplus Q_i$ and there are injective homomorphisms $\varphi_i : Z_i \to Z_{i+1}$

$$\varphi_i(y, q) = (e_i(y), t_i(y) + f_i(q)), \ y \in Y_i, q \in Q_i$$

for some non-zero homomorphisms $e_i : Y_i \to Y_{i+1}$, $t_i : Y_i \to Q_{i+1}$ and $f_i : Q_i \to Q_{i+1}$. Clearly, $Z = \varinjlim Z_i$.

On the other hand, $M = \varinjlim M_i$ for some nested finite-dimensional $\mathfrak{g}_i$-submodules $M_i \subset M$ such that $Y_i \subset M_i$. Moreover, $\dim \mathrm{Hom}_{\mathfrak{g}_i}(Y_i, M_i) = 1$ by Lemma 6.6. Therefore, M_i has a unique $\mathfrak{g}_i$-module decomposition $M_i = R_i \oplus Y_i$. The inclusions $\psi_i : M_i \to M_{i+1}$ are given by

$$\psi_i(r, y) = (p_i(r), s_i(r) + e_i(y)), \ y \in Y_i, r \in R_i$$

for some non-zero homomorphisms $p_i : R_i \to R_{i+1}$ and $s_i : R_i \to Y_{i+1}$.

Define $\tilde{M}_i := R_i \oplus Y_i \oplus Q_i$ and let $\zeta_i : \tilde{M}_i \to \tilde{M}_{i+1}$ be given by the formula

$$\zeta(r, y, q) = (p_i(r), s_i(r) + e_i(y), t_i(y) + f_i(q)).$$

Set $\tilde{M} := \varinjlim \tilde{M}_i$. It is easy to check that $\tilde{M}$ satisfies the conditions of the lemma. □

Lemma 6.10. *If $\mathrm{Hom}_{\mathfrak{g}}(I_\lambda, I_\mu) \neq 0$, then $|\mu| \leq |\lambda|$. If I is any injective object of $\widetilde{\mathrm{Tens}}_{\mathfrak{g}}$ and $0 \neq \varphi \in \mathrm{Hom}_{\mathfrak{g}}(I, I_\mu)$, then φ is surjective.*

Proof. The first statement follows immediately from Lemma 6.6.

To prove the second statement put $X = \mathrm{Ker}\varphi$, $Y = V_\mu$, $Z = \varphi^{-1}(Y)$ and $M = I_\mu$. Construct $\tilde{M}$ as in Lemma 6.9. By the injectivity of I, the injective homomorphism $Z \to \tilde{M}$ extends to a homomorphism $\tilde{M} \to I$. The latter induces a homomorphism $\eta : M = I_\mu \to I/X$.

Let now $\bar{\varphi} : I/X \to I_\mu$ denote the injective homomorphism induced by φ. Then it is obvious that $\bar{\varphi} \circ \eta(y) = y$ for any $y \in Y$. By Corollary 6.8, we have $\bar{\varphi} \circ \eta = \mathrm{Id}$. Hence $\bar{\varphi}$ is an isomorphism, i.e. φ is surjective. □

Proposition 6.11. *The Loewy length of I_λ equals $|\lambda| + 1$.*

Proof. By Lemma 6.6 we know that the Loewy length of I_λ is at most $|\lambda| + 1$. We prove equality by induction in $|\lambda|$. Fix $\mu \in \Theta$ such that $|\mu| = |\lambda| - 1$ and $\mathrm{Hom}_{\mathfrak{g}_i}(V_\mu^i, V_\lambda^{i+1}) \neq 0$. We claim that $\mathrm{Ext}^1(V_\mu, V_\lambda) \neq 0$. Indeed, consider non-zero homomorphisms $\varphi_i \in \mathrm{Hom}_{\mathfrak{g}_i}(V_\mu^i, V_\lambda^{i+1})$. Set $X = \varinjlim X_i$, where $X_i = V_\mu^i \oplus V_\lambda^i$, $q_i : X_i \to X_{i+1}$ is given by $q_i(x,y) = (e_i(x), \varphi_i(x) + f_i(y))$ for $x \in V_\mu$, $y \in V_\lambda$, and $e_i : V_\mu^i \to V_\mu^{i+1}$ and $f_i : V_\lambda^i \to V_\lambda^{i+1}$ denote the fixed inclusions. It is easy to see that X is a non-trivial extension of V_μ by V_λ.

This implies the existence of a non-zero homomorphism $I_\lambda \to I_\mu$. By Lemma 6.10, this homomorphism is surjective. Hence the Loewy length of I_λ is greater or equal to the Loewy length of I_μ plus 1. The statement follows. $\square$

The following theorem strengthens the claim of Corollary 6.3.

Theorem 6.12. *Let $M \in \mathrm{Int}_\mathfrak{g}$. Then $M \in \widetilde{\mathrm{Tens}}_\mathfrak{g}$ if and only if there exists a finite subset $\Theta_M \subset \Theta$ such that any simple subquotient of M is isomorphic to V_μ for $\mu \in \Theta_M$.*

Proof. Assume that $M \in \widetilde{\mathrm{Tens}}_\mathfrak{g}$. It is sufficient to prove the existence of Θ_M for a semisimple M since then the general case follows from Lemma 6.6. Without loss of generality we may assume that $M = \bigoplus_{j \in C} V_{\lambda_j}$, where V_{λ_j} are pairwise non-isomorphic. We claim that if C is infinite, then M^* does not have finite Loewy length. Indeed, M^* contains a submodule isomorphic to $\bigoplus_{j \in C} I_{\mu_j}$, where $V_{\mu_j} = (V_{\lambda_j})_*$. If C is infinite, then $|\mu_j| = |\lambda_j|$ is unbounded and the socle filtration of $\bigoplus_{j \in C} I_{\mu_j}$ is infinite. This contradiction shows that C is finite, i.e. that there exists a finite set Θ_M as required.

Now assume that M admits a finite set Θ_M as in the statement of the theorem. We claim first that if M' is a quotient of M and $\mathrm{Ext}^1_\mathfrak{g}(M', V_\lambda) \neq 0$ for some $\lambda \in \Theta$, then M has a subquotient isomorphic to V_μ for some μ with $|\mu| < |\lambda|$. Indeed, by extending the sequence $0 \to V_\lambda \to I_\lambda$ to a minimal injective resolution $0 \to V_\lambda \to I_\lambda \xrightarrow{i} I_\lambda^1 \to ...$, we see that there is a non-zero homomorphism $M' \xrightarrow{p} I_\lambda^1$. Furthermore, by the minimality of the resolution, we have $\mathrm{soc}(I_\lambda^1) \subset \mathrm{Im}\, i$. Hence by Lemma 6.6 every simple constituent of $\mathrm{soc}(I_\lambda^1)$ is of the form V_ν for $|\nu| < |\lambda|$. Since $(\mathrm{Im}\, p) \cap \mathrm{soc}(I_\lambda^1) \neq 0$, some simple constituent of $\mathrm{soc}(I_\lambda')$ is isomorphic to a subquotient of M' and thus of M.

We show now that M has finite Loewy length. Consider a weight $\lambda \in \Theta_M$ with minimal $|\lambda|$. The above argument shows that $\mathrm{Ext}^1_\mathfrak{g}(M', V_\lambda) = 0$ for any quotient M' of M. This implies that every subquotient of M isomorphic to V_λ is a quotient of M. Hence M admits a surjective homomorphism $\zeta : M \to M_\lambda$, where M_λ is isomorphic to a direct sum of copies of V_λ and $\Theta_{\mathrm{ker}\,\zeta} = \Theta_M \setminus \{\lambda\}$. By an induction argument we obtain that M has finite Loewy length. Therefore, $M \in \widetilde{\mathrm{Tens}}_\mathfrak{g}$ by Corollary 6.3 c). $\square$

Corollary 6.13. *A $\mathfrak{g}$-module $M \in \mathrm{Int}_\mathfrak{g}$ is an object of $\widetilde{\mathrm{Tens}}_\mathfrak{g}$ if and only if both M and $\Gamma_\mathfrak{g}(M^*)$ have finite Loewy length.*

Proof. In one direction the statement is trivial. We need to prove that, if $M \in \mathrm{Int}_\mathfrak{g}$ satisfies the above two conditions, then $M^* \in \mathrm{Int}_\mathfrak{g}$. For a semisimple M this follows directly from Theorem 6.12 (as we have already pointed out). The argument is completed by induction on the Loewy length. Let $M \in \mathrm{Int}_\mathfrak{g}$ have Loewy length k, and $\Gamma_\mathfrak{g}(M^*)$ have finite Loewy length. Consider the homomorphism

$\pi : M \to \operatorname{top}(M)$ onto the maximal semisimple quotient $\operatorname{top}(M)$ of M. Then $\Gamma_{\mathfrak{g}}(\operatorname{top}(M)^*) \subset \Gamma_{\mathfrak{g}}(M^*)$, hence $\operatorname{top}(M) \in \widetilde{\operatorname{Tens}}_{\mathfrak{g}}$, i.e. in particular $\operatorname{top}(M)^* \in \operatorname{Int}_{\mathfrak{g}}$. Therefore, there is an exact sequence

$$0 \to \operatorname{top}(M)^* \to \Gamma_{\mathfrak{g}}(M^*) \to \Gamma_{\mathfrak{g}}((\operatorname{Ker}\pi)^*) \to 0,$$

implying that $\Gamma_{\mathfrak{g}}((\operatorname{Ker}\pi)^*)$ has finite Loewy length. Since the Loewy length of $\operatorname{Ker}\pi$ equals $k-1$, the induction assumption allows us to conclude that $(\operatorname{Ker}\pi)^* \in \operatorname{Int}_{\mathfrak{g}}$. Hence $\Gamma_{\mathfrak{g}}(M^*) = M^*$. □

Corollary 6.14. $\widetilde{\operatorname{Tens}}_{\mathfrak{g}}$ *is a tensor category with respect to* $\otimes$.

Proof. It suffices to show that $\widetilde{\operatorname{Tens}}_{\mathfrak{g}}$ is closed with respect to $\otimes$. The fact that, if $M \in \widetilde{\operatorname{Tens}}_{\mathfrak{g}}$ and $M' \in \widetilde{\operatorname{Tens}}_{\mathfrak{g}}$ then $M \otimes M' \in \widetilde{\operatorname{Tens}}_{\mathfrak{g}}$, follows immediately from Theorem 6.12. □

The following theorem concerns the structure of injective modules in $\widetilde{\operatorname{Tens}}_{\mathfrak{g}}$.

Theorem 6.15. *Any injective module* $I \in \widetilde{\operatorname{Tens}}_{\mathfrak{g}}$ *has a finite filtration* $\{I_j\}$ *such that, for each* j, I_{j+1}/I_j *is isomorphic to a direct sum of copies of* I_{μ_j} *for some* $\mu_j \in \Theta$.

Proof. We use induction on the length of the filtration. Assume that

$$0 = I_0 \subset I_1 \subset ... \subset I_k$$

is already constructed. Let $\operatorname{soc}(I/I_k) = \bigoplus_{f \in \mathcal{F}} Y_f$ for a family $\{Y_f\}_{f \in \mathcal{F}}$ of simple modules Y_f (there are only finitely many non-isomorphic modules among $\{Y_f\}_{f \in \mathcal{F}}$). Denoting by p the projection $I \to I/I_k$, set $X_f := p^{-1}(Y_f)$. By Lemma 6.9, there exists $\tilde{Y}_f \in \operatorname{Int}_{\mathfrak{g}}$ such that $I_k \subset X_f \subset \tilde{Y}_f$ and $\tilde{Y}_f/I_k \simeq I_{\mu_f}$, $\mu_f \in \Theta$ being the highest weight of Y_f. The inclusion $X_f \subset I$ induces a homomorphism $\psi_f : \tilde{Y}_f \to I$. Let $\bar{\psi}_f : \tilde{Y}_f/I_k \stackrel{\sim}{\to} I_{\mu_f} \to I/I_k$ the corresponding homomorphism of quotients. Then $\bar{\psi} := \bigoplus_{f \in \mathcal{F}} \bar{\psi}_f : \bigoplus_{f \in \mathcal{F}} I_{\mu_f} \to I$ is injective since its restriction to $\operatorname{soc}(\bigoplus_{f \in \mathcal{F}} I_{\mu_f})$ is an isomorphism. This shows that if $I_{k+1} := p^{-1}(\bar{\psi}(\bigoplus_{f \in \mathcal{F}} I_{\mu_f}))$, there is an isomorphism $I_{k+1}/I_k \simeq \bigoplus_{f \in \mathcal{F}} I_{\mu_f}$.

The filtration $\{I_j\}$ terminates at a finite step as I has finite Loewy length. □

Example 6.16.

Let $\mathfrak{g} = sl(\infty), o(\infty), sp(\infty)$ and let M be a countable direct sum of copies of V, i.e. $M = \bigoplus_{f \in \mathcal{F}} V_f$, $\operatorname{card}\mathcal{F} = \beth_0$. Then $(M_*)^*$ can be identified with the set of all infinite matrices $\{b_{ij}\}_{i,j \in \mathbb{Z}_{>0}}$, the action of $\mathfrak{g}$ being left multiplication. The socle $\operatorname{soc}((M_*)^*)$ is the space of matrices F_r with finitely many non-zero rows and is isomorphic to $\bigoplus_{g \in 2^{\mathcal{F}}} V_g$. (Note that the module $\prod_{f \in \mathcal{F}} V_f$ considered in Example 6.2 d) is a submodule of $(M_*)^*$ and has the same socle as $(M_*)^*$). We thus obtain the diagram

$$\begin{array}{ccc} \bigoplus_{g \in 2^{\mathcal{F}}} V_g & \subset & (M_*)^* \\ \cup & & \cup \\ M & \subset & I_M \end{array},$$

I_M being an injective hull of M. Furthermore, I_M is the largest submodule of $(M_*)^*$ such that $\mathfrak{g} \cdot I_M = M$. A direct computation shows that I_M coincides with the space of all matrices with finite rows (i.e. each row has finitely many non-zero entries).

Note that $I_M \neq \bigoplus_{f \in \mathcal{F}} (I_{\varepsilon_1})_f$ ($\varepsilon_1 \in \Theta$ is the highest weight of V). In fact I_M has the following filtration as in Theorem 6.15: $0 \subset \bigoplus_{f \in \mathcal{F}} (I_{\varepsilon_1})_f \subset I_M$. Here $I_M / \bigoplus_{f \in \mathcal{F}} (I_{\varepsilon_1})_f$ is a trivial module of cardinality $2^{\mathcal{F}}$ which is interpreted as a direct sum of $2^{\mathcal{F}}$ copies of I_0.

For any $k \in \mathbb{Z}_{>0}$ we now define $\widetilde{\mathrm{Tens}}_{\mathfrak{g}}^k$ be the subcategory of modules whose simple quotients are isomorphic to V_μ with $|\mu| \leq k$. Theorem 6.12 and Corollary 6.3 a) imply the following.

Corollary 6.17. *The category $\widetilde{\mathrm{Tens}}_{\mathfrak{g}}^k$ is closed under direct products and direct sums.*

Corollary 6.18. *a) The category $\widetilde{\mathrm{Tens}}_{\mathfrak{g}}$ equals the direct limit $\varinjlim \widetilde{\mathrm{Tens}}_{\mathfrak{g}}^k$.*

b) If $\{M_f\}_{f \in \mathcal{F}}$ is an infinite family of objects of $\widetilde{\mathrm{Tens}}_{\mathfrak{g}}$, then $\prod_{f \in \mathcal{F}} M_f \in \widetilde{\mathrm{Tens}}_{\mathfrak{g}}$ (equivalently, $\bigoplus_{f \in \mathcal{F}} \in \widetilde{\mathrm{Tens}}_{\mathfrak{g}}$) if and only if there is k such that $M_f \in \widetilde{\mathrm{Tens}}_{\mathfrak{g}}^k$ for all $f \in \mathcal{F}$.

Proof. a) follows directly from Theorem 6.12.

Consider now $\prod_{f \in \mathcal{F}} M_f$. If $M_f \in \widetilde{\mathrm{Tens}}_{\mathfrak{g}}^k$ for some k, then $\prod_{f \in \mathcal{F}} M_f \in \widetilde{\mathrm{Tens}}_{\mathfrak{g}}^k$ (and thus also $\bigoplus_{f \in \mathcal{F}} M_f \in \widetilde{\mathrm{Tens}}_{\mathfrak{g}}^k$) by Corollary 6.3 a). If no such k exists, then $\bigoplus_{f \in \mathcal{F}} M_f \notin \widetilde{\mathrm{Tens}}_{\mathfrak{g}}$ by Theorem 6.12, hence also $\prod_{f \in \mathcal{F}} M_f \notin \widetilde{\mathrm{Tens}}_{\mathfrak{g}}$. $\qquad\square$

Corollary 6.19. *Every object in $\widetilde{\mathrm{Tens}}_{\mathfrak{g}}$ has a finite injective resolution.*

We now introduce the following partial order on Θ: we set $\mu \preceq \lambda$ if for any sufficiently large i there exists $j > i$ such that $\mathrm{Hom}_{\mathfrak{g}_i}(V_\mu^i, V_\lambda^j) \neq 0$. If $\mu \preceq \lambda$, then $l(\lambda, \mu)$ denotes the length of a maximal chain $\mu \prec \mu_1 \prec \cdots \prec \lambda$ in Θ.

Lemma 6.20. $\mathrm{Ext}_{\mathfrak{g}}^1(V_\mu, V_\lambda) \neq 0$ *if and only if $\mu \prec \lambda$. If $\mu \prec \lambda$, then $\dim \mathrm{Ext}_{\mathfrak{g}}^1(V_\mu, V_\lambda) = \beth_1$.*

Proof. Assume that there is a non-trivial extension

$$(8) \qquad\qquad 0 \to V_\lambda \to X \to V_\mu \to 0.$$

We will show that $\mu \prec \lambda$. Let, on the contrary, $\mathrm{Hom}_{\mathfrak{g}_i}(V_\mu^i, V_\lambda^j) = 0$ for all $j > i$. Then $\mathrm{Hom}_{\mathfrak{g}_i}(V_\mu^i, V_\lambda) = 0$. Since $\dim \mathrm{Hom}_{\mathfrak{g}_i}(V_\mu^i, V_\mu) = 1$, we have $\dim \mathrm{Hom}_{\mathfrak{g}_i}(V_\mu^i, X) = 1$. Let $\varphi : V_\mu^i \to X$ be a non-zero homomorphism. Then $U(\mathfrak{g}) \cdot \varphi(V_\mu^i) \simeq X$. Therefore, φ extends to a homomorphism of $\mathfrak{g}$-modules $V_\mu \to X$, and this yields a splitting of the exact sequence (8). Thus, $\mathrm{Ext}_{\mathfrak{g}}^1(V_\mu, V_\lambda) \neq 0$ implies $\mu \prec \lambda$.

Now let $\mu \prec \lambda$. Then there exists an infinite sequence $i_1, i_2, \ldots$ such that $\mathrm{Hom}_{\mathfrak{g}_{i_j}}(V_\mu^{i_j}, V_\lambda^{i_{j+1}}) \neq 0$ for all j. Consider a sequence of non-zero homomorphisms $\varphi_j \in \mathrm{Hom}_{\mathfrak{g}_{i_j}}(V_\mu^{i_j}, V_\lambda^{i_{j+1}})$ and set $Z_j := V_\mu^{i_j} \oplus V_\lambda^{i_j}$. Denote by e_j (respectively, f_j) the inclusion $V_\mu^{i_j} \to V_\mu^{i_{j+1}}$ (resp., $V_\lambda^{i_j} \to V_\lambda^{i_{j+1}}$). Define $\psi_j : Z_j \to Z_{j+1}$ by

$$\psi(x, y) = (e_j(x), \varphi_j(x) + f_j(y)), \ x \in V_\mu^{i_j}, y \in V_\lambda^{i_j}.$$

Consider $Z = \varinjlim Z_j$. It is an exercise to check that Z is an extension of V_μ by V_λ, and that it does not split if infinitely many $\varphi_j \neq 0$. Hence the dimension of $\text{Ext}^1_{\mathfrak{g}}(V_\mu, V_\lambda)$ is at least $\beth_1$. On the other hand, the dimension of $\text{Ext}^1_{\mathfrak{g}}(V_\mu, V_\lambda)$ is bounded by the multiplicity of V_μ in $\text{soc}^1(I_\lambda)/\text{soc}(I_\lambda)$. The dimension of $I_\mu = ((V_\mu)_*)^*$ is $\beth_1$, hence the dimension of $\text{Ext}^1_{\mathfrak{g}}(V_\mu, V_\lambda)$ is at most $\beth_1$.

To finish the proof just note that $\text{Ext}^1_{\mathfrak{g}}(V_\lambda, V_\lambda) = 0$ by Lemma 6.6. $\qquad\square$

Corollary 6.21. *The category $\widetilde{\text{Tens}}_{\mathfrak{g}}$ consists of a single block.*

Proof. According to Lemma 6.20, $\text{Ext}^1_{\mathfrak{g}}(\mathbb{C}, V_\mu) \neq 0$ for any $\mu \in \Theta$. $\qquad\square$

Proposition 6.22. *For $k \in \mathbb{Z}_{>0}$, set*
$$\Theta^k(\lambda) = \{\mu \prec \lambda \,|\, l(\lambda, \mu) \geq k+1\}.$$
Then
$$\text{soc}^k(I_\lambda)/\text{soc}^{k-1}(I_\lambda) = \bigoplus_{\mu \in \Theta^k(\lambda)} X^\mu \otimes V_\mu,$$
where each X^μ is a trivial $\mathfrak{g}$-module of dimension $\beth_1$.

Proof. For $k = 1$ the statement follows from Lemma 6.20. Now we proceed by induction on k. Note first that if V_μ is a simple constituent of $\text{soc}^k(I_\lambda)/\text{soc}^{k-1}(I_\lambda)$, then, by Lemma 6.20, $\mu \prec \chi$ for some simple constituent V_χ of $\text{soc}^{k-1}(I_\lambda)/\text{soc}^{k-2}(I_\lambda)$. By the induction assumption, $\chi \in \Theta^{k-1}(\lambda)$. In addition, it is clear that V_μ is a simple constituent of $\text{soc}^k(I_\lambda)/\text{soc}^{k-1}(I_\lambda)$ if and only if there exists a non-zero homomorphism $\varphi : I_\lambda \to I_\mu$, such that $\varphi(\text{soc}^{k-1}(I_\lambda)) = 0$. By Lemma 6.10, φ is surjective, so all simple constituents of $\text{soc}^1(I_\mu)/\text{soc}(I_\mu)$ are also simple constituents of $\text{soc}^k(I_\lambda)/\text{soc}^{k-1}(I_\lambda)$. This implies that V_μ is a simple constituent of $\text{soc}^k(I_\lambda)/\text{soc}^{k-1}(I_\lambda)$ if and only if there exists $\psi \in \Theta^{k-1}(\lambda)$ such that $\mu \in \Theta^1(\psi)$. Since $\mu \in \Theta^1(\psi)$ if and only if $\mu \in \Theta^k(\lambda)$, the statement follows. $\qquad\square$

Let $\text{Tens}_{\mathfrak{g}}$ be the full subcategory of $\widetilde{\text{Tens}}_{\mathfrak{g}}$ consisting of modules M whose cardinality $\text{card}M$ is bounded by $\beth_n$ for some n depending on M.

Theorem 6.23. *$\text{Tens}_{\mathfrak{g}}$ is the unique minimal abelian full subcategory of $\text{Int}_{\mathfrak{g}}$ which does not consist of trivial modules only and which is closed under $\otimes$ and *.*

Proof. Let $\mathcal{C}$ be a minimal abelian full subcategory of $\text{Int}_{\mathfrak{g}}$ which contains a non-trivial module M and is closed under $\otimes$ and *. We will show that $V \in \mathcal{C}$. Since $\text{End}_{\mathbb{C}}M$ is a $\mathfrak{g}$-submodule of $(M^* \otimes M)^*$ (through the map $\varphi(\psi \otimes m) = \psi(\varphi(m))$ for $m \in M$, $\psi \in M^*$, $\varphi \in \text{End}_{\mathbb{C}}M$), we have $\text{End}_{\mathbb{C}}M \in \mathcal{C}$. Furthermore, the adjoint module $\mathfrak{g}$ is a submodule of $\text{End}_{\mathbb{C}}M$. Hence $\mathfrak{g} \in \mathcal{C}$. Recall that $\mathfrak{g}$ is the socle of $V_* \otimes V$ for $sl(\infty)$, of $\Lambda^2(V)$ for $o(\infty)$, and of $S^2(V)$ for $sp(\infty)$. In all cases it is easy to see that $\mathfrak{g}^*$ contains a subquotient isomorphic to V. Therefore, $V \in \mathcal{C}$. In addition, $V_* = \text{soc}(V^*) \in \mathcal{C}$. Therefore, $T^{p,q} \in \mathcal{C}$ for all p, q, and $V_\lambda \in \mathcal{C}$ for all $\lambda \in \Theta$. Finally, by Corollary 6.7 a), any $M \in \widetilde{\text{Tens}}_{\mathfrak{g}}$ is a submodule of $(\text{soc}(M)_*)^*$, and the statement follows. $\qquad\square$

We conclude this paper with the remark that the category $\widetilde{\text{Tens}}_{\mathfrak{g}}$, for $\mathfrak{g} = sl(\infty), o(\infty), sp(\infty)$, is functorial with respect to any homomorphism of locally semisimple Lie algebras $\varphi : \mathfrak{g}' \to \mathfrak{g}$. By this we mean that any $M \in \widetilde{\text{Tens}}_{\mathfrak{g}}$ considered as a $\mathfrak{g}'$-module is an object of $\widetilde{\text{Tens}}_{\mathfrak{g}'}$.

To prove this, recall that the image of φ', being a locally semisimple subalgebra of $\mathfrak{g}$, is isomorphic to a direct sum of copies of $sl(\infty), o(\infty), sp(\infty)$ and of finite-dimensional simple Lie algebras, [DP2]. Furthermore, the result of [DP2] implies that as $\mathfrak{g}'$-modules both V and V_* have Loewy length at most 2 and that all non-trivial simple constituents of V and V_* are isomorphic to the natural and conatural representations $V_{\mathfrak{s}}$ and $(V_{\mathfrak{s}})_*$ for some simple direct summands $\mathfrak{s}$ of $\varphi(\mathfrak{g}')$ and that all non-trivial constituents occur with finite multiplicity. (The simple trivial representation may occur with up to countable multiplicity in both $\mathrm{soc}(V)$ and $V/\mathrm{soc}(V)$ (respectively, $\mathrm{soc}(V_*)$ and $V_*/\mathrm{soc}(V_*)$.) This allows us to conclude that any single simple object of $\widetilde{\mathrm{Tens}}_{\mathfrak{g}}$ is an object of $\widetilde{\mathrm{Tens}}_{\varphi(\mathfrak{g}')}$. Hence, by Theorem 6.12, any $M \in \widetilde{\mathrm{Tens}}_{\mathfrak{g}}$ is an object of $\widetilde{\mathrm{Tens}}_{\varphi(\mathfrak{g}')}$.

References

[Ba1] A. Baranov, Complex finitary simple Lie algebras, Archiv der Math. 71(1998), 1-6.

[Ba2] A. Baranov, Simple diagonal locally finite Lie algebras, Proc. London Math. Soc. (3) 77 (1998), 362-386.

[Ba3] A. Baranov, Finitary simple Lie algebras, J. Algebra 219(1999), 299-329.
[BZh] A. Baranov, A. Zhilinski, Diagonal direct limits of simple Lie algebras, Comm. Algebra 27 (1998), 2749-2766.

[DP1] I. Dimitrov, I. Penkov, Weight modules of direct limit Lie algebras, IMRN 1999, no. 5, 223-249.

[DP2] I. Dimitrov, I. Penkov, Locally semisimple and maximal subalgebras of the finitary Lie algebras gl(∞), sl(∞), so(∞), and sp(∞), Journ. Algebra 322 (2009), 2069-2081.

[GW] R. Goodman, N. Wallach, Representations and Invariants of the Classical Groups, Cambridge University Press, 1998.

[PS] I. Penkov, K. Styrkas, Tensor representations of infinite-dimensional root-reductive Lie algebras, in Developments and Trends in Infinite-Dimensional Lie Theory, Progress in Mathematics 288, Birkhauser, 2011, pp. 127-150.

[PStr] I. Penkov, H. Strade, Locally finite Lie algebras with root decomposition, Archiv der Math. 80 (2003), 478-485.

[W] C. Weibel, An introduction to homological algebra, Cambridge Studies in Advanced Mathematics, Cambridge University Press, 1994.

SCHOOL OF ENGINEERING AND SCIENCE, JACOBS UNIVERSITY BREMEN, CAMPUS RING 1, 28759 BREMEN, GERMANY
E-mail address: i.penkov@jacobs-university.de

DEPARTMENT OF MATHEMATICS, UNIVERSITY OF CALIFORNIA BERKELEY, BERKELEY CA 94720, USA
E-mail address: serganov@math.berkeley.edu

Contemporary Mathematics
Volume **557**, 2011

Binomial coefficients and Littlewood-Richardson coefficients for interpolation polynomials and Macdonald polynomials

Siddhartha Sahi

Dedicated to Gregg Zuckerman on his 60th birthday.

ABSTRACT. We establish a precise relationship between binomial coefficents and Littlewood-Richardson coefficients for interpolation polynomials and Macdonald polynomials, and obtain explicit formulas for both kinds of coefficients.

Introduction

Let $\mathbb{F} = \mathbb{Q}(q, t)$ denote the field of rational functions in q, t. In ([**18**], [**6**], [**19**], [**4**]) the author and F. Knop introduced two inhomogeneous polynomial bases

$$(0.1) \qquad \{G_\eta : \eta \in \mathcal{C}_n\} \subset \mathbb{F}[x_1, \ldots, x_n], \ \{R_\lambda : \lambda \in \mathcal{P}_n\} \subset \mathbb{F}[x_1, \ldots, x_n]^{S_n}$$

whose index sets are, respectively, compositions and partitions of length n:

$$\mathcal{C}_n := \{\eta = (\eta_1, \ldots, \eta_n) : \eta_i \in \mathbb{Z}_{\geq 0}\}, \ \mathcal{P}_n := \{\lambda \in \mathcal{C}_n : \lambda_1 \geq \lambda_2 \geq \ldots \geq \lambda_n\}.$$

R_λ and G_η are called interpolation polynomials and, as shown in ([**19**], [**4**]), their top degree terms are, respectively, the symmetric and nonsymmetric Macdonald polynomials of type A ([**12**], [**17**, **3**, **11**]).

In this paper we prove several new results about R_λ and G_η. We first introduce common notation to avoid having to state the results twice. Thus we write

$$\{h_v(x) : v \in L\} \subset \mathcal{R}$$

to denote *either* of the two situations in (0.1).

The index set L admits a partial order $\supseteq$, which, together with the "rank" function $|v| = v_1 + \cdots + v_n$, makes L into a graded poset ([**2**]). Furthermore there is a certain map $u \mapsto \overline{u} : L \to \mathbb{F}^n$ such that h_v is characterized as the unique polynomial in $\mathcal{R}$ of degree $|v|$ satisfying

$$(0.2) \qquad h_v(\overline{v}) = 1; \text{ and } h_v(\overline{u}) = 0 \text{ unless } u \supseteq v.$$

We refer the reader to sections 0.3 and 1.1 for precise definitions of $\supseteq$ and $u \mapsto \overline{u}$ in the symmetric and non-symmetric cases, respectively.

1991 *Mathematics Subject Classification.* Primary 33D52, 05E05; Secondary 05E15.

Key words and phrases. Macdonald polynomials, Interpolation polynomials, Littlewood-Richardson coefficients, Binomial coefficients.

0.1. Binomial coefficients. Our first result is a formula for the special values $h_v(\bar{u})$, which are called *binomial coefficients* in [**15, 21**]; we define

(0.3) $$b_{uv} := h_v(\bar{u})$$

We denote by $:\supset$ the covering relation of $\supseteq$; thus we have

$$u :\supset v \text{ iff } u \supseteq v \text{ and } |u| = |v| + 1$$

The b_{uv} are *explicitly* known if $u :\supset v$ (see [**1, 15**] and formulas (0.15),(0.18) below); to emphasize this fact we write

$$a_{uv} = \left\{ \begin{array}{ll} b_{uv} & \text{if } u :\supset v \\ 0 & \text{else} \end{array} \right. .$$

Consider the $L \times L$ matrices $A = (a_{uv})$, $B = (b_{uv})$, and the diagonal matrix $Z = (|\bar{u}|\,\delta_{uv})$, where we define for any n-tuple, e.g. for $y \in \mathbb{F}^n$

$$|y| := y_1 + \cdots + y_n$$

By (0.2), (0.3) B is unitriangular and hence invertible. We denote its inverse by

$$B^{-1} = (b'_{uv})$$

THEOREM 0.1. .

(1) *The following recursions characterize b_{uv} and b'_{uv}:*

(0.4)   (i) $b_{uu} = 1$, (ii) $(|\bar{u}| - |\bar{v}|)b_{uv} = \sum_{w:\supset v} b_{uw}(|\bar{w}| - |\bar{v}|)a_{wv}$.

(0.5)   (i) $b'_{uu} = 1$, (ii) $(|\bar{u}| - |\bar{v}|)b'_{uv} = \sum_{w \subset u} a_{uw}(|\bar{w}| - |\bar{u}|)b'_{wv}$.

(2) *The matrices A, B, Z satisfy the commutation relations*

(0.6)   (i) $[Z, B] = B[Z, A]$, (ii) $[Z, B^{-1}] = -[Z, A]B^{-1}$

(3) *Let $\mathfrak{C}_{uv} := \{\mathbf{w} = (w_0, w_1, \cdots, w_k) \mid w_0 = u, w_k = v, w_i :\supset w_{i+1}\}$; then*

(0.7) $b_{uv} = \sum_{\mathbf{w} \in \mathfrak{C}_{uv}} wt(\mathbf{w})$ *with* $wt(\mathbf{w}) = \prod_{i=0}^{k-1} \left[\frac{|\overline{w_i}| - |\overline{w_{i+1}}|}{|\overline{w_0}| - |\overline{w_{i+1}}|} a_{w_i, w_{i+1}} \right]$.

(0.8) $b'_{uv} = \sum_{\mathbf{w} \in \mathfrak{C}_{uv}} wt'(\mathbf{w})$ *with* $wt'(\mathbf{w}) = \prod_{i=0}^{k-1} \left[\frac{|\overline{w_{i+1}}| - |\overline{w_i}|}{|\overline{w_i}| - |\overline{w_k}|} a_{w_i, w_{i+1}} \right]$.

0.2. Littlewood Richardson coefficients. Our second result concerns the *Littlewood Richardson* coefficients $c_{uv} := c_{uv}(p)$, which are defined for each $p \in \mathcal{R}$ by the product expansion

(0.9) $$p(x) h_v(x) = \sum_u c_{uv} h_u(x).$$

THEOREM 0.2. *The following recursion characterizes $c_{uv} := c_{uv}(p)$:*
(0.10)
(i) $c_{uu} = p(\bar{u})$ (ii) $[|\bar{u}| - |\bar{v}|]c_{uv} = \sum_{w:\supset v} c_{uw}[|\bar{w}| - |\bar{v}|]a_{wv} - \sum_{w \subset u} [|\bar{u}| - |\bar{w}|]a_{uw}c_{wv}$

The matrices $C = C(p) = (c_{uv})$ and $D = D(p) = (p(\bar{u})\,\delta_{uv})$ satisfy:

(0.11)   (i) $C = B^{-1}DB$, (ii) $[Z, C] = [C, [Z, A]]$.

Of special interest are the Littlewood Richardson coefficients for h_w, which are defined as follows:

(0.12) $$c^u_{vw} := c_{uv}(h_w).$$

These can be expressed entirely in terms of binomial coefficients. Define

$$\mathfrak{C}^u_{vw}(z) = \mathfrak{C}_{uz} \times \mathfrak{C}_{zw} \times \mathfrak{C}_{zv}, \quad \mathfrak{C}^u_{vw} = \cup_z \mathfrak{C}^u_{vw}(z)$$

and for $\omega = \left(\mathbf{w}^1, \mathbf{w}^2, \mathbf{w}^3 \right) \in \mathfrak{C}^u_{vw}$ define

$$wt\left(\omega \right) = wt'\left(\mathbf{w}^1 \right) wt\left(\mathbf{w}^2 \right) wt\left(\mathbf{w}^3 \right) .$$

THEOREM 0.3. *The coefficient c^u_{vw} is given explicitly as follows:*

(0.13)
$$c^u_{vw} = \sum_z b'_{uz} b_{zw} b_{zv} = \sum_{\omega \in \mathfrak{C}^u_{vw}} wt\left(\omega \right)$$

0.3. The symmetric case. We now make the above results explicit in the symmetric case, and give an application of Theorem 0.3 to symmetric Macdonald polynomials.

DEFINITION 0.4. For $\lambda \in \mathcal{P}_n$ we define

$$\bar{\lambda} = \left(\bar{\lambda}_1, \cdots, \bar{\lambda}_n \right) \text{ where } \bar{\lambda}_i = q^{\lambda_i} t^{1-i}$$

For $\lambda, \mu \in \mathcal{P}_n$ we write $\lambda \supseteq \mu$ if $\lambda_i \leq \mu_i$ for all i, so that the diagram of λ contains that of μ. We write $\lambda :\supset \mu$ if $\lambda \supseteq \mu$ and $|\lambda| = |\mu| + 1$.

By [19], [4] for each $\lambda \in \mathcal{P}_n$ there exists a unique polynomial $R_\lambda(x)$ in $\mathbb{F}[x_1, \ldots, x_n]^{S_n}$ such that

$$\deg\left(R_\lambda \right) = |\lambda|, \ R_\lambda\left(\bar{\mu} \right) = \delta_{\lambda\mu} \text{ for } |\mu| \leq |\lambda|$$

DEFINITION 0.5. For $\lambda, \mu \in \mathcal{P}_n$ we define the symmetric binomial coefficient to be $b_{\lambda\mu} = R_\lambda\left(\bar{\mu} \right)$. If $\lambda :\supset \mu$ we write $a_{\lambda\mu} = b_{\lambda\mu}$.

Our result give an explicit formula for $b_{\lambda\mu}$. To state this formula we recall some standard notation related to partitions from [**12**].

The Young diagram of a partition λ is a left-justified array of boxes with λ_i boxes in row i. Transposing the diagram of λ gives the diagram of a new partition, usually denoted λ', such that λ'_j is the length of the jth column of the diagram of λ. If $s = (i, j)$ is the box in row i and column j; we define the *arm* and *leg* of s to be

$$a(s) = \lambda_i - j, \ l(s) = \lambda'_j - i.$$

and we define the (q, t)-*hooklengths* of λ as in [**12**, VI.8.1,1']:

(0.14)
$$c_\lambda(s) = 1 - q^{a(s)} t^{l(s)+1}, \ c_\lambda = \prod_{s \in \lambda} c_\lambda(s)$$
$$c'_\lambda(s) = 1 - q^{a(s)+1} t^{l(s)}, \ c'_\lambda = \prod_{s \in \lambda} c'_\lambda(s)$$

If $\lambda \supseteq \mu$ we write λ/μ for the "skew" diagram consisting of the boxes in λ which are not in μ. If $\lambda :\supset \mu$ then λ/μ consists of a single box.

A standard skew tableau of shape λ/μ is a labelling of the boxes of λ/μ by the numbers $1, 2, \cdots, k$ where $k = |\lambda| - |\mu|$, such that the labels increase from left to right along each row and from top to bottom along each column. We write $ST_{\lambda/\mu}$ for the set of such tableaux, which can also be regarded as sequences of partitions

$$\lambda = \lambda^0 :\supset \lambda^1 :\supset \cdots :\supset \lambda^k = \mu$$

where λ^i is obtained from λ by deleting the boxes with labels $1, \cdots, i$.

THEOREM 0.6. *If $\lambda :\supset \mu$ with $\lambda/\mu = (i, j)$, let R_i and C_j denote the (other) boxes in row i and column j, respectively, then we have*

(0.15)
$$a_{\lambda\mu} = t^{1-i} \prod_{s \in C_j} \frac{c_\lambda(s)}{c_\mu(s)} \prod_{s \in R_i} \frac{c'_\lambda(s)}{c'_\mu(s)}.$$

DEFINITION 0.7. *If* $T \in ST_{\lambda/\mu}$ *with* $T = \left(\lambda = \lambda^0 :\supset \lambda^1 :\supset \cdots :\supset \lambda^k = \mu \right)$ *we*
define

$$wt\,(T) = \prod_{i=0}^{k-1} \left[\frac{|\overline{\lambda^i}| - |\overline{\lambda^{i+1}}|}{|\overline{\lambda}| - |\overline{\lambda^{i+1}}|} a_{\lambda_i, \lambda_{i+1}} \right], \; wt'\,(T) = \prod_{i=0}^{k-1} \left[\frac{|\overline{\lambda^i}| - |\overline{\lambda^{i+1}}|}{|\overline{\mu}| - |\overline{\lambda^i}|} a_{\lambda_i, \lambda_{i+1}} \right]$$

THEOREM 0.8. *If* $\lambda \not\supseteq \mu$ *then* $b_{\lambda\mu} = 0$. *If* $\lambda \supseteq \mu$ *then we have*

$$b_{\lambda\mu} = \sum_{T \in ST_{\lambda/\mu}} wt\,(T)$$

Moreover if we define

$$b'_{\lambda\mu} := \sum_{T \in ST_{\lambda/\mu}} wt'\,(T)$$

then we have

$$\sum_\mu b_{\lambda\mu} b'_{\mu\nu} = \delta_{\lambda\nu}$$

DEFINITION 0.9. *For* $p\,(x) \in \mathbb{F}\,[x_1, \ldots, x_n]^{S_n}$ *we define its Littlewood-Richardson*
coefficients $c_{\lambda\mu} = c_{\lambda\mu}\,(p)$ *via the product expansion*

$$p\,(x)\,R_\mu\,(x) = \sum_\lambda c_{\lambda\mu} R_\lambda\,(x).$$

We also define

$$c^\lambda_{\mu\nu} = c_{\lambda\mu}\,(R_\nu) = c_{\lambda\nu}\,(R_\mu)$$

THEOREM 0.10. *The coefficients* $c_{\lambda\mu} = c_{\lambda\mu}\,(p)$ *are characterized as follows*

(i) $c_{\lambda\lambda} = p\,(\overline{\lambda})$ (ii) $[|\overline{\lambda}| - |\overline{\mu}|]c_{\lambda\mu} = \sum_{\nu:\supset\mu} c_{\lambda\nu}[|\overline{\nu}| - |\overline{\mu}|]a_{\nu\mu} - \sum_{\nu \subset \lambda} [|\overline{\lambda}| - |\overline{\nu}|]a_{\lambda\nu}c_{\nu\mu}$

Moreover we have

$$c^\lambda_{\mu\nu} = \sum_\kappa b'_{\lambda\kappa} b_{\kappa\mu} b_{\kappa\nu}$$

0.4. Macdonald polynomials. We now give an application of Theorem 0.10
to Macdonald polynomials.

Let $J_\lambda\,(x; q, t)$ be the "integral form" of the symmetric Macdonald polynomial
as in [**12**, VI.8.3]. The J_λ are orthogonal with respect to the (q, t)-inner product
$\langle ., . \rangle$ defined in [**12**, VI.1.5]. By [**12**, VI.8.7] we have

(0.16) $\langle J_\lambda, J_\mu \rangle = j_\lambda \delta_{\lambda\mu}$ where $j_\lambda = c_\lambda c'_\lambda$

Using Theorem 0.3 we can obtain an explicit formula for the scalar product $\langle J_\lambda, J_\mu J_\nu \rangle$.

DEFINITION 0.11. *For* λ, μ, ν *in* $\mathcal{P}_n$ *we define*

$$n\,(\lambda) = \sum_i (i - 1)\,\lambda_i = \sum_{(i,j) \in \lambda} (i - 1) = \sum_j \lambda'_j \left(\lambda'_j - 1 \right)/2$$
$$n\,(\lambda, \mu, \nu) = n\,(\lambda) - n\,(\mu) - n\,(\nu)$$

THEOREM 0.12. *We have*

(0.17) $\langle J_\lambda, J_\mu J_\nu \rangle = \begin{cases} c^\lambda_{\mu\nu} j_\mu j_\nu q^{-n(\lambda', \mu', \nu')} t^{2n(\lambda, \mu, \nu)} & \textit{if } |\lambda| = |\mu| + |\nu| \\ 0 & \textit{else} \end{cases}$

0.5. Remarks.

(1) The definitions and notations for the symmetric interpolation polynomials are slightly different in [**19**], [**4**], and [**15**]. The precise connection between these definitions is explained on P. 471 of [**21**].

(2) The nonsymmetric analog of Theorem 0.6 is contained in [**1**, Cor 4.2], and we give a concise reformulation. Suppose $\eta :\supset \gamma \in \mathcal{C}_n$, let $1 \leq i_1 < \ldots < i_k \leq n$ be the corresponding indices as in (1.1), and for $1 \leq j \leq n$ define constants $a_j, a'_j \in \mathbb{F}$ as follows:

$$a_j = \begin{cases} \overline{\gamma}_{i_l} & j \in [i_{l-1}, i_l) \\ q\overline{\gamma}_{i_1} & j \geq i_k \end{cases} , \quad a'_j = \begin{cases} \overline{\gamma}_{i_l} & j \in (i_{l-1}, i_l] \\ q\overline{\gamma}_{i_1} & j > i_k \end{cases} .$$

Then we have

(0.18) $$a_{\eta\gamma} = \frac{a_n - t^{1-n}}{1 - t} \cdot \prod_{j=1}^{n} \frac{a'_j - t\overline{\gamma}_j}{a_j - \overline{\gamma}_j}$$

The analogs of Theorems 0.8 and 0.10 are straightforward.

(3) The nonsymmetric analog of Theorem 0.12 involves three steps. We sketch the argument below and we leave the details to the interested reader.
   • The first step is to define the analog of the (q,t)-scalar product for nonsymmetric Macdonald polynomials. This involves a reinterpretation of the results of [**14**] along the lines of [**20**]. Note however that the natural scalar product is *Hermitian* (with $q^* = q^{-1}, t^* = t^{-1}$).
   • The second step is to define the integral form of the nonsymmetric Macdonald polynomials and compute its norm explicitly.
   • Finally one needs to compute the precise normalization constant relating the integral nonsymmetric Macdonald polynomial and the top term of the nonsymmetric interpolation polynomial.

(4) The results of this paper in the limiting case of Jack polynomials were obtained in [**22**].

1. Proofs of Theorems 0.1, 0.2, 0.3

1.1. Preliminaries. In this section we recall the definition of the partial order $\supseteq$ and the map $u \mapsto \overline{u}$ on the index set L. For $L = \mathcal{P}_n$ these are defined as in Definition 0.4.

For $L = \mathcal{C}_n$, the definition of $\supseteq$ is due to [**4**]. For γ, η in $\mathcal{C}_n$, we write $\eta :\supset \gamma$ if there are indices $1 \leq i_1 < \ldots < i_k \leq n$ such that

(1.1) $$\eta_i = \begin{cases} \gamma_{i_1} + 1 & \text{if } i = i_k \\ \gamma_{i_{j+1}} & \text{if } i = i_j, \ j < k \\ \gamma_i & \text{otherwise} \end{cases}$$

DEFINITION 1.1. [**4**] We define the partial order $\supseteq$ on $\mathcal{C}_n$ to be the transitive closure of $:\supset$; conversely $:\supset$ is the covering relation of $\supseteq$.

For $L = \mathcal{C}_n$ the definition of $\overline{u}$ is due to ([**6**], [**19**], [**4**]), and involves the permutation action of the symmetric group S_n on n-tuples (in $\mathcal{C}_n, \mathbb{F}^n, \mathbb{Z}^n$, etc.). The S_n-orbit of $\eta \in \mathcal{C}_n$ contains a unique partition that we denote η_+. The set $\{\sigma \in S_n : \sigma(\eta_+) = \eta\}$ contains a unique element of minimal length that we denote by σ_η. (Here, as usual, the length of a permutation σ is the number of σ-inversions, i.e. pairs of indices $1 \leq i < j \leq n$ such that $\sigma(i) > \sigma(j)$.)

DEFINITION 1.2. For η in $\mathcal{C}_n$ we define $\bar{\eta} \in \mathbb{F}^n$ to be

(1.2) $\bar{\eta} := \sigma_\eta \left(\overline{\eta_+} \right)$

REMARK 1.3. The restrictions of $(\supseteq, u \mapsto \bar{u})$ from $\mathcal{C}_n$ to $\mathcal{P}_n$ agree with the corresponding structures on $\mathcal{P}_n$.

Let L denote $\mathcal{C}_n$ or $\mathcal{P}_n$, and let $\mathcal{R}$ denote $\mathbb{F}[x_1, \ldots, x_n]$ or $\mathbb{F}[x_1, \ldots, x_n]^{S_n}$ accordingly. We recall that $|u| := u_1 + \cdots + u_n$, and for $d \in \mathbb{Z}_{\geq 0}$ we define

$$\mathcal{R}_d = \{p \in \mathcal{R} \mid \deg(p) \leq d\}, \; L_d = \{u \in L \mid |u| \leq d\}, \; \bar{L}_d = \{\bar{u} \mid u \in L_d\}$$

The following result is key to the definition of interpolation polynomials h_v.

PROPOSITION 1.4. [**19, 4**] *A polynomial in $\mathcal{R}_d$ is determined by its values on* $\bar{L}_d$.

We briefly sketch the argument. In the symmetric case the main idea goes back to [**18**] and arose in connection with author's joint work with B. Kostant [**8, 9**] on the Capelli identity. Evaluation gives a linear map $Ev : \mathcal{R}_d \to \mathbb{F}^{\bar{L}_d}$ and the proposition asserts that this is an isomorphism. We first note that both spaces have dimension $\#(L_d)$; this is obvious for $\mathbb{F}^{\bar{L}_d}$, while for $\mathcal{R}_d$ it follows by expressing a (symmetric) polynomial in terms of (symmetric) monomials. Therefore it suffices to prove that Ev is surjective, which can be carried out by induction on d.

Interpolation polynomials are images of delta functions under Ev^{-1}.

DEFINITION 1.5. $h_v(x)$ is the unique polynomial in $\mathcal{R}_{|v|}$ satisfying

$$h_v(\bar{u}) = \delta_{uv} \text{ for all } u \in L_{|v|}$$

The following "extra" vanishing result relates $h_v(x)$ and $\supseteq$.

PROPOSITION 1.6. [**6, 4**] *We have $h_v(\bar{u}) = 0$ unless $u \supseteq v$.*

1.2. Proofs. The proof of Theorem 0.1 depends on the following simple identity for $h_v(x)$.

PROPOSITION 1.7. *Let $|x|$ denote $x_1 + \cdots + x_n$, then we have*

(1.3) $(|x| - |\bar{v}|) h_v(x) = \sum_{w: \supseteq v} (|\bar{w}| - |\bar{v}|) a_{wv} h_w(x)$

PROOF. Both sides of (1.3) are polynomials of degree $d = |v| + 1$. By Proposition 1.4 it suffices to show that they agree on $\bar{L}_d$. Now let $x = \bar{u}$, then by formula (0.2) both sides vanish if $|u| < d$ and both become $(|\bar{u}| - |\bar{v}|) a_{uv}$ if $|u| = d$. □

PROOF OF THEOREM 0.1. We first prove (0.4). By formulas (0.2), (0.3) we get $b_{uu} = h_u(\bar{u}) = 1$, which is (0.4i). Next (0.4ii) follows from Proposition 1.7 by setting $x = \bar{u}$ in (1.3) and using formulas (0.2), (0.3). Finally (0.4) characterizes b_{uv} by induction on $|u| - |v|$.

Next note that (0.6i) is equivalent (0.4ii), and (0.6ii) is equivalent (0.5ii). Also (0.6ii) is equivalent to (0.6i) since

$$[Z, B^{-1}] = -B^{-1} (ZB - BZ) B = -B^{-1} [Z, B] B$$

This proves (0.6) and (0.5ii). Now (0.5i) is obvious, and (0.5) characterizes b'_{uv} by induction on $|u| - |v|$.

We next prove (0.7). Let $\bar{b}_{uv}$ temporarily denote the sum in (0.7), It suffices to verify that $\bar{b}_{uv}$ satisfies the recursion (0.4). Now (0.4i) holds since $\bar{b}_{uu}$ involves the single chain $\mathbf{w} = (u, u)$ whose weight is the empty product 1. For (0.4ii) we observe that

$$wt\left(\mathbf{w}\right) = wt\left(\bar{\mathbf{w}}\right) \frac{|\overline{w_{k-1}}| - |\overline{v}|}{|\overline{u}| - |\overline{v}|} a_{w_{k-1},v} \text{ where } \bar{\mathbf{w}} = (w_0, w_1, \cdots, w_{k-1})$$

Therefore collecting the terms in (0.7) with $w_{k-1} = w$, we get

$$\bar{b}_{uv} = \sum_{w : \supset v} \left[\sum_{\bar{\mathbf{w}} \in \mathfrak{C}_{uw}} wt\left(\bar{\mathbf{w}}\right)\right] \frac{|\overline{w}| - |\overline{v}|}{|\overline{u}| - |\overline{v}|} a_{wv} = \sum_{w : \supset v} \bar{b}_{uw} \frac{|\overline{w}| - |\overline{v}|}{|\overline{u}| - |\overline{v}|} a_{wv},$$

which is (0.4ii). Therefore $\bar{b}_{uv} = b_{uv}$ for all u, v. The proof of (0.8) is similar. □

PROOF OF THEOREM 0.2. We first prove (0.11). Substituting $x = \overline{w}$ in (0.9) we get

$$p\left(\overline{w}\right) h_v\left(\overline{w}\right) = \sum_u c_{uv} h_u\left(\overline{w}\right).$$

By (0.2,0.3) this becomes

$$d_{ww} b_{wv} = \sum_u b_{wu} c_{uv},$$

Hence we obtain the matrix identity $DB = BC$, which is equivalent to (0.11i).

To prove (0.11ii) we calculate as follows:

$$[Z, C] = \left[Z, B^{-1} DB\right] = \left[Z, B^{-1}\right] DB + B^{-1} [Z, D] B + B^{-1} D [Z, B]$$

The middle term vanishes since Z and D are both diagonal matrices. The first and last terms can be computed by formula (0.6) and we get

$$[Z, C] = -[Z, A] B^{-1} DB + B^{-1} DB [Z, A] = -[Z, A] C + C [Z, A] = [C, [Z, A]]$$

We now prove (0.10). Since B is unitriangular, (0.11i) implies that C and D have the same diagonal entries, which is (0.10i). Next (0.10ii) is equivalent to (0.11ii). Finally (0.11) characterizes c_{uv} by induction on $|u| - |v|$. □

PROOF OF THEOREM 0.3. For $p = h_w$, the diagonal matrix $D = D\left(h_w\right)$ has diagonal entries $d_{zz} = h_w\left(\bar{z}\right) = b_{zw}$. By formula (0.11ii) we have

$$c_{vw}^u = \sum_z b_{uz}' d_{zz} b_{zv} = \sum_z b_{uz}' b_{zw} b_{zv}$$

which is the first equality in (0.13). The second equality follows from (0.7), (0.8). □

2. Proofs of Theorems 0.6, 0.8, 0.10, 0.12

2.1. Preliminaries. In this section we recall some basic results on the symmetric interpolation polynomials $R_\lambda\left(x\right)$, which are needed for the proofs of Theorems 0.6, 0.8, 0.10, 0.12 below.

We write $\mathcal{R}_d^{(n)} := \left\{p \in \mathbb{F}\left[x_1, \ldots, x_n\right]^{S_n} \mid \deg\left(p\right) \leq d\right\}$ and define the symmetrized monomials

$$m_\lambda = \sum_{\sigma \in D(\lambda)} x_1^{\sigma_1} \cdots x_n^{\sigma_n} \text{ for } \lambda \in \mathcal{P}_n$$

where $D(\lambda)$ denotes the set of all distinct rearrangements of λ. Also define maps
$\omega : \mathcal{R}_d^{(n-1)} \to \mathcal{R}_d^{(n)}, \tau : \mathcal{R}_d^{(n)} \to \mathcal{R}_d^{(n)}, \tau' : \mathcal{R}_d^{(n-1)} \to \mathcal{R}_d^{(n-1)}, \upsilon : \mathcal{R}_d^{(n)} \to \mathcal{R}_{d+n}^{(n)}$

(2.1) $\qquad \omega\left(m_{\lambda_1,\dots,\lambda_{n-1}}\right) = m_{\lambda_1,\dots,\lambda_{n-1},0}$, extended by linearity

$$(\tau f)(x_1,\dots,x_n) = f\left(x_1 - t^{1-n},\dots,x_n - t^{1-n}\right)$$

$$(\tau' f)(x_1,\dots,x_{n-1}) = f\left(x_1 + t^{1-n},\dots,x_{n-1} + t^{1-n}\right)$$

$$(\upsilon f)(x) = f\left(q^{-1}x\right) \prod_{j=1}^{n} \frac{x_i - t^{1-n}}{\bar{\lambda}_i - t^{1-n}}$$

PROPOSITION 2.1. *If $\lambda \in \mathcal{P}_n$ with $|\lambda| = d$ there is a unique $R_\lambda(x) \in \mathcal{R}_d^{(n)}$ with*

(2.2) $\qquad R_\lambda(\bar{\mu}) = \delta_{\lambda\mu}$ *for all partitions μ with $|\mu| \le d$*

Moreover if $\lambda_n > 0$, then

(2.3) $\qquad R_\lambda = \upsilon(R_{\lambda-\varepsilon})$ *for $\varepsilon = (1,\cdots,1)$*

If $\lambda_n = 0$ then there is a unique $S(x) \in \mathcal{R}_{d-n}^{(n)}$ such that

(2.4) $\qquad R_\lambda = (\tau\omega\tau')(R_{\lambda^-}) + \upsilon(S)$ *for $\lambda^- = (\lambda_1,\dots,\lambda_{n-1})$*

This is proved in ([**19**], [**4**]). The function $S(x)$ is chosen by induction so that the right side of (2.4) vanishes for $\bar{\mu}$ if $|\mu| \le d$ and $\mu_n > 0$. One verifies that (2.3) and (2.4) serve to define $R_\lambda(x)$ by induction on $n + |\lambda|$.

As shown in ([**19**], [**4**]), the polynomials $R_\lambda(x)$ are eigenfunctions for certain difference operators. We recall the result below:

PROPOSITION 2.2. *Let D_1 be the operator defined by*

$$D_1 = \sum_k A_k(x)(1 - T_k).$$

where

$$A_k(x) = \left(1 - t^{1-n}x_k^{-1}\right) \prod_{l \ne k} \frac{x_k - tx_l}{x_k - x_l}$$

and T_k is the k-th q^{-1}-shift operator

$$T_k f(x_1,\dots,x_n) = f\left(x_1,\dots,q^{-1}x_k,\dots,x_n\right)$$

then we have

(2.5) $\qquad D_1 R_\mu(x) = \left[\sum_k \left(t^{k-1} - \bar{\mu}_k^{-1}\right)\right] R_\mu(x)$

2.2. Proofs.

PROOF OF THEOREM 0.6. We evaluate (2.5) at $x = \bar{\lambda}$ to get

$$\left[\sum_k \left(t^{k-1} - \bar{\mu}_k^{-1}\right)\right] R_\mu(\bar{\lambda}) = \left[\sum_k A_k(\bar{\lambda})\right] R_\mu(\bar{\lambda}) - \sum_k A_k(\bar{\lambda}) R_\mu\left(\overline{\lambda - \varepsilon_k}\right)$$

which we rewrite as follows:

(2.6) $\qquad \left[\sum_k A_k(\bar{\lambda}) - \sum_k \left(t^{k-1} - \bar{\mu}_k^{-1}\right)\right] R_\mu(\bar{\lambda}) = \sum_k A_k(\bar{\lambda}) R_\mu\left(\overline{\lambda - \varepsilon_k}\right)$

As in Lemma 3.5 of [**19**], we have

$$A_k(\bar{\lambda}) R_\mu\left(\overline{\lambda - \varepsilon_k}\right) = 0 \text{ if } k \ne i$$

and by Lemma 3.3 of [**19**], we have

$$\sum_k A_k(\bar{\lambda}) = \sum_k \left(t^{k-1} - \bar{\lambda}_k^{-1}\right)$$

Since $\bar{\mu}_k = \bar{\lambda}_k$ for $k \neq i$, (2.6) can be rewritten as

$$\left(\bar{\mu}_i^{-1} - \bar{\lambda}_i^{-1}\right) R_\mu\left(\bar{\lambda}\right) = A_i\left(\bar{\lambda}\right) = \left(1 - t^{1-n}\bar{\lambda}_i^{-1}\right) \prod_{l \neq i} \frac{\bar{\lambda}_i - t\bar{\lambda}_l}{\bar{\lambda}_i - \bar{\lambda}_l}$$

Substituting $\bar{\mu}_i = q^{-1}\bar{\lambda}_i$ and rewriting, we get

$$R_\mu\left(\bar{\lambda}\right) = \frac{\bar{\lambda}_i - t^{1-n}}{q-1} \prod_{l \neq i} \frac{\bar{\lambda}_i - t\bar{\lambda}_l}{\bar{\lambda}_i - \bar{\lambda}_l}$$

To complete the proof of (0.15) it suffices to verify the two identities

(2.7) $$\prod_{s \in C_j} \frac{c_\lambda(s)}{c_\mu(s)} = \prod_{l < i} \frac{\bar{\lambda}_i - t\bar{\lambda}_l}{\bar{\lambda}_i - \bar{\lambda}_l}$$

(2.8) $$t^{1-i} \prod_{s \in R_i} \frac{c'_\lambda(s)}{c'_\mu(s)} = \frac{\bar{\lambda}_i - t^{1-n}}{q-1} \prod_{l > i} \frac{\bar{\lambda}_i - t\bar{\lambda}_l}{\bar{\lambda}_i - \bar{\lambda}_l}.$$

Now C_j consists of boxes $\{(l,j) \mid l < i\}$. For $s = (l,j) \in C_j$ we have

$$a_\lambda(s) = a_\mu(s) = \lambda_l - j = \lambda_l - \lambda_i; \qquad l_\lambda(s) = l_\mu(s) + 1 = i - l.$$

$$\frac{c_\lambda(s)}{c_\mu(s)} = \frac{1 - q^{a_\lambda(s)}t^{l_\lambda(s)+1}}{1 - q^{a_\mu(s)}t^{l_\mu(s)+1}} = \frac{1 - q^{\lambda_l - \lambda_i}t^{i-l+1}}{1 - q^{\lambda_l - \lambda_i}t^{i-l}} = \frac{q^{\lambda_i}t^{1-i} - q^{\lambda_l}t^{2-l}}{q^{\lambda_i}t^{1-i} - q^{\lambda_l}t^{1-l}} = \frac{\bar{\lambda}_i - t\bar{\lambda}_l}{\bar{\lambda}_i - \bar{\lambda}_l}$$

which implies (2.7).

We now prove (2.8). Denote the left and right sides of (2.8) by $X(\lambda, i)$ and $Y(\lambda, i)$ respectively. First suppose $j = 1$. Then R_i is the empty set and $X(\lambda, i) = t^{1-i}$. Also we have $\lambda_i = 1$ and $\lambda_l = 0$ for $l > i$ therefore we get

$$Y(\lambda, i) = \frac{qt^{1-i} - t^{1-n}}{q-1} \prod_{l=i+1}^{n} \frac{qt^{1-i} - t^{2-l}}{qt^{1-i} - t^{1-l}} = t^{1-i} = X(\lambda, i)$$

Now suppose $j > 1$. Let k be the largest index such that $\lambda_k > 0$ and define

$$\lambda^* = (\lambda_1 - 1, \cdots, \lambda_k - 1, 0, \cdots, 0)$$

Note that necessarily $k \geq i$. Now we have

$$\frac{X(\lambda, i)}{X(\lambda^*, i)} = \frac{c'_\lambda(1, i)}{c'_\mu(1, i)} = \frac{1 - q^j t^{k-i}}{1 - q^{j-1} t^{k-i}}$$

Also for $l \leq k$ the ratios $\frac{\bar{\lambda}_i - t\bar{\lambda}_l}{\bar{\lambda}_i - \bar{\lambda}_l}$ are unchanged when we replace λ by λ^*. Thus

$$\frac{Y(\lambda, i)}{Y(\lambda^*, i)} = \frac{q^{\lambda_i}t^{1-i} - t^{1-n}}{q^{\lambda_i - 1}t^{1-i} - t^{1-n}} \prod_{l=k+1}^{n} \frac{q^{\lambda_i}t^{1-i} - t^{2-l}}{q^{\lambda_i}t^{1-i} - t^{1-l}} \prod_{l=k+1}^{n} \frac{q^{\lambda_i - 1}t^{1-i} - t^{1-l}}{q^{\lambda_i - 1}t^{1-i} - t^{2-l}}$$

$$= \frac{q^{\lambda_i}t^{1-i} - t^{2-(k+1)}}{q^{\lambda_i - 1}t^{1-i} - t^{2-(k+1)}} = \frac{1 - q^j t^{k-i}}{1 - q^{j-1} t^{k-i}} = \frac{X(\lambda, i)}{X(\lambda^*, i)}$$

and the identity $X(\lambda, i) = Y(\lambda, i)$ follows by induction on $|\lambda|$. $\square$

Theorems 0.8, 0.10 now follow from Theorems 0.1, 0.2, respectively. For the proof of Theorems 0.12 we need a preliminary result.

LEMMA 2.3. *Let k_λ be the coefficient of $m_\lambda(x)$ in $R_\lambda(x)$; then*

(2.9) $$k_\lambda = (-1)^{|\lambda|} t^{2n(\lambda)} q^{-n(\lambda')}/c'_\lambda.$$

PROOF. We proceed by induction on $n + |\lambda|$. The result is obvious for $n + |\lambda| = 0$, and so we may suppose $n + |\lambda| > 0$.

If $\lambda_n = 0$ then $R_\lambda = (\tau \omega \tau')(R_{\lambda^-}) + \upsilon(S)$ by formula (2.4). Now τ, τ' do not change the leading terms of a polynomial, ω maps m_{λ^-} to m_λ, and the coefficient of m_λ in $\upsilon(S)$ is 0. Therefore we deduce that $k_\lambda = k_{\lambda^-}$, and since the right side of (2.9) is unchanged under passage from λ to λ^- the equality (2.9) holds by induction.

If $\lambda_n > 0$ then let $\mu = \lambda - \varepsilon$. By formula (2.3) we deduce

$$k_\mu / k_\lambda = q^{|\mu|} \prod_{j=1}^{n} \left(\bar{\lambda}_i - t^{1-n} \right) = (-1)^n q^{|\mu|} t^{n(1-n)} \prod_{j=1}^{n} \left(1 - q^{\lambda_i} t^{n-i} \right)$$

Therefore by induction we get

$$k_\lambda = \frac{(-1)^{|\mu|} t^{2n(\mu)} q^{-n(\mu')} / c'_\mu}{(-1)^n q^{|\mu|} t^{n(1-n)} \prod_{j=1}^{n} \left(1 - q^{\lambda_i} t^{n-i} \right)} = (-1)^{|\lambda|} \frac{t^{2n(\mu) + n(n-1)} q^{-n(\mu') - |\mu|}}{c'_\mu \prod_{j=1}^{n} \left(1 - q^{\lambda_i} t^{n-i} \right)}$$

To complete the proof its suffices to verify the following identities for $\lambda = \mu + \varepsilon$

$$2n(\lambda) = 2n(\mu) + n(n-1), \ n(\lambda') = n(\mu') + |\mu|, \ c'_\lambda = c'_\mu \prod_{j=1}^{n} \left(1 - q^{\lambda_i} t^{n-i} \right)$$

whose (easy) verifications we leave to the reader. □

PROOF OF THEOREM 0.12. For two polynomials $p(x)$, $q(x)$ in $\mathbb{F}[x_1, \ldots, x_n]$ we write $p \sim_d q$ if $p - q$ has total degree $< d$.

As shown in [12], the coefficient of m_λ in $J_\lambda(x)$ is c_λ. Therefore if we define

(2.10) $$r_\lambda = c_\lambda / k_\lambda = (-1)^{|\lambda|} t^{-2n(\lambda)} q^{n(\lambda')} j_\lambda$$

then by Lemma 2.3 we get

$$J_\lambda(x) \sim_{|\lambda|} r_\lambda R_\lambda(x)$$

Therefore if $d = |\mu| + |\nu|$ then by Definition 0.9 we get

$$J_\mu J_\nu \sim_d r_\mu r_\nu R_\mu R_\nu = \sum_{|\lambda| \le d} r_\mu r_\nu c^\lambda_{\mu\nu} R_\lambda \sim_d \sum_{|\lambda| = d} r_\mu r_\nu r_\lambda^{-1} c^\lambda_{\mu\nu} J_\lambda.$$

Since the first and last polynomials are homogenous of degree d, they are equal. Therefore by (0.16) we get

$$\langle J_\lambda, J_\mu J_\nu \rangle = \begin{cases} r_\mu r_\nu r_\lambda^{-1} c^\lambda_{\mu\nu} j_\lambda & \text{if } |\lambda| = |\mu| + |\nu| \\ 0 & \text{else} \end{cases}.$$

To complete the proof it suffices to verify that

$$r_\mu r_\nu r_\lambda^{-1} j_\lambda = j_\mu j_\nu q^{-n(\lambda', \mu', \nu')} t^{2n(\lambda, \mu, \nu)} \text{ if } |\lambda| = |\mu| + |\nu|$$

which follows immediately from (2.10). □

References

[1] W. Baratta, *Pieri-Type Formulas for the Nonsymmetric Macdonald Polynomials*, IMRN (Internat. Math. Res. Notices) **15** (2009), 2829-2854.

[2] P. Doubilet, G.-C. Rota, and R. Stanley, *On the foundation of combinatorial theory (VI). The idea of generating functions*, in "Sixth Berkeley Symp. on Math. Stat. and Prob., Vol. 2: Probability Theory," pp. 267–318, Univ. of California, 1972.

[3] I. Cherednik, *Nonsymmetric Macdonald polynomials*, IMRN (Internat. Math. Res. Notices) **10** (1995), 483–515.

[4] F. Knop, *Symmetric and nonsymmetric quantum Capelli polynomials*, Comment. Math. Helv. **72** (1997), 84–100.

[5] F. Knop and S. Sahi, *A recursion and a combinatorial formula for Jack polynomials*, Invent. math. **128** (1997), 9–22.

[6] F. Knop and S. Sahi, *Difference equations and symmetric polynomials defined by their zeros*, IMRN (Internat. Math. Res. Notices) **10** (1996), 473–486.

[7] J. Kaneko, *Selberg integrals and hypergeometric functions associated with Jack polynomials*, SIAM J. Math. Anal. **24** (1993), 1086–1110.

[8] B. Kostant and S. Sahi, *The Capelli identity, tube domains, and the generalized Laplace transform*, Adv. Math. **106** (1991), 411–432.

[9] B. Kostant and S. Sahi, *Jordan algebras and Capelli identities*, Invent. math. **112** (1993), 657–664.

[10] M. Lassalle, *Une formule du binome generalisee pour les polynomes de Jack*, C.R. Acad. Sci. Paris Ser. I Math. **310** (1990), 253–256.

[11] I. G. Macdonald, *Commuting differential equations and zonal spherical functions*, in "Algebraic Groups, Utrecht 1986" (A.M. Cohen et al, Eds.), Lecture Notes in Math., Vol 1271, pp. 189-200, Springer-Verlag, Berlin/Heidelberg/New York, 1987.

[12] I. G. Macdonald, *Symmetric Functions and Hall Polynomials (2nd ed.)*, Oxford Univ. Press, Oxford, 1995.

[13] I. G. Macdonald, *Affine Hecke algebras and orthogonal polynomials*, Seminaire Bourbaki **797** (1994-95) Asterisque **237** (1996), 189-207.

[14] K. Mimachi and M. Noumi, *A reproducing kernel for nonsymmetric Macdonald polynomials*, Duke Math. Journal, **91** (1998), 621–634.

[15] A. Okounkov, *Binomial formula for Macdonald polynomials and its applications*, Math. Res. Lett. **4** (1997), 533–553.

[16] A. Okounkov and G. Olshanski, *Shifted Jack polynomials, binomial formula, and applications*, Math. Res. Lett. **4** (1997), 69–78.

[17] E. Opdam. *Harmonic analysis for certain representations of the graded Hecke algebra*, Acta Math. **175** (1995), 75–121.

[18] S. Sahi, *The spectrum of certain invariant differential operators associated to a Hermitian symmetric space*, in "Lie Theorey and Geometry", Progr. Math. **123**, Birkhauser, Boston, 1994, 569–576.

[19] S. Sahi. *Interpolation, integrality, and a generalization of Macdonald's polynomials*, IMRN (Internat. Math. Res. Notices) **10** (1996), 457–471.

[20] S. Sahi, *A new scalar product for nonsymmetric Jack polynomials*, IMRN (Internat. Math. Res. Notices) **20** (1996), 997–1004.

[21] S. Sahi, *The binomial formula for nonsymmetric Macdonald polynomials*, Duke Math. J. **94** (1998) 465–277.

[22] S. Sahi, *Binomial coefficients and Littlewood-Richardson coefficients for Jack polynomials*, IMRN (Internat. Math. Res. Notices) **7** (2011), 1597–1612.

DEPARTMENT OF MATHEMATICS, RUTGERS UNIVERSITY , NEW BRUNSWICK, NJ 08903, USA
E-mail address: sahi@math.rutgers.edu

Contemporary Mathematics
Volume **557**, 2011

Restriction of some representations of U(P,Q) to a symmetric subgroup

Birgit Speh

ABSTRACT. We consider the restriction of unitary representations $A_{\mathfrak{q}}$ of $U(p,q)$ with nontrivial $(\mathfrak{g}, K)$-cohomology which are cohomologically induced from a θ–stable parabolic subalgebra $\mathfrak{q}$ with Levi subgroup $L = U(1)^s U(p - s, q)$ to the symmetric subgroups $U(p, q - r)U(r)$ and $U(r)U(p - r, q)$. By considering a generalized bottom layer map with respect to non-compact subgroups we show that the restriction of $A_{\mathfrak{q}}$ to a symmetric subgroup $U(p, q - r)U(r)$ is a direct sum of irreducible representations each with finite multiplicity. For $U(p, p)$ we find examples of representations $A_{\mathfrak{q}}$ and of subgroups H_1 and H_2 which are isomorphic but not conjugate, so that the restriction of $A_{\mathfrak{q}}$ to one subgroup is a direct sum, whereas the restriction to the the other subgroup has continuous spectrum.

In the case r=1 we obtain information about the branching law in section 4.

We also relate in section 5. harmonic forms with coefficients in $A_{\mathfrak{q}}$ to harmonic forms with coefficients in the $U(p, q - 1)U(1)$ submodules generated by the minimal K–type.

Dedicated to Gregg Zuckerman

Introduction

Understanding a unitary representation π of a Lie groups G often involves understanding its restriction to suitable subgroups H. For example, when G is semisimple and K a maximal compact subgroup the collection of K-types and their multiplicities are an important invariant of a an irreducible representation and describe a good deal of its structure. Generally speaking, the more branching laws we know for a given irreducible representation, the better we understand the representation.

In recent years there has been much progress, both in abstract theory (for example [**7**], [**8**]), and in concrete examples of branching laws (see [**13**] and the references there), in particular for the branching laws for the restriction of a representation to a symmetric subgroup.

1991 *Mathematics Subject Classification.* 22E46.
partially supported by NSF grant DMS-0901024.

We consider in this paper a family of unitary representation $A_{\mathfrak{q}}$ of $U(p,q)$ whose infinitesimal character is the same as that of the trivial representation. These representations contain at least one K–type $\tau \otimes \mu$ where μ is a character of the compact subgroup U(q). These representations are cohomologically induced from a θ–stable parabolic subgroup with a Levi subgroup $L = U(1)^s U(p-s,q)$.

We discuss in this paper the restriction of $A_{\mathfrak{q}}$ to the subgroup $H = U(p, q - r)U(r)$. Using a generalization of the algebraic methods in [13] we show that the restriction of $A_{\mathfrak{q}}$ to H is a direct sum of irreducible unitary representations of H.

THEOREM 0.1. *The restriction of $A_{\mathfrak{q}}$ to $H = U(p, q - r)U(r)$ is direct sum of irreducible $(\mathfrak{h}, K \cap H)$-modules. If r=1 then each $(\mathfrak{h}, K \cap H)$–module occurs with multiplicity one.*

In section 4 we also determine part of the branching law for the restriction of $A_{\mathfrak{q}}$ to $H = U(p, q-1)U(1)$ and show that we can state this branching law in a form similar to the Blattner formula for the branching law of a irreducible representation to a maximal compact subgroup.

This result was also proved using completely different geometric techniques in [7] 4.7. There is also some overlap with results in [6].

In contrast we prove in the last

PROPOSITION 0.2. *If $A_{\mathfrak{q}}$ is not holomorphic or antiholomorphic then its restriction to $\bar{H} = U(r)U(p-r,q)$ is not a direct sum of irreducible representation,*

J. Harris and J.-S. Li showed in [4] that the restriction of $A_{\mathfrak{q}}$ to $U(p-r-s,q)$ has a discrete series representation $A_{\bar{\mathfrak{q}}}$ as a direct summand.

Combining these results we see that

COROLLARY 0.3. *Suppose that $p > 2$, $G = U(p,p)$, $H = U(1)U(p-1,q)$ and $H^\sharp = U(p, p-1)U(1)$. There exist families of unitary representations π with nontrivial $(\mathfrak{g}, K)$-cohomology which are H-admissible, but not $H^\sharp$-admissible.*

Since the representation $A_{\mathfrak{q}}$ has nontrivial $(\mathfrak{g}, K)$–cohomology in degree $R_G = \dim(\mathfrak{u} \cap \mathfrak{p})$ and is isomorphic to a representation in the discrete spectrum of a Shimura variety, these representations are also interesting from the point of view of automorphic forms and of the cohomology of discrete groups. See for example [14]. Understanding the restrictions of representations with nontrivial cohomology to subgroups is particular interest, since it is related to the problems of understanding the contribution of cycles defined by subgroups to the homology of a locally symmetric space and modular symbols. See for example [1], [4] [11]. If the restriction of $A_{\mathfrak{q}}$ to $\mathfrak{h}$ is a direct sum of irreducible $(\mathfrak{h}, K \cap H)$–modules then the "bottom H-type" i.e the $(\mathfrak{h}, K \cap H)$–module generated by the minimal K–type, and its $(\mathfrak{h}, K \cap H)$–cohomology carries information about the restriction of harmonic forms to subvarieties.

Now suppose that $H = U(p, q-1)U(1)$. We write the Cartan decomposition $\mathfrak{g} = \mathfrak{k} \oplus \mathbf{p}$. The Lie algebra $\mathfrak{h}$ is the fix point set of an involution σ of $\mathfrak{g}$. Let $\mathbf{s}$ be the σ-invariant complement to $\mathfrak{h}$ in $\mathfrak{g}$, $R_H = \dim \mathfrak{u} \cap \mathbf{p} \cap \mathfrak{h}$ and $J_H = R_G - R_H = \dim \mathfrak{u} \cap \mathbf{p} \cap \mathbf{s}$.

The bottom H–type of $A_{\mathfrak{q}}$ is an irreducible $(\mathfrak{h}, K \cap H)$–module $A_{\mathfrak{q}_H}(\lambda_H)$ and if $A_{\mathfrak{q}}$ is not holomorphic or anti holomorphic then it has the nontrivial cohomology with coefficients in a representation $V \subset \wedge^{J_H}\mathbf{s}$ in degree R_H, where $R_H < R_G$. In section 5. we determine the relationship between the harmonic form ω representing the cohomology of $A_{\mathfrak{q}}$ and the harmonic form representing the cohomology of the bottom H–type $A_{\mathfrak{q}_H}(\lambda_H)$.

THEOREM 0.4. *Let $p, q > 1$ and $H = U(p, q-1)U(1)$. Define the projection*

$$\mathcal{P}: \; A_{\mathfrak{q}} \otimes_{K \cap H} \wedge^{R_G}\mathbf{p}^* \to (A_{\mathfrak{q}_H}(\lambda_H) \otimes \wedge^{J_H}(\mathbf{s} \cap \mathbf{p})^*) \otimes_{K \cap H} \wedge^{R_H}\mathbf{p}_H^*.$$

Let $\omega \in A_{\mathfrak{q}} \otimes_K \wedge^{R_G}\mathbf{p}^*$ be a harmonic form representing a nontrivial $(\mathfrak{g}, K)$–cohomology class of $A_{\mathfrak{q}}$. Then there exists an irreducible representation $V \subset \wedge^{T_S}(\mathbf{s} \cap \mathbf{p})$ so that

$$
\begin{aligned}
\mathcal{P}(\omega) \;\; &\in \;\; (A_{\mathfrak{q}_H}(\lambda_H) \otimes V^*) \otimes_{K \cap H} \wedge^{R_H}\mathbf{p}_H^* \\
&\subset \;\; ((A_{\mathfrak{q}})_{|\mathfrak{h}} \otimes \wedge^{J_H}\mathbf{s}^*) \otimes_{K \cap H} \wedge^{R_H}\mathbf{p}_H^*
\end{aligned}
$$

is a harmonic form representing a nontrivial class in

$$H^{R_H}(\mathfrak{h}, K \cap H, A_{\mathfrak{q}_H}(\lambda_H) \otimes V^*).$$

We plan to discuss the applications of this theorem to the cohomology of discrete groups in a sequel to this note.

The paper is organized as follows

    I. : The restriction of cohomologically induced representations
    II.: Representations of K and $K \cap H$
    III.: The restriction of $A_{\mathfrak{q}}$ to $U(p, q-r)U(r)$.
    IV.: Branching Laws
    V.: Applications to $(\mathfrak{g}, K)$–cohomology
    VI.: The restriction of $A_{\mathfrak{q}}$ to $U(r)U(p-r, q)$.

1. The restriction of cohomologically induced representations.

In this section we introduce the notation and we recall the construction of the cohomologically induced representation $A_{\mathfrak{q}}$. After recalling some results from [**13**] we discuss bottom layer $(\mathfrak{h}, H \cap K)$–types.

1.1. Let G be a connected linear semisimple Lie group. We fix a maximal compact subgroup K and Cartan involution θ. Let H be a θ-stable connected semisimple subgroup with maximal compact subgroup $K_H = K \cap H$. We pick a fundamental Cartan subgroup $C_H = T_H \cdot A_H$ of H. It is contained in a fundamental Cartan subgroup $C = T \cdot A$ of G so that $T_H = T \cap H$ and $A_H = A \cap H$. We denote the Cartan decomposition by $\mathfrak{g} = \mathfrak{k} \oplus \mathbf{p}$.

We assume that the θ–stable parabolic subalgebras of $\mathfrak{g}$ and $\mathfrak{h}$ which define the representations are "well aligned", namely we fix x_o in $\mathfrak{t}_H$. Then $i\, x_o$ defines well aligned θ-stable parabolic subalgebras $\mathfrak{q} = \mathfrak{l} \oplus \mathfrak{u}$ and $\mathfrak{q}_H = \mathfrak{l}_H \oplus \mathfrak{u}_H = \mathfrak{q} \cap \mathfrak{h}$ of $\mathfrak{g}$, respectively $\mathfrak{h}$; for details see page 274 in [**5**].

We write L and L_H for the centralizer of x_0 in G and in H, respectively. For a unitary character λ of L we write λ_H for the restriction of λ to L_H. We will always consider representations of L and not of the metaplectic cover of L as some other authors.

Recall that the following complex A^* defines the cohomologically induced representations:

Let V be a representation of L. We consider $U(\mathfrak{g})$ as right $U(\mathfrak{q})$–module and write $V^\sharp = V \otimes \wedge^{\mathrm{top}}\mathfrak{u}$. Let $\mathbf{p}_L$ be a $L \cap K$–invariant complement of $\mathfrak{l} \cap \mathfrak{k}$ in $\mathfrak{l}$. We write $\mathbf{r}_G = \mathbf{p}_L \oplus \mathfrak{u}$. Since all the groups considered in the paper are connected we use the original definition of the Zuckerman functor [15] and do not use the Hecke algebra $R(\mathfrak{g}, K)$ as on page 167 in [5] to define the representations $A_{\mathfrak{q}}(V)$. Consider the complex

$$0 \to \mathrm{Hom}_{L\cap K}(U(\mathfrak{g}), \mathrm{Hom}(\wedge^0 \mathbf{r}_G, V^\sharp))_K \to$$
$$\to \mathrm{Hom}_{L\cap K}(U(\mathfrak{g}), \mathrm{Hom}(\wedge^1 \mathbf{r}_G, V^\sharp))_K \to$$
$$\to \mathrm{Hom}_{L\cap K}(U(\mathfrak{g}), \mathrm{Hom}(\wedge^2 \mathbf{r}_G, V^\sharp))_K \to \cdots .$$

Here the subscript K denotes the subspace of K–finite vectors. For $T(x, U(\cdot)) \in \mathrm{Hom}_{L\cap K}(U(\mathfrak{g}), \mathrm{Hom}_{\mathbb{C}}(\wedge^{n-1}\mathbf{r}_G, V^\sharp))_K$ we define

$$d\, T(x, U(X_1 \wedge X_2 \wedge \cdots \wedge X_n)) = \sum_{i=1}^n (-1)^i T(X_i x, U(X_1 \wedge X_2 \wedge \cdots \hat{X}_i \cdots \wedge X_n))$$

$$+ \sum_{i=1}^n (-1)^{i+1} T(x, X_i U(X_1 \wedge X_2 \wedge \cdots \hat{X}_i \cdots \wedge X_n))$$

$$+ \sum_{i<j} (-1)^{i+j} T(x, U(P_{\mathbf{r}_G}[X_i, X_j] \wedge X_1 \wedge X_2 \wedge \cdots \hat{X}_i \cdots \hat{X}_j \cdots \wedge X_n)),$$

where $x \in U(\mathfrak{g})$, $X_j \in \mathbf{r}_G$ and $P_{\mathbf{r}_G}$ is the projection onto $\mathbf{r}_G$ along $\mathfrak{l} \cap \mathfrak{k}$. Let χ be the infinitesimal character of V. If

$$\frac{2\langle \chi + \rho(\mathfrak{u}), \alpha\rangle}{|\alpha|^2} \notin \{0, -1, -2, -3, \dots\} \qquad \text{for} \quad \alpha \in \Delta(\mathfrak{u}),$$

where $\langle\ ,\ \rangle$ denotes the Killing form of $\mathfrak{g}$, then the cohomology is zero except in degree $S = S_G = \dim(\mathfrak{u} \cap \mathfrak{k})$. If V is irreducible this defines an irreducible $(U(\mathfrak{g}), K)$–module $\mathcal{R}^S_{(\mathfrak{g},K)}(V)$ in degree S (8.28 in [5]). By 5.24 in [5], (see also the remark/example on page 344) the infinitesimal character of $\mathcal{R}^S_{(\mathfrak{g},K)}(V)$ is $\chi + \rho(\mathfrak{u})$. If V is a character λ we write often write $A_{\mathfrak{q}}(\lambda)$ instead of $\mathcal{R}^S_{(\mathfrak{g},K)}(V)$.

If V is trivial the infinitesimal character of $A_{\mathfrak{q}}(V)$ is ρ_G and we write simply $A_{\mathfrak{q}}$. Two representations $A_{\mathfrak{q}}$ and $A_{\mathfrak{q}'}$ are equivalent if $\mathfrak{q}$ and $\mathfrak{q}'$ are conjugate under the compact Weyl group W_K.

1.2. We assume now that H is the connected component of the fix point set of an involution σ which commutes with the Cartan involution and let

$$\mathfrak{g} = \mathfrak{h} \oplus \mathfrak{s}$$

be the corresponding decomposition. As in [13] we consider additional complexes.

Let $\mathbb{C}_{\lambda_H}$ be the one dimensional representation of L_H defined by $\lambda \otimes \wedge^{\mathrm{top}}(\mathfrak{u} \cap \mathfrak{s})$. Then

$$\mathbb{C}_\lambda^\sharp = \mathbb{C}_\lambda \otimes \wedge^{\mathrm{top}}\mathfrak{u} = \mathbb{C}_{\lambda_H} \otimes \wedge^{\mathrm{top}}\mathfrak{u}^H = \mathbb{C}_{\lambda_H}^\sharp .$$

We define the complex L_H^* as follows

(1.1) $$(\mathrm{Hom}_{L \cap K \cap H}(U(\mathfrak{g}), \mathrm{Hom}(\wedge^i \mathbf{r}_H, \mathbb{C}_{\lambda_H}^\sharp))_{K \cap H}, d_H).$$

Here $\mathbf{r}_H = \mathbf{r}_G \cap \mathfrak{h}$ and d_H is defined analogously to the differential d previously defined.

Remark: In the notation of [**5**] this is the complex

$$P_{\mathfrak{h}, K \cap H}^{\mathfrak{g}, K \cap H}(\mathrm{Hom}_{L \cap K \cap H}(U(\mathfrak{h}), \mathrm{Hom}(\wedge^i \mathbf{r}_H, \mathbb{C}_{\lambda_H}^\sharp))_{K \cap H}), d_H).$$

As a left $U(\mathfrak{l}^H)$-module

$$U(\mathfrak{g}) = Q \otimes U(\mathfrak{h}),$$

where Q is the symmetric algebra $S(\mathbf{s})$. (See [**5**, 2.56].) We have

$$\mathrm{Hom}_{L \cap K \cap H}(U(\mathfrak{g}), \mathrm{Hom}(\wedge^i \mathbf{r}_H, \mathbb{C}_{\lambda_H}^\sharp))_{K \cap H} =$$
$$= \mathrm{Hom}_{L \cap K \cap H}(Q \otimes U(\mathfrak{h}), \mathrm{Hom}(\wedge^i \mathbf{r}_H, \mathbb{C}_{\lambda_H}^\sharp))_{K \cap H}$$
$$= \mathrm{Hom}_{L \cap K \cap H}(U(\mathfrak{h}), \mathrm{Hom}(\wedge^i \mathbf{r}_H, Q^* \otimes \mathbb{C}_{\lambda_H}^\sharp))_{K \cap H}.$$

$U(\mathfrak{g})$ acts on this space from the right and a quick check shows that d_H commutes with this action. Therefore we have an action of $U(\mathfrak{g})$ on the cohomology of the complex L_H^* .

Finally we consider the "large complex" L^*

$$(\mathrm{Hom}_{K \cap L \cap H}(U(\mathfrak{g}), \mathrm{Hom}(\wedge^i \mathbf{r}_G, \mathbb{C}_\lambda^\sharp))_{K \cap H}, d).$$

We consider the complex A^* defined in 1.1 using the forgetful functor as a subcomplex of L^*.

1.3. We have

$$\mathbf{r}_G = \mathbf{r}_H \oplus (\mathfrak{u} \cap \mathbf{s}) \oplus (\mathbf{p}_L \cap \mathbf{s}),$$

and so

$$\wedge^i \mathbf{r}_G = \oplus_{l+k=i} \wedge^k \mathbf{r}_H \otimes \wedge^l (\mathfrak{u} \cap \mathbf{s} \oplus \mathbf{p}_L \cap \mathbf{s}).$$

Using this and the observation that $(\mathfrak{u} \cap \mathbf{s}) \oplus (\mathbf{p}_L \cap \mathbf{s})$ is invariant under $L \cap K \cap H$, we can define a "pull back" map of forms

$$\mathbf{pb}_H^{i,j} : \ \mathrm{Hom}_{L \cap K \cap H}(U(\mathfrak{h}), \mathrm{Hom}(\wedge^i \mathbf{r}_H, \wedge^j (\mathfrak{u} \cap \mathbf{s} \oplus \mathbf{p}_L \cap \mathbf{s})^* \otimes (Q^* \otimes \mathbb{C}_{\lambda_H}^\sharp))_{K \cap H}$$
$$\to \mathrm{Hom}_{K \cap L \cap H}(U(\mathfrak{g}), \mathrm{Hom}(\wedge^{i+j} \mathbf{r}_G, \mathbb{C}_\lambda^\sharp))_{K \cap H}.$$

The pullback map commutes with the right action of $U(\mathfrak{h})$ and induces a map of complexes.

Thus restricting to the subcomplex of K–finite forms we get an $(\mathfrak{h}, K \cap H)$–map

(1.2) $$\mathcal{R}_{(\mathfrak{g},K)}^q(\lambda) \to \mathcal{R}_{(\mathfrak{h},H \cap K)}^q(\lambda_H)$$

If $q = S_H = S_G$ and if the image is nonzero we call the image a **bottom** $(\mathfrak{h}, K \cap H)$–**type of** $\mathcal{R}_{(\mathfrak{g},K)}^{S_G}(\lambda)$.

Remark: This construction follows the ideas of the construction of the bottom layer K–types in [**5**].

1.4. Since the parabolic subalgebras are well aligned we have

$$\mathfrak{u} \cap \mathbf{s} = (\mathfrak{u} \cap \mathfrak{k} \cap \mathbf{s}) \oplus (\mathfrak{u} \cap \mathbf{s} \cap \mathbf{p})$$

and both summands are invariant under the adjoint action of $\mathfrak{l} \cap \mathfrak{k} \cap \mathfrak{h}$. Let $J_H = \dim \mathfrak{u} \cap \mathfrak{k} \cap \mathbf{s}$ and consider

$$\mathrm{Hom}_{L \cap K \cap H}(U(\mathfrak{h}_K), \mathrm{Hom}(\wedge^i(\mathbf{r}_H \cap \mathfrak{k}) \otimes \wedge^{J_H}(\mathfrak{u} \cap \mathfrak{k} \cap \mathbf{s}), \mathbb{C}^{\sharp}_{\lambda_H}))_{K \cap H}.$$

Write λ_{K_H} for the character on $\wedge^{J_H}(\mathfrak{u} \cap \mathfrak{k} \cap \mathbf{s})^* \otimes \lambda_H$. Using the differential d_{K_H} we obtain again a complex. Its cohomology defines a representation

$$\mathcal{R}^{S_H}_{(H \cap K)}(\lambda_1 \lambda_{H_K})$$

of $K \cap H$ in cohomology of degree $S_H = \dim(\mathfrak{u} \cap \mathfrak{k} \cap \mathfrak{h})$. The same argument as in 1.3 shows that we have a $(H \cap K)$–map

$$\mathcal{R}^{S_H}_{(K)}(\lambda_K) \to \mathcal{R}^{S_H}_{(H \cap K)}(\lambda_{H_K}).$$

1.5. The diagram

$$
\begin{array}{ccc}
\mathcal{R}^{S_H}_{(\mathfrak{h}, H \cap K)}(\lambda_H) & \longleftarrow & \mathcal{R}^{S_H}_{(\mathfrak{g}, K)}(\lambda) \\
\downarrow & & \downarrow \\
\mathcal{R}^{S_H}_{(H \cap K)}(\lambda_{H_K}) & \longleftarrow & \mathcal{R}^{S_H}_{(K)}(\lambda_K)
\end{array}
$$

commutes and therefore

PROPOSITION 1.1. *Suppose that $S_H = S_G = S$ and that the map*

$$\mathcal{R}^S_{(K)}(\lambda_K) \to \mathcal{R}^S_{(H \cap K)}(\lambda_{H_K})$$

is not zero. Then

$$\mathcal{R}^S_{(\mathfrak{h}, H \cap K)}(\lambda_H)$$

is a bottom layer $(\mathfrak{h}, H \cap K)$–type.

We will show that if $G = U(p, q)$, $H = U(p, q - m)U(m)$ and $A_{\mathfrak{q}}$ is a coho-mologically induced representation from a θ–stable parabolic subgroup with Levi $L = U(1)^n U(p - n, q)$ then the assumption of 1.1 are satisfied.

I expect that, if $G = Sp(p, q)$, $H = Sp(p, q - m)Sp(m)$ or if G and H are the connected components of $SO(p, q)$ and $SO(p, q - m)SO(m)$ respectively, then the representations $A_{\mathfrak{q}}$ with lowest K–types of the form $\tau \otimes \mu$ for a character μ also satisfy the assumptions of the theorem.

2. Representations of K and $K \cap H$.

In this section we show, that under certain assumptions on G, H and $\mathfrak{q}$, the restriction of the minimal K-type of $A_{\mathfrak{q}}(\lambda)$ to $H \cap K$ is irreducible and equal to minimal $(K \cap H)$–type of a representation $A_{\mathfrak{q} \cap \mathfrak{h}}(\tilde{\lambda})$ for some character $\tilde{\lambda}$.

2.1. Assume now that $\mathfrak{q} = \mathfrak{l} \oplus \mathfrak{u}$ is a θ–stable parabolic subalgebra and that

$$K = K_1 \cdot K_2$$

$$H \cap K = K_1 \cdot (K_2 \cap H)$$

and

$$L \cap K = (K_1 \cap L) \cdot K_2.$$

We will refer this as **assumptions $\mathcal{A}$.**

Example:

(1) $G = U(p,q)$, $H = U(p, q-r)U(r)$, $r < q$ and $L = U(1)^{p-s}U(s,q)$.
(2) $G = SO(p,q)^0$, $H = SO(p, p-r)^0 SO(r)$ and $L = SO(2)^s SO(p-2s, q)^0$.
(3) $G = Sp(p,q)$, $H = Sp(p, q-r)Sp(r)$ and $L = T^s Sp(p-s, q)$

We have

$$H \cap L \cap K = (K_1 \cap L) \cdot (K_2 \cap H)$$

and

$$
\begin{aligned}
\mathfrak{u} \cap \mathfrak{k} \cap \mathfrak{h} &= \mathfrak{u} \cap \mathfrak{k}_1 \oplus \mathfrak{u} \cap \mathfrak{k}_2 \cap \mathfrak{h} && \text{since } \mathfrak{k}_1 \subset \mathfrak{h} \\
&= \mathfrak{u} \cap \mathfrak{k}_1 && \text{since } \mathfrak{k}_2 \subset \mathfrak{l} \\
&= \mathfrak{u} \cap \mathfrak{k}
\end{aligned}
$$

and so

$$S_H = S_G$$

and $\mathfrak{u} \cap \mathfrak{k} \cap \mathbf{s} = 0$. Furthermore

$$\mathfrak{l} \cap \mathfrak{k}_1 \cap \mathfrak{h} = \mathfrak{l} \cap \mathfrak{k}_1$$

and

$$\mathfrak{q} \cap \mathfrak{k}_1 \cap \mathfrak{h} = \mathfrak{q} \cap \mathfrak{k}_1.$$

Recall that as characters of $L \cap H \cap K$ we have $\lambda^\sharp = \lambda \otimes \wedge^{top}\mathfrak{u}$ and

$$\lambda_H = \lambda \otimes \wedge^{top}\mathfrak{u} \cap \mathbf{s} = \lambda \otimes \wedge^{top}\mathfrak{u} \cap \mathbf{s} \cap \mathbf{p}$$

Thus

$$\lambda \otimes \wedge^{top}\mathfrak{u} \cap \mathbf{p} = \lambda_H \otimes \wedge^{top}\mathfrak{u} \cap \mathbf{p} \cap \mathfrak{h}.$$

2.2. We denote the minimal K–types of $A_\mathfrak{q}(\lambda)$, respectively $(K \cap H)$–types of $A_{\mathfrak{q}\cap\mathfrak{h}}(\lambda_H)$, by $V_\mathfrak{q}(\lambda)$ respectively by $V_{\mathfrak{q}\cap\mathfrak{h}}(\lambda_H)$.

PROPOSITION 2.1. *Under the above assumptions the restriction of $V_\mathfrak{q}(\lambda)$ to $(H \cap K)$ is irreducible. It is isomorphic to $V_{\mathfrak{q}\cap\mathfrak{h}}(\lambda_H)$.*

Proof: The highest weight of $V_\mathfrak{q}(\lambda)$ is equal to

$$\lambda + 2\rho(\mathfrak{u} \cap \mathbf{p})$$

and that of $V_{\mathfrak{q}\cap\mathfrak{h}}(\lambda_H)$

$$
\begin{aligned}
\lambda_H + 2\rho(\mathfrak{u} \cap \mathbf{p} \cap \mathfrak{h}) &= \lambda \otimes \wedge^{top}\mathfrak{u} \cap \mathbf{s} + 2\rho(\mathfrak{u} \cap \mathbf{p} \cap \mathfrak{h}) \\
&= \lambda + 2\rho(\mathfrak{u} \cap \mathbf{s}) \cap \mathbf{p} + 2\rho(\mathfrak{u} \cap \mathbf{p} \cap \mathfrak{h}) \\
&= \lambda + 2\rho(\mathfrak{u} \cap \mathbf{p}).
\end{aligned}
$$

Thus the highest weights of the minimal K–types $V_{\mathfrak{q}}(\lambda)$ and $V_{\mathfrak{q}\cap\mathfrak{h}}(\lambda_H)$ are equal. Thus we consider $V_{\mathfrak{q}\cap\mathfrak{h}}(\lambda_H)$ as a $U(\mathfrak{h})$–submodule of $V_{\mathfrak{q}}(\lambda)$ generated by the highest weight vector.

Now $\mathfrak{q}\cap\mathfrak{k}$ is a parabolic subalgebra of $\mathfrak{k}$ and $\mathfrak{q}\cap\mathfrak{k}\cap\mathfrak{h}$ is a parabolic subalgebra of $\mathfrak{k}\cap\mathfrak{h}$ with Levi subalgebra of $\mathfrak{l}\cap\mathfrak{k}$, respectively $\mathfrak{l}\cap\mathfrak{k}\cap\mathfrak{h}$. The highest weight space of $V_{\mathfrak{q}}(\lambda)$ is invariant under $\mathfrak{l}\cap\mathfrak{k}$. As a $\mathfrak{k}\cap\mathfrak{u}$–module $V_{\mathfrak{q}}(\lambda)$ is generated by its highest weight space and since $\mathfrak{u}\cap\mathfrak{k}=\mathfrak{u}\cap\mathfrak{k}\cap\mathfrak{h}$ the $\mathfrak{k}\cap\mathfrak{h}\cap\mathfrak{u}$ –module generated by $\mathfrak{u}\cap\mathfrak{k}\cap\mathfrak{h}$ is equal to $V_{\mathfrak{q}}(\lambda)$. Hence the restriction of $V_{\mathfrak{q}}(\lambda)$ is irreducible. $\qquad\square$

Remark: Our assumptions are closely related to those of corollary 4.4 in [7]

3.   The restriction of $A_{\mathfrak{q}}$ to $U(p, q-r)U(r)$.

An irreducible $(\mathfrak{h}, K\cap H)$–modules π_H is called an H–**type of an irreducible** $(\mathfrak{g}, K)$–**module** π, if π is a direct sum of irreducible $(\mathfrak{h}, K\cap K)$–modules and π_H is a direct summand of π. It is called a minimal H–type of π, if it is a direct summand of the $U(\mathfrak{h})$–submodule generated by the minimal K–type $V_{\mathfrak{q}}$.

In this section we assume that $G = U(p,q)$, $H = U(p, q-r)U(r)$, $r < q$ and $L = U(1)^{p-s}U(s,q)$. Then $S_H = S_G = S$ and the restriction of the minimal K–type of $A_{\mathfrak{q}}$ to $H\cap K$ is irreducible. We show that this implies that $A_{\mathfrak{q}}$ is a direct sum of irreducible $(\mathfrak{h}, K\cap H)$–modules and we determine its minimal H–type.

3.1. Recall the definition of the bottom layer K–module $\mathcal{R}_K^{S_G}(\lambda)$ from section 1. We have bottom layer maps of $\mathfrak{k}$–modules.

$$\mathcal{B}(\lambda) : A_{\mathfrak{q}} \to \mathcal{R}_K^S(\lambda_0)$$

and

$$\mathcal{B}(\lambda_H) : A_{\mathfrak{q}\cap\mathfrak{h}}(\lambda_H) \to \mathcal{R}_{K\cap H}^S(\lambda_H),$$

where λ_0 is the trivial character of $L\cap K$. These maps are defined by the inclusion of complexes and hence of forms. See Theorem V.5.80 and its proof in [5]. For $A_{\mathfrak{q}}$, respectively $A_{\mathfrak{q}_H}(\lambda_H)$, the bottom layer is irreducible and contains only the minimal K–type, respectively $K\cap H$–type. (10.24 in [5]).

Now $K\cap H$ and K are again a symmetric pair, and

$$\mathbf{r}_G\cap\mathfrak{k}\cap\mathfrak{h} \oplus \mathbf{r}_G\cap\mathfrak{k}\cap\mathbf{s} = \mathbf{r}_G\cap\mathfrak{k}.$$

So have a pullback of of complexes

$$\mathbf{pb}_{H\cap K}^{i,j} : \ \mathrm{Hom}_{K\cap L\cap H}(U(\mathfrak{k}\cap\mathfrak{h}), \mathrm{Hom}(\wedge^i(\mathbf{r}_G\cap\mathfrak{k}\cap\mathfrak{h}), \wedge^j(\mathbf{r}_G\cap\mathfrak{k}\cap\mathbf{s})^*\mathbb{C}_{\lambda_H}^\sharp))_{K\cap H}$$
$$\to \mathrm{Hom}_{K\cap L\cap H}(U(\mathfrak{k}), \mathrm{Hom}(\wedge^{i+j}(\mathbf{r}_G\cap\mathfrak{k}), \mathbb{C}_{\lambda_0}^\sharp))_{K\cap H}.$$

Here the notation is in analogy with the case in 1.3, but now we take $G = K$. In particular

$$\mathbf{pb}_{H\cap K}^{i,0} : \ \mathrm{Hom}_{K\cap L\cap H}(U(\mathfrak{k}\cap\mathfrak{h}), \mathrm{Hom}(\wedge^i(\mathbf{r}_G\cap\mathfrak{k}\cap\mathfrak{h}), \wedge^0(\mathfrak{u}\cap\mathfrak{k}\cap\mathbf{s})^*\mathbb{C}_{\lambda_H}^\sharp))_{K\cap H}$$
$$\to \mathrm{Hom}_{K\cap L\cap H}(U(\mathfrak{k}), \mathrm{Hom}(\wedge^i(\mathbf{r}_G\cap\mathfrak{k}), \mathbb{C}_{\lambda_0}^\sharp))_{K\cap H}.$$

Using a forgetful functor we may consider

$$\mathrm{Hom}_{K\cap L}(U(\mathfrak{k}), \mathrm{Hom}(\wedge^i(\mathbf{r}_G\cap\mathfrak{k}), \mathbb{C}_{\lambda_0}^\sharp))_K$$

as a subspace, respectively subcomplex, of

$$\mathrm{Hom}_{K \cap L \cap H}(U(\mathfrak{k}), \mathrm{Hom}(\wedge^i(\mathbf{r}_G \cap \mathfrak{k} \cap \mathfrak{h}), \mathbb{C}^\sharp_{\lambda_0}))_{K \cap H}$$

$$= \mathrm{Hom}_{K \cap L \cap H}(U(\mathfrak{k} \cap \mathfrak{h}), \mathrm{Hom}(\wedge^i(\mathbf{r}_G \cap \mathfrak{k} \cap \mathfrak{h}), Q_H \otimes \mathbb{C}^\sharp_{\lambda_0}))_{K \cap H},$$

where Q_H is the symmetric algebra of the complement of $\mathfrak{h} \cap \mathfrak{k}$ in $\mathfrak{k}$. The cohomology of the complex

$$\mathrm{Hom}_{K \cap L \cap H}(U(\mathfrak{k} \cap \mathfrak{h}), \mathrm{Hom}(\wedge^i(\mathbf{r}_G \cap \mathfrak{k} \cap \mathfrak{h}), \mathbb{C}^\sharp_{\lambda_H}))_{K \cap H}$$

is isomorphic to $\mathcal{R}^S_{K \cap H}(\lambda_H)$, and the cohomology of the complex

$$\mathrm{Hom}_{K \cap L}(U(\mathfrak{k}), \mathrm{Hom}(\wedge^i(\mathbf{r}_G \cap \mathfrak{k}), \mathbb{C}^\sharp_{\lambda_0}))_K$$

is isomorphic to $\mathcal{R}^S_K(\lambda_0)$ and the class of the form

$$\mathrm{Hom}_{K \cap L}(U(\mathfrak{k} \cap \mathfrak{h}), \mathrm{Hom}(\wedge^S \mathfrak{k} \cap \mathfrak{n}), \mathbb{C}^\sharp_{\lambda_H}))_{K \cap H}$$

defines the highest weight vector in both representations. Thus the pullback of forms defines a nontrivial map

$$\mathcal{R}^S_K(\lambda_0)_{|K \cap H} \to \mathcal{R}^S_{K \cap H}(\lambda_H)$$

and so by 1.1 a nontrivial map in

$$\mathrm{Hom}_{\mathfrak{h}, K \cap H}(A_{\mathfrak{q}}, A_{\mathfrak{q} \cap \mathfrak{h}}(\lambda_H))$$

Thus the assumptions of Theorem I.1 are satisfied and so

THEOREM 3.1. *Assume that $G = U(p,q)$, $H = U(p, q-r)U(r)$, $r < q$ and that $\mathfrak{q}$ is a θ–stable parabolic subalgebra with $L = U(1)^{p-s}U(s,q)$. Then $A_{\mathfrak{q} \cap \mathfrak{h}}(\lambda_H)$ is a subrepresentation of the restriction of $A_{\mathfrak{q}}$ to H.*

By a theorem of Kobayashi (4.2.5 in [8]) this implies

COROLLARY 3.2. *Assume that $G = U(p,q)$, $H = U(p, q-r)U(r)$, $r < q$ and that $\mathfrak{q}$ is a θ stable parabolic subalgebra with $L = U(1)^{p-s}U(s,q)$. The $(\mathfrak{g}, K)$–module $A_{\mathfrak{q}}$ is infinitesimally H–admissible, i.e. it is a direct sum of irreducible $(\mathfrak{h}, K \cap H)$–modules.*

Remark : It is possible to prove the corollary directly using theorem 4.2 in [6]. Our method doesn't only prove that $(\mathfrak{g}, K)$–module $A_{\mathfrak{q}}$ is infinitesimally H-admissible, but it also provides an irreducible summand.

Recall that H is defined as the fix point set of an involution σ. Its partner in a pseudo dual pair is the group $\bar{H}$ which is the fix point set of the involution $\sigma\theta$ (for more detail see 4.2). A representation is infinitesimally H–admissible, iff it is infinitesimally $\bar{H}$–admissible. Thus the above argument also applies to $\bar{H}$. Since $\bar{H} = U(p,r)U(q-r)$, $r < q$ this case is already covered by the corollary 3.2.

4. Branching laws for the restrictions of $A_{\mathfrak{q}}$.

We assume in this section that $G = U(p,q)$ and $H = U(p, q-1)U(1)$ and that the representation $A_{\mathfrak{q}}$ satisfies the assumptions of section 3. We prove some branching laws for the restriction of the $(\mathfrak{g}, K)$–module $A_{\mathfrak{q}}$ to $(\mathfrak{h}, K \cap H)$.

4.1. Some θ–stable parabolic subalgebras of u(p,q). Put n=p+q. Fix a compact Cartan subalgebra $\mathfrak{c}$. We consider a root system of type A_{n-1} with positive roots $e_i - e_j$ with $i < j \leq n$. We pick as positive compact roots the roots

$$e_i - e_j, \qquad 1 \leq i,j \leq p \quad \text{or} \quad p+1 \leq i,j \leq n.$$

Fix two non negative integers s, r with $1 \leq r+s \leq p$ and let $L(r+s)$ be equal to the subgroup $U(1)^{r+s}U(p-r-s,q)$. Let $\mathfrak{l} \oplus \mathfrak{u}$ be the θ–stable parabolic subalgebra with roots $\Delta(\mathfrak{u},\mathfrak{c})$ of the nilpotent radical given by

$$
\begin{aligned}
\Delta(\mathfrak{u},\mathfrak{c}) \;=\; & \{e_i - e_j : 1 \leq i < j \leq r+s\} \\
& \cup \{e_i - e_j;\; 1 \leq i \leq r, \quad r+s+1 \leq j \leq p+q\} \\
& \cup \{-e_i + e_j;\; r+1 \leq i \leq r+s, \quad r+s+1 \leq j \leq p+q\}.
\end{aligned}
$$

The compact roots in $\Delta(\mathfrak{u},\mathfrak{c})$ are denoted by $\Delta(\mathfrak{u},\mathfrak{c})_c$, the noncompact ones by $\Delta(\mathfrak{u},\mathfrak{c})_{nc}$. We say that this parabolic subalgebra is of type r,s and denote it by $\mathfrak{q}(r,s)$.

There are r+s +1 K–conjugacy classes of θ–stable parabolic subalgebras of $\mathfrak{g}$ with same Levi part $\mathfrak{l}(d)$. A complete system of representatives is given by $\mathfrak{q}(r,s)$ where $r + s = d$.

We have

$$
\begin{aligned}
\rho(r,s) \;=\; & \sum_{\alpha \in \Delta(\mathfrak{u},\mathfrak{h})} \alpha \\
=\; & (p+q-1, p+q-3, \ldots, p+q-2r+1, -p-q+2s-1, \ldots \\
& \ldots, -p-q-1, s-r, \ldots s-r)
\end{aligned}
$$

and

$$
\begin{aligned}
\rho(r,s)_{nc} \;=\; & \sum_{\alpha \in \Delta(\mathfrak{u},\mathfrak{h})_{nc}} \alpha \\
=\; & (q, \ldots q, -q, \cdots -q, 0, \ldots, 0, s-r, \ldots s-r).
\end{aligned}
$$

A character λ of $L(r+s)$ can be identified with integers

$$m_1 \geq m_2 \geq \cdots \geq m_r \geq m_{r+1} \geq \cdots \geq m_{r+s}$$

where

$$\lambda = exp\left(m_1 e_1 + \ldots m_{r+s} e_{r+s} + \sum_{j \leq r+s} \frac{s-r}{2} e_j\right).$$

It defines a nonzero representation $A_{\mathfrak{q}}(\lambda)$ provided that the conditions 2.4.3 (page 24) of [**9**] are satisfied. We write $A_{\mathfrak{q}}$ for the representation with $\lambda = \rho(r,s)$. For more detail see [**9**]. Note that our parameter λ is denoted by $\lambda + \rho$ in [**5**].

The representations are holomorphic, respectively antiholomorphic, if s=0 respectively r=0.

4.2. Pseudo dual pairs. We consider the subgroup H of G defined by the involution

$$g \rightarrow MgM^{-1}$$

where M is the diagonal matrix with p+q-1 entries +1 followed by one entry -1 . Its "pseudo dual partner group" $\bar{H}$ is the fix point set of the involution

$$g \rightarrow M_0 g M_0^{-1},$$

where M_0 is the diagonal matrix with p entries $+1$ followed by q-1 entries -1 and one entry $+1$. Then $\bar{H}$ is isomorphic to $U(p,1)U(q-1)$.

The pair $(H, \bar{H})$ is often referred to an associate symmetric pair. See [13] for an explanation of the terminology "pseudo dual pair".

We have

$$L \cap H = U(1)^r U(p - r, q - 1) U(1)$$

and

$$L \cap \bar{H} = U(1)^r U(p - r, 1) U(q - 1).$$

Furthermore $\mathfrak{q}(r,s) \cap \mathfrak{h}$ and $\mathfrak{q}(r,s) \cap \bar{\mathfrak{h}}$ are again θ–stable parabolic subalgebras of $\mathfrak{h}$ respectively $\bar{\mathfrak{h}}$ of type r, s.

We write $2\rho(r, s, H)$, respectively $2\rho(r, s, \bar{H})$, for the sum of roots of $\mathfrak{c}$ in $\mathfrak{u} \cap \mathfrak{h}$, respectively $\mathfrak{u} \cap \bar{\mathfrak{h}}$. Then

$$\lambda_H = 2\rho(r,s) - 2\rho(r,s,H) = (1,1,\ldots 1, -1, \ldots, -1, 0, \ldots, 0, s - r)$$

and

$$\begin{aligned}
\lambda_{\bar{H}} &= 2\rho(r,s) - 2\rho(r,s,\bar{H}) \\
&= (q-1,\ldots,q-1,1-q,\ldots 1-q,0,\ldots 0, s-r, \ldots s-r, 0).
\end{aligned}$$

By section III $A_{\mathfrak{q} \cap \bar{\mathfrak{h}}}(\lambda_{\bar{H}})$ is a direct summand of the restriction of $A_{\mathfrak{q}}$ to $\bar{H}$. It is the product of a one dimensional representation of U(q-1) and an infinite dimensional irreducible representation U(p,1). Therefore all its $(K \cap \bar{H})$–types have multiplicity one.

If

$$L = U(1)^{r+s} U(p - r - s, q)$$

then the representation $A_{\mathfrak{q}}$ has a minimal K–type with highest weight

$$\mu_{\mathfrak{q}} = 2\rho(r,s)_{nc} = (q, \ldots q, -q, \cdots - q, 0, \ldots, 0, s - r, \ldots, s - r).$$

Recall that this also the highest weight of the minimal $(K \cap \bar{H})$–type of $A_{\mathfrak{q} \cap \bar{\mathfrak{h}}}(\lambda_{\bar{H}})$ as well as the minimal $(K \cap H)$–type of $A_{\mathfrak{q} \cap \mathfrak{h}}(\lambda_H)$.

4.3. The H–types of $A_{\mathfrak{q}}$.

LEMMA 4.1. *Every $(K \cap \bar{H})$–type of $A_{\mathfrak{q} \cap \bar{\mathfrak{h}}}(\lambda_{\bar{H}})$ has multiplicity one.*

PROOF. The $(K \cap \bar{H})$–types of an irreducible unitary representation of $U(p,1)$ have multiplicity one. $\square$

Since every irreducible $(\mathfrak{h}, H \cap K)$–submodule of $A_{\mathfrak{q}}$ contains at least one $(K \cap \bar{H})$–type of $A_{\mathfrak{q} \cap \bar{\mathfrak{h}}}(\lambda_{\bar{H}})$ we see that every irreducible $(\mathfrak{h}, H \cap K)$–submodule of $A_{\mathfrak{q}}$ occurs with multiplicity at most one.

LEMMA 4.2. *Every irreducible subrepresentation of $(A_{\mathfrak{q}})_{|H}$ is cohomologically induced from $\mathfrak{q} \cap \mathfrak{h}$.*

PROOF. By [10] all the irreducible subrepresentations of $(A_{\mathfrak{q}})_{|H}$ have the same associated variety. $\square$

The highest weight of a $(K \cap \bar{H})$–type of $A_{\mathfrak{q} \cap \bar{\mathfrak{h}}}(\lambda_{\bar{H}})$ defines a character μ_H of $L \cap H$ since it is of the form

$$\lambda_{\bar{H}} + \sum_{\alpha \in (\mathfrak{u} \cap \mathfrak{p} \cap \bar{\mathfrak{h}})} m_\alpha \alpha$$

with $0 \leq m_\alpha$. Hence $A_{\mathfrak{q} \cap \mathfrak{h}}(\mu_H)$ is nonzero and irreducible.

THEOREM 4.3. *(Branching law) Let μ_H be a character of $L \cap H$ obtained from a highest weight of $(K \cap \bar{H})$–type of $A_{\mathfrak{q} \cap \bar{\mathfrak{h}}}(\lambda_{\bar{H}})$. Then*

$$Hom_{\mathfrak{h}, K \cap H}(A_{\mathfrak{q} \cap \mathfrak{h}}(\mu_H), A_{\mathfrak{q}}) \leq 1$$

and every $(\mathfrak{h}, K \cap H)$–type of $A_{\mathfrak{q}}$ is of this form.

PROOF. A minimal $(K \cap H)$–type of a representation cohomologically induced from $\mathfrak{q} \cap \mathfrak{h}$ has a highest weight of the form

$$\nu = (a_1, \ldots, a_r, -a_{r+1}, \cdots - a_{r+s}, 0, \ldots, 0, b_1, \ldots b_1, b_q).$$

and the weights of its $(K \cap H)$–types are in a cone

$$\nu + \sum_{\alpha \in \mathfrak{u} \cap \mathfrak{p}} n_\alpha \alpha.$$

This cone intersects the $(K \cap H)$–types of $A_{\mathfrak{q} \cap \bar{\mathfrak{h}}}(\lambda_{\bar{H}})$ only if ν is the highest weight of a $(K \cap H)$–type of $A_{\mathfrak{q} \cap \bar{\mathfrak{h}}}(\lambda_{\bar{H}})$. $\qquad\square$

Conjecture: Under the assumptions of 4.3

$$Hom_{\mathfrak{h}, K \cap H}(A_{\mathfrak{q} \cap \mathfrak{h}}(\mu_H), A_{\mathfrak{q}}) = 1.$$

4.4. The following **example** supports this conjecture.

Let $G = U(3, q)$, $\mathfrak{q} = \mathfrak{q}(1, 1)$, $H = U(3, q-1)U(1)$ and $\bar{H} = U(3, 1)U(q-1)$. The minimal K–type of $A_{\mathfrak{q}}$ has highest weight

$$(q, -q, 0, \ldots, 0),$$

and $\lambda_H = (1/2, -1/2, 0, \ldots, 0)$. Using 5.108 in [**5**] we deduce that the multiplicity of a K–type of $A_{\mathfrak{q}}$ with highest weight

$$\mu_n = (q + n, -q - n, 0, \ldots, 0) \qquad 0 \leq n$$

equals one. The multiplicity of a $K \cap H$–type in $A_{\mathfrak{q}_H}(\lambda_H)$ and a $K \cap \bar{H}$–type in $A_{\mathfrak{q}_{\bar{H}}}(\lambda_{\bar{H}})$ with highest weight μ_n is also one. Furthermore the restriction of a K–type with highest weight μ_n to $K \cap H = K \cap \bar{H}$ is irreducible and every K–type of $A_{\mathfrak{q}}$, whose restriction to $K \cap H = K \cap \bar{H}$ contains a representation with highest weight μ_n, has highest weight

$$(q + n, -q - n, 0, k, 0, \ldots, 0, k) \qquad k \leq n.$$

Thus the $(K \cap H)$–type with highest μ_n has multiplicity n in the restriction of $A_{\mathfrak{q}}$ to $K \cap H$.

LEMMA 4.4. *Each $(K \cap \bar{H})$–type of $A_{\mathfrak{q}_H}(\lambda_{\bar{H}})$ with highest weight μ_n generates an irreducible $(\mathfrak{h}, K \cap H)$–submodules of $A_{\mathfrak{q}}$.*

PROOF. The result is true for μ_0 by the previous section. Now suppose that it is true for μ_m with $m < n$. The submodule generated by the representations of weight μ_m, $m < n$, contains the $K \cap \bar{H}$-type with highest weight μ_n exactly n-1 times and it also contains all representations of $K \cap H$ of highest weight μ_m, $m < n$. $\qquad\square$

All the $(K \cap H)$–types of $A_{\mathfrak{q}_H}(\lambda_{\bar{H}})$ are of the form

$$2\rho(\mathfrak{u} \cap \mathbf{p}) + (i, -j, 0, \ldots, 0, j - i) \qquad i, j \in \mathbb{N}.$$

Using that $SO(3) \subset H$ and that $\mathfrak{u} \cap \mathfrak{s}$ commutes with $SO(q-1)$ a slight modification of this lemma shows that all the $(K \cap H)$–types of $A_{\mathfrak{q}_H}(\lambda_{\bar{H}})$ generate irreducible $(\mathfrak{h}, K_H)$–submodules.

Hence we obtain the formulas in [10]

THEOREM 4.5. (**Branching theorem**) Let $G = U(3,q)$, $\mathfrak{q} = \mathfrak{q}(1,1)$, $H = U(3, q-1)U(1)$ and $\bar{H} = U(3,1)U(q-1)$. We have a 1-1 correspondence between $K_{\bar{H}}$–types of $A_{\mathfrak{q}_{\bar{H}}}$ and irreducible subrepresentations of the restriction of $A_{\mathfrak{q}}$ to H.

We can rephrase this branching law as a Blattner type formula (see [5] page 376).

COROLLARY 4.6. (**Blattner formula**) Let $G = U(3,q)$, $\mathfrak{q} = \mathfrak{q}(1,1)$, $H = U(3, q-1)U(1)$. Then for every irreducible $(\mathfrak{h}, K \cap H)$–module V

$$\text{dim Hom }_{\mathfrak{h},K \cap H}(V, A_{\mathfrak{q}}) =$$

$$\sum_{j=0}^{S}(-1)^{S-j} \sum_{n=0}^{\infty} \text{dim Hom }_{\mathfrak{l}, L \cap H \cap K}(H_j(\mathfrak{u} \cap \mathfrak{h}, V), S^n(\mathfrak{u} \cap \mathfrak{s}) \otimes \mathbb{C}_{2\rho(\mathfrak{u})})$$

PROOF. Note that for an irreducible $(\mathfrak{h}, K \cap H)$–module V by 8.11 of [5]

$$\text{Hom}_{\mathfrak{l}, L \cap H \cap K}(H_S(\mathfrak{u} \cap \mathfrak{h}, V), S^n(\mathfrak{u} \cap \mathfrak{s}) \otimes \mathbb{C}_{2\rho(\mathfrak{u})}) =$$
$$\text{Hom}_{\mathfrak{h}, K \cap H}(V, \mathcal{R}_H^S(S^n(\mathfrak{u} \cap \mathfrak{s}) \otimes \mathbb{C}_{2\rho(\mathfrak{u})-2\rho(\mathfrak{u} \cap \mathfrak{h})})).$$

The dimension of $\mathfrak{u} \cap \mathfrak{s}$ is 2 and the weights are of the form

$$\mu(i,j) = (i, -j, 0, \ldots, 0, j-i) \qquad i,j \in \mathbb{N}.$$

and

$$2\rho(\mathfrak{u}) - 2\rho(\mathfrak{u} \cap \mathfrak{h}) = 2\rho(\mathfrak{u} \cap \mathbf{p}) - 2\rho(\mathfrak{u} \cap \mathfrak{h} \cap \mathbf{p}) = \lambda_H.$$

Since

$$\mathcal{R}_H^k(\lambda_H \otimes \mu(i,j)) = 0 \quad \text{if } k \neq S$$

we deduce that

$$\sum_{j=0}^{S}(-1)^{S-j} \sum_{n=0}^{\infty} \text{dim Hom }_{\mathfrak{l}, L \cap H \cap K}(H_j(\mathfrak{u} \cap \mathfrak{h}, V), S^n(\mathfrak{u} \cap \mathfrak{s}) \otimes \mathbb{C}_{2\rho(\mathfrak{u})}) \neq 0$$

iff $V = \mathcal{R}_H^k(\lambda_H \otimes \mu(i,j))$. Furthermore

$$\text{dim Hom}_{\mathfrak{h}, K \cap H}(V = \mathcal{R}_H^k(\lambda_H \otimes \mu(i,j)), \mathcal{R}_H^S(S^n(\mathfrak{u} \cap \mathfrak{s}) \otimes \mathbb{C}_{2\rho(\mathfrak{u})-2\rho(\mathfrak{u} \cap \mathfrak{h})})) = 1$$

Hence the remarks before theorem 4.5 imply the Blattner formula.

$\square$

Remark: Complete branching laws for the restriction of discrete series representations of $U(p,q)$ to symmetric subgroups have been obtained by M. Duflo and J. Vargas under the assumption that the branching laws are discretely decomposable. [2]

5. Applications to $(\mathfrak{g}, K)$–cohomology.

We continue to assume that $G = U(p,q)$ and $H = U(p,q-1)U(1)$ and that L satisfies the assumptions of section 3. We show first that $H^*(\mathfrak{h}, K \cap H, A_{\mathfrak{q}\cap\mathfrak{h}}(\lambda_H) \otimes \wedge^{R_{\bar{H}}}\mathbf{s}^*) \neq 0$. We also derive a formula relating the differential forms representing $H^{R_G}(\mathfrak{g}, K, A_{\mathfrak{q}})$ to forms representing $H^{R_H}(\mathfrak{h}, K \cap H, A_{\mathfrak{q}\cap\mathfrak{h}}(\lambda_H) \otimes \wedge^{R_{\bar{H}}}\mathbf{s}^*)$.

5.1. The $(\mathfrak{h}, K \cap H)$–submodule $A_{\mathfrak{q}\cap\mathfrak{h}}(\lambda_H)$ of $\mathbf{A}_{\mathfrak{q}}$ containing the minimal K-type has infinitesimal character $\lambda_H + \rho_H$, where

$$\lambda_H = 2\rho(\mathfrak{u} \cap \mathbf{p}) - 2\rho(\mathfrak{u} \cap \mathbf{p}_\mathfrak{h}) = \sum_{\alpha \in \Sigma(\mathfrak{u}\cap\mathbf{p}\cap\mathbf{s})} \alpha.$$

The representation of H on $\mathbf{s}$ is a direct sum of of the representations

$$X \to gXu^{-1}$$

and

$$X \to uX^{tr}g^{-1}$$

where $g \in U(p,q-1)$ and $u \in U(1)$. These representations are dual to each other. The dimension of $\mathbf{s} \cap \mathfrak{u} \cap \mathbf{p}$ is equal to $R_{\bar{H}}$. Thus the representation of H on $\wedge^{R_{\bar{H}}}\mathbf{s}$ can be decomposed using the dual pair $U(1), U(p,q-1)$. In particular the tensor product of the $R_{\bar{H}}$'th fundamental representation with the character $u \to u^{-R_{\bar{H}}}$ of $U(1)$ is a direct summand and no other subrepresentation is a tensor product with this character. This representation V_R has the infinitesimal character λ_H. So

LEMMA 5.1. *Under the above assumptions*

$$H^{R_H}(\mathfrak{h}, K \cap H, A_{\mathfrak{q}\cap\mathfrak{h}}(\lambda_H) \otimes \wedge^{R_{\bar{H}}}\mathbf{s}^*) \neq 0.$$

In particular

$$H^{R_H}(\mathfrak{h}, K \cap H, A_{\mathfrak{q}\cap\mathfrak{h}}(\lambda_H) \otimes \wedge^{R_{\bar{H}}}\mathbf{s}^*) = H^{R_H}(\mathfrak{h}, K \cap H, A_{\mathfrak{q}\cap\mathfrak{h}}(\lambda_H) \otimes V_R^*).$$

PROOF. Note that the dual representation V_R^* to V_R is a subrepresentation of $\wedge^{R_{\bar{H}}}\mathbf{s}^*$. □

5.2. Restriction of harmonic forms with coefficients in $A_{\mathfrak{q}}$. Let $\omega \in A_{\mathfrak{q}} \otimes_K \wedge^{R_G}\mathbf{p}^*$ be a harmonic form representing the lowest non trivial $(\mathfrak{g}, K)$–cohomology class of $A_{\mathfrak{q}}$ where $R_G = \dim \mathfrak{u} \cap \mathbf{p}$.

We have $\mathbf{p} = \mathbf{p}_H \oplus \mathbf{p}_{\bar{H}}$ and thus

$$\wedge^{R_G}\mathbf{p} = \oplus_{i+j=R_G} \wedge^i \mathbf{p}_H \oplus \wedge^j \mathbf{p}_{\bar{H}}$$

and hence

$$A_{\mathfrak{q}} \otimes_{K\cap H} \wedge^{R_G}\mathbf{p}^*$$

is isomorphic to

$$\oplus_{i+j=R_G} A_{\mathfrak{q}} \otimes_{K\cap H} (\wedge^i\mathbf{p}_H^* \otimes \wedge^j\mathbf{p}_{\bar{H}}^*)$$

and hence to

$$\oplus_{i+j=R_G}(A_{\mathfrak{q}} \otimes \wedge^j\mathbf{p}_{\bar{H}}^*) \otimes_{K\cap H} \wedge^i\mathbf{p}_H^*.$$

As $(\mathfrak{h}, K \cap H)$–module

$$A_{\mathfrak{q}} = \oplus_k A_{\mathfrak{q}\cap\mathfrak{h}}(\lambda_H(k))$$

and so $A_{\mathfrak{q}} \otimes_{K \cap H} \wedge^{R_G} \mathbf{p}^*$ is isomorphic to

$$\oplus_k \oplus_{i+j=R_G} (A_{\mathfrak{q} \cap \mathfrak{h}}(\lambda_H(k)) \otimes \wedge^j \mathbf{p}_{\bar{H}}^*) \otimes_{K \cap H} \wedge^i \mathbf{p}_H^*.$$

Since $\mathbf{p}_{\bar{H}}$ is a $(K \cap H)$–submodule of the $\mathfrak{h}$–module $\mathbf{s}$ this is a subspace of

$$\oplus_k \oplus_{i+j=R_G} (A_{\mathfrak{q} \cap \mathfrak{h}}(\lambda_H(k)) \otimes \wedge^j \mathbf{s}^*) \otimes_{K \cap H} \wedge^i \mathbf{p}_H^*.$$

Recall that $R_H = \dim \mathfrak{u} \cap \mathfrak{h} \cap \mathbf{p}$, $R_{\bar{H}} = \dim \mathfrak{u} \cap \bar{\mathfrak{h}} \cap \mathbf{p}$ and $R_G = R_H + R_{\bar{H}}$. Since the minimal K–type of $A_{\mathfrak{q}}$ restricted to $H \cap K$ is irreducible and is the minimal $(K \cap H)$–type of $A_{\mathfrak{q} \cap \mathfrak{h}}(\lambda_H)$, ω defines a form

$$\omega_H \in A_{\mathfrak{q} \cap \mathfrak{h}}(\lambda_H) \otimes_{K \cap H} \wedge^{R_G} \mathbf{p}^*.$$

By construction the highest weight vector $v_{\mathfrak{q}} \in \wedge^{R_G} \mathbf{p}$ of the minimal K–type of $A_{\mathfrak{q}}$ is of the form

$$\omega_{\mathfrak{n} \cap \mathfrak{h} \cap \mathbf{p}} \otimes \omega_{\mathfrak{n} \cap \mathbf{s} \cap \mathbf{p}} \in \wedge^{R_H} \mathbf{p}_H \otimes \wedge^{R_{\bar{H}}} \mathbf{p}_{\bar{H}}$$

and so

$$\omega_H \in A_{\mathfrak{q} \cap \mathfrak{h}}(\lambda_H) \otimes \wedge^{R_{\bar{H}}} \mathbf{s}^* \otimes_{K \cap H} \wedge^{R_H} \mathbf{p}_H^*.$$

To conclude that ω_H represents a nontrivial class in

$$H^{R_H}(\mathfrak{h}, K \cap H, A_{\mathfrak{q}_H}(\lambda_H) \otimes \wedge^{R_{\bar{H}}} \mathbf{s}^*)$$

it suffices to show

LEMMA 5.2. *Let $V_R \subset \wedge^{R_{\bar{H}}} \mathbf{s}$ be the unique subrepresentation with infinitesimal character λ_H. Then*

$$\omega_{\mathfrak{n} \cap \mathbf{s} \cap p} \in V_R.$$

PROOF. Note that $\omega_{\mathfrak{n} \cap \mathbf{s} \cap p} \in \wedge^{R_{\bar{H}}} \mathfrak{n} \cap \mathbf{s} \cap \mathbf{p}$ and $U(1)$ operates on it by the character $u \to u^{-R_G}$. $\square$

Let $p, q > 1$ and $H = U(p, q-1)U(1)$. Write $R_G = R_H + J_H$. Define the projection

$$\mathcal{P} : A_{\mathfrak{q}} \otimes_{K \cap H} \wedge^{R_G} \mathbf{p}^* \to (A_{\mathfrak{q}_H}(\lambda_H) \otimes \wedge^{J_H} (\mathbf{s} \cap \mathbf{p})^*) \otimes_{K \cap H} \wedge^{R_H} \mathbf{p}_H^*.$$

We proved

THEOREM 5.3. *Let $\omega \in A_{\mathfrak{q}} \otimes_K \wedge^{R_G} \mathbf{p}^*$ be a harmonic form representing a non trivial $(\mathfrak{g}, K)$–cohomology class of $A_{\mathfrak{q}}$. Then*

$$\mathcal{P}(\omega) = \omega_H \in A_{\mathfrak{q}_H}(\lambda_H) \otimes \wedge^{J_H} \mathbf{s}^* \otimes_{K \cap H} \wedge^{R_H} \mathbf{p}_H^*$$

is a harmonic form representing a nontrivial class in

$$H^{R_H}(\mathfrak{h}, K \cap H, A_{\mathfrak{q}_H}(\lambda_H) \otimes \wedge^{J_H} \mathbf{s}^*).$$

We will discuss geometric applications of Theorem 5.3 in another article.

6.   The restriction of $A_\mathfrak{q}$ to $U(r)U(p-r,q)$.

We assume here that $G = U(p,q)$ with $p,q > 1$ and that H, L satisfy the assumptions $\mathcal{A}$. In a previous section we showed that the restriction of $A_\mathfrak{q}$ to $H = U(p,q-1)U(1)$ is a direct sum of irreducible $(\mathfrak{h}, K \cap H)$–modules with finite multiplicities. In this section we show that the restriction of $A_\mathfrak{q}$ with $\mathfrak{q} = \mathfrak{q}(r,s)$, $r > 0$ and $s > 0$ to $H^\sharp = U(1)U(p-1,q)$ doesn't decompose into direct sum of irreducible representations.

6.1. We use Theorem 4.2 in [**10**]. The rank of

$$
\begin{aligned}
K/K \cap H^\sharp &= (U(p)U(q))/U(1)U(p-1)U(q) \\
&= U(p)/U(1)U(p-1)
\end{aligned}
$$

is equal to the real rank of $U(1, p-1)$, hence equal to 1. Let $\mathfrak{t}_0$ be the Lie algebra of a maximal abelian subalgebra of $K/K \cap H^\sharp$. It is contained in $u(p)$. We extend it to a Cartan subalgebra $\mathfrak{t}$ of $u(p)$. The space $\sqrt{-1}\mathfrak{t}_0^*$ is spanned by a root of $(\mathfrak{t}, \mathfrak{u}(p))$.

On the other hand

$$
\mathbb{R}_+\langle \mathfrak{u} \cap \mathbf{p}\rangle = \sum_{\delta(\mathfrak{u} \cap \mathbf{p})} m_\alpha \alpha
$$

contains a root of $(\mathfrak{t}, \mathfrak{u}(p))$. Since all roots of $(\mathfrak{t}, \mathfrak{u}(p))$ are conjugate under the compact Weyl group we see that

$$
\sqrt{-1}\mathfrak{t}_0^* \cap \mathbb{R}_+\langle \mathfrak{u} \cap \mathbf{p}\rangle \neq 0.
$$

Thus we proved

PROPOSITION 6.1. *Suppose that $G = U(p,q)$ with $p,q > 1$, that $H^\sharp = U(1)U(p-1,q)$ and that $\mathfrak{q} = \mathfrak{q}(r,s)$ with $r > 0$, $s > 0$. The representation $A_\mathfrak{q}$ is not discretely decomposable as a $(\mathfrak{h}, K \cap H^\sharp)$–module.*

The same methods also prove the proposition for the restriction to $U(l)U(p-l,q)$.

We may summarize these results as follows.

THEOREM 6.2. *Suppose that $G = U(p,q)$ with $p > q > 1$ and that $\mathfrak{q} = \mathfrak{q}(r,s)$ with $r > 0$ and $s > 0$. Then the restriction of $A_\mathfrak{q}$ to*
  $U(p,q-1)U(1)$ is   a direct sum of irreducible modules,
  $U(p,1)U(q-1)$ is   a direct sum of irreducible modules
  $U(1)U(p-1,q)$ is   not a direct sum of irreducible modules
  $U(p-1)U(1,q)$ is   not a direct sum of irreducible modules.

Remark 1: J. Harris and J.-S. Li showed in [**4**] that the restriction of $A_\mathfrak{q}$ to $U(p-r-s,q)$ has a discrete series representation $A_{\tilde{\mathfrak{q}}}$ as a direct summand where $\tilde{L}$ is isomorphic to $U(1)^{p-r-s}U(q)$ and $\tilde{\mathfrak{q}} = \tilde{\mathfrak{q}}(r,s)$.

6.2. The rank of
$$U(p,q)/(U(1)U(p-1,q))$$
is equal to the real rank of $U(1, p + q - 1)$ which is equal to min(1,p+q -1). Thus
$$\operatorname{rank}(G/H^\sharp) = \operatorname{rank} K/K \cap H^\sharp.$$
Hence the discrete spectrum of $L^2(U(p,q)/U(1)U(p-1,q))$ is not empty [**3**].

The representations in the discrete spectrum appear with multiplicity one and are contragredient to those which are cohomologically induced from a θ–stable parabolic subgroup where L is isomorphic to $U(p-2,q)U(1)^2$ and $\mathfrak{q} = \mathfrak{q}(1,1)$.

6.3. In [**13**] we found a unitary representation of $SL(4,\mathbb{R})$ whose restriction to symplectic subgroups H depends on the conjugacy class of H. The restriction to an symplectic group in one conjugacy class is a direct sum of irreducible representations, whereas the restriction of a symplectic group in the other conjugacy class is a direct integral. The following example shows that this occurs frequently.

Suppose $G = U(p,p)$, $H = U(r)U(p-r,q)$ and $H^\sharp = U(p,p-r)U(r)$. These subgroups are isomorphic but not conjugate in G.

COROLLARY 6.3. *Suppose that* $p > 2$, $G = U(p,p)$, *and let* $H = U(1)U(p-1,q)$ *and* $H^\sharp = U(p,p-1)U(1)$. *There exist families of unitary representations* π *with nontrivial* $(\mathfrak{g}, K)$-*cohomology which are infinitesimally H-admissible, but not infinitesimally* $H^\sharp$-*admissible.*

References

[1] L.Clozel and T.N.Venkataramana, Restriction of the holomorphic cohomology of a Shimura variety to a smaller Shimura variety, Duke Math J. vol 95 (1) 51-106 (1998).

[2] M.Dulfo and J.Vargas, electronic communication.

[3] M.Flensted-Jensen, Ann. of Math. vol 111 (2), 253-311, (1980).

[4] M.Harris and J. Li, A Lefschetz property for subvarieties of Shimura varieties, J. Algebraic Geom. 7, no. 1, 77–122, (1998).

[5] A.W. Knapp and D. Vogan, Jr. *Cohomological induction and unitary representations,* Princeton Mathematical Series, 45. Princeton University Press, Princeton, NJ, 1995.

[6] T.Kobayashi, The restriction of $A_\mathfrak{q}(\lambda)$ to reductive subgroups, Proc. Acad. Japan Vol. 69, 7, (1993), p. 262.

[7] T.Kobayashi, Discrete decomposability of the restriction of $A_\mathfrak{q}(?)$ with respect to reductive subgroups and its applications. Invent. Math. 117 (1994), no. 2.

[8] T.Kobayashi, Restrictions of unitary representations of real Lie groups, in *Lie Theory, unitary representations and compactifications of symmetric spaces,* Jean-Phillippe Anker, Bent Orsted (Editors), Birkhäuser, Progress in Mathematics 229.

[9] T.Kobayashi, Singular Unitary Representations and Discrete Series Representations for Indefinite Stiefel Manifolds U(p,q,F)/U(p-m,q,F), Men. Amer.Math. Soc., Vol 462, (1992).

[10] T Kobayashi, Discrete decomposability of the restriction of $A_\mathfrak{q}(\lambda)$ with respect to reductive subgroups III., Restriction of Harish-Chandra modules and associated varieties, Invent. Math. vol. 131 (2) pp 229-256, (1998).

[11] T.Kobayashi and T.Oda, A vanishing theorem for modular symbols on locally symmetric spaces, Comment. Math. Helv. 45-70 (1998).

[12] T.Oshima, T. Matsuki, A description of discrete series for semisimple symmetric spaces, Adv. Stud. math. 4, 332-390 (1984).

[13] B. Ørsted and B.Speh, Branching Laws for Some Unitary Representations of SL(4,ℝ), SIGMA 4 (2008).

[14] B.Speh, Cohomology of discrete groups and representation theory, in *Geometry, Analysis and Topology of Discrete Groups*, 346–373, Adv. Lect. Math. (ALM), 6, Int. Press, Somerville, MA, (2008).

[15] D.Vogan, Jr.,*Representations of real reductive Lie groups,* Progress in Mathematics, 15. Birkhäuser, Boston, Mass., 1981.

DEPARTMENT OF MATHEMATICS, MALOTT HALL,, CORNELL UNIVERSITY, ITHACA, NY 14853, USA

E-mail address: speh@math.cornell.edu

Titles in This Series

For a complete list of titles in this series, visit the
AMS Bookstore at **www.ams.org/bookstore/**.